S0-AUH-164

INTERNATIONAL SERIES OF MONOGRAPHS ON
ELECTROMAGNETIC WAVES
EDITORS: A. L. CULLEN, V. A. FOCK AND J. R. WAIT

VOLUME 6

ELECTROMAGNETIC THEORY AND ANTENNAS

PART 1

OTHER TITLES IN THE SERIES ON

ELECTROMAGNETIC WAVES

Vol. 1 *Diffraction by Convex Surface*—FOCK

Vol. 2 *Ionospheric Sporadic E*—SMITH AND MATSUSHITA (Editors)

Vol. 3 *Electromagnetic Waves in Stratified Media*—WAIT

Vol. 4 *The Scattering of Electromagnetic Waves from Rough Surfaces*—BECKMANN AND SPIZZICHINO

Vol. 5 *Electromagnetic Scattering*—KERKER

"Symposium on Electromagnetic Theory and Antennas, Copenhagen, 1962.

ELECTROMAGNETIC THEORY AND ANTENNAS

Proceedings of a Symposium held at
Copenhagen, Denmark, June 1962

Edited by

E. C. JORDAN

PART 1

Sponsored by

COMMISSION VI OF URSI
TECHNICAL UNIVERSITY OF DENMARK
DANISH ACADEMY OF TECHNICAL SCIENCES
DANISH NATIONAL COMMITTEE OF URSI

SYMPOSIUM PUBLICATIONS DIVISION
PERGAMON PRESS
OXFORD · LONDON · NEW YORK · PARIS
1963

PERGAMON PRESS LTD.
Headington Hill Hall, Oxford
4 & 5 *Fitzroy Square, London, W.*1

PERGAMON PRESS INC.
122 *East* 55*th Street, New York* 22, *N.Y.*

GAUTHIER-VILLARS ED.
55 *Quai des Grands-Augustins, Paris* 6

PERGAMON PRESS G.m.b.H.
Kaiserstrasse 75, *Frankfurt am Main*

Distributed in the Western Hemisphere by
THE MACMILLAN COMPANY – NEW YORK
pursuant to a special arrangement with
PERGAMON PRESS LIMITED

Library of Congress Card No. 63-17037

Printed in Great Britain by Adlard & Son Ltd,
London & Dorking

Symposium Committees

ORGANIZING COMMITTEE

J. Rybner, Denmark; President

V. A. Fock, U.S.S.R.
N. E. Holmblad, Denmark
H. Lottrup Knudsen, Denmark; Secretary
J. Oskar, Nielsen, Denmark
Gunnar Pedersen, Denmark
K. M. Siegel, U.S.A.

COMMITTEE FOR THE TECHNICAL PROGRAM

James R. Wait, U.S.A.; Chairman

Executive Committee

H. Lottrup Knudsen, Denmark
N. Marcuvitz, U.S.A.
S. Silver, U.S.A.
George Sinclair, Canada

Corresponding Members

G. Barzilai, Italy
P. Beckmann, Czechoslovakia
K. Bochenek, Poland
H. Bremmer, Netherlands
J. Brown, England
G. Toraldo di Francia, Italy
G. Eckart, Germany
W. Franz, Germany
Z. Godzinski, Poland
Morris Kline, U.S.A.
Julien Loeb, France
P. Mattila, Finland
M. A. Miller, U.S.S.R.
K. Morita, Japan
A. A. Oliner, U.S.A.
J. A. Ortusi, France
C. L. Pekeris, Israel
O. E. H. Rydbeck, Sweden
K. M. Siegel, U.S.A.
A. Tonning, Norway
Francis J. Zucker, U.S.A.

Coordinators

R. C. Hansen, U.S.A.
J. B. Keller, U.S.A.
H. Lottrup Knudsen, Denmark
N. Marcuvitz, U.S.A.
A. A. Oliner, U.S.A.
K. M. Siegel, U.S.A.
S. Silver, U.S.A.
A. D. Wheelon, U.S.A.
James R. Wait, U.S.A.
Francis J. Zucker, U.S.A.

Preprints Committee

J. Bach Anderson, Denmark; Chairman
Jesper Hansen, Denmark
Mrs. Tove Larsen, Denmark

Arrangements Committee

K. Særmark, Denmark; Chairman

G. E. R. Bramslev, Denmark
M. Gronlund, Denmark
Sven Gylov, Denmark
Bruno Henriksen, Denmark
H. Lottrup Knudsen, Denmark
Mrs. Jorgen Rybner, Denmark

CONTENTS

PART 1

Section B. Anisotropic and Stratified Media

PART 2

Section C. Random Media and Partial Coherence

Section D. Surface Waves, Leaky Waves and Mode Propagation

Section E. Antenna Theory and Radiating Elements

Section F. Antenna Arrays and Data Processing

INTRODUCTION

FROM June 25 through June 30, 1962, a Symposium on Electromagnetic Theory and antennas was held at the Technical University of Denmark in Copenhagen. It was sponsored by the International Scientific Radio Union, The Technical University of Denmark, The Danish Academy of Technical Sciences, and the Danish National Committee of URSI.

This symposium has been preceded by three other international symposia with similar scopes, all being co-sponsored by the International Scientific Radio Union. The first of these was held at McGill University, Montreal, Canada, in 1953, and the following at the University of Michigan, Ann Arbor, Michigan, U.S.A., in 1955, and at the University of Toronto, Toronto, Canada, in 1959.

Support to the symposium was given by the International Scientific Radio Union and Pergamon Press Limited, Oxford, England, and by the following Danish foundations and firms: Aktieselskabet Brüel & Kjær, A/S Hellesens, Aktieselskabet Nordiske Kabel- og Traadfabriker, Carlsberg Bryggerierne, Dansk Siemens Aktieselskab, Det Store Nordiske Telegraf-Selskab A/S, DISA Elektronik, Division of Dansk Industri Syndikat A/S, Fyns Kommunale Telefonselskab, Industriselskabet Ferroperm A/S, Jydsk Telefon-Aktieselskab, Kjøbenhavns Telefon Aktieselskab, Philips A/S, Statens Teknisk-videnskabelige Fond, Thomas B. Thriges Fond, Tuborgs Bryggerier, and Vitrohm Elektronisk Fabrik Aktieselskab.

The organisers of the Symposium accepted 144 papers for presentation in the technical sessions. Of these 126 papers were actually presented at the symposium, and 121 papers were submitted for publication in the Proceedings. This volume contains 00 full-length papers and 00 summaries. In addition the book includes the welcoming address by M. Loeb, International Chairman of Commission VI of URSI, and the lecture by Professor Mogens Pihl on "The Scientific Works of L. V. Lorenz".

I should like to express the sincere thanks of the organizers to the many persons and institutions in Denmark and abroad, whose kind co-operation and support have contributed to the success of the symposium.

JØRGEN RYBNER

WELCOMING ADDRESS

JULIEN LOEB*

MESDAMES ET MESSIEURS,

Je tiens tout d'abord à remercier Monsieur E. Greve Knuth Winterfeldt Président de l'Université Technique du Danemark, pour l'admirable cadre qu'il a offert à nos séances, et pour ses paroles de bienvenue.

C'est une joie pour nous de séjourner dans ce pays, qui représente à nos yeux tout ce que la civilisation Européenne a de plus vénérable et de plus moderne à la fois.

Les vieilles pierres n'ont pas été, comme si souvent ailleurs, noircies par les siècles, si bien que les audacieuses réalisations de l'architecture actuelle (l'Aeroport, le Terminus SAS etc.) s'harmonisent tout à fait avec l'oeuvre du passé.

C'est aussi un privilège pour nous de tenir nos réunions dans cette ville où travaillèrent les grands savants Danois Bertholin, Roemer, Oersted, V. Lorentz et Pedersen.

Au nom de la VIème Commission de l'URSI, j'adresse mes plus vifs remerciements aux organisateurs de cette importante manifestation.

Je sais par expérience ce que représente l'organisation matérielle d'une semblable réunion, et combien nous devons être reconnaissants à M. K. Saemark, Président du Comité de réception, et à Madame J. Rybner qui a bien voulu organiser pour nos épouses les visites et les distractions.

Je remercie notre hôte Prof. Jorgens Rybner et ses collègues M.M. Knudsen, Holmblad, Nielsen, Pedersen, Fock et Siegel.

C'est avec un certain complexe de culpabilité que je pense au Comité pour le Programme technique: MM. J.R. Wait, Knudsen, Marcuwits, S. Silver (mon prédécesseur à la VIème Commission) et George Sinclair.

Ce sont de vieux compagnons de travail au sein de l'URSI. Vous pourrez apprécier au cours de cette semaine l'intérêt extrême du résultat de leur travail de préparation, le choix minutieux des sujets et des auteurs.

J'ai dit éprouver un complexe de culpabilité, car tout ce travail, accompli sous l'égide de l'URSI (VIème Commission) fut fait avec une participation plutôt symbolique de son Président.

Aussi, pour ne pas être accusé de venir ici comme "Ingénieur des Travaux Finis", je voudrais vous montrer comment les communications de ce Symposium s'inscrivent dans le domaine de notre Commission.

* Chairman of International Commission VI of URSI.

L'idéé de créer notre Commission est due à l'intuition de notre regretté Président, le Professeur B. Van der Pol. On pouvait se demander a priori ce que la théorie de l'information, celle des Circuits et l'Electromagnétisme pouvaient avoir en commun.

Puis, au fur et à mesure que les techniques se développaient — cela se fit très vite, sous nos yeux – les frontières entre ces domains devinrent plus floues. Un guide d'ondes relève-t-il de la théorie de la propagation d'ondes en présence de conditions aux limites déterminées, ou de la théorie des circuits caractérisés par les matrices bien connues?

La Théorie des Communications avait fait il y a environ 20 ans un progrès décisif lorsque l'on a pu traiter mathématiquement, avec Wiener, Kolmogoroff et Shannon, le signal comme une entité aléatoire. Voici maintenant que nous sommes amenés par la force des choses à calculer la propagation des ondes dans des milieux eux mêmes stochastiques. De tels calculs se sont montrés indispensables lorsque sur l'initiative du Professeur Van der Pol nous avons appliqué la Theorie de l'Information aux Radiocommunications réelles, soumises aux caprices de la propagation.

S'il avait fallu un point de rencontre supplémentaire, les recherches spatiales nous l'offrent d'une façon particulièrement impérative.

Toute tentative de télécommande, de télémesure, de localisation par Radar, met en jeu:

— La propagation dans les milieux (stochastiques bien entendu) formés de plasmas, contenant de la matière en mouvement, en présence de champs magnétiques.
— La diffraction, la réflexion diffuse ou spéculaire des rayonnements électromagnétiques sur des surfaces données.
— Les antennes à très large bande de fréquences.
— Le traitement de l'Information (filtrage des signaux, reconnaissance des formes, etc.).

Enfin, d'un point de vue plus pratique aux yeux d'un Président de Commission les travaux de ce Symposium constituent une préparation idéale pour la prochaine Assemblée Générale de l'URSI.

Notre tâche a été précisée et rendue plus lourde par les décisions du dernier Comité de Coordination de l'URSI.

Nous devons, le plus souvent possible, inviter les spécialistes mondialement connus, à présenter des rapports d'ensemble dans le cadre de questions scientifiques vivantes et importantes.

Où, plus qu'ici, pourrais-je trouver ce qu'il me faudra comme auteurs particulièrement qualifiés?

THE SCIENTIFIC ACHIEVEMENTS OF L. V. LORENZ

MOGENS PIHL

Institute of Theoretical Physics, University of Copenhagen

LUDVIG VALENTIN LORENZ (1829–91) was by training a civil engineer, graduated from the Technical University of Copenhagen, and by profession he was a teacher at the Danish Military Academy and at a teachers' training college in Copenhagen. During the last few years of his life he was financially supported by the newly established Carlsberg Foundation to such an extent that he could devote all his time to scientific research. He was one of the eminent physicists of the last century, although the content and range of some of his achievements were not fully realized and others not even known by his contemporaries.

Therefore, it is with pleasure and gratitude that a Dane is taking advantage of this symposium held here in Copenhagen to give you an idea of some of the most important contributions to physics made by this great scientist of ours. I hope that I shall be able to do so without leaving an impression of undue national pride.

The lack of recognition that some of Lorenz's scientific works met, was mainly due to himself, above all to his great difficulties in presenting his ideas and calculations in an accessible form, but it was also due to the fact that he published certain of his important papers in the Danish language only. His contemporaries in the scientific circles of Copenhagen no doubt realized that he was an eminent scientist and by the standard of his days his research work was supported accordingly. But although he was much respected, few understood what he actually did. This lack of deeper understanding on the part of his surroundings in combination with his own tendency to isolation and a rather outspoken critical attitude towards some of his colleagues, made this selftaught man something of a lonely philosopher, but apparently he cherished no feelings of resentment. In certain respects he appears to us congenial to his contemporary American colleague Joseph Willard Gibbs, although it must be admitted that he did not possess the classical simplicity so characteristic of the scientific works of Gibbs.

The one who best understood Lorenz was probably his younger contemporary C. Christiansen, professor of physics at the University of Copenhagen. In that capacity he was the teacher of the generations of physicists counting among others Martin Knudsen, distinguished for his experimental researches in the field of the kinetic theory of gases, and among the younger Niels Bohr. Niels Bohr's thesis for the doctorate, from 1911, deals with the electron theory of conductivities of metals for heat and electricity, the keystone of which theory always being the relation between the two specific conductivities and the temperature as discovered by Lorenz, and of which we shall speak later.

Christiansen, an original personality of independent mind, came into close contact with Lorenz and learnt much from him. And we, of later generations, have learnt much about Lorenz's work and life through Christiansen, although it must be admitted that he seemed not to have penetrated into all of Lorenz's achievements in physics. A very important contribution to our understanding of Lorenz's works has been given by H. Valentiner, a Danish mathematician who edited his collected works—*Oeuvres scientifiques de L. V. Lorenz*—a great task, indeed. Several mathematical mistakes and errors in his calculations were thereby corrected.

Lorenz's researches in physics mainly fall within the fields of *optics* and of *conductivities of metals* for heat and electricity, and they have been carried out with equal emphasis on theoretical and experimental methods. In his mastery of both of these techniques of physics he is surpassed only by very few. *Rayleigh*, before all others, could be mentioned.

Lorenz was also deeply interested in problems of *pure mathematics*, such as the distribution of prime numbers, but especially here he met a rather severe criticism from the Danish mathematicians, who did not appreciate the undoubted lightness—if not to say recklessness—with which he applied the methods of mathematics. I am convinced, however, that there are also points of great interest to be found in his mathematical works, but on this occasion I shall confine myself to his main achievements in *physics*. And although he was very interested also in *technology*, I shall discuss only one of his contributions in this field, his *theory of telephone cables*.

The basis on which Lorenz started his research in *optics*, was, of course, the *propagation of light waves* conceptualized in view of the *theory of elasticity*. But finally he got convinced of the incompatibility of the boundary conditions of the theory of elasticity with the *formulas of Fresnels* for reflection and refraction, and then he concentrated his efforts on finding a *phenomenological description* of the propagation of light waves, instead of speculation on the nature of light. Independent of the Irish physicist MacCullagh, who arrived at the same result, he showed in 1863 that the differential equation for the light vector $\mathbf{u}$ should be

$$- \text{curl curl } \mathbf{u} = \frac{1}{a^2} \frac{\partial^2 \mathbf{u}}{\partial t^2} \tag{A}$$

where the phase-velocity a is a constant, characteristic for the homogeneous optical medium in consideration. His confidence in this differential equation is, above all, based upon the fact that it leads to the existence of transversal waves in homogeneous media, which at the surface of discontinuity between two substances, 1 and 2, must satisfy the boundary conditions

$$\begin{aligned} u^1_{\text{tangential}} &= u^2_{\text{tangential}} \\ (\text{curl } \mathbf{u}^1)_{\text{tangential}} &= (\text{curl } \mathbf{u}^2)_{\text{tangential}}, \end{aligned} \qquad (A')$$

these conditions being equivalent to Fresnel's formulas. Here we meet for the first time the idea that such *boundary conditions* can be derived as *consequences of a fundamental differential equation*—a point which later on was emphasized by Heinrich Hertz in connection with Maxwell's equations of electrodynamics.

Lorenz then showed, that in case a is not considered as a constant, but as a *periodical function of space*, his wave equation leads to the theory of *double refraction*, when the periods of a are small as compared with the wavelength, which theory by use of the same mathematical technique as applied by Lorenz has been taken up in much later times and also is under discussion at this symposium. He also gives, on the same basis, a theory of the so-called *optical activity*, but I must admit that it has not been possible for me to penetrate into all the details of his calculations on this problem.

A further remarkable result of his optical researches on the basis of his fundamental differential equation (A) is his derivation—again difficult to follow in detail—of the well-known formula (Lorenz–Lorentz formula) for the so-called *refraction constant* R, according to which

$$R = \frac{n_\infty^2 - 1}{n_\infty^2 + 2} v = \text{constant (independent of temperature and pressure)}, \qquad (B)$$

where n_∞ is the index of refraction for an infinite wavelength and v the specific volume. In the semi-macroscopical model used by him, the molecules are considered as small bodies with a well-defined refraction index n_0, and he finds for the value of R

$$R = \frac{n_0^2 - 1}{n_0^2 + 2} v_0, \qquad (B')$$

v_0 being the proper specific volume occupied by the molecules in the volume v of the substance.

He also showed that for a mixture

$$R = \sum_i \frac{n_i^2 - 1}{n_i^2 + 2} v_i, \qquad (B'')$$

where the index $_i$ refers to the components of the mixture.

His first paper on the subject of refraction constant, in which he proves his formula and gives an experimental verification of it in case of H_2O, dates from 1869. In 1870 the famous Dutch physicist H. A. Lorentz, independently of our Lorenz, arrived at the same result on the basis of the electromagnetic theory of light. In our days the formula has been improved by a better approximation due to L. Onsager, but the Lorenz–Lorentz formula is still of great practical use, as you will know.

In 1890, the year before his death, Lorenz published in the *Proceedings of the Danish Academy of Science* a paper on the *diffraction of a plane light wave by a transparent sphere*, again using his fundamental differential equation (A) and the corresponding boundary conditions (A′), and thereby anticipating later calculations by Mie, Debye and others. Unfortunately, this very important paper was published in Danish only, and for this reason it has been without any influence upon the later development of this interesting problem, which always has been a challenge to mathematical physicists; before Lorenz the problem was treated by such men as Clebsch and Stokes.

It is in this paper we find the first *determination of the number of molecules in a given volume of air*, based upon the *scattering in the sky of sun light*. The first fairly accurate estimation of the order of magnitude of the number, now called Avogadro's Number, and usually ascribed to Rayleigh, who published it in 1899, nine years after the publication of Lorenz's paper, and, of course, independently of Lorenz.

The reasoning of Lorenz runs as follows: First he proved what essentially already had been shown by Rayleigh that the cross-section σ_{sc} for scattering of light of wavelength λ by a sphere with radius R and refraction index n_0 for $\lambda \gg R$ is inversely proportional to λ^4. He finds

$$\sigma_{sc} = \frac{128\,\pi^5 R^6}{3\,\lambda^4}\left(\frac{n_0^2-1}{n_0^2+2}\right)^2 \qquad \text{(C)}$$

If now the molecules of air, by use of the semi-macroscopical model referred to above, are considered as small spheres with radius R and a well-defined refraction index n_0, not to be confounded with the refraction index n of air, and further if we assume with Lorenz that the total scattering by the molecules in a certain element of volume is the sum of the scatterings by the individual molecules, which is true for gases, it follows, that the *extinction* or *attenuation* of the intensity of light through scattering along a x-axis is given by

$$-\frac{\mathrm{d}I}{\mathrm{d}x} = A\sigma_{sc}I \text{ or } I = I_0 e^{-hx}, \qquad h = A\sigma_{sc},$$

where A is the number of molecules in a unit of volume and h is the so-called *extinction* coefficient. As mentioned before, the refraction constant by air is given by (B′):

$$R = \frac{n^2-1}{n^2+2}v = \frac{n_0^2-1}{n_0^2+2}v_0$$

or

$$\frac{n^2-1}{n^2+2} = \frac{n_0^2-1}{n_0^2+2}A\frac{4\pi}{3}R^3.$$

where v is the specific volume of air and v_0 the proper volume occupied by the molecules in the volume v. The coefficient of extinction is then determined by

$$h = \frac{24\pi^3}{A\lambda^4}\left(\frac{n^2-1}{n^2+2}\right)^2, \tag{D}$$

whereby the unknown magnitudes n_0 and R, belonging to the individual molecules, have been eliminated. Or

$$h \approx \frac{32\pi^3(n-1)^2}{3A\lambda^4} \tag{D'}$$

because

$$n \approx 1 \text{ or } \frac{n^2-1}{n^2+2} \approx \frac{2}{3}(n-1).$$

This is exactly the value of h given by Rayleigh in his paper from 1899. From the experimentally known value of $h\lambda^4$ Lorenz found

$$A \approx 1{\cdot}63 \times 10^{22} \text{ mol/l.},$$

from which follows

$$\text{Avogadro's constant} \approx 3{\cdot}7 \times 10^{23}.$$

A more correct theory must, of course, take the *statistical fluctuations of the density of air* into consideration, as shown by Einstein in 1910, but this theory leads to the same expression for h, when applied to an ideal gas.

Most impressive of all Lorenz's achievements in optics is his *electromagnetic theory of light*, developed in a rather unknown paper from 1867, two years after Maxwell's famous paper on the same subject. There is no doubt that Lorenz did not know of Maxwell's theory, his approach being quite different from this—and by his contemporaries, including Maxwell himself, the two theories were considered even more different in nature than we do now.

Lorenz's starting-point was the well-known equations—here written in modern units—for quasi-stationary currents developed by Weber, Kirchhoff and others:

$$\mathbf{i} = -\sigma\left(\text{grad }\psi + \frac{\partial \mathbf{A}}{\partial t}\right) \tag{1}$$

$$\psi(\mathbf{x}, t) = \frac{1}{4\pi\epsilon_0}\int \frac{\rho(\mathbf{x}, t)}{|x - x'|}\, d\omega' \tag{2}$$

or

$$\Delta\psi = -\frac{\rho}{\epsilon_0}$$

$$\mathbf{A}(x, t) = \frac{\mu_0}{4\pi}\int \frac{\mathbf{i}(\mathbf{x}, t)}{|\mathbf{x} - \mathbf{x}'|}\, d\omega' \tag{3}$$

or

$$\Delta \mathbf{A} = -\mu_0 \mathbf{i}$$

$$\frac{\partial \rho}{\partial t} + \operatorname{div} \mathbf{i} = 0, \tag{4}$$

where **i** is the density of the current vector, σ the specific conductivity for electricity, ψ the scalar electric potential, **A** the vector potential for the magnetic field and ρ the density of electric charge; ϵ_0 and μ_0 are well-known constants. Lorenz did not introduce a dielectrical constant and a permeability, and he is concerned with the current-density vector **i** only, not with the electric and magnetic field intensities. In confining himself only to media, for which Ohm's law is valid, his theory is somewhat more restricted than Maxwell's but it easily admits of a generalization by which the two theories become equivalent.

He now assumes that the equation (1) should be kept, while (2) and (3) should be substituted by the *retarded potentials*

$$\psi(\mathbf{x}, t) = \frac{1}{4\pi\epsilon_0} \int \frac{\rho\left(\mathbf{x}, t - \frac{|\mathbf{x} - \mathbf{x}'|}{a}\right)}{|\mathbf{x} - \mathbf{x}'|} d\omega'$$

and

$$\mathbf{A} = \frac{\mu_0}{4\pi} \int \frac{\mathbf{i}\left(\mathbf{x}, t - \frac{|\mathbf{x} - \mathbf{x}'|}{a}\right)}{|\mathbf{x} - \mathbf{x}'|} d\omega',$$

where a is a velocity constant for propagations of actions from the distributions of charges and currents. *This concept of retarded potential is due to Lorenz* and was originally developed by him in connection with investigations on the theory of elasticity. As shown by him the above expressions for ψ and **A** are equivalent to

$$\Delta \psi - \frac{1}{a^2} \frac{\partial^2 \psi}{\partial t^2} = -\frac{\rho}{\epsilon_0} \tag{5}$$

and

$$\Delta \mathbf{A} - \frac{1}{a^2} \frac{\partial^2 \mathbf{A}}{\partial t^2} = -\mu_0 \mathbf{i}, \tag{6}$$

and he further shows that from the equation of continuity (4) follows

$$\operatorname{div} \mathbf{A} + \mu_0 \epsilon_0 \frac{\partial \psi}{\partial t} = 0. \tag{7}$$

Lorenz now makes the important observation that if a is chosen such that

$$\frac{1}{a^2} = \epsilon_0 \mu_0, \tag{8}$$

it follows from (1), (5), (6) and (7), that for *this choice, and for this only,*

$$-\operatorname{curl} \operatorname{curl} \mathbf{i} = \frac{1}{a^2} \frac{\partial^2 \mathbf{i}}{\partial t^2} + \mu_0 \frac{\partial \mathbf{i}}{\partial t}, \tag{E}$$

this differential equation for **i** *being the same as his fundamental equation* (*A*) *for the light vector* **u**, *apart from the term with* $\partial \mathbf{i}/\partial t$ *on the right side, which causes an absorption of* **i**. The value of *a* determined by (8) from the known values of ϵ_0 and μ_0 now being exactly the *velocity of light* immediately leads him to the *interpretation of light as propagation of the electric current-density vector*, whereby at the same time he gives an explanation of the absorption of light in good conductors.

I think it is not necessary to comment further on this great achievement, which took place almost 50 years after H. C. Ørsted's discovery of *electromagnetism*, and almost 200 years after Ole Rømer's *determination of the velocity of light* through observations of the eclipses of Jupiter's moons, the latter achievement, in turn, being preceded by Rasmus Bartholin's discovery a few years before of the *double refraction*. And when I now draw to your attention that the above mentioned C. Christiansen three years after the publication of this paper of Lorenz discovered the *anomalous dispersion of light*, and that we are looking forward to celebrate the 50th anniversary of Niels Bohr's interpretation of the *spectrum of hydrogen* next year, you will understand how tempting it would be for us to talk of a tradition for optics in Danish physics. However, these historical facts are not felt by us as testimonies of an unbroken tradition. If you ask our unbiased students or unsentimental young colleagues, they will not visualize these events as something which is bound together by the inspiring power of a tradition. They are conscious of them only as interesting, but rather accidental flashes in the history of physics in Denmark; apart from the last mentioned event, Niels Bohr's paper on the interpretation of the hydrogen spectrum, which, of course, did start a school or tradition of atomic physics, and not only in our own country. More generally speaking, it is an historical fact, which Denmark shares with many other of the old countries, that although Danish science, from the days of Tyge Brahe, our great astronomer, and up to the beginning of this century, may refer to several brilliant, but rather isolated achievements in science, we cannot until recently speak about a firmly established and living scientific tradition in our cultural life. Even among Danish physicists the works of Lorenz are hardly known. And when every Danish schoolboy has heard of Tyge Brahe it is more because of his fame and fascinating personality than of his scientific works, of which he will only know, that they have something to do with astronomy. Of course, he will also know that the velocity of light was discovered by Ole Rømer and electromagnetism by H. C. Ørsted, but nothing more about their scientific activities. On the other hand, he will know quite a lot about our great artists and statesmen.

I do not see any reason why this historical fact should be regretted. The minds of the best among scientists will always be turned more toward the future than the past, and they will not insist upon the cultivation of an unbroken tradition, which they cannot regard as such. A tradition is more than a mere collection of historical events. This concept implies the existence of a power of will to direct future activities from a given basis. It is interesting that the word "tradition" mostly is used in connection with the word "create". It is not only something which happens.

We now turn our attention to Lorenz's investigations in the field of *conductivities of metals for heat and electricity*; his most important contribution here is the well-known *Lorenz law*, according to which the *ratio between the two conductivities, λ and σ, for heat and electricity is proportional to the absolute or Kelvin-temperature T*:

$$\frac{\lambda}{\sigma} = CT, \tag{F}$$

where the constant C is *the same for all metals*. It had already been shown by Wiedemann and Franz, that λ/σ at the same temperature is the same for all metals, and Lorenz now proved through very accurate measurements of λ and σ at 0° Celsius and 100° Celsius that this ratio is proportional to the Kelvin temperature.

It is characteristic of his mastery of both theoretical and experimental methods that when in the determination of λ he met the problem of the *cooling-down of a body through heat convection by the air*, he solved this problem by calculation on the basis of the *equations of hydrodynamics* by Navier and Stokes. It is a very difficult problem in mathematical physics, but he succeeded in solving it by using quite modern methods of *similarity*. And having finished his computations he found them confirmed by experiments.

In the determination of the specific conductivities for electricity he made use of a well-known absolute method of his own invention, which later on was to be applied by himself and others, such as Rayleigh, to fix the *international unit of resistance*, 1 *ohm*.

The time at my disposal here does not permit me to go into further details concerning these investigations on conductivities of metals, for which Lorenz actually received much recognition. But there remains a posthumous paper of his of such great interest that it ought to be mentioned.

It is a paper in which Lorenz develops a *theory of currents in telephone-cables*. He starts with a deduction at the easily interpreted equations

$$\left.\begin{aligned} -\frac{\partial\psi}{\partial x} - l\frac{\partial I}{\partial t} &= rI \text{ (generalized Ohm's law)} \\ -\frac{\partial I}{\partial x} &= c\frac{\partial\psi}{\partial t} + g\psi \text{ (conservation of electric charge)} \end{aligned}\right\} \tag{G}$$

for the potential ψ and the current I along a cable in the direction of a x-axis. Here r, c and l are the *resistance*, *capacity* and *self-inductance* by unit of length. $g\psi$ is the *leakage* to the surroundings by unit of length and time, but in what now follows this small term will be neglected. He now tries to satisfy these equations by solutions of the form

$$I = I_0 e^{j(\omega t - \lambda x)}, \qquad \psi = \psi_0 e^{j(\omega t - \lambda x)}$$

where $j = \sqrt{-1}$ and $\lambda = \beta - \alpha j$ is a complex number, and it is easily seen that the value of α is given by

$$\alpha^2 = \frac{\omega c}{2} \frac{r^2}{l\omega + \sqrt{(r^2 + l^2\omega^2)}}.$$

α determines the *attenuation* of the current along the cable, which should be as small as possible. This can be done by making *r small and l large*, and in this case

$$\alpha \approx \frac{r}{2}\sqrt{\frac{c}{l}}$$

so that the attenuation is the same for all frequencies ω, no relative distortion of the transmitted signals then taking place.

Lorenz now proposes, that αl should be made large by using cables *wound with a covering of soft iron*. An idea proposed also by Heaviside, independently of Lorenz, and which later on was taken up by the Danish electrotechnician Krarup, who knew of Lorenz's theory. The link between Krarup and Lorenz was the Danish mathematician and telephone engineer J. L. W. V. Jensen—best known for his investigations into the theory of analytical functions, and like Lorenz he was a fascinating character of unusual fate. This improvement of cables by Lorenz is now known as *continuous loading*. Another method is the so-called *Pupinisation* of cables, where the inductance, as proposed by Pupin, is placed lumpwise along the cable.

My lecture is drawing to a close. I have not mentioned all of Lorenz's contributions to physics, but only those which seemed most important to me. I do not think I am exaggerating, when repeating what I said at the beginning of this lecture: he was one of the eminent physicists of the last century. I also said that some of his works were not fully understood or even known by his contemporaries, and that this lack of recognition mainly was due to himself, to his great difficulties in presenting his ideas and computations in an accessible form, and to the fact that he published some of his important papers in the Danish language only. In the history of science the question of recognition often is a rather petty one, and it is always very difficult to judge honestly, and so I think that we should not pay much attention to it. What really touches us and gives us inspiration when thinking of Lorenz is the story of a great character, who in solitude, though not without the respect and material support, too, of his surroundings was able to pursue his very original intentions in scientific inquiry with such a will-power and gift for research, that he could leave behind him this important work performed by one man. Since the days of Lorenz's physics and the methods of research applicable to physics have changed so violently that we can hardly imagine how a modern scientist, in a situation as solitary as that of Lorenz, would be able to carry through research of the same relative range and importance. A lonely working physicist of today is almost a self-contradiction. To my knowledge the young Einstein in his very productive years at the Patent Office at Berne at the beginning of this century was the last of the heroes—a very great one, indeed—of this old-world style.

And so the account of Lorenz's life and works, like so many other tales in history, recalls to some of us that romantic enthusiasm for the irrevocable past unduly called the "good old days", with which history inevitably is connected, and the excessive cultivation of which may involve a great danger of sentimentality among historians and, I admit, also to a historian of science. Instead of being sentimental we should rather look upon things in the light of the question: How much would Lorenz not have gained if the opportunity of collaboration which in our days is a matter of course, had been at the disposal of a man of his temperament and his need for isolation?

SECTION A

SCATTERING AND DIFFRACTION THEORY

A SURVEY OF SHORT WAVELENGTH DIFFRACTION THEORY†

JOSEPH B. KELLER

Courant Institute of Mathematical Sciences, New York University,
New York 3, N.Y.

ABSTRACT

A review of recent developments in the theory of diffraction of waves of short wavelength is presented. The concept of progagation along rays is emphasized. The local character of diffraction is also stressed. The relationship of these concepts of the work of various authors is described.

1. INTRODUCTION

In the last twenty years there has been an extensive development of the theory of propagation, scattering and diffraction of electromagnetic waves. It was stimulated by the invention of radar and of techniques for producing very short waves. Therefore the theoretical advances have primarily concerned short waves. Previous investigations of such waves had mainly aimed at providing an electromagnetic theory of various optical phenomena, on the one hand, and of accounting for the propagation of radiowaves around the earth, on the other. Those investigations provided some of the ideas which have played an important role in the recent developments.

From the physical viewpoint, probably the main conclusion which has emerged from this development is that the propagation of short waves can be satisfactorily described by means of rays. However, the usual rays of geometrical optics do not suffice for this purpose. Instead various types of new rays, which are called diffracted rays, must be introduced in addition. Some of them will be described below. Furthermore a quantitative description of the electromagnetic field can be obtained with the aid of the rays, as we shall see.

The second most important conclusion which is intimately related to the previous one, is that the processes which give rise to incident, reflected, refracted and diffracted rays are *local*. This means that the reflected and refracted fields near a point on an interface, the diffracted field near a point on

† This work was supported in part by the National Science Foundation under grant number G19671.

an edge and the diffracted field near a vertex, depend only upon the incident field and the properties of the medium in the neighborhood of the point. V. A. Fock has called this conclusion the "principle of the local field". It has been embodied in the geometrical theory of diffraction by introducing diffraction coefficients in analogy with reflection and transmission coefficients. These coefficients depend only upon the local geometry and the incident field near the point of diffraction and consequently they can be determined from the solutions of *canonical problems*. The latter are simplified problems in which occur only the local properties of the field and medium near the point of diffraction.

It follows from these two conclusions that the solutions of complex problems can be synthesized from the solutions of simpler problems. This possibility makes more important and more necessary the solution of the various canonical problems. On the other hand it makes less important and less necessary the direct solution of complex problems. Of course some such problems must be solved directly to test the validity of the preceding conclusions.

Mathematically, the new developments are based upon asymptotic expansions, valid for short wavelengths, of solutions of boundary-value problems for Maxwell's equations. Thus they have employed various techniques of asymptotic analysis such as asymptotic evaluation of integrals by the saddle point and stationary phase methods, asymptotic solutions of ordinary differential equations with and without turning points using the W.K.B. method, as well as the methods of Langer, Cherry, Olver and others, asymptotic evaluation of series by the Watson transformation, asymptotic expansions of special functions, boundary-layer analysis, etc. In addition new methods of asymptotic analysis have been devised, especially for partial differential equation problems, in the course of the new developments.

2. DIFFRACTION PHENOMENA

It is convenient to describe diffraction theory in terms of the following phenomena which it covers:

Propagation
Reflection and refraction
Diffraction by an edge
Diffraction by a vertex
Diffraction by a curved surface
Focusing at a caustic
Formation of a shadow boundary
Confluence of a caustic and a shadow boundary
Propagation of a surface wave

An attempt will be made to explain these phenomena by describing some of the theoretical results which have been obtained concerning them. Only time harmonic fields of angular frequency ω and propagation constant $k = \omega/c$ will be considered.

3. PROPAGATION

Sommerfeld and Runge in 1911 considered a scalar field $u(x)$ satisfying the reduced wave equation

$$(\Delta + k^2 n^2)\, u = 0. \tag{1}$$

Here $n(x)$ denotes the index of refraction of the propagation medium. They considered solutions of the form

$$u = A(x, k)\, e^{ikS(x)}. \tag{2}$$

They showed that for k large the phase function $S(x)$ must satisfy the eiconal equation

$$(\nabla S)^2 = n^2. \tag{3}$$

From (3) it follows that the orthogonal trajectories of the surfaces S = constant are the rays of geometrical optics. They also showed that the amplitude A varies along a ray in accordance with the principle of energy conservation in a tube of rays. Thus they showed how certain aspects of geometrical optics follow from (1) and how geometrical optics can be used to construct approximate solutions of (1).

R. K. Luneburg in 1944 applied similar considerations to the electromagnetic field. Both E and H were expressed in the form (2) with A denoting a vector, and (1) was replaced by Maxwell's equations. Luneburg represented A as an asymptotic series in k^{-1} in the form

$$A(x, k) \sim \sum_{m=0}^{\infty} A_m(x)\, (ik)^{-m}. \tag{4}$$

He showed that in isotropic media S still satisfied (3) and that E_0 and H_0, the leading coefficients in the amplitudes of E and H, were perpendicular to each other and to the rays. Thus the leading term in the field behaves locally like a plane wave. Furthermore the magnitudes of the coefficients E_0 and H_0 satisfied the energy conservation principle in a tube of rays. He also showed how the directions of E_0 and H_0 varied along a ray. In addition he obtained corresponding results for non-isotropic symmetric media. Some of the same results were obtained independently by F. G. Friedlander.

M. Kline pursued Luneburg's analysis and obtained a recursive system of ordinary differential equations along the rays for the higher coefficients E_m and H_m. Alternative derivations of some of Luneburg's results were given by H. Bremmer and E. T. Copson. Corresponding results for general hyperbolic second order partial differential equations were obtained by M. Kline, and for general hyperbolic systems of first-order equations by R. Courant and P. D. Lax, R. M. Lewis, and D. Ludwig.

4. REFLECTION AND REFRACTION

Luneburg considered the rays associated with the fields reflected and transmitted by a discontinuity in the medium. He showed that these rays satisfy the laws of reflection and refraction. He also showed that the reflected and transmitted fields E_0 and H_0 are related to the incident field by the Fresnel reflection and transmission coefficients. Thus the reflection and transmission process for the leading term is the same as for a local plane wave incident upon a plane interface. Independently of this work, on physical grounds, R. Spencer, H. Barker and R. S. Riblet and H. Primakoff and J. B. Keller used geometrical optics to determine fields reflected from curved surfaces. Then H. B. Keller and J. B. Keller used it to determine transmitted fields. Previously P. Debye had shown that the exact solution of the problem of plane wave scattering from a cylinder yielded the geometrical optics result for the reflected field, in the short wavelength limit. The same result had been obtained by B. van der Pol and H. Bremmer for scattering of a spherical wave from a sphere. Recently many other exact solutions have been shown to lead to the geometrical optics result for short wavelengths. In this symposium the papers of S. I. Rubinow, R. D. Kodis and C. W. Helstrom contain such results.

J. B. Keller, R. M. Lewis and B. D. Seckler employed expressions of the form (2) with A given by (4) to solve a large number of problems of reflection from objects of simple shapes. In some cases all the coefficients in the expansion were found explicitly. C. Schensted also used such expansions to find two terms in the field reflected from bodies of revolution with a plane wave incident along the symmetry axis. In this way he found that in the electromagnetic case, the geometrical optics field is the exact field for reflection by a paraboloid of revolution.

C. Chester and J. B. Keller considered reflection and transmission by a discontinuity in any derivative of the index of refraction. They found that the reflection coefficient is proportional to k^{-J} if the Jth derivative of $n(x)$ is the first discontinuous one. Their analysis, which is one-dimensional, is now being extended to the three-dimensional case.

5. DIFFRACTION BY AN EDGE

A. Sommerfeld's solution for plane wave diffraction by a half-plane or wedge at normal incidence contains a diffracted cylindrical wave emerging from the edge of the half-plane or wedge. For oblique incidence the emerging wave is conical. The corresponding diffracted rays make the same angle with the edge as do the incident rays, which is an instance of the law of edge diffraction. Such diffracted rays also result from A. Rubinowicz's treatment of diffraction by an aperture with a curved edge on the basis of the Kirchhoff theory. They are also contained in the subsequent analyses of this problem by N. G. van Kampen and J. B. Keller, R. M. Lewis and B. D. Secker. An appropriate modification of Fermat's principle also yields such rays. I have used them to analyze various diffraction problems involving edges, obtaining the appropriate diffraction coefficients from Sommerfeld's solution. The same results follow from Braunbek's modification of the Kirchhoff theory.

H. Levine obtained various results for slits, circular apertures and, together with T. T. Wu, for general convex apertures. His work is based upon asymptotic solution of the appropriate integral equations using various methods including that of Wiener and Hopf. This method was introduced into diffraction theory by V. A. Fock, J. Schwinger and E. T. Copson. It was used extensively by L. A. Wainstein and by A. Heins. Wainstein showed that the radiation from a semi-infinite parallel plate waveguide could be treated in terms of diffracted rays emerging from the two edges. R. F. Millar, T. B. Senior and P. Clemmow have also used the integral equation method to analyze edge diffraction. R. Buchal and J. B. Keller have applied boundary-layer methods to edge diffraction problems.

E. B. Hansen, in his paper at this symposium, has shown that the exact solutions for diffraction by an infinite strip and by a circular disc yield results which have been obtained by means of diffracted rays. Diffraction at the edge of a wedge with impedance boundary conditions on its surfaces, has been studied by T. B. Senior, and S. N. Karp and F. C. Karal. Their results yield new diffraction coefficients for such edges. S. N. Karp and J. B. Keller also obtained the diffraction coefficient for diffraction by a hard screen of a wave incident parallel to the screen.

6. DIFFRACTION BY A VERTEX

When a wave is diffracted by a cone, a spherical diffracted wave emerges from the vertex. The corresponding diffracted rays thus leave the tip in all directions. The field on these rays, which determines the diffraction coefficient, has been studied for a circular cone by C. Schensted, K. M. Siegel *et al.* and very extensively by L. B. Felsen. The only other cone for which the solution is known is the elliptic cone, and its degenerate case of the plane angular sector, treated by L. Kraus and L. Levine. However, they have not yet evaluated the diffraction coefficient.

7. DIFFRACTION BY A CURVED SURFACE

G. N. Watson, in attempting to clarify the work of Poincare and of Nicholson, showed in 1917 that diffracted waves travel along great circles on a sphere. This type of diffraction was extensively studied in the 1930's by B. van der Pol and H. Bremmer in connection with radio wave propagation around the earth. Later W. Franz rediscovered these diffracted waves on circular cylinders and called them "creeping waves". He and his collaborators analyzed them in detail for spheres and circular cylinders as well as for non-circular cylinders. F. G. Friedlander also studied the rays and wave fronts associated with such waves. J. B. Keller introduced a modified Fermat principle which yielded the diffracted rays. He also introduced the decay exponents and diffraction coefficients associated with them, which permitted the solutions to various problems to be constructed. This was done by B. R. Levy and J. B. Keller for diffraction by a number of smooth objects.

Exact solutions of several problems have confirmed the ray picture. Among these are S. O. Rice's solution for diffraction by a parabolic cylinder, B. R. Levy's and R. K. Ritt and N. D. Kazarinoff's solution for diffraction by an elliptic cylinder, and Levy and Keller's and Ritt and Kazarinoff's solution for diffraction by an ellipsoid of revolution.

Derivations of the ray results for non-circular cylinders were partly given by F. G. Friedlander and J. B. Keller in terms of a general type of asymptotic solution. T. T. Wu and S. Seshadri gave a more complete derivation using boundary-layer methods. N. A. Logan and K. S. Yee gave a derivation employing a uniform expansion based upon the work of V. A. Fock described in the next section. In their paper at this symposium V. A. Fock and L. A. Wainstein will present a very similar result.

8. CONFLUENCE OF A CAUSTIC AND A SHADOW BOUNDARY

Time does not permit me to discuss the special problems related to caustics and shadow boundaries. However, I would like to mention the results on the especially difficult problem of the analysis of the field where a caustic and a shadow boundary come together. This occurs at the line of tangency of a shadow boundary with the object which casts the shadow. This problem was analyzed by V. A. Fock who introduced certain universal functions to describe the field in this vicinity for a smooth object. Similar results were obtained by C. L. Pekeris and by S. O. Rice in special problems, but they did not point out the general significance of the results. Fock's functions have been tabulated and utilized extensively in various problems. N. A. Logan has made a detailed study of the properties of these and related functions, including an historical account of the solution of certain problems in which they occur.

9. CONCLUSION

It is obvious that I have omitted a great deal of important work in this brief outline, and this was inevitable. However, I hope that what I have described will show that the features of ray propagation and of the local character of diffraction phenomena are characteristic of a great many recent results in diffraction theory. It is essential that we infer general principles from all our results, and organize the results in terms of them. Otherwise our subject will degenerate into a collection of special solutions. This tendency is becoming especially ominous as computing machines invade diffraction theory.

ON THE TRANSVERSE DIFFUSION OF SHORT WAVES DIFFRACTED BY A CONVEX CYLINDER

V. Fock and L. Wainstein

ABSTRACT

The two-dimensional problem of diffraction of short waves on a convex cylinder of arbitrary section is reduced to the solution of a parabolic equation expressed in ray co-ordinates; this equation takes into account the transverse diffusion but neglects the longitudinal diffusion. The radius of curvature of the cylinder section is supposed to be large (compared with the wavelength) and slowly varying. The boundary conditions considered are of impedance type (with a special dependence of the impedance parameter on the curvature radius); the most important case thereof is that of a perfectly reflecting cylinder. Under the assumptions stated above an asymptotic solution of the diffraction problem is obtained for an arbitrary position of the source and of the observation point.

1. INTRODUCTION

It is well known that the field near the surface of a conducting convex body (with boundary conditions of impedance type) is expressible in terms of universal functions first introduced by Fock (1949). The original expressions were given for the penumbra region, but the investigation of the diffraction on a sphere and on a circular and an elliptic cylinder (Malyughinetz, 1961; Wainstein, 1961a and b) has shown that the same universal functions are valid in the deep shadow. This suggests the possibility of finding generalized expressions describing the diffraction on other surfaces with variable curvature and valid for any distances from the surface.

Such generalizations can be based, on the one hand, on the conception of diffracted rays (Keller, 1956) and, on the other hand, on the conception of transverse diffusion (Malyughinetz, 1959); the latter constitutes a physical interpretation of Leontovich's parabolic equation.

2. RAY COORDINATES

In the present communication we consider a two-dimensional diffraction problem, namely the diffraction of a cylindrical wave on a convex cylinder of arbitrary section. We use "ray coordinates" introduced by Malyughinetz (1961) and defined according to Figs. 1 and 2. Let ξ, η be ray coordinates for the observation point P and ξ', η' those for the (two-dimensional) source P'. The straight lines PT and $P'T'$ are tangent to the cylinder. The position of the tangent points T and T' is characterized by means of the two arcs

$$\eta = \widehat{T_0T} \quad \text{and} \quad \eta' = \widehat{T_0T'}, \tag{1}$$

where the point T_0 on the contour (the origin of the arc reckoning) is arbitrary. The position of the points P and P' is characterized by the two distances

$$\xi - \eta = PT \quad \text{and} \quad \eta' - \xi' = P'T'. \tag{2}$$

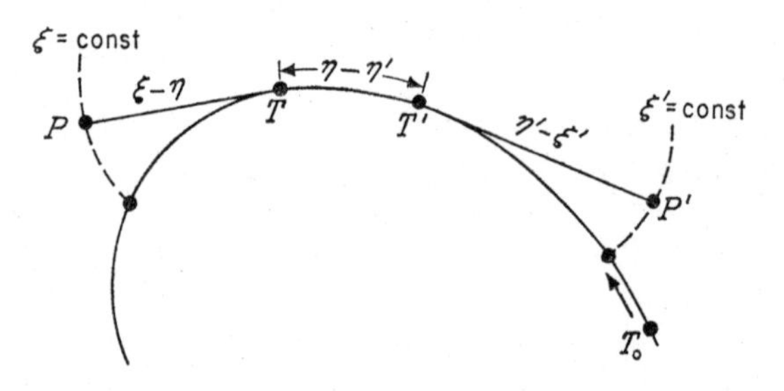

FIG. 1.

We choose the positive direction of the arcs so that $\xi > \xi'$; then if P is in the shadow region (Fig. 1) we have $\eta > \eta'$ and the total length $P'P$ (including two straight parts and a curved part $\eta - \eta'$ on the cylinder surface) will be equal to

$$(\eta' - \xi') + (\eta - \eta') + (\xi - \eta) = \xi - \xi'. \tag{3}$$

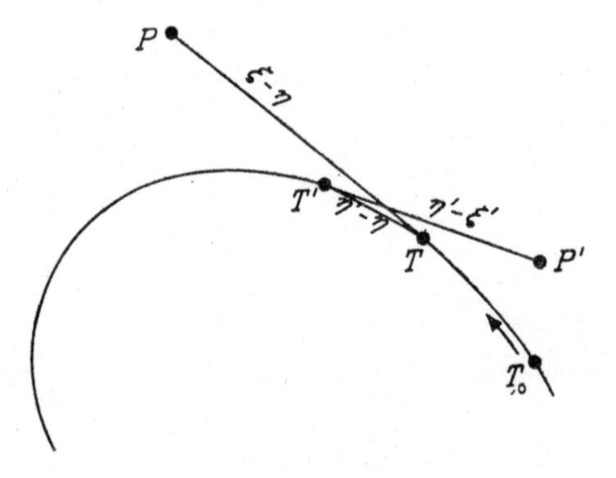

FIG. 2.

The expression for the square of the line-element in the plane of the normal section of the cylinder in terms of the ray coordinates is of the form

$$ds^2 = d\xi^2 + \frac{(\xi - \eta)^2}{\rho^2(\eta)}\, d\eta^2, \tag{4}$$

where $\rho(\eta)$ is the radius of curvature of the cylinder. The quantity

$$d\theta = \frac{d\eta}{\rho(\eta)} \tag{5}$$

is the increment of the angle θ between the tangent and a fixed direction. Since the equation of the cylinder is $\xi - \eta = 0$, the tangent line-element is $ds = d\xi$ and the line-element along the normal near the surface is

$$dn = (\xi - \eta)\, d\theta = \frac{\xi - \eta}{\rho(\eta)}\, d\eta. \tag{6}$$

Introducing the angle θ defined above we can form the expressions

$$\left.\begin{aligned} X &= \xi \cos\theta + \int_0^\theta \eta \sin\theta\, d\theta \\ Y &= \xi \sin\theta - \int_0^\theta \eta \cos\theta\, d\theta \end{aligned}\right\} \tag{7}$$

that can be interpreted as rectangular coordinates in the plane of the cylinder section.

The ray coordinates defined above can also be used in that part of the penumbra region where $\eta < \eta'$ (see Fig. 2).

3. DIFFERENTIAL EQUATIONS

We introduce the Green function Γ in the many-sheeted Riemann plane (Malyughinetz, 1961; Wainstein, 1961a and b). This is a function of two pairs of coordinates: the coordinates ξ, η of the observation point and the coordinates ξ', η' of the source. The Green function Γ must satisfy the time-independent wave equation

$$\Delta\Gamma + k^2\Gamma = 0 \tag{8}$$

with the boundary condition

$$\frac{\partial\Gamma}{\partial n} + ikg\Gamma = 0 \tag{9}$$

on the cylinder surface, and must fulfil the radiation condition at infinity. We take the incident cylindrical wave in the form

$$\Gamma^0 = i\pi H_0^{(1)}(kr), \tag{10}$$

r being the distance between P and P'. Then the singularity of Γ must be such that the difference $\Gamma - \Gamma^0$ as well as its derivatives is everywhere finite and continuous.

For a cylinder of infinite perimeter the function Γ reduces to the ordinary Green function; in the general case the latter is expressible in terms of Γ.

It is easily seen from (4) that the wave equation in ray coordinates assumes the form

$$\frac{1}{\xi-\eta}\frac{\partial}{\partial\xi}\left[(\xi-\eta)\frac{\partial\Gamma}{\partial\xi}\right]+\frac{\rho(\eta)}{\xi-\eta}\frac{\partial}{\partial\eta}\left[\frac{\rho(\eta)}{\xi-\eta}\frac{\partial\Gamma}{\partial\eta}\right]+k^2\Gamma=0. \quad (11)$$

It follows that the substitution

$$\Gamma=e^{ik(\xi-\xi')}\,W \quad (12)$$

reduces the differential equation (11) to the form

$$\left[\frac{1}{\xi-\eta}\frac{\partial}{\partial\xi}(\xi-\eta)\frac{\partial}{\partial\xi}+D^2+ikR\right]W=0, \quad (13)$$

where D and R denote the operators

$$D=\frac{\rho(\eta)}{\xi-\eta}\frac{\partial}{\partial\eta}, \qquad R=2\frac{\partial}{\partial\xi}+\frac{1}{\xi-\eta}. \quad (14)$$

The operator D (as well as D_1, D_2, and D_3 introduced later) is connected with transverse diffusion and may be called "diffusion operator". The operator R (as well as R_γ, R_β, and R_α, see later) is connected with the law of decrease of the amplitude of a cylindrical wave and may be called "ray amplitude operator" or, simpler, "ray operator". The second order differential operator in the first term of (13) is connected with longitudinal diffusion.

If both diffusion operators are neglected, equation (9) assumes the form

$$RW=0 \quad (15)$$

and has a solution

$$W=\frac{V_0(\eta)}{\sqrt{(\xi-\eta)}} \quad (16)$$

showing the law of decrease of the amplitude in a region where diffusion is absent (region GD, see Fig. 3 and Section 6 below).

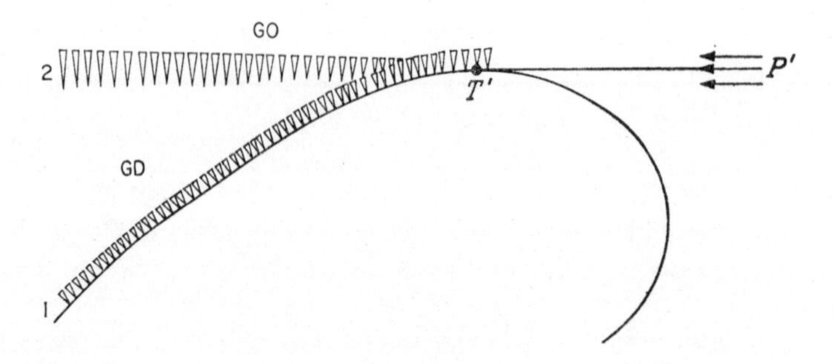

FIG. 3.

In order to estimate the relative importance of the two diffusion operators we introduce the dimensionless quantity

$$M(\eta) = \left[\frac{k\rho(\eta)}{2}\right]^{1/3} \tag{17}$$

and make a transformation from ξ, η to new independent variables

$$z = \frac{k}{2}\int_0^\eta \frac{d\bar{\eta}}{M^2(\bar{\eta})}, \qquad \gamma = \frac{k(\xi-\eta)}{2M^2(\eta)}. \tag{18}$$

Taking as the new dependent variable the function W_1 defined by

$$W = \frac{1}{\sqrt{[M(\eta)]}}\, W_1(z, \gamma) \tag{19}$$

we obtain for W_1 the equation

$$\left[L^2 + D_1^2 + \left(\chi + \frac{1}{\gamma}\right) D_1 + 2iR_\gamma\right] W_1 = 0, \tag{20}$$

where L, D_1 and R_γ are the operators

$$L = \frac{1}{M(\eta)}\,\gamma\,\frac{\partial}{\partial\gamma}, \qquad D_1 = \frac{\partial}{\partial z} - (1 + 2\chi\gamma)\frac{\partial}{\partial\gamma} - \frac{\chi}{2} \tag{21}$$

$$R_\gamma = 2\gamma^2\frac{\partial}{\partial\gamma} + \gamma \tag{22}$$

and χ denotes the function

$$\chi(z) = \frac{d \ln M(\eta)}{dz} = \frac{2M(\eta)}{k}\frac{dM(\eta)}{d\eta}. \tag{23}$$

The transformation from (13) to (20) is most easily performed with the help of the relations

$$\left.\begin{aligned} \frac{1}{\xi-\eta}\frac{\partial}{\partial\xi}(\xi-\eta)\frac{\partial}{\partial\xi} &= \frac{k^2}{4M^2\gamma^2}L^2 \\ \sqrt{M}\,.\,D^2\frac{1}{\sqrt{M}} &= \frac{k^2}{4M^2\gamma^2}\left[D_1^2 + \left(\chi + \frac{1}{\gamma}\right) D_1\right] \\ \sqrt{M}\,.\,kR\frac{1}{\sqrt{M}} &= \frac{k^2}{4M^2\gamma^2}2R_\gamma \end{aligned}\right\} \tag{24}$$

connecting the respective operators. The operator L corresponds to longitudinal and D_1 to transverse diffusion and R_γ is the new ray operator.

Now, if the radius of curvature $\rho(\eta)$ along the arc $T'T$ is large compared with the wavelength, the quantity M defined by (17) can be considered as large. Since the operator L involves M in the denominator, the L^2 term in (20) is small of the order $1/M^2$ compared with the other terms. Neglecting it (and thus neglecting longitudinal diffusion) we obtain the parabolic equation

$$\left[D_1^2 + \left(\chi + \frac{1}{\gamma}\right) D_1 + 2iR_\gamma\right] W_1 = 0 \tag{25}$$

that corresponds to the equation

$$(D^2 + ikR)\, W = 0 \tag{26}$$

in the original variables ξ, η.

A further transformation of variables is suitable for some purposes, especially for the study of the solution near the surface of the cylinder. We put

$$W_1 = \frac{1}{\sqrt{(1+\chi\gamma)}}\, W_2(z, \beta), \qquad \beta = \frac{\gamma}{1+\chi\gamma} \tag{27}$$

and introduce the notations

$$\left.\begin{aligned} D_2 &= \frac{\partial}{\partial z} - \frac{\partial}{\partial \beta} + 2\kappa R_\beta, \\ R_\beta &= 2\beta^2 \frac{\partial}{\partial \beta} + \beta \end{aligned}\right\} \tag{28}$$

where

$$\kappa = \kappa(z) = \frac{1}{4}\left(\chi^2 - \frac{d\chi}{dz}\right) = -\frac{M^3(\eta)}{k^2}\frac{d^2 M(\eta)}{d\eta^2}. \tag{29}$$

Then we have

$$\left.\begin{aligned} (1+\chi\gamma)^{1/2} D_1 (1+\chi\gamma)^{-1/2} &= D_2 \\ (1+\chi\gamma)^{1/2} R_\gamma (1+\chi\gamma)^{-1/2} &= R_\beta \end{aligned}\right\} \tag{30}$$

and eqn (25) transforms into

$$\left(D_2^2 + \frac{1}{\beta} D_2 + 2iR_\beta\right) W_2 = 0. \tag{31}$$

The boundary of the cylinder is given by the relation $\xi = \eta$ in the original coordinates and by the relations $\gamma = 0$ and $\beta = 0$ in the transformed coordinates.

By means of the substitution

$$\eta_2 = \int_0^\eta \frac{\xi - \bar{\eta}}{\rho(\bar{\eta})}\, d\bar{\eta} \tag{32}$$

and a suitable change of the dependent variable it is also possible to reduce the parabolic equation (26) to the simple form

$$\frac{\partial^2 F}{\partial \eta_2^2} + 2ik \frac{\partial F}{\partial \xi} = 0, \tag{33}$$

which is useful in some respects, e.g. for the investigation of the phase of the incident wave and of the Fresnel diffraction. But it is difficult to apply eqn (33) to the boundary value problem, since the equation of the boundary expressed in the variables ξ, η_2 is in general complicated.

4. CIRCULAR CYLINDER

As a preliminary stage we consider the solution of the equations stated above for the case of a circular cylinder. The quantity M is then a constant so that $\chi = 0$ and $\kappa = 0$. The transformation (27) becomes an identity and we have $\gamma = \beta$. Equation (25) or (31) takes the form

$$\left[\left(\frac{\partial}{\partial z} - \frac{\partial}{\partial \beta}\right)^2 + \frac{1}{\beta}\left(\frac{\partial}{\partial z} - \frac{\partial}{\partial \beta}\right) + 2iR_\beta\right] W_2 = 0. \tag{34}$$

By means of the substitution

$$W_2 = e^{-i2/3\beta^3} W_0(x, y), \qquad x = z + \beta, \qquad y = \beta^2 \tag{35}$$

eqn (34) reduces to the well-known form

$$\frac{\partial^2 W_0}{\partial y^2} + i \frac{\partial W_0}{\partial x} + y W_0 = 0. \tag{36}$$

The boundary condition (9) becomes

$$\frac{\partial \Gamma}{\partial y} + igM\Gamma = 0 \qquad \text{for } y = 0 \tag{37}$$

since we have, using (6),

$$k dn = M dy. \tag{38}$$

The same condition (37) holds for the function W in (12) and also for W_2 and W_0. Putting

$$q = igM \tag{39}$$

we may write this condition in the form

$$\frac{\partial W}{\partial y} + qW = 0 \qquad \text{for } y = 0. \tag{40}$$

Near the singularity of the Green function the parabolic equations do not hold and the requirement that $\Gamma - \Gamma^0$ should be finite at the singularity is to be replaced by another one. We must impose the condition that on the boundary of geometrical shadow (far from the singularity) the incident wave should be the same as that given by Γ^0.

The solution satisfying these conditions can be expressed in terms of the attenuation factor $\Psi(x, y, y', q)$ defined (for $x > 0$ and $y' > y$) by the integral

$$\Psi(x, y, y', q) = \frac{1}{2\pi i} \int\limits_C e^{itx}\, w(t - y') \left[v(t - y) - \right.$$

$$\left. - \frac{v'(t) - qv(t)}{w'(t) - qw(t)}\, w\,(t - y) \right] dt, \tag{41}$$

where $w(t)$ is the Airy function satisfying the differential equation

$$w''(t) = tw(t) \tag{42}$$

and having for $y \to \infty$ the asymptotic expression

$$w(t - y) = \frac{1}{y^{1/4}} \exp\left\{ i\left(\frac{2}{3}\, y^{3/2} - t\sqrt{y} + \frac{t^2}{4\sqrt{y}} + \frac{\pi}{4} \right) \right\}. \tag{43}$$

The Airy function $v(t)$ satisfies the same equation and is the imaginary part of $w(t)$ for real t. The contour C in (41) encloses the first quadrant of the complex t-plane, where the poles of the integrand are situated, and runs in the positive direction.

The solution for a circular cylinder can be written as

$$W = \frac{2\pi i}{M}\, e^{-i2/3(y^{3/2} + y'^{3/2})}\, \Psi(x - x', y, y', q). \tag{44}$$

The variables y and y' and their square roots $\sqrt{y}$ and $\sqrt{y'}$ are defined in accordance with (54) and (55) below.

Near the boundary of geometrical shadow the Ψ-function can be described as a Fresnel integral superimposed on a slowly varying function representing a weak background (Fock, 1950). Putting in accordance with (18) and (35)

$$z = x - \sqrt{y}, \qquad z' = x' + \sqrt{y'}, \qquad z - z' = \zeta \tag{45}$$

and introducing the positive quantity μ defined by

$$\mu^2 = \frac{\sqrt{yy'}}{\sqrt{y} + \sqrt{y'}} \tag{46}$$

we will have for the Fresnel integral term

$$\Psi = \frac{1}{2\pi i}\, \frac{1}{\sqrt[4]{yy'}}\, e^{i2/3(y^{3/2} + y'^{3/2})}\, \mu F(\mu\zeta) \tag{47}$$

and consequently

$$W = \frac{1}{M\sqrt[4]{yy'}}\, \mu F(\mu\zeta), \tag{48}$$

where

$$F(\tau) = e^{-i\tau^2} \int\limits_\tau^\infty e^{it^2}\, dt. \tag{49}$$

If μ is large, the terms neglected in (47) and (48) (the background) are small of the order $1/\mu$ with respect to the main term written out. (See also Belkina, 1957.)

The expression for the incident wave follows from (48) for large negative values of $\mu\zeta$ and is approximately equal to

$$W^0 = \frac{\mu}{M\sqrt[4]{yy'}} \sqrt{\pi}\, e^{i\pi/4}\, e^{-i\mu^2\zeta^2}. \tag{50}$$

It is easy to verify that for small values of ζ the quantity $\mu^2\zeta^2$ is nearly equal to the phase difference $k(\xi - \xi' - r)$. Thus the expression (44) satisfies the differential equation, the boundary condition and the condition on the shadow boundary. It is therefore the required solution.

5. CYLINDER WITH A VARIABLE RADIUS OF CURVATURE AND WITH $\kappa = 0$

Equation (34) holds rigorously not only in the case $\chi = 0$, but also in the case $\chi \neq 0$, $\kappa = 0$. The substitution resulting from (19) and (27) may be written as

$$W = \frac{1}{\sqrt{[N(\xi, \eta)]}} W_2, \tag{51}$$

where

$$N(\xi, \eta) = M(\eta) + (\xi - \eta) \frac{dM(\eta)}{d\eta} \tag{52}$$

is a linear function of ξ. If $\kappa = 0$, we have $N(\xi, \eta) = M(\xi)$.

We use the variables z and z' defined by (18) so that

$$z - z' = \zeta = \frac{k}{2} \int_{\eta'}^{\eta} \frac{d\bar{\eta}}{M^2(\bar{\eta})} \tag{53}$$

and also the variables β and β' where

$$\left.\begin{aligned} \beta &= \frac{k(\xi - \eta)}{2M(\eta)\, N(\xi, \eta)} = \frac{\gamma}{1 + \chi(z)\,\gamma}, & \gamma &= \frac{k(\xi - \eta)}{2M^2(\eta)} \\ \beta' &= \frac{k(\xi' - \eta')}{2M(\eta')\, N(\xi', \eta')} = \frac{\gamma'}{1 + \chi(z')\,\gamma'}, & \gamma' &= \frac{k(\xi' - \eta')}{2M^2(\eta')} \end{aligned}\right\} \tag{54}$$

Since we are considering the case when $\xi - \eta > 0$, $\xi' - \eta' < 0$, we will have $\gamma > 0$, $\gamma' < 0$ and for moderate values of γ and γ' also $\beta > 0$, $\beta' < 0$. We take the square roots $\sqrt{y}$ and $\sqrt{y'}$ in (44) always positive; therefore, we have

$$\sqrt{y} = \beta, \qquad \sqrt{y'} = -\beta'. \tag{55}$$

Introducing these values in (44) and putting for brevity

$$N = N(\xi, \eta), \qquad N' = N(\xi', \eta') \tag{56}$$

we can write as a generalization of the solution (44) for the case $\chi \neq 0$, $\kappa = 0$

$$W = \frac{2\pi i}{\sqrt{NN'}}\, e^{-i2/3(\beta^3-\beta'^3)}\, \Psi(\zeta + \beta - \beta', \beta^2, \beta'^2, q). \tag{57}$$

In the region where $\zeta = z - z'$ is small but β and $|\beta'| = -\beta'$ are large the function W can be approximately expressed in terms of Fresnel integrals, as in the case of a circular cylinder. We will have

$$W = \frac{1}{\sqrt{NN'}}\frac{1}{\sqrt[4]{yy'}}\,\mu F(\mu\zeta), \tag{58}$$

where μ is defined by

$$\frac{1}{\mu^2} = \frac{1}{\sqrt{y}} + \frac{1}{\sqrt{y'}} = \frac{1}{\beta} - \frac{1}{\beta'} = \frac{1}{\gamma} - \frac{1}{\gamma'} + \chi(z) - \chi(z'). \tag{59}$$

Since we are considering small values of ζ we can neglect the difference $\chi(z) - \chi(z')$ and also the difference between $M^2(\eta)$ and $M^2(\eta')$ in the definition (54) of γ and γ'. We obtain approximately

$$\frac{1}{\mu^2} = \frac{1}{\gamma} - \frac{1}{\gamma'} = \frac{2}{k}\, M(\eta)\, M(\eta')\, \frac{\xi - \xi'}{(\xi - \eta)(\eta' - \xi')}. \tag{60}$$

With this value of μ and the definition (53) of ζ it is easy to see that the square of the argument in the Fresnel integral is equal to

$$\mu^2\zeta^2 = ks = k(\xi - \xi' - z). \tag{61}$$

This gives the correct value of the phase of the incident wave. On the other hand, the relation

$$NN'\sqrt{yy'} = M(\eta)\, M(\eta')\,(-\gamma\gamma') = \frac{k^2}{4}\,\frac{(\xi - \eta)(\eta' - \xi')}{M(\eta)\, M(\eta')} \tag{62}$$

together with (60) shows that formula (58) may be written as

$$W = \sqrt{\left(\frac{2}{k(\xi - \xi')}\right)}\, F(\mu\zeta). \tag{63}$$

Using the expression for the Fresnel integral valid for large negative values of the argument we obtain for the incident wave

$$W^0 = \sqrt{\left(\frac{2\pi}{k(\xi - \xi')}\right)}\, e^{-iks+i\pi/4} \tag{64}$$

as it ought to be. It is possible to find for s an explicit expression such that W^0 is an exact solution of the parabolic equation (26) for the general case (not only for $\kappa = 0$).

We thus have shown that our solution (57) correctly represents the incident wave in a certain domain near the boundary between light and shadow.

The following remarks are to be made at this place.

Strictly, expression (57) satisfies the boundary condition (40) with constant q. Since q is related to g and M by means of (40), the constancy of q is a rather artificial requirement. For a perfect conductor, however, we have for the value of g in the original boundary condition either $g = \infty$ or $g = 0$ and these values give $q = \infty$ and $q = 0$ respectively, so that the solution (57) is then applicable.

Secondly, it is possible that $N(\xi, \eta)$ vanishes for finite values of ξ, so that β becomes infinite for finite distances from the cylinder. The corresponding curve in the ξ, η plane may be called diffractional caustic. There is no physical singularity near the diffractional caustic and if one uses the asymptotic expression (43) for the Airy function, one can verify that W passes smoothly across the diffractional caustic. Beyond the caustic the Airy function $w = w_1 = u + iv$ is to be replaced by $w_2 = u - iv$.

6. THE GENERAL CASE OF A CONVEX CYLINDER

The solution of the parabolic equation obtained in the preceding section, is a rigorous solution only in the case $\kappa \equiv 0$, but the formulae have been written so as to retain their meaning in the general case $\kappa \neq 0$ as well. It is therefore plausible that these formulae yield an approximate solution also if κ is different from zero but small.

Before investigating this fundamental question, we shall examine the physical aspect of the problem (Malyughinetz, 1959). Suppose a plane or a cylindrical wave from a remote source is falling on a convex cylinder. Then in the illuminated region the laws of geometrical optics apply. But there are two other regions (or zones) where the transverse diffusion plays a dominant part. One of them (zone 1) is close to the shadowed part of the cylinder surface and the other (zone 2) lies near the geometrical boundary between light and shadow, where the Fresnel diffraction takes place. Between the zones 1 and 2 a region GD is situated, where the notion of diffraction rays applies. In the regions GO and GD transverse diffusion is inessential.

If we take the solution in the form (57) then in the first zone β is small or finite. In the second zone β is large, $\zeta = z - z'$ is small and $\mu\zeta$ is small or finite.

Let us suppose κ to be very small and let us see whether the transition from the full parabolic equation (31) to the abridged form (34) is legitimate in the first zone. This transition amounts to the neglect of the term $2\kappa R_\beta$ in the operator

$$D_2 = \frac{\partial}{\partial z} - \frac{\partial}{\partial \beta} + 2\kappa R_\beta.$$

Now, if κ and β are small the term $2\ \kappa R_\beta W_2$ is obviously small compared with other terms in $D_2 W_2$; this term can be neglected even in the expression $1/(\beta)\ D_2 W_2$ for $\beta \to 0$. Because of the smallness of κ the term $2\ \kappa R_\beta W_2$ is also small for finite values of β. Thus eqn (34) is a good substitute for (31) everywhere in the first zone, including the immediate vicinity of the cylinder surface. (It is to be noted that this is not the case for the equation obtained from (25) by neglecting the terms with χ.)

Next, let us consider the GD region and the second zone where β is large. In the GD region the term $R_2 W_2$ is of the same order as $\partial W_2/\partial z$; this follows from the fact that W_2 is approximately proportional to $1/\sqrt{\beta}$ (the factor appearing in (43)). Thus in the GD region the term $2\kappa R_\beta$ is unimportant and can be neglected.

In the second zone, however, where the Fresnel diffraction takes place, the term $R_\beta W_2$ is of the order $\mu(\partial W_2/\partial z)$, so that $2\ \kappa R_\beta W_2$ can be neglected in $D_2 W_2$ only if $4\kappa\mu \ll 1$ which is the case only in a part of the second zone. Therefore, the second zone must be investigated separately.

The simplest way is to go back to eqn (25) and to neglect therein the terms in χ. Then eqn (25) takes the form (34) with β replaced by γ. The solution can be written thus

$$W = \frac{2\pi i}{\sqrt{[M(\eta)\ M(\eta')]}}\ \mathrm{e}^{-i2/3(\gamma^3 - \gamma'^3)}\ \Psi(\zeta + \gamma - \gamma', \gamma^2, \gamma'^2, q). \tag{65}$$

The asymptotic form of this solution in the Fresnel zone is according to (58)

$$W = \frac{1}{\sqrt{[M(\eta)\ M(\eta')]}}\ \frac{1}{\sqrt{(-\gamma\gamma')}}\ \mu F(\mu\zeta) \tag{66}$$

with μ defined in terms of γ and γ' according to (60). Using (66) one can verify that the terms in χ in the operator D_1 are small if χ is small so that the approximation (65) is self-consistent.

It is, however, possible to improve the approximation in the second zone by using in place of γ a slightly modified independent variable, namely

$$\alpha = \gamma\,\frac{1 + a\eta}{1 + a\xi} = \frac{\gamma}{1 + \chi^0\gamma}, \tag{67}$$

where a is a constant and

$$\chi^0 = \chi^0(z) = \frac{2M^2(\eta)}{k}\ \frac{a}{1 + a\eta} = \frac{\mathrm{d}}{\mathrm{d}z} \ln(1 + a\eta). \tag{68}$$

This transformation corresponds to the change of the dependent variables

$$W_1 = \frac{1}{\sqrt{(1 + \chi^0\gamma)}}\ W_3(z, \alpha). \tag{69}$$

The transformed operators are

$$(1+\chi^0\gamma)^{1/2} D_1(1+\chi^0\gamma)^{-1/2} = D_3 \tag{70}$$

$$(1+\chi^0\gamma)^{1/2} R_\gamma(1+\chi^0\gamma)^{-1/2} = R_\alpha, \tag{71}$$

where

$$D_3 = \frac{\partial}{\partial z} - \frac{\partial}{\partial \alpha} + (\chi^0 - \chi)\left(2\alpha\frac{\partial}{\partial \alpha} + \frac{1}{2}\right) \tag{72}$$

and

$$R_\alpha = 2\alpha^2\frac{\partial}{\partial \alpha} + \alpha. \tag{73}$$

We see that the transformation (67) does not introduce "long-range" terms of the type $2\kappa R_\alpha$ in the operator D_3 and at the same time replaces the coefficient in D_1 by the coefficient $\chi^0 - \chi$. The constant in (67) can be adjusted so as to make this coefficient vanish at a given point η_0 (between η and η', say). Comparing the expressions (23) and (68) for χ and χ^0 we see that the necessary condition is

$$\frac{a}{1+a\eta} = \frac{1}{M(\eta)}\frac{\mathrm{d}M(\eta)}{\mathrm{d}\eta} \qquad \text{for } \eta = \eta_0. \tag{74}$$

The solution takes the form

$$W = 2\pi i\sqrt{\left(\frac{1+a\eta}{(1+a\xi)\,M(\eta)}\,\frac{1+a\eta'}{(1+a\xi')\,M(\eta')}\right)}\,W_3, \tag{75}$$

where

$$W_3 = \mathrm{e}^{-i2/3(\alpha^3-\alpha'^3)}\,\Psi(\zeta+\alpha-\alpha', \alpha^2, \alpha'^2, q). \tag{76}$$

We observe that if $\kappa = 0$ and $M(\eta)$ is a linear function of η then (74) holds everywhere (and not only for $\eta = \eta_0$) and the transformation (67) is the same as (27). Thus in the case $\kappa = 0$ expressions (75) and (76) give the exact solution.

If the terms in $\partial/\partial\alpha$ and in $\chi^0 - \chi$ are neglected in the operator D_3 and if the term corresponding to $1/(\beta)\, D_2$ in (31) is also neglected, we obtain for W_3 an equation

$$\frac{\partial^2 W_3}{\partial z^2} + 2iR_\alpha W_3 = 0 \tag{77}$$

which is satisfied by the Fresnel integral.

The remarks at the end of Section 5 concerning infinite values and change of sign of the variable β are applicable to the variable α as well.

7. TRANSFORMATION GROUP OF THE PARABOLIC EQUATION

The results obtained can be considered in conjunction with the transformation group of the parabolic equation. We have according to (14) and (26)

$$(D^2 + ikR)\,W = 0, \tag{78}$$

where

$$D = \frac{\rho(\eta)}{\xi - \eta} \frac{\partial}{\partial \eta}, \qquad R = 2\frac{\partial}{\partial \xi} + \frac{1}{\xi - \eta}. \tag{79}$$

We make the transformation from ξ, η, W to new variables ξ^*, η^*, W^*, defined by

$$\xi^* = \frac{\xi}{1 + a\xi}, \qquad \eta^* = \frac{\eta}{1 + a\eta} \tag{80}$$

$$W(\xi, \eta) = \frac{1}{\sqrt{(1 + a\xi)}} W^*(\xi^*, \eta^*) \tag{81}$$

and we use the notation

$$\rho^*(\eta^*) = \frac{\rho(\eta)}{(1 + a\eta)^3}. \tag{82}$$

Then equation (78) transforms into

$$(D^{*2} + ikR^*) W^* = 0, \tag{83}$$

where

$$D^* = \frac{\rho^*(\eta^*)}{\xi^* - \eta^*} \frac{\partial}{\partial \eta^*}, \qquad R^* = 2\frac{\partial}{\partial \xi^*} + \frac{1}{\xi^* - \eta^*}. \tag{84}$$

Thus the transformed equation is of the same form as the original one with the only difference that the function $\rho(\eta)$ is replaced by $\rho^*(\eta^*)$. Therefore we have a family of cylinders depending on the parameter a and such that the solution of the diffraction problem for one of them yields the solution for the other members of the family. From this standpoint it is clear why the case $\kappa = 0$ considered in Section 5 is reducible to the case of a circular cylinder ($\rho =$ const.) considered in Section 4.

The variable z is not affected by the transformation because

$$M^*(\eta^*) = \frac{M(\eta)}{1 + a\eta} \tag{85}$$

and

$$\mathrm{d}z = \frac{k}{2} \frac{\mathrm{d}\eta}{M^2(\eta)} = \frac{k}{2} \frac{\mathrm{d}\eta^*}{M^{*2}(\eta^*)}. \tag{86}$$

The variable θ^*, however, is not identical with θ since we have

$$\eta^* \mathrm{d}\theta^* = \eta \mathrm{d}\theta. \tag{87}$$

We note that the quantity μ defined by

$$\frac{1}{\mu^2} = \frac{2}{k} M(\eta) M(\eta') \frac{\xi - \xi'}{(\xi - \eta)(\eta' - \xi')} \tag{88}$$

(see (60)) is invariant with respect to the transformation (80).

The transformation (80) with a suitable choice of the parameter a can also be used as a preliminary stage of the solution of the diffraction problem (before the introduction of the variables γ or β). The variable α used in Section 6 in order to improve the solution in the Fresnel zone is simply $\alpha = \gamma^*$.

8. CONCLUSION

In the course of investigating the diffraction of short waves by a convex cylinder with slowly varying curvature we were led to three expressions (65), (57), and (75)–(76) which will be referred to as the γ-, β-, and α-expressions. The γ-expression is a rather crude one, especially near the cylinder surface and gives a relative error of the order χ (or $k^{-1/3}$ when $k \to \infty$) where it is applicable. The β-expression is valid in the first zone (near the surface) in the *GD* region, and in a part of the second (Fresnel, see Fig. 3) zone. The α-expression is valid in the *GD* region and in the entire second zone. The approximation given by the β-expression is of the order κ or $k^{-2/3}$; for $\kappa=0$ the α- and the β-expressions coincide and give a rigorous solution of the parabolic equation. For a circular cylinder all three expressions coincide.

The results obtained show that the parabolic equation in ray coordinates permits one to find an asymptotic solution of diffraction problems in cases when a rigorous solution cannot be expressed in finite terms. Further possibilities of improving the solution are offered by the aplication of the transformation given in Section 7.

REFERENCES

BELKINA, M. G. and ASRILANT, P. A. (1957) *Numerical Results of the Theory of Radio-wave Diffraction around the Earth.* Soviet Radio, Moscow.

FOCK, V. A. (1949) The field of a vertical and a horizontal dipole raised above the earth's surface, *J. Exp. Theor. Phys.* (*ZETF*) **19,** No. 10, 916–929 (October).

FOCK, V. A. (1950) Fresnel diffraction by convex bodies, *Progr. of Phys. Sciences* (*Uspekhi*), **43,** No. 4, 587–599 (April).

KELLER, J. B. (1956) Diffraction by a convex cylinder, *I.R.E. Trans.* **AP-4,** No. 3, 312–321 (July).

MALYUGHINETZ, G. D. (1959) Development of ideas on diffraction phenomena, *Progr. of Phys. Sciences* (*Uspekhi*), **69,** No. 2, 321–334 (October).

MALYUGHINETZ, G. D. and WAINSTEIN, L. A. (1961) Transversal diffusion in the diffraction on impedance cylinder of large radius. Part I. Parabolic equation in ray coordinates. *Radioengineering a. Electronics* (*Radiotekhnika i Elecktronika*), **6,** No. 8, 1247–1258 (August).

WAINSTEIN, L. A. and FEDOROV, A. A. (1961a) Scattering of plane and cylindrical waves on elliptic cylinder and concept of diffracted rays, *Radioengineering a. Electronics* (*Radiotekhnika i Elektronika*), **6,** No. 1, 31–46 (January).

WAINSTEIN, L. A. and MALYUGHINETZ, G. D. (1961b) Transversal diffusion in the diffraction on impedance cylinder of large radius. Part II. Asymptotic laws of diffraction in polar coordinates, *Radioengineering a. Electronics* (*Radiotekhnika i Elektronika*), **6,** No. 9, 1489–1495 (September).

SCALAR DIFFRACTION BY A THIN OBLATE SPHEROID*

R. F. Goodrich,† N. D. Kazarinoff† and V. H. Weston‡

ABSTRACT

The pressure distribution u on the surface B of an oblate spheroid is found under the conditions: (1) B is illuminated by a harmonic point source of wavelength λ located on its axis; (2) if 2a and 2b are the axes of B, then $2\pi a/\lambda \gg 1$ and $[2\pi b^2/(\lambda a)]^{1/2} \ll 1$, namely, B is nearly a disc; (3) u satisfies either Dirichlet's or Neumann's condition on B. The distribution u is derived from the solution of Helmholtz's equation by using asymptotic theories of differential equations with turning points. It is interpreted in terms of traveling waves which are reflected at the rim of B. The reflection coefficients are determined.

1. INTRODUCTION

In this study we determine the surface distribution induced on an oblate spheroid B by an incident acoustic wave. We take the source to lie on the axis of symmetry and suppose either Dirichlet or Neumann boundary conditions to hold on B. The essential restriction on the problem for application of our method is that the spheroid be nearly a disc whose radius is large compared with the wavelength of the incident radiation.

The precise restrictions are that if (ξ, η, ϕ) are oblate spheroidal coordinates, c is half the inter-focal distance, $k = 2\pi/\lambda$ is the wavenumber, and if $\xi = \xi_0$ on B, then

$$kc \equiv \gamma \gg 1, \qquad \gamma^{1/2}\,\xi_0 \ll 1. \tag{1.1}$$

That is to say, the radius of our near-disc is large compared to one wavelength λ, and the radius of curvature at the rim is much smaller than λ.

Our analysis is based upon two separate theories. The first is the theory of the asymptotic solutions of the separated differential equations arising from (2.1) below under the conditions (1.1). This theory has been developed by McKelvey (1955) and Kazarinoff (1957). The second is an integral representation of the Green's function and a generalization of the Watson transform developed by Kazarinoff and Ritt (1959).

* The research in this paper was supported by the Air Force Cambridge Research Laboratories under Air Force Contract AF19(604)–6655.

† The University of Michigan, Radiation Laboratory, Ann Arbor, Michigan, U.S.A.

‡ Formerly with The University of Michigan Radiation Laboratory; now with the Conductron Corporation, Ann Arbor, Michigan, U.S.A.

2. REPRESENTATION OF THE FIELD

In oblate spheroidal coordinates, Helmholtz's equation becomes

$$[(\xi^2 + 1)\, u_\xi]_\xi + [(1 - \eta^2)\, u_\eta]_\eta + \gamma^2(\eta^2 + \xi^2)\, u \equiv (-L_\xi - L_\eta)\, u = 0. \quad (2.1)$$

We assume that B is "illuminated" by a point source at $(\Xi, 1)$ $(\Xi > \xi_0)$. We shall derive approximations for u on the surface of B (that is, for $\xi = \xi_0$) in the cases $u(\xi_0, \eta) = 0$ and $\partial u(\xi_0, \eta)/\partial \xi = 0$ (the Dirichlet and Neumann problems, respectively). The basis of our work is the contour integral representation for u (Kazarinoff and Ritt, 1959, pp. 280–283). It is

$$u(\xi_0, \eta, \Xi, 1) = \lim_{s \to 0^+} \frac{1}{2\pi i} \int_\Gamma G(\xi_0, \Xi, \nu)\, \tilde{G}(\eta, 1, -\nu)\, d\nu, \qquad (2.2)$$

where G and $\tilde{G}$ are the resolvent Green's functions for the operators L_ξ and L_η, respectively, and Γ is a straight line path in the ν-plane that is a distance proportional to s above the real axis. The integral represents u if the parameter γ in (2.1) is chosen to be

$$\gamma = c(k - is) \qquad (s > 0). \qquad (2.3)$$

If ϕ_2 is that solution (up to a constant multiple) of $(L_\xi - \nu)\, \phi = 0$ which is in $L_2[\xi_0, \infty)$ then

$$\frac{\partial}{\partial \xi} G(\xi_0, \Xi, \nu) = \frac{-\phi_2(\Xi, \nu)}{(\xi_0^2 + 1)\, \phi_2(\xi_0, \nu)} \quad \text{(in the Dirichlet problem)},$$

$$G(\xi_0, \Xi, \nu) = \frac{\phi_2(\Xi, \nu)}{(\xi_0^2 + 1)\, \phi_2'(\xi_0, \nu)} \quad \text{(in the Neumann problem)}. \qquad (2.4)$$

If $\psi_2(\eta, \nu)$ is that solution (up to a constant multiple) of $(L_\eta + \nu)\, \psi = 0$ which is regular at $\eta = -1$ and ψ_1 is defined by the condition

$$\psi_1(\eta) = \psi_2(-\eta), \qquad (2.5)$$

then

$$\tilde{G}(\eta, 1, -\nu) = \frac{\psi_2(\eta)\, \psi_1(1)}{-(1 - \eta^2)\, W(\psi_1, \psi_2, \eta)}, \qquad (2.6)$$

where $W(\psi_1, \psi_2)$ is the Wronskian of ψ_1 and ψ_2.

Our approximations for u in the case where $|\eta| \leq \epsilon$ (ϵ positive and sufficiently small) are based upon an evaluation of the early terms in the residue series for the integral in (2.2), the residue series associated with the poles of $G(\xi, \Xi, \nu)$. It is clear from (2.4) that the residue series is determined either by the zeros of $\phi_2(\xi_0, \nu)$ (considered as a function of ν) or of $\phi_2'(\xi_0, \nu)$ depending upon the boundary condition for u at ξ_0. The zeros of $\phi_2(\nu)$ and $\phi_2'(\nu)$ will be determined in Section 3 along with the residue contribution of $G(\xi_0, \Xi, \nu)$ at these zeros.

3. THE RADIAL EIGENVALUES

If we adopt the notation $\lambda = -i\gamma$ and $\sigma = -\lambda^{-1}\nu$, then the function

$$w_2 = (\xi^2 + 1)^{1/2} \phi_2 \tag{3.1}$$

is that solution of the differential equation

$$\frac{d^2 w}{d\xi^2} - \left[\frac{\lambda^2 \xi^2}{\xi^2 + 1} + \frac{\lambda\sigma}{(\xi^2 + 1)} + \frac{1}{(\xi^2 + 1)^2}\right] w = 0 \tag{3.2}$$

which is in $L_2[\xi_0, \infty)$. We seek the behavior of $w_2(\xi_0, \nu)$ when $|\xi_0| \ll |\gamma|^{-1/2}$ and $|\gamma| \gg 1$. Thus we study the asymptotic behavior of solutions of (3.2) in a neighborhood of the origin when $|\lambda| \gg 1$. The origin is a turning point of second order for the differential equation, and the asymptotic behavior of its solutions in a neighborhood of the origin may be determined by an application of the theory given by McKelvey (1955). An almost identical application has been carried out by Goodrich and Kazarinoff (1962b) in their investigation of scalar diffraction by a thin elliptic cylinder. We have therefore omitted the derivation of the results given in the remainder of this section.

Let

$$\left\{\begin{aligned} &\phi(\xi) = \xi[1 + \xi^2]^{-1/2} \\ &\zeta(\xi) = -2i\gamma[(1 + \xi^2)^{1/2} - 1],\ \zeta_0 = \zeta(\xi_0) = -i\gamma\xi_0^2[1 + O(\xi_0^2)], \\ &\Psi(\xi) = \{[(1 + \xi^2)^{1/2} - 1]\,(1 + \xi^2)\}^{1/4}\,\xi^{-1/2},\ \Psi(0) = 2^{-1/4}, \\ &\Theta(\xi) = -\frac{\sigma}{4}\ln\left[\frac{(\xi^2 + 1)^{1/2} + 1}{2}\right], \\ &\mu_0 = \cosh\Theta, \quad \text{and} \\ &\mu_1 = \phi^{-1}(\xi)\sinh\Theta. \end{aligned}\right. \tag{3.3}$$

Then if

$$W_{-1}(\xi) = W_{\sigma/4,\,1/4}(e^{\pi i}\xi), \tag{3.4}$$

where $W_{k,\,m}(z)$ is Whittaker's confluent hypergeometric function,

$$w_2(\xi, \nu) \sim \left[\mu_0 + \frac{i\mu_1}{\gamma}\frac{d}{d\xi}\right] \Psi\zeta^{-1/4}\, W_{-1}(\zeta) \qquad (|\gamma|) \to +\infty,\ |\arg\gamma| < \epsilon). \tag{3.5}$$

If $|\zeta|$ is large and if $-5\pi/2 + \epsilon \leq \arg\zeta \leq \pi/2 - \epsilon$,

$$W_{-1}(\xi) = (e^{\pi i}\zeta)^{\sigma/4}\, e^{\zeta/2}\,[1 + O(\zeta^{-1})]. \tag{3.6}$$

In our case, $\arg\zeta = -\pi/2 + \arg\gamma \sim -\pi/2$, so that we may use (3.6). The relationship

$$W_{-1}(\zeta) = \frac{e^{3\pi i/4}\,\Gamma(-1/2)}{\Gamma\left(\frac{1-\sigma}{4}\right)} M_{-\sigma/4,\,1/4}(\xi) + \frac{e^{\pi i/4}\,\Gamma(1/2)}{\Gamma\left(\frac{3-\sigma}{4}\right)} M_{-\sigma/4,\,-1/4}(\zeta) \tag{3.7}$$

serves to determine the behavior of W_{-1} when ζ is small.

The zeroes ν_n^D and ν_n^N of $\phi_2(\xi_0, \nu)$ and $d\phi_2(\xi_0, \nu)/d\xi$, respectively, are found by replacing the quantities appearing in the right-hand member of (3.5) by the final one or two terms of their power series in ζ_0 or ξ_0 and carrying out appropriate computations. Doing so, we have proved the following theorem.

THEOREM 1. *Under the hypotheses* (1.1), *that solutions of the differential equation* $(L_\xi - \nu)\,\phi = 0$ *(see* (2.1) *which lies in* $L_2\,[\xi_0, \infty)$ *if* $i\gamma < 0$, *has, considered as a function of* ν, *a finite number of zeros described by the formula*

$$\nu_n^D = i\gamma\,[4n + 3 + 4\epsilon_n^D + O((\epsilon_n^D)^2)] \tag{3.8a}$$

where

$$\epsilon_n^D \sim \frac{-2}{\pi}\,e^{\pi i/4}\,\frac{\Gamma(n + 3/2)}{n!}\,\gamma^{1/2}\,\xi_0. \tag{3.8b}$$

These are the only zeros of the form $O(\gamma)$. *Under the same hypotheses the derivative of this solution has a finite number of zeros of the form*

$$\nu_n^N = i\gamma\,[4n + 1 + 4\,\epsilon_n^N + O((\epsilon_n^N)^2)], \tag{3.9a}$$

where

$$\epsilon_n^N \sim \frac{(4n + 1)\,\Gamma(n + 1/2)}{2\pi n!}\,e^{\pi i/4}\,\gamma^{1/2}\,\xi_0. \tag{3.9b}$$

How large n may be once γ is chosen we do not know. These eigenvalues coincide with those for the "thin" elliptic cylinder (see Goodrich and Kazarinoff, 1962b). Also,

$$\begin{aligned} \frac{\partial}{\partial \nu}\,\phi_2(\xi_0, \nu_n^D) &\sim 2^{-9/4}\,\pi^{1/2}\,\gamma^{-1}\,n!\,(-1)^n\,e^{3\pi i/4}, \\ \frac{\partial}{\partial \nu}\,\phi_2'(\xi_0, \nu_n^N) &\sim 2^{-5/4}\,\pi^{1/2}\,n!\,(-1)^n\,\gamma^{-1/2}. \end{aligned} \tag{3.10}$$

Lastly, we observe that if $|\zeta|$ is large, and if $-5\pi/2 + \epsilon \le \pi/2 - \epsilon$ (which it is in our case),

$$w_2(\xi, \nu) \sim 2^{(2\sigma-1)/4}\,\gamma^{(\sigma-1)/4}\exp[i\gamma + (\sigma + 1)\,\pi i/8 - i\gamma(1 + \xi^2)^{1/2}] \left(\frac{1 + \xi^2}{\xi^2}\right)^{1/4} \left(\frac{(1 + \xi^2)^{1/2} - 1}{(1 + \xi^2)^{1/2} + 1}\right)^{\sigma/4}.$$

Thus if Ξ is large,

$$w_2(\Xi, \nu_n^N) \sim 2^{2n+5/4}\,\gamma^{n+1/2}\,e^{-i\gamma\Xi + i\gamma + (n+1)\pi i/2}, \tag{3.11}$$

and

$$w_2(\Xi, \nu_n^N) \sim 2^{2n+1/4}\,\gamma^n\,e^{-i\gamma\Xi + i\gamma + (n+1/2)\,\pi i/2}. \tag{3.12}$$

4. ANGULAR SOLUTIONS FOR η NEAR -1

The angular resolvent Green's function is given in terms of that solution ψ_2 of the differential equation

$$-\,[(1 - \eta^2)\,\psi_\eta]_\eta - (\gamma^2\eta^2 - \nu)\,\psi = 0(\nu = i\gamma\sigma) \tag{4.1}$$

which is regular at -1. Thus it is natural to begin by determining the asymptotic behavior of ψ_2 as $|\gamma| \to \infty$ for η close to -1. In Section 5, we derive the structure of ψ_2 for $|\eta|$ small. The point -1 is a regular singular point of (4.1), but for the values of ν_n of ν, the functional dependence of ψ_2 on η and γ for $|\gamma| \to \infty$ cannot be determined by ordinary power series methods. However, it can be found by using the theory of differential equations with turning points; the particular algorithm we need is one of Kazarinoff's (1957).

Let

$$V = (1 - \eta^2)^{1/2}\,\psi, \quad z = 1 + \eta, \text{ and } \lambda = 2^{-1/2}\,\gamma. \tag{4.2}$$

Then (4.1) takes the canonical form,

$$\frac{d^2V}{dz^2} + \left[\lambda^2 \frac{2(1-z)^2}{z(2-z)} + \lambda \frac{e^{-\pi i/2}\, 2^{1/2}\,\sigma}{z(2-z)} + \frac{1}{z^2(2-z)^2}\right] V = 0, \tag{4.3}$$

used in (Kazarinoff, 1957). For the application of Kazarinoff's theory, we adopt the following notation:

$$\left\{\begin{aligned}
\phi &= 2^{1/2}\,(1-z)\,[z(2-z)]^{-1/2} = -2^{1/2}\,\eta(1-\eta^2)^{-1/2} \\
\Phi &= \int_0^z \phi(t)\,dt = 2^{1/2}\,(1-\eta^2)^{1/2} \\
\xi &= \lambda\Phi = \gamma(1-\eta^2)^{1/2} \\
\Psi &= [\phi\Phi]^{-1/2} = (-2\eta)^{-1/2} \\
\theta &= \frac{-i\sigma}{4} \ln\left(\frac{1+(1-\eta^2)^{1/2}}{1-(1-\eta^2)^{1/2}}\right) \\
\mu_0 &= \cos\theta \\
\mu_1 &= \phi^{-1}\sin\theta.
\end{aligned}\right. \tag{4.4}$$

The regular solutions of (4.1) and (4.3) are asymptotically related to $J_0(\xi)$, where $J_0(\xi)$ is the Bessel function of order zero. In particular, by (Kazarinoff, 1957),

$$V_2 \sim \left[\mu_0 + \frac{2^{1/2}}{\gamma}\,\mu_1\,\frac{d}{d\eta}\right] \Psi\xi J_0(\xi), \qquad (V_2 = (1-\eta^2)^{1/2}\,\psi_2). \tag{4.5}$$

If $|\xi|$ is large and $|\arg\xi| < \pi$ (in our case $\arg\xi \sim 0$),

$$J_0(\xi) = \left(\frac{2}{\pi\xi}\right)^{1/2} \cos(\xi - \pi/4)\,[1 + O(\xi^{-1})];$$

thus if $|\gamma(1-\eta^2)^{1/2}|$ is large and $-1 < \eta \leq -\epsilon < 0$,

$$V_2(\eta) \sim \left[\frac{\gamma(1-\eta^2)^{1/2}}{-\pi\eta}\right]^{1/2} \cos(\gamma(1-\eta^2)^{1/2} + \theta(\eta) - \pi/4). \tag{4.6}$$

If $|\gamma(1-\eta^2)|^{1/2}$ is small and $-1 \leq \eta \leq -\epsilon < 0$,

$$V_2(\eta) \sim \left\{\cos\left[\frac{i\sigma}{2}(1-\eta^2)^{1/2}\right] + \frac{(1-\eta^2)^{1/2}}{\gamma\eta}\sin\left[\frac{i\sigma}{2}(1-\eta^2)^{1/2}\right]\frac{d}{d\eta}\right\}$$

$$\frac{\gamma(1-\eta^2)^{1/2}}{(-2\eta)^{1/2}} J_0[\gamma(1-\eta^2)^{1/2}], \tag{4.7}$$

or (if $|\gamma(1-\eta^2)^{1/2}| \ll 1$)

$$V_2(\eta) \sim \gamma(-2\eta)^{-1/2}(1-\eta^2)^{1/2}. \tag{4.8}$$

5. ANGULAR SOLUTIONS FOR η NEAR 0

The differential equations (4.1) has a second order turning point at the origin. The structure of its solutions for $|\gamma|$ large may therefore be determined from McKelvey's theory (McKelvey, 1955). In this section we adopt the notation

$$\begin{aligned}
\phi &= \eta(1-\eta^2)^{-1/2} \\
\Phi &= \int_0^\eta \phi(t)\,dt = [1-(1-\eta^2)^{1/2}] \\
\xi &= -2i\gamma\Phi \\
\Psi &= \Phi^{1/4}\phi^{-1/2} \\
\tilde{\theta} &= \frac{\sigma}{4}\ln\left(\frac{1+(1-\eta^2)^{1/2}}{2}\right) \\
\mu_0 &= \cosh\tilde{\theta} \\
\mu_1 &= \phi^{-1}\sinh\tilde{\theta}.
\end{aligned} \tag{5.1}$$

We note that if $\eta < 0$,

$$\arg\phi = \pi \text{ and } \arg\Phi = 2\pi.$$

Thus,

$$\Psi = [(1-\eta^2)\,.\,|1-(1-\eta^2)^{1/2}|\,.\,|\eta|^{-2}]^{1/4}. \tag{5.2}$$

The differential equation satisfied by the function V_2 of Section 3 is

$$\frac{d^2V}{d\eta^2} + \left[\frac{\gamma^2\eta^2}{1-\eta^2} + \frac{e^{-\pi i/2}\gamma\sigma}{1-\eta^2} + \frac{1}{(1-\eta^2)^2}\right]V = 0. \tag{5.3}$$

It has linearly independent solutions $\tilde{V}_l(l = 1, 2)$ that have a reasonably simple structure when $|\xi|$ is large and $\eta < 0$; namely, if

$$W_l(\xi) = W_{(-1)^l\sigma/4,\,1/4}(\xi e^{-l\pi i}), \tag{5.4}$$

where $W_{k,m}(z)$ is Whittaker's confluent hypergeometric function, then by (McKelvey, 1955)

$$\tilde{V}_l \sim \left[\mu_0 - \frac{\mu_1}{i\gamma}\frac{d}{d\eta}\right]\Psi(\eta)\,\xi^{-1/4}\,W_l(\xi e^{-l\pi i}). \tag{5.5}$$

This is so because

$$W_l(\xi) = [1 + O(\xi^{-1})] \exp[(-1)^{l+1} \xi/2 + (-1)^l (\sigma/4) \ln (\xi\, e^{-l\pi i})]$$

$$(\epsilon + (l - 3/2)\pi \leq \arg \xi \leq (l + 3/2)\pi - \epsilon). \quad (5.6)$$

We need to determine which linear combination of $\tilde{V}_{+1}$ and $\tilde{V}_{+2}$ is the solution V_2 of Section 3.

To do this we first recall that γ has a small imaginary part; see (2.3). We are concerned primarily with the case $s = 0$, for in the end the residue contributions of the integral (2.2) must be evaluated for $s = 0$. But for $s = 0$ and $\eta < 0$, $\arg \xi = 3\pi/2$. Thus if $|\xi|$ is large and $\eta < 0$, neither W_1 nor W_2, and hence neither $\tilde{V}_1$ nor $\tilde{V}_2$ is exponentially small. Indeed, both are oscillatory, and points of $(-1, 0)$ exist for which $\tilde{V}_1$ and $\tilde{V}_2$ are of the same order of magnitude. On the other hand, if we choose

$$a_1 = \pi^{-1/2}\, 2^{(2\sigma-3)/4}\, \gamma^{(3+\sigma)/4} \exp[i\gamma - \pi i(3\sigma - 1/8)],$$

$$a_2 = \pi^{-1/2}\, 2^{-(2\sigma+3)/4}\, \gamma^{(3-\sigma)/4} \exp[-i\gamma + \pi i(5\sigma + 5/8)], \quad (5.7)$$

then

$$a_1\tilde{V}_1 + a_2\tilde{V}_2 \sim \left[\frac{\gamma(1-\eta^2)^{1/2}}{-\pi\eta}\right]^{1/2} \cos[\gamma(1-\eta^2)^{1/2} + \theta(\eta) - \pi/4], \quad (5.8)$$

where θ is defined by (4.4). In view of the representation (4.6), we would like to identify $a_1\tilde{V}_1 + a_2\tilde{V}_2$ with V_2. Since V_1 and V_2 are both oscillatory if $s = 0$, $\eta < 0$, and ξ is large and since asymptotic representations are unique. It follows from the representations (4.8) and (3.6) that if $s = 0$,

$$V_2 = (a_1\tilde{V}_1 + a_2\tilde{V}_2) + e_1(\gamma)\, \tilde{V}_1 + e_2(\gamma)\, \tilde{V}_2, \quad (5.9)$$

where

$$e_l/a_l = 0(1) \ (|\gamma| \to \infty;\ l = 1, 2).$$

We next observe that (Buchholz, 1953, pp. 26–27)

$$W_l(\xi) = \frac{e^{-l\pi i/4}\, \Gamma(-1/2)}{i^l \Gamma\left(\dfrac{1-\beta\sigma}{4}\right)} M_{\sigma/4,\,1/4}(\xi) + \frac{e^{l\pi i/4}\, \Gamma(1/2)}{i^l \Gamma\left(\dfrac{3-\beta\sigma}{4}\right)} M_{\sigma/4,\,-1/4}(\xi),$$

$$(\beta = (-1)^l).$$

The behavior of V_2 if ξ is small, that is if $|\gamma[1 - (1 - \eta^2)^{1/2}]|$ is small, is therefore described by the formula

$$V_2 = b_+V_+ + b_-V_-, \quad (5.10)$$

where V_+ and V_- are the solutions asymptotic to the right-hand member of

(5.5) with $M_{\sigma/4,\ 1/4}(\xi)$ and $M_{\sigma/4,\ -1/4}(\xi)$, respectively, replacing W_l and where

$$b_+ = e^{-3\pi i/4}\,\Gamma(-1/2)\left\{\frac{a_1}{\Gamma\left(\frac{1+\sigma}{4}\right)} + \frac{e^{-3\pi i/4}a_2}{\Gamma\left(\frac{1-\sigma}{4}\right)}\right\},$$

$$b_- = e^{-\pi i/4}\,\Gamma(1/2)\left\{\frac{a_1}{\Gamma\left(\frac{3+\sigma}{4}\right)} + \frac{e^{-i\pi/4}a_2}{\Gamma\left(\frac{3-\sigma}{4}\right)}\right\}. \tag{5.11}$$

Now,

$$\psi_1(\eta) = \psi_2(-\eta).$$

We therefore define

$$V_1(\eta) = V_2(-\eta). \tag{5.12}$$

Also (Buchholz, 1953, p. 11),

$$M_{k,\ \pm 1/4}(\xi e^{2\pi i}) = e^{\mp\pi i}\,M_{k,\ \pm 1/4}(\xi), \text{ and } \xi^{-1/4}\,(-\eta) = e^{-\pi i/2}\,\xi(\eta).$$

Therefore,

$$V_1(\eta) = b_- V_-(\eta) - b_+ V_+(\eta). \tag{5.13}$$

Lastly, we observe that by (4.5)

$$\psi_1(1) \sim 2^{-1/2}\,\gamma. \tag{5.14}$$

6. THE WRONSKIAN OF THE ANGULAR SOLUTIONS

We see from (1.7) that in order to obtain a representation for the angular Green's function, we need to have representations for $\psi_2(\eta)$, $\psi_1(1)$ and the Wronskian of ψ_1 and ψ_2. We have already found representations for $\psi_1(1)$ and $\psi_2(\eta)$ in Sections 4 and 5. It remains to evaluate the Wronskian $W(\psi_1, \psi_2)$ in order to approximate the Green's function $\tilde{G}$. To do this we note that

$$(1 - \eta^2)\,W(\psi_1, \psi_2; \eta) = W(V_1, V_2; \eta).$$

But

$$W(V_1, V_2; \eta) \equiv W(V_1, V_2; 0),$$

and

$$W(V_1, V_2; 0) = -2b_+b_-\,W(V_+, V_-; 0).$$

It is easy to compute $W(V_+, V_-; 0)$; see (McKelvey, 1955). The result is

$$W(V_+, V_-; 0) = 2(-i\gamma)^{1/2}\,W(M_{\sigma/4,\ 1/4}(\xi), M_{\sigma/4,\ -1/4}(\xi); \xi) = -\frac{(-i\gamma)^{1/2}}{2}$$

Therefore,

$$(1 - \eta^2)\,W(\psi_1, \psi_2; \eta) = (-i\gamma)^{1/2}\,b_+b_-.$$

It follows from (5.11) and (5.7) that

$$b_+b_- = \frac{2\pi a_1^2}{\Gamma\left(\frac{1+\sigma}{4}\right)\Gamma\left(\frac{3+\sigma}{4}\right)} + O(\gamma^{3/2}),$$

where

$$a_1^2 = \gamma^{3+2n}\, O(1) \qquad (n = 0, 1, \ldots)$$

in the Dirichlet problem, and

$$a_1^2 = \gamma^{2+2n}\, O(1) \qquad (n = 0, 1, \ldots).$$

in the Neumann problem. We therefore determine that in the Dirichlet problem,

$$(1-\eta^2)\, W(\psi_1, \psi_2) = \frac{e^{2i\gamma+\pi i/2}\,\gamma^2}{2^{1/2}\,\pi R_n^D}(1 - R_n^D\, e^{-2i\gamma}), \tag{6.1}$$

where

$$R_n^D = \frac{(-1)^n\, i(2n+1)!\, e^{i\pi/4}}{2^{6n+4}\,\pi^{1/2}\,\gamma^{2n+3/2}}. \tag{6.2}$$

In the Neumann problem,

$$(1-\eta^2)\, W(\psi_1, \psi_2) = \frac{e^{2i\gamma-\pi i/2}\,\gamma^2}{2^{1/2}\,\pi R_n^N}(1 - R_n^N\, e^{-2i\gamma}), \tag{6.3}$$

where

$$R_n^N = \frac{(-1)^n\, i(2n)!\, e^{-i\pi/4}}{2^{6n+1}\,\pi^{1/2}\,\gamma^{2n+1/2}}. \tag{6.4}$$

The factors R_n^N, R_n^D appearing in these expressions will play the role of reflection coefficients. Unlike the reflection coefficients that occur in the theory of scalar diffraction by a thin prolate spheroid (Goodrich and Kazarinoff, 1962a), these reflection coefficients are only very weakly dependent upon the eccentricity of the generating ellipse, the behavior exhibited by the corresponding coefficients in the case of scalar diffraction by a thin elliptic cylinder. The phenomenon is altogether reasonable in view of the fact that changes in the eccentricity of an oblate spheroid that is nearly a disc should only slightly effect the scattered field for "broadside" incidence of the incoming energy.

7. THE SURFACE DISTRIBUTION

We now approximate the integral (2.2) as a residue series, the terms of which we asymptotically evaluate by means of the formulas (2.4), (2.6), (3.8)–(3.12), (5.14), and (6.1)–(6.4) and the various approximations to $\psi_2(\eta)$ given in the previous sections.

If $u(\xi_0, \eta; \Xi, 1) = 0$, then

$$\frac{\partial u(\xi_0, \eta; \Xi, 1)}{\partial \xi} \sim \frac{e^{-i\gamma(\Xi+1)}}{\Xi} \sum_{n=0}^{N} \frac{i^{-n}\, 2^{-4n-5/4}\,(2n+1)!}{(1 - R_n^D\, e^{-2i\gamma})\, n!\, \gamma^{n+1}} \psi_2(\eta, \nu_n^D), \tag{7.1}$$

where R_n^D is defined by (6.2).

There are three particular cases of interest. If $|\gamma^{1/2}\eta| \ll 1$ and $\eta < 0$,

$$\frac{\partial u(\xi_0, \eta; \Xi, 1)}{\partial \xi} \sim \frac{e^{-i\gamma\Xi}}{\Xi} (1 - \eta^2)^{-1/4} \sum_{n=0}^{N} T_{N,0}^D [1 - R_n^D\, e^{-2i\gamma}]^{-1}, \tag{7.2}$$

where

$$T_{n,0}^D = e^{-i\pi/4}\, \gamma^{1/2}\, (-1)^n\, 2^{-n}\, \frac{(2n+1)!}{n!\, \Gamma(n+3/2)}.$$

If $\gamma(1 - \eta^2)^{1/2} \gg 1$ and $-1 < \eta \leq -\epsilon < 0$,

$$\frac{\partial u}{\partial \xi}(\xi_0, \eta; \Xi, 1) \sim \frac{e^{-i\gamma\Xi}}{\Xi} (-\eta)^{-1/2} (1 - \eta^2)^{-1/4} \sum_{n=0}^{N} T_n^D [1 - R_n^D\, e^{-2i\gamma}]^{-1}\, .$$

$$\cdot\, [e^{-i\gamma(1-(1-\eta^2)^{1/2})} [f(\eta)]^{n+3/4} - i\, e^{-i\gamma(1+(1-\eta^2)^{1/2})} [f(\eta)]^{-n-3/4}], \tag{7.3}$$

where

$$T_n^D = i^{-n-1/2}\, 2^{-4n-9/4}\, \pi^{-1/2}\, \frac{(2n+1)!}{n!}\, \gamma^{-n}$$

and

$$f(\eta) = \frac{1 + (1 - \eta^2)^{1/2}}{1 - (1 - \eta^2)^{1/2}}.$$

If $\gamma(1 - \eta^2)^{1/2} \ll 1$ and $\eta < 0$,

$$\frac{\partial u(\xi_0, \eta; \Xi, 1)}{\partial \xi} \sim \frac{e^{-i\gamma\Xi}}{\Xi} (-\eta)^{-1/2} \sum_{n=0}^{N} T_{n,-1}^D\, e^{-i\gamma} [1 - R_n^D\, e^{-2i\gamma}]^{-1}, \tag{7.4}$$

where

$$T_{n,-1}^D = i^{1/2}\, 2^{1/2}\, \pi^{1/2}\, T_n^D.$$

If $\partial u(\xi_0, \eta; \Xi, 1)/\partial \xi = 0$, then

$$u(\xi_0, \eta; \Xi, 1) \sim \frac{e^{-i\gamma(\Xi+1)}}{\Xi} \sum_{n=0}^{N} \frac{i^{n-1/2}\, 2^{-4n+1/2}\, (2n)!}{n!(1 - R_n^N\, e^{-2i\gamma})\, \gamma^{n+1}} \psi_2(\eta), \tag{7.5}$$

where R_n^N is defined by (6.4).

In particular, if $|\gamma^{1/2}\eta| \ll 1$ and $\eta < 0$,

$$u(\xi_0, \eta; \Xi, 1) \sim \frac{e^{-i\gamma\Xi}}{\Xi} (1 - \eta^2)^{-1/4} \sum_{n=0}^{M} T_{n,0}^N [1 - R_n^N\, e^{-2i\gamma}]^{-1}, \tag{7.6}$$

where

$$T_{n,0}^N = \frac{(-1)^{n+1}\, (2n)!\, i}{2^{2n}\, (n!)^2}.$$

If $\gamma(1-\eta^2)^{1/2} \gg 1$ and $-1 < \eta \leq -\epsilon < 0$,

$$u(\xi_0, \eta; \Xi, 1) \sim \frac{e^{-i\gamma\Xi}}{\Xi} (-\eta)^{-1/2} (1-\eta^2)^{-1/4} \sum_{n=0}^{M} T_n^N [1 - R_n^N e^{-2i\gamma}]^{-1} \cdot$$

$$\cdot [e^{-i\gamma(1-(1-\eta^2)^{1/2})} [f(\eta)]^{n+1/4} - i e^{-i\gamma(1+(1-\eta^2)^{1/2})} [f(\eta)]^{-n-1/4}], \quad (7.7)$$

where

$$T_n^N = i^{n-1/2} 2^{-4n-1/2} \pi^{-1/2} \frac{(2n)!}{n!} \gamma^{-n-1/2}.$$

If $|\gamma(1-\eta^2)^{1/2}| \ll 1$ and $\eta < 0$,

$$u(\xi_0, \eta; \Xi, 1) \sim \frac{e^{-i\gamma\Xi}}{\Xi} (-\eta)^{-1/2} \sum_{n=0}^{M} T_{n,-1}^N e^{-i\gamma} [1 - R_n^N e^{-2i\gamma}]^{-1}, \quad (7.8)$$

where

$$T_{n,-1}^N = \frac{i^n (2n)!}{n!\, 2^{4n} \gamma^n}.$$

The above results can be interpreted in complete analogy with those in (Goodrich and Kazarinoff, 1962a and b). We therefore omit a detailed discussion here. Briefly, a model to which the above results correspond is as follows. The quantity $\gamma[1-(1-\eta^2)^{1/2}]/k$ is approximately the geodesic distance l from a point $(\xi_0, 0, \phi)$ on the shadow boundary to a point (ξ_0, η, ϕ) in the shadow, γ/k is approximately one quarter of the perimeter of the generating ellipse, call it L, and $\gamma[1+(1-\eta^2)^{1/2}]/k$ is approximately the geodesic distance $2L - l$ from the point $(\xi_0, 0, \phi + \pi)$ to (ξ_0, η, ϕ). We assume that the waves incident on the shadow boundary, together with any energy coming from surface waves on the lit portion of the spheroid, give rise to traveling waves in various modes, one for each value of n in the above results. We shall consider a single mode only. A traveling wave is launched at the shadow boundary with amplitude T. Hence at some point away from the caustic, but in the shadow, there will be a field

$$\frac{e^{-i\gamma\Xi}}{\Xi} [T e^{ikl} + T e^{ik(2L-l)}] (1-\eta^2)^{-1/4},$$

where the factor involving Ξ results from the path from the source to the shadow boundary. There are additional contributions corresponding to waves that have traveled back and forth on B, being reflected with a reflection coefficient R each time they reach the rim (shadow boundary). In the figure, we have illustrated the first three possible paths originating from a point $(\xi_0, 0, \phi)$ and terminating at (ξ_0, η, ϕ). Corresponding to these and the still longer paths, the contributions to the surface field are

$$T e^{-ikl} \; TR e^{-ik(4L-l)}, \; TR^2 e^{-ik(4L+l)},$$

etc. where we have suppressed the factor $e^{-i\gamma\Xi}/\Xi$. Those terms corresponding to waves launched at the diametrically opposite point are

$$T\,e^{-ik(2L-l)},\ TR\,e^{-ik(2L+l)},\ TR^2\,e^{-ik(6L-l)},$$

etc. On adding all these terms, we find that the field at (ξ_0, η, ϕ) is of the form

$$\frac{e^{-i\gamma\Xi}}{\Xi}\,T\,[e^{-ikl} + e^{-ik(2L-l)}]\,(1-\eta^2)^{-1/4}\,(1-\mathrm{R}e^{-2ikL})^{-1}.$$

This is almost the form of (7.3) and (7.7).

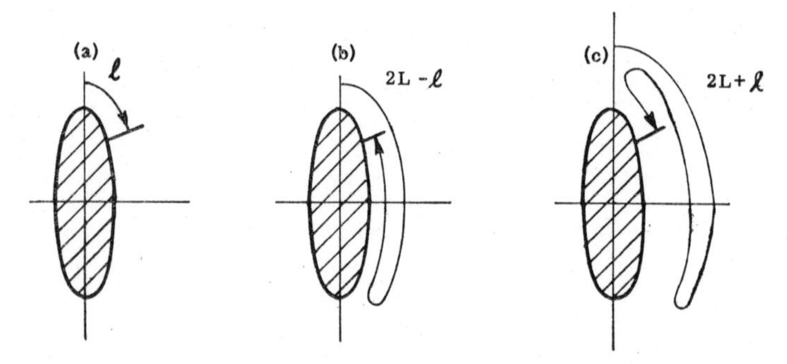

The additional factors involving η that appear in (7.3) and (7.7) arise because the spheroid is not actually a disc and has nonzero curvature. The coefficient of the $e^{-i\gamma(2L-l)}$-term is not 1 because of a phase change the traveling waves undergo as they pass through the caustic.

Near the caustic $(\xi_0, -1, \phi)$, of course, the field behavior is different in that there is a geometric focusing of the field. However, the formulas (7.4) and (7.8) exhibit the expected phase.

REFERENCES

BUCHHOLZ, H. (1953) Die konfluente hypergeometrische Funktion. Springer-Verlag, Berlin.

GOODRICH, R. F. and KAZARINOFF, N. D. (1963) Scalar diffraction by prolate spheroids whose eccentricities are almost one, *Proc. Camb. Philos. Soc.*, **59**, 167–183.

GOODRICH, R. F. and KAZARINOFF, N. D. (1963) Scalar diffraction by elliptic cylinders whose eccentricities are almost one, *Mich. Math. J.*, **10**, No. 2 (to appear).

KAZARINOFF, N. D. (1957) Asymptotic solutions with respect to a parameter of ordinary differential equations having a regular singular point, *Mich. Math. J.* **4**, 207–220.

KAZARINOFF, N. D. and RITT, R. K. (1959) On the theory of scalar diffraction and its application to the prolate spheroid, *Ann. of Phys.* **6**, 277–299.

MCKELVEY, R. W. (1955) The solutions of second order linear differential equations about a turning point of order two. *Trans. Amer. Math. Soc.* **79,** 103–123.

DIFFRACTION OF A SCALAR WAVE BY A PLANE CIRCULAR DISK†

E. B. HANSEN‡

ABSTRACT

The problem of diffraction of a scalar spherical wave by a circular disk is considered in the case when the wavelength is very small compared with the radius of the disk. By means of separating the variables, an exact series expression in terms of spheroidal functions of complex order is obtained. The expression is expanded asymptotically for the wavelength tending towards zero and the asymptotic representation for the field in the geometrical optics shadow region is found. This representation agrees completely with that predicted by the geometric theory of diffraction; thus it provides a verification of that theory.

SUMMARY

In the paper which is summarized here we consider the problem of diffraction of a scalar wave by a plane, circular disk in the case when the wavelength is very small compared with the radius of the disk. The disk may be perfectly soft or perfectly hard. For the sake of definiteness we here consider it to be perfectly hard.

From the previous literature on diffraction by plane obstacles we particularly mention the work of Keller (1951) since it is the purpose of the present investigation to demonstrate the validity of the expression for the field derived from his geometrical theory of diffraction by an argument which uses the classical method of separation of the variables.

The disk is shown in Fig. 1 together with a Cartesian coordinate system in which the source (which is a unipole source) has the coordinates $(x_0, 0, z_0)$. We introduce oblate spheroidal coordinates ξ, η and ϕ defined by

$$\begin{Bmatrix} x \\ y \end{Bmatrix} = a(1+\xi^2)^{1/2}(1-\eta^2)^{1/2}\begin{Bmatrix} \cos\phi \\ \sin\phi \end{Bmatrix}$$
$$z = a\xi\eta \tag{1}$$
$$0 \leqq \xi < \infty;\ -1 \leqq \eta \leqq +1;\ 0 \leqq \phi \leqq 2\pi.$$

In this system the source coordinates are denoted by $(\xi, \eta, \phi) = (\xi_0, \eta_0, 0)$. If the incident field is given by $(4\pi R)^{-1} \exp(ik_1R)$ (k_1 being the propagation

† The present contribution describes part of an investigation on high-frequency diffraction by infinite strips and circular disks which was carried out while the author was visiting the Courant Institute of Mathematical Sciences, New York University, New York. A full description of the work is given in *Journal of Mathematics and Physics*, **71**, No. 4, 229–245, Dec. 1962.

‡ Laboratory of Electromagnetic Theory, The Technical University of Denmark, Copenhagen, Denmark.

constant of the medium surrounding the disk and R the distance from the source) the equation of the problem (with $k = k_1 a$) is

$$\frac{\partial}{\partial \xi}\left((1+\xi^2)\frac{\partial u}{\partial \xi}\right) + \frac{\partial}{\partial \eta}\left((1-\eta^2)\frac{\partial u}{\partial \eta}\right) + \frac{\xi^2+\eta^2}{(1+\xi^2)(1-\eta^2)}\frac{\partial^2 u}{\partial \phi^2} + k^2(\xi^2+\eta^2)\,u = -\frac{1}{a}\,\delta(\xi-\xi_0)\,\delta(\eta-\eta_0)\,\delta(\phi). \tag{2}$$

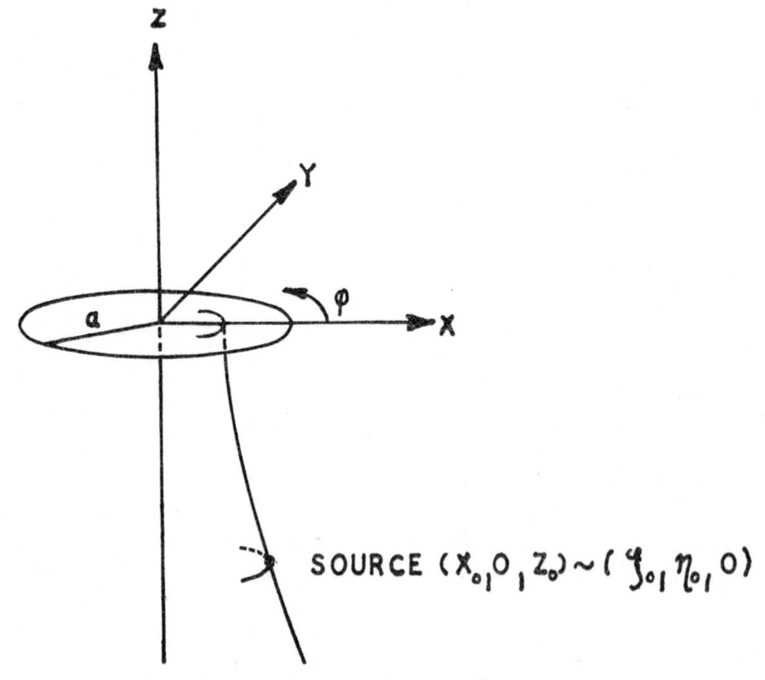

FIG. 1. Definition of the coordinate systems.

Following the method of Sommerfeld (1947) we try to express the field $u(\xi, \eta, \phi)$ as a series

$$u = \sum_{(m,n)=(0,1)}^{\infty} V_{m,n}^{(1)}(\xi)\, X_{m,n}(\eta) \cos m\phi, \tag{3}$$

where $V_{m,n}^{(1)}(\xi)$ satisfies the conditions

$$V_{m,n}^{(1)}(\xi) \sim \frac{e^{ik\xi}}{\xi} \text{ for } \xi \to \infty$$

and

$$(\mathrm{d}V_{m,n}^{(1)}(\xi)/\mathrm{d}\xi)_{\xi=0} = 0. \tag{4}$$

The functions $V^{(1)}_{m,n}(\xi)$ shall then satisfy the equation

$$\frac{d}{d\xi}\left((\xi^2+1)\frac{dV^{(1)}_{m,n}}{d\xi}\right)+\left(k^2\xi^2+A_{m,n}-\frac{m^2}{1+\xi^2}\right)V^{(1)}_{m,n}=0. \tag{5}$$

This equation together with (4) constitute an eigenvalue problem with the eigenvalues $A_{m,n}$. The functions $\chi_{m,n}(\eta)$ satisfy the equation which results from (5) when ξ is substituted by $i\eta$. Among the solutions of this equation we shall reserve the notation $\chi_{m,n}(\eta)$ for a solution which is finite at $\eta=+1$. Standard procedures then lead to the following expression for u:

$$u(\xi,\eta,\phi)=\frac{1}{a\pi}\sum_{(m,n)=(0,1)}^{\infty}\frac{\chi_{m,n}(\eta_>)\,\chi_{m,n}(-\eta_<)\,V^{(1)}_{m,n}(\xi_0)\,V^{(1)}_{m,n}(\xi)\cos m\phi}{(1+\delta_{0,n})\int\limits_0^{\infty}(V^{(1)}_{m,n}(\xi))^2\,d\xi\;W_{m,n}} \tag{6}$$

where $W_{m,n}$ is the Wronskian of the two functions $\chi_{m,n}(\eta_>)$ and $\chi_{m,n}(-\eta_<)$; $\eta_>$ ($\eta_<$) being the bigger (smaller) of η and η_0.

The crucial point in the determination of the expression (6) for $u(\xi,\eta,\phi)$ is the solution of the eigenvalue problem defined by (4) and (5). We wish to solve this eigenvalue problem asymptotically for $k\to\infty$. In order to do so we set

$$V^{(1)}_{m,n}(\xi)=\left(1+\frac{x^2}{4}\right)^{-1/4}\left(1+\frac{x^2}{2}\right)^{-1/2}w(x), \tag{7}$$

where

$$x=2^{1/2}((1+\xi^2)^{1/2}-1)^{1/2}$$

and we assume that

$$A_{m,n}\sim-2i\alpha_{m,n}k+0(k^\circ). \tag{8}$$

The equation for $w(x)$ is then

$$w''(x)+\left(k^2x^2-\frac{8i\alpha_{m,n}k}{4+x^2}+0(k^\circ)\right)w(x)=0. \tag{9}$$

This equation resembles the equation for the parabolic cylinder functions $D_\nu(y)$ as defined in Erdelyi (1953). We therefore set:

$$w(x)\sim D_{\alpha_{m,n}-\frac{1}{2}}(y)\sum_{s=0}^{\infty}f_s(x)\,k^{-s}+D'_{\alpha_{m,n}-\frac{1}{2}}(y)\sum_{s=1}^{\infty}g_s(x)\,k^{-s-1/2} \tag{10}$$

where $y=(2k)^{1/2}\,e^{-i\frac{\pi}{4}}\,x$, and where $f_s(x)$ and $g_s(x)$ are functions to be determined. This may be done by methods developed by Olver (1954) and we find the following asymptotic expression for an eigenfunction $V^{(1)}_{m,n}(\xi)$:

$$\begin{aligned}V^{(1)}_{m,n}(\xi)\sim{}&(1+(1+\xi^2)^{1/2})^{-1/4}(1+\xi^2)^{-1/4}\times\\&\times\left\{D_{2(n-1)}(y)\cosh\left((n-\tfrac{3}{4})\ln\left(\frac{1+(1+\xi^2)^{1/2}}{2}\right)\right)\right.\\&\left.+\frac{2}{y}D'_{2(n-1)}(y)\sinh\left((n-\tfrac{3}{4})\ln\left(\frac{1+(1+\xi^2)^{1/2}}{2}\right)\right)\right\}\times\\&\times\{1+0(k^{-1})\}\end{aligned} \tag{11}$$

The equation for $\chi_{m,n}(\eta)$ may be handled in much the same way. However, since the regular singularities at $\eta = \pm 1$ are in the interval of interest, several different asymptotic representations must be used each valid in one of the intervals $-1 \leqq \eta < 0$, $-1 < \eta < 1$ and $0 < \eta \leqq 1$. They are all of a type similar to (11) but we shall not show the explicit expressions here.

When the asymptotic solutions are inserted in (5) a series expression for the function $u(\xi, \eta, \phi)$ is obtained of which all the terms are of the same (namely the zeroth) order of k. If we limit ourselves to the interesting case when the source is on the axis of the disk ($\eta_0 = -1$) the series is a sum of two geometric series which are convergent inside the geometric shadow region.

When $k(1-\eta^2)^{1/2} \gg 1$, i.e. when the point of observation is well off the axis the sum of the series is expressed by

$$u \sim \frac{e^{ik_1 r_0}}{4\pi r_0} \frac{e^{i[k_1 r_1 + \pi/4]}}{2\left(2\pi k_1 \frac{r_1}{a}\left(1 + \frac{r_1}{a} \sin \theta_1\right)\right)^{1/2}} \left\{\operatorname{cosec} \frac{\theta_1 - a}{2} - \sec \frac{\theta_1 + a}{2}\right\}$$
$$+ \frac{e^{ik_1 r_0}}{4\pi r_0} \frac{e^{i[k_1 r_2 - \pi/4]}}{2\left(2\pi k_1 \frac{r_2}{a}\left(-1 - \frac{r_2}{a} \sin \theta_2\right)\right)^{1/2}}$$
$$\times \left\{\operatorname{cosec} \frac{\theta_2 - a}{2} - \sec \frac{\theta_2 + a}{2}\right\} \qquad (12)$$

Here u is expressed in terms of the geometric quantities shown in Fig. 2. This expression is the same as the one which follows from formulas (12) and (13) in Keller (1957).

When the point of observation is very close to the axis of the disk the sum of the series becomes

$$u \sim \left\{\frac{e^{ik_1 r_0}}{4\pi r_0} \frac{\left(1 + \frac{a}{r_0}\right)^{1/2}}{1 + \frac{r_1}{r_0}}\right\} \frac{1}{2}\left(\frac{a \cos \delta}{r_1}\right)^{1/2} \times$$
$$\times \left[\sec\left(\frac{\delta}{2} + \frac{\pi}{4}\right) + \operatorname{cosec}\left(\frac{\delta}{2} + \frac{\pi}{4}\right)\right] e^{ik_1 r_1} J_0(k\rho \cos \delta) \qquad (13)$$

where $J_0(k\rho \cos \delta)$ is the Bessel function of order zero; δ and ρ are shown in Fig. 2. In the case of plane wave incidence ($u^i = e^{ik_1 z}$) the curly bracket in front of the expression should be substituted by unity. Then (13) agrees with formula (40) of Keller (1957) as may be seen by application of Babinets principle.

In conclusion it should be mentioned that an investigation of scalar diffraction by a thin oblate spheroid and—in the limiting case—a circular disk has recently been carried out by Goodrich and Kazarinoff (to be published). An outline of that investigation was presented at this symposium in a contribution by Goodrich, Kazarinoff and Weston (1963). Although Goodrich and Kazarinoff focus their attention on the behavior of the field on the spheroid while this paper only discusses the field away from the disk, some results are coincident. To the extent that this is the case the agreement is complete.

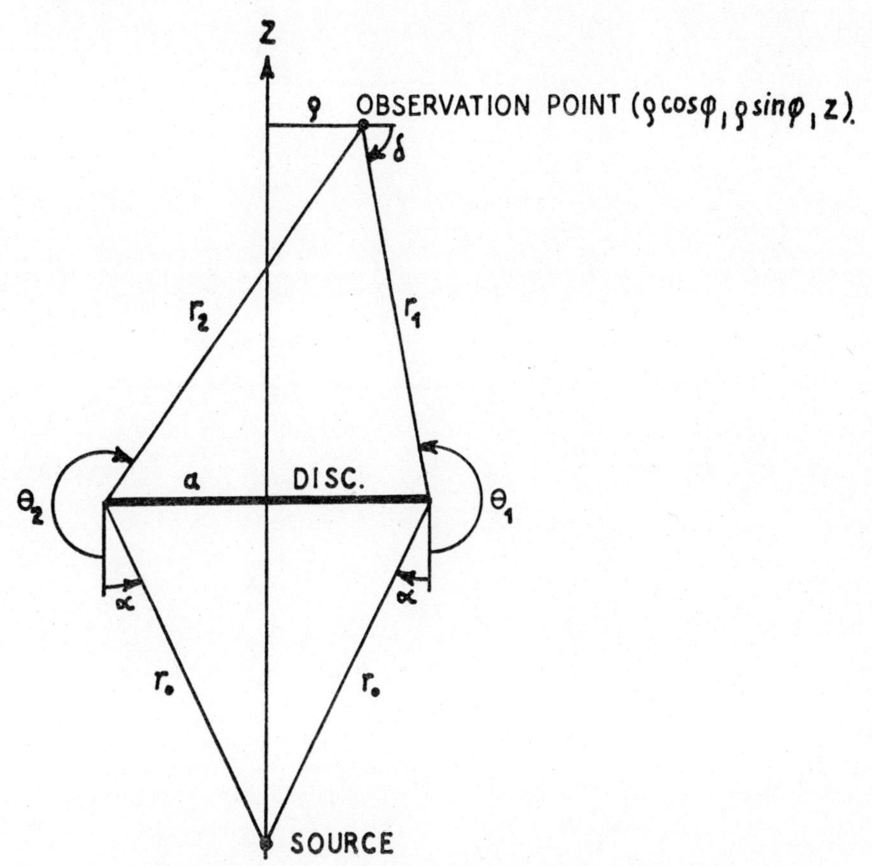

FIG. 2. Definition of quantities occurring in (12) and (13). Note that r_1 is the shorter and r_2 is the longer of the two diffracted rays through the observation point.

REFERENCES

ERDELYI, A. *et al.* (1953) *Higher Transcendental Functions*, vol. II. New York, p. 117.

GOODRICH, R. F., KAZARINOFF, N. D. and WESTON, V. H. (1963) These proceedings, pp. 27–38.

GOODRICH, R. F. and KAZARINOFF, N. D. To be published.

KELLER, J. B. (1957) *J. Appl. Phys.* **28**, 426–444.

OLVER, F. W. J. (1954) *Phil. Trans. Roy. Soc. London*, Ser. A **247**, 307.

SOMMERFELD, A. (1947) *Vorlesungen über theoretische Physik* Wiesbaden, VI, p. 216.

DIFFRACTION OF RADIO WAVES BY SEVERAL SMOOTH SURFACES

K. FURUTSU†

National Bureau of Standards, Boulder, Colorado

ABSTRACT

The diffraction of electromagnetic waves by several smooth surfaces is treated on a rather general condition that the radius of curvature of the diffracting surfaces are sufficiently large compared with the wavelength. The unified formulas of field strength are obtained in the form of multiple residue series and valid for a wide range of diffraction angles including the region near the line of sight. The several supplementary formulas are also prepared when the convergence is poor. The formulas are available for the multiple diffraction by mountains of finite radius of curvature and arbitrary electrical properties, in which the earth surface can also be included as a spherical mountain. When the radius of curvature is sufficiently small, as in the case of ridge diffraction, the Kirchoff approximation term appears as the leading term of the formula. Some results of comparison of the theory with experiment are displayed in the case of two-ridge diffraction.

1. STATEMENT OF PROBLEM AND THE RESULT IN THE SIMPLEST CASE OF ONE DIFFRACTING SURFACE

Many authors have discussed the diffraction of electromagnetic waves by a single object such as sphere, cylinder, paraboloid, ellipsoid, plane, etc. But there are many difficulties when trying to calculate the diffraction loss by actual mountains or hills, because they usually cannot be seen as a single smooth diffracting obstacle, but more or less have irregularities. Strictly speaking, the actual diffraction of radio waves by mountains is a problem in multiple diffraction. But it is, of course, impossible to derive the formula of diffraction loss for such complicated obstacles, and usually different models have been adopted for the diffracting mountains according to the different range of frequency. However, if we could assume that every diffracting mountain has a smooth surface whose radius of curvature is sufficiently large compared with the wavelength, the problem can be treated on fairly general conditions, and we can get the result in a unified form, independently of the number of diffracting mountains.

In the simplest case of one diffracting surface, the result can be deduced for the ordinary van der Pol and Bremmer formula for the diffraction by a large spherical surface: on referring to Fig. 1, the attenuation coefficient A,

† On leave from Radio Research Laboratories, Kokubunji, Tokyo.

which is the ratio of the field strength to that in free space, is given by

$$A = \sqrt{(2\pi k_1 r)}\,(2/k_1 a)^{1/3} \sum_s (t_s - q^2)^{-1}$$
$$\times \exp[-i\{(r/a)\,(k_1 a/2)^{1/3}\, t_s + \pi/4\}]\, f_s(y_1)\, f_s(y_2). \quad (1.1)$$

Here t_s's are the roots of

$$W'(t) - qW(t) = 0,$$
$$W(-t) = \sqrt{(\pi/3)}\, e^{-i2\pi/3}\, t^{1/2}\, H^{(2)}_{1/3}\,(\tfrac{2}{3}t^{3/2}), \quad (1.2)$$

and

$$f_s(y) = W(t_s - y)/W(t_s), \qquad y_i = (2/k_1 a)^{1/3}\, k_1 h_i, \qquad (h_i \ll a)$$
$$iq = (k_1 a/2)^{1/3} \times \begin{cases} k_1\sqrt{(k_2^2 - k_1^2)}/k_2^2, & \text{Ver. Pol.} \\ \sqrt{(k_2^2 - k_1^2)}/k_1, & \text{Hori. Pol.} \end{cases} \quad (1.3)$$

For $y \gg 1$, the height gain function $f_s(y)$ takes the asymptotic form

$$f_s(y) \simeq y^{-1/4} \exp[-i(\tfrac{2}{3}y^{3/2} - t_s y^{1/2} + \tfrac{1}{4}y^{-1/2}t_s^2 + \pi/4)]/W(t_s), \quad (1.4)$$

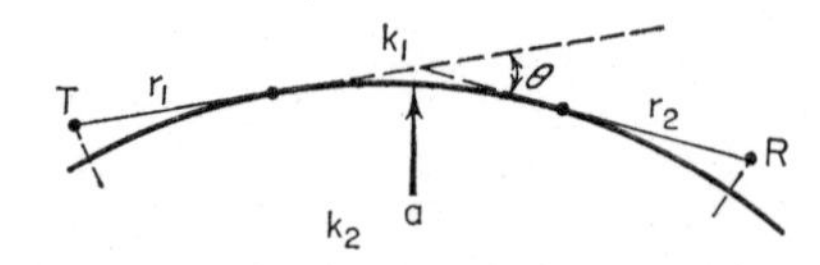

FIG. 1. The spherical diffracting surface and the notations for eqn (1.1).

where the argument of the exponential function is expanded with respect to t_s and taken up to the square of t_s^2. Thus, when the height of one or both of the transmitter and receiver take sufficiently large values as $y_1, y_2 \gg 1$, the attenuation coefficient A of (1.1) takes the following forms:

(i) $y_1 \gg 1, \quad y_2 \lesssim 1$

$$A = \exp[-i(k_1 a/3)\,(r_1/a)^3]\, 2\sqrt{\pi}\sqrt{(r/r_1)} \sum_s (t_s - q^2)^{-1}\, W(t_s)^{-1}$$
$$\times \exp[-i\{\theta(k_1 a/2)^{1/3}\, t_s + \pi/2\}]\, f_s(y_2),$$
$$\theta = (r - r_1)/a, \qquad r_1 = \sqrt{(2ah_1)}, \qquad r_2 = \sqrt{(2ah_2)}. \quad (1.5)$$

(ii) $y_1 \gg 1, \quad y_2 \gg 1$

$$A \simeq \exp\left[-i\frac{k_1 a}{3}\left\{\left(\frac{r_1}{a}\right)^3 + \left(\frac{r_2}{a}\right)^3\right\}\right] \sqrt{\left(\frac{r}{k_1 r_1 r_2}\right)}$$
$$\times \sum_s 2\sqrt{(2\pi)}\,(k_1 a/2)^{1/3}\,(t_s - q^2)^{-1}\, W(t_s)^{-2}$$
$$\times \exp[-i\{\theta(k_1 a/2)^{1/3}\, t_s + (2k_1)^{-1}\,(k_1 a/2)^{2/3}\,(1/r_1 + 1/r_2)\, t_s^2 + 3\pi/4\}],$$
$$\theta = (r - r_1 - r_2)/a. \quad (1.6)$$

In the case (ii), it is a convergent series for the whole range of θ. On the other hand, in the case (i), the series diverges for negative values of θ, and hence it must be analytically continued, as will be stated later. Thus the analytically continued function A of θ is regular on the whole range of θ, including negative values. However, in the case (ii), if the square term of t_s^2 were neglected in the exponential function, the function A of θ would have a pole at $\theta = 0$, even though it is analytically continued. Thus the square term of t_s^2 cannot be neglected in the case (ii). The convergence of series of these formulas is good for large diffraction angles. But, otherwise, it becomes poor. However, there is some general rule to overcome this poor convergence, as will be treated in the following section.

Since the above results have been developed according to the original van der Pol and Bremmer formula given in residue series, they could be valid only when both the transmitter and receiver are not far from the surface, and it is not immediately clear whether they are also valid even when calculating the field far from the surface.

But it can be proven (Furutsu, 1956; Wait and Conda, 1959) that the results (1.5) and (1.6) are still correct even when the transmitter and receiver are at great distance from the diffracting surface compared with the radius of curvature, if we reinterpret $r_1[\doteqdot\sqrt{(2ah_1)}]$ and $r_2[\doteqdot\sqrt{(2ah_2)}]$ as the lengths of the parts of wave path from the transmitter and receiver to the first contracting points on the diffracting surface, respectively, as illustrated in Fig. 1; the important part of the diffracting surface which decisively contributes to the diffracting waves is the very small part of the surface in the vicinity of wave path, and the other part is not important, provided that the radius of curvature of surface is sufficiently large compared with the wavelength. We shall return to this subject again in Section 3.

2, METHODS FOR THE CASE OF POOR CONVERGENCE

For the kinds of series like (1.5) and (1.6) there are some general rules to overcome the difficulty when the convergence of series is poor. Taking into account

$$\operatorname*{Res}_{t=t_s}\left[\frac{(t-q^2)\,W(t)}{W'(t)-qW(t)}\right] = 1, \tag{2.1}$$

we have for the arbitrary series

$$\sum_{s=0}^{\infty} a(t_s), \tag{2.2}$$

$$\sum_s a(t_s) = \sum_s \frac{1}{2\pi i}\int_{C_s} \frac{(t-q^2)\,W(t)}{W'(t)-qW(t)}\,a(t)\,\mathrm{d}t, \tag{2.3}$$

where C_s is the infinitesimal contour integration path around the sth pole.

Hence, if the integrand thus formulated does not have any pole besides the pole t_s's, as in the case (i) of (1.5), the sum of the contour paths is equivalent to the contour path C around the set of poles, as illustrated in Fig. 2. Further, if the integrand tends to zero sufficiently rapidly at infinity, we can further deform it to the path $C_1 + C_2$ in Fig. 2; these paths are proven to be the best paths for numerical integration in the meaning that the integrand decreases most rapidly on these paths. Furthermore, the integral converges for negative values of θ, even when the original series diverged, as in (1.5). Thus

$$\sum_s \int_{C_s} = \int_C = \int_{C_1+C_2} . \tag{2.4}$$

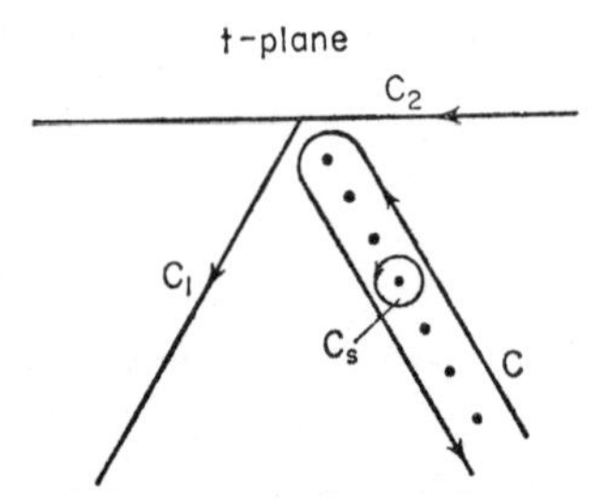

FIG. 2. The infinitesimal contour path c_s and the integration paths c_1 and c_2 for eqns (2.3) and (2.4).

On the other hand, however, if the integrand has extra poles, like those of the function $W(t)$, besides the necessary poles of t_s, the foregoing method cannot be used and another integrand must be sought.

From the Wronskian identity, we see that

$$\{qv(t_s) - v'(t_s)\}\, W(t_s) = 1,$$

$$v(-t) = \tfrac{1}{2}\sqrt{(\pi/3)}\, \{e^{-i\pi/6}\, t^{1/2}\, H^{(2)}_{1/3}\, (\tfrac{2}{3}t^{3/2}) + e^{i\pi/6}\, t^{1/2}\, H^{(1)}_{1/3}(\tfrac{2}{3}t^{3/2})\}. \tag{2.5}$$

Hence, in principle, we could multiply this function to the integrand to any power as we wish. For example, by multiplying it once, the equation (2.3) is replaced by

$$\sum_s a(t_s) = \frac{-1}{2\pi i} \sum_s \int_{C_s} \left(\frac{v'(t) - qv(t)}{W'(t) - qW(t)}\right) (t - q^2)\, W(t)^2\, a(t)\, dt. \tag{2.6}$$

Thus, even if the old integrand had the undesirable extra poles of the function $W(t)$, the new integrand would not have them. Thus we have the contour integration path C, and, further, the path $C_1 + C_2$ if the integrand decreases sufficiently rapidly at infinity. This situation actually occurs in the series of (1.6).

Here the question may occur whether it is possible to multiply the square of the function (2.5) to the integrand and to deform the integration path to the path $C_1 + C_2$. The answer is no because, though the integrand has no extra pole besides t_s's, it diverges at infinity on the way deforming the path from the C to the path $C_1 + C_2$. Hence, generally, there exists only one integrand for which the integration path $C_1 + C_2$ is available.

Usually the Kirchoff approximation terms appear as the leading terms of these integrals, and are obtained from the asymptotic forms of the integrands for large magnitude of t; for instance,

$$\frac{-1}{2\pi i}\left\{\frac{v'(t) - qv(t)}{W'(t) - qW(t)}\right\} \sim -\frac{1}{4\pi} \tag{2.7}$$

on the path C_1 or any path of $(0, \infty\,\mathrm{e}^{-i\beta})$ in the range $\pi > \beta > \pi/3$, and it tends to zero as $|t| \to \infty$ on the path C_2.

Using these asymptotic forms in the integrand of A of the case (ii), it becomes

$$A \simeq \exp[\quad]\sqrt{\left(\frac{r}{k_1 r_1 r_2}\right)}\,(k_1 a/2)^{1/3}\,\frac{1}{\sqrt{(2\pi)}}\int\limits_0^{\infty \mathrm{e}^{-i\pi/2}} \mathrm{d}t$$

$$\exp[-\,i\{\theta(k_1 a/2)^{1/3}\,t + (2k_1)^{-1}\,(1/r_1 + 1/r_2)\,(k_1 a/2)^{2/3}\,t^2\}]$$

$$= \exp[\quad]\,\frac{1}{\sqrt{\pi}}\,\mathrm{e}^{i\zeta^2}\int\limits_{\zeta \mathrm{e}^{i\pi/4}}^{\infty} \mathrm{e}^{-t^2}\,\mathrm{d}t, \tag{2.8}$$

$$\zeta = \theta\sqrt{[k_1 r_1 r_2/2(r_1 + r_2)]},$$

with

$$\exp[\quad] = \exp[-\,i(k_1 a/3)\,\{(r_1/a)^3 + (r_2/a)^3\}],$$

The result of exact evaluation takes the form

$$A = \exp[\quad]\left\{\frac{1}{\sqrt{\pi}}\,\mathrm{e}^{i(\xi/\eta)^2}\int\limits_{(\xi/\eta)\mathrm{e}^{i\pi/4}}^{\infty} \mathrm{e}^{-t^2}\,\mathrm{d}t - \eta G(\xi)\right\},$$

$$\xi = \theta(k_1 a/2)^{1/3}, \qquad \eta = (k_1 a/2)^{1/3}\sqrt{[(2/k_1)\,(1/r_1 + 1/r_2)]}. \tag{2.9}$$

Here, the numerical values of the function $G(\xi)$ have been calculated by Logan (1959) for the surface of perfect conductor, and by Wait and Conda (1959) for a wide range of surface impedance. It may be remarked that the similar numerical integration method for the cases of poor convergence has been adopted more previously by Fock (1946) and Rice.

3. EQUATION FORMULATION

In this section, the equations are formulated to obtain the field in the general case in which several mountains exist along the wave path. In order to survey the subject from a wider point of view and also to facilitate the equation formulation, it is convenient to represent the six components of electromagnetic field in a unified form (Furutsu, 1952; Marcuvitz, 1962).

Using the usual notations in Gaussian units, Maxwell's equations are given by

$$\begin{aligned} \text{rot } H - i(\omega/c)\,\epsilon E &= 4\pi I/c, \\ \text{rot } E + i(\omega/c)\,\mu H &= 0. \end{aligned} \tag{3.1}$$

Introducing the matrices

$$s_1 = \begin{pmatrix} 0 & 0 & 0 \\ 0 & 0 & i \\ 0 & -i & 0 \end{pmatrix}, \quad s_2 = \begin{pmatrix} 0 & 0 & -i \\ 0 & 0 & 0 \\ i & 0 & 0 \end{pmatrix}, \quad s_3 = \begin{pmatrix} 0 & i & 0 \\ -i & 0 & 0 \\ 0 & 0 & 0 \end{pmatrix}, \tag{3.2}$$

the operator "rot" as a three-rowed matrix is expressed by

$$\text{rot} = \sum_{i=1}^{3} s_i \partial_i \equiv (s\partial), \quad \partial_j = i\frac{\partial}{\partial x_j} \quad (j = 1, 2, 3). \tag{3.3}$$

Thus, eqn (3.1) can be expressed in the form

$$(s\partial)\begin{pmatrix} 0 & 1 \\ 1 & 0 \end{pmatrix}\begin{pmatrix} iE \\ H \end{pmatrix} - \frac{\omega}{c}\begin{pmatrix} \epsilon & 0 \\ 0 & \mu \end{pmatrix}\begin{pmatrix} iE \\ H \end{pmatrix} = \begin{pmatrix} \frac{4\pi}{c} I \\ 0 \end{pmatrix}, \tag{3.4}$$

which takes the form

$$[\rho_1(s\partial) - k]\,\psi = j. \tag{3.5}$$

Here,

$$\psi = \begin{pmatrix} iE \\ H \end{pmatrix}, \quad j = \begin{pmatrix} \frac{4\pi}{c} I \\ 0 \end{pmatrix}, \quad k = \frac{\omega}{2c}\{(1 + \rho_3)\,\epsilon + (1 - \rho_3)\,\mu\}, \tag{3.6}$$

and ρ_i's are the two-rowed Pauli's matrices

$$\rho_1 = \begin{pmatrix} 0 & 1 \\ 1 & 0 \end{pmatrix}, \quad \rho_3 = \begin{pmatrix} 0 & -i \\ i & 0 \end{pmatrix}, \quad \rho_3 = \begin{pmatrix} 1 & 0 \\ 0 & -1 \end{pmatrix}. \tag{3.7}$$

Equation (3.5) finally takes the form

$$\left[\sum_{i=1}^{3} \gamma_i \partial_i - k\right] \psi = j, \qquad \gamma_i = \rho_1 s_i. \tag{3.8}$$

Here, γ_i's are antisymmetrical Hermitian matrices and thus

$$\gamma_i^* = \gamma_i^T = -\gamma_i, \tag{3.9}$$

where γ_i^* and γ_i^T are the complex conjugate and transposed matrices of γ_i, respectively. Further, s_i's and ρ_i's have the following algebraic relations which are often very convenient to use:

$$\begin{aligned} &s_i s_j s_k + s_k s_j s_i = \delta_{ij} s_k + \delta_{jk} s_i, \\ &s_i s_j - s_j s_i = i^{-1} \sum_{k=1}^{3} \epsilon_{ijk} s_k, \quad s_1^2 + s_2^2 + s_3^2 = 2, \\ &\rho_1^2 = \rho_2^2 = \rho_3^2 = 1, \quad \rho_1\rho_2 = -\rho_2\rho_1 = i\rho_3, \quad \text{etc.} \end{aligned} \tag{3.10}$$

Here ϵ_{ijk} is the antisymmetrical tensor of $\epsilon_{123} = 1$ and it is $+1$ or -1 according to whether the number of permutations necessary to change the order of subscripts i, j, k into the order 1, 2, 3 is even or odd.

Green's theorem in this representation takes the following form: let ψ'' and ψ' be arbitrary continuous ψ-vectors in the space Σ, then

$$\psi''[(\gamma\partial) - k]\,\psi' - \psi''[-(\gamma\overleftarrow{\partial}) - k]\,\psi' = i \sum_{j=1}^{3} \frac{\partial}{\partial x_j}(\psi''\gamma_j\psi'),$$

gives, on integrating both sides in the space Σ,

$$\int_\Sigma \{\psi''[(\gamma\partial) - k]\,\psi' - \psi''[-(\gamma\overleftarrow{\partial}) - k]\,\psi'\}\,dv = -i\,[\psi''(s)\,(\eta\gamma)\,\psi'(s)]. \tag{3.11}$$

Here

$$[\psi''(s)\,(n\gamma)\,\psi'(s)] = \int_S \psi''(n\gamma)\,\psi'\,ds = \int_S [H' \times E'' + E' \times H'']_n\,ds \tag{3.12}$$

and n is the unit vector inward, normal to the surface S enclosing the space Σ, and E', H' and E'', H'' are the electric and magnetic fields of ψ' and ψ'' respectively.

It may be remarked that the integrand of the surface integral of (3.12) is continuous in whole space, even on boundaries of discontinuous medium, because it contains only the tangential components of E–M field.

For later convenience, the Green function $\psi(x, x')$ is introduced: it satisfies the given conditions in space and at boundaries and also

$$\begin{aligned} &[(\gamma\partial) - k]\,\psi(x, x') = -i\delta(x - x'), \\ &\psi(x, x')\,[-(\gamma\overleftarrow{\partial}') - k] = -i\delta(x - x'), \end{aligned} \tag{3.13}$$

where $\overleftarrow{\partial}'$ operates on the coordinates x' on the left side. The latter equation of (3.13), being valid even in anisotropic medium, can be proven by the use of Green's theorem (3.11): let $\overline{\psi}(x', x)$ be the solution of

$$\overline{\psi}(x', x)\,[-(\gamma\overleftarrow{\partial}) - k] = -i\delta(x' - x)$$

with the same boundary conditions as those of $\psi(x, x')$. Now in (3.11), let $\psi'(x) = \psi(x, x'')$ and $\psi''(x) = \bar{\psi}(x', x)$ and take for the space Σ, the whole space. Then

$$\bar{\psi}(x', x'') - \psi(x', x'') = [\bar{\psi}(x', s)\,(n\gamma)\,\psi(s, x'')]_{s=\infty} = 0,$$

which gives the necessary proof.

The solution of (3.13) in a homogeneous medium can be obtained as follows: put

$$\psi(x, x') = [(\gamma\partial) + \{(\gamma\partial)^2 - \partial^2 + k_0^2\}\,k^{-1}]\,\phi(x, x'), \tag{3.14}$$

then, on using the relation

$$[(\gamma\partial) - k]\,[(\gamma\partial) + \{(\gamma\partial)^2 - \partial^2 + k_0^2\}\,k^{-1}] = \partial^2 - k_0^2,$$
$$k_0^2 = k\rho_1 k\rho_1 = (\omega/c)^2\,\epsilon\mu \tag{3.15}$$

which is readily proven by the use of (3.10), it follows from (3.13) that $\phi(x, x')$ satisfies

$$[\partial^2 - k_0^2]\,\phi(x, x') = -\,i\delta(x - x'),$$

or

$$\phi(x, x') = \frac{-i}{4\pi\,|x - x'|}\,e^{-ik_0|x-x'|}. \tag{3.16}$$

Now we can formulate the necessary equations for the general case of diffraction by several surfaces. We shall first begin with the simplest case of one diffracting surface like Fig. 3: in the Green theorem (3.11), let

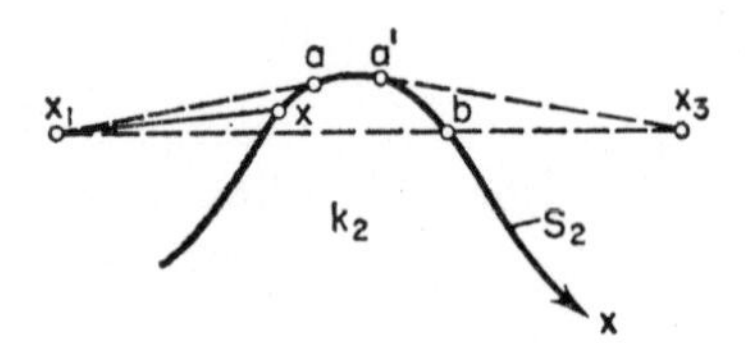

FIG. 3. The form of diffracting surface and the notations for eqn (3.19).

$\psi'(x) = \psi_0(x, x_1)$ the solution of (3.13) in free space, and $\psi''(x) = \psi_2(x_3, x)$ the solution to be obtained for the diffracting surface having the medium constant k_2, and for the space Σ take the whole upper space over the diffracting surface, then we have for arbitrary points x_3 and x_1 in Σ,

$$\psi_2(x_3, x_1) = \psi_0(x_3, x_1) + [\psi_2(x_3, s_2)\,(n_2\gamma)\,\psi_0(s_2, x_1)],$$
$$n_2 = n(s_2). \tag{3.17}$$

Here s_2 stands for the diffracting surface of k_2, and the outward propagating condition at infinity for both ψ_2 and ψ_0 is taken into account in (3.17).

In the same way if we put $\psi'(x) = \psi_2(x, x_1)$ and $\psi''(x) = \psi_0(x_3, x)$ in

(3.11), we would have

$$\psi_2(x_3, x_1) = \psi_0(x_3, x_1) - [\psi_0(x_3, s_2)\,(n_2\gamma)\,\psi_2(s_2, x_1)]. \qquad (3.18)$$

On the other hand we could use the expression (3.18) for $\psi_2(x_3, s_2)$ in the integrand of (3.17):

$$\psi_2(x_3, x_1) = \psi_0(x_3, x_1) - [\psi_0(x_3, s_2)\,(n_2\gamma)\,\psi_2(s_2, s_2')\,(n_2'\gamma)\,\psi_0(s_2', x_1)],$$

$$n_2 = n(s_2), \qquad n_2' = n(s_2'). \qquad (3.19)$$

Here, the relation

$$[\psi_0(x_3, s_2)\,(n_2\gamma)\,\psi_0(s_2, x_1)] \equiv 0$$

is taken into account, and s_2' stands for another surface integral of (3.12) on the same surface as s_2.

The second term of (3.18) on the right side is evidently the reflected waves from the surface. The same is true of the 2nd term of (3.19). Furthermore, there is a clear one-to-one correspondence between the terms of the integrand of (3.18) and the respective parts of wave path, i.e. the term $\psi_2(s_2, x_1)$ corresponds to the path from the point x_1 to the diffracting surface s_2 and the term $\psi_0(x_3, s_2)$ corresponds to the propagation from the surface to the point x_3. This correspondence is best in the integrand of (3.19): $\psi_0(s_2', x_1)$ corresponds to the propagation in free space from the point x_1 to the diffracting surface, and $\psi_2(s_2, s_2')$ corresponds to the propagation along the diffracting surface, and $\psi_0(x_3, s_2)$ corresponds to the propagation, again in free space, from the surface to the point x_3.

We shall first consider the integrand of (3.18). Taking the coordinate x along the diffracting surface, the phase of the integrand changes with x as in Fig. 4. From the point of view of the phase stationary method of approxi-

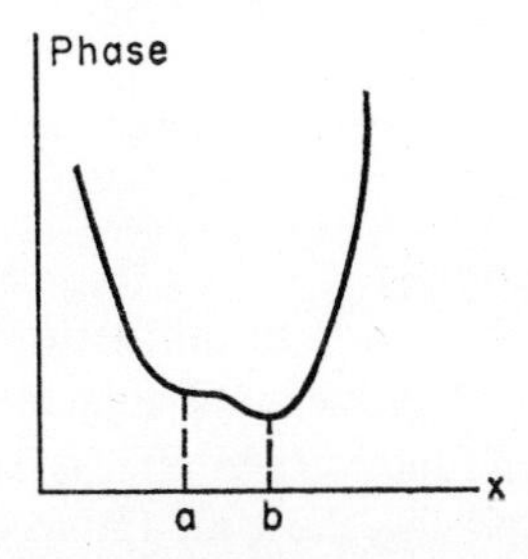

FIG. 4. The phase of the integrand of (3.18) as a function of x.

mation, the surface integral is decisively determined by integration over the small domain of s_2 in the vicinity of wave path, where the phase is almost stationary. In this small domain of diffracting surface, the surface could be expressed as a surface of 2nd degree, having the same radius of curvature as a sphere, cylinder, parabolid, etc., and therefore it implies that we could solve the wave equation exactly, in the range of the important part of diffracting surface. In fact, their Green functions usually happen to be the same in this small range, independently of the kind of surfaces adopted, provided that the radius of curvature of the diffracting surface is defined along the wave path.

It is now very easy to proceed to the general case of diffraction by two or more diffracting surfaces. Let $\psi_{n,\ldots,2}(x_{n+1}, x_1)$ be the solution of (3.13) to be obtained in the case where diffracting surfaces of the numbers from 2 to n exist along the wave path. The problem is equivalent to getting the recurrence formula for the solution $\psi_{n,\ldots,2}$ i.e. the solution $\psi_{n+1,\ldots,2}$ in terms of the solution $\psi_{n,\ldots,2}$ in which the $n+1$th diffracting obstacle is not present.

The way of formulating the integral equation is just the same as in the case of one diffracting surface: in eqns (3.18) and (3.19) we could replace ψ_0 by $\psi_{n,\ldots,2}$ and ψ_2 by $\psi_{n+1,\ldots,2}$ and the space Σ by the whole space surrounding the $n+1$th obstacle, and thus

$$\psi_{n+1,\ldots,2}(x_{n+2}, x_1) = \psi_{n,\ldots,2}(x_{n+2}, x_1) + [\psi_{n+1,\ldots,2}(x_{n+2}, s_{n+1})(n_{n+1}\gamma)\,\psi_{n,\ldots,2}(s_{n+1}, x_1)], \quad (3.20)$$

or

$$\psi_{n+1,\ldots,2}(x_{n+2}, x_1) = \psi_{n,\ldots,2}(x_{n+2}, x_1) - [\psi_{n,\ldots,2}(x_{n+2}, s'_{n+1})(n'_{n+1}\gamma)\,\psi_{n+1,\ldots,2}(s'_{n+1}, s_{n+1}) \times (n_{n+1}\gamma)\,\psi_{n,\ldots,2}(s_{n+1}, x_1)]. \quad (3.21)$$

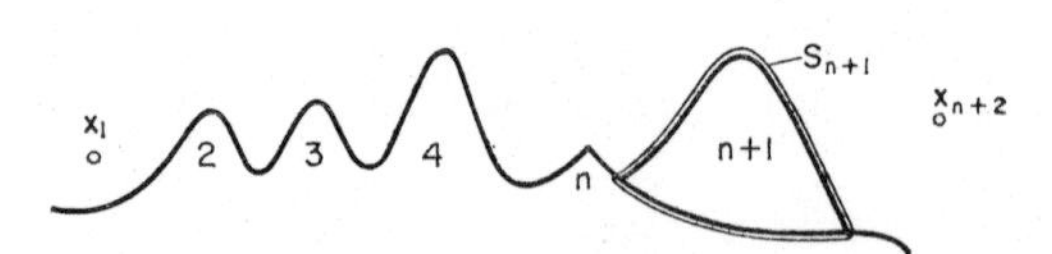

FIG. 5. The form of terrain and the notations for eqns (3.21) and (3.23).

Here, as in (3.13), the term $\psi_{n+1,\ldots,2}(s'_{n+1}, s_{n+1})$ in (3.21) corresponds to the propagation along the $n+1$-th diffracting surface s_{n+1} and therefore, from the point of view of the phase stationary method of approximation, we may replace it by the Green function $\psi^{(n+1)}(s'_{n+1}, s_{n+1})$ for a surface having the same surface impedance and also the same radius of curvature as the $n+1$-th diffracting surface in the vicinity of wave path. On the other hand, the term $\psi_{n,\ldots,2}(x_{n+2}, s_{n+1})$ in (3.21) corresponds to the propagation from the surface s_{n+1} to the point x_{n+2} in the case in which the $n+1$-th obstacle was not present. Hence, if the point x_{n+2} is sufficiently far from other diffracting surfaces as in Fig. 5, the effect of reflections from their surfaces to the term $\psi_{n,\ldots,2}(x_{n+2}, s_{n+1})$ will be very small and thus the latter could be replaced by $\psi_0(x_{n+2}, s_{n+1})$, the solution in free space.

As a result, we could put in (3.21)

$$\psi_{n+1,\ldots,2}(s'_{n+1}, s_{n+1}) \simeq \psi^{(n+1)}(s'_{n+1}, s_{n+1}),$$
$$\psi_{n,\ldots,2}(x_{n+2}, s'_{n+1}) \simeq \psi_0(x_{n+2}, s'_{n+1}), \quad (3.22)$$

and thus we have

$$\psi_{n+1,\ldots,2}(x_{n+2}, x_1) \simeq \psi_{n,\ldots,2}(x_{n+2}, x_1)$$
$$-[\psi_0(x_{n+2}, s'_{n+1})\,(n'_{n+1}\gamma)\,\psi^{(n+1)}(s'_{n+1}, s_{n+1})\,(n_{n+1}\gamma)\,\psi_{n,\ldots,2}(s_{n+1}, x_1)]. \quad (3.23)$$

In the same way, when the point x_{n+2} is on or close to the surface s_{n+1}, we may put in (3.20)

$$\psi_{n+1,\ldots,2}(x_{n+2}, s_{n+1}) \simeq \psi^{(n+1)}(x_{n+2}, s_{n+1}) \quad (3.24)$$

and have

$$\psi_{n+1,\ldots,2}(x_{n+2}, x_1) \simeq \psi_{n,\ldots,2}(x_{n+2}, x_1)$$
$$+[\psi^{(n+1)}(x_{n+2}, s_{n+1})\,(n_{n+1}, \gamma)\,\psi_{n,\ldots,2}(s_{n+1}, x_1)]. \quad (3.25)$$

Equations (3.23) and (3.25) give us the required recurrence formulas for $\psi_{n,\ldots,2}$.

The actual evaluation has been performed on the assumption that the diffracting obstacles distribute so that the optical wave lies in a plane in the whole range of the wave path (Furutsu, 1956). The principle of evaluation is just the same as the phase stationary method of approximation. However, it is remarked that the phase is not stationary at one point but in some range or domain, and therefore we must take into account not only the change of the phase term but also the exact change of the whole integrand.

4. GENERAL FORMULA FOR DIFFRACTION BY TWO OR MORE SURFACES

As illustrated in Figs. 6 and 7, the general case is considered here where the waves propagate over several mountains having the numbers $2, 3, \ldots, n$ along the wave path from the point x_1 to x_{n+1}, and the radii of curvature (along the wave path) $a_2, a_3, \ldots, a_n$, and the propagation constants $k_2, k_3, \ldots, k_n$ respectively. The diffraction angles of the respective mountains will be denoted by $\theta_2, \theta_3, \ldots, \theta_n$, and thus $d_m = a_m\theta_m (m = 2, 3, \ldots, n)$ will be the distance of part of the wave path contacting the mth mountain.

On the assumption that the whole wave path lies in a plane profile, the result of evaluation gives the following expressions for the attenuation coefficient A which is just a generalizations of (1.5) and (1.6): using the notation t_m for the set of roots of eqn (1.2) in the case of $q = q_m$, they are (Furutsu, 1956):

(i) $(a_m/r_{m,\, m\pm1})^{3/4}(k_1a_m)^{-1/4} \ll 1 \quad (2 \leqslant m \leqslant n)$

$$A = \{(r_{n+1,\, n} + d_n + \ldots + r_{32} + d_2 + r_{21}) / k_1^{n-1} r_{n+1,\, n} r_{m,\, n-1} \ldots r_{32} r_{21}\}^{1/2}$$
$$\times \sum_{t_n, \ldots, t_3, t_2} T(r_{n+1,\, n})_{0,\, t_n} T(\xi_n)_{t_n} \ldots T(\xi_3)_{t_3} T(r_{32})_{t_3, t_2} \times T(\xi_2)_{t_2} T(r_{21})_{t_2,\, 0}. \quad (4.1)$$

(ii) $(a_2/r_{21})^{3/4}(k_1a_2)^{-1/4} \gtrsim 1, \quad (a_m/r_{m,\, m\pm1})^{3/4}(k_1a_m)^{-1/4} \ll 1 \; (3 \leqslant m \leqslant n)$

$$A = \{(r_{n+1,\, n} + d_n + \ldots + r_{32} + d_2)/k_1^{n-2} r_{n+1,\, n} r_{n,\, n-1} \ldots r_{43} r_{32}\}^{1/2}$$
$$\times \sum_{t_n, \ldots, t_3, t_2} T(r_{n+1,\, n})_{0,\, t_n} T(\xi_n)_{t_n} \ldots T(r_{43})_{t_4,\, t_3} T(\xi_3)_{t_3} \times T(r_{32})_{t\; t_3,\, 2} T(\xi_2, z_1)_{t_2}. \quad (4.2)$$

(iii) $(a_2/r_{21})^{3/4}(k_1 a_2)^{-1/4} \gtrsim 1, \quad (a_n/r_{n+1,\, n})^{3/4}(k_1a_n)^{-1/4} \gtrsim 1$ and

$(a_m/r_{m,\, m\pm1})^{3/4}(k_1a_m)^{-1/4} \ll 1 \quad (3 \leqslant m \leqslant n-1)$

$$A = \{(d_n + r_{n,\, n-1} + \ldots + d_3 + r_{32} + d_2) / k_1^{n-3} r_{n,\, n-1} r_{n-1,\, n-2} \ldots r_{43} r_{32}\}^{1/2}$$
$$\times \sum_{t_n, \ldots, t_3, t_2} T(\xi_n, z_{n+1})_{t_n} T(r_{n,\, n-1})_{t_n,\, t_{n-1}} T(\xi_{n-1})_{t_{n-1}} \ldots \times T(\xi_3)_{t_3} T(r_{32})_{t_3,\, t_2} T(\xi_2, z_1)_{t2}. \quad (4.3)$$

Here, $r_{m,\, n}$ is the distance between the mth and nth mountains and

$$\xi_m = \theta_m(k_1 a_m/2)^{1/3},$$
$$T(\xi)_{t_m} = 2\sqrt{2\pi}(k_1a_m/2)^{1/3}(t_m - q_m^2)^{-1} W(t_m)^{-2} e^{-i(\xi t_m + 3\pi/4)},$$
$$T(r)_{t_m,\, t_n} = \exp[-i(2k_1 r)^{-1}\{(k_1 a_m/2)^{1/3} t_m - (k_1a_n/2)^{1/3} t_n\}^2],$$
$$T(\xi, z)_{t_m} = 2\sqrt{\pi}(t_m - q_m^2)^{-1} W(t_m)^{-1} f_{t_m}(z, a_m) e^{-i(\xi t_m + \pi/2)}. \quad (4.4)$$

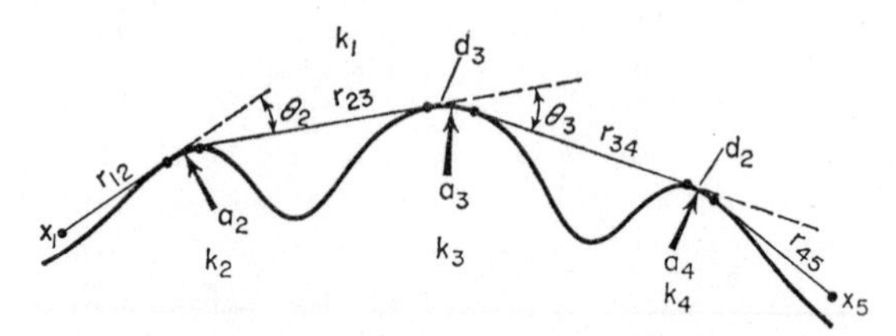

FIG. 6. The form of diffracting surfaces and the notations for eqn (4.1).

In the case (i), both the transmitter and receiver are sufficiently apart from the diffracting mountains (Fig. 6). The formula (4.1) in this case consists of two kinds of terms one of which, $T(\xi_m)_{t_m}$, depends only on the diffraction angle, the surface impedance and the radius of curvature of the mth mountain, but does not depend on those quantities of other diffracting mountains.

On the other hand, the other kind of term $T(r_{m,\,n})_{t_m,\,t_n}$ depends on the radii of curvature of the m and nth mountains and their electrical properties through t_m and t_n and also the propagation distance between them, but does not depend on other quantities. Therefore it works as the coupling term between the m and nth mountains. There is no other kind of term.

This fact facilitates the understanding of this formula considerably; there is a clear one-to-one correspondence between the terms in the formula and the respective parts of the wave path: for instance $T(r_{21})_{t_2,\,0}$ corresponds to the wave path from the point x_1 to the 2nd diffracting surface, the term $T(\xi_2)_{t_2}$ corresponds to the wave path along the same diffracting surface, and $T(r_{32})_{t_3,\,t_2}$ corresponds to the propagation in free space from the 2nd to the 3rd diffracting surface, etc.

As already stated, this formula is essentially the generalization of formula (1.6) for one diffracting surface, which was derived from the original van der Pol and Bremmer formula on the restricted condition, i.e. both the transmitter and the receiver are not far from the diffracting surface compared with the radius of curvature. But here it is derived from this general formula as the special case of one diffracting surface.

In the case (ii), either the transmitter or receiver say the point x_1 is on or near the diffracting surface (Fig. 7). The only difference from that of the

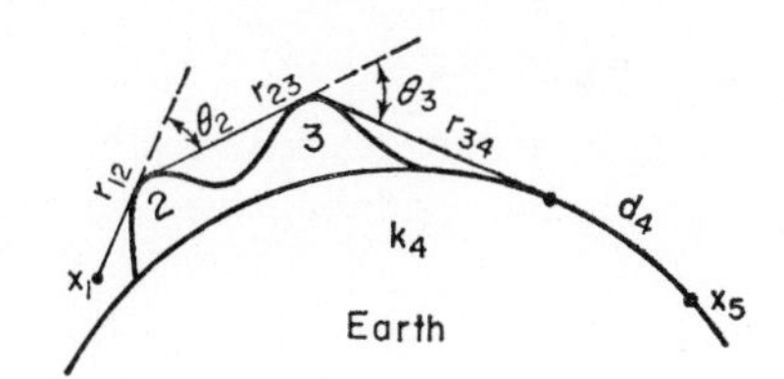

FIG. 7. The form of terrain and the notations for eqn (4.2).

case (i) is the last two terms; they are replaced here by the term $T(\xi_2, z_1)$ and the others are just the same. Here, $f_{t_m}(z, a_m)$ is the ordinarily defined height gain function. Again, the special case of this formula agrees with (1.5).

In the case (iii), both the transmitter and receiver are on or near the respective diffracting surfaces. The change to be made in the preceding formulas is so evident that it is not necessary to mention it here explicitly.

Sometimes it may be necessary to take into account the contributions of the waves which are reflected from the surfaces between the diffracting mountains. Mathematically speaking, these reflecting points on the surfaces are just the phase stationary points of the integrand, and we could get the result from these formulas by multiplying the reflection coefficients of the surfaces and also adjusting the diffraction angles so that they correspond to the reflected wave paths.

5. SUMMATION OF THE SERIES

Finally the problem of convergence of the formulas (4.1)–(4.3) remains. There are two kinds of series in the above formulas; one is of the form

$$\sum_{t_m} T(r_{m+1,\ m})_{t_{m+1},\ t_m}\ T(\xi_m)_{t_m}\ T(r_{m,\ m-1})_{t_m,\ t_{m-1}} \tag{5.1}$$

and its convergence is poor when the ξ_m is near or smaller than 1. Now the treatment for the case of poor convergence is exactly the same as in Section 2. Using the same method, we have the result

$$T(r_{m+1,\ m})_{t_{m+1},\ 0}\ T(r_{m,\ m-1})_{0,\ t_{m-1}} \sqrt{2}(k_1 a_m/2)^{1/3} F(\xi_m - \xi',\ \eta_m, q_m). \tag{5.2}$$

Here, on referring to eqn (2.9)

$$F(\xi, \eta, q) = \frac{1}{\sqrt{\pi}.\eta}\, e^{i(\xi/\eta)^2} \int\limits_{(\xi/\eta)e^{i\pi/4}}^{\infty} e^{-t^2}\, dt - G(\xi),$$

$$\eta_m = (k_1 a_m/2)^{1/3} \sqrt{[(2/k_1)\ (1/r_{m+1,\ m} + 1/r_{m,\ m-1})]},$$

$$\xi'_m = (k_1 a_m/2)^{1/3} \{(k_1 r_{m+1,\ m})^{-1} (k_1 a_{m+1}/2)^{1/3}\ t_{m+1} + (k_1 r_{m,\ m-1})^{-1} (k_1\ a_{m-1}/2)^{1/3}\ t_{m-1}\}, \tag{5.3}$$

and ξ'_m depends on the preceding and succeeding diffracting surfaces through $t_{m\pm 1}$ and thus it works as the only coupling variable between the diffracting surfaces.

When the diffraction angle of the mth surface is small and/or the radius of curvature is sufficiently small, the 2nd correction term becomes very small in most cases and can be neglected compared with the 1st leading term; in the case of $\xi_m \ll 1$, the series (5.1) takes a simple form

$$\sum_{t_m} \simeq T(r_{m+1,\ m})_{t_{m+1},\ 0}\ T(r_{m,\ m-1})_{0,\ t_{m-1}}\ \tfrac{1}{2}\mathscr{E}((\xi_m - \xi'_m)\ \eta_m \cdot e^{i\pi/4}) \times \sqrt{[k_1/(1/r_{m+1,\ m} + 1/r_{m,\ m-1})]},\ \xi_m \ll 1,$$

$$\mathscr{E}(z) = \frac{2}{\sqrt{\pi}}\, e^{z^2} \int\limits_{2}^{\infty} e^{-t^2}\, dt. \tag{5.3a}$$

Another special case is the case in which the diffraction angles of $m \pm 1$-th mountains are sufficiently large as $\xi_{m\pm 1}/\eta_{m\pm 1} \gg 1$. It follows then that $\xi'_m/\eta_m \ll 1$ and thus ξ'_m in (5.3) can be neglected:

$$\sum_{t_m} = T(r_{m+1,\ m})_{t_{m+1},\ 0}\ T(r_{m,\ m-1})_{0,\ t_{m-1}} \sqrt{2}.(k_1 a_m/2)^{1/3}\ F(\xi_m, \eta_m, q_m). \tag{5.3b}$$

Similarly, on the condition of $\xi_m/\eta_m \gg 1$ and $\xi_l/\eta_l \gtrsim -1$ $(l \neq m)$,

$$\sqrt{2}.(k_1 a_m/2)^{1/3} F(\xi_m - \xi_m', \eta_m, q_m)$$
$$\simeq \sqrt{2}.(k_1 a_m/2)^{1/3} \left[\frac{1}{2\sqrt{\pi\xi_m}} e^{-i\pi/4} - G(\xi_m)\right] \equiv M(\xi_m, a_m),$$
$$\xi_m \gg |\xi_m'|. \qquad (5.3c)$$

Here, the function $M(\xi_m, a_m)$ is introduced for later convenience.

The other kind of series is

$$\sum_{t_2} T(r_{32})_{t_3,\, t_2} T(\xi_2, z_1)_{t_2} \simeq T(r_{32})_{t_3,\, 0} \sum_{t_2} T(\xi_2 - \xi_2', z_1)_{t_2}, \qquad (5.4)$$

which always occurs with the height gain function of the receiver or transmitter. Using the same method as in Section 2, it takes the form, on the condition that $\xi_l/\eta_l \gg 1$ $(l \neq 2)$ and $\xi_2/\eta_2 \gtrsim -1$,

$$\sum_{t_2} T(\xi_2 - \xi_2', z_1)_{t_2} \simeq g(\xi_2)\{1 + i(k_1/k_2')\, z_1\}, \qquad (2/k, a_2)^{1/3} k_1 z_1 \ll 1,$$

$$k_1/k_2' = \begin{cases} k_1\sqrt{(k_2^2 - k_1^2)}/k_2^2, & \text{Ver. Pol.} \\ \sqrt{(k_2^2 - k_1^2)}/k_1, & \text{Horiz. Pol.} \end{cases} \qquad (5.5)$$

Here, z_1 is the height of the point x_1 from the surface, and the function $g(\xi)$ has been numerically calculated by Logan (1957) for the case of a perfect conductor and by Wait and Conda (1959) for a wide range of surface impedance.

The special applications of the supplemental formulas will be as follows:

(i) $\xi_m/\eta_m \gg 1 \qquad (n \geq m \geq 2)$

This is the case where the diffraction angle of every diffracting surface is sufficiently large. On using (5.3c) in the formula (4.1), the attenuation coefficient A takes the form

$$A = \{(r_{n+1,\,n} + d_n + \ldots + r_{32} + d_2 + r_{21})/ k_1^{n-1} r_{n+1,\,n} r_{n,\,n-1} \ldots r_{32} r_{21}\}^{1/2}$$
$$\times M(\xi_n, a_n) M(\xi_{n-1}, a_{n-1}) \ldots M(\xi_2, a_2), \qquad (5.6a)$$

or, when either the transmitter or receiver or both are on or near the diffracting surface, the formulas (4.2–3) respectively give, on using (5.5),

$$A = \{(r_{n+1,\,n} + d_n + \ldots + r_{32} + d_2)/ k_1^{n-2} r_{n+1,\,n} r_{n,\,n-1} \ldots r_{43} r_{32}\}^{1/2}$$
$$\times M(\xi_n, a_n) \ldots M(\xi_3, a_3)\, g(\xi_3) f(z_1, a_2), \qquad (5.6b)$$

$$A = \{(d_n + r_{n,\,n-1} + \ldots + r_{32} + d_z)/k_1^{n-3} r_{n,\,n-1} \ldots r_{43} r_{32}\}^{1/2}$$
$$\times f(z_{n+1}, a_n)\, g(\xi_n) M(\xi_{n-1}, a_{n-1}) \ldots M(\xi_3, a_3)\, g(\xi_2) f(z_1, a_2). \qquad (5.6c)$$

These results are just the so-called multiplication rule in diffraction.

Another application that may be of interest will be when one of the diffracting surfaces does not work as a diffracting obstacle but simply as a reflecting surface, as in Fig. 8. In this case, the coupling between the preceding and succeeding surfaces becomes very serious. From the leading term of the obtained formula (5.3a), we have

(ii) $\xi_m/\eta_m \ll -1$

$$\sum_{t_m} T(r_{m+1,\ m})_{t_{m+1},\ t_m}\, T(\xi_m)t_m\, T(r_{m,\ m-1})_{t_m,\ t_{m-1}}$$
$$\simeq T(r_{m+1,\ m})_{t_{m+1},\ 0}\, T(r_{m,\ m-1})_{0,\ t_{m-1}} \sqrt{[k_1/(1/r_{m+1,\ m} + 1/r_{m,\ m-1})]}$$
$$\times \exp[i(\xi_m - \xi_m')^2/\eta_m^2].$$
$$= T(r_{m+1,\ m} + r_{m,\ m-1})_{t_{m+1},\ t_{m-1}}$$
$$\times \exp[i\{(\xi_m/\eta_m)^2 - \theta_m(r_{m,\ m-1}/(r_{m+1,\ m} + r_{m,\ m-1}))(k_1 a_{m+1}/2)^{1/3}\, t_{m+1}$$
$$- \theta_m(r_{m+1,\ m}/(r_{m+1,\ m} + r_{m,\ m-1}))(k_1 a_{m-1}/2)^{1/3}\, t_{m-1}\}],$$
$$\xi_m/\eta_m \ll -1. \tag{5.7}$$

When this result is substituted in the formulas (4.1–3), they take just those values when there is no diffracting surface at all between these $m \pm 1$-th diffracting surfaces: the phase terms of

$$\theta_m\{r_{m,\ m\mp1}/(r_{m+1,\ m} + r_{m,\ m-1})\}\, (k_1 a_{m\pm1}/2)^{1/3}\, t_{m\pm1}$$

are respectively combined with the terms of $T(\xi_{m\pm1})_{t_{m\pm1}}$ to give the new diffraction angles $\theta_{m\pm1}'$, due to the omission of mth surface (Fig. 8).

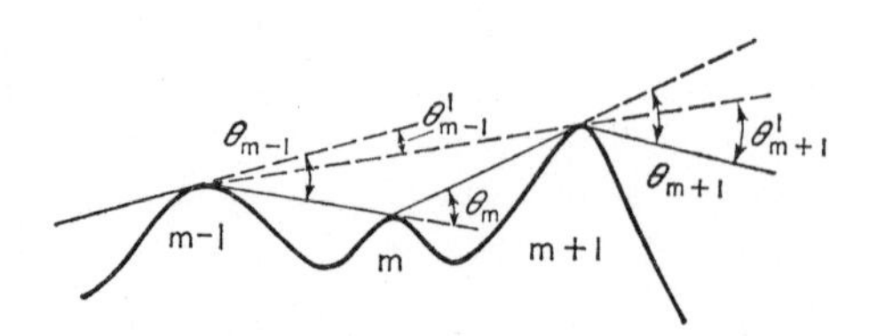

FIG. 8. The situation in the case of eqn (5.7).

In just the same way, when the convergence of the double series

$$\sum_{t_m,\ t_{m+1}} T(r_{m+2,\ m+1})_{t_{m+2},\ t_{m+1}}\, T(\xi_{m+1})_{t_{m+1}}\, T(r_{m+1,\ m})_{t_{m+1},\ t_m}$$
$$\times T(\xi_m)_{t_m}\, T(r_{m,\ m-1})_{t_m,\ t_{m-1}} \tag{5.8}$$

is poor with respect to both t_m and t_{m+1}, we have the leading term (Fig. 9)

$$T(r_{m+2,\ m+1})_{t_{m+2},\ 0}\, T(r_{m,\ m-1})_{0,\ t_{m-1}}$$
$$\times \{k_1^2\, r_{m+2,\ m+1}\, r_{m+1,\ m}\, r_{m,\ m-1}/(r_{m+2,\ m+1} + r_{m+1,\ m} + r_{m,\ m-1})\}^{1/2}$$
$$\times \tfrac{1}{4}\{\mathscr{E}(\zeta_{m+1}\, e^{i\pi/4}, \beta_{m+1}/\zeta_{m+1}) + \mathscr{E}(\zeta_m\, e^{i\pi/4}, \beta_m/\zeta_m)\}, \tag{5.9}$$

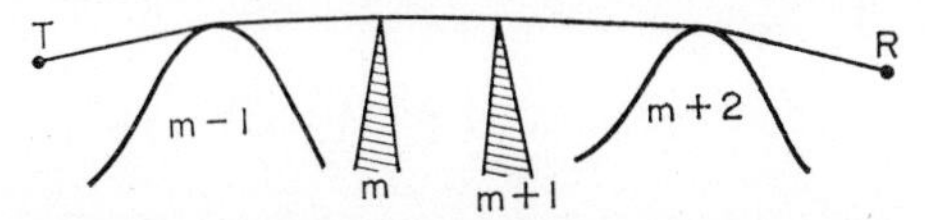

FIG. 9. The form of diffracting surfaces for eqns (5.8) and (5.9).

which is valid for $\eta_m \ll 1$, $\eta_{m+1} \ll 1$. Here,

$$\beta_m = \{\xi_m - (k_1 r_{m,\, m-1})^{-1} (k_1 a_m/2)^{1/3} (k_1 a_{m-1}/2)^{1/3}\, t_{m-1}\}/\eta_m,$$
$$\zeta_m = \{\gamma \beta_m + \beta_{m+1}\} \times [(r_{m+2,\, m+1} + r_{m+1,\, m})(r_{m+1,\, m} + r_{m,\, m-1}) / r_{m+1,\, m}(r_{m+2,\, m+1} + r_{m+1,\, m} + r_{m,\, m-1})]^{1/2},$$
$$\gamma = \{r_{m+2,\, m+1}\, r_{m,\, m-1}/(r_{m+1,\, m} + r_{m,\, m-1})(r_{m+2,\, m+1} + r_{m+1,\, m})\}^{1/2},$$
etc.,

$$\mathscr{E}(z, \eta) = \left(\frac{2}{\sqrt{\pi}}\right)^2 e^{(1+\eta^2)Z^2} \int_z^\infty dt_2\, e^{-t_2^2} \int_{\eta t_2}^\infty e^{-t_1^2}\, dt_1. \tag{5.10}$$

The result of (5.9) corresponds to (5.3a) and can be applied to the formulas (4.1–3) on the condition of $\xi_m \ll 1$, $\xi_{m+1} \ll 1$.

When $\xi_{m-1}/\eta_{m-1} \gg 1$ and $\xi_{m+2}/\eta_{m+2} \gg 1$, the terms of t_{m-1} and t_{m+2} in (5.10) can be neglected as in (5.3b) and it follows that

$$\zeta_m = \{(\theta_{m+1} + \theta_m)\, r_{m,\, m-1} + \theta_{m+1}\, r_{m+1,\, m}\} \times \{k_1 r_{m+2,\, m+1}/2(r_{m+1,\, m} + r_{m,\, m-1})(r_{m+2,\, m+1} + r_{m+1,\, m} + r_{m,\, m-1})\}^{1/2},$$
$$\beta_m = \theta_m \sqrt{[k_1/2(1/r_{m+1,\, m} + 1/r_{m,\, m-1})]}, \text{ etc.} \tag{5.11}$$

In the case in which (5.11) can be used, the coupling between the m- and $m+1$-th surfaces and the other surfaces is lost, and thus they behave independently in the formulas (4.1–3) for the attenuation coefficient A.

6. DIFFRACTION BY TWO RIDGES

The special case of $n = 3$ and ξ_2, $\xi_3 \ll 1$, in which (5.9) and (5.11) are valid, corresponds to the case of diffraction by two ridges, as illustrated in Fig.

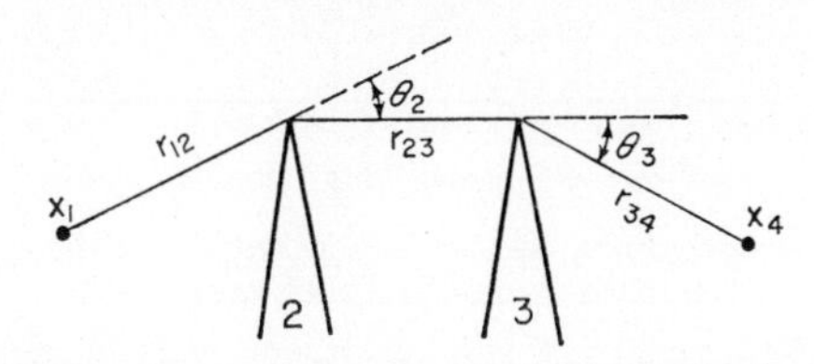

FIG. 10. Two ridges and the notations for eqns (6.1) and (6.2).

10. The straight application gives the result

$$A = \tfrac{1}{4}[\mathscr{E}(\zeta_2\, e^{i\pi/4}, \beta_2/\zeta_2) + \mathscr{E}(\zeta_3\, e^{i\pi/4}, \beta_3/\zeta_3)]. \tag{6.1}$$

Here

$$\zeta_2 = \{(\theta_2 + \theta_3)\, r_{21} + \theta_3 r_{32}\} \sqrt{[k_1 r_{43}/2(r_{32} + r_{21})\,(r_{43} + r_{32} + r_{21})]},$$
$$\zeta_3 = \{(\theta_2 + \theta_3)\, r_{43} + \theta_2 r_{32}\} \sqrt{[k_1 r_{21}/2(r_{43} + r_{32})\,(r_{43} + r_{32} + r_{21})]},$$
$$\beta_2 = \theta_2 \sqrt{[k_1/2(1/r_{32} + 1/r_{21})]}, \quad \beta_3 = \theta_3 \sqrt{[k_1/2\,(1/r_{32} + 1/r_{43})]}. \tag{6.2}$$

Equation (6.1) corresponds to the Fresnel integral in the case of a single ridge. Indeed, putting $r_{32} \to 0$, it follows that

$$\zeta_3 = \zeta_2 = (\theta_3 + \theta_2) \sqrt{[k_1 r_{43} r_{21}/2\,(r_{43} + r_{21})]},$$

and thus, according to the definition of $\mathscr{E}(z, n)$ in (5.10) and $\mathscr{E}(z)$ in (5.3a) eqn (6.1) becomes

$$A = \tfrac{1}{2}\mathscr{E}(\zeta_2\, e^{i\pi/4}). \tag{6.3}$$

Equation (6.1) is given in terms of the function $\mathscr{E}(a, b/a)$ whose precise analytical treatment has been studied in another paper (Furutsu, 1955) and also in the Appendix of this paper, the results of which are briefly described for convenience of computation. It is a many valued function and the follow-

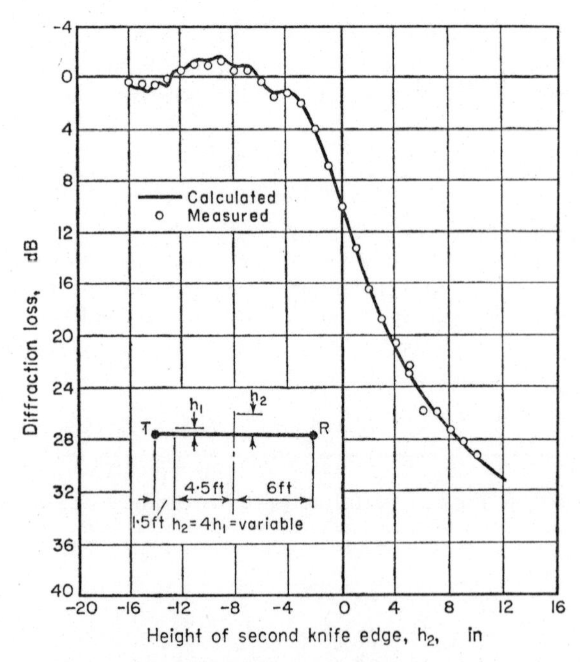

FIG. 11. A few comparisons of experimental and theoretical values for two-ridge diffraction.

(*a*) Knife edges at variable heights with transmitter and two edges in line.

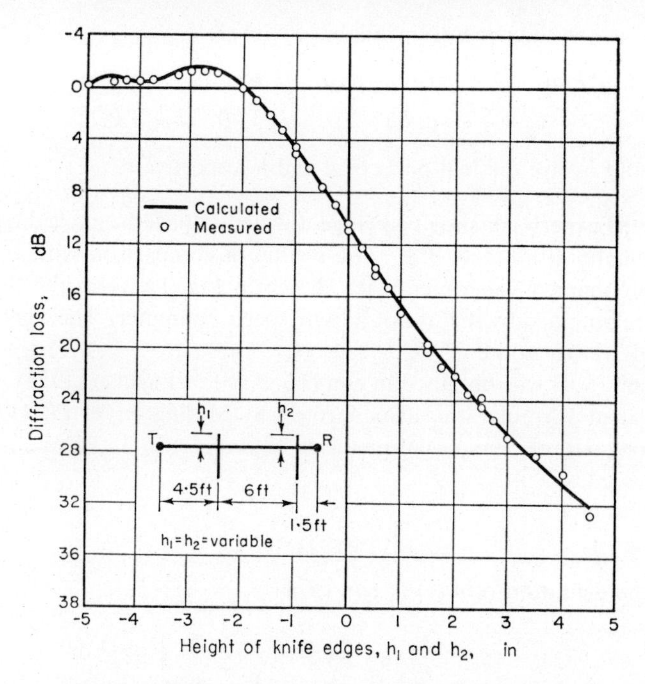

FIG. 11. (*b*) Knife edges at equal variable heights, frequency 24.21 Gc/s.

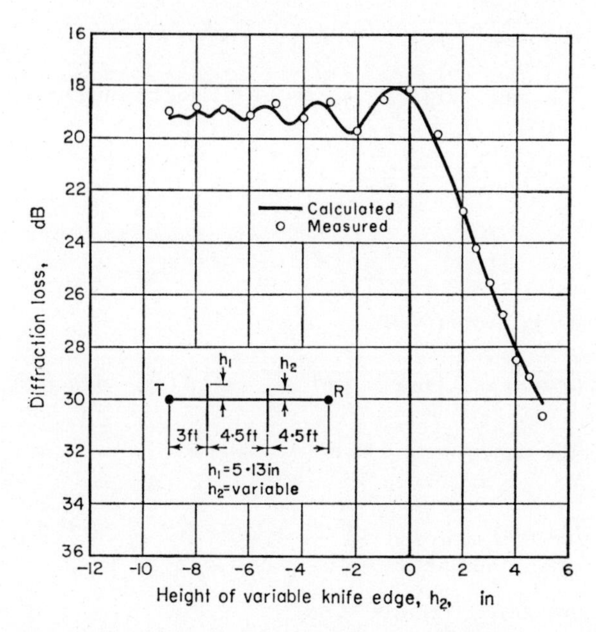

FIG. 11. (*c*) First knife edge fixed, second variable in height, frequency 24.14 Gc/s.

ing relations are conveniently available for (6.1):

$$\mathscr{E}(a, b/a) = 2e^{b^2}\,\mathscr{E}(a) - \mathscr{E}(a, -b/a), \quad a_R > 0, \quad b_R \leqslant 0,$$
$$= -\mathscr{E}(-a, b/(-a)), \quad a_R \leqslant 0, \quad b_R > 0. \tag{6.4}$$

Here, a_R and b_R are the real parts of a and b respectively.

The model experiments for two ridge diffraction have been tried by Decker and his collaborators in N.B.S. The results of comparison with the theory and the preliminary experiments are shown in Fig. 11. Here the theoretical values were obtained by the use of a high speed computer. The agreement is surprisingly rather good.

The result (6.1) was obtained in eqn (12.2.15) of Furutsu (1956). Recently the equivalent formula was also derived by Millington *et al.* (1962) and comparisons with the empirical methods of Bullington and Epsten-Peterson were discussed.

APPENDIX

From the definition of $\mathscr{E}(z)$ in eqn (5.3a),

$$\mathscr{E}(z) = e^{z^2} - \mathscr{E}(z), \quad \bar{\mathscr{E}}(z) = \frac{2}{\sqrt{\pi}} e^{z^2} \int_0^z e^{-t^2}\, dt. \tag{A.1}$$

Here

$$\bar{\mathscr{E}}(z) = \sum_{n=1}^{\infty} C_n, \qquad C_n = \frac{z^2}{n - 1/2} C_{n-1}, \qquad C_1 = \frac{2}{\sqrt{\pi}} z. \tag{A.2}$$

For $|z| \gg 1$ and $|\arg(z)| < 3\pi/4$, $\mathscr{E}(z)$ takes the asymptotic form

$$\mathscr{E}(z) \sim \sum_{n=1}^{\infty} g_n, \qquad |z| \gg 1, \qquad |\arg(z)| < 3\pi/4, \tag{A.3}$$

where

$$g_n = -\frac{n - 3/2}{z^2} g_{n-1}, \qquad g_1 = \frac{1}{\sqrt{\pi}.z}. \tag{A.4}$$

In the same way, from (5.10),

$$\mathscr{E}(a, b/a) = \frac{2}{\pi}\left(\tan^{-1}\frac{a}{b}\right) e^{a^2+b^2} - e^{b^2}\,\bar{\mathscr{E}}(a) + \bar{\mathscr{E}}(a, b/a). \tag{A.5}$$

Here $\bar{\mathscr{E}}(a)$ is the same function as in (A.1) and

$$\bar{\mathscr{E}}(a, b/a) = \left(\frac{2}{\sqrt{\pi}}\right)^2 e^{a^2+b^2} \int_0^a dt_2\, e^{-t_2^2} \int_0^{(b/a)t_2} e^{-t_1^2}\, dt_1,$$

$$\mathscr{E}(0, b/a) = \frac{2}{\pi} \tan^{-1}\frac{a}{b}. \tag{A.6}$$

The function $\bar{\mathscr{E}}(a, b/a)$ can be expanded in the absolutely convergent series of

$$\bar{\mathscr{E}}(a, b/a) = \sum_{n=1}^{\infty} h_n. \tag{A.7}$$

Here

$$h_n = \frac{1}{n}\{(a^2+b^2)\,h_{n-1} + d_n\}, \quad d_n = \frac{1}{n-1/2}\,b^2\,d_{n-1}, \quad h_1 = d_1 = \frac{2}{\pi}\,ab. \tag{A.8}$$

For $|a^2+b^2|^{1/2} \gg 1$ and $|a/b| < 1$ and $|\arg(a)|$, $|\arg(b)| < 3\pi/4$, the function $\mathscr{E}(a, b/a)$ takes the asymptotic form

$$\mathscr{E}(a, b/a) \sim \frac{1}{\sqrt{\pi}}\left(\frac{a}{a^2+b^2}\right)\sum_{n=0}^{\infty} j_n,$$

$$|a^2+b^2|^{1/2} \gg 1, \quad |a/b| < 1, \quad |\arg(a)|, \quad |\arg(b)| < 3\pi/4. \tag{A.9}$$

Here

$$\begin{cases} j_{2n+1} = Bj_{2n} - nAj_{2n-1} - k_n, \quad j_0 = \mathscr{E}(b), \\ j_{2n} \;\;= Bj_{2n-1} - (n-1/2)\,Aj_{2n-2} \end{cases}$$

$$\begin{aligned} k_n &= -(n-1/2)\,(a^2+b^2)^{-1}\,k_{n-1}, \\ k_0 &= \pi^{-1/2}\,b\,(a^2+b^2)^{-1}, \\ A &= a^2(a^2+b^2)^{-2}, \quad B = b^2(a^2+b^2)^{-1}. \end{aligned} \tag{A.10}$$

For the case of $|a/b| > 1$, the following relation is available:

$$\mathscr{E}(a, b/a) = \mathscr{E}(a)\,\mathscr{E}(b) - \mathscr{E}(b, a/b). \tag{A.11}$$

REFERENCES

FOCK, V. A. (1946) *Diffraction of Radio Waves around the Earth's Surface.* Academy of Sciences, U.S.S.R.

FURUTSU, K. (1952) On the group velocity, wave path and their relations to the Poynting vector of E-M field in an absorbing medium. *J. Phys. Soc. Japan* **8,** No. 4.

FURUTSU, K. (1955) Propagation of electromagnetic waves over a flat earth across two boundaries. *J. Radio Res. Lab.* **2,** 239–279.

FURUTSU, K. (1956) On the multiple diffraction of electromagnetic waves by spherical mountains. *J. Radio Res. Lab.* **3,** 331–390.

LOGAN, N. A. (1959) Fresnel diffraction by convex surfaces. Paper presented at Washington meeting of International Scientific Radio Union, 4–7 May 1959.

MARCUVITZ, N. (1962) *Abstract Operator Methods in Electromagnetic Waves.* The University of Wisconsin Press, p. 109.

MILLINGTON, G., HEWITT, R. and IMMIRZI, F. S. (1962) The Fresnel surface integral, I.E.E. Monograph No. 508E; Double knife-edge diffraction in field strength predictions, Monograph No. 507E.

WAIT, J. R. and CONDA, M. (1959) Diffraction of electromagnetic waves by smooth obstacles for grazing angles. *J. Res. of NBS* **63D,** 181–197.

THE SCATTERING OF A PLANE ELECTROMAGNETIC WAVE BY A FINITE CONE†

C. C. ROGERS,‡ J. K. SCHINDLER,§ and F. V. SCHULTZ§

ABSTRACT

The problem considered is that of determining, as an electromagnetic boundary-value problem, the scattered electromagnetic field set up by a single-frequency plane electromagnetic wave striking, nose-on, a perfectly conducting cone of finite height. The base of the cone is assumed to consist of a spherical cap with center at the apex of the cone. The object is to calculate the radar cross-section in the Rayleigh and resonance regions. The scattered field components are expressed as infinite series of vector eigenfunctions and the coefficients of these series determine the boundary conditions. A few values of the radar cross-section are calculated in the Rayleigh and near-resonance regions and these are compared with measured values obtained by Keys and with approximate values calculated by Siegel and Keller.

1. INTRODUCTION

The problem undertaken herein is that of the exact solution for the scattering of an incident plane electromagnetic wave by a finite-sized perfectly conducting cone. We consider here only "nose-on" incidence (see Fig. 1) and, in order to retain a spherical system throughout, the endcap of the cone is assumed to be a spherical sector, centered at the apex of the cone. Time variations are given by $e^{i\omega t}$ and spherical coordinates are used.

It is known that if Φ is a solution of the homogeneous scalar Helmholtz-equation, then the functions, **l**, **m**, and **n**, defined by

$$\left.\begin{aligned} \mathbf{l} &= \operatorname{grad} \Phi, \\ \mathbf{m} &= \operatorname{curl} \Phi \mathbf{a}_r, \\ \mathbf{n} &= \frac{1}{k} \operatorname{curl} \mathbf{m}, \end{aligned}\right\} \tag{1}$$

are solutions of the homogeneous vector Helmholtz-equation and form the basis for its general solution. Here **r** is the unit radial-vector in spherical co-

† The research reported herein was sponsored by the Electronics Research Directorate, Air Force Cambridge Research Laboratories, under Contract AF 19(604)-4051.

‡ Formerly at School of Electrical Engineering, Purdue University, Lafayette, Indiana, U.S.A. Now of Rose Polytechnic Institute, Terre Haute, Indiana, U.S.A.

§ School of Electrical Engineering, Purdue University, Lafayette, Indiana, U.S.A.

ordinates and $k = 2\pi/\lambda$. Since, for the case under consideration,

$$\operatorname{div} \mathbf{E} = \operatorname{div} \mathbf{H} = 0,$$

and since $\operatorname{div} \mathbf{l} \neq 0$, only **m** and **n** may be used in the expansions of the field quantities. Now

$$\Phi^n_{\substack{e\\o}\nu m}(r, \theta, \phi) = z^n_\nu(kr)\, P^m_\nu(\cos\theta) \begin{bmatrix}\cos m\phi\\ \sin m\phi\end{bmatrix} \tag{2}$$

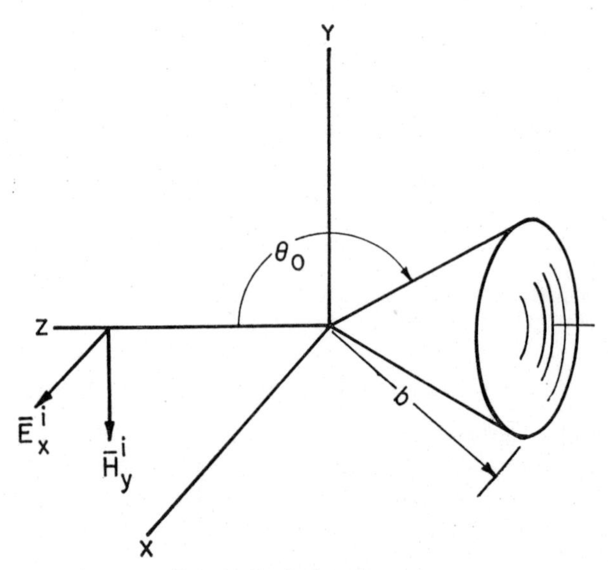

FIG. 1. Physical configuration.

where we let "e" (even) or "o" (odd) indicate the use of $\cos m\phi$ or $\sin m\phi$, respectively. Here $z^n_\nu(kr)$ is a spherical Bessel or Hankel function of order ν, and n may take on the values 1, 2, 3, or 4, where these numbers represent Bessel functions of the first and second kind, and Hankel functions of the first and second kind, respectively. $P^m_\nu(\cos\theta)$ is the associated Legendre function of degree ν and order m.

As discussed by Rogers and Schultz (1960) it is desirable to subdivide space as shown in Fig. 2. In Region II the complete field is split into two component fields, the incident field and the scattered field. In Region I the complete field is dealt with, without subdividing it into incident and scattered components. Because the radiation condition must be satisfied for the scattered field, Hankel functions of the second kind are used to represent the scattered field in Region II, since they possess the required behavior as $r \to \infty$. In Region I, the finite energy condition at the tip of the cone forces the use of Bessel functions of the first kind.

When one considers the behavior of the associated Legendre functions, the situation forces the use of functions of integral degree (i.e. polynomials) for all $r > b$, since in this region the fields exist and are finite throughout the complete θ-domain, and any associated Legendre function of non-integral degree becomes infinite at either $\theta = 0$ or π. For $r \leqq b$, $\theta = \pi$ is not in the domain of interest, and consequently non-integral degree associated Legendre functions may be used. In addition, as will be seen later, the proper selection of the degree is used for the satisfaction of the boundary conditions at the surface of the cone.

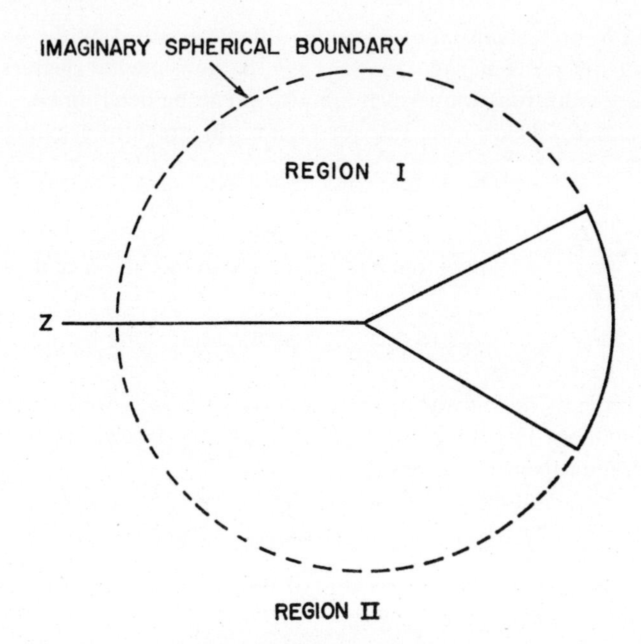

FIG. 2. Space sectionalizing.

2. FIELD EXPANSIONS

We now consider the expansions for the electric fields.

In Region II the incident electric-field may be expressed (Stratton, 1941) thus

$$\mathbf{E}^{i}_{\text{II}} = a_x \, e^{ikz} = \mathbf{a}_x \, e^{ikr \cos \theta} = \sum_n (\gamma_n \mathbf{m}^{(1)}_{\text{oln}} + \Gamma_n \mathbf{n}^{(1)}_{\text{eln}}), \tag{3}$$

where

$$\gamma_n = i^n \frac{2n+1}{n(n+1)}, \qquad \Gamma_n = -(i)^{n+1} \frac{2n+1}{n(n+1)},$$

and $\mathbf{a}_x$ is a unit vector in the x-direction. In this case, the summation is over all of the integers, n, from one to infinity. Consideration of the ϕ-variation of the incident field led to the choice of the odd **m** and even **n**-functions for the expansion, and the ϕ-variation also limits us to $m = 1$. As a result, we use odd **m** and even **n**-functions with $m = 1$ for all expansions of the electric fields.

In Region I we are not necessarily interested in the separate incident and scattered fields, and assume an expansion of the total field only, as mentioned previously. Consideration of the prior arguments about the radial and spherical functions leads to an expansion of the form

$$\mathbf{E}_{\mathrm{I}}^{t} = \sum_{\nu} a_{\nu}\mathbf{m}_{\mathrm{o1}\nu}^{(1)} + \sum_{\mu} b_{\mu}\mathbf{n}_{\mathrm{e1}\mu}^{(1)}, \tag{4}$$

where a_ν and b_μ are expansion coefficients to be determined by the boundary conditions of the problem, and μ and ν are the non-integral degrees of the associated Legendre functions which are also yet to be determined.

For the scattered field in the exterior region, the previous arguments lead to the choice

$$\mathbf{E}_{\mathrm{II}}^{s} = \sum_{n} (c_n\mathbf{m}_{\mathrm{o1}n}^{(4)} + d_n\mathbf{n}_{\mathrm{e1}n}^{(4)}), \tag{5}$$

where c_n and d_n are constants to be determined. Here we have selected $z_n^{(4)}(kr) = h_n^{(2)}(kr)$, the Hankel function of the second kind, since it possesses an asymptotic form

$$\lim_{kr\to\infty} h_n^{(2)}(kr) \to \frac{1}{kr}\, i^{(n+1)}\, \mathrm{e}^{-ikr}, \tag{6}$$

and thus represents an outward-traveling wave at infinity and satisfies the radiation condition. Hereafter, since we use only the Hankel function of the second kind, the superscript is omitted.

3. THE PROBLEM SOLUTION

Equations (4) and (5) contain six sets of unknown constants which must be determined: μ, ν, a_ν, b_μ, c_n, and d_n.

Through the choice of functions, we have insured the satisfaction of the finite energy condition at the tip of the cone and the radiation condition at infinity. There remain, then, the following boundary conditions:

$$[\mathbf{E}_{\mathrm{I}}^{t}]_{r,\,\phi} = 0 \text{ at } \theta = \theta_0, \quad r \leqq b; \tag{7}$$

$$[\mathbf{E}_{\mathrm{II}}^{i} + \mathbf{E}_{\mathrm{II}}^{s}]_{\theta,\,\phi} = 0 \text{ at } r = b, \quad \theta_0 \leqq \theta \leqq \pi; \tag{8}$$

$$\left.\begin{aligned} [\mathbf{E}_{\mathrm{II}}^{i} + \mathbf{E}_{\mathrm{II}}^{s}]_{\theta,\,\phi} &= [\mathbf{E}_{\mathrm{I}}^{t}]_{\theta,\,\phi} \quad (9)\\ [\mathbf{H}_{\mathrm{II}}^{i} + \mathbf{H}_{\mathrm{II}}^{s}]_{\theta,\,\phi} &= [\mathbf{H}_{\mathrm{I}}^{t}]_{\theta,\,\phi} \quad (10) \end{aligned}\right\} \text{ at } r = b, \quad 0 \leqq \theta < \theta_0;$$

the finite energy condition at the edge of the cone,

$$r \to b, \qquad \theta \to \theta_0, \tag{11}$$

where b is the radius of the spherical end-cap and θ_0 is the exterior apex-angle.

The third and fourth conditions insure the continuity of the field components across the imaginary spherical boundary. For future reference, the field quantities are expanded in their entirety below:

$$\begin{aligned}\mathbf{E}_{\mathrm{I}}^{t} = &\left[\sum_{\mu} b_{\mu}\frac{\mu(\mu+1)}{kr} j_{\mu}(kr)\, P_{\mu}^{1}(\cos\theta)\right]\cos\phi\, \mathbf{a}_{r}\\ &+\left[\sum_{\nu} a_{\nu} j_{\nu}(kr)\frac{P_{\nu}^{1}(\cos\theta)}{\sin\theta}+\sum_{\mu} b_{\mu} j_{\mu}'(kr)\frac{\mathrm{d}P_{\mu}^{1}}{\mathrm{d}\theta}\right]\cos\phi\,\mathbf{a}_{\theta}\\ &-\left[\sum_{\nu} a_{\nu} j_{\nu}(kr)\frac{\mathrm{d}P_{\nu}^{1}}{\mathrm{d}\theta}+\sum_{\mu} b_{\mu} j_{\mu}'(kr)\frac{P_{\mu}^{1}(\cos\theta)}{\sin\theta}\right]\sin\phi\,\mathbf{a}_{\phi};\end{aligned} \tag{12}$$

$$\begin{aligned}\mathbf{E}_{\mathrm{II}}^{i} = \sum_{n}\Bigg\{&\left[\Gamma_{n}\frac{n(n+1)}{kr} j_{n}(kr)\, P_{n}^{1}(\cos\theta)\right]\cos\phi\,\mathbf{a}_{r}\\ &+\left[\gamma_{n} j_{n}(kr)\frac{P_{n}^{1}(\cos\theta)}{\sin\theta}+\Gamma_{n} j_{n}'(kr)\frac{\mathrm{d}P_{n}^{1}}{\mathrm{d}\theta}\right]\cos\phi\,\mathbf{a}_{\theta}\\ &-\left[\gamma_{n} j_{n}(kr)\frac{\mathrm{d}P_{n}^{1}}{\mathrm{d}\theta}+\Gamma_{n} j_{n}'(kr)\frac{P_{n}^{1}(\cos\theta)}{\sin\theta}\right]\sin\phi\,\mathbf{a}_{\phi}\Bigg\};\end{aligned} \tag{13}$$

$$\begin{aligned}\mathbf{E}_{\mathrm{II}}^{s} = \sum_{n}\Bigg\{&\left[d_{n}\frac{n(n+1)}{kr} h_{n}(kr)\, P_{n}^{1}(\cos\theta)\right]\cos\phi\,\mathbf{a}_{r}\\ &+\left[c_{n} h_{n}(kr)\frac{P_{n}^{1}(\cos\theta)}{\sin\theta}+d_{n} h_{n}'(kr)\frac{\mathrm{d}P_{n}^{1}}{\mathrm{d}\theta}\right]\cos\phi\,\mathbf{a}_{\theta}\\ &-\left[c_{n} h_{n}(kr)\frac{\mathrm{d}P_{n}^{1}}{\mathrm{d}\theta}+d_{n} h_{n}'(kr)\frac{P_{n}^{1}(\cos\theta)}{\sin\theta}\right]\sin\phi\,\mathbf{a}_{\phi}\Bigg\};\end{aligned} \tag{14}$$

$$\begin{aligned}\mathbf{H}_{\mathrm{I}}^{t} = \frac{i}{\eta}\Bigg\{&\left[\sum_{\nu} a_{\nu}\frac{\nu(\nu+1)}{kr} j_{\nu}(kr)\, P_{\nu}^{1}(\cos\theta)\right]\sin\phi\,\mathbf{a}_{r}\\ &+\left[\sum_{\nu} a_{\nu} j_{\nu}'(kr)\frac{\mathrm{d}P_{\nu}^{1}}{\mathrm{d}\theta}-\sum_{\mu} b_{\mu} j_{\mu}(kr)\frac{P_{\mu}^{1}(\cos\theta)}{\sin\theta}\right]\sin\phi\,\mathbf{a}_{\theta}\\ &+\left[\sum_{\nu} a_{\nu} j_{\nu}'(kr)\frac{P_{\nu}^{1}(\cos\theta)}{\sin\theta}-\sum_{\mu} b_{\mu} j_{\mu}(kr)\frac{\mathrm{d}P_{\mu}^{1}}{\mathrm{d}\theta}\right]\cos\phi\,\mathbf{a}_{\phi}\Bigg\};\end{aligned} \tag{15}$$

$$\begin{aligned}\mathbf{H}_{\mathrm{II}}^{i} = \frac{i}{\eta}\sum_{n}\Bigg\{&\left[\gamma_{n}\frac{n(n+1)}{kr} j_{n}(kr)\, P_{n}^{1}(\cos\theta)\right]\sin\phi\,\mathbf{a}_{r}\\ &+\left[\gamma_{n} j_{n}'(kr)\frac{\mathrm{d}P_{n}^{1}}{\mathrm{d}\theta}-\Gamma_{n} j_{n}(kr)\frac{P_{n}^{1}(\cos\theta)}{\sin\theta}\right]\sin\phi\,\mathbf{a}_{\theta}\\ &+\left[\gamma_{n} j_{n}'(kr)\frac{P_{n}^{1}(\cos\theta)}{\sin\theta}-\Gamma_{n} j_{n}(kr)\frac{\mathrm{d}P_{n}^{1}}{\mathrm{d}\theta}\right]\cos\phi\,\mathbf{a}_{\phi}\Bigg\};\end{aligned} \tag{16}$$

$$\mathbf{H}_{\mathrm{II}}^{s} = \frac{i}{\eta} \sum_{n} \left\{ \left[c_n \frac{n(n+1)}{kr} h_n(kr) P_n^1(\cos\theta) \right] \sin\phi \mathbf{a}_r \right.$$
$$+ \left[c_n h_n'(kr) \frac{\mathrm{d}P_n^1}{\mathrm{d}\theta} - d_n h_n(kr) \frac{P_n^1(\cos\theta)}{\sin\theta} \right] \sin\phi \mathbf{a}_\theta \tag{17}$$
$$\left. + \left[c_n h_n'(kr) \frac{P_n^1(\cos\theta)}{\sin\theta} - d_n h_n(kr) \frac{\mathrm{d}P_n^1}{\mathrm{d}\theta} \right] \cos\phi \mathbf{a} \right\}.$$

We now apply the boundary conditions at the surface of the cone. To satisfy (7), we equate the r-component of $\mathbf{E}_{\mathrm{I}}^{t}$ to zero at $\theta = \theta_0$,

$$\sum_{\mu} b_\mu \frac{\mu(\mu+1)}{kr} j_\mu(kr) P_\mu^1(\cos\theta_0) = 0, \tag{18}$$

and thus set

$$P_\mu^1(\cos\theta_0) = 0. \tag{19}$$

This equation thus determines the values of μ. Equating the ϕ-component of $\mathbf{E}_{\mathrm{I}}^{t}$ to zero gives

$$\sum_{\nu} a_\nu j_\nu(kr) \left. \frac{\mathrm{d}P_\nu^1}{\mathrm{d}\theta} \right|_{\theta=\theta_0} + \sum_{\mu} b_\mu j_\mu'(kr) \frac{P_\mu^1(\cos\theta_0)}{\sin\theta_0} = 0. \tag{20}$$

Since $P_\mu^1(\cos\theta_0) = 0$, we set

$$\left. \frac{\mathrm{d}P_\nu^1}{\mathrm{d}\theta} \right|_{\theta=\theta_0} = 0 \tag{21}$$

and thus determine the values of ν.

From (8), (13), and (14), we have for $r = b$, $\theta_0 \leqq \theta \leqq \pi$,

$$\sum_{n} \left\{ [\gamma_n j_n(kb) + c_n h_n(kb)] \frac{P_n^1(\cos\theta)}{\sin\theta} + [\Gamma_n j_n'(kb) + d_n h_n'(kb)] \frac{\mathrm{d}P_n^1}{\mathrm{d}\theta} \right\} = 0 \tag{22}$$

for the θ-component, and, for the ϕ-component

$$\sum_{n} \left\{ [\gamma_n j_n(kb) + c_n h_n(kb)] \frac{\mathrm{d}P_n^1}{\mathrm{d}\theta} + [\Gamma_n j_n'(kb) + d_n h_n'(kb)] \frac{P_n^1(\cos\theta)}{\sin\theta} \right\} = 0. \tag{23}$$

These two equations contain the unknowns c_n and d_n and apply over a portion of the θ-domain. In order to obtain equations involving only one set of coefficients, we first multiply (22) by $\sin\theta$ and then differentiate with respect to θ; there results

$$\sum_{n} \left\{ [\gamma_n j_n(kb) + c_n h_n(kb)] \frac{\mathrm{d}P_n^1}{\mathrm{d}\theta} + [\Gamma_n j_n'(kb) + d_n h_n'(kb)] \frac{\mathrm{d}}{\mathrm{d}\theta} \left(\sin\theta \frac{\mathrm{d}P_n^1}{\mathrm{d}\theta} \right) \right\} = 0. \tag{24}$$

The subtraction of (24) from (23) yields

$$\sum_n [\Gamma_n j_n'(kb) + d_n h_n'(kb)] \left\{ -\frac{\mathrm{d}}{\mathrm{d}\theta}\left(\sin\theta \frac{\mathrm{d}P_n^1}{\mathrm{d}\theta}\right) + \frac{P_n^1(\cos\theta)}{\sin\theta}\right\} = 0. \quad (25)$$

Furthermore, since the associated Legendre equation may be expressed as

$$\frac{\mathrm{d}}{\mathrm{d}\theta}\left(\sin\theta \frac{\mathrm{d}P_\nu^m}{\mathrm{d}\theta}\right) - \frac{m^2}{\sin\theta} P_\nu^m(\cos\theta) = -\nu(\nu+1)\sin\theta\, P_\nu^m(\cos\theta); \quad (26)$$

and because, in this particular case, $\nu = n$, $m = 1$, (25) may be written

$$\sum_n n(n+1)\,[\Gamma_n j_n'(kb) + d_n h_n'(kb)]\, P_n^1(\cos\theta) = 0. \quad (27)$$

Let us signify the first N-terms of the resulting series as $S_N(\theta)$, i.e.

$$S_N(\theta) = \sum_{n=1}^{N} n(n+1)\,[\Gamma_n j_n'(kb) + d_n h_n'(kb)]\, P_n^1(\cos\theta) \quad (28)$$

$$S_\infty(\theta) = 0, \qquad \theta_0 \leqq \theta \leqq \pi.$$

In a similar manner, we multiply (23) by $\sin\theta$, differentiate with respect to θ, and subtract the resulting equation from (22). There results

$$T_N(\theta) = \sum_{n=1}^{N} n(n+1)\,[\gamma_n j_n(kb) + c_n h_n(kb)]\, P_n^1(\cos\theta) \quad (29)$$

$$T_\infty(\theta) = 0, \qquad \theta_0 \leqq \theta \leqq \pi.$$

We have thus obtained two equations involving the unknown coefficients c_n and d_n for a portion of the range of θ. Next we apply (9). From (12) through (14) we have for the θ-component, where $r = b$, $0 \leqq \theta < \theta_0$:

$$\sum_\nu a_\nu j_\nu(kb) \frac{P_\nu^1(\cos\theta)}{\sin\theta} + \sum_\mu b_\mu j_\mu'(kb) \frac{\mathrm{d}P_\mu^1}{\mathrm{d}\theta}$$

$$= \sum_n \left\{ [\gamma_n j_n(kb) + c_n h_n(kb)] \frac{P_n^1(\cos\theta)}{\sin\theta} + [\Gamma_n j_n'(kb) + d_n h_n'(kb)] \frac{\mathrm{d}P_n^1}{\mathrm{d}\theta}\right\} \quad (30)$$

and for the ϕ-component,

$$\sum_\nu a_\nu j_\nu(kb) \frac{\mathrm{d}P_\nu^1}{\mathrm{d}\theta} + \sum_\mu b_\mu j_\mu'(kb) \frac{P_\mu^1(\cos\theta)}{\sin\theta}$$

$$= \sum_n \left\{ [\gamma_n j_n(kb) + c_n h_n(kb)] \frac{\mathrm{d}P_n^1}{\mathrm{d}\theta} + [\Gamma_n j_n'(kb) + d_n h_n'(kb)] \frac{P_n^1(\cos\theta)}{\sin\theta}\right\}. \quad (31)$$

In a manner exactly analogous to that used in obtaining equation (27), we may first multiply (31) by $\sin\theta$, differentiate with respect to θ, and subtract the resulting equation from (30). There results

$$\sum_{\nu} a_\nu j_\nu(kb)\, \nu(\nu+1)\, P_\nu^1(\cos\theta)$$
$$= \sum_n [\gamma_n j_n(kb) + c_n h_n(kb)]\, n(n+1)\, P_n^1(\cos\theta) \tag{32}$$
$$= T_\infty(\theta); \qquad 0 \leqq \theta < \theta_0.$$

Performing the same operations on (30) and subtracting the results from (31) yields

$$\sum_{\mu} b_\mu j_\mu'(kb)\, \mu(\mu+1)\, P_\mu^1(\cos\theta)$$
$$= \sum_n [\Gamma_n j_n'(kb) + d_n h_n'(kb)]\, n(n+1)\, P_n^1(\cos\theta) \tag{33}$$
$$= S_\infty(\theta); \qquad 0 \leqq \theta < \theta_0.$$

Considering (27) and (33), let us define a function $f(\theta)$ as follows:

$$f(\theta) = \begin{cases} \sum_{\mu} b_\mu j_\mu'(kb)\, \mu(\mu+1)\, P_\mu^1(\cos\theta), & 0 \leqq \theta < \theta_0; \\ 0, \qquad \theta_0 \leqq \theta \leqq \pi. \end{cases} \tag{34}$$

We now think of $f(\theta)$ as a defined function which we wish to represent as accurately as possible by a finite series, $S_N(\theta)$. To do this we minimize the weighted mean square error to find the coefficients d_n in terms of the b_μ-coefficients.

Let $\epsilon(\theta) = f(\theta) - S_N(\theta)$ represent the error, and then form the mean square error weighted by an amount $\sin\theta$, thus

$$M = \frac{1}{\pi} \int_0^\pi \epsilon^2(\theta) \sin\theta \, d\theta. \tag{35}$$

Since this weighting factor is always positive in the range of integration it does not destroy the primary significance of M, but only causes the error in the center of the range to be weighted more heavily than that at the end points.

In order to minimize the mean square error with respect to a particular coefficient d_m, we set

$$\frac{\partial M}{\partial d_m} = 0. \tag{36}$$

If we now insert the expression for $S_N(\theta)$ into the equation and move the portion containing that series to the right-hand side, the orthogonality of the associated Legendre functions produces

$$\int_0^{\pi} f(\theta)\, P_m^1(\cos\theta)\sin\theta\, d\theta$$

$$= [\Gamma_m j_m'(kb) + d_m h_m'(kb)]\, m(m+1) \int_0^{\pi} \sin\theta\, [P_m^1(\cos\theta)]^2\, d\theta. \qquad (37)$$

By inserting the expression for $f(\theta)$ and evaluating the integral on the right side, one obtains

$$\sum_{\mu} b_\mu j_\mu'(kb)\, \mu(\mu+1) \int_0^{\theta_0} \sin\theta\, P_\mu^1(\cos\theta)\, P_m^1(\cos\theta)\, d\theta$$

$$= [\Gamma_m j_m'(kb) + d_m h_m(kb)]\, \frac{2[m(m+1)]^2}{2m+1}. \qquad (38)$$

One may evaluate the integral on the left (Schelkunoff, 1948) obtaining for it

$$\frac{\sin\theta_0}{m(m+1) - \mu(\mu+1)}\, P_m^1(\cos\theta_0) \left.\frac{dP_\mu^1}{d\theta}\right|_{\theta=\theta_0}. \qquad (39)$$

By replacing the integral in (38) by (39) and then solving for the coefficient d_m, one obtains

$$d_m = \frac{2m+1}{m(m+1)} \frac{j_m'(kb)}{h_m'(kb)} \left[\frac{1}{2}\sin\theta_0\, P_m^1(\cos\theta_0)\right.$$

$$\left. \times \sum_{\mu} b_\mu \frac{j_\mu'(kb)}{j_m'(kb)} \frac{\mu(\mu+1)}{m(m+1)} \frac{1}{m(m+1) - \mu(\mu+1)} \left.\frac{dP_\mu^1}{d\theta}\right|_{\theta=\theta_0} + i^{m+1}\right]. \qquad (40)$$

In a similar manner, we may use (29) and (32) to obtain

$$c_m = \frac{2m+1}{m(m+1)} \frac{j_m(kb)}{h_m(kb)} \left[\frac{1}{2}\sin\theta_0 \left.\frac{dP_m^1}{d\theta}\right|_{\theta=\theta_0}\right.$$

$$\left. \times \sum_{\nu} a_\nu \frac{j_\nu(kb)}{j_m(kb)} \frac{\nu(\nu+1)}{m(m+1)} \frac{P_\nu^1(\cos\theta_0)}{\nu(\nu+1) - m(m+1)} - i^m\right]. \qquad (41)$$

We next apply the same technique to the tangential magnetic fields over the imaginary spherical boundary. By minimizing the errors with respect to a_ν and b_μ, one may obtain expressions for these coefficients in terms of the coefficients c_n and d_n, which, upon substitution into eqns (40) and (41), yield an infinite set of equations for the coefficients c_n and d_n. The reverse substitution also produces a similar set of equations for the coefficients a_ν and b_μ. We finally obtain:

$$\sum_{\nu}\sum_{n} c_n({}_c\xi_\nu^{nm} - {}_cK_n\delta_{mn}) = -\sum_{\nu}\sum_{n} ({}_c\psi_\nu^{nm} - {}_c\chi_n\delta_{mn}), \qquad m = 1, 2, \ldots \quad (42)$$

where

$$ {}_c\xi_\nu^{nm} = \frac{1}{2j_\nu'(kb)\, B_\nu} \frac{n(n+1)}{m(m+1)} \frac{j_\nu(kb)}{h_m(kb)} h_n'(kb) $$

$$ \times \frac{\sin^2\theta_0\, [P_\nu^1(\cos\theta_0)]^2}{[\nu(\nu+1) - n(n+1)]\,[\nu(\nu+1) - m(m+1)]} \left[\frac{\mathrm{d}P_m^1}{\mathrm{d}\theta} \frac{\mathrm{d}P_n^1}{\mathrm{d}\theta} \bigg|_{\theta=\theta_0} \right], $$

$$ B_\nu = \int_0^{\theta_0} [P_\nu^1(\cos\theta)]^2 \sin\theta \, \mathrm{d}\theta, $$

$$ {}_c\psi_\nu^{nm} = i^n \frac{2n+1}{n(n+1)} \frac{j_n'(kb)}{h_n'(kb)}\, {}_c\xi_\nu^{nm}, $$

$$ {}_c\chi_n = i^n \frac{j_n(kb)}{h_n(kb)}, $$

$$ {}_cK_n = \frac{n(n+1)}{2n+1}; $$

and

$$ \sum_\mu \sum_n d_n({}_d\xi_\mu^{nm} - {}_dK_n\delta_{mn}) = \sum_\mu \sum_n ({}_d\psi_\mu^{nm} - {}_d\chi_n\delta_{mn}), \qquad m = 1, 2, \ldots \quad (43) $$

where

$$ {}_d\xi_\mu^{nm} = \frac{1}{2j_\mu(kb)\, B_\mu} \frac{n(n+1)}{m(m+1)} \frac{j_\mu'(kb)}{h_m'(kb)} h_n(kb) $$

$$ \times \frac{\sin^2\theta_0 P_n^1(\cos\theta_0)\, P_m^1(\cos\theta_0)}{[n(n+1) - \mu(\mu+1)]\,[m(m+1) - \mu(\mu+1)]} \left[\frac{\mathrm{d}P_\mu^1}{\mathrm{d}\theta} \bigg|_{\theta=\theta_0} \right]^2, $$

$$ {}_d\psi_\mu^{nm} = i^{n+1} \frac{2n+1}{n(n+1)} \frac{j_n(kb)}{h_n(kb)}\, {}_d\xi_\mu^{nm}, $$

$$ {}_d\chi_n = i^{n+1} \frac{j_n'(kb)}{h_n'(kb)}, $$

$$ {}_dK_n = \frac{n(n+1)}{(2n+1)}. $$

Similar equations may be obtained for a_ν and b_μ, and these are listed by Rogers and Schultz (1960).

4. NUMERICAL RESULTS

In order to calculate the coefficients a_ν, b_μ, c_n and d_n it is necessary to know the values of ν and μ, of the integrals B_ν and B_μ, which are defined between (42) and (43), and of $P_\nu^1(\cos\theta_0)$ and $(\mathrm{d}P_\mu^1/\mathrm{d}\theta)|_{\theta=\theta_0}$. Each of these constants is dependent upon θ_0, shown in Fig. 1. All of the numerical results presented herein were calculated for $\theta_0 = 165$ deg (total included angle of cone = 30 deg).

As shown by Rogers and Schultz (1960), the calculated value of the radar cross-section is extremely sensitive to small errors in the values of ν and μ which are used. Approximate values for ν and μ have been calculated by Siegel *et al.* (1953) and by the Institute of Numerical Analysis of the University of California, which are listed by Siegel (1953). Because of the need for high accuracy it has been necessary to recalculate some of these quantities.

The infinite series expansion for the associated Legendre function and its θ-derivative were used along with the upper bound estimate of the remainder, calculated by Rogers and Schultz (1960), to obtain the values of the functions. A program for the Burroughs Datatron 205 Computer was developed to vary the degree of the functions until a change in the least significant figure of the degree caused a sign change in the value of the function. A double precision routine which carried 17-significant figures was used in the summation of the series.

A tabulation of the results of these calculations will be published in a forthcoming report.

A series expansion for the integral B_μ was used in obtaining numerical values for this quantity. These values will also be published in the aforementioned report. The values of B_ν were taken from Siegel. The values of $P_\nu^1(\cos\theta_0)$ and $(dP_\mu^1/d\theta)_{\theta=\theta_0}$ were computed from infinite series-expansions for the functions.

These results were used in (42) and (43) to obtain the values of scattering coefficients, c_n and d_n. Using the definition of the monostatic radar cross-section

$$\sigma = \lim_{r\to\infty}\left[4\pi r^2\left|\frac{S_{II}^s}{S_{II}^i}\right|\right], \tag{44}$$

one can show that for the finite cone considered herein

$$\sigma = \frac{\lambda^2}{4\pi}\left|\sum_n i^n n(n+1)(c_n - id_n)\right|^2. \tag{45}$$

A few calculations of σ have been made for values of kb in the Rayleigh region and just below the resonance region, and these are listed below, along with some measured values contained in a private communication from Keys and calculated values obtained by Siegel (1959) and by Keller (1961 and private communication). The quantity a occurring in Table 1 is the radius of the circular edge.

The reasons for the discrepancies are being investigated. It is to be noted that the first three values of $\sigma/\pi a^2$ calculated by the method discussed herein, obey very closely the required λ^{-4}-variation. Also, it should be mentioned that the values calculated in the present report are for a cone with a spherical cap

TABLE 1

kb	ka	$\sigma/\pi a^2$			
		Keys	Siegel	Keller	Rogers, Schindler, Schultz
0·1	0·0259		$35 \cdot 2 \times 10^{-7}$		$7 \cdot 24 \times 10^{-7}$
0·2	0·0518				$1 \cdot 16 \times 10^{-5}$
0·5	0·1295				$4 \cdot 70 \times 10^{-4}$
3·6	0·932	1·41			0·589
5·4	1·40	2·95		2·80	1·65

for a base, whereas the other calculated and measured results listed in Table 1 are for a cone with a flat base.

The finite energy condition at the tip of the cone has been used in selecting the eigenfunctions used in the field expansions, but the finite energy-requirement at the edge of the cone has not been invoked. It is necessary, therefore,

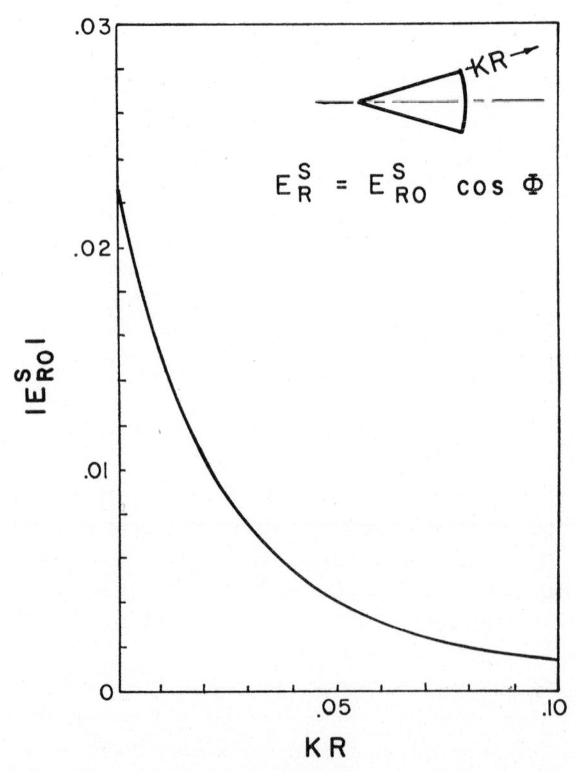

FIG. 3. The scattered radial electric-field near the cone edge.

to investigate the behavior of the solution at this edge. To do this, numerical methods were used. In order to reduce the work involved, a very small cone was considered ($kb = 0{\cdot}1$) and the approach to the edge was taken along a radial line $r \geqq b$, $\Theta = \Theta_0$, $\phi =$ constant. E_r^s, E_Θ^s, and E_ϕ^s were calculated along the line, the results for E_r^s being shown in Figs. 3 and 4. In making the calculations, three, four, five, six, and seven sets of equations were successively used to calculate a corresponding number of scattering coefficients,

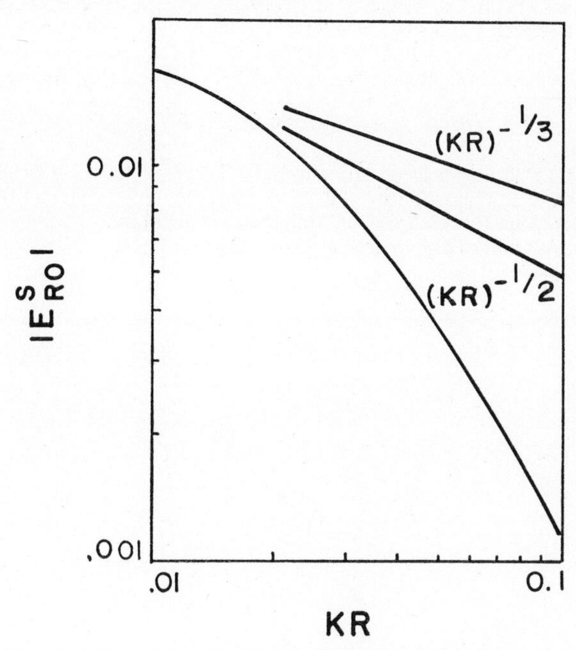

FIG. 4. Order of the edge singularity in the radial electric-field near the cone edge.

c_n and d_n. In every case the magnitudes of the coefficients computed by using only three equations lie within 2 per cent of the magnitudes of those computed by using seven equations. Also, an examination of $|c_1/c_2|$ and $|d_1/d_2|$ yielded factors of 200 and 140, respectively, indicating very rapid convergence of the series, as may be expected for such a small value of kb.

Although one could not expect to obtain a true singularity at the edge of the cone by using only a finite number of terms, Fig. 3 indicates that such a singularity is being approached. Furthermore, Fig. 4 shows that the singularities not only lie well within the $(kr)^{-1}$ limit imposed by the finite energy condition, but also closely approximate the $(kr)^{-1/3}$ singularity predicted by Meixner (1954) and Rogers and Schultz (1960) for a 90-deg edge. As a further check, the seven terms in the expansion of E_r were investigated individually and it was found that all were of the same order of magnitude and six of them of the same algebraic sign, indicating a strong likelihood that the series was diverging at this expected singularity.

REFERENCES

KELLER, J. B. (1961) Backscattering from a finite cone—comparison of theory and experiment, *I.R.E. Trans.* **AP–9,** 411.

MEIXNER, JOSEF (1954) The behavior of electromagnetic fields at edges, New York University, Research Report No. EM-72.

ROGERS, C. C. and SCHULTZ, F. V. (1960) The scattering of a plane electromagnetic wave by a finite cone, Scientific Report No. 1, School of Electrical Engineering, Purdue University.

SCHELKUNOFF, S. A. (1948) *Applied Mathematics for Engineers and Scientists.* D. Van Nostrand.

SIEGEL, K. M. (1959) Far-field scattering from bodies of revolution, *Appl. Sci. Res.* **B7,** 293.

SIEGEL, K. M., *et al.* (1951–3) Studies in radar cross-sections II, III, and IV, UMM 82, 87, 92, Willow Run Research Center, University of Michigan.

STRATTON, J. A. (1941) *Electromagnetic Theory.* McGraw-Hill.

SCATTERING BY NONSPHERICAL PARTICLES WHOSE SIZE IS OF THE ORDER OF THE WAVELENGTH†

J. M. GREENBERG, L. LIBELO, A. LIND and R. T. WANG

Rensselaer Polytechnic Institute, Troy, New York

ABSTRACT

Some experimental and theoretical total cross-sections for the scattering of electromagnetic radiation by a variety of nonspherical dielectric particles are presented. Comparisons among these results are made. It is found that, for radiation propagated normal to the symmetry axis, the total cross-sections for finite circular cylinders of rather small elongation (2 : 1) are strikingly similar to those of infinite cylinders. It is also found that for radiation propagated along the direction of the axis there is a resonance which appears to be related to the wave-guide properties. A long wave length expansion by Stevenson is applied to spheroids and is found to give qualitatively incorrect results at least up to one term beyond the Rayleigh approximation.

I. INTRODUCTION

On the basis of the most generally accepted theoretical interpretation of the astronomical data on interstellar extinction and polarization we are mainly interested in the total (extinction) cross-sections for arbitrarily oriented nonspherical particles whose sizes are in the range $0 < ka \lesssim 10$ ($ka = 2\pi a/\lambda$, $a =$ characteristic linear dimension, $\lambda =$ wavelength of the radiation) and whose indices of refraction are of the form $m \approx 1{\cdot}4 - i\delta$ where $\delta \ll 1$. As is well known, this range spans the most inaccessible region for obtaining scattering information; namely, the so-called resonance region which lies between the range of applicability of the Rayleigh approximation and the geometrical optics approximation.

Because of the rather formidable theoretical difficulties presented by the scattering problems of interest it has been necessary to attempt a simultaneous experimental and theoretical approach to the study of the scattering properties of mathematically simple-shaped bodies. The hope is that at least semi-empirical approximations may be indicated which are sufficiently accurate to be useful. Furthermore the experimental results will also serve as a check on various long wavelength and short wavelength expansions.

† Supported by the National Science Foundation and the National Aeronautics and Space Administration.

We shall present some of our recent work along these lines: In Section 2, a brief outline of the experimental method and some results on the total cross-sections of several types of particles. In Section 3, some results of exact calculation of the scattering by infinitely long cylinders. In Section 4, a discussion of an attempt (Stevenson, 1953) to obtain an extension of the Rayleigh (1897) approximation.

2. MICROWAVE ANALOG METHOD

The experimental study has been made using 3 cm microwave radiation. The details of the method have been presented elsewhere (Greenberg *et al.*, 1961). However, at least a summary account is useful here in order to clarify the nature of the results and also to indicate the need for some modifications of the published method under certain conditions.

Total cross-sections are found by using the relationship between the total cross-section and the complex forward-scattering amplitude

$$\sigma_t = (4\pi/k^2)\,\mathrm{Re}\{\mathbf{q} \cdot \mathbf{f}_\mathbf{q}(0)\},$$

where $\mathbf{q}$ is a unit vector along the direction of polarization and $\mathbf{f}_\mathbf{q}$ is the complex-scattering amplitude. The magnitude and phase of the scattering amplitude are determined from the variation of the detected signal at the receiver antenna as the particle is moved perpendicular to the direction of the incident radiation. A comparison of the variation produced by a given scattering target is made with similar variations produced by a set of standard spheres for which the scattering has been computed by Mie theory. As a particle is moved the signal goes through a maximum and minimum. The difference between these gives the magnitude of the scattering amplitude and their positions define the phase, both of these being normalized to the results for the standards. Because the wavelength is fixed, the dependence of the total cross-section on ka is obtained by considering a set of particles of identical shape and refractive properties but of varying sizes. If the particles are not too large the small angle-scattering amplitude variations may be neglected within the range of angles involved in laterally displacing the particles. However we have made measurements of the small angle (0°–5°) scattering amplitudes and have disclosed significant differences between those of differently oriented nonspherical particles whose sizes correspond to the scattering region at and beyond the first main resonance peak. For example, if the transmitter and receiver are adjusted to vertical polarization then, in the plane perpendicular to the polarization, the scattering amplitude for a prolate spheroid whose symmetry axis is along the polarization direction decreases more slowly than the scattering amplitude for the same spheroid, with its symmetry axis perpendicular to the polarization. The present experimental method thus generally is limited to particles bounded by an upper size limit. The lower size limit is determined by considerations of noise (transmitter phase stability, mechanical rigidity of the transmitter mounting).

In Figs. 1–6 we present some total cross-sections obtained for oblate and prolate spheroids. The particles with index of refraction $m = 1 \cdot 613$ are machined from lucite. The other particles are molded from polystyrene

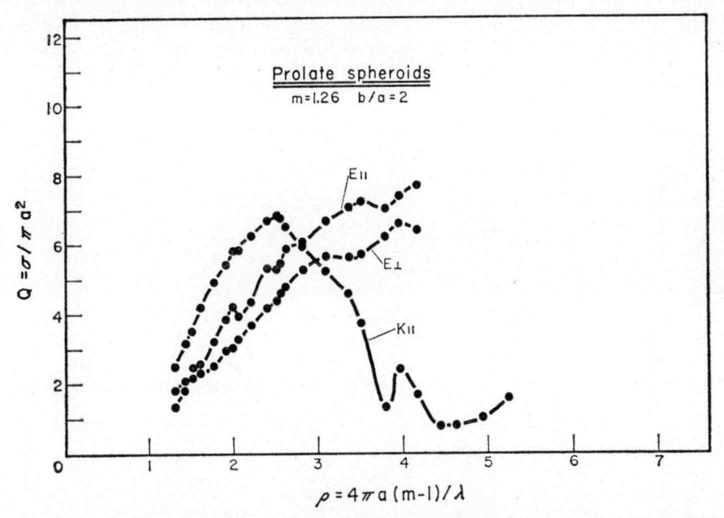

FIG. 1. Total cross-sections for prolate spheroids of elongation $b/a=2$ and index of refraction $m=1 \cdot 26$ at three orthogonal orientations: $k_{\parallel}$ =symmetry axis along the direction of propagation, $E_{\parallel}$ =symmetry axis parallel to the electric vector, $E_{\perp}$=symmetry axis perpendicular to the electric vector and to the propagation vector.

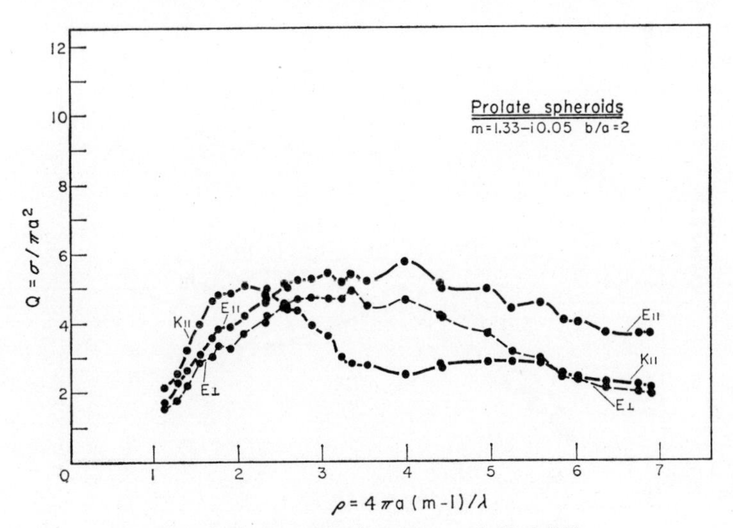

FIG. 2. Same as Fig. 1 except $m=1 \cdot 33–0 \cdot 05i$.

expandable beads. The imaginary part of the index of refraction is produced by admixing a small quantity of finely divided carbon. Our experiments have so far been performed for three orthogonal orientations of the particles: Electric vector parallel to the symmetry axis ($E_{\parallel}$), magnetic vector parallel to the axis ($E_{\perp}$), propagation vector parallel to the axis ($k_{\parallel}$). Because both the magnitude and phase of the forward-scattering amplitude are determined,

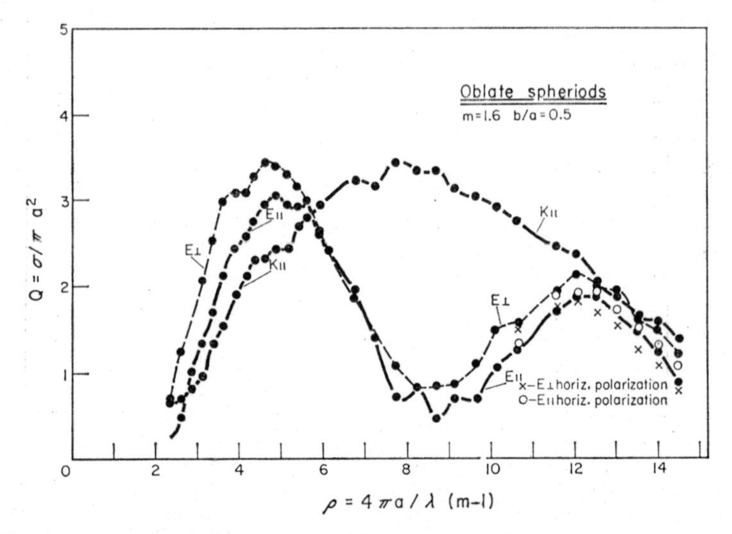

FIG. 3. Total cross-sections for oblate spheroids of elongation $b/a = 1/2$ and index of refraction $m = 1 \cdot 61$ at three orthogonal orientations.

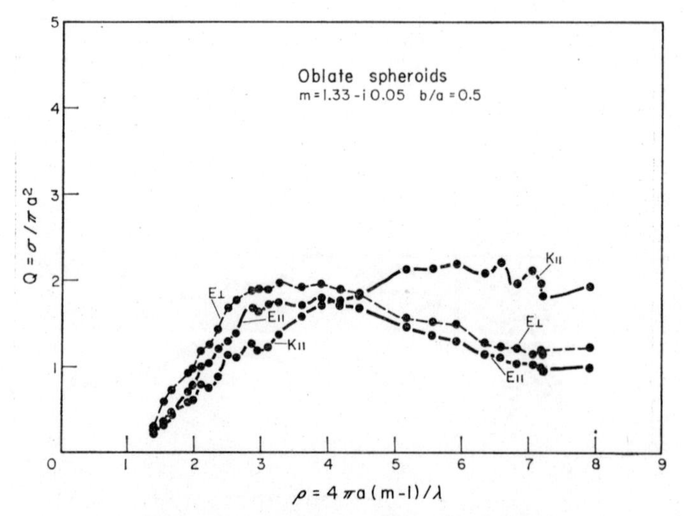

FIG. 4. Same as Fig. 3 except $m = 1 \cdot 33 - 0 \cdot 05i$.

the total cross-sections for an *arbitrarily* oriented particle of cylindrical symmetry can be obtained from data for the particle axis lying in two perpendicular planes: (1) plane containing E and k, (2) plane containing H and k. Thus additional data for perhaps only four more properly selected orientations should fairly complete the general picture.

In Figs. 1–4 we have plotted the ratio of the total cross-section to the geometrical area along the symmetry axis vs. $2ka(m - 1) = \rho$. For prolate spheroids ($b/a = 2$) it is to be expected, on the basis of a low index of refraction approximation (Greenberg, 1960) that the first major peak should occur at $\rho \simeq 2$ for $k_{\parallel}$ and at $\rho \simeq 4$ for $E_{\parallel}$(or $E_{\perp}$). It is also to be expected that, at their respective peaks, $Q_{k_{\parallel}} \simeq Q_{E_{\parallel}}/2$. As can be seen in Figs. 1 and 2, the former is roughly borne out, although with a tendency for the peak for $Q_{k_{\parallel}}$ to occur at a value of ρ somewhat greater than $\rho = 2$. While the magnitude of $Q_{E_{\parallel}}$ at the peak is quite reasonable, it is not completely clear why the magnitude of $Q_{k_{\parallel}}$ at its peak is so much larger than was to be expected. Apparently the approximate theory referred to above is rather more limited in application to nonspherical bodies than it is to spheres where the comparison is quite good (van de Hulst, 1957). Although the curves in Fig. 2 have been plotted up to $\rho = 7$ we know that the values of $Q_{E_{\parallel}}$ and $Q_{E_{\perp}}$ are quite incorrect for the larger sizes in the sense that they should actually be generally almost equal† beyond $\rho = 4$. On the other hand the values of $Q_{k_{\parallel}}$ are almost certainly good up to $\rho = 4$ and a little beyond. The details of this analysis are based on our observations of the small-angle scattering distributions and will be published elsewhere. The results shown in Fig. 3 for oblate spheroids are probably reliable up to about $\rho = 8$. An indication of the magnitude of the errors in $Q_{E_{\perp}}$ and $Q_{E_{\parallel}}$ for larger values of ρ is given by the points which were taken with the receiver and transmitter rotated through 90°. This rotates the planes in which the effect of small-angle scattering changes is demonstrated. We note that this effect shows up as an *apparent* decrease of $Q_{E_{\perp}}$ and increase of $Q_{E_{\parallel}}$ for sufficiently large values of ρ. For the lower index of refraction oblate spheroids in Fig. 4 this effect, which is primarily a function of size, sets in at somewhat smaller values of ρ and it is estimated that the values are reliable only up to $\rho \sim 5$. A rather sensitive method of determining the reliability of our results is illustrated in Figs. 5 and 6, where some complex forward-scattering amplitudes (times $4k^2/a^2$) are plotted on a polar diagram. Any serious discrepancies show up clearly in the deviations from a reasonable spiral which should asymptotically approach the value 2 in Figs. 5 and 6. These curves are generally similar to those for equivalent spheres. For $m = 1{\cdot}61$ (Fig. 5) the rather large swing beyond the value 2 has already been noted in the discussion of the large magnitudes of $Q_{k_{\parallel}}$ at its major resonance peaks. For complex m (Fig. 6) the damping of the spiral is analogous to the Mie theory results for spheres,

† See discussion in Section 3 on the results of exact calculations on infinite cylinders.

although the amount of tangling of the curve appears to be greater than one would expect for such a modest real part of the index of refraction.

A rather simple scattering situation for which there exists as yet no very satisfying theoretical approach is exemplified in Fig. 7. We see here the extinction efficiencies, $\sigma/\pi a^2$, for finite circular cylinders vs. index of re-

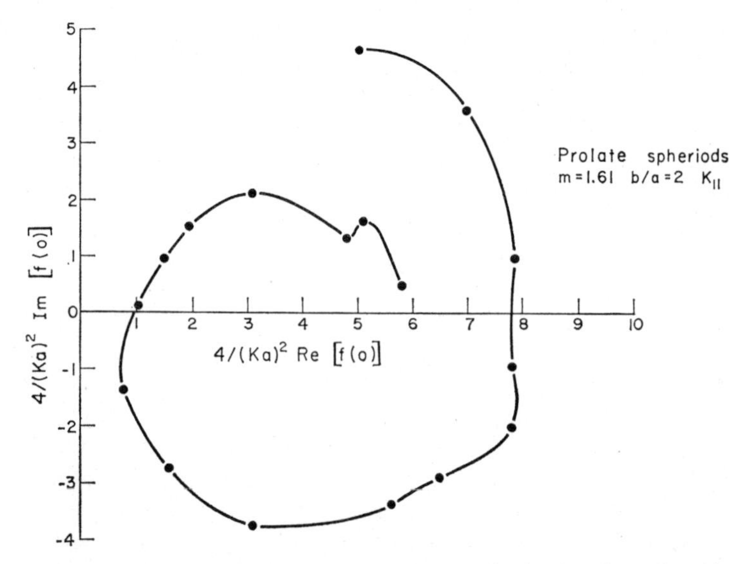

FIG. 5. Plot of $4/(ka)^2$ times forward-scattering amplitude of prolate spheroids of index of refraction $m=1 \cdot 61$ in the complex plane. The abscissa is the real part and the ordinate is the imaginary part.

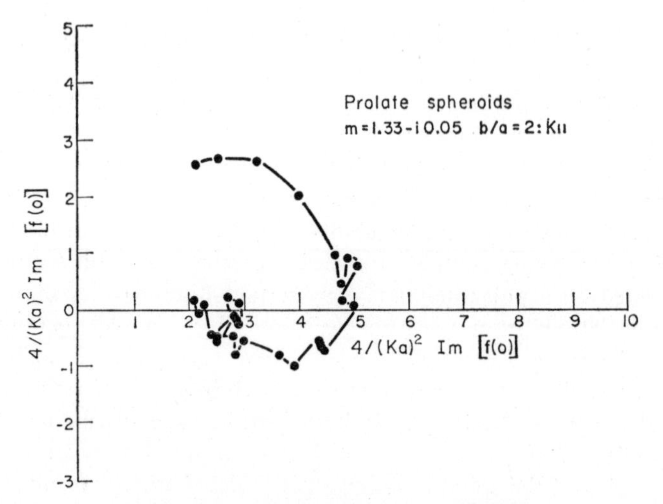

FIG. 6. Same as Fig. 5 except $m=1 \cdot 33-0 \cdot 05i$.

fraction times circumference for radiation propagated along the axis. In Fig. 7a the length ($2b$) is 3 in. and in Fig. 7b the length is 5 in. The resonance in both cases occurs at $(m^2 - 1)^{1/2} ka \approx 1 \cdot 7$. It is to be noted that this value

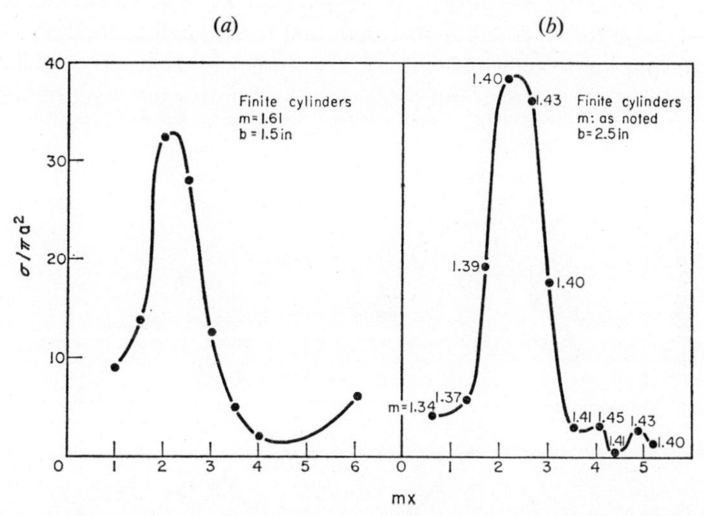

FIG. 7. Total cross-sections vs. index of refraction times circumference of finite cylinders for radiation propagated parallel to the axis: (a) Length of cylinders $2b=3$ in., index of refraction $m=1 \cdot 61$. (b) Length of cylinders $2b=5$ in., index of refraction $m \cong 1 \cdot 4$.

is smaller by a factor of about $(1 \cdot 4)^{-1}$ from the value $2 \cdot 4$ which defines the cutoff wavelength for lowest order mode propagation in an infinite dielectric cylinder. Work is in progress to determine by how much the length must be increased in order to reach this value.

3. INFINITE CIRCULAR CYLINDERS

The exact computation of the scattering by cylinders serves several purposes: (1) Comparison with analogous experimental results, (2) determination of qualitative features of polarization of radiation by aligned cylinders, (3) comparison with approximate theories for infinite cylinders and with approximate theories for other types of elongated particles. For these reasons we have programmed a computation for the IBM 7090 which results in the complete scattering characteristics (complex angular-scattering amplitudes) for circular cylinders with either real or complex indices of refraction. Extensive numerical data have been obtained and we shall present here a few illustrative examples.

An interesting and significant comparison between theory and experiment is shown in Fig. 8. It can be seen that the experimental total cross-sections for finite cylinders only twice as long as they are wide are strikingly similar in detail to the total cross-sections for infinite cylinders in the resonance region. Note particularly the ripples in both sets of curves around the first major peak for the two orientations—$E_{\parallel}$ and $E_{\perp}$. The fact that the experimental curve for $E_{\parallel}$ is still higher than that for $E_{\perp}$ well beyond $\rho = 4$ is probably a spurious result which is due to the aforementioned effects of differences in angular-scattering distributions at small angles for the two orientations.

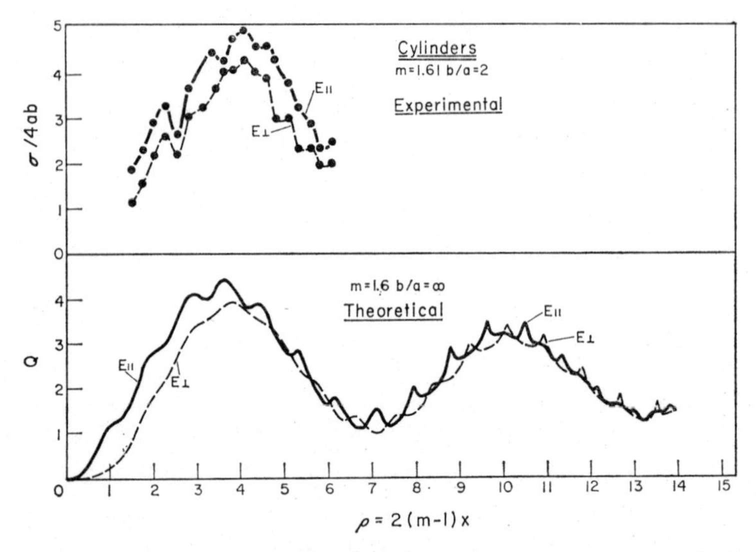

FIG. 8. Total cross-sections for cylinders whose symmetry axes are perpendicular to the propagation direction. Upper curves are experimental results for finite cylinders of elongation $b/a=2$ and index of refraction $m=1\cdot61$. Lower curves are exact theoretical results for infinitely long cylinders of index of refraction $m=1\cdot60$.

The increasing degree of complexity of the size dependence of the total cross-sections with increasing real index of refraction is shown by comparing Figs. 8–10. One of the apparently basic characteristics which is preserved is that just beyond the first major resonance the polarization of infinite cylinders, $1 - E_{\parallel}/E_{\perp}$, oscillates and on the average is quite small. This is perhaps not too surprising when it is realized that the ray approximation appears to become valid in this region and that therefore polarization effects become increasingly negligible. The small ripples in the curves have not been satisfactorily explained. The tentative explanation of the frequency of ripples in the total cross-sections for spheres by van de Hulst (1957) is roughly applicable to the cylinders, but a great difficulty in accepting this explanation is the problem of accounting for their magnitudes.

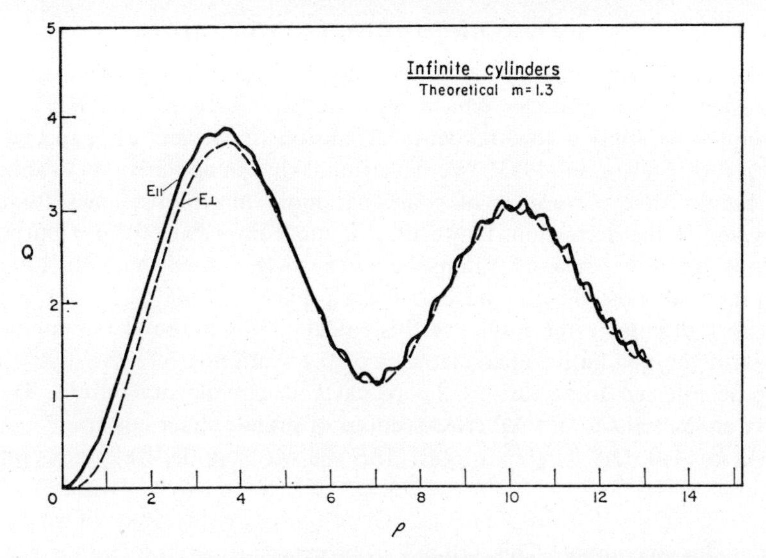

FIG. 9. Mie series total cross-sections of infinitely long cylinders with index of refraction $m=1\cdot30$.

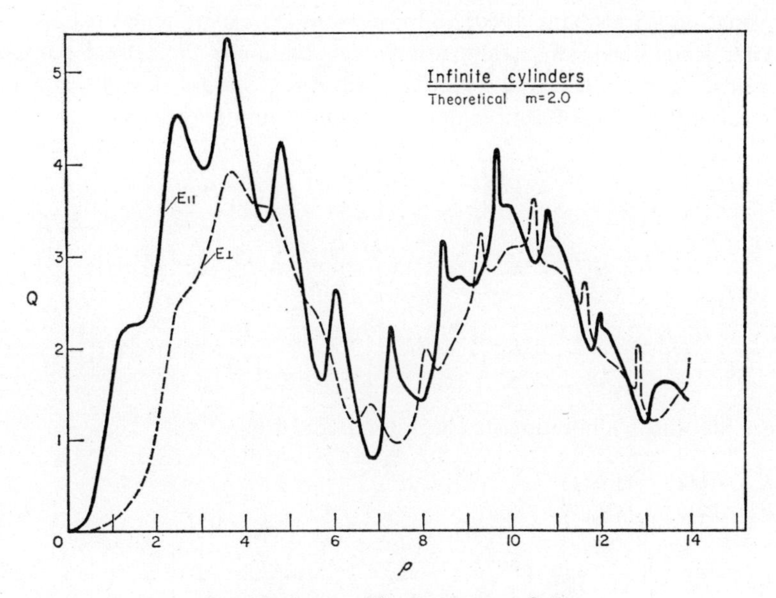

FIG. 10. Same as Fig. 9 except $m=2\cdot00$.

4. A LOW-FREQUENCY EXPANSION

Because of the experimental difficulties involved in obtaining accurate cross-sections for particles which approach the Rayleigh region we have attempted to apply a low-frequency expansion (in powers of $\chi = ka$) developed by Stevenson (1953). We have found that in applications to spheres the Stevenson expansion is of somewhat more limited usefulness than is indicated in the derivation. However, it is more important for our purposes to consider the expansion for nonspherical particles for which no other analytical methods exist. This we have done for prolate spheroids.

Before discussing the results of this calculation it is useful to summarize a few of the qualitative characteristics of the scattering by spheroids which may be inferred from various theoretical and experimental results. Qualitative characteristics of total cross-sections of prolate spheroids are:

(1) As $ka \to 0$, $Q_{k_\parallel} = Q_{E_\perp} < Q_{E_\parallel}$. This follows from the Rayleigh approximation.

(2) $Q_{E_\parallel} > Q_{E_\perp}$, $0 < ka \lesssim (ka)_{mm}$, where $(ka)_{mm}$ is the value of ka at the first major maximum. This follows from experimental results (see Figs. 1 and 2) and from analogy with infinite cylinders (see Figs. 8–10).

(3) There exists a value of $ka = (ka)_1 < (ka)_{mm}$ such that for $ka < (ka)_1$ $Q_{E_\parallel} > Q_{k_\parallel} > Q_{E_\perp}$.

(4) For $(ka)_{mm} > ka > (ka)_1$, $Q_{k_\parallel} > Q_{E_\parallel} > Q_{E_\perp}$. This, and (3) follow from the fact that eventually (in this range of ka) $Q_{k_\parallel}$ becomes larger than both $Q_{E_\parallel}$ and $Q_{E_\perp}$. This is readily inferred from the low index of refraction approximation (Greenberg, 1960) and also from the experimental results.

After some algebraic manipulations we obtain the three total scattering efficiencies for a dielectric spheroid (dielectric constant $\epsilon = m^2$, magnetic permeability $\mu = 1$) of semi-major axis c and semi-minor axis a,

$$Q^s_{k_\parallel} = a^{-2}\frac{8}{3}k^4K_1^2\left\{1 + 2k^2\left[\frac{L_1(k_\parallel)}{K_1} - \frac{1}{150}(4a^2 + c^2)\right]\right\} \tag{1}$$

$$Q^s_{E_\perp} = a^{-2}\frac{8}{3}k^4K_1^2\left\{1 + 2k^2\left[\frac{L_1(E_\perp)}{K_1} - \frac{1}{150}(4a^2 + c^2)\right]\right\} \tag{2}$$

$$Q^s_{E_\parallel} = a^{-2}\frac{8}{3}k^4K_3^2\left\{1 + 2k^2\left[\frac{L_3(E_\parallel)}{K_3} - \frac{1}{150}(3c^2 + 2a^2)\right]\right\} \tag{3}$$

where Stevenson's notation has been used, and where

$$K_1 = \tfrac{2}{3}(\epsilon - 1)\,f_1(\epsilon)$$
$$K_3 = \tfrac{2}{3}(\epsilon - 1)\,f_3(\epsilon)$$

$$15\,L_1 = f_1(\epsilon)\left\{(\epsilon - 1)\,l_1\left[\frac{5a^2 - c^2}{3} - a^2(l^2 + m^2) - c^2n^2\right] + \right.$$

$$+ \epsilon(a^2mn_2 - c^2nm_2)\Big\} + [f_1(\epsilon)]^2 l_1 \Big\{(\epsilon - 1)\Big[(\epsilon - 2)\, I +$$

$$+ \epsilon a^2 I_a - \frac{4}{c}\Big] + \epsilon^2 \frac{(a^2 + c^2)}{a^2c}\Big\} + f_1(\epsilon)\frac{(a^2c)}{2}$$

$$\Big\{(I_a - I_c)\, \tfrac{1}{2}(c^2 - a^2)^2\, (mn_2 - nm_2) - \frac{\epsilon}{a^2c}$$

$$(a^2 - c^2)\,(mn_2 + nm_2)\Big\} +$$

$$+ [f_1(\epsilon)]^2 \frac{(a^2c)}{2}(I_a - I_c)\, l_1\, .$$

$$\cdot \Big\{(\epsilon - 1)\Big[a^2 I_a - c^2 I_c + \tfrac{1}{2}\epsilon(a^2 + c^2)\,(I_a - I_c) + \epsilon(\epsilon - 2)\frac{(a^2 - c^2)}{a^2c}\Big\}$$

$$15\, L_3 = f_3(\epsilon)\Big\{(\epsilon - 1)\, n_1 \Big[\frac{6c^2 - 2a^2}{3} - a^2(l^2 + m^2) - c^2n^2\Big] +$$

$$+ \epsilon(c^2lm_2 - a^2ml_2)\Big\} + [f_3(\epsilon)]^2$$

$$n_1\Big\{(\epsilon - 1)\Big[(\epsilon - 2)\, I + \epsilon c^2 I_c - \frac{4c}{a^2}\Big] + 2\epsilon^2/c\Big\}.$$

The direction cosines l, m, etc., are defined as follows for the three orientations:

	$k_{\parallel}$	$E_{\perp}\,(H_{\parallel})$	$E_{\parallel}$
$l, m, n,$	$(0, 0, -1)$	$(0, -1, 0)$	$(0, -1, 0)$
l_1, m_1, n_1	$(1, 0, 0)$	$(1, 0, 0)$	$(0, 0, 1)$
l_2, m_2, n_2	$(0, -1, 0)$	$(0, 0, 1)$	$(-1, 0, 0)$

The Rayleigh approximation gives only the first term on the right of eqns (1)–(3) and we see that in this limit $Q_{k_{\parallel}} = Q_{E_{\perp}}$ and $Q_{E_{\perp}} < Q_{E_{\parallel}}$. The difference between $Q^s_{k_{\parallel}}$ and $Q^s_{E_{\perp}}$ occurs through the difference $L_1(k_{\parallel}) - L_1(E_{\perp})$ which is given by

$$L_1(k_{\parallel}) - L_1(E_{\perp}) = \frac{f_1(\epsilon)}{15}(\epsilon - 1)\,(a^2 - c^2) < 0. \tag{4}$$

This result implies that $Q^s_{k_{\parallel}} < Q^s_{E_{\perp}}$ up to terms in the expansion of the order (ka)† and is in contradiction to conditions (3) and (4). In order to consider other qualitative characteristics which are not clearly discernible from eqns (1)–(3) we have chosen a particular spheroid of elongation $c/a = 2$.

† See discussion in Section 3 on the results of exact calculations on infinite cylinders.

For $1 \leqq m \leqq (2)^{1/2}$ we find that $Q^s_{k\parallel}$ will, for some small value of ka, exceed $Q^s_{E\parallel}$. This agrees with the *first* part of condition (4) but it also implies that, for this value of ka, $Q^s_{E\perp} > Q^s_{E\parallel}$ which should only occur for larger values of ka; namely for values of ka beyond the first major resonance. For selected values of $m > (2)^{1/2}$ ($1 \cdot 68 \leqq m \leqq 2 \cdot 39$, and $m \gg 1$) we find in *all* cases that $Q^s_{k\parallel} < Q_{E\parallel}$ for *all* values of ka. This violates condition (4). We are thus forced to the conclusion that the Stevenson expansion of the scattering to one term in ka beyond the Rayleigh term is insufficient to describe the correct qualitative behavior of the total cross-sections for spheroids even for small ka. This does not preclude the possibility that some other low-frequency expansion will not be more successful.

5. CONCLUSIONS

The microwave analog method has made it possible to study the scattering by nonspherical particles in a size range which bridges most of the gap between the ray optics and the Rayleigh regions. The difficulties in using the present phasing method for the determination of the cross-sections occurs for large particles (if $m \lesssim 1 \cdot 4$). Because the scattering amplitude for such particles is quite large and is also strongly directed forward, we shall in the future eliminate the particle motion and obtain our results by a null method at precisely zero degrees. This will extend the range of reliable data as far in ρ as we wish. For small values of $\rho (ka < 1)$ it still appears to be necessary to develop accurate formulas, probably as reasonably converging expansions in ka.

ACKNOWLEDGEMENTS

We should like to express our appreciation to the U.S. Naval Ordnance Laboratory for granting us the use of their high speed digital computer (IBM 7090) in carrying out the computations for the infinite cylinders.

REFERENCES

GREENBERG, J. M. (1960) *J.A.P.* **31,** 82.
GREENBERG, J. M., PEDERSEN, N. E. and PEDERSEN, J. C. (1961) *J.A.P.* **32,** 233.
RAYLEIGH, LORD (1897) *Phil. Mag.* **44,** 28.
STEVENSON, A. F. (1953) *J.A.P.* **24,** 1134, 1143.
VAN DE HULST, H. C. (1957) *Light Scattering by Small Particles*, fig. 32, p. 117. John Wiley, New York.
VAN DE HULST, H. C. (1957) *Light Scattering by Small Particles*, p. 374. John Wiley.

SCATTERING AND DIFFRACTION OF TRANSIENT PLANE ELECTROMAGNETIC WAVES

EDWARD M. KENNAUGH

Antenna Laboratory, Department of Electrical Engineering,
The Ohio State University, Columbus 10, Ohio

SUMMARY

The general solution to the scattering of a transient plane wave by an object is related to the impulse response waveform. This is the electromagnetic waveform scattered in various directions by an obstacle when illuminated with a delta-function plane wave pulse. The scattered waveform for any incident transient plane wave can be constructed from the impulse response waveform, using the convolution principle. The scattered field for a monochromatic incident wave can also be derived using transform methods. Certain properties of the impulse response waveform for a finite obstacle recommend its use as a mathematical and conceptual tool. First, it is found that the waveform is of finite duration, in most cases enduring no longer than 4 or $5T$, where T is the time for a pulse to travel the major linear dimension of the obstacle. Secondly, it is found that the waveform changes in a regular manner with scattering angle, so that interpolation is possible in many instances to obtain the waveform for intermediate scattering angles. Finally, it appears that the waveform may be related directly to certain physical properties of the scattering obstacle, ultimately permitting its direct interpretation or prediction by simple means.

The purpose of this paper is to present the impulse response waveform for various simple objects, as determined from a Fourier synthesis of known monochromatic plane wave scattering solutions, and to approximate by several methods the impulse response waveform for other objects. To illustrate the form of the impulse response waveform derived by Fourier synthesis methods, the scattered waveform at various directions from metallic and dielectric spheres are presented for a periodic train of short single-cycle plane wave pulses, proceeding to the limit of extremely short pulse duration and long pulse separation. The salient features of the impulse response waveform can be derived by this means for pulse widths as long as $0 \cdot 4T$ and pulse spacings as short as $10T$. For example, Fig. 1 presents one cycle of the far-zone scattered waveforms for a perfectly conducting sphere illuminated by pulses of $0 \cdot 392T$ duration and $12 \cdot 8T$ spacing. These are shown as "snapshots" of the relative electric field amplitude at various scattering angles, with the sphere and waveforms in the same scale of distance. The waveforms at each scattering angle contain the initial spike predicted by geometrical optics as

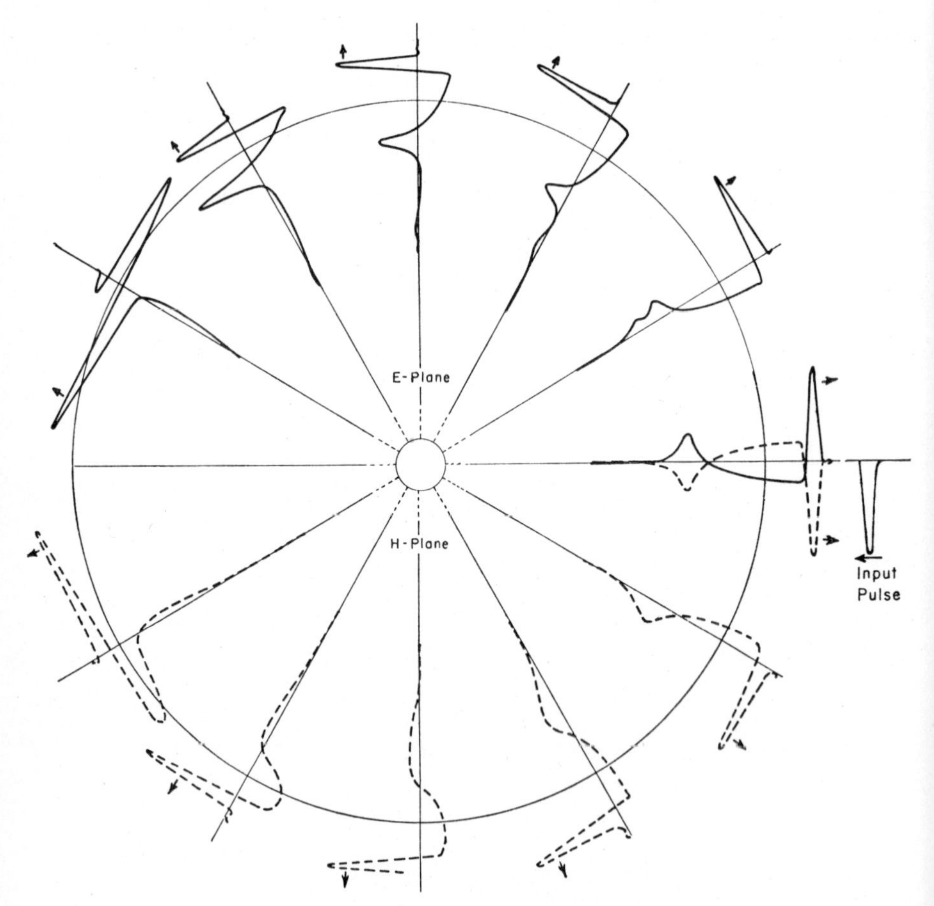

FIG. 1. Scattering of short electromagnetic pulses by a perfectly conducting sphere.

well as secondary peaks which are predicted by the geometrical theory of diffraction or the "creeping" wave concept (Keller, 1958). In addition, those features of the waveform which indicate the scatterer properties in the low and middle frequency range can be given qualitative interpretation.

Proceeding from the derivation and analysis of the impulse response waveform for obstacles such that a rigorous solution to the scattering problem exists, the derivation of approximate impulse response waveforms in other cases is considered. It is shown how conventional high and low-frequency approximate methods can be applied to determine salient features of the scattered waveform. For example, the coefficients in a power series expansion of the scattered field as a function of frequency (for which the Rayleigh coefficient would be the leading term) are directly proportional to the various moment integrals of the impulse response waveform (Kennaugh and Cosgriff, 1958). Similarly, the coefficients of an asymptotic series expansion of the scattered field (in which the geometrical optics term would be the leading term) are related directly to discontinuities in various time derivatives of the waveform. It is thus possible to combine the features of low and high frequency approximations in a single model from which a scattering approximation at any frequency can be made.

Next, the use of approximate or exact solutions for the impulse response waveform to construct an approximate solution for scattering by a composite body is discussed. The method is illustrated and applied to conducting bodies whose surfaces are composed of conical and spherical sections. It is shown how known solutions for each of the segments can be combined and adjusted to derive the approximate impulse response waveform. Theoretical results obtained in specific cases have been compared with experimentally determined values with excellent agreement (Kennaugh and Moffatt, 1962).

REFERENCES

KELLER, J. B. (1958) A geometrical theory of diffraction, *Proc. Symp. on Applied Mathematics*, **8**, 27–52. McGraw-Hill, New York.

KENNAUGH, E. M. and COSGRIFF, R. L. (1958) The use of impulse response in electromagnetic scattering problems, *I.R.E. National Convention Record*, Part 1, pp. 72–77.

KENNAUGH, E. M. and MOFFATT, D. L. (1962) On the axial echo area of the cone-sphere shape, *Proc. I.R.E.* **50**, No. 2 (February).

PLANE WAVE DIFFRACTION BY A STRIP†

RALPH E. KLEINMAN

The University of Michigan Radiation Laboratory Ann Arbor, Michigan, U. S. A.

ABSTRACT

An exact integral representation is presented of the field diffracted by a soft (Dirichlet boundary condition) infinitely long strip of width $2a$ in the presence of a plane wave. The result is obtained as a limiting case of the corresponding problem with a line source excitation after recasting the line source solution in a form suitable for passage to the limit. Some simplification in the form of the solution obtains in the case of normal incidence and these results are given.

1. INTRODUCTION

Although the exact solution for the problem of diffraction of plane or cylindrical waves by an infinitely long strip of finite width was presented over fifty years ago (Sieger, 1908), the infinite series of Mathieu functions in which these solutions are expressed have proven to be sufficiently unwieldy so as to prevent a thorough analysis of the behavior of the scattered field. An integral representation of these solutions comparable to Sommerfeld's result for the half plane (Sommerfeld, 1896) has long been sought. Recently, an exact

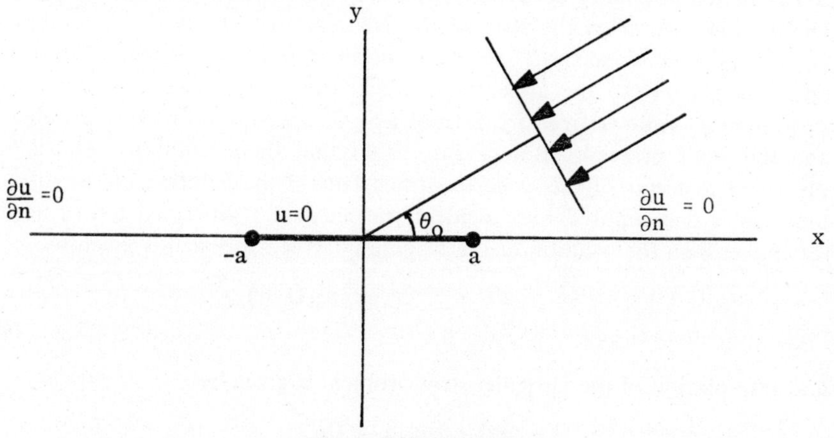

FIG. 1. Strip geometry.

† The research reported in this paper was sponsored by Air Force Cambridge Research Laboratories under Contract AF-19(604)-6655.

integral expression for the problem of diffraction of cylindrical waves by a soft (Dirichlet boundary condition) strip has been found (Timman and Kleinman, 1960; Kleinman, 1961). In the present paper, this result is extended to the case of plane wave excitation.

The integral expressions for the line source case are of such complexity that a direct determination of the asymptotic behavior for the source located far from both strip and field point is not trivial. However, by making use of Weber's two-dimensional analogue of the Helmholtz integral representation of the field (Baker and Copson, 1950), it is possible to circumvent the asymptotic difficulty and obtain the plane wave result.

2. STATEMENT OF THE PROBLEM

The precise problem under consideration (see Fig. 1) is the determination of a two-dimensional wave function $u(x, y, \phi_0)$ such that

(a) $(\nabla^2 + k^2)\, u = 0,$

(b) $u\,\big|_{y=0} = 0, \qquad |\,x\,| < a, \qquad \left.\dfrac{\partial u}{\partial y}\right|_{y=0} = 0, \qquad |\,x\,| > a$

(c) $u = e^{ikr\cos(\phi-\phi_0)} + e^{ikr\cos(\phi+\phi_0)} + u_{\text{dif}},$

where u_{dif} satisfies a Sommerfeld radiation condition.

(d) u is everywhere finite.

The harmonic time dependence e^{ikct} is assumed so $e^{ikr\cos(\phi-\phi_0)}$ represents an incoming plane wave, k is the wave number $2\pi/\lambda$, and c is the velocity of propagation.

The mixed boundary condition, (b), is considered rather than treating the Dirichlet condition on the strip or the Neumann condition on the complementary screen as separate problems because it is convenient to restrict attention entirely to the upper half plane. This is not a serious limitation since it easily follows from the formulation of aperture diffraction problems and Babinet's principle (Bouwkamp, 1954), that these solutions can all be expressed in terms of the same function. Thus if the solution of the mixed problem is decomposed into incident, reflected and diffracted terms, as in (c) above, then the solution of the Neumann screen problem is given by

$$\begin{aligned} u_N &= e^{ikr\cos(\phi-\phi_0)} + e^{ikr\cos(\phi+\phi_0)} + \tfrac{1}{2}u_{\text{dif}}(x, y),\ y > 0 \\ &= -\tfrac{1}{2}u_{\text{dif}}(x, -y),\ y < 0, \end{aligned} \tag{1}$$

and the solution of the Dirichlet strip problem is given by

$$\begin{aligned} u_D &= e^{ikr\cos(\phi-\phi_0)} + \tfrac{1}{2}u_{\text{dif}}(x, y),\ y > 0 \\ &= e^{ikr\cos(\phi-\phi_0)} + \tfrac{1}{2}u_{\text{dif}}(x, -y),\ y < 0, \end{aligned} \tag{2}$$

where u_{dif} is the same function throughout and need be defined only for non-negative values of y. Here ϕ_0 denotes the direction from which the plane wave is incident and r, ϕ are the usual polar coordinates.

3. SOLUTION

To obtain the solution of the mixed problem for plane wave incidence from the corresponding solution for line source excitation it is convenient to introduce elliptic coordinates. Thus we set

$$\begin{aligned} x &= a \cosh \mu \cos \theta, \qquad \mu \geqslant 0, \; -\pi \leqslant \theta \leqslant \pi \\ y &= a \sinh \mu \sin \theta. \end{aligned} \tag{3}$$

The coordinate surfaces are confocal ellipses and hyperbolae with the limiting cases $\mu = 0$ and $\theta = 0$, π corresponding to segments of the real axis as shown in Fig. 2.

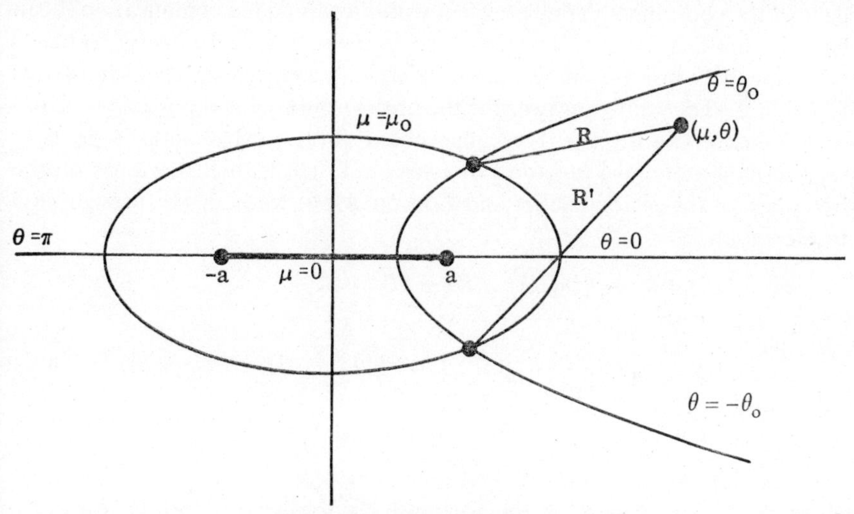

FIG. 2. Elliptic coordinates.

In these coordinates the distances R and R' are

$$R = \sqrt{[(x - x_0)^2 + (y - y_0)^2]} = a \left| \cosh(\mu + i\theta) - \cosh(\mu_0 + i\theta_0) \right|$$

and

$$R' = \sqrt{[(x - x_0)^2 + (y + y_0)^2]} = a \left| \cosh(\mu + i\theta) - \cosh(\mu_0 - i\theta_0) \right|. \tag{4}$$

With the assumed time dependence, Hankel functions of the second kind represent outgoing waves and the line source is normalized to be

$$-\pi i H_0^{(2)}(kR).$$

The solution (denoted by $\Phi(\mu, \theta, \mu_0, \theta_0)$) of the mixed problem for this excitation is found to be (Timman and Kleinman, 1960),

$$\Phi(\mu, \theta, \mu_0, \theta_0) = \int_{-\mu+i\theta}^{\mu+i\theta} d\alpha \int_{-\mu_0+i\theta_0}^{\mu_0+i\theta_0} d\beta J(\mu, \theta, \alpha)\, J(\mu_0, \theta_0, \beta)\, F(\alpha, \beta) \tag{5}$$

where

$$\mu, \mu_0 \geqslant 0,\; 0 \leqslant \theta, \theta_0 \leqslant \pi,$$

the short hand for the Bessel function

$$J(\mu, \theta, \alpha) \equiv J_0(\sqrt{\{ka[\cosh(\mu + i\theta) - \cosh \alpha]\, [\cosh(\mu - i\theta) - \cosh \alpha]\}}) \tag{6}$$

is employed, and $F(\alpha, \beta)$ is a complicated function which shall be defined shortly. For the present it is necessary to know only that $F(\alpha, \beta)$ is independent of the coordinate variables. Even without this added complication it can be seen that this form of Φ offers difficulties in any direct attempt to determine the behavior for large values of μ_0. An asymptotic analysis of this function is under way, however for the present purpose it is possible to avoid such considerations. The two-dimensional form of Helmholtz's integral representation of wave functions in a region, in terms of their values on the boundary of the region (Baker and Copson, 1950) leads in the present case to the equation

$$\Phi(\mu, \theta, \mu_0, \theta_0) = -i\pi[H_0^{(2)}(kR) + H_0^{(2)}(kR')]$$

$$+\frac{i}{2} \int_0^\pi H_0^{(2)}(ka \mid \cosh(\mu + i\theta) - \cos \theta_1 \mid) \frac{\partial}{\partial \mu_1} \Phi(\mu_1, \theta_1, \mu_0, \theta_0) \bigg|_{\mu_1=0} d\theta_1. \tag{7}$$

This relation would be an integral equation were Φ unknown, but with Φ given by (5), it provides a means for present purposes by which to transform the result into a more useful form. A significant check would be provided by a direct demonstration that (7) is an identity when Φ is given by (5) but this is not a trivial calculation and will be reported elsewhere.

Since a plane wave can be expressed as a limit of a cylindrical wave as follows

$$e^{ikr\cos(\phi-\theta_0)} = e^{ika(\cosh \mu \cos \theta \cos \theta_0 + \sinh \mu \sin \theta \sin \theta_0)}$$

$$= \lim_{\mu_0 \to \infty} \sqrt{\left(\frac{\pi\, ka \cosh \mu_0}{2}\right)}\, e^{i(ka \cosh \mu_0 - \pi/4)}\, H_0^{(2)}(kR), \tag{8}$$

the solution of the mixed boundary value problem for plane wave incidence can be expressed as the limit of solution of the line source problem in the following manner,

$$u(\mu, \theta, \theta_0) = \lim_{\mu_0 \to \infty} \frac{1}{-\pi i} \sqrt{\left(\frac{\pi ka \cosh \mu_0}{2}\right)} e^{i(ka \cosh \mu_0 - \pi/4)} \Phi(\mu, \theta, \mu_0, \theta_0) \tag{9}$$

where the factor $(1/-\pi i)$ arises from the particular normalization of the source chosen.

The reciprocity relation existing between source and field point permits the identity (7) to be rewritten in even more convenient form. Since interchanging μ with μ_0 and θ with θ_0 must leave Φ unaltered, (7) can be rewritten

$$\Phi(\mu, \theta, \mu_0, \theta_0) = -i\pi[H_0^{(2)}(kR) + H_0^{(2)}(kR')]$$
$$+ \frac{i}{2} \int_0^{\pi} H_0^{(2)}(ka \mid \cosh(\mu_0 + i\theta_0) - \cos \theta_1 \mid) \frac{\partial}{\partial \mu_1} \Phi(\mu_1, \theta_1, \mu, \theta) \bigg|_{\mu_1 = 0} d\theta_1. \tag{10}$$

Substituting (10) in (9) and performing the indicated limit operation, which is now easily done, leads to the plane wave solution in elliptic coordinates

$$u(\mu, \theta, \theta_0) = e^{ika \cosh \mu \cos \theta \cos \theta_0} [e^{ika \sinh \mu \sin \theta \sin \theta_0} + e^{-ika \sinh \mu \sin \theta \sin \theta_0}]$$
$$- \frac{1}{2\pi} \int_0^{\pi} e^{ika \cos \theta_1 \cos \theta_0} \frac{\partial}{\partial \mu_1} \Phi(\mu_1, \theta_1, \mu, \theta) \bigg|_{\mu_1 = 0} d\theta_1. \tag{11}$$

Note that in the limit, the elliptic variable θ_0 becomes the ordinary polar angle ϕ_0. With (6) it is seen that $J(\mu_1, \theta_1, \pm\mu_1 + i\theta_1) = 1$, thus

$$\frac{\partial \Phi}{\partial \mu_1}(\mu_1, \theta_1, \mu, \theta) \big|_{\mu_1 = 0} = 2 \int_{-\mu + i\theta}^{\mu + i\theta} J(\mu, \theta, a) F(i\theta_1, a) \, da \tag{12}$$

and it is only necessary to define F to completely specify the solution. This is (Timman and Kleinman, 1960),

$$F(i\theta_1, a) = -\frac{\sinh a \sin \theta_1}{2(\cosh a - \cos \theta_1)^2} \int_{-\gamma(i\theta_1, a)}^{\gamma(i\theta_1, a)} \frac{e^{-ika \cos \omega [\pm(\cosh a - \cos \theta_1)]}}{\sin^2 \omega} d\omega \tag{13}$$

where

$$\gamma(i\theta_1, a) = i \log \frac{1 - \cosh(a + i\theta_1)}{\pm(\cosh a - \cos \theta_1)} \tag{14}$$

and the positive sign is to be used when $\theta < \theta_1$ and the negative sign when $\theta > \theta_1$. The path of integration in (12) is taken to be a straight line in the

α-plane and the branch cuts of the logarithm in (14) are defined so that the log is real when $\cosh \alpha$ is real, i.e. when $\alpha = i\theta$, and so as not to intersect the path of integration in the α-plane.

4. DISCUSSION

The function F is investigated more thoroughly elsewhere (Timman and Kleinman, 1960) where it is shown, with the help of integral representations of the difference

$$H_0^{(2)}(kR) - H_0^{(2)}(kR'),$$

to be sufficiently well behaved despite its apparent singular behavior when $\theta = \theta_1$. Curiously it is the difference of the two Hankel functions which is important here whereas in (7) the sum plays a vital role. Integral representations of $H_0^{(2)}(kR) + H_0^{(2)}(kR')$, comparable to those developed for $H_0^{(2)}(kR) - H_0^{(2)}(kR')$, have been found and are expected to facilitate analysis in the independent boundary value problem where a Neumann condition is specified on the strip, and/or a Dirichlet condition on the complementary screen.

As is readily apparent the solution presented here will have to be considerably simplified if it is to be comparable with the elegant and useful form of the half plane solution. One immediate simplification is evident in the case of normal incidence, $\theta_0 = \pi/2$. Combining (11), (12), (13) and (14) in this case yields

$$u(\mu, \theta, \pi/2) = e^{ika \sinh \mu \sin \theta} + e^{-ika \sinh \mu \sin \theta}$$

$$+ \frac{1}{2\pi} \int_0^{\pi} d\theta_1 \int_{-\mu+i\theta}^{\mu+i\theta} d\alpha \left\{ \frac{J(\mu, \theta, \alpha) \sinh \alpha \sin \theta_1}{(\cosh \alpha - \cos \theta_1)^2} \int_{-i \log \frac{1-\cosh(\alpha+i\theta_1)}{\pm(\cosh \alpha - \cos \theta_1)}}^{i \log \frac{1-\cosh(\alpha+i\theta_1)}{\pm(\cosh \alpha - \cos \theta_1)}} \frac{e^{-ika \cos \omega [\pm(\cosh \alpha - \cos \theta_1)]}}{\sin^2 \omega} d\omega \right\}. \quad (15)$$

Attempts to achieve further simplification will be the subject of future investigations. These hold promise of success since some preliminary results in an attempt to reduce (15) to useful computational form are quite encouraging.

REFERENCES

BAKER, B. B. and COPSON, E. T. (1950) *The Mathematical Theory of Huygen's Principle.* Oxford University Press, pp. 48–52.

BOUWKAMP, C. J. (1954) Diffraction theory, *Rep. Prog. Phys.* **17,** 35–100.

KLEINMAN, R. E. (1961) Integral representations of solutions of the Helmholtz equation with application to diffraction by a strip. Dissertation, Delft, Netherlands.

SIEGER, B. (1908) Die beugung einerebenen elektrischen welle an einem schirm von elliptischen querschnitt, *Ann. Phys.* **27**, 626.

SOMMERFELD, A. (1896) Mathematische theorie der diffraction, *Math. Ann.* **47**, 317.

TIMMAN, R. and KLEINMAN R. E. (1960) Integral representation for the field diffracted by a strip URSI General Assembly, London. To be published in the *Proceedings*.

SCATTERING BY A WIDE GRATING

R. F. MILLAR

Department of Mathematics, Royal Military College of Canada,
Kingston, Ontario, Canada

ABSTRACT

Scalar plane-wave scattering by a finite grating of arbitrary cylinders is treated. The scattered field representation is an integral over plane waves, involving an unknown scattering function. The grating is considered as part of an infinite grating with known properties. A plane-wave superposition principle yields an integral equation relating the functions for the isolated finite grating and infinite grating. An asymptotic series solution, valid when the width suffices (except at Rayleigh wavelengths and certain angles of observation), is derived. Only the leading term seems significant for gratings wider than one wavelength; this incorporates end effects through functions associated with semi-infinite gratings.

SUMMARY

This paper describes a method for solution of the problem of scattering of a plane wave by a wide, finite grating of identical cylinders. No restrictions are placed on the size, uniform spacing, or physical properties of the elements. Because the analysis is relatively straightforward, much of this summary will be an exposition of the method of formulation of the problem.

In the past ten years or so, the concept of an angular spectrum of plane waves has been employed in the analysis of several two-dimensional diffraction problems (see, for example, Clemmow, 1950, 1951). The basis for this analysis is that an essentially arbitrary solution of the reduced wave equation may be expressed as an integral over plane waves, the integration being with respect to the (complex) angle of incidence. Application of boundary conditions then leads to integral equations for the undetermined amplitude of the waves.

More recently, Twersky (1956, 1961) has employed related notions in certain multiple-scattering problems. But rather than relate the unknown functions through the boundary conditions, Twersky associates the unknown "many-body scattering functions" of individual scatterers for plane-wave incidence, with the corresponding functions (assumed known) for isolated scatterers, by a plane-wave superposition principle. This leads to certain integral relationships between the scattering functions. (These scattering functions are, for real angles of observation, essentially the polar diagrams of the far-fields of the scatterers.)

A point of view quite similar to that of Twersky has been adopted by the author to study scattering by many identical objects. However, the analysis proceeds rather differently, and a few words on this topic are worth while at the present point.

It is customary to consider the individual scatterers as fundamental elements in a system. But it has been found that some problems may be more amenable to analysis when considered as two-body problems (even though at least one of the two bodies contains more than one scatterer), provided the scattering properties of the bodies, together or in isolation, are known. By way of example, the semi-infinite grating of identical cylinders may be considered to be one-half an infinite grating whose scattering properties are, in turn, known in terms of the scattering properties of an isolated cylinder. Alternatively, decomposition into the end element, and the remaining infinity of elements is possible; here the properties of the end element in isolation are given. Of these two possibilities, the first has proved until now to be the more encouraging, and the analysis of the semi-infinite grating by this method has been described elsewhere (Millar, 1961).

Although some work has been done in connection with the finite grating (see, for example, Twersky, 1952), a completely satisfactory analysis, which takes into account end effects and other interaction phenomena (especially when many elements are involved), still seems to be lacking. The present work is an attempt to provide a step in this direction.

The basic notions involved in the analysis are described above. Here, too, there appear to be two possible methods of attack. In the first method, the grating of N elements may be regarded as having been constructed in N stages, each stage consisting of the addition of one element. At each stage, it is necessary to solve a two-body problem; for example, at the nth stage, the two bodies whose scattering properties are known are the $n-1$ scatterers already present and the one scatterer added at this stage. This leads to the solution for n bodies, and thus the solution for N scatterers requires the solution of $N-1$ two-body problems. Although this might be feasible for small values of N, it seems less likely that the method would be useful if N were large, where the second procedure, to be described below, is most suitable.

Consider then, a finite grating of $M+N+1$ identical elements C_ν: $\nu = -M, -M+1, \ldots, N$, as comprising the central portion of an infinite grating. This infinite grating is then assumed to be composed of three bodies, namely: the cylinders C_ν for $-\infty < \nu < -M$, $-M \leqslant \nu \leqslant N$, $N < \nu < \infty$, respectively. If the corresponding scattering functions for these three bodies as parts of an infinite grating are denoted by p_-, p, p_+, and if P represents the scattering function for the finite grating in isolation, then the superposition principle may be employed to relate the four functions, Since the three functions p_-, p and p_+ refer to the (periodic) infinite grating.

each may be expressed in terms of p_0, where p_0 is the scattering function for the cylinder C_0 in the infinite grating. If p_0 is known, then the superposition relation connects the desired function P with the known p_0.

It is next shown that it is natural to split P into a sum of two parts Q and R, say. When these are inserted into the relation involving P, a pair of simultaneous integral relationships is obtained between Q, R and p_0. These relationships are observed to be very similar to the type which arise in the study of scattering by the (isolated) semi-infinite gratings consisting of elements C_ν: $-M \leqslant \nu < \infty$ and C_ν: $-\infty < \nu \leqslant N$. It is shown that Q and R are equal to the functions for the corresponding semi-infinite gratings, plus correction functions U and V, respectively. The equations satisfied by U and V are then determined, and it is observed that, if certain angles of incidence and observation are excluded, asymptotic series for U and V in inverse powers of a parameter (assumed to be large) are readily generated. The large parameter involved in the expansions is $\eta = (N + M)\,ka$ where a is the spacing between consecutive cylinders, while k is the propagation constant of the incident plane wave. Thus η is approximately 2π times the width of the grating in wavelengths. The first terms in the series for U and V are found to be of order $\eta^{-3/2}$, and are probably of little importance if $\eta \gtrsim 5$. Thus the results are appropriate for wide gratings, optical gratings in particular.

The physical interpretation of the various terms in the resultant asymptotic series for P is quite simple. The terms which do not depend on an inverse power of η may be written as a sum of three parts. The first is just the function p, introduced above, which relates to scattering by $C_{-M}, \ldots, C_N$ in the infinite grating, and which therefore ignores end effects. The remaining two parts contribute end corrections, and involve scattering functions related to the two semi-infinite gratings C_ν: $-M \leqslant \nu < \infty$ and C_ν: $-\infty < \nu \leqslant N$. Thus end effects are included even in a first approximation. On the other hand, the terms which depend on inverse powers of η arise through multiple scattering by the two edges of the grating.

The close analogy with certain treatments of diffraction by a wide strip is apparent; the half-planes which enter into these latter analyses are here replaced by semi-infinite gratings. The fact that the first terms in the asymptotic series for U and V are of order $\eta^{-3/2}$ is also consistent with corresponding (E-polarization) results for the wide strip (see, for example, Millar, 1958).

The excluded angles of incidence correspond to the Rayleigh wavelengths of the grating. It is here that interaction effects are greatest. The excepted angles of observation are related to certain phenomena peculiar to semi-infinite gratings. In each case, exclusion seems necessary because, for these angles, what may be a singularity in an integrand then lies on the contour of integration. (Similar problems arise for grazing incidence on a strip.) A more careful analysis should overcome this limitation, and it is hoped that a uniform asymptotic expansion, valid for all angles of incidence and observation, will be forthcoming.

REFERENCES

CLEMMOW, P. C. (1950) A note on the diffraction of a cylindrical wave by a perfectly-conducting half-plane, *Quart. J. Mech.* **3,** Part 3, 377–384.

CLEMMOW, P. C. (1951) A method for the exact solution of a class of two-dimensional diffraction problems, *Proc. Roy. Soc.* **A 205,** 286–308.

MILLAR, R. F. (1958) Diffraction by a wide slit and complementary strip, *Proc. Camb. Phil.* 1961 *Soc.* **54,** Part 4, 479–511.

MILLAR, R. F. (1961) Plane wave spectra in multiple scattering theory; paper presented at URSI-IRE Spring meeting, Washington, D.C. (to be published).

TWERSKY, V. (1952) On a multiple scattering theory of the finite grating and the Wood anomalies, *J. Appl. Phys.* **23,** No. 10, 1099–1118 (October).

TWERSKY, V. (1956) On the scattering of waves by an infinite grating, *I.R.E. Trans.* **AP-4,** 330–345.

TWERSKY, V. (1961) On scattering of waves by two objects; Electronic Defense Laboratories, Mountain View, California; Engineering Report EDL-E60, March 6, 1961. See also pp. 361–389 in *Electromagnetic Waves*, edited by Rudolph E. Langer, Madison: The University of Wisconsin Press, 1962.

A BOUNDARY WAVE THEORY OF DIFFRACTION AT AN APERTURE†

E. WOLF,‡ E. MARCHAND§ and K. MIYAMOTO||

SUMMARY

In well-known investigations Maggi and Rubinowicz (Maggi, 1888; Rubinowicz, 1917, 1953) showed that within the accuracy of Kirchhoff's diffraction theory, the field $U_K(P)$ arising from the diffraction of a monochromatic spherical or a plane scalar wave at an aperture in an opaque screen may be expressed in the form

$$U_K(P) = U_K{}^{(B)}(P) + U^{(G)}(P), \tag{1}$$

where $U_K{}^{(B)}$ represents a combined effect of disturbances propagated from each point of the boundary of the aperture (boundary wave), and $U^{(G)}$ represents the field predicted by geometrical optics. In the present paper an account will be given of a number of recent investigations (Miyamoto and Wolf, 1962; Marchand and Wolf, 1962) concerned with generalizations of the Maggi–Rubinowicz theory both within and outside the domain of the Kirchhoff theory.

The existence of the boundary wave is intimately related to the existence of a certain potential of the diffracted wave field. To show this consider a Green's type integral representation of the solution of the Helmholtz equation:

$$U(P) = \iint_S \mathbf{V}(P, Q) \,.\, \mathbf{n}\, \mathrm{d}S, \tag{2}$$

where

$$\mathbf{V}(P, Q) = G(P, Q)\, \nabla_Q U(Q) - U(Q)\, \nabla_Q G(P, Q). \tag{3}$$

Q is a typical point on a surface S surrounding P, n is the unit outward normal to S and G is any "principal solution" of the Helmholtz equation. An elementary calculation shows that

$$\mathrm{div}\,_Q \mathbf{V}(P, Q) = 0. \tag{4}$$

† The research described in this paper was supported in part by the U.S. Air Force under a contract monitored by the Air Force Office of Scientific Research.

‡ Department of Physics and Astronomy, University of Rochester, Rochester, N.Y., U.S.A.

§ Research Laboratories, Eastman Kodak Company, Rochester, N.Y., U.S.A.

|| Nippon Kokagu, K.K. (Japan Optical Industrial Co. Ltd.), Tokyo, Japan.

Since (4) holds for points Q on any arbitrary closed surface S surrounding P this implies that there exists a potential $\mathbf{W}(P, Q)$ such that

$$\mathbf{V}(P, Q) = \text{curl}_Q \mathbf{W}(P, Q). \tag{5}$$

Closed expressions for the potential $\mathbf{W}$ will be given, for any field U which obeys the Sommerfeld radiation condition at infinity, in an appropriate angular segment of space. The potential $\mathbf{W}$ considered as a function of Q has in general some singularities on S. It may be shown that in many cases of practical interest these singularities are discrete points, finite in number. Only such cases will be considered here.

Because of many well known difficulties which one encounters when one tries to solve aperture diffraction problems rigorously one often uses (2) in an approximate form. One extends the integral over the aperture A only and one replaces the unknown exact boundary values of U or its normal derivative $\partial U/\partial n$ by $U^{(i)}$ or $\partial U^{(i)}/\partial n$ where $U^{(i)}$ denotes the unperturbed incident field. If this approximation is made it then follows by using (5) and Stokes theorem, that the diffracted field may be represented in the form

$$U(P) = U^{(B)}(P) + \sum_j F_j(P), \tag{6}$$

where

$$U^{(B)}(P) = \int_\Gamma \mathbf{W}(P, Q) \cdot \mathbf{l} \, d\mathbf{l} \tag{7}$$

and the integral is taken along the rim Γ of the aperture. Thus $U^{(B)}$ represents a boundary wave. Further $F_j(P)$ represents the contribution from a singularity Q_j of the vector potential and is expressed in the form of an integral similar to (7), taken along the boundary of a vanishingly small circle surrounding the singularity. The summation in (6) is taken over all the singularities Q_j in A.

The representation (6) corresponding to the following choices of the principal solution G will be discussed more fully:

$$\text{(I) } G \equiv G_K, \qquad G_K = \frac{1}{4\pi} \frac{\exp iks}{s},$$

$$\text{(II) } G \equiv G_{RI}, \qquad G_{RI} = \frac{1}{4\pi} \left\{ \frac{\exp iks}{s} - \frac{\exp iks^*}{s^*} \right\},$$

$$\text{(III) } G \equiv G_{RII}, \qquad G_{RII} = \frac{1}{4\pi} \left\{ \frac{\exp iks}{s} + \frac{\exp iks^*}{s^*} \right\},$$

where $s = PQ$, $s^* = P^*Q$, P^* being the image of P in the plane of the aperture.† In particular it will be shown that when the wave incident upon the

† With the choice $G = G_K$ the integral representation (2) reduces to the Kirchoff solution; with the choice $G_{R\alpha}$ (α = I, II) it reduces to the Rayleigh type solution, α = I corresponding to the case of Dirichlet's boundary conditions and α = II to the case of Neumann's boundary conditions.

aperture is plane or spherical, the contribution $\sum_j F_j(P)$ in (6) from the singularities of the potential in the aperture is in all these three cases exactly equal to the field $U^{(G)}$ specified by geometrical optics; and moreover that the corresponding boundary waves $U_K^{(B)}$, $U_{RI}^{(B)}$ and $U_{RII}^{(B)}$ differ from each other only in the form of an inclination factor.

This theory gives a new insight into the physical process of diffraction at an aperture when the linear dimensions of the aperture and the distance of the field point P from the aperture are both large compared with the wavelength, and when the angles of diffraction are not too large. Since the representation (6) involves only one-dimensional integrals it may lead to simplifications in the practical calculations of diffracted fields.

Finally it may be mentioned that some of our results were recently rederived by A. Rubinowicz (1962) using different methods; and also that K. Miyamoto (1962) extended some of the results to non-monochromatic wave fields. These investigations of Rubinowicz and Miyamoto are, however, restricted to the domain of the Kirchhoff diffraction theory.

REFERENCES

MAGGI, G. A. (1888) *Ann. di Mat.* **16,** 21. RUBINOWICZ, A. (1917) *Ann. der Phys.* (4), **53,** 257; (1953) *Acta Phys. Polonica*, **12,** 225. For comprehensive account of these researches see RUBINOWICZ, A., *Die Beugunswelle in der Kirchhoffschen Theorie der Beugung* (Warszawa, Polska Akademia Nauk, 1957). Other accounts will be found in BAKER, B. B. and COPSON, E. T., *The Mathematical Theory of Huygens' Principle* (Oxford, Clarendon Press, 2nd ed., 1950, p. 74) and BORN, M. and WOLF, E., *Principles of Optics* (London and New York, Pergamon Press, 1959, § 8.9).

MARCHAND, E. and WOLF, E. (1962) *J. Opt. Soc. Amer.* **52,** 761.

MIYAMOTO, K. and WOLF, E. (1962) *J. Opt. Soc. Amer.* **52,** 615.

MIYAMOTO, K. and WOLF, E. (1962) *J. Opt. Soc. Amer.* **52,** 626.

MIYAMOTO, K. (1962) *Proc. Phys. Soc.* **79,** 617.

RUBINOWICZ, A. (1962) *J. Opt. Soc. Amer.* **52,** 717.

RUBINOWICZ, A. (1962) *Acta Phys. Polonica*, **21,** 61.

DIFFRACTION WAVE IN CASE OF AN ARBITRARY INCIDENT FIELD IN THE ELECTROMAGNETIC KIRCHHOFF THEORY

A. RUBINOWICZ
Polish Academy of Sciences

SUMMARY †

When a spherical scalar wave is diffracted at an aperture, the resulting wave motion may be represented, within the Kirchhoff theory, as the sum of a geometric optical wave and a diffracted wave in agreement with the views of Thomas Young about the origin of diffraction phenomena. A decomposition of this type was obtained also in the case of an electromagnetic Kirchhoff wave motion (Kottler, 1923; Laporte and Meixner, 1958) arising from the diffraction of an electric dipole wave.

Recently Miyamoto and Wolf (1962a and b) have shown that even in case of an arbitrary incident wave a similar splitting of the scalar Kirchhoff wave motion is possible. The aim of this paper is to prove that a strictly similar decomposition is also obtainable for an electromagnetic Kirchhoff wave motion.

To show this, we apply a geometric method which was used (Rubinowicz, 1962a and b) to derive the Miyamoto–Wolf formula for a diffracted scalar wave. By applying the electromagnetic reciprocity theorems to the Larmor principle, we can show that in a region R of space bounded by a closed surface S, any electromagnetic field $\mathbf{E}$, $\mathbf{H}$ regular throughout this region may be represented in the form

$$\mathbf{E}(P) = \int_S \hat{V}_e(P, Q)\, \mathbf{n}\, \mathrm{d}S_Q, \qquad \mathbf{H}(P) = \int_S \hat{V}_m(P, Q)\, \mathbf{n}\, \mathrm{d}S_Q. \tag{1}$$

Here $\hat{V}_e(P, Q)$ and $\hat{V}_m(P, Q)$ are certain tensor fields. The components of the "electric" tensor field $\hat{V}_e(P, Q)$ are defined by

$$V_{e|li}(P, Q) = \frac{c}{4\pi}\, \epsilon_{ijk}(E_{e|lj}\, H_k + H_{e|lj}\, E_k). \tag{2}$$

Here E_k, H_k denote the kth components of the given electromagnetic field $\mathbf{E}(Q)$, $\mathbf{H}(Q)$. $E_{e|lj}$, $H_{e|lj}$ are at an integration point Q, the jth components

† The complete paper has been publishcd in full in *Acta Physica Polonica* (**21**, 451, 1962).

of the electromagnetic field $\mathbf{E}_{e|l}(P, Q)$, $\mathbf{H}_{e|l}(P, Q)$ due to an electric dipole situated in the observation point P. Its dipole moment is given by a current element which is equal to the unit vector $\mathbf{i}_l$ in the direction of the positive lth axis†.

The corresponding expression for the components of the magnetic tensor field $\hat{V}_m(P, Q)$ may be obtained by replacing in (2) the index e by m.

As both the electromagnetic fields $\mathbf{E}(Q)$, $\mathbf{H}(Q)$ and $\mathbf{E}_{e|l}(P, Q)$, $\mathbf{H}_{e|l}(P, Q)$ are (as functions of Q) solutions of the Maxwell equations, the tensor divergences of the fields $\hat{V}_e(P, Q)$ and $\hat{V}_m(P, Q)$ are zero so that

$$\frac{\partial}{\partial x_i} V_{e|li} = 0.$$

Hence they can be represented by tensor curls of some tensor potentials $\hat{W}_e(P, Q)$, $\hat{W}_m(P, Q)$:

$$\hat{V}_e(P, Q) = \mathrm{Curl}_Q\, \hat{W}_e(P, Q), \qquad \hat{V}_m(P, Q) = \mathrm{Curl}_Q\, \hat{W}_m(P, Q). \tag{3}$$

This means that

$$V_{e|li} = \epsilon_{ijk} \frac{\partial}{\partial x_j} W_{e|lk} \qquad \text{and} \qquad V_{m|li} = \epsilon_{ijk} \frac{\partial}{\partial x_j} W_{m|lk}.$$

For such a tensorial curl there exists in analogy to the vectorial Stokes' theorem, a tensorial integral theorem,

$$\int_\sigma \mathrm{Curl}_Q\, \hat{W}_e(P, Q)\, \mathbf{n}\, \mathrm{d}S_Q = \int_B \hat{W}_e(P, Q)\, \mathrm{d}\mathbf{s}_Q \tag{4}$$

where B is the boundary of an open surface σ.

To obtain an expression for the tensor potential $\hat{W}_e(P, Q)$ we take a fictitious open surface σ bounded by a curve Γ and denote by P, some point outside of σ (Fig. 1). Next let L be the surface of the geometric shadow of σ generated by a light source at P. Then the surfaces σ and L bound a region R of space which has the shape of an infinite conical frustum. L is its lateral area and σ its covering surface.

Let us now apply the Larmor principle (1) to the region R and to an electromagnetic field which is regular in R and satisfies the electromagnetic Sommerfeld radiation conditions at infinity. For the point P in (1) we now take the vertex of the lateral area L. As P is outside the space R we have

$$\int_{L+\sigma} \hat{V}_e(P, Q)\, \mathbf{n}\, \mathrm{d}S_Q = 0. \tag{5}$$

† We remark that $\mathbf{E}_{e|l}(P, Q)$, $\mathbf{H}_{e|l}(P, Q)$ are represented by the formulae for the electromagnetic field of an electric dipole, but have not the dimensions of such a field. We have namely replaced, in the formulae for the electric dipole radiation, the current element by unit vectors without any dimensions. Nevertheless we shall call them electric or magnetic fields for the sake of convenience. For the same reason we shall call also the unit vector $\mathbf{i}_l$ a current element.

To transform the integral over L, we consider a strip on L bounded by a line element $d\mathbf{s}_Q$ of Γ and by two rulings A and A' of L passing through the endpoints of $d\mathbf{s}_Q$. If Q denotes a point on $d\mathbf{s}_Q$ we have in a point Q' of such a strip

$$\mathbf{n}\, dS_{Q'} = d\mathbf{r}_{Q'} \times d\mathbf{s}_Q\, (r_{Q'}/r_Q) \tag{6}$$

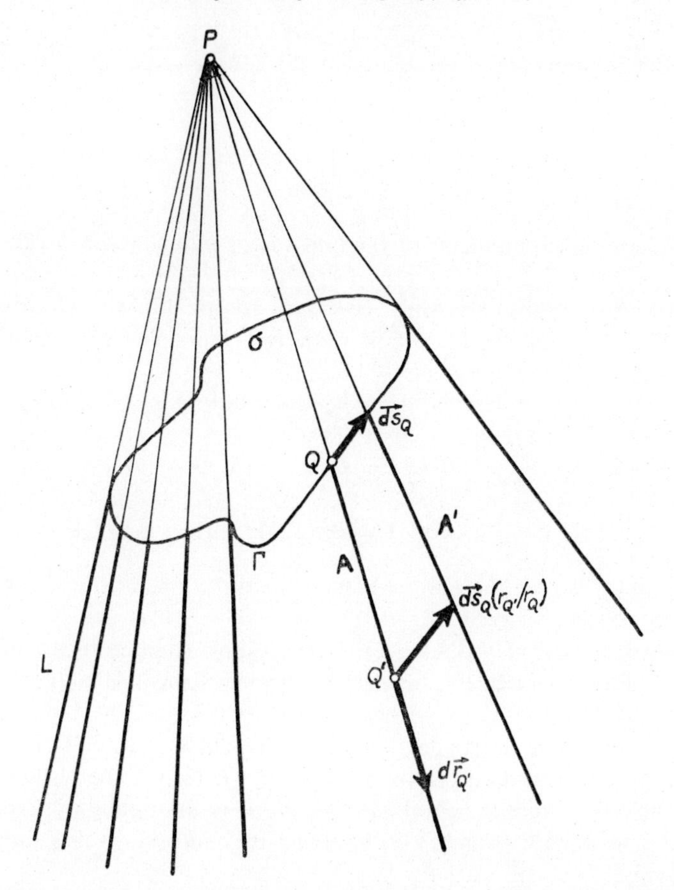

FIG. 1. Conical frustum bounded by the covering surface σ and the lateral area L. Point P is the vertex of the cone belonging to L, Γ is the boundary curve of σ and L. A vector surface element on L is given by $\mathbf{n}\ dS_{Q'} = d\mathbf{r}_{Q'} \times d\mathbf{s}_{Q'}\ (r_{Q'}/r_Q)$.

where $d\mathbf{s}_Q\ (r_Q'/r_Q)$ is a line element parallel to $d\mathbf{s}_Q$ between the two rulings A and A'. The surface element (6) has therefore the form of a parallelogram, two sides of which are given by $d\mathbf{r}_Q'$ and $d\mathbf{s}_Q\ (r_{Q'}/r_Q)$. Using (6) we obtain for the integral over L in (5) the expression

$$\int_L \hat{V}_e(P, Q')\, \mathbf{n}\, dS_{Q'} = \int_L (r_{Q'}/r_Q)\, \hat{V}_e(P, Q')(d\mathbf{r}_{Q'} \times d\mathbf{s}_Q). \tag{7}$$

If we carry out the integration in (7) over the rulings first and then over the boundary curve Γ, i.e. first over $\mathbf{d r}_{Q'}$ and then over $\mathbf{d s}_Q$, we can write the integral (7) in the form

$$\int_L \hat{V}_e(P, Q')\, \mathbf{n}\, dS_{Q'} = -\int_\Gamma \hat{W}_e(P, Q)\, \mathbf{ds}_Q \tag{8a}$$

where the components of the tensor $\hat{W}_e(P, Q)$ are given by

$$W_{e|li}(P, Q) = -\frac{1}{r_Q} \int_{r_Q}^{+\infty} \epsilon_{ijk}\, V_{e|lj}\, i_{r|k}\, r_{Q'}\, dr_{Q'}. \tag{8b}$$

Here $i_{r|k}$ are the components of the unit vector $\mathbf{i}_r$ having the direction of the ruling.

To transform (8b) it is only necessary to use (2), the electromagnetic reciprocity theorems for the dipole radiations and some vector algebra. We then obtain the expression

$$\left.\begin{aligned} W_{e|li}(P, Q) = -\frac{c}{4\pi}\frac{1}{r_Q}\mathbf{j}_l \int_{r_Q}^{+\infty} \{&\mathbf{E}_e(P, Q';\ \mathbf{H} \times (\mathbf{i}_r \times \mathbf{j}_i)) - \\ &- \mathbf{E}_m(P, Q';\ \mathbf{E} \times (\mathbf{i}_r \times \mathbf{j}_i))\}\, r_{Q'}\, dr_{Q'}. \end{aligned}\right\} \tag{9}$$

Here $\mathbf{j}_l$ and $\mathbf{j}_i$ denote the unit vectors in the directions of the lth and ith co-ordinate axes respectively. $\mathbf{E}_e(P, Q';\ \mathbf{H} \times (\mathbf{i}_r \times \mathbf{j}_i))$ is the electric field at the observation point P due to an electric dipole situated in the integration point Q' and generated by the electric current element $\mathbf{H} \times (\mathbf{i}_r \times \mathbf{j}_i)$. Similarly $\mathbf{E}_m(P, Q';\ \mathbf{E} \times (\mathbf{i}_r \times \mathbf{j}_i))$ denotes the electric field at P of a magnetic dipole generated by the magnetic current element $\mathbf{E} \times (\mathbf{i}_r \times \mathbf{j}_i)$ in Q'.†

Now we can very easily show that $\hat{W}_e(P, Q)$ (8b) is the tensor potential of the "electric" tensor field $\hat{V}_e(P, Q)$. As Γ is the boundary curve of the surface σ we obtain from (5) on applying the tensorial Stokes' theorem (4) to (8a)

$$\int_\sigma \hat{V}_e(P, Q)\, \mathbf{n}\, dS_Q = \int_\sigma \mathrm{Curl}_Q\, \hat{W}_e(P, Q)\, \mathbf{n}\, dS_Q.$$

This relation holds for any surface σ. Hence we obtain from it immediately the first relation in (3) for the tensor field $\hat{V}_e(P, Q)$, with the expression (9) for the tensor potential $\hat{W}_e(P, Q)$.

By similar considerations we can derive the tensor potential $\hat{W}_m(P, Q)$ for the tensor field $\hat{V}_m(P, Q)$.

The electromagnetic Kirchhoff "ansatz" implies that in (1) we integrate

† Also $\mathbf{E}_e$ and $\mathbf{E}_m$ do not have here the dimensions of an electric field.

over an open surface A covering the diffraction aperture. With regard to (3) it has also the form

$$\left.\begin{aligned} \mathbf{E}_K(P) &= \int_A \operatorname{Curl}_Q \hat{W}_e(P, Q)\, \mathbf{n}\, \mathrm{d}S_Q, \\ \mathbf{H}_K(P) &= \int_A \operatorname{Curl}_Q \hat{W}_m(P, Q)\, \mathbf{n}\, \mathrm{d}S_Q. \end{aligned}\right\} \tag{10}$$

Using Stokes' tensorial theory (4) we can reduce the surface integrals (10) to line integrals over the boundary B of A:

$$\mathbf{E}_d(P) = \int_B \hat{W}_e(P, Q)\, \mathrm{d}\mathbf{s}_Q, \qquad \mathbf{H}_d(P) = \int_B \hat{W}_m(P, Q)\, \mathrm{d}\mathbf{s}_Q, \tag{11}$$

and to integrals over the singularities† of the tensor potentials on A. We can regard the wave motion (11) as a diffracted wave as far as it consists of contributions of the different line elements $\mathrm{d}\mathbf{s}_Q$ of the diffracting edge B. If we suppose that the whole Kirchhoff wave motion (10) is given by a diffracted and a geometric–optical wave, then we must assume that the latter wave motion is given by the integrals over the singularities of $\hat{W}_e(P, Q)$ and $\hat{W}_m(P, Q)$ on the surface A. But we must remark that the definition (11) of the diffracted wave does not in general agree with Young's views. For Young supposed that this wave is caused by a scattering of the incident wave on the different line elements $\mathrm{d}\mathbf{s}_Q$ of the diffracting edge. According to Young it could therefore depend only on the local behaviour of the incident electromagnetic field at the line elements $\mathrm{d}\mathbf{s}_Q$. As may be seen from (9) the diffracted wave (11) depends, however, on the values of the incident field at all points situated on a half-ray which begins in $\mathrm{d}\mathbf{s}_Q$ and the continuation of which passes through P. Only in special cases, such as spherical waves, can the integrals in (11) be calculated (Kottler, 1923; Laporte and Meixner, 1958) and give diffracted waves in agreement with Young's views.

The procedure used above can be applied also to Kottler's (1923) formulation of the electromagnetic Huygens principle, as was shown in a paper published in the *Acta Physica Polonica* (Rubinowicz 1962*c*). As was shown by Karczewski (1963) we then obtain from the line integrals in Kottler's formulae contributions to the diffracted wave depending only on the local behaviour of the incident elecromagnetic field. But Kottler's surface integrals give contributions to the diffracted wave only of the Miyamoto–Wolf type considered above.

† If we use in the integrands of the Kirchhoff ansatz (10) the in A regular tensor fields $\hat{V}_e(P, Q)$ and $\hat{V}_m(P, Q)$ instead of $\operatorname{Curl}_Q \hat{W}_e(P, Q)$ and $\operatorname{Curl}_Q \hat{W}_m(P, Q)$ respectively, we have to integrate over the whole surface A without any exception. In case there are present on A diffuse singularities of $\hat{W}_e(P, Q)$ or $\hat{W}_m(P, Q)$ we obtain, applying Stoke's theorem (4) to (10), therefore not only loop integrals surrounding the singularities in question, but also surface integrals over these singularities themselves. If any singularity on A is localized in a point or in a curve, the application of Stokes' theorem (4) yields loop integrals only.

REFERENCES

KARCZEWSKI, B. (1963) *J. Opt. Soc. Amer.*, in press.
KOTTLER, F. (1923) *Ann. Phys.* [Leipzig] **71**, 457.
LAPORTE, O. and MEIXNER, J. (1958) *Z. Phys.* **153**, 129.
MIYAMOTO, K. and WOLF, E. (1962*a*) *J. Opt. Soc. Amer.* **52**, 615.
MIYAMOTO, K. and WOLF, E. (1962*b*) *J. Opt. Soc. Amer.* **52**, 626.
RUBINOWICZ, A. (1962*a*) *J. Opt. Soc. Amer.* **52**, 717.
RUBINOWICZ, A. (1962*b*) *Acta Phys. Polon.* **21**, 61.
RUBINOWICZ, A. (1962c) *Acta Phys. Polon.* **21**, 451.

AN ASYMPTOTIC EXPANSION OF ELECTRIC VECTOR FIELDS WITH COMPLEX PHASE FUNCTION

S. POGORZELSKI†

SUMMARY

An electric field **E**, harmonic in time, is subjected in a source-free region to the known vector equation

$$\nabla^2 \mathbf{E} + k_0^2 \mathbf{E} = 0 \tag{1}$$

where $k_0 = 2\pi/\lambda_0$ relates to the empty space.

Let us introduce a vector amplitude function $\mathbf{A}(P)$ and a phase function $L(P)$ by assuming

$$\mathbf{E}(P) = \mathbf{A}(P)\, e^{ik_0 L(P)}. \tag{2}$$

Substituting (2) into (1) we obtain

$$(ik_0)^2 [(\nabla L)^2 - 1]\, \mathbf{A} + ik_0[2(\nabla L \,.\, \nabla)\, \mathbf{A} + (\nabla^2 L)\, \mathbf{A}] + \nabla^2 \mathbf{A} = 0. \tag{3}$$

In order properly to define the phase function $L(P)$, the additional relation

$$ik_0[(\nabla L)^2 - 1] + \nabla^2 L = 0 \tag{4}$$

is imposed.

In accordance with (3) the amplitude **A** must now satisfy

$$2ik_0(\nabla L \,.\, \nabla)\, \mathbf{A} + \nabla^2 \mathbf{A} = 0. \tag{5}$$

The equations (4) and (5) are to be solved by introducing the asymptotic series

$$L \sim \sum_{m=0}^{\infty} \frac{L_m}{(ik_0)^m}; \qquad \mathbf{A} \sim \sum_{m=0}^{\infty} \frac{\mathbf{A}_m}{(ik_0)^m}. \tag{6}$$

Both functions L and **A** are generally complex. It should be noted that the phase function does not depend on the amplitude **A**. Therefore we are able to find the phase function without regard to the vector character of the field. It can be hoped that the phase function should contain interesting scalar properties of the field.

Let us begin to compute with L_m coefficients. The coefficient L_0 satisfies the known eiconal equation

$$(\nabla L_0)^2 = 1 \tag{7}$$

† Przemystowy Instytut Telekomunikacji, Warszawa, Poligonowa 30, Poland.

with the initial condition that $L_0 = f(u, v)$ on the initial surface $\mathbf{r}_0 = \mathbf{r}_0(u, v)$. Complex functions $f(u, v)$ are allowed, thereafter the eiconal L_0 may be complex.

The complex integral surface of eqn (8) satisfying imposed initial conditions can be written as

$$\mathbf{r}(u, v, s) = \mathbf{r}_0(u, v) + [s - f(u, v)]\, \mathbf{Q}(u, v) \tag{8}$$

where $s = L_0$

$$\mathbf{Q} = \frac{\dfrac{\partial \mathbf{r}_0}{\partial v} \times \mathbf{N}}{\mathbf{N}^2} \frac{\partial f}{\partial u} + \frac{\mathbf{N} \times \dfrac{\partial \mathbf{r}_0}{\partial u}}{\mathbf{N}^2} \frac{\partial f}{\partial v} \pm \frac{\mathbf{N}}{\mathbf{N}^2} \sqrt{\left[\mathbf{N}^2 - \left(\frac{\partial f}{\partial v}\frac{\partial \mathbf{r}_0}{\partial u} - \frac{\partial f}{\partial u}\frac{\partial \mathbf{r}_0}{\partial v}\right)^2\right]}$$

$$\mathbf{N} = \frac{\partial \mathbf{r}_0}{\partial u} \times \frac{\partial \mathbf{r}_0}{\partial v}.$$

Keeping in (8) u = const. and v = const. the equation for a complex ray is obtained. The parameter s can be interpreted as the arc length of the complex ray. One can introduce also real and imaginary rays as lines of $\nabla \mathrm{Re}[s]$ and $\nabla \mathrm{Jm}[s]$ respectively.

The remaining phase coefficients L_m satisfy linear differential equations along complex rays. The equations have the form:

$$\frac{\partial L_m}{\partial s} = - \sum_{\substack{p+q=m \\ p \neq m,\, q \neq m}} \nabla L_p \,.\, \nabla L_q + \nabla^2 L_{m-1}. \tag{9}$$

In order to integrate the right side of (9), it is necessary to investigate metric properties of the u, v, s space as given by (8). It may be shown that the covariant components of the metric tensor are quadratic polynomials in s and the contravariant components are rational functions of s. So the right side of (9) is a rational function of s and the integration is always possible. For instance we get

$$L_1 = \tfrac{1}{2} \ln \sqrt{\left(\frac{g(s_0)}{g(s)}\right)}, \tag{10}$$

where s_0 is the initial value of s on the initial surface and g is the determinant of the metric tensor in the space of the variables u, v, s.

Amplitude coefficients are now taken into consideration. The coefficient $\mathbf{A}_0$ satisfies the simple relation

$$\frac{\partial \mathbf{A}_0}{\partial s} = 0. \tag{11}$$

Thereafter $\mathbf{A}_0$ is constant along complex rays. It may be shown that the amplitude $\mathbf{A}_0$ is orthogonal to complex rays

$$\mathbf{A}_0 \,.\, \nabla L_0 = 0. \tag{12}$$

However, $\mathbf{A}_0$ can have components along real and imaginary rays. The higher coefficients $\mathbf{A}_m$ also satisfy linear differential equations of the form

$$\frac{\partial \mathbf{A}_m}{\partial s} = - \sum_{\substack{p+q=m \\ q \neq m}} (\nabla L_p \,.\, \nabla)\, \mathbf{A}q - \tfrac{1}{2}\nabla^2\, \mathbf{A}_{m-1}. \tag{13}$$

The right side of (13) is a rational function of s, so we can always get explicit expressions for the coefficients $\mathbf{A}_m$.

The simplest illustration of our analysis is offered by a surface wave which can propagate along a dielectric coat extended over a flat perfect conductor. It may be shown that the electric field of this surface wave has the form

$$\mathbf{E} = \mathbf{A}_0\, e^{ik_0 L_0}$$

where $\mathbf{A}_0$ and L_0 are functions as already defined. L_0 is in this case a complex function.

FORWARD AND BACKWARD SCATTERING FROM A PENETRABLE SPHERE AT SHORT WAVELENGTHS

S. I. Rubinow†

SUMMARY

In a previous work,‡ the problem of a scalar plane wave incident on a penetrable sphere was considered in the short wavelength limit for which the wavelength is assumed to be small compared to the radius of the sphere. The wave function and its derivatives are assumed to be continuous at the surface of the sphere. These boundary conditions are appropriate to the quantum mechanical scattering from a square-well potential, although, it is convenient to think of this problem as a scalar analog of the scattering of short wavelength electromagnetic waves (light) from water droplets. By means of the Poisson summation formula, a representation of the scattering amplitude is introduced which is particularly appropriate in this limit and which requires only the evaluation of certain integrals. Some of these integrals may be evaluated by the method of steepest descent and yield the contribution of the geometrical optics field. This includes the bow field, which is associated with a particular type of saddle point. The remainder of the integrals may be evaluated by the method of residues and lead to the contribution of the diffracted field. This "diffracted ray" field is well known from other recent investigations in diffraction theory. Thus, all the results that were obtained are in accordance with modern geometrical diffraction theory.

However, the representation that is employed breaks down in the neighborhood of the forward and backward scattering directions. Mathematically, this failure is associated with the infinite behavior of the generalized Legendre function that appears on the analysis, in the neighborhood of the forward and backward directions. Geometrically, this is associated with the fact that these particular directions are caustics of the ray system that characterizes the scattered field. We will show how it is possible to recast the representation for the scattering amplitude so that the neighborhood of these caustic directions may be properly examined. As an example, the total scattering cross-section may be obtained directly from the forward scattering by means of the cross-section theorem.

† Stevens Institute of Technology, Castle Point Station, Hoboken, New Jersey, U.S.A.

‡ S. I. Rubinow (1961) *Annals of Physics*, **14**, 305.

APPLICATION OF BOUNDARY LAYER TECHNIQUES TO THE CALCULATION OF SHADOW REGION DIFFRACTION FIELDS

W. P. BROWN, Jr.

Hughes Research Laboratories, Malibu, California

SUMMARY

It is shown that the nature of the diffraction fields in the shadow region of a smooth obstacle can be deduced directly from the asymptotic form of the reduced wave equation and the boundary conditions on the surface of the scatterer. The appropriate decomposition of the operator $\nabla^2 + k^2$ is obtained by applying the boundary layer techniques frequently employed in fluid flow problems and recently applied to diffraction theory by Wu (1957, 1958) and Keller (1959). The results which are obtained by this procedure provide a deductive justification for the functional form of the diffraction fields which J. B. Keller employs in his geometrical theory of diffraction (1956). Keller obtained his results by generalizing the known results for the cylinder to the case of a surface of non-constant curvature.

The asymptotic solution of the reduced wave equation is expressed as the product of a rapidly varying function and a slowly varying function. In this formulation it is recognized that in addition to finiteness in the vicinity of the caustic a satisfactory solution of the diffraction problem must be capable of satisfying a boundary condition on the surface of the caustic. In the simplest case the boundary condition is either of the Dirichlet or Neumann type. The Dirichlet case is solved explicitly and the extension to other types of boundary conditions is indicated.

REFERENCES

KELLER, J. B. (1956) Diffraction by a convex cylinder, *I.R.E. Trans. on Antennas and Prop.*, special suppl. *Proceedings of the Symposium on Electromagnetic Wave Theory*, 312–321.

KELLER, J. B. and BUCHAL, R. N. (1959) Boundary layer problems in diffraction theory, Research Report No. EM-131, New York University, Institute of Mathematical Sciences, Div. of Electromagnetic Research.

WU, T. T. (1957) The electromagnetic theory of light. I, Scientific Report No. 9, Cruft Laboratory, Harvard University.

WU, T. T. and SESHADRI, S. R. (1958) The electromagnetic theory of light. II, Scientific Report No. 22, Cruft Laboratory, Harvard University.

THE RADAR CROSS-SECTION OF A CONDUCTING CYLINDER WITH DIELECTRIC SLEEVE AT THE OPTICAL LIMIT†

RALPH D. KODIS

By means of the Poisson summation formula it is possible to express the back-scattered amplitude for a plane electromagnetic wave, e^{-ik_0x}, incident upon a conducting cylinder with a dielectric sleeve, as a continuous spectrum of radial eigenfunctions (Kodis, 1959). If the frequency of the incident radiation is high enough so that, within limits, the Debye approximation is valid for all Hankel functions which appear in this integral representation, the integrand can be developed in a Taylor's series. The expansion that results is a sum of integrals, each of which makes its principal contribution at a point of stationary phase, subject to certain restrictions on the parameter values. When these conditions are satisfied, the integrals can be evaluated asymptotically and the result put in the form:

$$\left.\begin{aligned}\psi^{sc}(x) \sim \left(\frac{\beta_0}{2k_0x}\right)^{1/2} e^{ik_0x} \Bigg[& r\, e^{-i2\beta_0} \\ & + \sum_{s=1}^{\infty} \left(1 + \frac{s\delta}{n(1-\delta)}\right)^{-1/2} (-1)^s\, t\, r_n^{s-1}\, t_n\, e^{i2\beta_0(sn\delta-1)} \\ & + 2 \sum_{m=1}^{\infty} \sum_{s=1}^{\infty} \left(\frac{1}{\cos\phi_i}\right. \\ & \left. + \frac{s}{n}\left\{\frac{1}{(1-\delta)\cos\phi_r} - \frac{1}{\cos\phi_n}\right\}\right)_{sm}^{-1/2} (-1)^s\, (TR_n^{s-1}T_n)_{sm} \\ & .\, \exp i2\beta_0 \left(sn\left\{\cos\phi_n - (1-\delta)\cos\phi_r\right\} - \cos\phi_i\right)_{sm} \Bigg]\end{aligned}\right\} \quad (1)$$

In this expression:

$$\beta_0 = k_0 b = 2\pi b/\lambda_0,$$

where b is the outer radius of the dielectric sleeve and λ_0 is the wavelength in empty space;

$$1/n = \lambda/\lambda_0,$$

† This research has been sponsored in part by the Office of Naval Research and the David Taylor Model Basin, and in part by the Electronics Research Directorate of the Air Force, Cambridge Research Laboratories.

where n in the refractive index and λ is the wavelength in the dielectric; and

$$1 - \delta = 1 - (d/b) = a/b,$$

where d is the thickness and a the inner radius of the sleeve.

The remaining quantities of formula (1) are most easily understood in relation to a physical analogue by means of which each term is associated with an optical ray. The first term has been derived previously and corresponds to the central ray reflected from the outer dielectric boundary. The remaining terms, which constitute the main result of this paper, correspond to rays reflected s times at the conducting cylinder. The first group of these contributions is made by the central ray after $s - 1$ internal reflections at the dielectric boundary. The second group corresponds to rays incident upon the sleeve at larger angles, which enables them to circle the cylinder axis

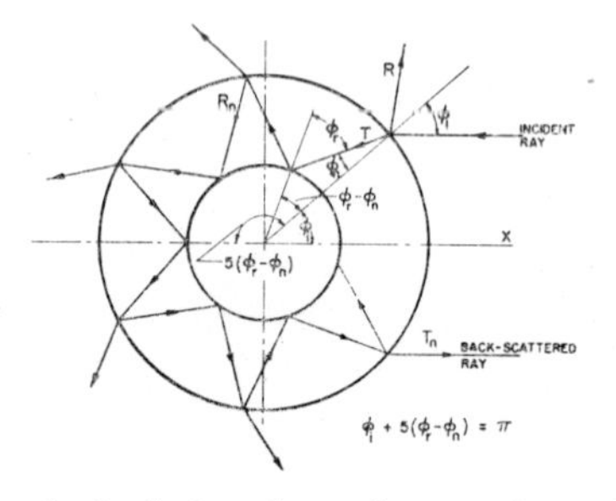

FIG. 1. Optical analogue for $m = 1$, $s = 5$.

$m \geq 1$ times before they emerge from the dielectric to contribute to the back-scattered amplitude. From symmetry it is clear that there are two each of these rays.

The amplitude of each ray is proportional, in the first place, to its Fresnel coefficients. For the central ray these are: $r = (1 - n)/(1 + n)$ at the external reflection; $r_n = (n - 1)/(n + 1)$ at each of $s - 1$ internal reflections; $t = 2/(n + 1)$ and $t_n = 2n/(n + 1)$ at each crossing of the outer dielectric boundary. For off-center rays which circle the origin one or more times the more general Fresnel coefficients, R_n, T and T_n, are required. These depend in the usual way upon the angles of incidence and refraction as well as on the index, n. All other amplitude factors are associated with the change in intensity brought about by the spreading of the incident ray bundles. They are monotonically decreasing functions of s.

The relative phase of each term in (1) corresponds to the path length of each ray in the dielectric, plus π radians for each of the s reflections at the conducting core. The geometrical counterparts of ϕ_i, ϕ_n and ϕ_r, which determine these paths, are shown in the sketch of a typical case (Fig. 1: $m = 1$, $s = 5$). The angles clearly satisfy the equation,

$$\phi_i + s(\phi_r - \phi_n) = m\pi, \tag{2}$$

which is precisely the condition that determines the point of stationary phase for each integral in the original asymptotic representation.

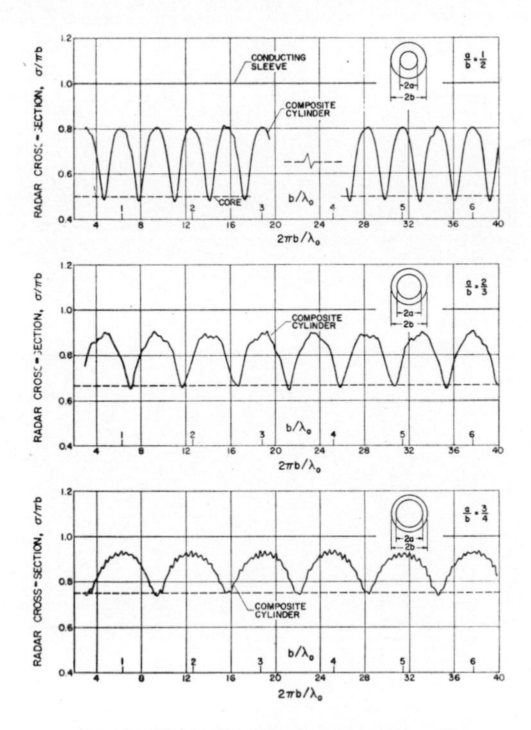

FIG. 2. Normalized back scatter ($n = 2$).

Since $\phi_n \leq \phi_r < \frac{1}{2}\pi$ and $0 \leq \phi_i < \frac{1}{2}\pi$, eqn (2) also tells us that $s \geq 2m$. A few constructions make it clear that the smaller values of s at every m require near-grazing incidence on either the outer or the inner cylinder, which is equivalent to saying that the limit of validity for the Debye approximation has been reached. On the other hand, it can be shown that if $\alpha = nk_0a < \beta_0$, i.e. $n(1 - \delta) < 1$, additional stationary phase points may exist. The corresponding rays are those which are characteristic of a solid dielectric but which do not intersect the inner core. For these rays $\sin \phi_i > n(1 - \delta)$ and the stationary phase equation is

$$\phi_i + s\left(\frac{\pi}{2} - \phi_n\right) = m\pi. \tag{3}$$

Since ϕ_i, the angle of incidence, must decrease with increasing s, only a finite number of such rays is possible with a given m. The contribution of a symmetrical pair to the scattered field is

$$\left.\begin{aligned} \psi^{sc}_{sm}(x) \sim 2\left[\frac{1}{\cos\phi_i} - \frac{s}{n}\frac{1}{\cos\phi_n}\right]^{-1/2}_{sm} & (-i)^s (T R_n^{s-1} T_n)_{sm} \,. \\ & . \exp i2\beta_0 (sn\cos\phi_n - \cos\phi_i)_{sm}, \end{aligned}\right\} \quad (4)$$

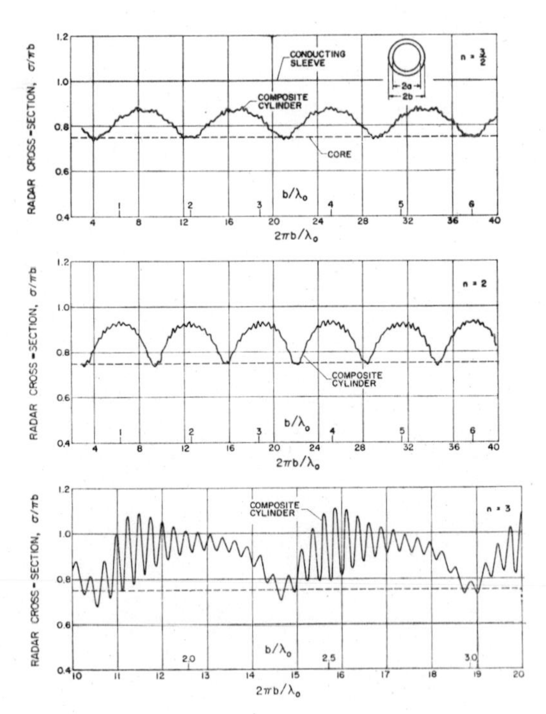

FIG. 3. Normalized back scatter $\left(\delta = \frac{1}{4}, \frac{a}{b} = \frac{3}{4}\right)$.

in which the various factors have the same significance as in eqn (1). In this context, however, s is the number of path segments traversed by the ray in the dielectric, and it should be noted that the radical may be either real or imaginary and can become large if $n \cos\phi_n/\cos\phi_i \sim s$. These features of (4), together with the phase factor $(-i)^s$ are probably due to the focusing properties of a solid dielectric.

If n, the index of refraction, is not too high, the reflection coefficients, r_n and R_n, are considerably smaller than one and the series converges reasonably well. Radar cross-sections can then be calculated as a function of frequency from the formula

$$\frac{\sigma}{\pi b} = \frac{2x}{b} \mid \psi^{sc}(x) \mid^2. \tag{5}$$

It is only necessary to solve the stationary phase equation for each ray and to do the sums indicated by formula (1). These calculations have been carried

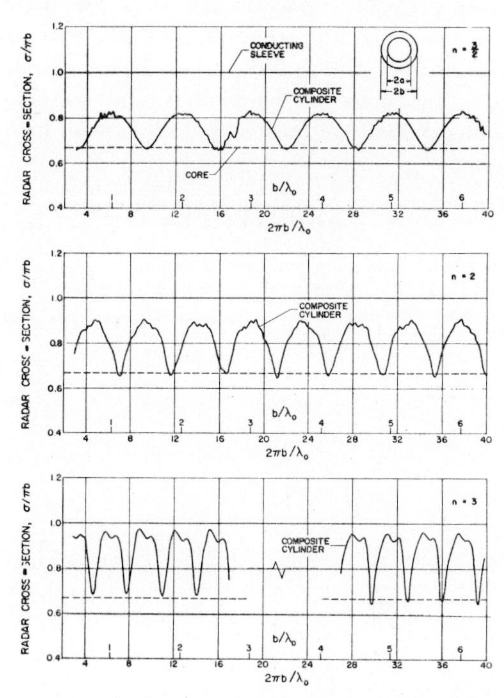

FIG. 4. Normalized back scatter $\left(\delta = \frac{1}{3}, \frac{a}{b} = \frac{2}{3}\right)$.

out on the I.B.M. 7070 at Brown University's Computing Center. Since the inclusion of the dielectric rays described by (4) would have led to additional programming difficulties, the computations were restricted to situations in which such rays could not contribute. A wide range of parameters is accessible even with this limitation.

The results, shown on Figs. 2–4, exhibit some rather interesting features. The fine structure found by Van de Hulst (1957) in a different way for a solid dielectric of low refractive index is clearly in evidence. The major variation of the cross-section with frequency is periodic, the maxima tending toward the value for a perfectly conducting cylinder of radius b as n increases. In general, the minima remain at the core value regardless of the index.

As the magnitude of the product, $n(1 - \delta)$, is increased there is a tendency for the valleys of the curve to fill while the maximum grows. At the largest value used, however, the curve breaks into rapid oscillations reminiscent of the Gibb's phenomenon encountered whenever a discontinuity is approximated by a Fourier-type series. The true cross-section in this case is probably close to that of the perfectly conducting sleeve at all frequencies outside a set of narrow bands in which it drops abruptly to the core value. It would be interesting to see whether this behavior can be observed experimentally. Unfortunately, Tang's choice of parameters and variables (Tang, 1957) makes it impossible to compare our computations with his results except at a few isolated points.

REFERENCES

KODIS, R. D. (1959) *I.R.E. Trans. Ant. and Prop.* **7**, S468.

VAN DE HULST, H. C. (1957) *Light Scattering by Small Particles*, p. 177, 1st ed. John Wiley, New York.

TANG, C. C. (1957) *J. Apply. Phys.* **28**, 628.

SCATTERING FROM A CYLINDER COATED WITH A DIELECTRIC MATERIAL

CARL W. HELSTROM

Mathematics Department, Westinghouse Research Laboratories,
Pittsburgh 35, Pennsylvania

ABSTRACT

A perfectly conducting cylinder of many wavelengths' radius is coated with a thin layer of dielectric material. The scattering of incident plane electromagnetic waves from it is resolved into a specular part and a diffracted part similar to the fields scattered by a bare cylinder, except that the former is modified by a reflection factor and the amplitudes and eigenvalues of the components of the latter are altered. For thick enough coatings the waves moving around the cylinder resemble those trapped in a plane dielectric slab bounded on one side by a perfect conductor. For certain frequencies of the radiation, resonances appear in the scattering cross-section; the most favorable conditions for detecting them are investigated.

1. THE GENERAL SOLUTION

An infinitely long, perfectly conducting, circular cylinder of radius a is coated with a dielectric material of uniform thickness τ, the radius of whose outer surface is $b = a + \tau$. Plane electromagnetic radiation of frequency $f = \omega/2\pi$ falls normally upon it. To be calculated is the amplitude of the scattered radiation far from the cylinder.

Tang (1957) has expressed the scattered field as a series involving Bessel and Hankel functions of integral order, and has presented numerical and experimental results for radiation of a wavelength λ comparable with the radius of the cylinder. When the incident wavelength is much smaller than the radius, the series converges too slowly to be useful for computation. Kodis (1961) has worked out the shortwave scattering in terms of the difference between the scattered fields of the coated cylinder and of a perfectly conducting cylinder of radius $b = a + \tau$, an approach enabling him to treat the dielectric coating as a perturbation.

The field scattered by a perfectly conducting cylinder consists, for $\lambda \ll b$, of a specular part and a diffracted part. The former arises from reflection at the surface, the latter from the passage around the cylinder of the so-called "creeping waves". The creeping waves correspond to normal modes of the field whose eigenvalues are related to the zeros of the Hankel function or its derivative as functions of the order (Franz, 1957). When the coating of dielectric material is very thin, this structure of the scattered field should be preserved, except for changes in the intensity and the eigenvalues of the creeping-wave modes. Assuming this is so, we shall demonstrate how the scattering by the coated cylinder can be analyzed from this standpoint, an aspect not treated by Kodis (1961).

We shall assume that the coating is much thinner than the cylinder, $\tau = b - a \ll b$, but the thickness τ and the wavelength λ of the radiation may be of comparable magnitudes. We take the radii a and b to be much greater than the wavelength λ.

The dielectric material has a capacitivity ε and a permeability μ; the corresponding quantities for free space are ε_0 and μ_0. For the most part we shall consider the coating to be nonconducting, so that ε is real. On the capacitivity and the permeability we place the minor restriction, needed later, that

$$|\mu\varepsilon - \mu_0\varepsilon_0|/\mu_0\varepsilon_0 \gg (\lambda/b)^{2/3}.$$

All fields have the time dependence $e^{-i\omega t}$. The propagation constants are $K = 2\pi/\lambda = \omega\sqrt{(\mu_0\varepsilon_0)}$ in free space and $k = \omega\sqrt{(\mu\varepsilon)}$ in the dielectric. We consider separately a TM-field, in which the magnetic intensity **H** is parallel to the axis of the cylinder, and a TE-field, in which the electric intensity **E** is parallel to the axis. Any other incident polarization can be expressed as a linear combination of these two types of field.

In the TM-field u represents the axial component of **H** inside the dielectric, and u_0 represents the same in free space. In the TE-field, u and u_0 represent the axial component of **E**. These functions u and u_0 satisfy the Helmholtz equations

$$\nabla^2 u + k^2 u = 0, \quad a < r < b; \qquad \nabla^2 u_0 + K^2 u_0 = 0, \quad r > b. \tag{1}$$

The radial co-ordinate is r, the azimuthal is ϕ. At the interface $r = b$ the boundary conditions are

$$\begin{aligned} u &= u_0, \qquad \frac{1}{\varepsilon}\frac{\partial u}{\partial r} = \frac{1}{\varepsilon_0}\frac{\partial u_0}{\partial r} \quad \text{(TM-field)} \\ u &= u_0, \qquad \frac{1}{\mu}\frac{\partial u}{\partial r} = \frac{1}{\mu_0}\frac{\partial u_0}{\partial r} \quad \text{(TE-field),} \end{aligned} \tag{2}$$

At the conducting surface $r = a$, $\partial u/\partial r = 0$ for the TM-field and $u = 0$ for the TE-field. The incident field is

$$u_{\text{inc}} = \mathrm{e}^{-iKx} = \mathrm{e}^{-iKr\cos\phi}. \tag{3}$$

In setting up our expressions we follow Franz's (1957) treatment of scattering by a bare conducting cylinder. The total field outside the cylinder is given by

$$u_0(r, \phi) = \frac{1}{2i}\int_C \mathrm{d}\nu\, \mathrm{e}^{-i\nu\pi/2}\frac{\cos\nu(\pi - \phi)}{\sin\nu\pi}[H_\nu^{(2)}(Kr) - F_j(\nu)H_\nu^{(1)}(Kr)], \tag{4}$$

where C is a contour surrounding the positive real ν-axis and bisecting the origin. Here

$$F_j(\nu) = \frac{H_\nu'^{(2)}(Kb) - R_j(\nu)H_\nu^{(2)}(Kb)}{H_\nu'^{(1)}(Kb) - R_j(\nu)H_\nu^{(1)}(Kb)} \tag{5}$$

where $j = 1$ for the TM-field and $j = 2$ for the TE-field, and

$$R_1(\nu) = \frac{Z}{Z_0}\frac{J_\nu'(ka)H_\nu'^{(1)}(kb) - H_\nu'^{(1)}(ka)J_\nu'(kb)}{J_\nu'(ka)H_\nu^{(1)}(kb) - H_\nu'^{(1)}(ka)J_\nu(kb)}$$

$$R_2(\nu) = \frac{Z_0}{Z}\frac{J_\nu(ka)H_\nu'^{(1)}(kb) - H_\nu^{(1)}(ka)J_\nu'(kb)}{J_\nu(ka)H_\nu^{(1)}(kb) - H_\nu^{(1)}(ka)J_\nu(kb)} \tag{6}$$

$$Z = \sqrt{(\mu/\varepsilon)}, \qquad Z_0 = \sqrt{(\mu_0/\varepsilon_0)}.$$

The integrand in (4) has poles at $\nu = n$, n all integers. Since the residue of $1/\sin \nu\pi$ is $(-1)^n/\pi$ at $\nu = n$, application of Cauchy's residue theorem to (4) reduces it to the series solution (Tang, 1957). The relation

$$J_\nu(x) = \tfrac{1}{2}[H_\nu^{(1)}(x) + H_\nu^{(2)}(x)] \tag{7}$$

is used here. The expression for the field inside the coating will not be needed, and we forbear writing it down.

To obtain a series that converges rapidly when $\lambda \ll b$, we deform the contour of integration in (4) until it surrounds the poles of $F_j(\nu)$. As with the solution for the bare conducting cylinder, the resulting residue series is useful only for points of observation in the geometrical shadow of the cylinder. Before deforming the contour we therefore make the substitution suggested by Franz (1957, p. 43),

$$\cos \nu(\pi - \phi) = e^{i\nu\pi} \cos \nu\phi - i\, e^{i\nu\phi} \sin \nu\pi. \tag{8}$$

We obtain

$$u_0(r, \phi) = \frac{1}{2i}\int_C d\nu\, e^{i\nu\pi/2} \frac{\cos \nu\phi}{\sin \nu\pi}[H_\nu^{(2)}(Kr) - F_j(\nu)H_\nu^{(1)}(Kr)]$$

$$- \frac{1}{2}\int_C d\nu\, e^{i\nu(\phi - \pi/2)}[H_\nu^{(2)}(Kr) - F_j(\nu)H_\nu^{(1)}(Kr)]. \tag{9}$$

2. THE SPECULAR FIELD

The second integral in (9) can, as for the bare conducting cylinder, be shown to contain the incident field u_{inc} and a field u_{sp} that can be attributed to specular reflection from the surface $r = b$ (Franz, 1957, pp. 47–50). For

$Kb \gg 1$ we substitute the following asymptotic form of the Hankel function (Erdélyi, 1953, vol. 2, p. 86, eqn. (11)):

$$H_\nu^{(1)}(z) \sim \sqrt{\frac{2}{\pi}}(z^2-\nu^2)^{-1/4} \exp i\Psi(z,\nu),$$
$$\Psi(z,\nu) = \sqrt{(z^2-\nu^2)} - \nu \cos^{-1}(\nu/z) - \pi/4, \tag{10}$$

with the corresponding form for $H_\nu^{(2)}(z)$ obtained by changing i to $-i$. The integral can then be evaluated by the method of steepest descents.

The representation of the specular field at a great distance from the cylinder arises from a saddlepoint at $\nu = Kb \sin(\phi/2)$. The first term of the asymptotic expansion for the specular field is, at a great distance r from the cylinder,

$$u_{sp} \sim Z_{sp}(\phi)\sqrt{\left(\frac{2}{i\pi Kr}\right)}\, e^{iKr}, \qquad Kr \gg 1 \tag{11}$$

with the specular scattering amplitude $Z_{sp}(\phi)$ given by

$$Z_{sp}(\phi) \cong \frac{R_j(Kb \sin \phi/2) + i \cos \phi/2}{R_j(Kb \sin \phi/2) - i \cos \phi/2}\sqrt{\left(\frac{i\pi Kb \cos \phi/2}{4}\right)}\, e^{-2iKb\cos\phi/2} \tag{12}$$

where for $\tau \ll a$,

$$R_1(\nu) = -\frac{Z}{Z_0}\Psi_z(kb,\nu)\tan[k\tau\, \Psi_z(kb,\nu)],$$
$$R_2(\nu) = \frac{Z_0}{Z}\Psi_z(kb,\nu)\cot[k\tau\, \Psi_z(kb,\nu)], \tag{13}$$
$$\Psi_z(kb,\nu) = \left(1-\frac{\nu^2}{k^2b^2}\right)^{1/2}.$$

These expressions for $R_j(\nu)$ were obtained by using the asymptotic form of (10) in (6), after replacing the Bessel functions by Hankel functions by means of (7). To obtain further terms in the asymptotic series for the specular field, one must use additional terms of the expansion for the Hankel function and refine the saddlepoint integration.

The specular scattering amplitude $Z_{sp}(\phi)$ in (13) is the same as for a bare conducting cylinder, except for the fraction at the beginning. That part of the expression is equal to the reflection coefficient for plane radiation striking a perfectly conducting plane with a dielectric coating of thickness τ, when the angle of incidence is $\phi/2$. This is to be expected, for when Kb is very large, the specular field can be computed by a Kirchhoff approximation, in which the far field is obtained by integrating field values on the surface found by replacing the surface at each point by a tangent plane. The field observed at azimuth ϕ arises mostly from points on the surface near azimuth $\phi/2$.

3. THE DIFFRACTED FIELD

In the rest of the paper we fix our attention on the first integral of (9), which contains the diffracted field arising from waves that move around the back of the cylinder. The contour C is deformed into another that surrounds the poles of $F_j(\nu)$ in the upper half of the ν-plane, and the integral is evaluated by the residue theorem. After the Wronskian for the Hankel functions is used, the diffracted field far from the cylinder ($Kr \gg 1$) can be written

$$u_d(r) \sim Z_j(\phi) \sqrt{\left(\frac{2}{i\pi Kr}\right)} e^{iKr}, \qquad j = 1, 2 \tag{14}$$

$$Z_j(\phi) = \sum_{s=1}^{\infty} b_s \frac{\cos \nu_s \phi}{\sin \nu_s \pi} \tag{15}$$

where the coefficients b_s in the scattering amplitude $Z_j(\phi)$ are given by

$$b_s^{-1} = \frac{iKb}{4} H_\nu^{(1)}(Kb) \frac{\partial}{\partial \nu} [H_\nu'^{(1)}(Kb) - R_j(\nu) H_\nu^{(1)}(Kb)] \Big|_{\nu=\nu_s}, \tag{16}$$

and the eigenvalues (eits) ν_s are the roots of the equation

$$H_\nu'^{(1)}(Kb) - R_j(\nu) H_\nu^{(1)}(Kb) = 0, \qquad \nu = \nu_s, \tag{17}$$

with $R_j(\nu)$ again given by (13).

When $Kb \gg 1$ and ν_s is not too close to kb, the scattering amplitude can be rewritten in terms of Airy functions by using the first term of the asymptotic series given by Olver (1954):

$$H_\nu^{(1)}(Kb) \sim 2\sqrt{\frac{2}{i}} [q/(\nu^2 - K^2 b^2)]^{1/4} Ai(-q) \tag{18}$$

$$\xi = \frac{2}{3} q^{3/2} = -i\rho \tag{19}$$

$$\rho = Kb(\sigma \cosh \sigma - \sinh \sigma) \tag{20}$$

$$\nu = Kb \cosh \sigma, \tag{21}$$

where for fixed Kb, $q = q(\nu)$ is a function of ν.† To terms of relative order $(Kb)^{-2/3}$,

$$b_s^{-1} = -2q^{1/2}[Ai(-q)]^2 \frac{\partial}{\partial \nu} [L(q) - \Gamma_j(\nu)] \Big|_{\nu = \nu_s} \tag{22}$$

with

$$L(q) - \Gamma_j(\nu) = 0, \qquad \nu = \nu_s, \qquad q = q(\nu_s) \tag{23}$$

† Justification of these approximations and details of calculations can be found in a report (Helstrom, 1962).

where

$$L(q) = iq^{1/2} \frac{Ai'(-q)}{Ai\,(-q)}, \tag{24}$$

$$\Gamma_j(v) = -R_j(v) \operatorname{cosech} \sigma. \tag{25}$$

Let the eits for the bare conducting cylinder be $\tilde{v}_p$ in the TE-mode, $\tilde{v}'_p$ in the TM-mode,

$$H^{(1)}_{\tilde{v}_p}(Kb) = 0, \qquad H'^{(1)}_{\tilde{v}'_p}(Kb) = 0. \tag{26}$$

For $Kb \gg 1$, these eits are given approximately by (Franz, 1957)

$$\begin{aligned} \tilde{v}_p &\doteq Kb + (Kb/2)^{1/3} q_p \, e^{i\pi/3} \\ \tilde{v}'_p &\doteq Kb + (Kb/2)^{1/3} q'_p \, e^{i\pi/3} \\ Ai(-q_p) = 0,& \qquad Ai'(-q'_p) = 0, \\ 0 < q'_1 &< q_1 < q'_2 < q_2 \dots . \end{aligned} \tag{27}$$

Consider an eit v_s for the TM-mode on the coated cylinder. As the thickness τ of the coating increases from zero, v_s leaves $\tilde{v}'_s$ and moves down along a sinuous curve like that shown in Fig. 1. It passes successively very close to the eits $\tilde{v}_{s-1}$, $\tilde{v}'_{s-1}$, $\tilde{v}_{s-2}$, $\tilde{v}'_{s-2}$, and so on. Finally it goes past $\tilde{v}'_1$ and leaves the neighborhood of these eits, approaching the Rl v-axis asymptotically as it heads toward the point $v = kb$. An eit v_s for the TE-mode behaves in the same way, but starts at $\tilde{v}_s$.

As long as the eit v_s lies near the eits for the bare cylinder, we can replace v by Kb in $R_j(v)$ and assume $\sigma \ll 1$ in (20), (21). For most values of the thickness the eit v_s is given approximately by

$$\begin{aligned} &v_s \doteq \tilde{v}_p - [R_j(Kb)]^{-1} \\ &(s-p-1)\pi < \tau\sqrt{(k^2-K^2)} < (s-p)\pi \qquad \text{(TM)} \\ &(s-p-\tfrac{1}{2})\pi < \tau\sqrt{(k^2-K^2)} < (s-p+\tfrac{1}{2})\pi \qquad \text{(TE)} \\ &|R_j(Kb)| \gtrsim (2/Kb)^{1/3}. \end{aligned} \tag{28}$$

The waves near the cylinder then resemble the creeping waves on a bare cylinder in the TE-mode, and the coefficient b_s in the scattering amplitude is approximately the same as for a bare cylinder,

$$b_s \doteq \tfrac{1}{2} \, e^{5i\pi/6} (Kb/2)^{1/3} [Ai'(-q_p)]^{-2}, \qquad v_s \cong \tilde{v}_p. \tag{29}$$

For such thicknesses that $|R_j(Kb)| \lesssim (2/Kb)^{1/3}$, the eit v_s is near one of the TM-eigenvalues $\tilde{v}'_p$.

The terms in $Z_j(\phi)$ corresponding to these creeping waves contribute little to the scattering because the imaginary part of v_s is so large that the waves are rapidly attenuated as they move around the back of the cylinder. However, when $\tau\sqrt{(k^2 - K^2)}$ increases beyond $(s-1)\pi$ for the TM-mode or

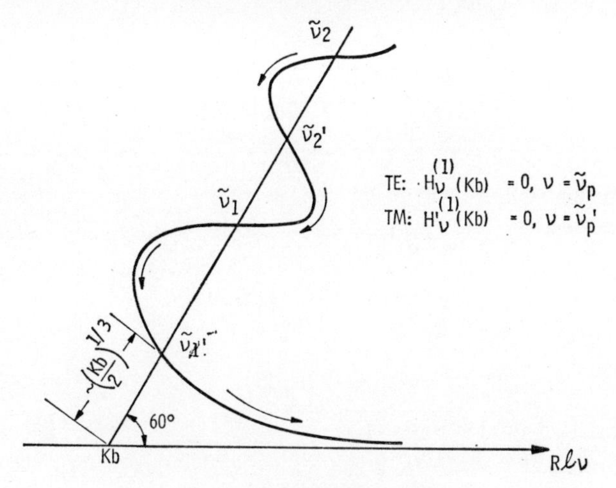

FIG. 1. Locus of eigenvalues.

$(s - \frac{1}{2})\pi$ for the TE-mode, the eit ν_s passes by $\tilde{\nu}_1'$ and begins its asymptotic approach to the Rl ν-axis. The imaginary part of ν_s decreases rapidly, and the waves are no longer much attenuated as they move around the cylinder. They now resemble the waves trapped in a plane dielectric slab with a perfect conductor on one face. Elliott (1955) pointed out the existence of similar waves on the circular cylinder.

In this case the quantity $\xi = \frac{2}{3}q^{3/2}$ in (19) acquires a large negative imaginary part, and we can use asymptotic expansions for the Airy function given by Olver (1954):

$$Ai(-q) \sim \pi^{-1/2}q^{-1/4}\left[\cos\left(\xi - \frac{\pi}{4}\right) + \frac{5}{72\xi}\sin\left(\xi - \frac{\pi}{4}\right)\right]$$

$$Ai'(-q) \sim \pi^{-1/2}q^{1/4}\left[\sin\left(\xi - \frac{\pi}{4}\right) + \frac{7}{72\xi}\cos\left(\xi - \frac{\pi}{4}\right)\right] \tag{30}$$

$$\xi = \tfrac{2}{3}q^{3/2} = -i\rho$$

to write for $L(q)$ of (24)

$$L(q) \sim 1 - \frac{1}{6\rho} - 2i\,e^{-2\rho}\left(1 - \frac{5}{36\rho}\right), \qquad |\rho| \gg 1. \tag{31}$$

We expand ρ, σ, and ν about real numbers ρ_0, σ_0, ν_0 satisfying

$$L_0(\rho_0) = 1 - \frac{1}{6\rho_0} = \Gamma_j(\nu_0) \tag{32}$$

$$\rho_0 = Kb(\sigma_0 \cosh \sigma_0 - \sinh \sigma_0) \tag{33}$$

$$\nu_0 = Kb \cosh \sigma_0 \tag{34}$$

with

$$\rho \doteq \rho_0 + (\sigma - \sigma_0)Kb\sigma_0 \sinh \sigma_0 \tag{35}$$

$$\nu \doteq \nu_0 + (\sigma - \sigma_0)Kb \sinh \sigma_0, \tag{36}$$

$$\therefore \rho - \rho_0 \doteq \sigma_0(\nu - \nu_0). \tag{37}$$

The eit ν_s in this region is given by

$$\nu_s \doteq \nu_0 + i \operatorname{Im} \nu_s \tag{38}$$

$$\operatorname{Im} \nu_s \doteq 2\, e^{-2\rho_0}\left(1 - \frac{5}{36\rho_0}\right)\left(G_j + \sigma_0 \frac{\partial L_0}{\partial \rho_0}\right)^{-1} \tag{39}$$

where

$$G_j = -\partial\Gamma_j/\partial\nu\big|_{\nu=\nu_0} \tag{40}$$

$$G_1 = \frac{\Gamma_1(\nu_0)\cosh\sigma_0}{Kb\sinh^2\sigma_0}\left[1 + \frac{\varepsilon_0^2 \tan^2\Phi(1 + 2\Phi \operatorname{cosec} 2\Phi)}{\varepsilon^2[\Gamma_1(\nu_0)]^2}\right]$$

$$G_2 = \frac{\Gamma_2(\nu_0)\cosh\sigma_0}{Kb\sinh^2\sigma_0}\left[1 + \frac{\mu_0^2 \cot^2\Phi(1 - 2\Phi \operatorname{cosec} 2\Phi)}{\mu^2[\Gamma_2(\nu_0)]^2}\right] \tag{41}$$

$$\Phi = \tau\sqrt{(k^2 - K^2\cosh^2\sigma_0)}.$$

The associated coefficient b_s in the scattering amplitude is, with these approximations,

$$b_s \doteq -i\pi \operatorname{Im} \nu_s. \tag{42}$$

Electromagnetic waves travelling through a plane dielectric slab bounded on one side by a perfect conductor are reflected back and forth between the faces, their directions making an angle θ with the normal to the faces. The fields contain a factor $\exp i(k_1 x - \omega t)$, $k_1 = k \sin\theta$, where x is the direction of propagation along the slab. Such fields can exist only when the constant k_1 and the angle θ are given by

$$\Gamma_j(k_1 b) = \Gamma_j(kb \sin\theta) = 1, \tag{43}$$

with $j = 1$ for TM-waves, $j = 2$ for TE-waves. (The length b cancels out of this equation when (13) is combined with (21) in (25) to form $\Gamma_j(k_1 b)$.) The quantity $k_1 b$ corresponds to the eit ν_s, and (32) reduces to (43) for $Kb \gg 1$, $\rho_0 \gg 1$. The imaginary part of ν_s, given by (39), represents the attenuation of the waves due to the curvature of the coating on the cylinder. As ρ_0 increases, that imaginary part becomes very small.

4. THE RESONANCES

With (42) and (15) a typical term in the scattering amplitude $Z_j(\phi)$ is, when $\operatorname{Im} \nu_s \ll 1$,

$$Z_s(\phi) = -\frac{i\pi \operatorname{Im} \nu_s \cos \nu_s\phi}{\sin(\pi \operatorname{Rl} \nu_s + i\pi \operatorname{Im} \nu_s)}. \tag{44}$$

As the frequency $\omega/2\pi$ increases, the quantity $\text{Rl}\ \nu_s \cong Kb = \omega b/c$ passes successively through integral values, at each of which the scattering amplitude exhibits a resonance. The amplitude at the peaks of the resonances may be compared with the specular part $Z_{\text{sp}}(\phi)$ of the scattering amplitude as given in (12). The widths of the resonances are mostly very small, and the Q's are very large,

$$Q = \frac{\text{Rl}\ \nu_s}{2\ \text{Im}\ \nu_s} \cong \frac{1}{4}\ \text{Rl}\ \nu_s\, e^{2\rho_0} \qquad G_j \gg 1. \tag{45}$$

The best chance of detecting the resonances seems to be when the eit ν_s has just moved away from $\tilde{\nu}_1'$ toward the Rl ν-axis. It is still close enough to $\nu = Kb$ that the parameters σ and σ_0 in (20), (21), (33), and (34) are small, and we can make a power series expansion,

$$\rho_0 \doteq \tfrac{1}{3}Kb\sigma_0{}^3, \qquad \nu_0 \doteq Kb(1 + \tfrac{1}{2}\sigma_0^2) \tag{46}$$

In (41) the phase Φ has for the TM-mode passed through $(s-1)\pi$, and $\tan\Phi$ is small enough that we can neglect the term with $(\varepsilon_0/\varepsilon)^2$ in G_1. Similarly $\cot\Phi \ll 1$ for G_2. Then we can put approximately, by (32),

$$G_j \doteq \left(1 - \frac{1}{6\rho_0}\right) \Big/ Kb\sigma_0^2 \tag{47}$$

and by (39), since $\partial L_0/\partial\rho_0 = 1/6\rho_0{}^2$,

$$\text{Im}\ \nu_s \doteq 2Kb\sigma_0^2\, e^{-2\rho_0}\left(1 - \frac{17}{36\rho_0}\right) \tag{48}$$

with σ_0 to be determined from (25), (32):

$$1 - \frac{1}{6\rho_0} = \Gamma_j(\nu_0) \doteq -R_j(Kb)/\sigma_0. \tag{49}$$

This is solved by trial and error, using (46). Finally, Im ν_s is obtained from (48).

As an example we have plotted in Fig. 2 the maxima and minima of the partial backscattering amplitude $|Z_s(0)|$ for frequencies $f = \omega/2\pi$ near $f_1 = c/\lambda_1$, for a cylinder of radius $b = 100\ \lambda_1$ coated with a dielectric having $\varepsilon/\varepsilon_0 = 2$, $\mu/\mu_0 = 1$ and thickness $\tau = 0.07\lambda_1$. In the region $0.9f_1 < f < 1.1f_1$ there are about 127 resonances. The quantity $\log_{10} Q$ for these resonances is plotted as a dashed curve on Fig. 2. As the frequency f increases, the peaks of $|Z_s(0)|$ approach 1, but their widths become extremely small, and the valleys between them lie nearly at zero.

The most favorable condition for detecting these resonances seems to be that π Im ν_s is of the order of 1. We have therefore solved the equation $\pi\text{Im}\ \nu_s = 1$ to get a relation between σ_0 and Kb, which through (49), (13)

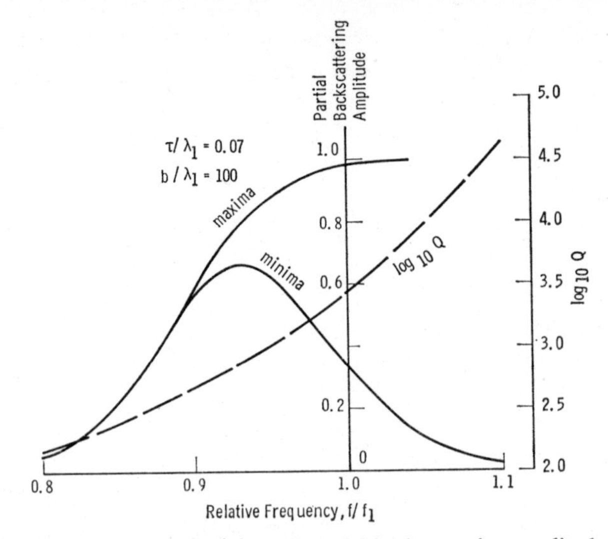

FIG. 2. Maxima and minima of partial backscattering amplitude.

gives a relation between τ/λ and b/λ, where λ is the wavelength of the radiation. This in turn gives the ratio b/λ as a function of the ratio τ/b, which is plotted in Fig. 3 for materials of dielectric constants 2 and 16. From this one can determine the wavelength λ of radiation to be used to hunt for evidence of the fields diffracted by a cylinder of given radius, coated with a material of given thickness. The diffracted waves will show up by their interference with the specularly reflected waves.

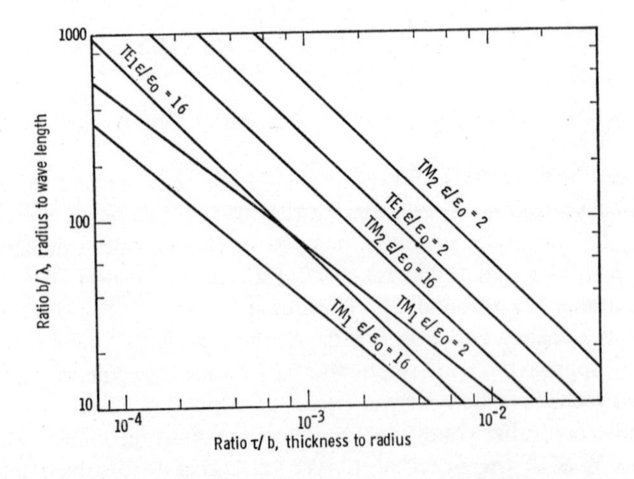

FIG. 3. Conditions for detection of resonances.

5. THE EFFECT OF CONDUCTIVITY

If the dielectric material is conductive, the waves will suffer an additional attenuation as they move around the cylinder. The amount of this attenuation can be estimated as though the dielectric were in the form of a plane slab. We solve (43) for k_1 after replacing the capacitivity ε everywhere by $\varepsilon + i\gamma/\omega$, where γ is the conductivity of the material. This will cause k_1 to have a positive imaginary part given when $\gamma/\varepsilon\omega \ll 1$ by

$$\frac{\operatorname{Im} k_1}{k_1} \doteq \frac{\gamma}{2\varepsilon\omega}\frac{\sec^2\theta}{KbG_j}\beta_j(\theta) \tag{50}$$

$$\begin{aligned} \beta_1(\theta) &= 2\Phi \operatorname{cosec} 2\Phi - \cos 2\theta \qquad \text{(TM)} \\ \beta_2(\theta) &= 1 - 2\Phi \operatorname{cosec} 2\Phi \qquad \text{(TE)} \\ \Phi &= k\tau\cos\theta, \qquad k_1 = k\sin\theta, \end{aligned} \tag{51}$$

where G_j is given by (41) with $\Gamma_j = 1$.

If the value of $\operatorname{Im} k_1 b$ as given by (50) is smaller than the value $\operatorname{Im} \nu_s$ for the cylinder as given by (39), the conductivity will not dominate the curvature in attenuating the waves. We therefore form their ratio, which is, in the most interesting region where (48) holds,

$$\begin{aligned} \frac{\operatorname{Im} k_1 b}{\operatorname{Im} \nu_s} &\cong \frac{\gamma}{2\varepsilon\omega}\frac{e^{2\rho_0}}{n^2-1} && (\text{TM}_1) \\ &\cong \frac{\gamma}{4\varepsilon\omega}\left(\frac{\varepsilon_0}{\varepsilon}\right)\frac{n^2}{\sqrt{(n^2-1)}}\left(\frac{e^{2\rho_0}}{\sigma_0}\right)(s-1)\pi,\ s>1 && (\text{TM}_s) \\ &\cong \frac{\gamma}{4\varepsilon\omega}\left(\frac{\mu_0}{\mu}\right)\frac{n^2}{\sqrt{(n^2-1)}}\left(\frac{e^{2\rho_0}}{\sigma_0}\right)(s-\tfrac{1}{2})\pi && (\text{TE}_s) \end{aligned} \tag{52}$$

where $n = k/K$ is the index of refraction of the coating. Now for the curves in Fig. 3, the value of ρ_0 ranges from about 2 to about 4, and σ_0 lies between 0.1 and 0.6. For a dielectric material of resistivity 10^{10} ohm-cm and with radiation of 1000 mc,

$$\gamma/\varepsilon\omega = 1.8 \times 10^{-7}(\varepsilon_0/\varepsilon).$$

Thus the condition that $\operatorname{Im} k_1 b$ be much less than $\operatorname{Im} \nu_s$ does not seem difficult to attain. As the frequency increases, however, $\operatorname{Im} \nu_s$ will at some point decrease below $\operatorname{Im} k_1 b$, after which the resonances will disappear and the amplitudes of the diffracted waves will be uniformly small.

REFERENCES

ELLIOTT, R. S. (1955) Azimuthal surface waves on circular cylinders, *J. Appl. Phys.* **26**, No. 4, 368–376 (April).

ERDÉLYI, A., *et al.* (1953) *Higher Transcendental Functions.* McGraw-Hill, New York.

FRANZ, W. (1957) *Theorie der Beugung elektromagnetischer Wellen.* Springer-Verlag, Berlin.

HELSTROM, C. W. (1962) The mathematics of scattering from a dielectric-coated cylinder, Westinghouse Res. Rep't 62–106–521–R2 (May 18).

KODIS, R. D. (1961) On the theory of diffraction by a composite cylinder, *J. Res. Nat. Bur. Stand.* **65D**, No. 1, 19–33 (Jan., Feb.).

OLVER, F. W. J. (1954) The asymptotic expansion of Bessel functions of large order, *Phil. Trans. Roy. Soc.* **247**, No. A930, 328–368 (Dec. 28).

TANG, C. C. H. (1957) Backscattering from dielectric-coated infinite cylindrical obstacles, *J. Appl. Phys.* **28**, No. 5, 628–633 (May).

SCATTERING BY A FINITE CYLINDER

RICHARD B. KIEBURTZ

Department of Electrical Engineering, New York University, New York 53, New York

ABSTRACT

A high-frequency asymptotic solution is constructed for the scattering of a cylindrically symmetric electromagnetic wave by a solid, conducting, right circular cylinder of finite length. The solution is found from the spatial Fourier transform of the tangential electric field on the surface formed by the infinite extension of the cylinder. The form of this transform function is deduced by applying boundary conditions on the field near the ends of the conducting cylinder and at large distances from it, and by employing the concept of a pulse solution.

1. INTRODUCTION

Although there now exist approximate solutions to many electromagnetic scattering problems, providing results often in good agreement with experiment, relatively few of such problems can be solved rigorously as boundary-value problems. The development of asymptotic methods of solution of such problems provides a means of extending the methods of classical analysis. This paper treats such an extension for the problem of scattering by a solid, conducting, right circular cylinder of small radius and finite length.

The problem of scattering by a cylinder is closely related to the cylindrical antenna problem first formulated by Hallén (1938), except that the excitation is simpler, since no finite generator need be assumed. Hallén formulated the problem in terms of an integral equation for the current on the antenna. The boundary condition at the end of the antenna caused difficulty if the ends were supposed solid; this difficulty was obviated in later treatments by assuming hollow cylinders with thin walls. If the radius is sufficiently small the presence or absence of conducting caps on the ends is unimportant, although this is not the case as the radius becomes an appreciable fraction of a wavelength.

Since Hallén's first treatment, many other authors have applied various methods to obtain solutions to the integral equation. Among these, Storer (1951) was the first to apply a variational method to this problem, King and Middleton (1946) have used an iteration method, Williams (1956) has used an extension of the Wiener–Hopf technique and has considered scattering of a plane wave by a cylindrical tube of finite length, and Wu (1961) has applied this technique to the antenna problem. Chen and Keller (1962) have used an asymptotic method, based on a construction of traveling waves multiply-reflected by the surface discontinuities, to obtain a solution to the antenna problem.

In this treatment, however, we shall follow the suggestions of Luneberg (1947–8) and Kline (1955) to develop an asymptotic solution. Basically, the main idea is the following. If one considers the pulse solution to a diffraction problem, then the geometrical optics wave front will undergo discontinuities, at least in its derivatives, at points in time when the wave front encounters points or lines on the boundary surface which are discontinuous, or have discontinuous derivatives. Luneberg and Kline point out that the discontinuities in the temporal description of the pulse solution are equivalent in the time-harmonic solution to a series involving inverse powers of the frequency, multiplied by a phase constant. The leading inverse powers are determined by the order of the discontinuity in the temporal description by means of an integral transform, and may be fractional.

In applying such a method to boundaries with multiple discontinuities, from which a wave may be multiply scattered, it is more convenient to proceed from a spatial, rather than a temporal representation of the pulse solution. Application of an integral transform to the spatial representation will give an expansion in inverse powers of the wavenumber. The effects of multiple scattering will be accounted for implicitly in such a representation. However, since one cannot know *a priori*, the complete spatial representation for an unsolved problem, one must make use of all of the information which is known. The data which are known from the boundary conditions are the locations of the surface discontinuities and the asymptotic behavior of the field components in the vicinity of each discontinuity. We shall see that we can use this information to construct an asymptotic solution.

2. FORMULATION OF THE PROBLEM

Consider a perfectly conducting, solid, right circular cylinder of radius a and length $2b$, located at the center of a cylindrical coordinate system as shown in Fig. 1. Let the radius be restricted by an inequality $ka \ll 1$, where k is the wavenumber of the incident radiation. We shall consider only the scattering of a cylindrically symmetric wave, in which case the fields can be derived from a single scalar potential. Let

$$\begin{aligned} \mathbf{E} &= -j\omega\mu_0 \nabla x \mathbf{\Pi}_h \\ \mathbf{H} &= k^2\mathbf{\Pi}_h + \nabla\nabla\cdot\mathbf{\Pi}_h \qquad (1) \\ \mathbf{\Pi}_h &= \mathbf{a}_\Phi u(r, z)\, e^{j\omega t}\,. \end{aligned}$$

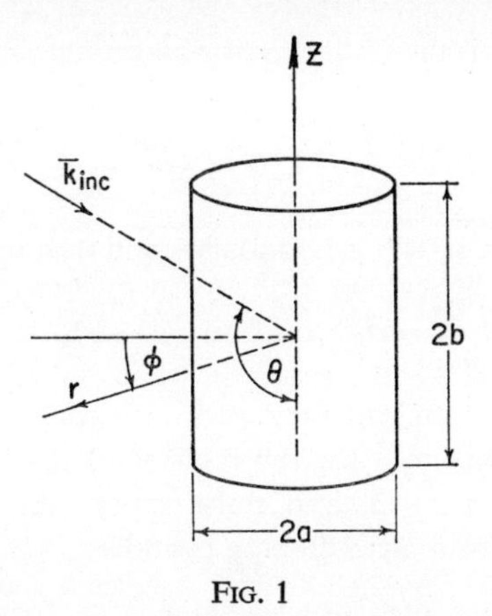

FIG. 1

The potential function for the incident wave will be

$$u_{\text{inc.}}(r, z) = j\sqrt{\left(\frac{\varepsilon_0}{\mu_0}\right)}\frac{E_0}{k^2}J_1(kr \sin\theta)\, e^{-jkz\cos\theta}\,. \tag{2}$$

Let us express the scattered field in terms of the tangential electric field on the infinite extension of the surface of the cylinder, $r = a$. Define a reflected potential as that which would be scattered if the cylindrical surface $r = a$ was replaced by an infinitely long perfect conductor. This reflected potential is

$$u_{\text{refl.}}(r, z) = -j\sqrt{\left(\frac{\varepsilon_0}{\mu_0}\right)}\frac{E_0}{k^2}\frac{J_0(ka\sin\theta)}{H_0^{(2)}(ka\sin\theta)}H_1^{(2)}(kr\sin\theta)\,. \tag{3}$$

The total potential in the two regions separated by the surface $r = a$ is then

$$u = \begin{cases} u_{\text{inc.}} + u_{\text{refl.}} + u_1 & (r > a)\,, \\ u_2 & (r < a)\,. \end{cases} \tag{4}$$

The functions u_1 and u_2, representing the diffracted field, both satisfy a wave equation

$$\frac{\partial}{\partial r}\frac{1}{r}\frac{\partial}{\partial r}(ru) + \frac{\partial^2 u}{\partial z^2} + k^2 u = 0\,. \tag{5}$$

Now let e_z designate the tangential component of electric field on $r = a$. This may be considered a source for u_1 and u_2, and we may write

$$u_{1,2}(r, z) = \mp\frac{j\omega}{k^2}\int_{-\infty}^{\infty} e_z(z')G_{1,2}(r, z|a, z')\,dz' \qquad (r \gtrless a)\,. \tag{6}$$

The functions $G_{1,2}$ are Green's functions for the infinitely long conducting cylinder, evaluated when the radial source coordinate is a, and satisfy a wave equation

$$\frac{\partial}{\partial r}\frac{1}{r}\frac{\partial}{\partial r}(rG) + \frac{\partial^2 G}{\partial z^2} + k^2 G = -\frac{\delta(r-a)}{r}\,\delta(z-z')\,. \tag{7}$$

Both Green's functions satisfy a boundary condition on the cylinder

$$\frac{\partial}{\partial r}(rG) = 0 \quad \text{on} \quad r = a\,.$$

In addition, G_1, which is defined for $r \geq a$, satisfies a radiation condition for large r. The Green's function for the interior of the cylinder, G_2, must be bounded everywhere in $r \leq a$, and must satisfy the additional boundary condition on the ends of the conducting cylinder,

$$\frac{\partial G_2}{\partial z} = 0 \quad \text{on} \quad |z| = b\,.$$

We shall use a Fourier integral representation for the Green's functions,

$$G_1(r,z|a,z') = \frac{1}{2\pi}\int_{-\infty}^{\infty} f_1(r,\lambda)\,\mathrm{e}^{-j\lambda(z-z')}\,\mathrm{d}\lambda\,, \tag{8a}$$

$$G_2(r,z|a,z') = \frac{1}{2\pi}\int_{-\infty}^{\infty} f_2(r,\lambda)\,(\mathrm{e}^{-j\lambda(z-z')} + \mathrm{e}^{-j\lambda(z+z'-2b)})\,\mathrm{d}\lambda \tag{8b}$$

for z, $z' > b$, where

$$f_1(r,\lambda) = f(a,\lambda)\,\frac{H_1^{(2)}[\sqrt{(k^2-\lambda^2)}\,r]}{H_0^{(2)}[\sqrt{(k^2-\lambda^2)}\,a]}$$

$$f_2(r,\lambda) = -f(a,\lambda)\,\frac{J_1[\sqrt{(k^2-\lambda^2)}\,r]}{J_0[\sqrt{(k^2-\lambda^2)}\,a]}$$

and $f(a, \lambda)$ is a common factor which is an even function of λ for λ real. The branches of the radical are chosen such that G_1 satisfies a radiation condition for large r. The Green's function G_2 for z, $z' < -b$ has the same form as (8b) except that b is replaced by $-b$.

As a consequence of the formulation, the tangential component of electric field is continuous across $r = a$ on $|z| > b$ and is zero on the conducting cylinder. By requiring continuity of the normal electric field and tangential magnetic field across $r = a$ for $|z| > b$, one arrives at a differential equation satisfied by the analytic continuation of the potential u onto the surface $r = a$.

Thus if

$$u_0(z) = \lim_{\delta\to 0}\,[u_2(r-\delta,z) - u_1(r+\delta,z)]\,, \tag{9}$$

we find that

$$\left[\frac{\mathrm{d}^2}{\mathrm{d}z^2} + k^2\right]u_0(z) = \frac{1}{j\omega\mu_0}\left[\frac{\partial}{\partial r}(E_{z\,\text{inc.}} - E_{z\,\text{refl.}})\right]_{r=a}. \tag{10}$$

A solution of this equation relates u_0 to the incident field amplitude. It also contains a complementary part which must satisfy a radiation condition for large $|z|$. This solution is

$$u_0(z) = U\,\mathrm{e}^{-jkz\cos\theta} + \begin{cases} C^+\,\mathrm{e}^{-jkz} & (z > b) \\ C^-\,\mathrm{e}^{+jkz} & (z < b) \end{cases} \tag{11}$$

where

$$U = j\sqrt{\left(\frac{\varepsilon_0}{\mu_0}\right)}\,\frac{E_0}{k^2\sin\theta}\left[J_1(ka\sin\theta) - \frac{J_0(ka\sin\theta)}{H_0^{(2)}(ka\sin\theta)}\,H_1^{(2)}(ka\sin\theta)\right].$$

Upon substituting the integral form (6) for u_1 and u_2, and using (9) and (11), we arrive at an integral equation for the tangential electric field distribution on the surface $r = a$,

$$\int_{-\infty}^{\infty} e_z(z')[G_1(a,z|a,z') + G_2(a,z|a,z')]\,\mathrm{d}z' = U\,\mathrm{e}^{-jkz\cos\theta} + \begin{cases} C^+\,\mathrm{e}^{-jkz} & (z > b)\,, \\ C^-\,\mathrm{e}^{jkz} & (z < b)\,. \end{cases} \tag{12}$$

3. CONSTRUCTION OF AN ASYMPTOTIC SOLUTION

To obtain an asymptotic solution to the integral equation (12) we shall take the Fourier transform of both sides, then attempt to construct an asymptotic series solution for the transform of e_z. The function u_1 can then be expressed as an inversion integral, and its behavior for large r can be obtained by conventional methods of asymptotic evaluation of integrals.

In order to take the Fourier transform of (12) we must continue the definition of $u_0(z)$ into the entire range $-\infty < z < \infty$. We can define u_0 over this range as

$$u_0(z) = U\,\mathrm{e}^{-jkz\cos\theta} + u^+(z)\,\mathrm{e}^{-jk|z+b|} + u^-(z)\,\mathrm{e}^{-jk|z-b|} \tag{13}$$

where the functions u^+ and u^- give the amplitudes of the scattered waves emanating from the ends of the cylinder at $z = -b$ and $z = b$, respectively. The sum of these wave functions in the region $z > b$ is $C^+\,\mathrm{e}^{-jkz}$ and in $z < -b$ is $C^-\,\mathrm{e}^{jkz}$. At large distances from the ends, however, we find that the field scattered parallel to the axis must decrease as $|z|^{-2}$, and so we conclude that $C^+ = C^- = 0$.

We can obtain an expression for the Fourier transform of e_z by taking separately the Fourier sine and cosine transforms of (12), with the range of definition extended by (13), and then combine these transforms to obtain the exponential transform. In view of the fact that e_z vanishes for $-b < z < b$ we find it convenient to write a related transform, of a function obtained by shifting the coordinate of e_z, to close this "gap." The representation of this function by a transform then has a slightly different form for each of the regions $z > b$ and $z < -b$. This is necessitated by the fact that the Green's function G_2 has a different representation for these two regions. The result valid for $z > b$ is

$$F(\lambda) = \int_0^\infty e_z(z' + b)\, e^{j\lambda z'}\, dz' + \int_{-\infty}^0 e_z(z' - b)\, e^{j\lambda z'}\, dz'$$

$$= \frac{1}{f(a,\lambda)} \left[j \frac{Ug_1(a,\lambda)\, e^{-j\lambda b}}{\lambda - k\cos\theta} + j \frac{Ug_2(a,\lambda)\, e^{j\lambda b}}{\lambda + k\cos\theta} \right. \tag{14}$$

$$+ g_1(a,\lambda)\, e^{-j\lambda b}\{L^+(K^+) + L^-(K^-)\}$$

$$\left. + g_2(a,\lambda)\, e^{j\lambda b}\{L^+(-K^-) + L^-(-K^+)\} \right]$$

where

$$g_1 = \frac{2H_0^{(2)}(\kappa a)[J_1(\kappa a)H_0^{(2)}(\kappa a) - J_0(\kappa a)H_1^{(2)}(\kappa a)]}{H_1^{(2)}(\kappa a)[2J_1(\kappa a)H_0^{(2)}(\kappa a) - J_0(\kappa a)H_1^{(2)}(\kappa a)]}$$

$$g_2 = -\frac{2H_0^{(2)}(\kappa a)[J_1(\kappa a)H_0^{(2)}(\kappa a)]}{H_1^{(2)}(\kappa a)[2J_1(\kappa a)H_0^{(2)}(\kappa a) - J_0(\kappa a)H_1^{(2)}(\kappa a)]}$$

$$\kappa = \sqrt{(k^2 - \lambda^2)},$$

and the transforms L^+ and L^- are

$$L^+(K) = e^{-jkb} \int_{-\infty}^\infty u^+(z')\, e^{jKz'}\, dz'$$

$$L^-(K) = e^{-jkb} \int_{-\infty}^\infty u^-(z')\, e^{jKz'}\, dz'.$$

The arguments of L^+ and L^- appearing in (14) are linear functions of λ with the origin shifted,

$$K^+ = \lambda - k$$

$$K^- = \lambda + k.$$

To obtain the transform of the field in $z < -b$, we replace b by $-b$ on the right hand side of (14). This is the function $F(\lambda)$ which is found when the form of G_2 appropriate to the region $z < -b$ is substituted into (12).

Now we must determine the form of the functions L^+ and L^-. To do this we shall employ the boundary conditions on e_z in the vicinity of the ends of the conducting cylinder and at large distances from the cylinder.

First it is helpful to examine the dependence of the scattered potential on the transform $F(\lambda)$. This is obtained from (6), by substituting (8a) for G_1 and the inverse transform for e_z,

$$e_z(z) = \begin{cases} \dfrac{1}{2\pi}\displaystyle\int_{-\infty}^{\infty} F(\lambda')\, \mathrm{e}^{-j\lambda'(z-b)}\, \mathrm{d}\lambda' & (z > b)\,, \\ \dfrac{1}{2\pi}\displaystyle\int_{-\infty}^{\infty} F(\lambda')\, \mathrm{e}^{-j\lambda'(z+b)}\, \mathrm{d}\lambda' & (z < -b)\,, \end{cases}$$

where the form of $F(\lambda)$ appropriate to each region is used. This gives

$$u_1(r, z) = \frac{1}{2\pi}\int_{-\infty}^{\infty} f(a, \lambda)\, \frac{H_1^{(2)}(\kappa r)}{H_0^{(2)}(\kappa a)}\, F(\lambda)\, \mathrm{e}^{-j\lambda(z \mp b)}\, \mathrm{d}\lambda\,, \qquad \begin{pmatrix} z > b \\ z < -b \end{pmatrix}. \tag{15}$$

These two forms are merely alternate representations of u_1. Now the total scattered field, which is $u_1 - u_{\mathrm{refl}}$, cannot contain any cylindrical waves, since the scatterer is of finite dimensions. Thus u_1 must contain a cylindrical wave term which cancels the cylindrical wave u_{refl}. From (15) one sees that cylindrical wave terms will occur if there are simple poles of $F(\lambda)$. If one takes $z > b$, the path of integration of (15) can be closed in the lower half plane. In order to obtain a cylindrical wave with z-dependence agreeing with that of u_{refl}, the pole at $\lambda = k \cos\theta$ should be taken to lie within the contour. We may assume that the wavenumber k has a small negative imaginary part, thus shifting the desired pole slightly below the real λ-axis, in order to avoid ambiguity in the contour of integration.

The residue in the pole appearing in the first term on the right side of (14) is not sufficient to cancel the cylindrical wave, however, so there must be additional poles in the terms L^+ and L^-. Since $L^+(K^+)$ and $L^-(K^-)$ are the transforms of u^- and u^+, the waves scattered from the cylinder ends, we see that it is these terms which must contribute poles at $\lambda = k\cos\theta$. Also we can see that this pole must produce a cylindrical wave component of u^+ for $z < -b$ and u^- for $z > b$. Therefore, L^+ and L^- must contain exponential factors so that the closure of the contour of the inversion integral shifts from one half-plane to the other as z passes the points $z = -b$ and b respectively. We therefore write

$$\begin{aligned} L^+(K) &= \frac{\mathrm{e}^{-j(K+k)b}}{K + k - k\cos\theta}\, h^+(K)\,, \\ L^-(K) &= \frac{\mathrm{e}^{-j(K-k)b}}{K - k - k\cos\theta}\, h^-(K)\,. \end{aligned} \tag{16}$$

The functions h^+ and h^- contain no poles. The pole of L^+ we take to be slightly above the contour of the inversion integral and of L^- we take slightly below, in order that cylindrical waves are produced only in the region $|z| > b$ in the spatial domain.

When we evaluate the cylindrical wave contributions of u_1 from (15) and set the amplitude of the wave whose phase constant is $\mathrm{e}^{-jkz\cos\theta}$ equal to $-u_{\mathrm{refl}}$ to cancel the scattered cylindrical wave in each of the regions $z > b$

and $z < -b$, we obtain two conditions on the amplitudes h^+ and h^-. These relate the residues in the poles of L^+ and L^- to known amplitudes,

$$j\,\mathrm{e}^{jkb\cos\theta}h^-(k\cos\theta + k) = U - j\sqrt{\left(\frac{E_0}{\mu_0}\right)}\frac{E_0}{k^2}\frac{J_0(ka\sin\theta)}{g_1(a, k\cos\theta)} \tag{17}$$

$$-j\,\mathrm{e}^{-jkb\cos\theta}h^+(k\cos\theta - k) = U - j\sqrt{\left(\frac{E_0}{\mu_0}\right)}\frac{E_0}{k^2}\frac{J_0(ka\sin\theta)}{g_1(a, k\cos\theta)}. \tag{18}$$

There is also a pole found at $\lambda = -k\cos\theta$ in various terms of $F(\lambda)$. This pole never lies within the contour of integration of the inversion integral, hence never contributes a cylindrical wave to u_1.

Next we shall determine the form of h^+ and h^-. First, note that in the integral formula (15) for u_1, any singularity of $F(\lambda)$ lying within the path of integration at some point $\lambda = p$ will give rise to a term in u_1 of the form $f_p(r, z)\,\mathrm{e}^{-jp(z+b)}$. In particular, at the top end of the cylinder, $z = b$, the phase delay will be $2pb$. This phase delay cannot be arbitrary, but can only have certain allowable values which depend on the cylinder length. For the cylindrically symmetric case considered here, the phase delays are due to traveling waves which are scattered by the surface discontinuities at the ends of the conducting cylinder and travel along the cylinder to the other end, whereupon they undergo further scattering. Since the phase velocity of these traveling waves may be expected to be the velocity of light in the medium external to the cylinder, one may expect the allowable values of the phase delay to be integer multiples of $2kb$. Indeed no constants of physical significance appear in the specification of the problem other than the cylinder dimensions and the wavenumber, so that there is no justification for supposing that any physical significant quantity can depend on constants other than these.

The singularities of $F(\lambda)$ other than its simple poles must therefore be located at points $\lambda = nk$, where n is a positive or negative integer or zero. Negative values of n give rise to terms having a negative phase delay, or advanced phase. Since the traveling waves on the cylinder are not unidirectional, but are scattered from both ends, the inclusion of both advanced and retarded potentials in the solution violates no physical principle. The advanced phase may be identified with waves traveling in the negative z-direction.

Now that the locations of the singularities of $F(\lambda)$ in the complex λ-plane are known, it remains to determine their order. This can be done by application of the boundary condition at the surface discontinuity at the end of the conducting cylinder, and by utilizing the concept of a pulse solution.

The behavior of the fields at distances arbitrarily close to the circular edge of the end of the conducting cylinder must obey the same boundary condition as the field close to the vertex of a right-angled wedge. This boundary condition, which is applicable at small distances compared with either the wavelength or the radius of the cylinder, requires that the potential function u must tend to zero with distance ρ from the edge as fast as $\rho^{2/3}$ (Meixner, 1948). Since the spatial behavior at small distances from singular points determines the behavior of the transform of large argument, one finds that $h^+(K)$ and $h^-(K)$ must be $O(K^{-2/3})$ as $|K| \to \infty$.

In the pulse solution for a wedge, it has been shown that when a wave exhibiting a singularity in time strikes the vertex of a wedge, it undergoes a reduction in the order of the time singularity (Keller, 1962). In the transform representation, the asymptotic behavior of the transform of the scattered field is of a higher negative power than is that of the incident field. But for the simple case of a pulse striking a wedge, the one-dimensional spatial Fourier transform and the temporal Fourier transform are closely related; they differ in their asymptotic behavior by a power of unity. Thus the spatial transform for each multiply scattered wave will also show the increase in order of asymptotic behavior. The fractional increase in the negative power of the asymptotic form is determined by the edge condition on the fields, and so is related to the wedge angle.

These results for the pulse solution are easily generalized to the case of harmonic time dependence. Application of these results to the problem under consideration tells us that the functions $h^+(K)$ and $h^-(K)$ should consist of a series of terms whose asymptotic behavior for large $|K|$ is as $K^{-2/3}$, $K^{-4/3}$, K^{-2}, etc.

The singularities of these terms in the complex K-plane can be of no more negative power than $(K - nk)^{-1/2}$ however, since the scattered fields must satisfy a radiation condition for large positive or negative values of z in the spatial representation.

By considering all of the information as to the locations of the singularities and their order, and the asymptotic behavior in the transform plane, it is possible to construct a series representation for h^+ and h^-. These functions have the form

$$\begin{aligned} h^+(K) &= \alpha_1\psi^+(K) + \alpha_2\psi^+(K-k)\psi^-(K-k) \\ &\quad + \alpha_3\psi^+(K-2k)\psi^+(K)\psi^-(K+2k) + \dots \\ h^-(K) &= \beta_1\psi^-(K) + \beta_2\psi^-(K+k)\psi^+(K-k) \\ &\quad + \beta_3\psi^-(K+2k)\psi^-(K)\psi^+(K-2k) + \dots, \end{aligned} \tag{19}$$

where

$$\begin{aligned} \psi^+(K) &= K^{-1/2}[K^{1/2} + (K-2k)^{1/2}]^{-1/3} \\ \psi^-(K) &= K^{-1/2}[K^{1/2} + (K+2k)^{1/2}]^{1/3} \end{aligned}$$

and the branch cuts are as shown in Fig. 2. The choice of branch cuts is dictated by the consideration that $h^+(K^+)$ is a factor of the transform of u^+ and hence, for $z > -b$, when the contour of integration of the inversion

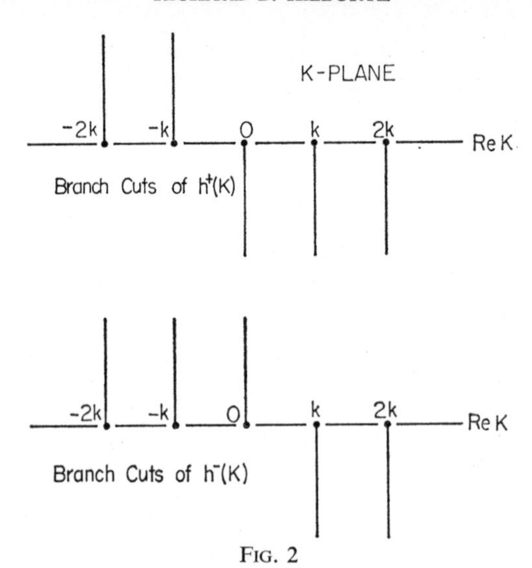

FIG. 2

integral is closed in the lower half λ-plane (or K-plane), the integral must contribute only terms with retarded phase corresponding to propagation in the positive z-direction. A similar argument applies to the contribution for $z > -b$, when waves of advanced phase must result, and to the contributions to u^- when it is evaluated for z greater or less than b.

All that now remains in the formal solution for the transform function $F(\lambda)$ is to obtain a set of equations determining the coefficients α_i and β_i. Equations (17) and (18) provide two such conditions. To obtain further equations, we again turn to the behavior of the potential near the ends of the cylinder. Since the total potential must vanish at the edge of the conducting cylinder on the surface $r = a$, we set z equal in turn to $b - \varepsilon$ and $-b + \varepsilon$, and take the limit as ε vanishes

$$\lim_{\varepsilon \to 0} [u_1(a, \pm b \mp \varepsilon) + u_{\text{refl.}}(a, \pm b \mp \varepsilon)] = O(\varepsilon^{2/3}) . \tag{20}$$

Upon substituting (15) into (20), for each end of the cylinder, and evaluating the integrals, we obtain two more equations in the coefficients α_i and β_i.

To obtain further equations in the coefficients, we must investigate the behavior of derivatives of u_1 near the cylinder ends. On the surface $r = a$, a series expansion for $u(a, z) - u_{\text{inc}}(a, z)$ for $z = b + \varepsilon$ is

$$u - u_{\text{inc.}} = a_1 \varepsilon^{2/3} + a_2 \varepsilon^{4/3} + a_3 \varepsilon^2 + \ldots . \tag{21}$$

Let us define a fractional derivative of a function regular in, at least a half-neighborhood of some point, by an integral operator. For non-negative powers, let this operator be

$$\frac{d^s}{dz^s} z^n = \frac{\Gamma(n+1)}{\Gamma(n-r)} \int_0^z \zeta^n \zeta^{-1-s} \, d\zeta = \frac{\Gamma(n+1)}{\Gamma(n-r+1)} z^{n-s} \tag{22}$$

where $0 < s < 1$. If we apply this operator to take a derivative of order 2/3 of (21), we find that the first term becomes infinite logarithmically as the order of the derivatives becomes 2/3. The finite part of the derivative is

$$\text{F.P.} \frac{\mathrm{d}^{2/3}}{\mathrm{d}z^{2/3}} u(a, z)\bigg|_{z=b+\varepsilon} = a_2 \frac{\Gamma(\frac{7}{3})}{\Gamma(\frac{5}{3})} \varepsilon^{2/3} + a_3 \frac{\Gamma(3)}{\Gamma(\frac{7}{3})} \varepsilon^{4/3} + \ldots . \tag{23}$$

Now the fractional derivatives of $u_{\text{inc.}}$, $u_{\text{refl.}}$, and the integral form of u_1 can also be taken, provided the definition is extended to include exponential functions. A consistent definition is

$$\frac{\mathrm{d}^s}{\mathrm{d}z^s} \mathrm{e}^{-\tau z} = -\frac{\Gamma(1)}{\Gamma(-s)} \int_{-\infty}^{z} \mathrm{e}^{-\tau\zeta}(z-\zeta)^{-1-s}\,\mathrm{d}\zeta = (-\tau)^s \mathrm{e}^{-\tau z} . \tag{24}$$

If the fractional derivative of 2/3 order is now taken of the potential $u - u_{\text{inc.}}$ evaluated on $r = a$, $z = \pm b \mp \varepsilon$, then the finite part of this derivative must tend to zero as $0(\varepsilon^{2/3})$. Applying this condition at both cylinder ends gives two more equations in coefficients α_i and β_i. The fractional derivatives of $u - u_{\text{inc.}}$ at the cylinder ends can be repeated, and by setting the finite part equal to $0(\varepsilon^{2/3})$ each time as infinite a set of algebraic equations in the unknown coefficients is evolved.

The final step in obtaining a solution for the transform function $F(\lambda)$ is the solution of the system of algebraic equations. While the convergence of such a system is difficult to show, it is not hard to show that truncation of the infinite set yields an asymptotic solution. The coefficients α_i and β_i can be shown to be of the order of increasing negative powers of $2kb$,

$$\alpha_{i+2} \sim (2kb)^{2/3}\alpha_i$$

$$\beta_{i+2} \sim (2kb)^{-2/3}\beta_i,$$

thus truncation yields a solution which is asymptotic to the exact solution in the high-frequency limit.

4. CONCLUSIONS

A method has been developed for construction of an asymptotic solution to the boundary-value problem posed by the scattering of a cylindrically symmetric electromagnetic wave by a perfectly conducting, right circular cylinder of finite length. The gist of the method is that the spatial Fourier transform of the tangential electric field, on the infinite extension of the cylinder, can be constructed by employing boundary conditions on the field near the ends of the cylinder and at large distances from it.

The positions of singularities in the transform plane can be deduced from a knowledge of the phase velocity of traveling waves on the surface of the cylinder. The order of these singularities is deduced by application of the concept of a pulse solution and from the boundary condition satisfied by the field near the ends of the conducting cylinder. Finally, a set of equations in the unknown coefficients of a series of terms of the transform representation are obtained, by employing the boundary condition at the cylinder ends and taking finite parts of integrals.

The method yields a solution which is asymptotic to the exact solution in the high-frequency limit. Furthermore it can be extended to a large number of other problems, such as scattering by right-angled wedges, thick plates and strips, cylinders of larger radius with noncylindrically symmetric excitation, and other high-frequency scattering problems amenable to attack by transform methods.

REFERENCES

CHEN, Y. M. and KELLER, J. B. (1962) Current on and input impedance of a cylindrical antenna, *N.B.S. J. Research*, **66D** (Radio Propagation), No. 1, 15–21 (Jan.–Feb. 1962).

KELLER, J. B. (1962) Diffraction by polygonal cylinders, pp. 129–137 in *Electromagnetic Waves*, proceedings of a symposium conducted by the Mathematics Research Center, U.S. Army, at the University of Wisconsin, edited by R. Langer, University of Wisconsin Press, Madison, Wis.

KING, R. W. P. (1956) *The Theory of Linear Antennas*, Harvard University Press, Cambridge, Mass.

KING, R. W. P. and MIDDLETON, D. (1946) The cylindrical antenna: current and impedance, *Quart. Appl. Math.* **3**, 302–335.

KLINE, M. (1955) Asymptotic solutions of Maxwell's equations involving fractional powers of the frequency, *Comm. Pure and Appl. Math.* **8**, 595–614.

LUNEBERG, R. K. (1947–8) Propagation of electromagnetic waves, NYU Lecture Notes.

MEIXNER, J. (1948) Strenge Theorie der Beugung elektromagnetischer Wellen en der vollkommen, leitenden Kreisscheibe, *Z. Naturforschung*, **3a**, 506–518.

STORER, J. E. (1951) Solution to thin wire antenna problems by variational methods, doctoral dissertation, Harvard University (June 1951). For a summary of this work, see King, (1956), pp. 237–259.

WILLIAMS, W. E. (1956) Diffraction by a cylinder of finite length, *Proc. Cambridge Phil. Soc.* **52**, 322–335.

WU, T. T. (1961) Theory of the dipole antenna and the two-wire transmission line, *J. Math. Phys.* **2**, No. 4, 550–574 (July–Aug. 1961).

THE DIFFRACTION FIELDS OF A NON-UNIFORM CIRCULAR APERTURE

S. CORNBLEET†

ABSTRACT

The application of symmetry conditions to the scalar diffraction integral reduces it to a finite transform of the aperture amplitude distribution function. Normal circular symmetry gives rise to the zero-order Hankel transform. Anti-symmetry gives rise to an "odd" transform, and other conditions give rise to higher order transforms.

To suit these transforms, and general amplitude distribution functions, this function is expanded in a series of circle polynomials. This expansion and the resultant far-field diffraction patterns are obtained for several distributions.

Complex distributions can be transformed by the same means and hence conditions involving radial phase change, can be considered. The radiation patterns of defocused systems, or the near field and focal distributions of apertures, can be obtained by this method.

INTRODUCTION

This paper gives a practical computational procedure for obtaining the far-field diffraction pattern of a circular aperture in which the field distribution is non-uniform both in phase and amplitude. With the application of symmetry conditions to the phase distribution, the diffraction integral is reduced to an integral transform of the amplitude distribution. Three symmetry conditions of phase are considered, the first of which, normal circular symmetry gives rise to the zero-order Hankel transform. The second phase distribution is that in which the aperture is divided by a diameter into anti-phase halves. This gives rise to a zero-order transform, here called the Lommel transform, which is odd if the Hankel transform is considered to be even. This transform is defined in terms of the Lommel–Weber functions. The third phase distribution is cyclic in character and gives rise to higher order Hankel transforms, the order of the transform being the number of cycles of phase variation in a circular path centered on the origin.

The usual radial amplitude functions $(1 - r^2)^p$ which transform simply by the zero-order Hankel transform are shown to be simply transformable by the zero-order Lommel transform. They do not, however, transform simply by the higher order transforms, nor are they particularly suitable for

† The General Electric Company Limited, Applied Electronics Laboratories, Stanmore, Middlesex.

the general description of aperture amplitude distributions arising in practice.

To meet this case, an expansion of the amplitude distribution is derived in terms of the orthogonal circle polynomials, which transform simply by any of the transforms given above. By this means complex amplitude functions can also be transformed and hence conditions involving radial variation of the phase can be considered. The major difference between the far, the near-field diffraction patterns and the field distribution at a focus lies in the inclusion of such a radial phase factor. Thus a change of definition of the variables used can then be made to allow the same transform and expansion method to be used for these fields. Among other things this permits the use of this method for the pattern synthesis problem. With the phase modes at our disposal and using the principle of superposition, any aperture distribution can be analyzed into terms which are transformable by this method.

1. THE DIFFRACTION INTEGRAL

For the circular aperture, the far-field diffraction pattern can be expressed as (Silver, 1949)

$$g(u, \phi) = \int_0^{2\pi} \int_0^1 F(r, \phi') \, e^{jur \cos \overline{\phi - \phi'}} r \, dr \, d\phi' \tag{1}$$

where $$u = \frac{2\pi a \sin \theta}{\lambda}$$

a is the radius of the aperture,

r, ϕ' are normalized polar coordinates in the aperture with respect to the center as origin,

and omitting constants of multiplication.

If symmetry conditions are applied to the function $F(r, \phi')$ this integral can be separated into radial and angular components, with the result that the function $g(u, \phi)$ becomes the integral transform of a radial amplitude distribution $f(r)$. This paper deals with three such symmetry conditions and their corresponding integral transforms.

2. THE INTEGRAL TRANSFORMS

2.1. *The Zero-Order Hankel Transform*

If the phase of the aperture illumination is constant, and without loss of generality, zero, and if the amplitude distribution is circularly symmetrical, $F(r, \phi')$ in (1) can be put equal to a real function of r alone, $f(r)$, which is finite in the range $0 \leqslant r \leqslant 1$ and zero outside.

Then $$g(u, \phi) = \int_0^1 f(r) r \, dr \int_0^{2\pi} e^{jur \cos \overline{\phi - \phi'}} \, d\phi'. \tag{2}$$

Since the aperture distribution and the diffraction pattern will have circular symmetry, the integration over ϕ' can be performed separately to give the circularly symmetric diffraction pattern

$$g(u) = 2\pi \int_0^1 f(r) J_0(ur) r \, dr, \tag{3}$$

which is one form of the zero-order Hankel transform of $f(r)$.

2.2. *The Zero-Order Lommel Transform*

Let the opposite halves of the circular aperture be in anti-phase. That is

$$F(r, \phi') = \begin{cases} f(r)\, e^{-j\pi/2} & 0 < \phi' \leqslant \pi \\ f(r)\, e^{+j\pi/2} & \pi < \phi' \leqslant 2\pi, \end{cases}$$

then

$$g(u,\phi) = e^{-j\pi/2} \int_0^{\pi} \int_0^1 f(r)\, e^{jur \cos \overline{\phi - \phi'}} r \, dr \, d\phi' + e^{j\pi/2} \int_0^{\pi} \int_0^1 f(r)\, e^{-jur \cos \overline{\phi - \phi'}} r \, dr \, d\phi'. \tag{4}$$

In the far-field the plane of interest is that at right angles to the diameter which divides the aperture into anti-phase halves, i.e. the plane $\phi = \pm\pi/2$. This gives

$$g\left(u, \pm\frac{\pi}{2}\right) = \pm 2 \int_0^1 f(r) r \, dr \int_0^{\pi} \sin(ur \sin \phi') \, d\phi'. \tag{5}$$

The integration over ϕ' in (5) bears the same relation to the Bessel integral in (3) as does the sine transform to the cosine transform in Fourier analysis. It defines the Lommel–Weber function $\Omega_0(ur)$ (Jahnke and Emde, 1945). Thus an 'odd' transform, called here the Lommel transform, is given by

$$g\left(u, \pm\frac{\pi}{2}\right) = \pm 2\pi \int_0^1 f(r) \Omega_0(ur) r \, dr. \tag{6}$$

It is simple to show that, for the plane $\phi = 0$ and π, $g(u, \phi)$ is zero.

A condition of practical importance arises when the two halves are not completely in anti-phase. The transform then contains both odd and even terms. Thus if the two halves are $\pm\alpha$ degrees in phase (4) can be rewritten

$$g(u, \phi) = e^{-j\alpha} \int_0^{\pi} \int_0^1 f(r)\, e^{jur \cos \overline{\phi - \phi'}} r \, dr \, d\phi' + e^{-j\alpha} \int_0^{\pi} \int_0^1 f(r)\, e^{-jur \cos \overline{\phi - \phi'}} r \, dr \, d\phi',$$

which for the plane $\phi = \pm\,\pi/2$ reduces to

$$g\left(u, \pm\frac{\pi}{2}\right) = 2\pi \cos\alpha \int_0^1 f(r) J_0(ur) r \, dr \pm 2\pi \sin\alpha \int_0^1 f(r) \Omega_0(ur) r \, dr. \tag{7}$$

2.3. *Higher Order Hankel Transforms*

2.3.1. *Cyclic phase-variation.* If the phase varies in a cyclic manner with p-cycles of variation in one circuit of the aperture, $F(r, \phi')$ can be written

$$F(r, \phi') = f(r)\, e^{-jp\phi'} \qquad p, \text{ positive integer}$$

then

$$g(u, \phi) = \int_0^1 f(r) r\, dr \int_0^{2\pi} e^{j(ur \cos \overline{\phi - \phi'} - p\phi')}\, d\phi' .$$

This gives (Bickley, 1953)

$$g\left(u, \frac{\pi}{2}\right) = 2\pi \int_0^1 f(r) J_p(ur) r\, dr . \tag{8}$$

Further

$$g(u, \phi) \equiv e^{-jp\phi} g(u, 0) ,$$

or

$$g(u, \phi) \equiv e^{jp[\pi/2 - \phi]} g\left(u, \frac{\pi}{2}\right)$$

i.e. $g(u, \phi)$ itself has a cyclic phase variation (as is physically self-evident).

2.3.3. *Sinusoidal phase-variation.* If a sinusoidal phase-variation with ϕ' has amplitude α, as shown diagrammatically in Fig. 1, higher order transforms again arise. The phase can be described by the function $e^{-j\alpha \sin \phi'}$. Thus for the plane $\phi = \pm \pi/2$

$$g\left(u, \pm \frac{\pi}{2}\right) = \int_0^1 \int_0^{2\pi} f(r)\, e^{j[\pm ur \sin \phi] - j\alpha \sin \phi'} r\, dr\, d\phi'$$

$$= 2\pi \int_0^1 f(r) J_0(\pm ur - \alpha) r\, dr . \tag{9}$$

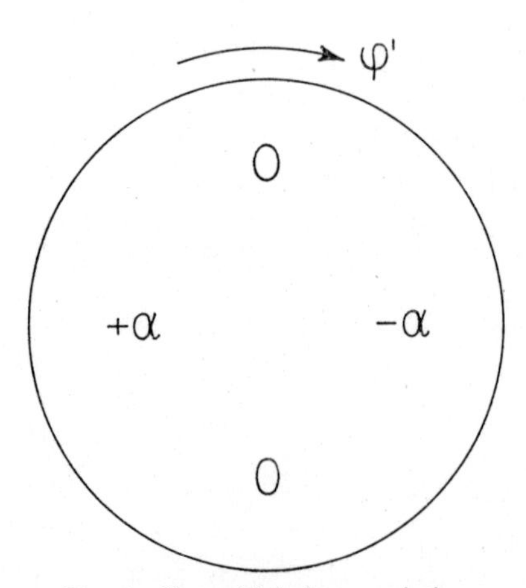

FIG. 1. Sinusoidal phase-variation.

Using the addition formula for Bessel functions (Watson, 1948), this gives

$$\frac{1}{2\pi} g\left(u, \pm\frac{\pi}{2}\right) = J_0(\alpha)\int_0^1 f(r)J_0(ur)r\,\mathrm{d}r + 2\sum_{m=1}^{\infty}(\pm 1)^m J_m(\alpha)\int_0^1 f(r)J_m(ur)r\,\mathrm{d}r. \quad (10)$$

3. EXPANSIONS OF THE AMPLITUDE FUNCTION

In the following we define for simplicity

$$\bar{g}(u, \phi) = \frac{g}{2\pi}(u, \phi).$$

3.1. *The Form* $(1-r^2)^p$

The simplest form of the radial amplitude function $f(r)$ which is transformable by the zero-order Hankel transform, is the form $(1-r^2)^p p = 0, 1, 2, 3$ (Lansraux, 1953). This is given in the relation

$$\int_0^1 (1-r^2)^p J_0(ur)r\,\mathrm{d}r = \frac{2^p p! J_{p+1}(u)}{u^{p+1}} = \frac{\Lambda_{p+1}(u)}{2(p+1)} \quad (11)$$

where the Λ function defined by this equation is tabulated (Jahnke and Emde, 1945). Analogous results are obtained in transforming these functions by the zero-order Lommel transform. Using the series expansion of $\Omega_0(ur)$ one can derive the following:

(a) $\underline{f(r) = 1}$ $\quad \bar{g}\left(u, \frac{\pi}{2}\right) = \frac{1}{u}\left[\Omega_1(u) + \frac{2}{\pi}\right]$ — Fig. 2

(b) $\underline{f(r) = 1 - r^2}$ $\quad \bar{g}\left(u, \frac{\pi}{2}\right) = \frac{2\Omega_2(u)}{u^2} + \frac{4}{3\pi u}$

(c) $\underline{f(r) = (1 - r^2)^2}$ $\quad \bar{g}\left(u, \frac{\pi}{2}\right) = \frac{8\Omega_3(u)}{u^3} + \frac{16}{3\pi u^3} + \frac{16}{3.5\pi u}$.

Several amplitude distributions of the kind found in practice (Fig. 3) can be transformed by expansions in terms of these functions as illustrated by the following:

(d) $\underline{f(r) = (1 - kr^2)^p}$ p integral $k < 1$,

$$\bar{g}(u) = \int_0^1 (1 - kr^2)^p J_0(ur)r\,\mathrm{d}r.$$

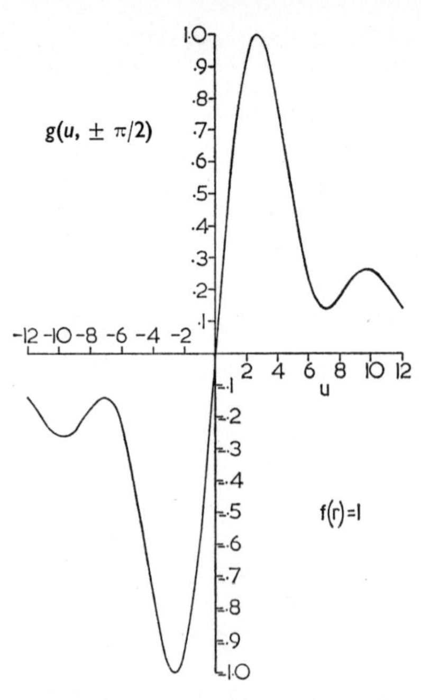

FIG. 2. Diffraction pattern of uniform amplitude distribution in anti-phased aperture.

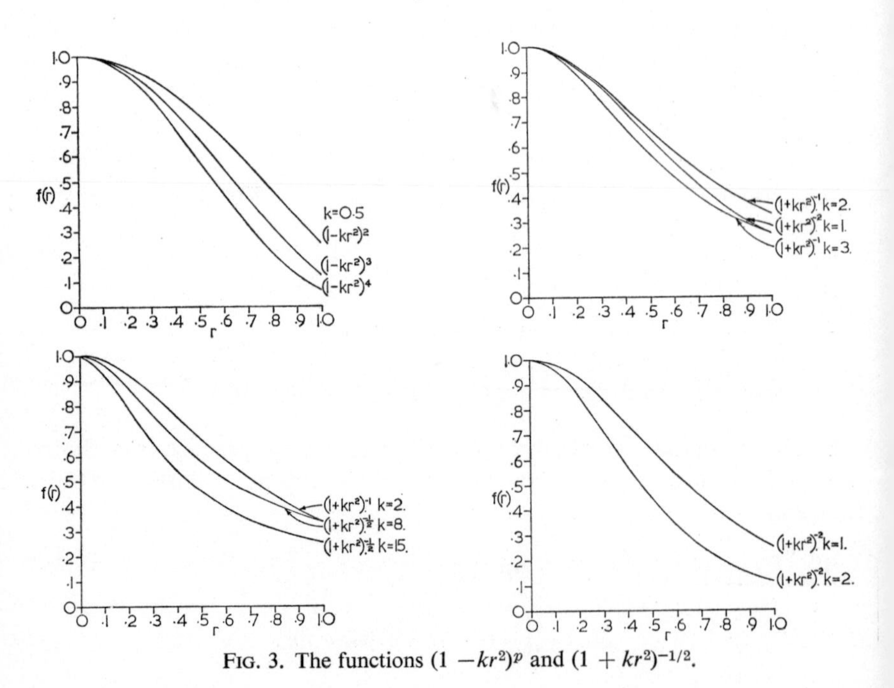

FIG. 3. The functions $(1 - kr^2)^p$ and $(1 + kr^2)^{-1/2}$.

Expanding the bracket by the binomial theorem

$$\bar{g}(u) = \int_0^1 \sum_{m=0}^{p} \frac{p!}{m!(p-m)!} (1-k)^{p-m} k^m (1-r^2)^p J_0(ur) r \, dr$$

$$= \sum_{m=0}^{p} \frac{p!(1-k)^{p-m} k^m}{m!(p-m)!} \frac{\Lambda_{m+1}(u)}{2(m+1)}. \tag{12}$$

(e) $f(r) = (1+kr^2)^{-1/2}$ $k \geqslant 1$,

$$f(r) = [1 + k - k(1-r^2)]^{-1/2}$$

$$= \frac{1}{(1+k)^{1/2}} \left[1 - \frac{k}{1+k}(1-r^2)\right]^{-1/2}.$$

Expanding by the binomial theorem and transforming term by term

$$\bar{g}(u) = \frac{1}{(1+k)^{1/2}} \left[\frac{J_1(u)}{u} + \frac{k}{1+k}\frac{J_2(u)}{u^2} + 3\left(\frac{k}{1+k}\right)^2 \frac{J_3(u)}{u^3} \cdots\right].$$

(f) $f(r) = (1+kr^2)^{-n}$ n integral > 0, $k \geqslant 1$.

similarly gives

$$\bar{g}(u) = \frac{1}{(1+k)^n} \left[\frac{J_1(u)}{u} + \frac{2nk}{1+k}\frac{J_2(u)}{u^2} + 4n(n+1)^2\left(\frac{k}{1+k}\right)^2 \frac{J_3(u)}{u^3} + \ldots\right].$$

Similar results can be derived for the zero-order Lommel transforms for these series.

However higher order transforms of the $(1-r^2)^p$ functions do not take on this simple form. Further since these functions are non-orthogonal they can rarely be used to approximate to a general amplitude distribution that might be obtained in practice.

3.2. *Series Expansion*

We require a set of orthogonal polynomials whose integral transforms of the above kind are readily available in analytical or tabulated forms. One such set, which has hitherto been used in the theory of diffraction images, is the set of Zernike polynomials $R_n^m(r)$ (Linfoot 1955).

These have the following properties, proofs of which can be found in the reference given

$$\int_0^1 R_n^m(r) R_p^m(r) r \, dr = \begin{cases} 0, & n \neq p \\ \dfrac{1}{2(n+1)}, & n = p \end{cases}. \tag{13}$$

For the Hankel transform of order m we have

$$\int_0^1 R_n^m(r) J_m(ur) r \, dr = (-1)^{(n-m)/2} \frac{J_{n+1}(u)}{u}. \tag{14}$$

Thus for the zero-order Hankel transform

$$\int_0^1 R_n^0(r)J_0(ur)r\,dr = (-1)^{n/2}\frac{J_{n+1}(u)}{u} \text{ if } n \text{ is even}$$
$$= 0 \text{ if } n \text{ is odd} \tag{15}$$

and

$$R_n^0(r) = P_{n/2}(2r^2-1) \qquad n, \text{ even}$$
$$= 0 \qquad n, \text{ odd},$$

where P_n is the Legendre polynomial.

It can be shown that

$$\Omega_s(x) = -\int_{-\infty}^{\infty}\frac{J_s(u)\,du}{u-x} \tag{16}$$

that is $\Omega_s(x)$ is the (inverse) Hilbert transform of $J_s(u)$. Hence, it can be shown that

$$\int_0^1 R_n^m(r)\Omega_m(ur)r\,dr = (-1)^{(n-m)/2}\left[\frac{\Omega_{n+1}(u)}{u}+\frac{2\cos^2 n\pi/2}{(n+1)u\pi}\right] \quad (n-m) \text{ even}$$

for the mth-order Lommel transform,

and $$\int_0^1 R_n^0(r)\Omega_0(ur)r\,dr = (-1)^{n/2}\left[\frac{\Omega_{n+1}(u)}{u}+\frac{2}{(n+1)\pi u}\right] \quad n \text{ even} \tag{17}$$

for the zero-order transform.

3.2.1. *Zero-order transforms.* Letting

$$f(r) = \sum_{t=0,2,4,6\ldots}^{\infty} a_t R_t^0(r) \tag{18}$$

then

$$\int_0^1 f(r)R_s^0(r)r\,dr = \sum_{t=0}^{\infty}\int_0^1 a_t R_s^0(r)R_t^0(r)r\,dr$$
$$= \frac{a_s}{2(s+1)} \quad \text{from (13).} \tag{19}$$

Then for the zero-order Hankel transform

$$\bar{g}(u) = \int_0^1 f(r)J_0(ur)r\,dr$$
$$= \sum_{s=0}^{\infty}\int_0^1 a_s R_s^0(r)J_0(ur)r\,dr$$

giving

$$\underline{\bar{g}(u) = \sum_{s=0,2,4,6\ldots}^{\infty} a_s(-1)^{s/2}\frac{J_{s+1}(u)}{u}} \tag{20}$$

Similarly, for the zero-order Lommel transform

$$\bar{g}\left(u, \pm\frac{\pi}{2}\right) = \pm\int_0^1 f(r)\Omega_0(ur)r\,\mathrm{d}r$$

$$= \pm\sum_{s=0}^{\infty} a_s(-1)^{s/2}\left[\frac{\Omega_{s+1}(u)}{u} + \frac{2}{\pi u(s+1)}\right]. \tag{21}$$

The coefficients a_s from (19) are given by

$$a_s = 2(s+1)\int_0^1 f(r)R_s^0(r)r\,\mathrm{d}r\,. \tag{22}$$

The first six of these coefficients are given in the table below.

TABLE 1

$$a_0 = 2\int_0^1 f(r)r\,\mathrm{d}r$$

$$a_2 = 6\int_0^1 f(\ )(2r^2-1)r\,\mathrm{d}r$$

$$a_4 = 10\int_0^1 f(r)(6r^4-6r^2+1)r\,\mathrm{d}r$$

$$a_6 = 14\int_0^1 f(r)(20r^6-30r^4+12r^2-1)r\,\mathrm{d}r$$

$$a_8 = 18\int_0^1 f(r)(70r^8-140r^6+90r^4-20r^2+1)r\,\mathrm{d}r$$

$$a_{10} = 22\int_0^1 f(r)(252r^{10}-630^8+560r^6-210r^4+30r^2-1)r\,\mathrm{d}r$$

3.2.2. *Higher order transforms.* Similarly, for the higher order transforms, $f(r)$ can be expanded in the appropriate series of Zernike polynomials, e.g.

$$f(r) = \Sigma\, a_t R_t^p(r)$$

for the pth-order Hankel transform.

In this instance t takes even values if p is even and odd values if p is odd. From the orthogonality of R_t^p we have

$$a_s = 2(s+1)\int_0^1 f(r)R_s^p(r)r\,\mathrm{d}r \tag{23}$$

and from (14)

$$\bar{g}(u) = \Sigma\, a_s(-1)^{(s-p)/2}\frac{J_{s+1}(u)}{u}. \tag{24}$$

3.2.3. *Sum of the coefficients.* Given

$$f(r) = \sum_{s=0,2,4\ldots} a_s R_s^0(r) \qquad \text{as in (18)}$$

and

$$\bar{g}(u) = \sum_{s=0,2,4\ldots} b_s \frac{J_{s+1}(u)}{u} \quad \text{with} \quad b_s = (-1)^{s/2} a_s = R_s^0(0) a_s$$

then

$$\sum_{s=0,2,4} b_s = \sum_{s=0,2,4\ldots} a_s R_s^0(0) = f(0), \quad \text{and similarly} \quad \sum_{s=0,2,4\ldots} a_s = f(1),$$

i.e. the sum of the coefficients $|a_s|$ equals the value of the illumination function at the periphery and, with alternating sign, at the origin.

Also since the Zernike polynomials are alternately odd and even functions of $(1-2r^2)$, and if $r \to \sqrt{(1-r^2)}$ then $(1-2r^2) \to (2r^2-1)$, then if

$$\bar{g}_1(u) = \int_0^1 f(r) J_0(ur) r \, dr = a_0 \frac{J_1(u)}{u} - a_2 \frac{J_3(u)}{u} + a_4 \frac{J_5(u)}{u} - \ldots$$

then

$$\bar{g}_2(u) = \int_0^1 f[\sqrt{(1-r^2)}] J_0(ur) r \, dr = a_0 \frac{J_1(u)}{u} + a_2 \frac{J_3(u)}{u} + a_4 \frac{J_5(u)}{u} + \ldots.$$

4. ZERO-ORDER TRANSFORMS OF PARTICULAR FUNCTIONS

We derive in this section the zero-order transforms of some particular functions which are of interest in diffraction theory. Since the process consists essentially of obtaining the coefficients a_s which are the same for the zero-order Lommel transform and the zero-order Hankel transform (equations (18) to (22)), we have confined the examples given to the Hankel transforms. Further, since $\Omega_0(u)$ is the Hilbert transform of $J_0(u)$, it follows that the Lommel transforms can be obtained directly from the Hankel transforms by this transformation.

4.1. $f(r) = \text{constant} = 1$ say

Since $$1 = R_0^0(r)$$

then $$a_0 = 2\int_0^1 r \, dr = 1$$

$$a_2 = a_4 = a_6 \text{ etc.} = 0 \quad \text{from (13).}$$

Therefore $$\bar{g}(u) = a_0 \frac{J_1(u)}{u} = \frac{J_1(u)}{u}.$$

4.2. $\underline{f(r) = 1 - r}$

$$\bar{g}(u) = \frac{J_1(u)}{u} - \int_0^1 J_0(ur)r^2\,dr$$
$$= \frac{J_1(u)}{u} - I$$

where
$$I = \sum_{s=0,2,4\ldots}^{\infty} a_s(-1)^{s/2}\frac{J_{s+1}(u)}{u}.$$

From Table 1
$$a_0 = 2\int_0^1 r^2\,dr = \tfrac{2}{3}$$
$$a_2 = 6\int_0^1 (2r^4 - r^2)\,dr = 6\cdot\frac{1}{1.3.5} \quad \text{etc.,}$$

i.e.
$$a_s = (-1)^{s/2+1}\frac{2}{(s-1)(s+3)}$$

giving
$$\bar{g}(u) = \tfrac{1}{3}J_1\frac{(u)}{u} + \frac{1}{u}\sum_{s=2,4,6,\ldots}\frac{2}{(s-1)(s+3)}J_{s+1}(u)$$
$$= \frac{1}{u}\left\{\frac{J_1(u)}{3} + \tfrac{2}{5}J_3(u) + \tfrac{2}{21}J_5(u) + \tfrac{2}{45}J_7(u)\ldots\right\} \tag{25}$$

See Fig. 4.

4.3. $\underline{f(r) = r^n}$

This key function can be used for any representation of $f(r)$ capable of a series expansion or a good polynomial approximation.

Thus:

$$a_0 = 2\int_0^1 r^n r\,dr = \frac{2}{n+2}$$
$$a_2 = 6\int_0^1 r^{n+1}(2r^2 - 1)\,dr = \frac{6n}{(n+2)(n+4)} \quad \text{etc.}$$
$$a_s = \frac{2(s+1)\cdot n(n-2)(n-4)\ldots(n-s+2)}{(n+2)(n+4)\ldots(n+s+2)}$$
$$= 2\cdot(s+1)\cdot\frac{1}{2}\frac{((n/2)!)^2}{(n/2+s/2+1)!(n/2-s/2)!} \quad \text{(Erdelyi, 1954a)†}$$
$$\therefore\ \bar{g}(u) = \sum_{s=0,2,4,6\ldots}\frac{(-1)^{s/2}(s+1)((n/2)!)^2}{((n+s)/2+1)!((n-s)/2)!}\frac{J_{s+1}(u)}{u}. \tag{26}$$

† This disagrees with the result given in the reference which can be shown to be in error by the substitution $n = 0$.

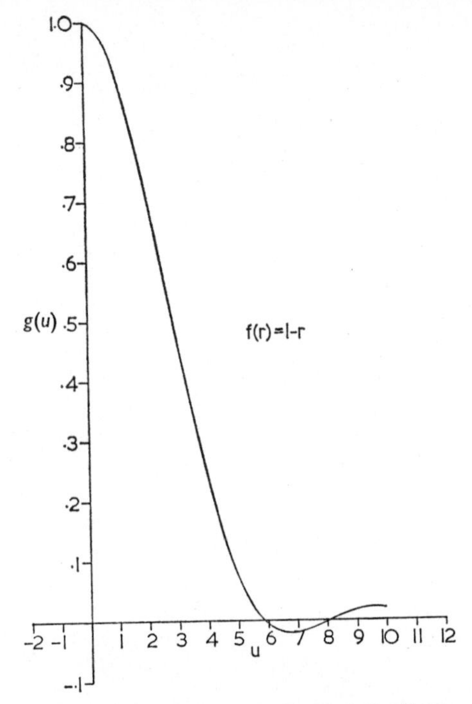

FIG. 4. Diffraction pattern of gabled distribution.

The factorials, if non-integral, can be replaced by their gamma function representation,

e.g. $$\frac{n}{2}! \equiv \Gamma\left(\frac{n}{2}+1\right).$$

This transform is also given (Erdelyi, 1954b) in the form

$$\bar{g}(u) = \frac{1}{n+2}\,{}_1F_2\left\{\frac{n}{2}+1; 1, \frac{n}{2}+2; -\frac{u^2}{4}\right\}_{n\geqslant 0}$$

where ${}_1F_2$ is a generalized hypergeometric function.

It is of interest to compare results by using the Zernike expansion for the quadratic forms of the previous subsection. Thus

4.4. $f(r) = (1-r^2)^p$

$$a_s = 2(s+1)\int_0^1 (1-r^2)pR_s^0(r)r\,\mathrm{d}r$$

$$= \frac{(-1)^{s/2}(s+1)(p!)^2}{(p+s/2+1)!(p-s/2)!} \qquad \text{(Erdelyi, 1954c)}$$

$$\therefore\quad \bar{g}(u) = \sum_{s=0,2,4,6\ldots}^{2p} \frac{(s+1)(p!)^2}{(p+s/2+1)!(p-s/2)!}\,\frac{J_{s+1}(u)}{u}. \qquad (27)$$

The representation of $g(u)$ in this case can be used for non-integral values of p by replacing the factorials with their gamma function equivalents. By convention the series terminates when a factorial becomes a negative integer.

The series can be shown to be equivalent to the standard result in (12) using the reduction formulas for Bessel functions.

4.5. $f(r) = \dfrac{1}{\sqrt{(1-r^2)}}$

$$a_s = 2(s+1)\int_0^1 (1-r^2)^{-1/2} R_s^0(r) r\,dr$$

$$= 2 \text{ for all } s \text{ (Erdelyi, 1954c).}$$

$$\therefore \quad \bar{g}(u) = \frac{2J_1(u)}{u} - \frac{2J_3(u)}{u} + \frac{2J_5(u)}{u} \dots$$

$$= \frac{\sin u}{u} \qquad \text{(Erdelyi, 1954d).} \tag{28}$$

4.6. *Functions with a Taylor Series Expansion*

From subsection 4.3 any amplitude function, capable of a series expansion or a polynomial representation can be transformed by the Zernike polynomial method to give a diffraction pattern represented by a series of Bessel functions.

That is if

$$f(r) = \sum_{n=0}^{\infty} \frac{r^n}{n!} f^{(n)}(0)$$

then

$$\bar{g}(u) = \sum_{n=0}^{\infty} \frac{f^{(n)}(0)}{n!} \sum_{s=0,2,4\dots}^{\infty} \frac{(-1)^{s/2}(s+1)((n/2)!)^2}{((n+s)/2+1)!((n-s)/2)!} \frac{J_{s+1}(u)}{u} \tag{29}$$

4.7. $f(r) = \cos \pi r/2$

$$\bar{g}(u) = \frac{J_1(u)}{u}\left[1 - \frac{\pi^2}{2!2!4} + \frac{\pi^4(2!)^4}{2!3!4!2^4} - \frac{\pi^6(3!)^2}{3!4!6!2^6} \dots\right]$$

$$+ \frac{3J_3(u)}{u}\left[\frac{\pi^2}{2!3!4!} - \frac{\pi^4(2!)^4}{4!4!2^4} + \frac{\pi^6(3!)^2}{2!5!6!2^6} \dots\right] \text{ etc.}$$

The series converge sufficiently rapidly and give (to 5 decimal places)

$$\bar{g}(u) = 0.46267\frac{J_1(u)}{u} + 0.49904\frac{J_3(u)}{u} + 0.03733\frac{J_5(u)}{u} + 0.00095\frac{J_7(u)}{u}. \tag{30}$$

4.8. $f(r) = \cos^2 \pi r/2$

Similarly using $\cos 2\theta = 2\cos^2 \theta - 1$ (to 3 decimal places)

$$\bar{g}(u) = 0.297 \frac{J_1(u)}{u} + 0.473 \frac{J_3(u)}{u} + 0.205 \frac{J_5(u)}{u} + 0.022 \frac{J_7(u)}{u} + 0.002 \frac{J_9(u)}{u}. \tag{31}$$

(Fig. 5)

but the convergence is much slower and limits the continuation of this procedure.

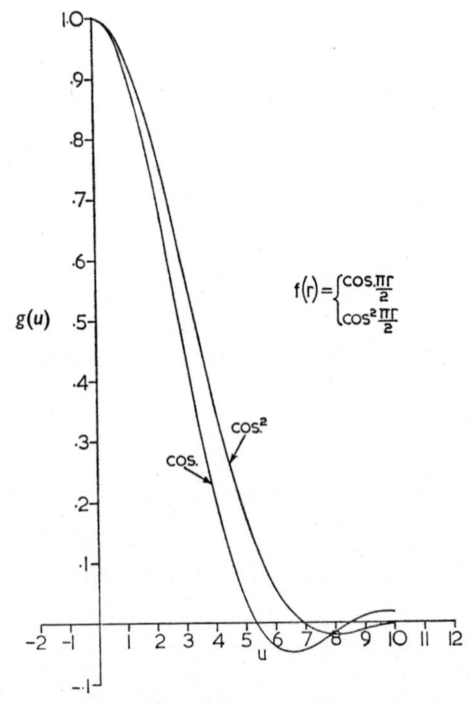

FIG. 5. Diffraction patterns of cosine and cosine² distributions

Similarly one can derive the transforms of

$$f(r) = \cos \frac{m\pi r}{2}$$

and

$$f(r) = \cos^2 \frac{m\pi r}{2}.$$

4.9. $f(r) = e^{-2br^2}$

For the exponential distribution we have

$$a_0 = 2\int_0^1 e^{-2br^2} r\,dr = \frac{1}{2b}[1 - e^{-2b}]$$

$$a_2 = 6\int_0^1 e^{-2br^2}(2r^2 - 1)r\,dr = -\frac{3}{2b}\left[e^{-2b}\left(1+\frac{1}{b}\right) + \left(1 - \frac{1}{b}\right)\right].$$

Using the recurrence properties of Legendre polynomials, we can derive the recurrence relation

$$\frac{a_s}{s+1} = \frac{a_{s-4}}{s-3} + \frac{a_{s-2}}{b} \qquad (32)$$

with a_0 and a_2 given above.

Table 2 gives these coefficients for values of $b = \frac{1}{2}$, 1, and -1.

From the properties of the coefficients given in subsection 3.2.3 it follows that the terms tend to zero and hence each bracket contains a continuously improving rational approximation to the appropriate power of e. It can be shown in fact that the relation in (32) is related to the simple continued fraction of Gauss for the exponential function

TABLE 2

	$b = \frac{1}{2}$	$b = 1$	$b = -1$
a_0	$1 - e^{-1}$	$\frac{1}{2}(1 - e^{-2}$	$\frac{1}{2}(e^2 - 1)$
a_2	$3(1 - 3e^{-1})$	$\frac{3}{2}(\ \ - 2e^{-2})$	$\frac{3}{2}(2)$
a_4	$5(7 - 19e^{-1})$	$\frac{5}{2}(1 - 7e^{-2})$	$\frac{5}{2}(e^2 - 7)$
a_6	$7(71 - 193e^{-1})$	$\frac{7}{2}(5 - 37e^{-2})$	$\frac{7}{2}(37 - 5e^2)$
a_8	$9(1001 - 2721e^{-1}$	$\frac{9}{2}(36 - 266e^{-2})$	etc.
a_{10}	$11(18089 - 49171e^{-1})$	$\frac{11}{2}(392 - 2431e^{-2})$	

The diffraction patterns obtained from these series are shown in Fig. 6.

5. HIGHER ORDER TRANSFORMS—CONSTANT AMPLITUDE

Using the method of subsection 3.2.2 we can obtain the integrals

$$g(u) = 2\pi\int_0^1 J_n(ur)r\,dr.$$

This gives for example

$$\int_0^1 J_1(ur)r\,dr = \frac{2.2}{1.3}\frac{J_2(u)}{u} + \frac{2.4}{3.5}\frac{J_4(u)}{u} + \ldots + \frac{2(s+1)}{s(s+2)}\frac{J_{s+1}(u)}{u} \quad (s,\ \text{odd}), \quad (33)$$

and

$$\int_0^1 J_2(ur)r\,\mathrm{d}r = \frac{6}{4}\frac{J_3(u)}{u} + \frac{10}{12}\frac{J_5(u)}{u} + \ldots + \frac{4(s+1)}{s(s+2)}\frac{J_{s+1}(u)}{u} \ldots \quad (s\text{, even}), \tag{34}$$

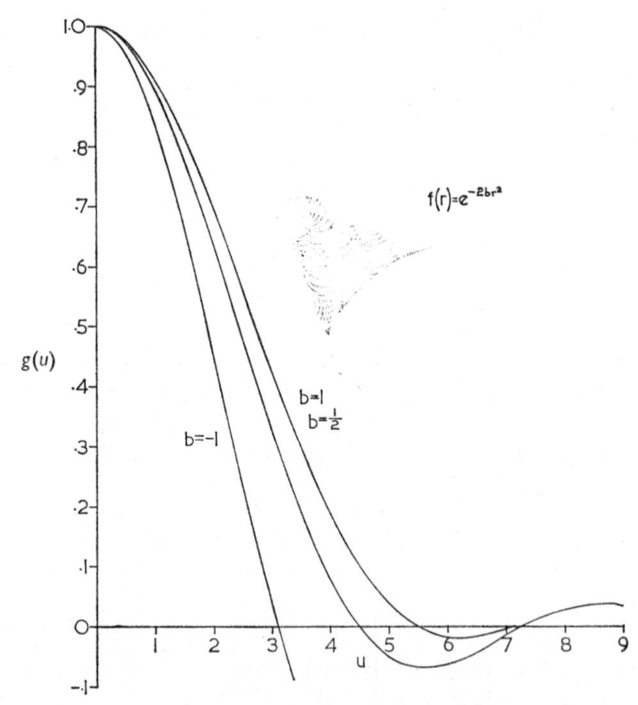

FIG. 6. Diffraction patterns of exponential distributions.

6. RADIAL PHASE-VARIATION

If in (1) $F(r, \phi')$ be a complex function of r, this would be a representation of a ϕ'-independent amplitude function with a radial variation in phase, i.e.

$$F(r, \phi') = f(r)\,\mathrm{e}^{-j\beta(r)}$$

where $f(r)$ and $\beta(r)$ are real functions of r. The factor $\mathrm{e}^{-j\beta(r)}$ being ϕ'-independent can be taken outside the angular integral in (2) to give

$$\bar{g}(u) = \int_0^1 f(r)\,\mathrm{e}^{-j\beta(r)} J_0(ur)r\,\mathrm{d}r.$$

The complex amplitude given by $f(r)\,\mathrm{e}^{-j\beta(r)}$ can be expressed as a series of Zernike polynomials and dealt with by the same methods as in the previous subsection.

The distribution of most interest is that in which $\beta(r) = v\, r^2/2$ giving

$$g(u) = 2\pi \int_0^1 f(r) J_0(ur)\, e^{-jvr^2/2} r\, dr \tag{35}$$

where v is a constant. This integral has more than one interpretation. In the far-field it is the diffraction pattern of an aperture with amplitude distribution $f(r)$ and with a quadratic phase error of the type that would be obtained by defocusing the system. In the near-field at a distance R the integral represents the field distribution of a circularly symmetrical *in-phase* aperture illumination, but with $v = ka^2/R$. The integral can also represent the three dimensional distribution of the field at a focus of the system (Born and Wolf, 1959).† For this case

$$v = \frac{ka^2}{f^2} z$$

$$u = \frac{ka}{f} \sqrt{(x^2 + y^2)} = \frac{2\pi a \sin\theta}{\lambda} \quad \text{as before,}$$

where (x, y, z) are coordinate axes at the focus with z along the axis of the circular aperture and f is the focal length.

Ming-Kuei Hu (1960) has shown how the normal treatment of this integral using the Lommel functions, which themselves are Bessel function series, can be extended for the case where $f(r)$ is of the form $(1 - r^2)^p$.

Proceeding, however, in the manner of the previous subsections and considering

$$f(r)\, e^{-jvr^2/2}$$

as a complex amplitude function with circular symmetry, the integral can be evaluated again as

$$g(u) = 2\pi \sum_{s=0,2,4,6\ldots} a_s (-1)^{s/2} \frac{J_{s+1}(u)}{u}$$

with

$$a_s = 2(s+1) \int_0^1 f(r)\, e^{-jvr^2/2} R_s^0(r) r\, dr. \tag{36}$$

6.1. *Uniform Illumination*

Putting $f(r) = 1$ we obtain

$$a_0 = \frac{2}{jv}[1 - e^{-jv/2}] = \frac{\sin v/2}{v/2} - j\left(\frac{1 - \cos v/2}{v/2}\right) = \cos \overset{*}{T} - j \sin \overset{*}{T} \tag{37}$$

† Note: this $u \cdot v$-notation is reversed from that given in the reference for consistency with the previous work.

$$a_2 = \frac{6}{jv}[-1 - e^{-jv/2} + 2a_0]$$

$$= 3 \cos \overset{*}{T} - \frac{12}{v} \sin \overset{*}{T} - j\left[\frac{12}{v} \cos \overset{*}{T} + 3 \sin \overset{*}{T} - \frac{12}{v}\right] \quad (38)$$

and

$$a_s = (s + 1)\left[\frac{a_{s-4}}{s - 3} - \frac{4ja_{s-2}}{v}\right]. \quad (39)$$

In the above $\cos \overset{*}{T}$ and $\sin \overset{*}{T}$ refer respectively to the finite Fourier cosine and sine transforms of, in this case unity.

6.2. *Tapered Illumination*

For any amplitude distribution $f(r)$ we have

$$a_0 = 2\int_0^1 f(r)\, e^{-jvr^2/2} r\, dr = \int_0^1 f(\sqrt{t})\, e^{-jvt/2}\, dt \quad (40)$$

$$\equiv \cos \overset{*}{T} - j \sin \overset{*}{T},$$

that is the finite Fourier transform of the same function but with the power of r halved.

6.3. $f(r) = 1 - r^2$

$$a_0 = \cos \overset{*}{T} - j \sin \overset{*}{T} = \frac{1 - \cos v/2}{(v/2)^2} - j\,\frac{v/2 - \sin v/2}{(v/2)^2}$$

where $\overset{*}{T}$ is the finite Fourier transform of $1 - r$.

6.4. $f(r) = \pi r^2/2$

$$a_0 \equiv \cos \overset{*}{T} - j \sin \overset{*}{T} = \frac{2\pi \cos v/2}{\pi^2 - v^2} + 2j\,\frac{v - \pi \sin v/2}{\pi^2 - v^2}$$

where $\overset{*}{T}$ is the transform of $\cos \pi r/2$.

6.5. $f(r) = (1 - r^2)^p$

The result of subsection 6.1 and 6.2 can be generalized for these polynomials. We define

$$\frac{a_{s,p}}{2(s + 1)} \equiv \int_0^1 (1 - r^2)^p R_s^0(r)\, e^{-jvr^2/2} r\, dr,$$

the required coefficients. Then integrating once by parts and using the Legendre reduction formula we have

$$\frac{a_{s,p}}{2(s+1)} = \frac{(-1)^{s/2}}{jv} - \frac{2p}{jv} a_{s,p-1} + \frac{2}{jv} [a_{s-2,p} + a_{s-6,p} + a_{s-10,p} \ldots],$$

where

$$\frac{a_{s,0}}{s+1} = \frac{a_{s-4,0}}{s-3} - \frac{4j}{v} a_{s-2,0} \quad \text{as in (39)}$$

$a_{2,0}$ is given by (38) and

$a_{0,p} \equiv \cos \overset{*}{T} - j \sin \overset{*}{T}$ where $\overset{*}{T}$ is the transform of $(1 - r)^p$.

From the properties of $R_s^0(r)$ the above series terminates for any even value of s.

REFERENCES

BICKLEY, W. G. (1953) *Bessel Functions and Formulae*, p. XXXIX. Cambridge University Press.

BORN, M. and WOLF, E. (1959) *Principles of Optics*, p. 436. Pergamon Press.

ERDELYI, *et al.* (1954) *Tables of Integral Transforms*, McGraw-Hill.
  (a) Vol. 2, No. 22, p. 278.
  (b) Vol. 1, No. 2, p. 326.
  (c) Vol. 2, No. 23, p. 278.
  (d) Vol. 2, No. 5, p. 7.

JAHNKE and EMDE (1954) *Tables of Functions*, p. 211, Dover.

LANSRAUX, G. (1953) Conditions functionelles de la diffraction instrumentale, *Cahier de Physique*, No. 45, p. 29.

LINFOOT, E. H. (1955) *Recent Advances in Optics*, p. 51, Oxford University Press.

MING-KUEI HU (1960) Some new methods of analysis and synthesis of near zone fields, *I.R.E. International Convention Record*, Pt. 1, p. 13.

SILVER, S. (1949) *Microwave Antenna Theory and Design*, M.I.T. Series Vol. 12, p. 173, McGraw-Hill.

WATSON, G. N. (1948) *Bessel Functions*, p. 143, McMillan.

DIFFRACTION ON A BROAD APERTURE IN A BROAD WAVEGUIDE

B. Z. KATSENELENBAUM

ABSTRACT

When waves are diffracted by inhomogeneities in waveguides having a large cross-section compared to the wavelength the optical and waveguide properties of the field are simultaneously revealed. Proposed here is a method for studying such phenomena, using the simple case of a two-dimensional scalar problem concerned with the field in a waveguide with a transverse slot; the wavelength $2\pi/\mathbf{k}$ is small compared to the width of the waveguide b and the slot width L. A wave traveling along the waveguide loses a portion of its energy; half of the lost energy escapes through the slot, and the other half is transformed into parasitic waves which fill (with respect to the wave number) a broad spectrum. For the boundary condition of the first kind the relative loss of energy is equal to $2\mu^3/3$, where $\mu = (\pi L/\mathbf{k}b^2)^{1/2}$. For the boundary condition of the second kind the losses are proportional to the first power of the smallness parameter. Similar results are obtained for diffractional losses at the mirror placed in a bend of a broad waveguide. Diffractional losses in such a bend have been also determined for a circular waveguide with the H_{01}-mode.

This method is based on the established fact that given the value of the field at the slot one can determine all quantitative characteristics of the resultant diffraction pattern which are of importance. This field can be found from elementary optical considerations which take into account the results of Sommerfeld's solution to the problem of diffraction on a half-plane.

1. INTRODUCTION

In broad waveguides, i.e. waveguides having large cross-sectional dimensions compared to the wavelength, the laws underlying wave propagation are of intermediate nature between the usual waveguide laws and optical laws. On the one hand, the field in the waveguide is close in structure to that of the plane wave, and therefore many of the proposed waveguide elements embody the optical principles in the reflection and refraction of plane waves (Comte, Carfort, Ponthus, Paris, 1957). On the other hand, in the cases under consideration the division into waveguide waves still has a physical meaning. For only one of the waves is useful while the rest are parasitic. It is this classification that is important whereas a detailed distribution of energy among the various parasitic waves is of no interest here. The waveguide elements should not be the cause of any considerable transformation of the useful wave into the parasitic waves, i.e. they must not give rise to large losses by transformation. The electromagnetic computation of such systems is reduced to asymptotic problems of the theory of diffraction, which may be solved by combining the theory of diffraction for open systems and the classical waveguide theory. Proposed below is a mathematical method (Katsenelenbaum, 1962) for the solution of such asymptotic problems. It is applied here to a scalar problem concerned with a wave traveling along a broad plane waveguide which has a broad slot in one of its walls. In conclusion it will be shown that the solution of this problem can be extended to the computation of the mirror in a broad waveguide bend.

Apparently, some elements of the proposed method can be useful for the elaboration of a general theory of geometric optics of broad waveguides (Katsenelenbaum, 1961).

2. FIELD AT THE SLOT

Sought for is the solution $u(x, z)$ of the wave equation

$$\frac{\partial^2 u}{\partial x^2} + \frac{\partial^2 u}{\partial z^2} + k^2 u = 0 \tag{1}$$

satisfying on the straight line $x = b$ and on two half-straight lines $x = 0$, $z < 0$ and $x = 0$, $z > L$, the boundary condition

$$u\big|_c = 0. \tag{2}$$

Coming from the left, from the region $z < 0$, $0 < x < b$, is the wave

$$u = \sin \alpha_1 x \cdot e^{-ih_1 z}, \qquad \alpha_m = \frac{\pi m}{b}, \qquad h_m = \sqrt{(k^2 - \alpha_m^2)} \tag{3}$$

No other wave comes in from infinity. The wavelength is small compared to the width of the waveguide and slot width, $kb \gg 1$, $kL \gg 1$.

Under these conditions the field in the waveguide is close to that of the incident wave. If we write the field on the right of the slot in the form

$$u(x, z) = \sum_{m=1}^{\infty} A_m \sin \alpha_m x \cdot e^{-ih_m z} + \sin \alpha_1 x \cdot e^{-ih_1 z} \tag{4}$$

then the quantities A_m which characterize the perturbation of the field will be small. The idea of the proposed method is to express these quantities in terms of the field at the waveguide slot, i.e. through function $u(0, z)$.

As we are presently to see, the field at the slot contains a small factor. Besides, a small factor also stands before the integral which expresses A_m in terms of $u(0, z)$. Thus, the sought-for amplitudes A_m are expressed as the product of a small numerical factor and the integral at the slot. The order of smallness of A_m is determined by this small factor. Consequently, a slight error in the field u at the slot will bring about a relatively slight error in A_m.

It is this feature of the formula which gives A_m in terms of u (see formula (5) below) that is turned to account by the proposed method; it enables us to substitute in (5) the approximate value of the field at the slot which is easy to find.

It is worth noting that things would be different if A_m were expressed in terms of the field at the mouth of the right-hand waveguide, i.e. for $z = L$, $0 < x < b$. Though u enters into the respective formulas also as an integrand term, they contain no small factor and the integrals themselves appear to have much smaller values than the integrand functions. To compute A_m from these formulas one would have to know very accurately the field at the mouth of the right-hand waveguide, as is ordinarily the case when expressions involving a difference between two great value quantities are used. In particular, one would have to take into account the boundary effects near the edges, and it would be impossible for instance merely to follow the Huyghen–Kirchhoff's principle in determining the field at the mouth of the right-hand waveguide from the field of the incident wave (3) at the mouth of the left-hand waveguide. On the other hand, in the proposed scheme of computation of A_m in the term of highest order with respect to the smallness parameter, the effect of the edges on the field at the slot can be disregarded altogether.

A_m is expressed in terms of $u(0, z)$ by ordinary methods of waveguide excitation theory. Since the field on the left of the slot comprises a single incident wave and a system of outgoing waves, while on the right it has the form (4), it is easy to obtain

$$A_m = \frac{-i}{h_m b} \int_0^L u(0, z) \frac{\partial v^m}{\partial x} \, dz, \qquad v^m = \sin \alpha_m x \cdot e^{-ih_m z} . \tag{5}$$

We shall now proceed to compute an approximate evaluation of the field $u(0, z)$. Under a certain condition which we shall presently obtain, the field at the slot is independent of the structure of the field far away from the slot, being determined solely by the values of u and $\partial u/\partial z$ at the portion of the plane $z = 0$ adjacent to the slot. This can be shown by expressing $u(0, z)$, according to Kirchhoff's formula, through the integral of the values of u and $\partial u/\partial n$ over a closed contour, for instance coinciding with the dash-line curve in Fig. 1. Essential in this integral appears to be only the region of integration along the line $z = 0$ in the proximity of the point $x = 0$. All other regions contribute little to the value of the field either because of the presence of a rapidly oscillating factor in the integrand, or due to a reciprocal compensation of the integrand terms proportional to u and $\partial u/\partial n$. The slot is illuminated by the incident wave coming in from the left. Not all of the wave is effective, but only one or two of the first Fresnel zones are effective, rather than the entire wave front. The size of that portion depends on the distance of the point at which the field is computed from the slot edge $z 0, = x = 0$. For the farthest point, i.e. for $z = L$, the size of the significant portion of the front is

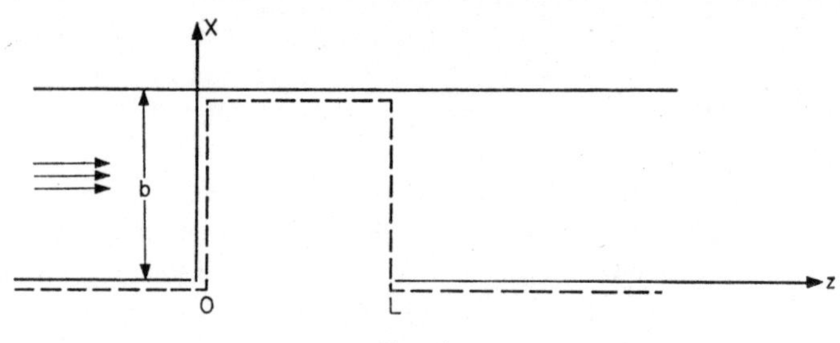

FIG. 1

of order $(L/k)^{1/2}$. For the field at the slot to be local, that portion of the front must be small compared to b, so that $(L/k)^{1/2} \ll b$. Let us introduce a dimensional parameter

$$\mu = \left(\frac{\pi L}{kb^2}\right)^{1/2}. \tag{6}$$

Then the condition for the applicability of further computations is

$$\mu \ll 1. \tag{7}$$

The field on the line $z > 0$, $x = 0$ is determined by the field on a small portion of the line $z = 0$, $x > 0$. The field on that portion can be regarded as equal to the field of the incident wave (3). Afterwards the field on the slot is found either by computing Kirchhoff's integral, or, what is somewhat simpler, from Sommerfeld's solution to the problem of the diffraction of an incident plane wave

$$u = \frac{1}{2i} e^{i\alpha_1 x} \cdot e^{-ih_1 z} \tag{8}$$

on the half-plane $x = 0$, $z < 0$. Function (8) is obtained from (3) by isolating the term corresponding to an incident plane wave.

Making use of the solution for the field formed when wave (8) falls on the half-plane at a very small glancing angle we obtain on the lower slot

$$u = -i^{3/2}\mu \sqrt{\left(\frac{z}{2L}\right)}\, e^{-ikz} (a) \quad \frac{\partial u}{\partial x} = \frac{\pi}{2b} e^{-ikz} (b). \tag{9}$$

Condition (7) permits h_1 in the exponent to be replaced by k.

The computation of the amplitudes A_m by means of (5) and (9) is a simple matter. The amplitude of the principal wave is

$$1 + A_1 = 1 - \frac{\sqrt{2}}{3} i^{1/2} \mu^3. \tag{10}$$

The relative energy loss by transformation, i.e. quantity $1 - |1 + A_1|^2$, is thus equal to $2\mu^3/3$.

3. PARASITIC WAVES

Let us now consider the distribution of this energy among the various parasitic waves. The energy carried away in a horizontal waveguide to the left can be shown to be higher in order of magnitude than μ^3. We shall determine the energy carried away to the right. It can be easily found that

$$A_m = -i^{1/2} \frac{1}{\sqrt{2}} \cdot \frac{k}{h_m} m\mu^3 \int_0^1 \sqrt{t} \cdot e^{-i(k-h_m)L \cdot t} \, dt \tag{11}$$

$$(k - h_m)L = \frac{2}{1 + h_m/k} \frac{\pi}{2} m^2\mu^2 . \tag{12}$$

These two formulas give the amplitude of the parasitic wave as a function of the wavenumber m. With m rising the energy of each wave $|A_m|^2 . \, h_m/k$ grows from values of order μ^6 at $m \sim 1$ up to a value of order μ^4 at $m \sim 1/\mu$, and then declines. Large m, of order $1/\mu^2$, at which the difference of the factor h_m/k from unity should be taken into account, corresponds to a very small energy.† Therefore the ratio h_m/k in (11) and (12) can be substituted with unity.

The set of waves with amplitudes (11) whose variation with the number is very slow can be described as a broad spectral line, although strictly speaking the spectrum of that line (with respect to m) is a line spectrum rather than continuous. The width of the line is $\Delta m \sim 1/\mu$. The relative energy over the line, i.e. the sum $\sum |A_m|^2 \cdot h_m/k$, can be computed according to Euler's summation formula as the integral over dm. The quantity $|A_m|^2$ decreasing rapidly as m grows, we can express the energy as

$$\int_0^\infty |A_m|^2 \cdot dm.$$

This integral can be computed exactly; it is equal to $\mu^3/3$.

Thus half of the energy lost by transformation is carried by parasitic waves to the right-hand waveguide. It is easy to see that the remaining portion of the energy escapes through the slot in accordance to the law of conservation of energy. This portion equal to

$$-\text{Im} \int_0^L u \frac{\partial u^*}{\partial x} \, dz$$

can be easily computed from (10). This integral divided by the incident wave energy is, in fact, equal to $\mu^3/3$.

The character of diffraction at a large aperture in a broad waveguide is determined as shown above, by the structure of the incident wave field in the near vicinity of the aperture boundary in a region of order μb. Since with the

† If at any frequency $h_m = 0$ for a certain m, the respective wave is not excited at all.

boundary condition (2) the incident wave field in that portion of the front is small, the field on the slot is small too (9), and the losses are proportional to the third power of the smallness parameter μ.

With the boundary condition $\partial u/\partial n = 0$ the major portion of the energy likewise passes to the right-hand waveguide in the form of a wave of the same number as the incident wave, yet the loss will be higher, viz. of order μ. In this case the field on the slot is

$$u = \tfrac{1}{2}\,\mathrm{e}^{-ikz}, \frac{\partial u}{\partial x} = -i^{1/2}\sqrt{\left(\frac{k}{2\pi z}\right)}\,\mathrm{e}^{-ikz}. \tag{13}$$

The relative decrease of the incident wave energy is equal to $2\mu/\pi$ in the case of a waveguide wave for which $u(x, 0) = \cos \pi x/b$, and is half as low in the case of the principal wave, i.e. a wave for which $u(x, 0) = 1$. Half of the energy is again carried away through the slot, the other half being transmitted to parasitic waves going to the right-hand waveguide. The parasitic wave amplitudes are

$$A_m = i^{3/2}\frac{\mu}{\pi\sqrt{2}}\int_0^1 \mathrm{e}^{-i(\pi/2)\mu^2 m^2 t}\frac{\mathrm{d}t}{\sqrt{t}}. \tag{14}$$

They decrease monotonically as m grows. This corresponds to a broad spectral line located likewise at $m \sim 1/\mu$ and having a width $\Delta m \sim 1/\mu$, yet in this case its maximum is of order μ^2, being located at $m \sim 1$, while its area is proportional to μ.

4. A MIRROR IN THE BEND OF A WAVEGUIDE

With the results obtained the problem of diffraction losses in a waveguide cross (Fig. 2) is practically solved as well. The presence of a second slot in the wall of a horizontal waveguide under condition (7) will not change the field on the first slot. The diffractional losses of the principal wave will be two times as high, and the area of the spectral line formed by parasitic waves in the right-hand waveguide will likewise increase twice. There will also take place a redistribution of energy between even and odd parasitic waves. The fields (9) or (13) in vertical waveguides will produce a narrow spectral line located at $m \sim 1/\mu^2$, its width being equal to $\Delta m \sim 1$. Its height will be of order μ^3 under boundary condition (2) and of order μ under boundary condition $\partial u/\partial n = 0$, i.e. one order of magnitude higher than the height of the broad spectral line formed by parasitic waves in the right-hand waveguide.

From Fig. 2 we can easily go over the case of a mirror in a waveguide bend (Fig. 3), if we assume that there are two incident waves in the waveguide cross coming with the respective signs, one from left the other from above. It must be taken into consideration that the parasitic waves formed in the

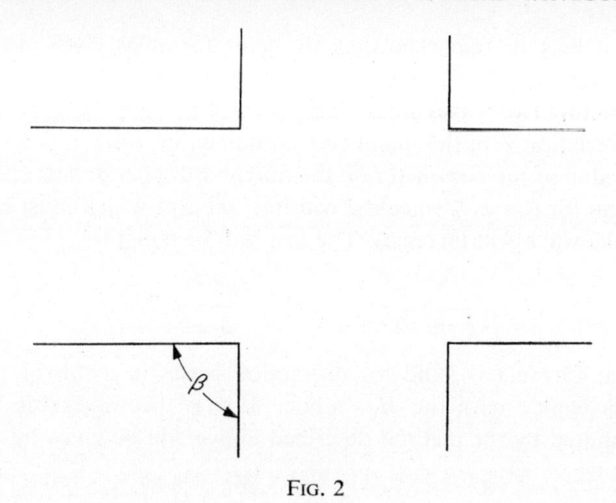

FIG. 2

lower waveguide (Fig. 2) from the two incident waves are located in different regions of the spectrum (with respect to m), and so they do not interfere.

Thus we obtain the following result for the boundary condition (2). The diffraction losses in the mirror are $4\mu^3/3$. A quarter of this loss ($\mu^3/3$) passes to the left-hand waveguide as a narrow line at $m \sim 1/\mu^2$, with $\Delta m \sim 1$ and the height of order μ^3. To the lower waveguide goes an energy equal to μ^3, two lines being formed here; one of them is similar to that in the left-hand waveguide with energy $\mu^3/3$, while the other is a broad line ($\Delta m \sim 1/\mu$) having the height of order μ^4 and containing half of the entire energy lost by transformation.

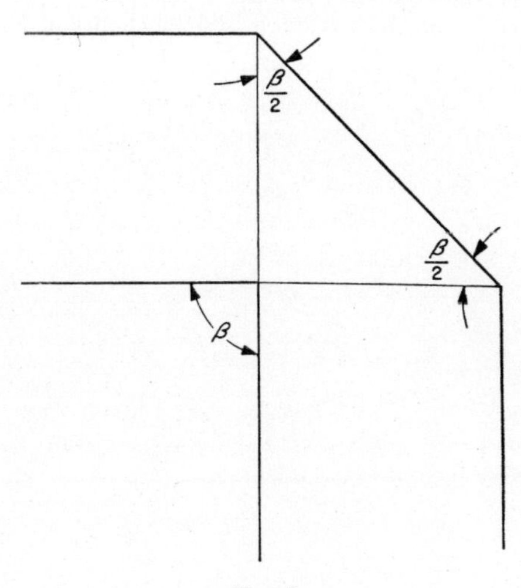

FIG. 3

Finally, it may be remarked that all of the formulas given in this section remain valid if angle β in Figs. 2 or 3 is different from $\pi/2$, i.e. in the case of a bend connecting two waveguides which are not at right angles to each other. The only variation is in the quantity L contained in μ (6), $L = b/\sin\beta$. With the same value of the parameter kb the mirror will operate less effectively for $\beta \neq \pi/2$ than for $\beta = \pi/2$, since the width of the slot which must be "jumped over" by the wave will increase. The loss will be equal to

$$\frac{C}{(kb \sin\beta)^{3/2}}, \qquad C = \frac{4\pi^{3/2}}{3} = 7,4\,. \tag{15}$$

A similar formula is valid for diffraction losses in a mirror placed in a circular waveguide with the H_{01} mode. If a is the waveguide radius, the losses computed by the method described above will be given by

$$\frac{C}{(ka \sin\beta)^{3/2}}, \qquad C = \frac{4\cdot 3{,}83^2}{9\pi^2}\cdot\Gamma^2\left(\frac{1}{4}\right) = 8,7\,. \tag{16}$$

Recently this system with $\beta = \pi/2$ has been experimentally investigated by Marcatili (1961).

REFERENCES

COMTE, G., CARFORT, F., PONTHUS, A. and PARIS, J. (1957) Utilisation de guide d'ondes circulaires pour la transmission à grand distance d'onde centimetrique et millimetrique, *Gables et Transm.* **11**, No. 4, 342.

KATSENELENBAUM, B. Z. (1961) *Theory of Irregular Waveguides with Slowly Varying Parameters*, U.S.S.R. Academy of Sciences Press, Moscow.

KATSENELENBAUM, B. Z. (1962) Diffraction at a Large Aperture in a Broad Waveguide, *Doklady Akademii Nauk S.S.S.R.*, **144**, No. 2, 322–324.

MARCATILI, E. (1961) A circular-electric hybrid junction and some channel-dropping filters, *B.S.T.J*, **40**, No. 1, 185.

SOME RECENT DEVELOPMENTS IN SCATTERING AND DIFFRACTION THEORY

ALBERT E. HEINS†

SUMMARY

THIS paper is concerned with the role of the theory of analytic functions of a complex variable in the solution of boundary-value problems in scattering and diffraction theory. Two recent developments will be discussed, both of them having been described in detail by MacCamy and Heins (1958. 1960), Benkard (1961) and the present speaker (1958, 1960).

The first development is concerned with the study of axially symmetric geometries such as, for example, the circular disk. Whether we employ symmetric excitation is immaterial since we can show that the problem may be reduced in the general case to a sequence of axially symmetric ones. In the scalar problem, that is, the one for which the wave function or its normal derivative vanishes on the disk, we discuss a method which employs the wave function (or its radial derivative) on the axis of symmetry, to produce a Fredholm integral equation of the second kind which is particularly useful for a study of the Rayleigh limit. This is accomplished by using a representation known to Poisson (Temple, 1958) in conjunction with a conventional Helmholtz representation, and leads to the possibility of calculating the Rayleigh solution to whatever accuracy one may desire by a routine procedure.

During the past year, Benkard has examined the vector problem, that is the electromagnetic case for a non-normally incident plane wave excitation. He finds that the vector problem may be decomposed into a series of problems related to those discussed by MacCamy and Heins. His results produce the various components of the electromagnetic field out to order (ak) (1958) (a = radius, k = wave number), again in the Rayleigh limit. In the two scalar problems and the vector problem, no mention or use is made of any special functions—that is spheroidal harmonics, and the final details are a standard iterative procedure.

A second development deals with a recent discussion given by MacCamy and Heins (1958) for the celebrated half-plane problem of Sommerfeld. Here essential use is made of the fact that the half-plane upon which Neumann or Dirichlet boundary conditions are supplied, is simultaneously a branch cut

† Department of Mathematics, University of Michigan, Ann Arbor, Michigan, U.S.A.

of an analytic function. This analytic function coincides with the wave function in the "aperture". Through an application of the Cauchy formula, it is possible to produce the analytic function in terms of a known function for plane wave, line or point source excitation. There is a delicate question of the growth of the analytic function which is reminiscent of a procedure employed in the method of Wiener and Hopf. It is interesting to observe that the Sommerfeld representation is by-passed going from the Cauchy type of the theory of functions of a complex variable to the familiar Fresnel integral type.

Time permitting, various ramifications of these methods will be discussed.

REFERENCES

HEINS, A. E. and MACCAMY, R. C. (1958) Axially symmetric solutions of elliptic differential equations. Technical Report No. 24 AF OSR, Contract No. AF49(638)–227, Carnegie Institute of Technology, November (to appear shortly).

HEINS, A. E. and MACCAMY, R. C. (1960) On the scattering of waves by a disk, *Zeitschrift fur Angewandte Mathematik und Physik* **11**, 249–264.

BENKARD, J. P. (1961) On an axially symmetric vector boundary-value problem with special reference to the circular disk. Dissertation, University of Michigan.

TEMPLE, G. (1958) Whittaker's work on the integral representation of harmonic functions. *Proc. Edinburgh Mathematical Soc.* **11**, 11–24.

HEINS, A. E. and MACCAMY, R. C. (1958) A function-theoretic solution of certain integral equations (I), *Quarterly Journal of Mathematics* (second series) **9**, 132–143.

REFLECTION AT THE JUNCTION OF AN INHOMOGENEOUSLY LOADED WAVEGUIDE —A QUASI-STATIC APPROACH

L. LEWIN

ABSTRACT

The problem of reflection in a waveguide, a length of which is partially loaded with dielectric, is formulated and solved to the quasi-static approximation by a method which can be further extended to more complicated configurations, e.g. the addition of a diaphragm, waveguide step, etc. The integral-equation method used can also be extended to higher-order mode approximations.

The electric field at the junction is found to not exhibit any singularities at the sharp edge of the dielectric insert.

Accurate approximate formulas for the eigenvalues in the loaded guide are obtained.

The formulas are compared numerically in a particular example with a pair of variational solutions, and the magnetic field formulation for the junction impedance is shown to be superior over the range of terminating reactances.

Very accurate results should ensue from combining the variational procedure with the integral-equation solution.

INTRODUCTION

The problem under consideration is the nature of the field, in rectangular waveguide, at the discontinuity where an empty section of guide joins a guide partially filled with a dielectric.

Propagation in such a partially filled waveguide has been treated by a number of authors[1,2], and various expressions have been obtained for the eigenfunctions and eigenvalues of mode propagation in the cases in which the dielectric extends across one or the other of the two axes determining the guide cross-section.

Angulo[3], and Collin and Vaillancourt[4], using the variational procedure and the Rayleigh–Ritz method respectively, have investigated the reflection at the junction between such unloaded and loaded waveguides. The emphasis has been on the equivalent circuit parameters rather than the field configuration, though in the case of the Rayleigh–Ritz method, using a finite, and in practice, small number of modes, some such information is implicit in the solution.

Similar methods to these can be used in other waveguide problems, e.g. reflection at diaphragms; but a much more powerful approach has been the quasi-static analysis of Schwinger[5], possibly combined with the Schwarz–Christoffel transformation. An alternative quasi-static method by Lewin[6] uses a formulation involving the solution of a singular integral equation. There is some affinity with Schwinger's analysis, but a different range of problems can be tackled. These methods have the advantage of being rigorous at the low frequency limit, and of being capable of ready extension to the higher frequencies by the retention of one or more dynamic terms in the equations. In particular, the solution has the correct form of singularity at sharp edges or corners, and can give much more accurate information on the field pattern than is obtainable from the modal solutions. Taken in conjunction with a variational formula it can then give an extremely accurate value for the circuit parameters.

The present paper is concerned with an extension of the singular integral equation method to the junction between homogeneously and inhomogeneously loaded waveguides. It is applicable to the case of a multi-layered loading consisting of any number of layers or electric constants, in either or both guides, subject to the layering being completely across either the broad or the narrow side of the guide. An inductive or capacitive insert can also be accommodated, or alternatively a two-to-one change in guide dimensions. One of the guides may be loaded with a transversely magnetised ferrite. The detailed changes necessary to include the above complications are discussed, in a different context, in ref. 6. Here we shall be concerned only with the restricted problem of formulating and solving the basic quasi-static equations for the particular case of a junction in which the loaded guide is partially filled in the H-plane by a single layer dielectric slab adjacent to one wall. The extension to other configurations will be seen to be quite straightforward, using the methods outlined in ref. 6. *Para passu*, some interesting formulas for the eigenvalues in the loaded waveguide are obtained.

1. MODAL FORMULATION

The configuration is sketched in Fig. 1. The guide, of width a, extends from $-\infty < z < 0$, region 1, and is air-filled; and from $0 < z < \infty$, region 2, it contains a dielectric slab. Region 2 is filled from $0 < y < c$, region 2a, with material of relative permittivity ε. For $c < y < a$, region 2b, the guide is empty. A wave $e^{-jk'z} \sin(\pi y/a)$ is incident from the left, and sets up a reflected wave and a train of evanescent higher-order modes,

$$ {}_1E_x = (e^{-jk'z} + R\, e^{jk'z})\sin(\pi y/a) + \sum_{2}^{\infty} R_n \exp[\Gamma_n z]\sin(n\pi y/a). \tag{1} $$

Here

$$ k' = 2\pi/\lambda_g = 2\pi(1 - \lambda^2/4a^2)^{1/2}/\lambda $$

$$ R = \text{voltage reflection coefficient} $$

$$ \Gamma_n = (n^2\pi^2/a^2 - k^2)^{1/2} \sim n\pi/a \text{ for large } n $$

$$ k = 2\pi/\lambda \text{ where } \lambda \text{ is the free-space wavelength.} $$

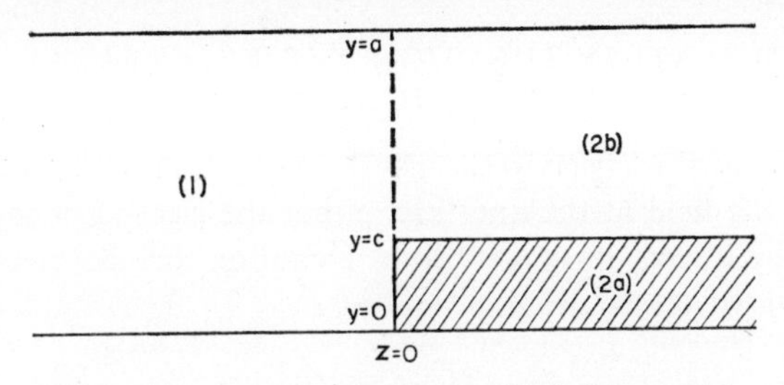

FIG. 1. Junction of loaded and unloaded waveguide.

The magnetic field in the y-direction is obtained by a differentiation. Denoting by Z_0 the impedance of free space, we get

$$kZ_0\,{}_1H_y = k'(e^{-jk'z} - R\,e^{jk'z})\sin(\pi y/a) + j\sum_2^\infty \Gamma_n R_n \exp[\Gamma_n z]\sin(n\pi y/a) \quad (2)$$

In region 2 the field takes a more complicated form. We assume that only the dominant mode can propagate, and that it is terminated by an impedance sufficiently far from the origin to not affect the higher-order modes. This is not an essential limitation, but a consideration of several propagating modes complicates the analysis unnecessarily.

The reflection coefficient of this impedance, referred in phase to the origin, will be denoted by r, so that $(1 - r)/(1 + r)$ is the relative admittance, Y, as seen at the origin, referred to the admittance of the dominant mode. The modes form a complete and orthogonal set[2] and the form of the nth mode will here be denoted by $\phi_n(\pi y/a)$, and normalized so that

$$\int_0^a \phi_n^2(\pi y/a)\,dy = \tfrac{1}{2}a. \quad (3)$$

In the absence of dielectric, $\phi_n \equiv \sin(n\pi y/a)$ and, as will be shown later, this is also the form approached by ϕ_n for high mode number in any case.

We can now write the field in region 2 as

$$_2E_x = \frac{T}{1+r}(e^{-j\gamma z} + r\,e^{j\gamma z})\phi_1(\pi y/a) + \sum_2^\infty T_n\phi_n(\pi y/a)\,e^{-\gamma_n z} \quad (4)$$

$$kZ_0\,{}_2H_y = \frac{T\gamma}{1+r}(e^{-j\gamma z} - r\,e^{j\gamma z})\phi_1(\pi y/a) - \sum_2^\infty T_n\gamma_n\phi_n(\pi y/a)\,e^{-\gamma_n z}. \quad (5)$$

Here, the γ's are the eigenvalues of the ϕ_n and, as will be shown later, $\gamma_n \sim n\pi/a$ for large n. The form of the coefficient of the first term has been written in a way to show explicitly the effect of the termination to the dominant mode. Thus T alone is the transmission coefficient which would exist in an unterminated (infinite) guide.

2. FORMULATION THROUGH THE ELECTRIC FIELD AT THE JUNCTION

In order to proceed we need to express the mode transfer coefficients R_n and T_n in terms of the field at the aperture, either the electric or magnetic field. We shall use the former in this section. Denoting this as yet undetermined field by $E(\pi y/a)$ we get, by Fourier analysis,

$$1 + R = (2/a)\int_0^a E(\pi y/a)\sin(\pi y/a)\,\mathrm{d}y\,.$$

Integrate by parts, discarding the integrated term which vanishes at the limits. Also change variable from y to θ where $\theta = \pi y/a$, so that the limits for θ are $0, \pi$. We find

$$1 + R = (2/\pi)\int_0^\pi E'(\theta)\cos\theta\,\mathrm{d}\theta\,. \tag{6}$$

Similarly

$$R_n = (2/\pi n)\int_0^\pi E'(\theta)\cos n\theta\,\mathrm{d}\theta\,. \tag{7}$$

Since the ϕ_n are orthogonal and are normalized in the same way as the sine function, a similar analysis can be applied to find the T_n. A slight complication arises because of the need to integrate by parts. We introduce the function $\bar{\phi}_n$, analogous to the cosine, by the formula

$$\int \phi_n(\theta)\,\mathrm{d}\theta = -\bar{\phi}_n(\theta)/n\,. \tag{8}$$

It satisfies the asymptotic relation $\bar{\phi}_n(\theta) \to \cos n\theta$ for large n.

In terms of this function

$$T = (2/\pi)\int_0^\pi \bar{\phi}_1(\theta)E'(\theta)\,\mathrm{d}\theta \tag{9}$$

$$T_n = (2/\pi n)\int_0^\pi \bar{\phi}_n(\theta)E'(\theta)\,\mathrm{d}\theta\,. \tag{10}$$

We now substitute these values of R_n and T_n in (2) and (4) and equate the resulting expressions in order to satisfy the requirement of equal tangential magnetic fields at the boundary. In so doing we introduce a current variable $\phi = \pi y/a$, on a par to the integration variable θ. We put, as above, $Y = (1 - r)/(1 + r)$ for the admittance of the dominant mode termination. An integral equation for $E'(\theta)$ results, in which the kernel is an infinite series in $\sin n\phi \cos n\theta$, and $\phi_n(\phi)\bar{\phi}_n(\theta)$, and involving the eigenvalues Γ_n and γ_n, and the mode number n,

$$k'(1-R)\sin\phi + j\sum_{2}^{\infty}(2\Gamma_n/\pi n)\sin n\phi\int_0^{\pi}E'(\theta)\cos n\theta\,\mathrm{d}\theta =$$

$$=(2\gamma Y/\pi)\phi_1(\phi)\int_0^{\pi}E'(\theta)\bar{\phi}_1(\theta)\,\mathrm{d}\theta - j\sum_{2}^{\infty}(2\gamma_n/\pi n)\phi_n(\phi)\int_0^{\pi}E'(\theta)\bar{\phi}_n(\theta)\,\mathrm{d}\theta. \quad (11)$$

In order to present this equation in a quasi-static form we replace Γ_n by its asymptotic value $n\pi/a$ plus a small difference term. The main-term trigonometric series can then be summed by adding and subtracting the first term of the series and using the result

$$\sum_{1}^{\infty}\sin n\phi\cos n\theta = \frac{\frac{1}{2}\sin\phi}{\cos\theta-\cos\phi}. \quad (12)$$

Similarly, the series involving $\bar{\phi}_n(\theta)\phi_n(\phi)(\gamma_n/n)$ can be replaced by a dominant term $\cos n\theta\sin n\phi(\pi/a)$ and a small difference term, bearing in mind that $\phi_n(\phi)\to\sin n\phi$ and $\gamma_n\sim n\pi/a$ for large n.

The resulting equation is

$$\sin\phi[j\tfrac{1}{2}k'a(1-R)+\pi(1+R)] - j(\gamma aY/\pi)\phi_1(\phi)\int_0^{\pi}E'(\theta)\bar{\phi}_1(\theta)\,\mathrm{d}\theta =$$

$$=\int_0^{\pi}\frac{\sin\phi E'(\theta)\,\mathrm{d}\theta}{\cos\theta-\cos\phi}+\pi S \quad (13)$$

where S is the correction to the quasi-static approximation,

$$S=\frac{1}{\pi}\sum_{2}^{\infty}\left(\frac{a\Gamma_n}{n\pi}-1\right)\int_0^{\pi}E'(\theta)\cos n\theta\sin n\phi\,\mathrm{d}\theta +$$

$$+\frac{1}{\pi}\sum_{2}^{\infty}\int_0^{\pi}E'(\theta)\left[\frac{a\gamma_n}{n\pi}\phi_n(\phi)\bar{\phi}_n(\theta)-\sin n\phi\cos n\theta\right]\mathrm{d}\theta. \quad (14)$$

We now introduce the following notation and changes of variable

$$\left.\begin{aligned}\tfrac{1}{2}jk'a(1-R)+\pi(1+R)&=\pi A\\ -j(\gamma aY/\pi)\int_0^{\pi}E'(\theta)\bar{\phi}_1(\theta)\,\mathrm{d}\theta&=\pi B\end{aligned}\right\} \quad (15)$$

are two constants (independent of ϕ).

$$x=\cos\theta \qquad y=\cos\phi \qquad E'(\theta)\,\mathrm{d}\theta=F(x)\,\mathrm{d}x \quad (16)$$

(this use of x and y as subsidiary variables will not be confused with their earlier use as coordinates. The function $F(x)$ replaces $E'(\theta)$ as unknown function).

Finally, write $f(y)=\phi_1(\phi)\operatorname{cosec}\phi$, a function which equals unity when the dielectric slab completely fills region 2, and which in any case is finite at $\phi=0,\pi$, despite the presence of the term $\operatorname{cosec}\phi$.

The integral eqn (13) becomes

$$\frac{1}{\pi}\int_{-1}^{1}\frac{F(x)\,dx}{x-y}=\frac{S(y)}{(1-y^2)^{1/2}}-A-Bf(y) \tag{17}$$

and has the solution (see ref. 6)

$$F(x)=\frac{1}{\pi}\int_{-1}^{1}\left(\frac{1-y^2}{1-x^2}\right)^{1/2}\frac{A+Bf(y)-S(y)(1-y^2)^{-1/2}}{y-x}\,dy. \tag{18}$$

The first term is immediately integrable, and (18) becomes, on returning to the original variables θ, ϕ and $E'(\theta)$,

$$E'(\theta)=A\cos\theta-\frac{B}{\pi}\int_{0}^{\pi}\frac{\phi_1(\phi)\sin\phi\,d\phi}{\cos\phi-\cos\theta}+\frac{1}{\pi}\int_{0}^{\pi}\frac{S(\phi)\sin\phi\,d\phi}{\cos\phi-\cos\theta}. \tag{19}$$

(The general solution to (18) would add an arbitrary constant to (19), but its value is zero because of the vanishing of the electric field at the guide walls.)

If we ignore the third term we have the quasi-static solution. If, in the expression for S, we retain the first few terms, then the quasi-static form becomes augmented with the correction for the first few modes. Each term retained introduces a constant, analogous to the integral for B in eqn (15). The final elimination of these constants involves determinants of order increasing with the number of correction terms retained. The increased accuracy is obtained at the cost of complication in the form of the results, but no new principle is introduced by the retention of terms from S. Since the object here is to illustrate the method rather than produce refinements in the answers, the term in S will henceforth be ignored, although it is quite straightforward to retain some early terms in (19). (An example involving the retention of such terms is given in ref. 6. See also the result given in the Appendix.)

We therefore write, for the quasi-static solution,

$$E'(\theta)=A\cos\theta-(B/\pi)\psi(\theta) \tag{20}$$

with

$$\psi(\theta)=\int_{0}^{\pi}\frac{\phi_1(\phi)\sin\phi\,d\phi}{\cos\phi-\cos\theta}. \tag{21}$$

The two terms in (20) are already a vast improvement over the simple form $E(\theta)=\sin\theta$, such as was used by Angulo[3] for his variational solution; and in fact, as will be shown, eqn (20) already leads directly to an expression in a very similar form for the junction impedance.

The term in $\psi(\theta)$ correctly accounts for the essential features of the electric field at the slab. The integral is a principal one, so the only possible singularities are at discontinuities of $\phi_1(\phi)$ or at the end points. Both ϕ_1 and its derivative are continuous throughout the range. Hence, there is no singularity arising at the sharp corner of the dielectric slab. At $\theta=0$ the denominator vanishes as ϕ^2 near $\phi = 0$, but so, too, does the numerator, hence $\psi(\theta)$ is finite throughout the range, and $E(\theta)$, from (20), is finite and continuous everywhere.

In order to proceed we substitute back for $E'(\theta)$ in (15) and (16), and eliminate A and B. The admittance at the junction becomes, after some simplification of terms,

$$\frac{1-R}{1+R} = \frac{\gamma Y}{k'}$$

$$\times \left\{ \frac{-\left(\frac{4}{\pi^3}\right) \int_0^\pi \psi(\theta) \cos\theta \, d\theta \int_0^\pi \bar{\Phi}_1(\phi) \cos\phi \, d\phi}{1 - j\frac{\gamma a Y}{\pi^3}\left[\int_0^\pi \psi(\theta)\bar{\Phi}_1(\theta)\, d\theta - \frac{2}{\pi}\int_0^\pi \psi(\theta)\cos\theta\, d\theta \int_0^\pi \bar{\Phi}_1(\phi)\cos\phi\, d\phi\right]} \right\}. \quad (22)$$

The integrals in this expression simplify considerably. We use the expansion (12) in order to obtain the following expression for $\psi(\theta)$

$$\psi(\theta) = -\pi \sum_1^\infty a_n \cos n\theta \quad (23)$$

where

$$a_n = (2/\pi) \int_0^\pi \phi_1(\phi) \sin n\phi \, d\phi \, . \quad (24)$$

Hence

$$\int_0^\pi \psi(\theta) \cos\theta \, d\theta = -\tfrac{1}{2}\pi^2 a_1 \quad (25)$$

$$\int_0^\pi \psi(\theta)\bar{\Phi}_1(\theta)\, d\theta = -\tfrac{1}{2}\pi^2 \sum_1^\infty a_n^2/n \quad (26)$$

and

$$\int_0^\pi \bar{\Phi}_1(\phi) \cos\phi \, d\phi = \tfrac{1}{2}\pi a_1 \, ,$$

the latter two on integration by parts. Substituting into (22)

$$\frac{1-R}{1+R} = \frac{\gamma Y}{k'} \frac{a_1^2}{1 + (j\gamma a Y/2\pi) \sum_2^\infty a_n^2/n} \quad (27)$$

a form very reminiscent of the variational expression obtained by using $\phi_1(\phi)$ as a trial function for the *magnetic* field. This aspect will be discussed later. Meanwhile we note that the initial factor $\gamma Y/k'$ in (27) is the expression to be expected on the simplest assumption that the empty guide is joined directly to a guide of characteristic admittance proportional to γ, and terminated in Y.

If (27) is inverted so as to give impedances rather than admittances, the equivalent circuit of Fig. 2 is derived. The turns ratio N^2 is simply $1/a_1^2$, and the normalized series reactance in the output is

$$(\gamma a/2\pi) \sum_{2}^{\infty} a_n^2/n \, .$$

These expressions will be further discussed in a later section.

Finally, integrating (20) for the electric field gives

$$E(\theta) = A \sin \theta + (B/\pi) \int_0^{\pi} \phi_1(\phi) \log \left| \frac{\sin(\frac{1}{2}\theta + \frac{1}{2}\phi)}{\sin(\frac{1}{2}\theta - \frac{1}{2}\phi)} \right| d\phi \tag{28}$$

where A and B are related by

$$B/A = - \frac{(j\gamma a Y/2\pi) a_1}{1 + (j\gamma a Y/2\pi) \sum_{1}^{\infty} a_n^2/n} \, . \tag{29}$$

Thus the form of the electric field depends on the termination to the dominant mode in the second part of the guide. In the particular case in which it is open-circuited, $Y = 0$ and hence $B = 0$. The electric field in this case (to the quasi-static approximation) is therefore the simple sinusoid of the empty guide.

This variation of the field pattern with termination impedance is quite different from the behavior of the field at, say, an inductive iris in rectangular waveguide, where the field pattern, though not its magnitude, is invariant.

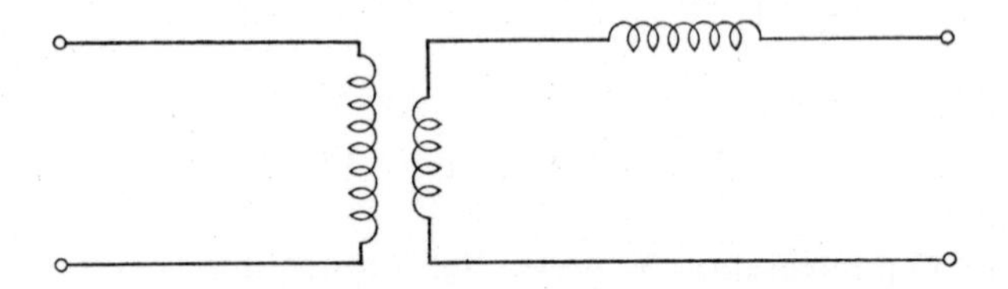

FIG. 2. Equivalent circuit corresponding to eqn (27).

3. FORMULATION THROUGH THE MAGNETIC FIELD AT THE JUNCTION

An alternative to the procedure of Section 2 is to use the magnetic field at the junction as the basic unknown in the problem. Instead of 6–10 we find (absorbing the factor kZ_0 into the symbol for magnetic field)

$$\left.\begin{aligned} k'(1-R) &= (2/\pi)\int_0^\pi H(\theta)\sin\theta\,\mathrm{d}\theta \\ jR_n\Gamma_n &= (2/\pi)\int_0^\pi H(\theta)\sin n\theta\,\mathrm{d}\theta \\ \gamma YT &= (2/\pi)\int_0^\pi H(\theta)\phi_1(\theta)\,\mathrm{d}\theta \\ -jT_n\gamma_n &= (2/\pi)\int_0^\pi H(\theta)\phi_n(\theta)\,\mathrm{d}\theta \end{aligned}\right\}. \tag{30}$$

Analogously to (11) we find the following integral equation for $H(\theta)$

$$(1+R)\sin\phi + \sum_2^\infty(-2j/\pi\Gamma_n)\sin n\phi\int_0^\pi H(\theta)\sin n\theta\,\mathrm{d}\theta =$$

$$= (2/\pi\gamma Y)\phi_1(\phi)\int_0^\pi H(\theta)\phi_1(\theta)\,\mathrm{d}\theta + \sum_2^\infty(2j/\pi\gamma_n)\phi_n(\phi)\int_0^\pi H(\theta)\phi_n(\theta)\,\mathrm{d}\theta\,. \tag{31}$$

To obtain the quasi-static equation we proceed as in Section 2, leading to

$$C\sin\phi + D\phi_1(\phi) = j(4a/\pi^2)\sum_1^\infty\int_0^\pi H(\theta)\frac{\sin n\theta\sin n\phi}{n}\,\mathrm{d}\theta + \bar{S} \tag{32}$$

where

$$\begin{aligned} C &= (1+R) + j(2k'a/\pi)(1-R) \\ D &= -(2/\pi\gamma Y)\int_0^\pi H(\theta)\phi_1(\theta)\,\mathrm{d}\theta \end{aligned} \tag{33}$$

and $\bar{S}$ is a correction term analogous to S in (13), and which will be ignored henceforth in the present treatment. In order to solve (32) for $H(\theta)$ differentiate both sides with respect to ϕ, sum the infinite series via (12), and invert the resulting integral equation as in ref. 6. The differentiation is necessary in order to get the kernel in the required form, but in the process any constant term is thereby removed. It is effectively restored on inversion of the integral equation, whose solution in fact involves an arbitrary constant. Denoting it by C' we get

$$H(\theta) = (-j/2a\sin\theta)\left\{C' + \int_0^\pi \frac{\sin^2\phi\,\mathrm{d}\phi}{\cos\phi - \cos\theta}[C\cos\phi + D\phi_1'(\phi)]\right\} \tag{34}$$

and C' is determined by the requirement that $H(\theta)$ remain finite at $\theta = 0, \pi$. This gives

$$C' = \int_0^\pi \cos\phi[C\cos\phi + D\phi_1'(\phi)]\,\mathrm{d}\phi$$

and inserting this into (34) gives finally

$$H(\theta) = (-j/2a)\left\{\pi C \sin\theta + D\sin\theta \int_0^\pi \frac{\phi_1'(\phi)\,\mathrm{d}\phi}{\cos\phi - \cos\theta}\right\}. \tag{35}$$

Using this expression in (30) and (33) enables C and D to be eliminated, resulting eventually in an expression for the junction impedance

$$\frac{1+R}{1-R} = \frac{k'}{\gamma Y}\frac{a_1^2}{1-(j\pi/2\gamma a Y)\sum_2^\infty n a_n^2} \tag{36}$$

with a_n as in (24).

This result is different in form from (27). Although both are "quasi-static" formulas they agree identically only in the extreme limit of very low frequency. For the wavelength occurs in the eigenvalue γ, and through it, into the coefficients a_n.

The numerical differences between the two results, both of which are equally "valid", is discussed in Section 6.

Equation (36) resembles closely the form obtained by the variational procedure using a trial function $\phi_1(\phi)$ for the *electric* field at the junction. This aspect will now be pursued in more detail.

4. VARIATIONAL EXPRESSIONS FOR THE JUNCTION IMPEDANCE

A variational equation for the admittance at the junction can be obtained from (11) by multiplying both sides by $E(\phi)$, integrating over the guide cross-section, and dividing by $\frac{1}{2}\pi k'(1+R)^2$. The result of eqn (6) is used to eliminate $(1+R)$, and integration by parts is used to restore E from E' in the formulas, the differential form not being required here. After some simplification we get

$$\frac{1-R}{1+R} = \frac{1}{k'\left[\int_0^\pi E(\theta)\sin\theta\,\mathrm{d}\theta\right]^2}\left\{\gamma Y\left[\int_0^\pi E(\theta)\phi_1(\theta)\,\mathrm{d}\theta\right]^2 - \right.$$

$$\left. - j\sum_2^\infty \Gamma_n\left[\int_0^\pi E(\theta)\sin n\theta\,\mathrm{d}\theta\right]^2 - j\sum_2^\infty \gamma_n\left[\int_0^\pi E(\theta)\phi_n(\theta)\,\mathrm{d}\theta\right]^2\right\}. \tag{37}$$

In this equation take $E(\theta) = \phi_1(\theta)$ as a trial function. Then the last summation disappears on account of the orthogonality of the ϕ_n and, introducing the coefficients a_n of (24), we find

$$\frac{1-R}{1+R} = \frac{\gamma Y}{k' a_1^2}\left[1-(j/\gamma Y)\sum_2^\infty \Gamma_n a_n^2\right]. \tag{38}$$

Although this expression is of variational form, it is not as it stands an absolute minimum or maximum on account of the appearance of complex terms. This can be avoided by taking Y as a pure reactance, $Y = jB$ say, and the modified expression is now an absolute minimum. If Γ_n be replaced by its asymptotic value $n\pi/a$ the bracketted form in (38) can be written

$$\left[1 - 2(\pi/2a\gamma B) \sum_2^\infty n a_n^2\right]$$

and is identical apart from the appearance of a factor 2, to the corresponding term in (36). Hence (36) departs from the possible extreme value in the permissible direction, i.e. to a greater value than the absolute minimum and *may*, therefore, give more accurate results than the particular variational formulation given above.

Of course $\phi_1(\theta)$ is not necessarily the best of the available simple forms to try. Angulo[3] preferred the mode form of the empty guide, $\sin\theta$, but no figures are available for comparison with the results from using $\phi_1(\theta)$ instead.

Two points of interest emerge. The first is that (38) is obtained from an *electric*-field variational formulation, but is compared to (36) arising from an integral-equation formulation via the *magnetic* field. Secondly the absence of the factor 2 in (36) can be traced to approximating $\phi_n(\theta)$ by $\sin n\theta$ in the kernel of the integral equation; failure to do this would give rise to the doubled term of (38). Since $\phi_1(\theta)$ is exactly orthogonal to $\phi_n(\theta)$ but only approximately so to $\sin n\theta$, an extra contribution is obtained and (36) results.

A similar comparison can be made by taking a variational formulation via the magnetic field, and the result can be compared with the integral-equation solution based on the electric field, eqn (27) In this case the variational expression is found to be

$$\frac{1-R}{1+R} = \frac{\gamma Y}{k'} \frac{a_1^2}{1 + j\gamma Y \sum_2^\infty a_n^2/\Gamma_n} \tag{39}$$

and this is an absolute maximum when Y is taken of the form jB. If the asymptotic approximation $\Gamma_n \sim n\pi/a$ is made, the denominator can then be written

$$\left[1 - 2(\gamma a B/2\pi) \sum_2^\infty a_n^2/n\right],$$

and again a factor 2 has appeared as compared to (27). Since (39) is an absolute maximum, (27), with a larger denominator, will be away from the extreme value in the direction which would permit it to be a more accurate formula. Actually the numerical discrepancy is even greater than this indicates, because of the approximation $(a/\pi)\Gamma_n \sim n$ in the above argument. This approximation is worst for $n = 2$ giving (in the case $\lambda = a\sqrt{2}$) a factor 2 instead of the correct value $\sqrt{2}$. Since the term in a_2^2 is the dominating one of the series, and Γ_n appears as a reciprocal; the relevant terms differ by a factor of nearly 3. The opposite effect appears in the comparison of (36) and (38), the difference being reduced to a factor $\sqrt{2}$.

The various formulas will be compared numerically in an example in Section 6.

5. THE EIGENVALUES

The functions ϕ_n must satisfy the equations

$$\begin{aligned} \partial^2\phi_n/\partial y^2 + (k^2\varepsilon - \gamma_n^2)\phi_n &= 0 \qquad 0 < y < c \\ \partial^2\phi_n/\partial y^2 + (k^2 - \gamma_n^2)\phi_n &= 0 \qquad c < y < a\,. \end{aligned} \tag{40}$$

Also ϕ_n and its derivative must be continuous at $y = c$ and ϕ_n must be zero at $y = 0, a$. If A is a normalizing factor then

$$\begin{aligned} \phi_n(y) &= A\frac{\sin(\alpha\pi y/a)}{\sin(\alpha\pi c/a)} \qquad 0 < y < c \\ &= A\frac{\sin[\beta\pi(1 - y/a)]}{\sin[\beta\pi(1 - c/a)]} \qquad c < y < a\,, \end{aligned} \tag{41}$$

where

$$\left.\begin{aligned} \alpha^2\pi^2/a^2 &= k^2\varepsilon + \gamma_n^2 \\ \beta^2\pi^2/a^2 &= k^2 + \gamma_n^2 \end{aligned}\right\} \tag{42}$$

and

$$\alpha\cot(\alpha\pi c/a) + \beta\cot[\beta\pi(1 - c/a)] = 0\,. \tag{43}$$

Equations (42) and (43), on elimination of α and β (which are functions of n, k and ε) determine the eigenvalues γ_n. β may be imaginary, giving rise to hyperbolic functions in (41). From (42) we see that when γ_n is large the effect of the terms $k^2\varepsilon$ and k^2 are swamped. Hence α and β will be asymptotically equal, and the eigenfunctions and eigenvalues will go over to the form appropriate to empty guide. The value of n at which this starts to happen will, of course, depend on ε and λ/a. A more quantitative form of this statement is given hereafter.

For the record the normalizing constant A is given by

$$1 = A^2\left\{\frac{c}{a}\,\frac{1 - \sin(2\alpha\pi c/a)/(2\alpha\pi c/a)}{\sin^2(\alpha\pi c/a)} + \right.$$
$$\left. + \frac{a-c}{a}\,\frac{1 - \sin[2\beta\pi(1 - c/a)]/2\beta\pi(1 - c/a)}{\sin^2[\beta\pi(1 - c/a)]}\right\}. \tag{44}$$

An extremely useful and surprisingly accurate approximation will now be derived for α, β and γ_n, and which is particularly valuable in estimating the number and cut-off positions of the higher-order modes.

We introduce a parameter p such that on eliminating γ_n^2 from (42) we get

$$\alpha^2 - \beta^2 = (\varepsilon - 1)\, 4a^2/\lambda^2 = 4p^2, \text{ say.} \tag{45}$$

We also introduce a further parameter q which measures the departure from a half-filled guide condition

$$q = 2c/a - 1\,. \tag{46}$$

If, finally we write $\alpha + \beta = 2v$, $\alpha - \beta = 2w$, eqns (45) and (43) become

$$\begin{aligned} vw &= p^2 \\ \sin[\pi(v + qp^2/v)] &= (p^2/v^2)\sin[\pi(qv + p^2/v)]\,. \end{aligned} \tag{47}$$

It can be seen immediately from the form of this equation that when v is large it takes on the asymptotic value n, where n is a large integer. This is also the exact value it takes when $p = 0$. An expansion for v would therefore be of the form

$$v = n + A_n p^2 + B_n p^4 + \ldots \tag{48}$$

where the coefficients A_n, B_n, etc. can be found by substituting into (47). An expression which constantly recurs, and is denoted by M, is given by

$$M = 1 - \cos n\pi \sin(n\pi q)/n\pi q\,. \tag{49}$$

In terms of it, it is found that

$$A_n = -qM/n$$

$$B_n = qA_n[M - \tfrac{1}{2} + \cos n\pi \cos n\pi q]/n^2 + \cos n\pi \cos(n\pi q)/n^3$$

and so on.

These values can be substituted back into (48) for v and (47) for w, and hence back to α and β. The resulting expressions are squared and terms up to $1/n^2$ retained, so that the following asymptotic expressions are found

$$\begin{aligned} \alpha^2 \sim\ & n^2 + 2p^2(1 - qM) + p^4 \\ & \times [3q^2M(1 - M) + (1 - q^2M)(1 + 2\cos n\pi \cos n\pi q)]/n^2 \\ \beta^2 \sim\ & n^2 - 2p^2(1 + qM) + p^4 \\ & \times [3q^2M(1 - M) + (1 - q^2M)(1 + 2\cos n\pi \cos n\pi q)]/n^2\,. \end{aligned} \tag{50}$$

Equation (42) now readily gives γ_n from either of these results.

6. NUMERICAL RESULTS

Although eqn (50) is derived for large n the results can in fact be used down to $n = 1$. Equation (48) gives only real values of v and hence it might be thought that $\beta = v - w$ must always appear real. However, the truncated series for β^2 in (50) can in fact go negative, yielding imaginary values† of β. The location of this change of sign of β^2 is a useful check on the accuracy of the equation, since at $\beta = 0$, $\alpha = 2p$ and (43) takes the simple form

$$\pi p \cot[\pi p(1 + q)] + 1/(1 - q) = 0 \tag{51}$$

and is readily solved for p from tables.

For example with $q = 0$ (slab half-way across guide) eqn (51) has the solution $p = 0.646$. Equation (50) with $n = 1$, $q = 0$ gives $p^4 + 2p^2 - 1 = 0$, $p = [\sqrt{(2)} - 1]^{1/2} = 0.644$ which is very close to the correct value. With $q = \frac{1}{2}$ (slab three-quarter way across guide) (51) gives $p = 0.521$ whilst (50) has the solution 0.520, again exceedingly close.

The remainder of this section will be concerned with a slab half-way across the guide ($q = 0$) and composed of material with $\varepsilon = 2$. The free space wavelength is taken to be related to a by $\lambda = a\sqrt{2}$, the center of the usual rectangular waveguide band. Then $p = \sqrt{\frac{1}{2}} = 0.707$ and the values of α and β from (50) are

$$\alpha = \sqrt{7/2} = 1.3229$$

$$\beta = j\tfrac{1}{2} \quad = j0.5\,.$$

These compare with the exact values

$$\alpha = 1.3288$$

$$\beta = j0.4840$$

obtained by numerical solution of (42) and (43).

The four formulas to be compared are the electric and magnetic field integral-equation solutions (27) and (36), and the extremal solutions (38) and (39). They each involve the same factor γ/k', so we shall concentrate on the remaining terms only, also replacing the termination Y by jB so as to exhibit explicitly the maximum and minimum nature of the variational solution.

Denoting by Y with a suitable subscript, (k'/γ) times the admittance we get for the above example

$$Y_E \quad = 0.935B/(1 - 0.0246B)$$

$$Y_H \quad = 1.070(B - 0.0446)$$

$$Y_{\max} = 0.935B/(1 - 0.0695B)$$

$$Y_{\min} = 1.070(B - 0.0605)\,.$$

† The justification for the procedure which leads to this result is that eqns (42) and (43) really determine α^2 and β^2 rather than α and β—as may be seen by series expansion of the cotangent functions—so that it is negative values of β^2 rather than imaginary values of v and w which are involved. The latter only appear as intermediate variables in the calculation, and could have been dispensed with if an alternative route, e.g. an assumed asymptotic expansion for β^2 had been chosen.

Figure 3a shows these four expressions plotted as functions of B from $B = -5$ to $+5$. It is seen that both Y_E and Y_H lie close to $Y_{\min}$ over the positive range, and that Y_E which was shown to be less than $Y_{\max}$, is in fact also slightly less than $Y_{\min}$ over part of this range. Y_H just exceeds $Y_{\max}$ by a slight amount in a short range around $B = 1$. Y_E would appear to give a better approximation for negative B, and Y_H for positive B.

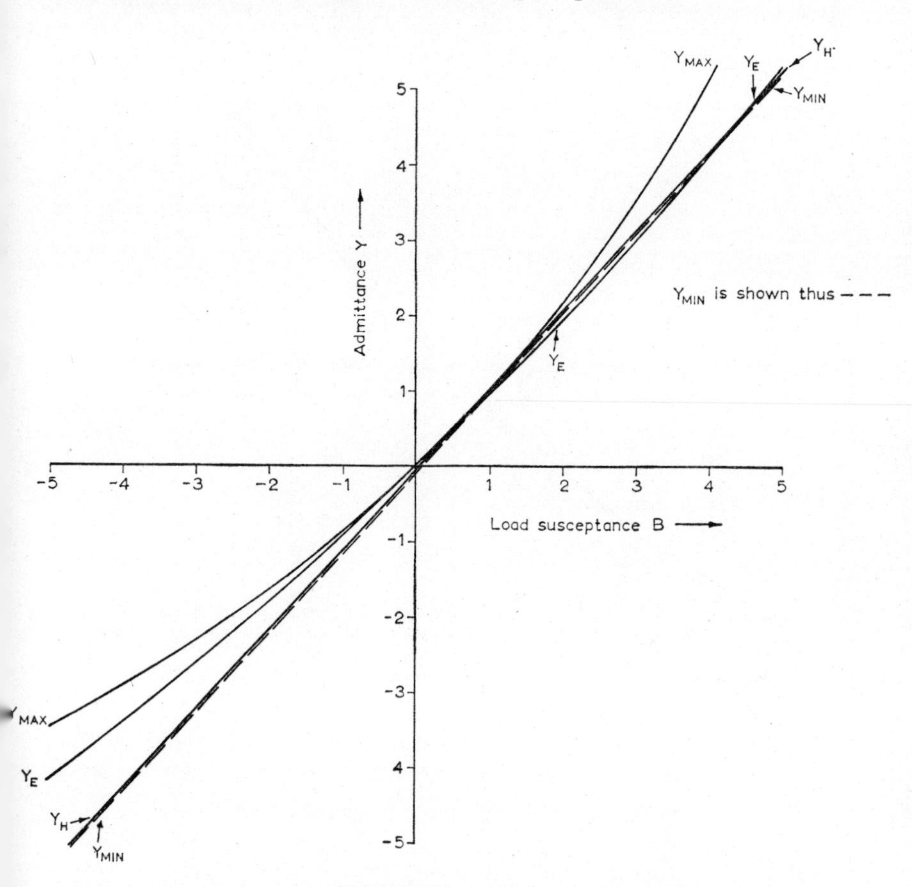

FIG. 3a. Plot of input admittance vs. termination susceptance.

On account of the resonances in the denominator of $Y_{\max}$ and Y_E the plot as an admittance is unsuitable for large values of B, and Fig. 3b shows a plot of $Z = 1/Y$ against $X = 1/B$ from $X = -5$ to $+5$. The range of B and X from 0.2 to 5 is common to both graphs. $Y_{\min}$, of course, has now become $Z_{\max}$. From the graph it would appear that Z_H is a good approximation over the whole region, despite the fact that it just falls below $Z_{\min}$ in a range around $X = 1$ where all the curves are closely bunched together.

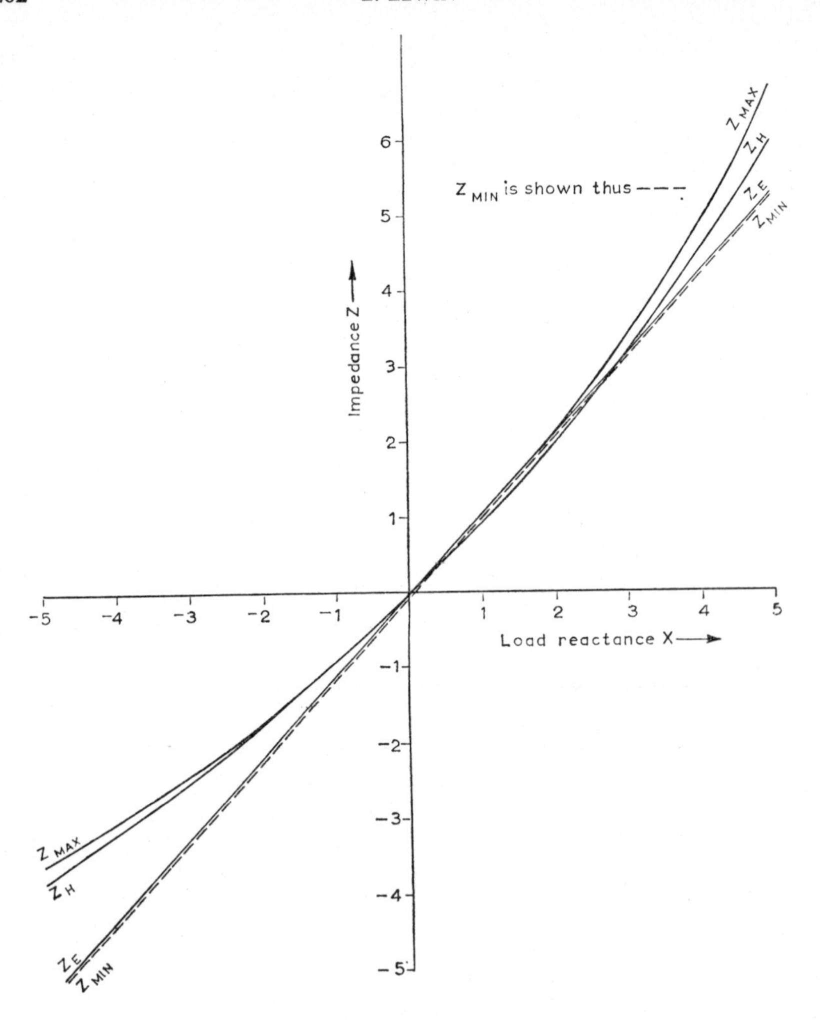

FIG. 3b. Plot of input impedance vs. termination reactance.

CONCLUSIONS

The first order quasi-static solution has been obtained for a simple configuration by a method which can be extended to more complicated arrangements or higher-order solutions. In the case of the example, involving changes across the *H*-plane, the magnetic field formulation appears to yield a result superior both to the electric field formulation and the two simple variational solutions, so far as the impedance is concerned.

The electromagnetic field is free from singularities at sharp dielectric edges, but its form at the junction is found to be a function of the guide termination.

The combination of the integral-equation solution with the variational procedure should give an extremely accurate value for the circuit parameters.

REFERENCES

1. CHAMBERS, L. G. 1953. Propagation in waveguides filled longitudinally with two or more dielectrics, *Brit. J. Appl. Phys.* **4**, 39–45, Feb.
2. COLLIN, R. E. 1960. *Field Theory of Guided Waves*, pp. 224–256. McGraw-Hill New York, N.Y.
3. ANGULO, C. M. 1957. Discontinuities in a rectangular waveguide partially filled with dielectric, *I.R.E. Trans. M.T.T.*, 68–74, Jan.
4. COLLIN, R. E. and VAILLANCOURT, R. M. 1957. Application of the Rayleigh–Ritz method to dielectric steps in waveguides, *I.R.E. Trans. M.T.T.*, 177–84, July.
5. J. Schwinger's work is reported in N. Marcuvitz *Waveguide Handbook*, p. 147. M.I.T. Rad. Lab. Ser., McGraw-Hill, New York, N.Y. 1951.
6. LEWIN, L. 1961. On the resolution of a class of waveguide discontinuity problems by the use of singular integral equations, *I.R.E. Trans. M.T.T.*, 321–32, July.

APPENDIX

A second-order formula for the admittance can be obtained from the integral-equation (13) by retaining terms in Γ_2, γ_2 and ϕ_2, and making the previous quasi-static approximation in the higher-order terms. The result will be quoted here only for the electric field formulation. Defining the following quantities

$$a_n = \frac{2}{\pi}\int_0^{\pi} \phi_1(\theta)\sin n\theta \, d\theta$$

$$b_n = \frac{2}{\pi}\int_0^{\pi} \phi_2(\theta)\sin n\theta \, d\theta$$

$$A = \frac{1}{\gamma Y} + j\left[\frac{a_2^2}{\Gamma_2} + \frac{a}{2\pi}\sum_3^{\infty} a_n^2/n\right]$$

$$B = -j2\left[\frac{a_2 b_2}{\Gamma_2} + \frac{a}{2\pi}\sum_3^{\infty} a_n b_n/n\right]$$

$$C = 4j\left[\frac{1}{\gamma_2} + \frac{b_2^2}{\Gamma_2} + \frac{a}{2\pi}\sum_3^{\infty} b_n^2/n\right]$$

then

$$\frac{1-R}{1+R} = \frac{1}{k'}\,\frac{Ca_1^2 + 2Ba_1b_1 + Ab_1^2}{AC - B^2}.$$

This reduces to (27) when $\Gamma_2 = \gamma_2 = 2\pi/a$ and $b_n = 0$ except for $b_2 = 1$.

Functionally it can be put in the form

$$\frac{1-R}{1+R} = \frac{1}{k'}\,\frac{N^2\gamma Y - jp}{1 + jq\gamma Y} \text{ with } N^2, p \text{ and } q \text{ independent of } Y,$$

so it combines both the series and parallel representation of the previous treatments.

SYSTEMATIC IMPROVEMENT OF QUASISTATIC CALCULATIONS

SAMUEL N. KARP

Courant Institute of Mathematical Sciences, New York University, New York, U.S.A.

SUMMARY

In the theory of diffraction, or more generally, wave propagation in variable media, the boundary-value problems may be solved explicitly in a limited number of special cases, by the method of separation of variables. (Also certain two-part boundary-value problems can be solved by Wiener–Hopf methods or related but more general methods.) The solutions given by the separation method are usually not rapidly convergent at high frequency and a great deal of effort has been expended in order to obtain their rigorous high frequency expressions. A geometrical theory of diffraction pioneered by Keller(1) which leans on some of the results, has shown itself valid in predicting other such results which were obtained by asymptotic expansion of the rigorous solutions. That theory therefore offers a highly reliable tool for the general class of problems which can't be treated at all by the separation method or other exact methods. Thus, the situation at high frequencies is quite satisfactory. Experience in special cases shows that the high frequency results are good even at moderately low frequency, e.g. in diffraction by a circular cylinder they almost agree with exact results for wavelengths of the order of magnitude of the perimeter.

On the other hand, another type of expansion is one based on a power series in inverse powers of wavelength, or more generally, on methods based on perturbations about the electrostatic or magnetostatic regimes, i.e. based on the solution of Laplace's equation. It is well known that such quasi static series have a finite radius of convergence, which is usually too small to overlap the high frequency region. Although the gap is easily bridged, when separability exists, there are few separable cases, and in these cases the low frequency method is unnecessary.

As to the virtues of the low-frequency method, they can be said to be the following: (a) There are many more separable problems. (b) There are many more problems explicitly solvable by other means, e.g. one can obtain the Green's function for a finite conical shell. (c) Conformal mapping methods are available in two dimensions, and correspondingly, source sink methods in three dimensions. (d) Excellent variational methods are available which give upper and lower bounds in the case of problems that can't be solved explicitly.

We remarked earlier that the bridge between low frequency and high frequency methods is the eigenfunction method. Another and closely related method, which is always available even for non-separable geometries, is the integral equation method. This method, which has been studied by Müller[(2)], has been shown by him to lead to existence theorems, but it has unfortunately not been exploited for the purpose of obtaining interpretable results. It is interesting that the eigenfunctions of the integral equation, in the case of diffraction by a smooth conducting obstacle, can be extended to the interior and exterior region of the obstacle. They provide analogues of the separation of variables solution. The extended functions can be characterized as solutions of the wave equation, regular and outgoing in the exterior of the obstacle, and regular in the interior also, and obeying at the boundary an eigenvalue-jump-condition of the type

$$\left[\frac{\partial U_n(P)}{\partial \nu}\right] = \lambda_n U_n(P) .$$

The λ_n are the eigenvalues of the integral equation. For example, for a circular boundary of radius a, the functions $U_n(P)$ are given by

$$U_n(P)\begin{cases} = J_n(kr)H_n^{(1)}(ka)\, \mathrm{e}^{\pm in\Theta} & r < a\,, \\ = J_n(ka)H_n^{(1)}(kr)\, \mathrm{e}^{\pm in\Theta} & r > a\,, \end{cases}$$

and

$$\lambda_n = (J_n(ka)H_n(ka))^{-1} .$$

In general, the solution of the boundary integral equation, and also, the eigenfunction expansion which may be derived therefrom, can probably be expected, at high frequency, to converge at a similar rate to that obtaining in the case of the circle, and probably deserves similar study. On the other hand, the Fredholm solution of the usual integral equation also fails to be useful at low frequency, since it obviously leads to the Fredholm solution for the electrostatic problem (which is not a closed form) even when a closed form is known on other grounds.

One purpose of the present paper is to explain a simple method for obtaining a bridge between static and high frequency methods. The method is conveniently based on an integral equation formulation. Briefly expressed, the idea is that one obtains an underlying integral equation for the problem at hand, for example, the Green's function integral equation. But then, use of the static Green's function (or static Neumann's function, or tensor Green's function, as the case may require) allows one to convert the original equation to a new Fredholm equation of the second kind. Of course, it is clear that by iteration of this Fredholm equation a series is generated which is rapidly convergent for low frequency and this leads to the systematic equivalent of the usual power series solution (or, in some two dimensional problems, to a series in powers of k and $\log k$). But, of course, a power series solution ceases to converge as the frequency is increased. The next step is to construct the solution by the Fredholm method, as a quotient of two series, the denominator being a function of the frequency parameter alone. This form of solution is always convergent for real k, the zeros of the denominator at complex resonant frequencies being responsible for the divergence of the power series. It is important to note that the successive terms in numerator and denominator can be obtained *iteratively*. It turns out to be possible to express them in terms of the quantities occurring in the usual Neumann series. When power series are introduced, then we find simple expressions for the coefficients of the numerator and denominator series in terms of the coefficients of the power series usually obtained by application of the static or quasi-static method. This means that given any number of terms of the low frequency expansion, one can construct therefrom a sequence of fractions which converges at all frequencies, while continuing to be asymptotic at low frequencies. The denominator allows for the analytical expression of resonances, which may be understood to occur whenever a complex root approaches fairly close to the real axis. This technique is actually not restricted to static approximations, and is also of use in multiple scattering, for example. But the formulation described above shows how one may conveniently generate all the terms of the low frequency expansion and, at the same time, it shows how to *convert* that expansion, which can be obtained by other methods, into a convergent expansion. In the case of diffraction theory, it is of interest to recall that the first few elements of this new static expression can be interpreted in terms of electrical properties of the obstacle, i.e. such quantities as capacitance and polarizibility. Even when one cannot obtain explicit solutions, the use of variational methods for harmonic functions is known to give good results, including upper and lower bounds, for such quantities.

ACKNOWLEDGEMENTS

The research reported in this document has been sponsored by the Electronics Research Directorate of the Air Force Cambridge Research Laboratories, Office of Aerospace Research (USAF) Bedford, Mass. under Contract No. AF 19(604)5238.

REFERENCES

1. KELLER, J. B. (1956) Diffraction by a convex cylinder, *Trans. I.R.E.* **AP-4**, 312.
2. MÜLLER, CLAUS, *Grundprobleme der Mathematischen Theorie Elektromagnetischer Schwingungen*, Berlin 1957.

REFLECTION AT INCIDENCE OF AN H_{mn}-WAVE AT JUNCTION OF CIRCULAR WAVEGUIDE AND CONICAL HORN

GERHARD PIEFKE

Central Labs., Siemens & Halske AG, Munich, Germany.

ABSTRACT

Formulas are derived for calculating the reflection and transmission coefficients with incidence of an H_{mn}-mode from a circular waveguide to an adjoining conical horn of infinite length. Curves show in particular the reflection coefficient for horn aperture angles up to 90° and incidence of an H_{11}-mode, if it is the only mode capable of propagation in the waveguide. With these curves the magnitude of the reflection coefficient decreases with increasing frequency initially very rapidly and remains thereafter approximately constant. With increasing aperture angle the magnitude increases. The phase is π at the cutoff frequency of the H_{11}-mode, decreases thereafter with increasing frequency, increases subsequently beyond π and finally even beyond $3\pi/2$. The equivalent circuit of the horn is accordingly a parallel connection of resistance, inductance, and capacitance. With incidence of an H_{11}-mode and very small horn aperture angles, the reflection of the H_{11}- and E_{11}-modes as well as the transmission of the E_{11}-mode are proportional to the horn aperture angle to a first approximation.

LIST OF PRINCIPAL SYMBOLS

List of principal symbols

ρ, ϕ, z cylinder coordinates

r, θ, ϕ spherical coordinates

a inner radius of the waveguide

r_0 place of the interface between waveguide and horn with spherical coordinates (see Fig. 2)

z_0 place of the interface between waveguide and horn with cylindrical coordinates (see Fig. 2)

ρ_0 half the inner diameter of the horn (see Fig. 2)

θ_0 half the aperture angle of the horn

μ_0 permeability of space

ε_0 permittivity of space

$Z_0 = \sqrt{(\mu_0/\varepsilon_0)}$ field impedance of space

$\beta_0 = 2\pi/\lambda_0$ phase constant of a planar wave in space

λ_0 wavelength of a planar wave in space

λ_c cutoff wavelength of the H_{11}-mode in the circular waveguide
J_M Bessel function of order M
J'_M Derivation of the Bessel function with respect to the argument
$H_\eta^{(2)}$ Hankel function of second kind and order η
$H_\zeta^{(2)}$ Hankel function of second kind and order ζ
$y_{nH} = n$th, $y_{pH} = p$th — not disappearing root of $J'_M = 0$; $n,p = 1,2,3, \ldots$
$y_{nE} = n$th, $y_{pE} = p$th — not disappearing root of $J_M = 0$; $n,p = 1,2,3, \ldots$

$$\beta_n = \begin{cases} \beta_0 \sqrt{\left[1 - \left(\dfrac{y_n}{\beta_0 \rho_0}\right)^2\right]} & \text{for } y_n < \beta_0 \rho_0 , \\ -j\beta_0 \sqrt{\left[\left(\dfrac{y_n}{\beta_0 \rho_0}\right)^2 - 1\right]} & \text{for } y_n > \beta_0 \rho_0 , \quad n = 1,2,3, \ldots \end{cases}$$

β_{nH} results from β_n, if $y_n = y_{nH}$ and $\rho_0 = a$,
β_{nE} results from β_n, if $y_n = y_{nE}$ and $\rho_0 = a$.
$Z_{nH} = Z_0 \beta_0 / \beta_{nH}$ field impedance of the H_{Mn}-mode in the waveguide ($n = 1,2,3, \ldots$)
$Z_{nE} = Z_0 \beta_{nE} / \beta_0$ field impedance of the E_{Mn}-mode in the waveguide ($n = 1,2,3, \ldots$)
$Z_{pH}^{(2)} = Z_{pH}^{(2)}(r)$ field impedance of the H_{Mp}-mode propagating in the direction of increasing r in the horn at the place r ($p = 1,2,3, \ldots$)
$Z_{pE}^{(2)} = Z_{pE}^{(2)}(r)$ field impedance of the E_{Mp}-mode propagating in the direction of increasing r in the horn at the place r ($p = 1,2,3, \ldots$)
r_{nH} reflection coefficients of the H_{Mn}-modes ($n = 1,2,3, \ldots$)
r_{nE} reflection coefficients of the E_{Mn}-modes ($n = 1,2,3, \ldots$)
$r_{nH0} = r_{nH} \exp j2\beta_{nH} r_0 \cos \theta_0$ reflection coefficients of the H_{Mn}-modes at the point $z = z_0$ ($n = 1,2,3, \ldots$)
$r_{nE0} = r_{nE} \exp j2\beta_{nE} r_0 \cos \theta_0$ reflection coefficients of the E_{Mn}-modes at the point $z = z_0$ ($n = 1,2,3, \ldots$)
$t_{pH} = \sqrt{(2/\pi)} d_{pH}$ transmission coefficients of the H_{Mp}-modes ($p = 1,2,3, \ldots$)
$t_{pE} = \sqrt{(2/\pi)} d_{pE}$ transmission coefficients of the E_{Mp}-modes ($p = 1,2,3, \ldots$)

1. INTRODUCTION. MATHEMATICAL METHOD

This paper gives a general basis, sufficiently accurate for all practical purposes, for the calculation of the reflection and transmission coefficients with the incidence of an *arbitrary* H_{mn}-mode, from a circular waveguide to a rotationally symmetrical flare-out of the latter. For this purpose equations and formulas, which also hold for large horn aperture angles, are derived for the reflection and transmission coefficients in the case of incidence of an H_{mn}-mode with a given $m = M$ and $n = N$ from a cylindrical waveguide to a conical horn of infinite length (Fig. 1).

Because of the rotationally symmetrical guide configuration (Figs. 1 and 2) only H_{Mn}- and E_{Mn}-modes ($n = 1,2,3, \dots$) are excited at the junction of

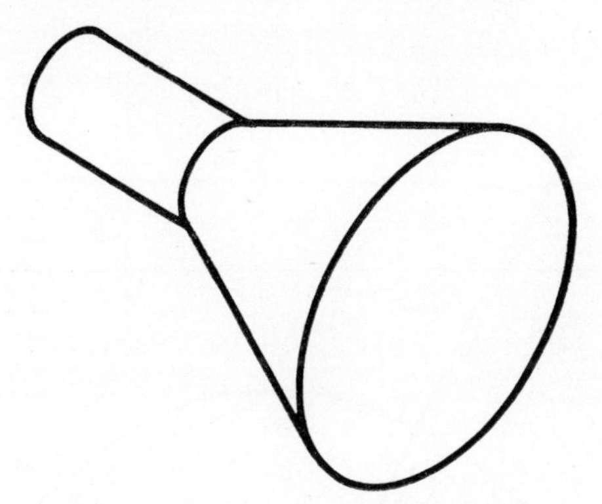

FIG. 1. Junction of circular waveguide and conical horn.

waveguide and horn. The modes in the conical horn are spherical. Unlike waveguide modes they have no cutoff frequency. In the far field the phase constant of the modes in the conical horn is identical with that of a planar wave in space.

For simplified calculation $\tan\,\theta = \sin\,\theta = \theta$ is put in the differential equation of the spherical function. This leads to Bessel functions instead of spherical functions. For the H_{11}-mode the difference between the spherical function and the Bessel function is less than 1 per cent and hence negligible

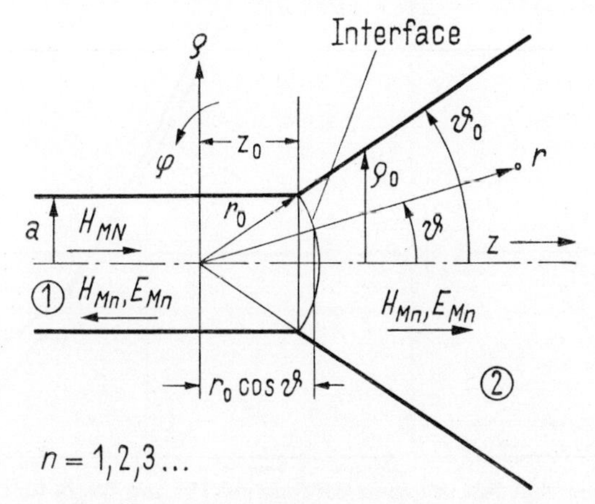

FIG. 2. Coordinates and modes with the incidence of an H_{MN}-mode on the junction of a circular waveguide and a conical horn.

even with a horn aperture angle of 180° (junction of the waveguide and a plane shield). For the other modes there results as a rule a difference as in the Figs. 3 and 4 where for a horn aperture angle of 180° the spherical and Bessel functions corresponding to the E_{11}- and H_{12}-modes are plotted as well as their derivatives. Up to a horn aperture angle of 120° the mean deviation

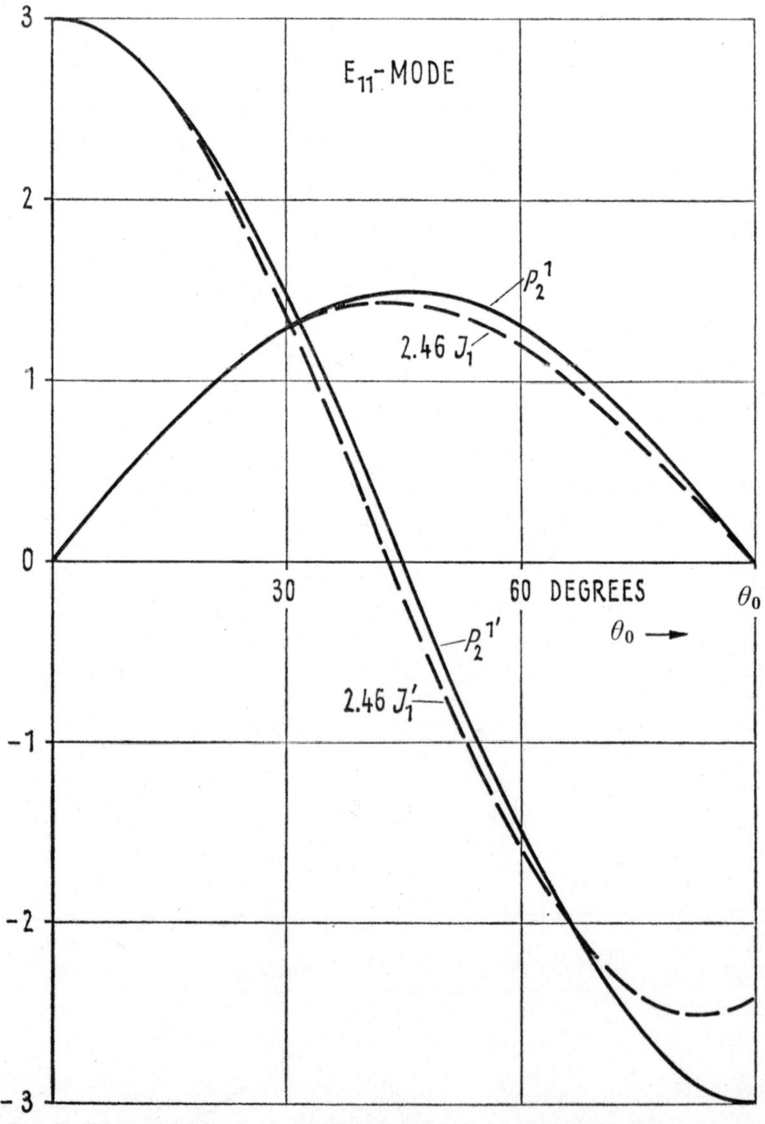

FIG. 3. Comparison between spherical function P_2^1 and Bessel function J_1 and their derivatives for the E_{11}-mode with a half-aperture angle $\theta_0 = 90°$ of the conical horn.

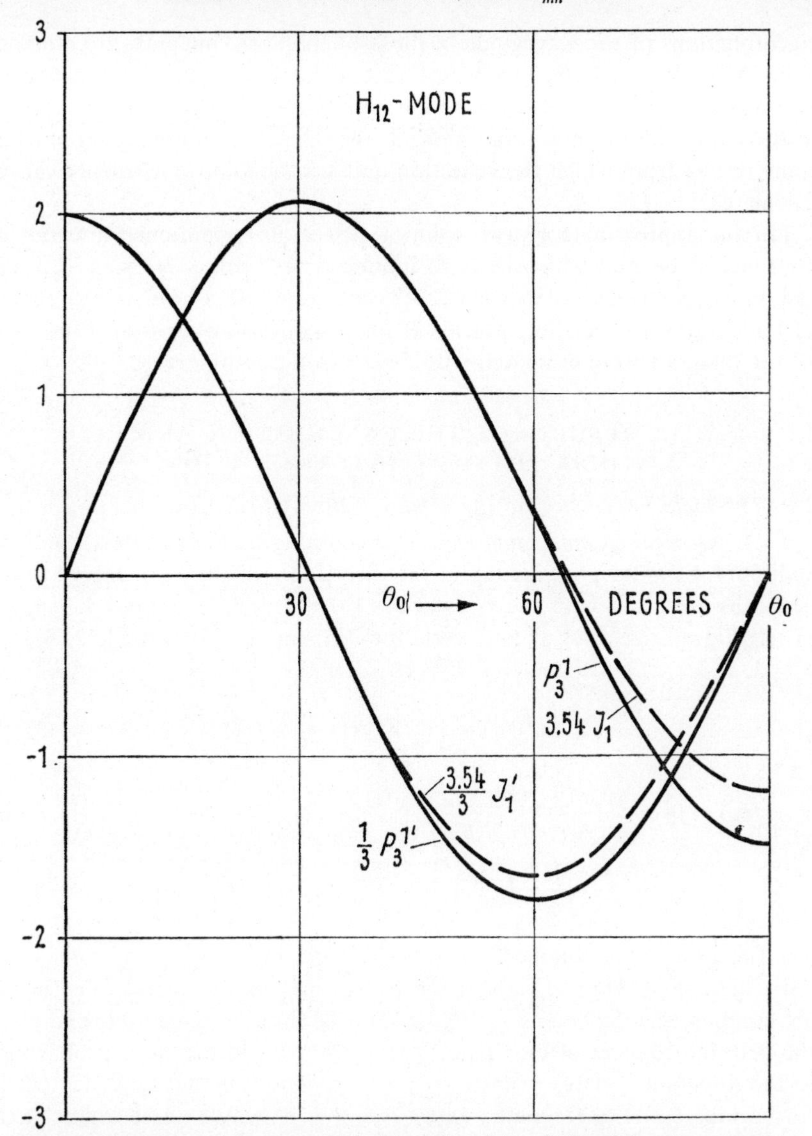

FIG. 4. Comparison between spherical function $P_3{}^1$ and Bessel function J_1 and their derivatives for the H_{12}-mode with a half-aperture angle $\theta_0 = 90°$ of the conical horn.

of the Bessel function from the spherical function as referred to the maximum of the latter is less than 2 per cent.

Because of $\tan\theta = \sin\theta = \theta$ it is sufficient to take either the plane $z = z_0$ or the spherical calotte (Fig. 2) as interface between waveguide and horn. Here the calotte has been taken as interface. In that case, expanding the eigenfunctions of the waveguide in those of the horn, integrals are obtained whose integrand includes as a factor an exponential function, not a Hankel function as would have been the case with use of a plane interface. After introducing the continuity conditions at the interface, an infinite set of equations results from which the reflection and transmission coefficients can be calculated.

For the approximate general solution of the aforementioned integrals an approach is devised which leads to infinite series with powers of θ_0. The resultant error is again less than 2 per cent even with a horn aperture angle of 120°. For the special numerical calculation of the curves in the Figs. 11 to 14 the integrals were evaluated with an electronic computer.

2. RESULTS OF THE CALCULATION AND THEIR PHYSICAL INTERPRETATION

2.1. *The Field Impedances of the Modes in the Conical Horn*

2.1.1. *General formulas and curves.* According to Piefke (1954) and (35) and (36) for modes propagating in the direction r in the horn the wave impedances are given with the H_{Mp}-modes ($M = 0, 1, 2, 3, \ldots ; p = 1, 2, 3, \ldots$) by $Z_{pH}^{(2)}(r) = -E_{\phi H}^{(2)}/H_{\theta H}^{(2)}$ and with the E_{Mp}-modes ($M = 0, 1, 2, 3, \ldots ;$ $p = 1, 2, 3, \ldots$) by $Z_{pE}^{(2)}(r) = -E_{\phi E}^{(2)}/H_{\theta H}^{(2)}$, i.e.

$$Z_{pH}^{(2)}(r) = \frac{Z_0}{f(\eta)}, \quad Z_{pE}^{(2)}(r) = Z_0 f(\zeta), \tag{1), (2}$$

$$f(\eta) = j\,\frac{\mathrm{d}/[\mathrm{d}(\beta_0 r)][\sqrt{(\beta_0 r)}H_\eta^{(2)}(\beta_0 r)]}{\sqrt{(\beta_0 r)}H_\eta^{(2)}(\beta_0 r)} = -j\left[\frac{\eta - \frac{1}{2}}{\beta_0 r} - \frac{H_{\eta-1}^{(2)}(\beta_0 r)}{H_\eta^{(2)}(\beta_0 r)}\right], \tag{3}$$

$$\eta = \sqrt{\left[\frac{1}{4} + \left(\frac{y_{pH}}{\theta_0}\right)^2\right]}, \quad \zeta = \sqrt{\left[\frac{1}{4} + \left(\frac{y_{pE}}{\theta_0}\right)^2\right]}. \tag{4), (5}$$

The function $f(\zeta)$ is obtained from $f(\eta)$ by replacing η by ζ.

Because of the Hankel functions the wave impedances of the horn modes are complex and depend on r. Figures 5 to 10 show by magnitude and phase the field impedances of the H_{11}-, E_{11}-, H_{12}-modes in the horn propagating in the direction r at the point $r = r_0$ as a function of the frequency (λ_c/λ_0) for various θ_0. The frequency range is $1 \leqq \lambda_c/\lambda_0 \leqq 2.1$, i.e. it begins at the cutoff frequency of the H_{11}-mode in the waveguide and ends at that of the E_{11}-mode in the waveguide. With the H_{mp}-modes the equivalent circuit of the wave impedance is a parallel connection of resistance and inductance, and with the E_{mp}-modes one of resistance and capacitance. In the far field ($r \to \infty$) all wave impedances change into those of space.

2.1.2. *Approximations for small horn aperture angles.* Let us assume $\sin\theta_0 \approx \theta_0$ and hence $\beta_0 r \approx \beta_0 \rho_0/\theta_0$ and besides

$$\eta, \zeta, \beta_0 r \gg 1 \tag{6}$$

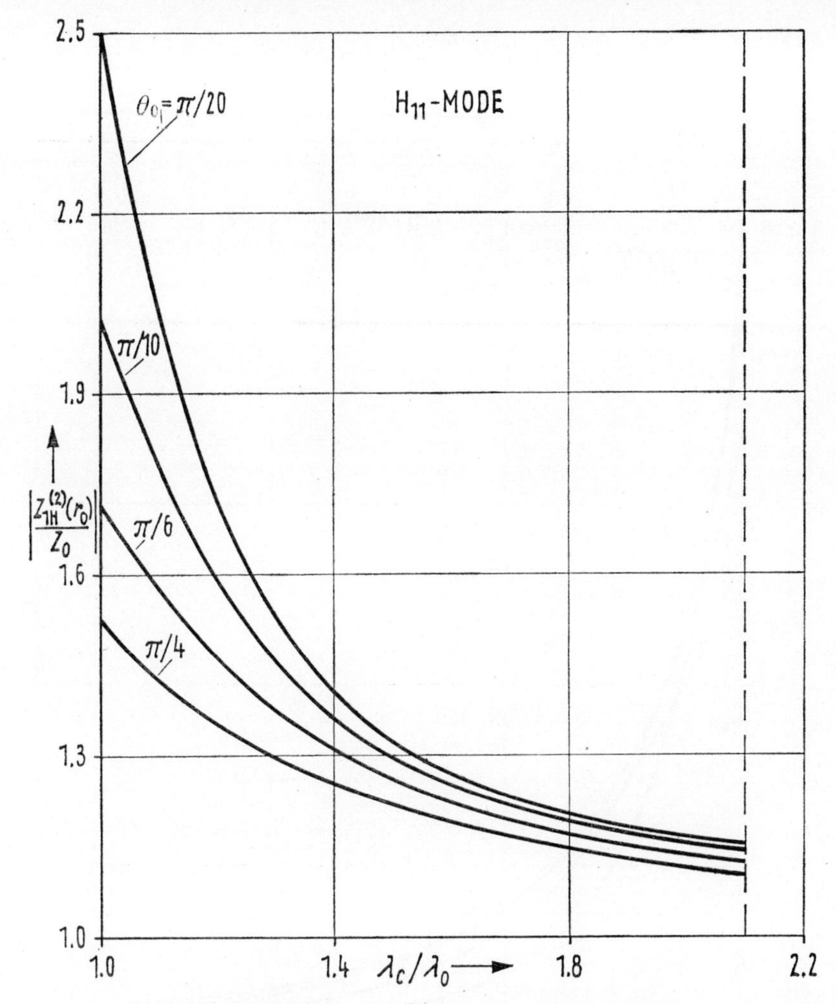

FIG. 5. Magnitude of $Z^{(2)}{}_{1H}(r_0)/Z_0$ as a function of λ_c/λ_0.

and thus according to (4) and (5) $\eta \approx y_{pH}/\theta_0$ and $\zeta \approx y_{pE}/\theta_0$. If the Hankel functions in (3) are replaced by the first terms of the Debye series (Magnus, Oberhettinger, 1948) the following approximations result for the wave impedances in the horn:

$$\text{(1)}\quad \left.\begin{matrix} \dfrac{Z_{pH}(\rho_0)}{Z_{pH}^{(2)}(r)} \\[2ex] \dfrac{Z_{pE}^{(2)}(r)}{Z_{pE}(\rho_0)} \end{matrix}\right\} = 1 + \frac{(\beta_0\rho_0)^4}{8(\beta_p\rho_0)^6}\left[1 + 4\left(\frac{y_p}{\beta_0\rho_0}\right)^2\right]\theta_0^2 - j\,\frac{y_p^2}{2(\beta_p\rho_0)^3}\,\theta_0\,, \tag{7}$$

if
$$y_p < \beta_0\rho_0\,, \quad \frac{\beta_0\rho_0}{y_p} - 1 > 3\left(\frac{\theta_0}{y_p}\right)^{2/3}. \qquad (8), (9)$$

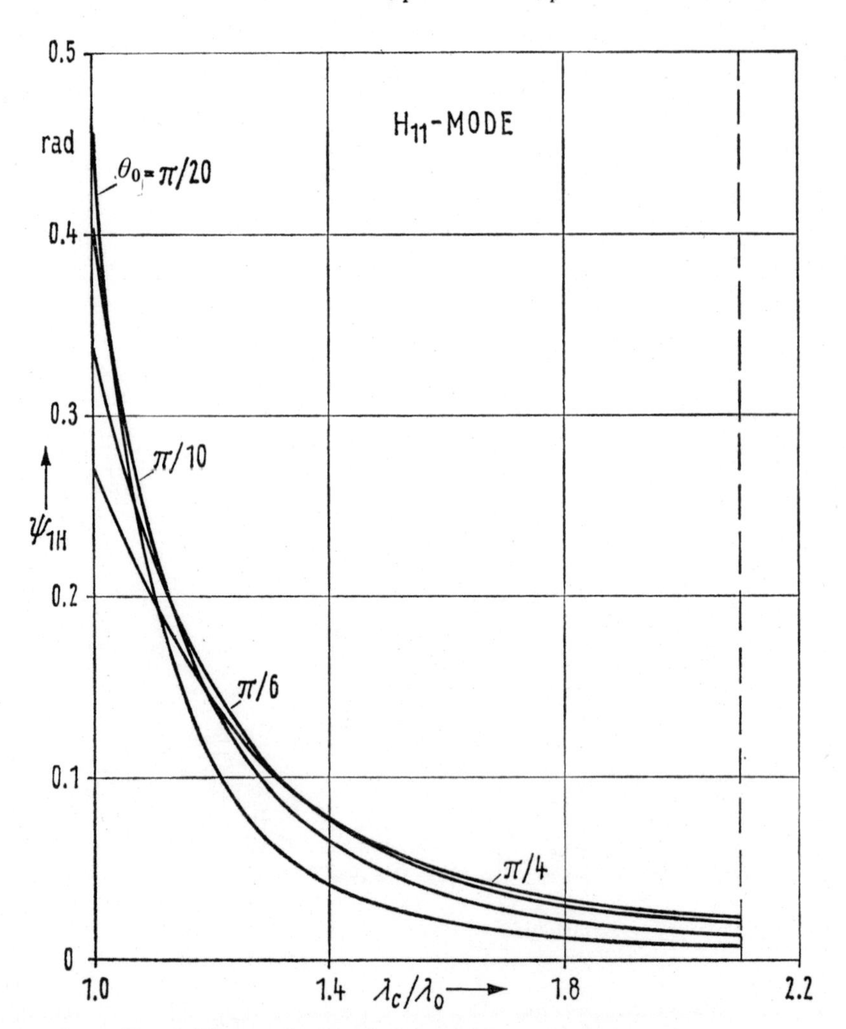

FIG. 6. Phase angle ψ_{1H} of $Z^{(2)}{}_{1H}(r_0)/Z_0$ as a function of λ_c/λ_0.

(2)
$$Z^{(2)}_{pH}(r) = Z_{pH}(\rho_0)\,, \quad Z^{(2)}_{pE}(r) = Z_{pE}(\rho_0)\,, \qquad (10), (11)$$

if
$$y_p > \beta_0\rho_0\,, \quad \left.\begin{matrix} |y_p^2 - (\beta_0\rho_0)^2|^{1/2} \\ |y_p^2 - (\beta_0\rho_0)^2|^{3/2} y_p^{-2} \end{matrix}\right\} \gg \theta_0\,. \qquad (12), (13)$$

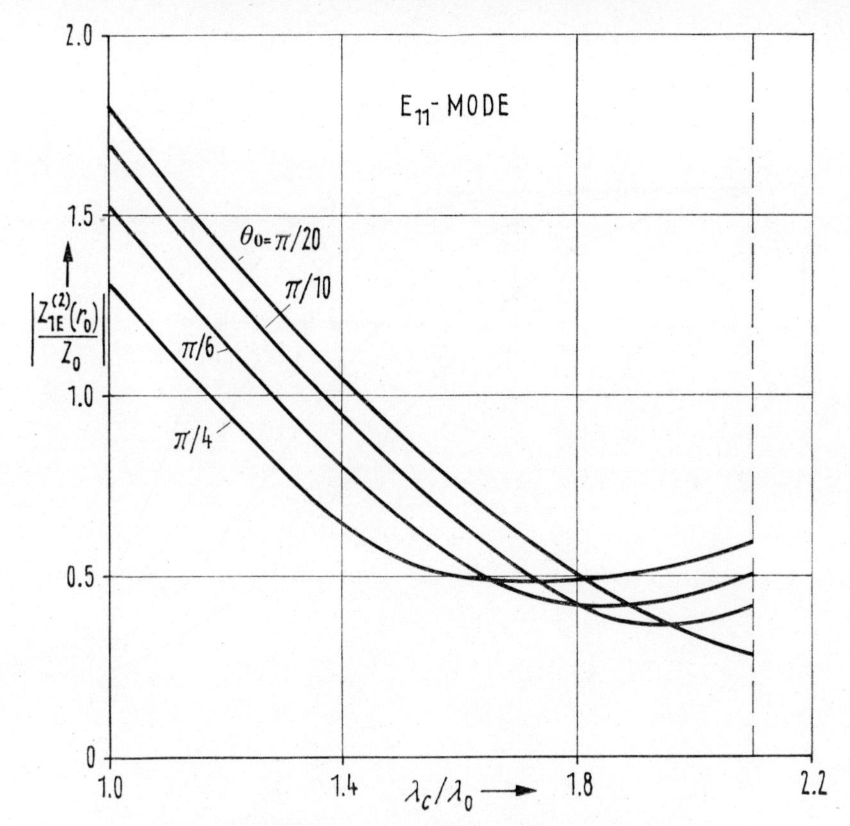

FIG. 7. Magnitude of $Z^{(2)}{}_{1E}(r_0)/Z_0$ as a function of λ_c/λ_0.

(3)
$$\left.\begin{aligned} &\frac{Z^{(2)}_{pH}(r)}{Z_0} \\ &\frac{Z_0}{Z^{(2)}_{pE}(r)} \end{aligned}\right\} = 1.089\sqrt[3]{\left(\frac{\beta_0\rho_0}{\theta_0}\right)}\,e^{j\pi/6}\,, \tag{14}$$

if
$$y_p \approx \beta_0\rho_0\,, \quad |y_p - \beta_0\rho_0| \ll \theta_0\sqrt[3]{\left(\frac{\beta_0\rho_0}{\theta_0}\right)}\,. \qquad (15), (16)$$

$Z_{pH}(\rho_0)$ results from Z_{nH} by putting $n = p$ and $a = \rho_0$. $Z_{pE}(\rho_0)$ is found from Z_{nE} by putting $n = p$ and $a = \rho_0$. β_p results from β_n, by putting $n = p$. In calculating $Z^{(2)}_{pH}(r)$ one puts $y_p = y_{pH}$, and in calculating $Z^{(2)}_{pE}(r)$ one puts $y_p = y_{pE}$. Equation (7) corresponds to the wave impedance in the guide with propagating waves, (10) and (11) to that with statically damped fields, (14) to that at the cutoff frequency.

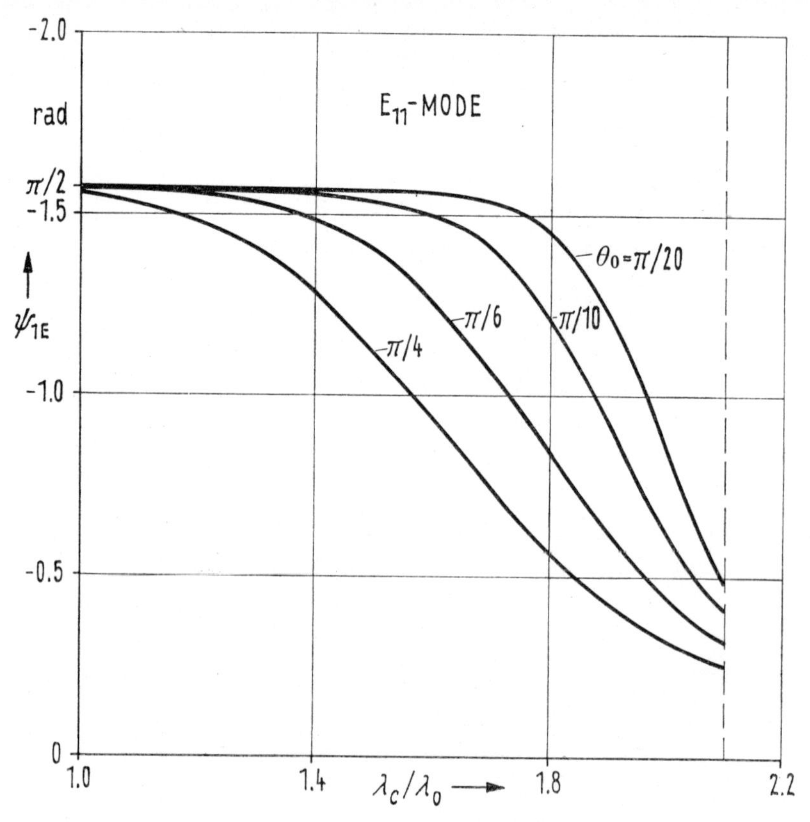

FIG. 8. Phase angle ψ_{1E} of $Z^{(2)}{}_{1E}(r_0)/Z_0$ as a function of λ_c/λ_0.

2.2. *Reflection Coefficients, in particular of the H_{11}-mode*

2.2.1. *Reflection coefficient with incidence of an H_{11}-mode.* For the general case or the incidence of an H_{mn}-mode the reflection and transmission coefficients are calculated from (49) or (50). Let us here consider as an example the incidence of an H_{11}-mode under the assumption that only this mode can exist in the waveguide. Only the reflection coefficient of the H_{11}-mode is of interest in such case, since the other excited modes will not transport any energy in the waveguide. With incidence of an H_{11}-mode the reflection coefficient r_{1H0} is the ratio of the electrical field strength of the reflected H_{11}-mode to that of the incident H_{11}-mode at the plane junction of waveguide and horn ($z = z_0$ in Fig. 2).

Figures 11 and 12 show in terms of magnitude and phase the first to the fourth approximation and the estimated ultimate solution of r_{1H0} for $1 \leqq \lambda_c/\lambda_0 \leqq 2.1$ with $\theta_0 = \pi/6$. The curves were calculated from (50). The first approximation means then the calculation of r_{1H} and accordingly of

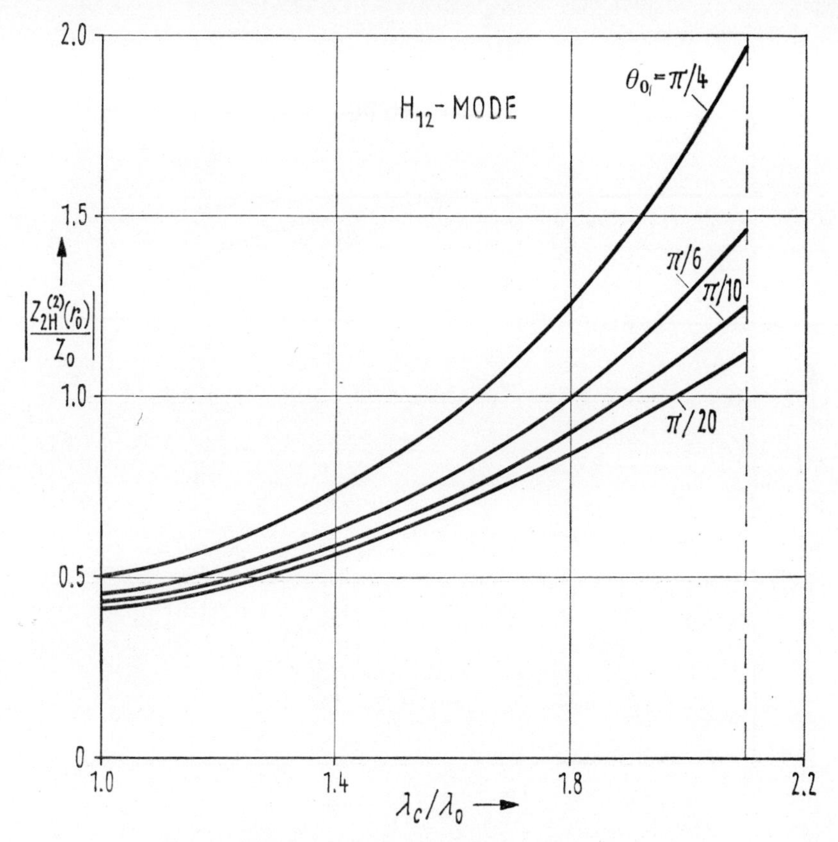

FIG. 9. Magnitude of $Z^{(2)}{}_{2H}(r_0)/Z_0$ as a function of λ_c/λ_0.

r_{1H0} from (50) by putting all $r_{nH} = r_{nE} = 0$ apart from r_{1H}. With the second approximation all $r_{nH} = r_{nE}$ are put equal to zero, except r_{1H} and r_{1E}. With the third approximation one puts correspondingly all $r_{nH} = r_{nE} = 0$, except r_{1H}, r_{1E}, and r_{2H}, etc. According to Figs. 11 and 12 the results present the remarkable fact that the fourth approximation deviates no longer strongly from the third and that besides the sequence of approximations alternates so that the ultimate value can be well estimated. This alternation of the sequence of approximations which is not encountered with the incidence of an H_{0n}-mode (Piefke 1961), is probably due to the simultaneous occurrence of H- and E-modes. The degree of approximation, of course, increases very rapidly as θ_0 decreases.

Figures 13 and 14 show magnitude and phase of r_{1H0} as a function of frequency ($1 \leqq \lambda_c/\lambda_0 \leqq 2.1$) for horn aperture angles of 0° to 90° ($0 \ldots \theta_0 \ldots \pi/4$). The curves were derived corresponding to the diagrams 11 and 12 by calculation of the individual approximations and subsequent estimation of

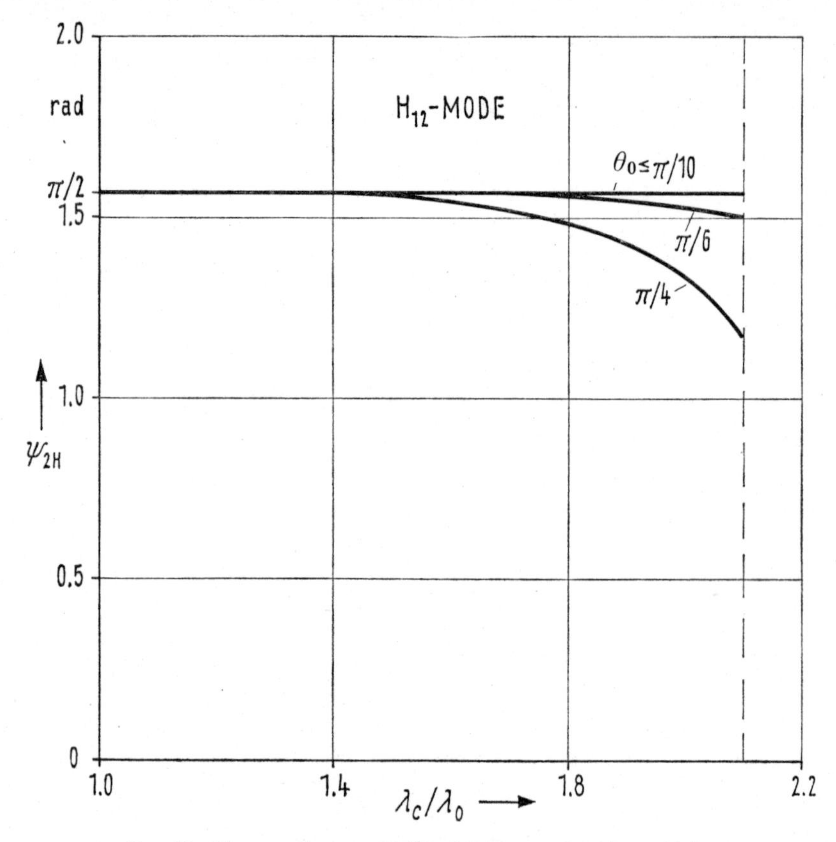

FIG. 10. Phase angle ψ_{2H} of $Z^{(2)}{}_{2H}(r_0)/Z_0$ as a function of λ_c/λ_0.

FIGS. 5–10. Relative wave impedances $Z^{(2)}{}_{pH}(r_0)/Z_0 = |Z^{(2)}{}_{pH}(r_0)/Z_0|\, e^{j\psi_{pH}}$ and $Z^{(2)}{}_{1E}(r_0)/Z_0 = |Z^{(2)}{}_{1E}(r_0)/Z_0|\, e^{j\psi_{1E}}$ of the H_{1p}-modes ($p = 1, 2$) and E_{11}-mode propagating of the horn in direction r as a function of frequency (λ_c/λ_0), where λ_0 wavelength of the planar wave in space, λ_c cutoff wavelength of the H_{11}-mode in the waveguide.

the ultimate shape. At the cutoff frequency of the H_{11}-mode in the guide ($\lambda_c/\lambda_0 = 1$) there is accordingly the reflection coefficient $r_{1H0} = -1$, the magnitude unity and the phase π. Besides, this value can also be exactly read direct from (50). With increasing frequency the magnitude decreases very rapidly and remains thereafter approximately constant. The phase decreases first with increasing frequency, becomes thus less than π, rises then again, exceeds π and finally even $3\pi/2$. Remarkable are the common intersections of the phase curves in Fig. 14. The equivalent circuit is accordingly a line corresponding to the waveguide with the wave impedance Z_{1H} which is terminated by a parallel connection of resistance, inductance, and capacitance (see Fig. 13).

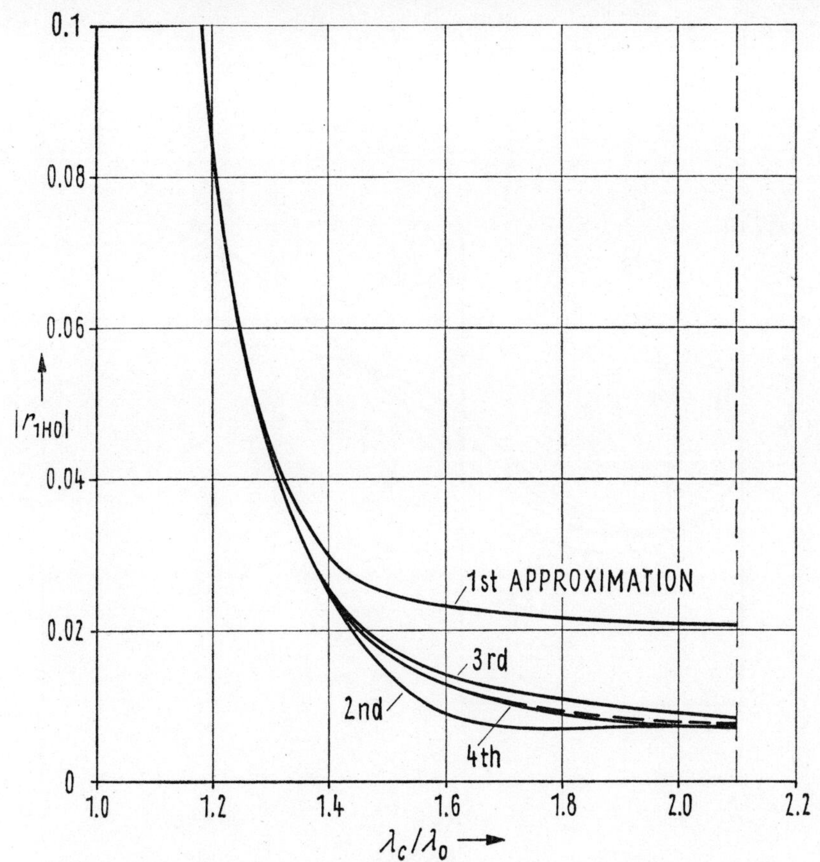

FIG. 11. Approximations of the magnitude of r_{1H0} as a function of λ_c/λ_0 with $\theta_0 = \pi/6$.

2.2.2. *Approximations for the reflection coefficients of the H_{11}-mode and E_{11}-mode with incidence of an H_{11}-mode and very small horn aperture angle.* If the method described at the end of subsection 3.2.2 is used for calculating the integrals and all terms with θ_0^2 and higher powers are neglected, the first approximation for r_{1H0} from (50) becomes with $r_0 = a/\theta_0$ and $\cos\theta_0 = 1 - \theta_0^2/2$

$$r_{1H0} \approx -\frac{\dfrac{Z_{1H}}{Z_{1H}^{(2)}(r_0)} - 1 + j\dfrac{\theta_0}{\beta_{1H}a(y_{1H}^2 - 1)}}{\dfrac{Z_{1H}}{Z_{1H}^{(2)}(r_0)} + 1 + j\dfrac{\theta_0}{\beta_{1H}a(y_{1H}^2 - 1)}}\, e^{-j0.634\beta_{1H}a\theta_0}. \tag{17}$$

With incidence of an H_{01}-mode the approximation formula for the reflection coefficient of the H_{01}-mode corresponding to (17) contains apart from the exponential function only the field impedances, even if the terms with θ_0^2 are also considered (Piefke 1961).

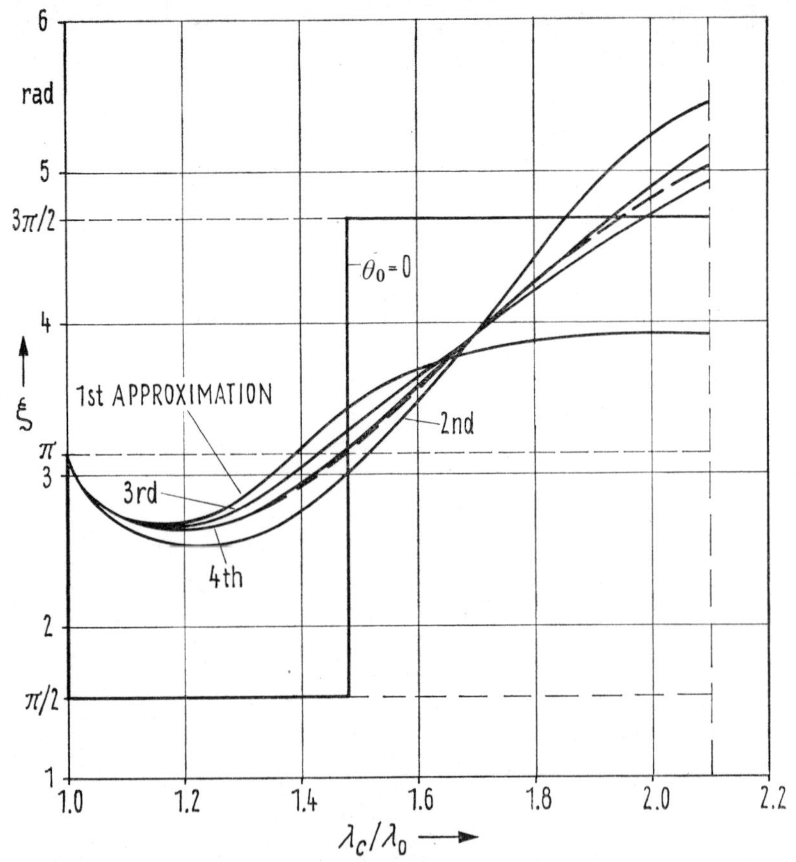

FIG. 12. Approximations of the phase angle ξ of r_{1H0} as a function of λ_c/λ_0 with $\theta_0 = \pi/6$.

FIGS. 11 and 12. 1 ... 4 Approximation of the reflection coefficient $r_{1H0} = |r_{1H0}|\, e^{j\xi}$ of the H_{11}-mode with incidence of an H_{11}-mode at junction of circular waveguide and conical horn of infinite length as a function of frequency (λ_c/λ_0) for $\theta_0 = \pi/6$. ---- estimated ultimate solution. For the meaning of λ_0 and λ_c refer to Figs. 5 to 10.

Under the assumption of (8) and (9) eqn (7) with $\rho_0 = a$ may be inserted in (17). There results then

$$r_{1H0} \approx \frac{|7.45 - (\beta_0 a)^2|\theta_0}{\sqrt{\{[4.78(\beta_{1H} a)^3]^2 + \theta_0^2[7.45 - (\beta_0 a)^2]^2\}}}\, e^{j\xi}, \tag{18}$$

$$\xi \approx \frac{\pi}{2} - 0.634\beta_{1\mathrm{H}} a\theta_0 + \arctan\left(\frac{11.53 + 0.915(\beta_0 a)^2 + 0.508(\beta_0 a)^4}{(\beta_{1\mathrm{H}} a)^3\,[7.45 - (\beta_0 a)^2]}\theta_0\right). \tag{19}$$

Equations (18) and (19) agree with Schnetzler, (1960) if the term with θ_0^2 below the root is neglected and the phase is equated to that at $\theta_0 = 0$.

To illustrate the degree of approximation according to (18) and (19), r_{1H0}

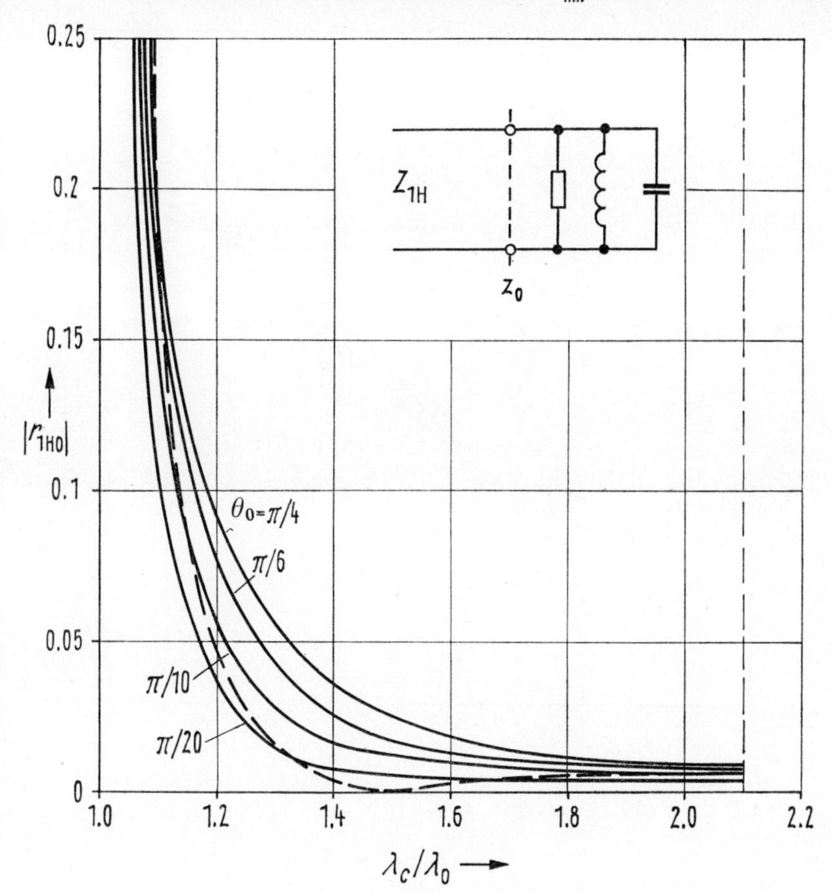

FIG. 13. Magnitude of r_{1H0} as a function of λ_c/λ_0 and equivalent circuit of the junction of waveguide and horn.

was calculated herefrom for $\theta_0 = \pi/20$ and entered in the diagrams 13 and 14.

An approximation for r_{1E0} is gained from (50) by putting all $r_{nH} = r_{nE} = 0$, except r_{1H} and r_{1E}. If the method described at the end of subsection 3.2.2 is used again and only the first approximations are considered, there results with $r_0 \approx a/\theta_0$ and $\cos\theta_0 = 1 - \theta_0^2/2$

$$r_{1E0} \approx 2\frac{A_{11E}^*}{E_{11E}}\frac{\dfrac{Z_{1H}}{Z_{1H}^{(2)}(r_0)} - \dfrac{Z_{1E}^{(2)}(r_0)}{Z_{1H}}\left[1+\left(\dfrac{y_{1H}}{\beta_{1H}a}\right)^2\right] + j\,\dfrac{0.418\theta_0}{\beta_{1H}a}}{\dfrac{\beta_{1E}}{\beta_0}\left(1+\dfrac{Z_{1H}}{Z_{1H}^{(2)}(r_0)}+j\,\dfrac{0.418\theta_0}{\beta_{1H}a}\right)\left(1+\dfrac{Z_{1E}^{(2)}(r_0)}{Z_{1E}}-j\,\dfrac{\theta_0}{\beta_{1E}a}\right)} \times$$

$$\times \exp j\left(\frac{2\beta_{1E}a}{\theta_0} - \beta_{1E}a\theta_0\right), \qquad (20)$$

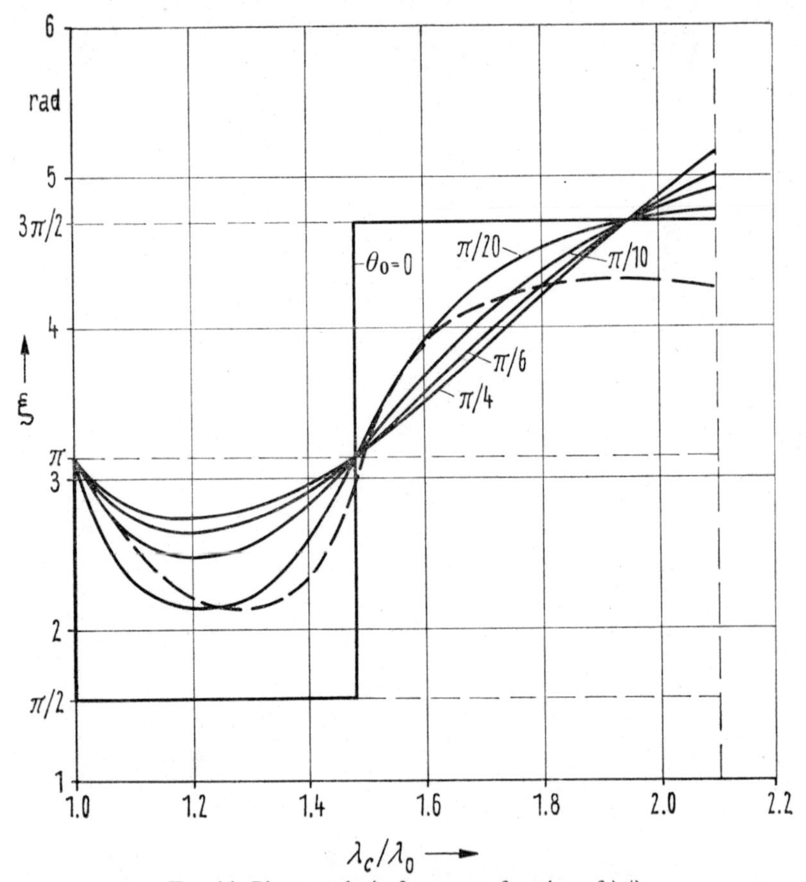

FIG. 14. Phase angle ξ of r_{1H0} as a function of λ_c/λ_0.

FIGS. 13 and 14. Reflection coefficient $r_{1H0} = |r_{1H0}| e^{j\xi}$ of the H_{11}-mode with incidence of an H_{11}-mode at junction of circular waveguide and conical horn of infinite length as a function of the frequency (λ_c/λ_0) and related equivalent circuit. ---- Approximation according to (18) and (19) for $\theta_0 = \pi/20$. Concerning the meaning of λ_0 and λ_c see Figs. 5 to 10.

$$\frac{A^*_{11E}}{E_{11E}} \approx j\,\frac{2\theta_0\beta_{1H}aJ_1(y_{1H})}{y_{1E}(y^2_{1E} - y^2_{1H})J_0(y_{1E})}\,\exp -j\,\frac{1}{\theta_0}\,(\beta_{1H}a + \beta_{1E}a)\,. \tag{21}$$

With $Z^{(2)}_{1H}(r_0) \approx Z_{1H}$ and $Z^{(2)}_{1E}(r_0) \approx Z_{1E}$, neglection of the terms with θ_0 in (20) and insertion of the numerical values for y_{1H}, y_{1E}, $J_1(y_{1H})$, $J_0(y_{1E})$ there results from (20) and (21)

$$r_{1E0} \approx -j\,0.0335\theta_0\beta_0 a\,\frac{\beta_{1H}}{\beta_{1E}}\left(1 - \frac{\beta_{1H}\beta_{1E}}{\beta_0^2}\left[1 + \left(\frac{y_{1H}}{\beta_{1H}a}\right)^2\right]\right) \times$$

$$\times \exp -j\left(\frac{\beta_{1H}a}{\theta_0} - \frac{\beta_{1E}a}{\theta_0} + \beta_{1E}a\theta_0\right). \tag{22}$$

2.3. *Approximation for the Transmission Coefficients of the H_{11}-mode and E_{11}-mode with Incidence of an H_{11}-mode and very small Horn Aperture Angle*

The first approximation for the transmission coefficient $t_{1H} = d_{1H}\sqrt{(2/\pi)}$ of the H_{11}-mode is gained from (49) by putting all $r_{nH} = r_{nE} = d_{pH} = d_{pE} = 0$ apart from r_{1H} and d_{1H}. If only the first approximations are considered as obtained by the method described at the end of subsection 3.2.2, there results

$$t_{1H} \approx 2\sqrt{\left(\frac{2}{\pi}\right)}\frac{\theta_0}{\sin\theta_0}\frac{(1 - j\,0.183\beta_{1H}a\theta_0)\exp -j\dfrac{\beta_{1H}a}{\theta_0}}{\left(\dfrac{Z_{1H}}{Z_{1H}^{(2)}(r_0)} + 1 + j\dfrac{0.418}{\beta_{1H}a}\theta_0\right)\sqrt{(\beta_0 r_0)}H_\eta^{(2)}(\beta_0 r_0)}. \tag{23}$$

After inserting (7) with $\rho_0 = a$ in (23) and replacing the Hankel function by the first approximation of the Debye series (Magnus, Oberhettinger, 1948) one obtains

$$|t_{1H}| = \sqrt{\left(\frac{Z_0}{Z_{1H}}\right)}\frac{4.78(\beta_{1H}a)^3}{\sqrt{\{[4.78(\beta_{1H}a)^3]^2 + \theta_0^2[7.45 - (\beta_0 a)^2]^2\}}}. \tag{24}$$

Equation (24) results also from the power balance, if according to (69) and (72) there is established $(P_{H11})_w/P_{H11} = |t_{1H}|^2 Z_{1H}/Z_0 = 1 - |r_{1HO}|^2$ and (18) inserted for $|r_{1HO}|$.

Corresponding to the calculation of t_{1H} there results

$$t_{1E} \approx 2\frac{A_{11E}^*}{H_E}\frac{\dfrac{Z_{1H}}{Z_{1H}^{(2)}(r_0)} + \dfrac{\beta_{1H}\beta_{1E}}{\beta_0^2}\left[1 + \left(\dfrac{y_{1H}}{\beta_{1H}a}\right)^2\right] - j\theta_0\dfrac{\beta_{1H}a}{(\beta_0 a)^2}\left[1 + \left(\dfrac{y_{1H}}{\beta_{1H}a}\right)^2 - 0.418\left(\dfrac{\beta_0}{\beta_{1H}}\right)^2\right]}{\dfrac{\beta_{1E}}{\beta_0}\left(1 + \dfrac{Z_{1H}}{Z_{1H}^{(2)}(r_0)} + j\dfrac{0.418\theta_0}{\beta_{1H}a}\right)\left(1 + \dfrac{Z_{1E}^{(2)}(r_0)}{Z_{1E}} - j\dfrac{\theta_0}{\beta_{1E}a}\right)}, \tag{25}$$

$$\frac{A_{11E}^*}{H_E} \approx j\sqrt{\left(\frac{2}{\pi}\right)}\frac{\theta_0}{\sin\theta_0}\frac{2\theta_0\beta_{1H}aJ_1(y_{1H})\exp -j(\beta_{1H}a/\theta_0)}{y_{1E}(y_{1E}^2 - y_{1H}^2)J_0(y_{1E})\sqrt{(\beta_0 r_0)}H_\zeta^{(2)}(\beta_0 r_0)}. \tag{26}$$

With $Z_{1H}^{(2)}(r_0) \approx Z_{1H}$, $Z_{1E}^{(2)}(r_0) \approx Z_{1E}$, $r_0 \approx a/\theta_0$, neglecting of the terms with θ_0 in (25), insertion of the numerical values for y_{1H}, y_{1E}, $J_1(y_{1H})$, $J_0(y_{1E})$ and replacing of the Hankel function by the first approximation of the Debye series (Magnus, Oberhettinger, 1948) one obtains from (25) and (26) with a real β_{1E} (E_{11}-mode in the guide is capable of propagation)

$$|t_{1E}| \approx 0.0335\theta_0\beta_0 a\frac{\beta_{1H}}{\beta_{1E}}\sqrt{\left(\frac{Z_{1E}}{Z_0}\right)}\left(1 + \frac{\beta_{1H}\beta_{1E}}{\beta_0^2}\left[1 + \left(\frac{y_{1H}}{\beta_{1H}a}\right)^2\right]\right). \tag{27}$$

Observe the similarity of the transmission coefficient of the E_{11}-mode with the reflection coefficient of the E_{11}-mode.

The ratio of the active power $(P_{E11})_w$ of the E_{11}-mode to the active power $(P_{H11})_w$ of the H_{11}-mode in the horn results with $|t_{1H}| \approx \sqrt{(Z_0/Z_{1H})}$ and (27) from (72) and (74) as

$$\frac{(P_{E11})_w}{(P_{H11})_w} = 3.3 \cdot 10^{-3} \theta_0^2 (\beta_0 a)^2 \frac{\beta_{1H}}{\beta_{1E}} \left(1 + \frac{\beta_{1H}\beta_{1E}}{\beta_0^2}\left[1 + \left(\frac{y_{1H}}{\beta_{1H} a}\right)^2\right]\right)^2 . \tag{28}$$

3. MATHEMATICAL DISCUSSION OF THE PROBLEM

3.1. *The Modes in Waveguide and Conical Horn*

3.1.1. *The waveguide modes.* With omission of the factor $e^{j\omega t}$ the following field strength components exist in the waveguide with incidence of the H_{MN}-mode as referred to the electrical field strength E_0 and the magnetic field strength $H_0 = E_0/Z_0$:

(1) H_{Mn}-modes ($E_{zH}^{(1)} = 0$)

$$\frac{E_{\rho H}^{(1)}}{E_0} = -\frac{a}{\rho} M \sin(M\phi)\left[J_M\left(y_{NH}\frac{\rho}{a}\right)e^{-j\beta_{NH}z} + \right.$$
$$\left. + \sum_n r_{nH} J_M\left(y_{nH}\frac{\rho}{a}\right)e^{j\beta_{nH}z}\right],$$

$$\frac{E_{\phi H}^{(1)}}{E_0} = -\cos(M\phi)\left[y_{NH}J'_M\left(y_{NH}\frac{\rho}{a}\right)e^{-j\beta_{NH}z} + \right.$$
$$\left. + \sum_n r_{nH} y_{nH} J'_M\left(y_{nH}\frac{\rho}{a}\right)e^{j\beta_{nH}z}\right],$$

$$\frac{H_{\rho H}^{(1)}}{H_0} = \cos(M\phi)\left[y_{NH}\frac{Z_0}{Z_{NH}}J'_M\left(y_{NH}\frac{\rho}{a}\right)e^{-j\beta_{NH}z} - \right.$$
$$\left. - \sum_n r_{nH} y_{nH}\frac{Z_0}{Z_{nH}} J'_M\left(y_{nH}\frac{\rho}{a}\right)e^{j\beta_{nH}z}\right],$$

$$\frac{H_{\phi H}^{(1)}}{H_0} = -\frac{a}{\rho} M \sin(M\phi)\left[\frac{Z_0}{Z_{NH}}J_M\left(y_{NH}\frac{\rho}{a}\right)e^{-j\beta_{NH}z} - \right.$$
$$\left. - \sum_n r_{nH}\frac{Z_0}{Z_{nH}} J_M\left(y_{nH}\frac{\rho}{a}\right)e^{j\beta_{nH}z}\right],$$

$$\frac{H_{zH}^{(1)}}{H_0} = \frac{j}{\beta_0 a}\cos(M\phi)\left[y_{NH}^2 J_M\left(y_{NH}\frac{\rho}{a}\right)e^{-j\beta_{NH}z} + \right.$$
$$\left. + \sum_n r_{nH} y_{nH}^2 J_M\left(y_{nH}\frac{\rho}{a}\right)e^{j\beta_{nH}z}\right].$$

$$n = 1,2,3,\ldots . \tag{29}$$

(2) E_{Mn}-modes ($H_{zE}^{(1)} = 0$)

$$\frac{H_{\rho E}^{(1)}}{H_0} = \frac{a}{\rho} M \cos(M\phi) \sum_n r_{nE} J_M\left(y_{nE}\frac{\rho}{a}\right) e^{j\beta_{nE}z},$$

$$\frac{H_{\phi E}^{(1)}}{H_0} = -\sin(M\phi) \sum_n r_{nE} y_{nE} J'_M\left(y_{nE}\frac{\rho}{a}\right) e^{j\beta_{nE}z},$$

$$\frac{E_{\rho E}^{(1)}}{E_0} = \sin(M\phi) \sum_n r_{nE} y_{nE} \frac{Z_{nE}}{Z_0} J'_M\left(y_{nE}\frac{\rho}{a}\right) e^{j\beta_{nE}z},$$

$$\frac{E_{\phi E}^{(1)}}{E_0} = \frac{a}{\rho} M \cos(M\phi) \sum_n r_{nE} \frac{Z_{nE}}{Z_0} J_M\left(y_{nE}\frac{\rho}{a}\right) e^{j\beta_{nE}z},$$

$$\frac{E_{zE}^{(1)}}{E_0} = -\frac{j}{\beta_0 a} \sin(M\phi) \sum_n r_{nE} y_{nE}^2 J_M\left(y_{nE}\frac{\rho}{a}\right) e^{j\beta_{nE}z}.$$

$$n = 1,2,3, \ldots . \qquad (30)$$

The superscript $^{(1)}$ in (29) and (30) means that modes in medium 1 (waveguide) are involved. The subscripts H and E show that the field strengths belong to the H_{Mn}-mode and E_{Mn}-mode, respectively.

3.1.2. *The modes in the conical horn.* According to Piefke (1954) the modes in a conical horn result from the vector potential

$$A = \begin{Bmatrix} \sin \\ \cos \end{Bmatrix} (\mu\phi) K_\nu^\mu(\cos\theta) \cdot \sqrt{(\beta_0 r)} Z_{\nu+1/2}(\beta_0 r) \qquad (31)$$

where $K_\nu^\mu(\cos\theta)$ is a linear combination of two spherical functions and $Z_{\nu+1/2}(\beta_0 r)$ a linear combination of two cylindrical functions of the order $\nu + 1/2$. The quantities ν and μ must be determined from the boundary conditions (Piefke, 1954). In the horn only $\mu = M = 0,1,2, \ldots$ is possible. The H-modes result from $E = \text{rot}\, A$, the E-modes from $H = \text{rot A}$.

Putting $\tan\theta = \sin\theta = \theta$ in the differential equation (11) in the annex of (Piefke, 1954) there results with $\mu = M$

$$K_\nu^\mu(\cos\theta) \approx J_M(\sqrt{[\nu(\nu+1)]}\theta), \qquad M = 0,1,2, \ldots . \qquad (32)$$

In analogy to the waveguide there must then hold because of the boundary conditions

$$\sqrt{[\nu(\nu+1)]}\theta_0 = \begin{cases} y_{pH} & \text{with } H\text{–modes} \\ y_{pE} & \text{with } E\text{–modes} \end{cases}, \qquad (33)$$

$$\nu + \tfrac{1}{2} = \begin{cases} \eta = \sqrt{\left[\frac{1}{4} + \left(\frac{y_{pH}}{\theta_0}\right)^2\right]} & \text{with } H\text{–modes} \\ \zeta = \sqrt{\left[\frac{1}{4} + \left(\frac{y_{pE}}{\theta_0}\right)^2\right]} & \text{with } E\text{–modes} \end{cases}. \qquad (34)$$

The ν resulting from (34) for the argument of the Bessel function can, of course, not be identical with the ν in the spherical function K_ν^μ since the latter must be determined from the boundary conditions for the spherical function. See in this respect the Figs. 3 and 4.

For distinction from the waveguide the H_{Mn}-modes and the E_{Mn}-modes in the horn are denoted H_{Mp}-modes and E_{Mp}-modes, respectively. The modes propagate in the horn merely in the direction of increasing r. For the cylindrical function the Hankel function of second kind is therefore written. According to Piefke (1954) there exist then with use of (32) and omission of the factor $e^{j\omega t}$ the following field strengths in the horn as referred to the electrical field strength E_0 and the magnetic field strength $H_0 = E_0/Z_0$:

(1) H_{Mp}–modes ($E_{rH}^{(2)} = 0$)

$$\frac{E_{\theta H}^{(2)}}{E_0} = -\frac{a}{r\sin\theta} M\sin(M\phi)\sum_p d_{pH} J_M\left(y_{pH}\frac{\theta}{\theta_0}\right)\sqrt{(\beta_0 r)}H_\eta^{(2)}(\beta_0 r),$$

$$\frac{E_{\phi H}^{(2)}}{E_0} = -\frac{a}{\theta_0 r}\cos(M\phi)\sum_p d_{pH} y_{pH} J'_M\left(y_{pH}\frac{\theta}{\theta_0}\right)\sqrt{(\beta_0 r)}H_\eta^{(2)}(\beta_0 r),$$

$$\frac{H_{\theta H}^{(2)}}{H_0} = \frac{a}{\theta_0 r}\cos(M\phi)\sum_p d_{pH} y_{pH}\frac{Z_0}{Z_{pH}^{(2)}} J'_M\left(y_{pH}\frac{\theta}{\theta_0}\right)\sqrt{(\beta_0 r)}H_\eta^{(2)}(\beta_0 r),$$

$$\frac{H_{\phi H}^{(2)}}{H_0} = -\frac{a}{r\sin\theta} M\sin(M\phi)\sum_p d_{pH}\frac{Z_0}{Z_{pH}^{(2)}} J_M\left(y_{pH}\frac{\theta}{\theta_0}\right)\sqrt{(\beta_0 r)}H_\eta^{(2)}(\beta_0 r),$$

$$\frac{H_{rH}^{(2)}}{H_0} = \frac{j\beta_0 a}{(\theta_0\beta_0 r)^2}\cos(M\phi)\sum_p d_{pH} y_{pH}^2 J_M\left(y_{pH}\frac{\theta}{\theta_0}\right)\sqrt{(\beta_0 r)}H_\eta^{(2)}(\beta_0 r).$$

$$p = 1,2,3,\ldots. \quad (35)$$

(2) E_{Mp}–modes ($H_{rE}^{(2)} = 0$)

$$\frac{H_{\theta E}^{(2)}}{H_0} = \frac{a}{r\sin\theta} M\cos(M\phi)\sum_p d_{pE} J_M\left(y_{pE}\frac{\theta}{\theta_0}\right)\cdot\sqrt{(\beta_0 r)}H_\zeta^{(2)}(\beta_0 r),$$

$$\frac{H_{\phi E}^{(2)}}{H_0} = -\frac{a}{\theta_0 r}\sin(M\phi)\sum_p d_{pE} y_{pE} J'_M\left(y_{pE}\frac{\theta}{\theta_0}\right)\cdot\sqrt{(\beta_0 r)}H_\zeta^{(2)}(\beta_0 r),$$

$$\frac{E_{\theta E}^{(2)}}{E_0} = -\frac{a}{\theta_0 r}\sin(M\phi)\sum_p d_{pE} y_{pE}\frac{Z_{pE}^{(2)}}{Z_0} J'_M\left(y_{pE}\frac{\theta}{\theta_0}\right)\cdot\sqrt{(\beta_0 r)}H_\zeta^{(2)}(\beta_0 r),$$

$$\frac{E_{\phi E}^{(2)}}{E_0} = -\frac{a}{r\sin\theta} M\cos(M\phi)\sum_p d_{pE}\frac{Z_{pE}^{(2)}}{Z_0} J_M\left(y_{pE}\frac{\theta}{\theta_0}\right)\cdot\sqrt{(\beta_0 r)}H_\zeta^{(2)}(\beta_0 r),$$

$$\frac{E_{rE}^{(2)}}{E_0} = -\frac{j\beta_0 a}{(\theta_0\beta_0 r)^2}\sin(M\phi)\sum_p d_{pE} y_{pE}^2 J_M\left(y_{pE}\frac{\theta}{\theta_0}\right)\cdot\sqrt{(\beta_0 r)}H_\zeta^{(2)}(\beta_0 r).$$

$$p = 1,2,3,\ldots. \quad (36)$$

In (35) and (36) the superscript (2) denotes that modes in medium 2 (conical horn) are involved. The subscripts H and E state again, that the field strengths belong to the H_{Mp}-mode and E_{Mp}-mode, respectively. The quantities $Z_{pH}^{(2)}$ and $Z_{pE}^{(2)}$ are evident from (1) and (2).

3.2. *The set of Equations for Calculating the Reflection and Transmission Coefficients*

3.2.1. *The field components in spherical coordinates.* If in cylinder coordinates the vector U has the components U_ρ, U_ϕ, U_z it has in spherical coordinates the components

$$U_r = U_\rho \sin\theta + U_z \cos\theta, \quad U_\theta = U_\rho \cos\theta - U_z \sin\theta \text{ and } U\phi = U\phi. \quad (37)$$

Let us assume i unit vector in direction θ, k unit vector in direction ϕ. The electrical field strength E_t tangential to a plane $r = \text{const.}$ and the tangential magnetic field strength H_t are then

$$E_t = iE_\theta + kE_\phi, \quad H_t = iH_\theta + kH_\phi.$$

After inserting the field strengths from (29) and (30) in (37) and subsequent insertion of (37) in (38) one obtains for the waveguide

$$\begin{aligned} E_t^{(1)} &= i[(E_{\rho H}^{(1)} + E_{\rho E}^{(1)}) \cos\theta - E_{zE}^{(1)} \sin\theta] + k[E_{\phi H}^{(1)} + E_{\phi E}^{(1)}], \\ H_t^{(1)} &= i[(H_{\rho H}^{(1)} + H_{\rho E}^{(1)}) \cos\theta - H_{zH}^{(1)} \sin\theta] + k[H_{\phi H}^{(1)} + H_{\phi E}^{(1)}]. \end{aligned} \quad (39)$$

The corresponding field strengths in the conical horn are given by

$$\begin{aligned} E_t^{(2)} &= i[E_{\theta H}^{(2)} + E_{\theta E}^{(2)}] + k[E_{\phi H}^{(2)} + E_{\phi E}^{(2)}], \\ H_t^{(2)} &= i[H_{\theta H}^{(2)} + H_{\theta E}^{(2)}] + k[H_{\phi H}^{(2} + H_{\phi E}^{(2)}]. \end{aligned} \quad (40)$$

With $\sin\theta = \theta$ there results after inserting (35) and (36) into (40) for $r = r_0$,

$$E_t^{(2)} = -E_0 \frac{a}{r_0} \sqrt{(\beta_0 r_0)} \sum_p \Big[d_{pH} e_{MpH} H_\eta^{(2)}(\beta_0 r_0) + $$

$$+ d_{pE} \frac{Z_{pE}^{(2)}(r_0)}{Z_0} e_{MpE} H_\zeta^{(2)}(\beta_0 r_0) \Big],$$

$$H_t^{(2)} = E_0 \frac{a}{r_0} \sqrt{(\beta_0 r_0)} \sum_p \left[\frac{d_{pH}}{Z_{pH}^{(2)}(r_0)} h_{MpH} H_\eta^{(2)}(\beta_0 r_0) + \frac{d_{pE}}{Z_0} h_{MpE} H_\zeta^{(2)}(\beta_0 r_0) \right],$$

$$p = 1,2,3,\ldots. \quad (41)$$

where

$$\begin{aligned} e_{MpH} &= i \frac{M}{\theta} \sin(M\phi) J_M\left(y_{pH} \frac{\theta}{\theta_0}\right) + k \frac{y_{pH}}{\theta_0} \cos(M\phi) J'_M\left(y_{pH} \frac{\theta}{\theta_0}\right), \\ e_{MpE} &= i \frac{y_{pE}}{\theta_0} \sin(M\phi) J'_M\left(y_{pE} \frac{\theta}{\theta_0}\right) + k \frac{M}{\theta} \cos(M\phi) J_M\left(y_{pE} \frac{\theta}{\theta_0}\right), \\ h_{MpH} &= i \frac{y_{pH}}{\theta_0} \cos(M\phi) J'_M\left(y_{pH} \frac{\theta}{\theta_0}\right) - k \frac{M}{\theta} \sin(M\phi) J_M\left(y_{pH} \frac{\theta}{\theta_0}\right), \end{aligned} \quad (42)$$

$$h_{MpE} = i\,\frac{M}{\theta}\cos(M\phi)J_M\left(y_{pE}\frac{\theta}{\theta_0}\right) - k\,\frac{y_{pE}}{\theta_0}\sin(M\phi)J'_M\left(y_{pE}\frac{\theta}{\theta_0}\right).$$

There hold the orthogonality relations

$$\int_0^{2\pi}\int_0^{\theta_0} e_{MpH}e_{MqH}\theta\,d\theta\,d\phi = \int_0^{2\pi}\int_0^{\theta_0} h_{MpH}h_{MqH}\theta\,d\theta\,d\phi =$$

$$= \begin{cases} 0 & \text{for } p \neq q\,, \\ \dfrac{\pi}{\delta_M}\,y_{pH}^2[J_M(y_{pH})]^2\left[1-\left(\dfrac{M}{y_{pH}}\right)^2\right] = \dfrac{1}{P_{pH}^2} & \text{for } p = q\,, \end{cases} \tag{43}$$

$$\int_0^{2\pi}\int_0^{\theta_0} e_{MpE}e_{MqE}\theta\,d\theta\,d\phi = \int_0^{2\pi}\int_0^{\theta_0} h_{MpE}h_{MqE}\theta\,d\theta\,d\phi =$$

$$= \begin{cases} 0 & \text{for } p \neq q\,, \\ \dfrac{\pi}{\delta_M}\,y_{pE}^2[J_{M+1}(y_{pE})]^2 = \dfrac{1}{P_{pE}^2} & \text{for } p = q\,, \end{cases} \tag{44}$$

$$M = 0,1,2,\ldots\,,$$

$$\delta_M = \begin{cases} 1 & \text{for } M = 0 \\ 2 & \text{for } M \neq 0 \end{cases}. \tag{45}$$

According to the expansion theorem there is

$$E_t^{(1)} = \sum_p (c_{pH}P_{pH}e_{MpH} + c_{pE}P_{pE}e_{MpE})\,,$$
$$H_t^{(1)} = \sum_p (\zeta_{pH}P_{pH}h_{MpH} + \zeta_{pE}P_{pE}h_{MpE})\,, \tag{46}$$
$$p = 1,2,3,\ldots$$

where

$$c_{pH} = P_{pH}\int_0^{2\pi}\int_0^{\theta_0} E_t^{(1)}e_{MpH}\theta\,d\theta\,d\phi\,,$$
$$c_{pE} = P_{pE}\int_0^{2\pi}\int_0^{\theta_0} E_t^{(1)}e_{MpE}\theta\,d\theta\,d\phi\,, \tag{47}$$

$$\zeta_{pH} = P_{pH}\int_0^{2\pi}\int_0^{\theta_0} H_t^{(1)}h_{MpH}\theta\,d\theta\,d\phi\,,$$
$$\zeta_{pE} = P_{pE}\int_0^{2\pi}\int_0^{\theta_0} H_t^{(1)}h_{MpE}\theta\,d\theta\,d\phi\,. \tag{47}$$

3.2.2. *The set of equations.* After inserting (41) and (46) into the continuity conditions $E_t^{(1)} = E_t^{(2)}$ and $H_t^{(1)} = H_t^{(2)}$ one obtains

$$
\begin{aligned}
c_{pH}P_{pH} &= -\, d_{pH}E_0 \frac{a}{r_0} \sqrt{(\beta_0 r_0)} H_\eta^{(2)}(\beta_0 r_0)\,, \\
c_{pE}P_{pE} &= -\, d_{pE}E_0 \frac{a}{r_0} \frac{Z_{pE}^{(2)}(r_0)}{Z_0} \sqrt{(\beta_0 r_0)} H_\zeta^{(2)}(\beta_0 r_0)\,, \\
\zeta_{pH}P_{pH} &= d_{pH} \frac{E_0}{Z_{pH}^{(2)}(r_0)} \frac{a}{r_0} \sqrt{(\beta_0 r_0)} H_\eta^{(2)}(\beta_0 r_0)\,, \\
\zeta_{pE}P_{pE} &= d_{pE} \frac{E_0}{Z_0} \frac{a}{r_0} \sqrt{(\beta_0 r_0)} H_\zeta^{(2)}(\beta_0 r_0)\,.
\end{aligned}
\tag{48}
$$

The field strengths from (29) and (30) are now inserted in (39) with use of $z = r_0 \cos\theta$. Subsequently (39) is inserted into (47) and (48). With $\rho = a\theta/\theta_0$ the result is then

$$
\begin{aligned}
A_{pNH}^* + \sum_n (r_{nH}A_{pnH} - r_{nE}P_{pnH}) &= \frac{\sin\theta_0}{\theta_0} \frac{\sqrt{(\beta_0 r_0)}}{P_{pH}^2} H_\eta^{(2)}(\beta_0 r_0) d_{pH}\,, \\
Q_{pNH}^* - \sum_n (r_{nH}Q_{pnH} - r_{nE}E_{pnH}) &= \frac{\sin\theta_0}{\theta_0} \frac{\sqrt{(\beta_0 r_0)}}{P_{pH}^2} H_\eta^{(2)}(\beta_0 r_0) \frac{Z_0}{Z_{pH}^{(2)}(r_0)} d_{pH}\,, \\
A_{pNE}^* + \sum_n (r_{nH}A_{pnE} - r_{nE}P_{pnE}) &= \frac{\sin\theta_0}{\theta_0} \frac{\sqrt{(\beta_0 r_0)}}{P_{pE}^2} H_\zeta^{(2)}(\beta_0 r_0) \frac{Z_{pE}^{(2)}(r_0)}{Z_0} d_{pE}\,. \\
Q_{pNE}^* - \sum_n (r_{nH}Q_{pnE} - r_{nE}E_{pnE}) &= \frac{\sin\theta_0}{\theta_0} \frac{\sqrt{(\beta_0 r_0)}}{P_{pE}^2} H_\zeta^{(2)}(\beta_0 r_0) d_{pE}\,. \\
& n,p = 1,2,3,\ldots\,.
\end{aligned}
\tag{49}
$$

After elimination of the transmission coefficients d_{pH} and d_{pE} there results from (49) the set of equations

$$
\begin{aligned}
\frac{Z_{pH}^{(2)}(r_0)}{Z_0} Q_{pNH}^* - A_{pNH}^* &= \sum_n (r_{nH}\alpha_{pnH} - r_{nE}\beta_{pnH})\,, \\
\frac{Z_{pE}^{(2)}(r_0)}{Z_0} Q_{pNE}^* - A_{pNE}^* &= \sum_n (r_{nH}\alpha_{pnE} - r_{nE}\beta_{pnE})\,. \\
& n,p = 1,2,3,\ldots\,.
\end{aligned}
\tag{50}
$$

Equations (49) and (50) contain the following auxiliary quantities

$$
P_{pnH} = \frac{Z_{nE}}{Z_0} B_{pnH} + j \frac{y_{nE}^2}{\beta_0 a\theta_0} C_{pnH}\,, \tag{51}
$$

$$
Q_{pnH} = \frac{Z_0}{Z_{nH}} D_{pnH} + j \frac{y_{nH}^2}{\beta_0 a\theta_0} F_{pnH}\,, \tag{25}
$$

$$\alpha_{pnH} = A_{pnH} + \frac{Z_{pH}^{(2)}(r_0)}{Z_0} Q_{pnH} , \tag{53}$$

$$\beta_{pnH} = P_{pnH} + \frac{Z_{pH}^{(2)}(r_0)}{Z_0} E_{pnH} . \tag{54}$$

P_{pnE} results from P_{pnH} if B_{pnH} and C_{pnH} are replaced by B_{pnE} and C_{pnE}, respectively. Q_{pnE} results from Q_{pnH} if D_{pnH} and F_{pnH} are replaced by D_{pnE} and F_{pnE}, respectively. α_{pnE} results from α_{pnH} if A_{pnH} and $Z_{pH}^{(2)}(r_0) \cdot Q_{pnH}$ are replaced by A_{pnE} and $Z_{pE}^{(2)}(r_0) \cdot Q_{pnE}$, respectively. β_{pnE} results from β_{pnH} if P_{pnH} and $Z_{pH}^{(2)}(r_0) \cdot E_{pnH}$ are replaced by P_{pnE} and $Z_{pE}^{(2)}(r_0) \cdot E_{pnE}$, respectively. There denote

$$A_{pnH} = \pi \int_0^{\theta_0} \left(\frac{M^2}{\theta^2} J_{MH} \bar{J}_{MH} \cos\theta + \frac{y_{nH} y_{pH}}{\theta_0^2} J'_{MH} \bar{J}'_{MH} \right) \mathrm{e}^{-2s_H \cos\theta} \theta \,\mathrm{d}\theta , \tag{55}$$

$$B_{pnH} = \frac{\pi M}{\theta_0} \int_0^{\theta_0} \left(\frac{y_{nE}}{\theta} J'_{ME} \bar{J}_{MH} \cos\theta + \frac{y_{pH}}{\theta} J_{ME} \bar{J}'_{MH} \right) \mathrm{e}^{-2s_E \cos\theta} \theta \,\mathrm{d}\theta , \tag{56}$$

$$C_{pnH} = \pi M \int_0^{\theta_0} J_{ME} \bar{J}_{MH} \, \mathrm{e}^{-2s_E \cos\theta} \theta \,\mathrm{d}\theta , \tag{57}$$

$$A_{pnE} = \frac{\pi M}{\theta_0} \int_0^{\theta_0} \left(\frac{y_{nH}}{\theta} J'_{MH} \bar{J}_{ME} + \frac{y_{pE}}{\theta} J_{MH} \bar{J}'_{ME} \cos\theta \right) \mathrm{e}^{-2s_H \cos\theta} \theta \,\mathrm{d}\theta , \tag{58}$$

$$B_{pnE} = \pi \int_0^{\theta_0} \left(\frac{M^2}{\theta^2} J_{ME} \bar{J}_{ME} + \frac{y_{nE} y_{pE}}{\theta_0^2} J'_{ME} \bar{J}'_{ME} \cos\theta \right) \mathrm{e}^{-2s_E \cos\theta} \theta \,\mathrm{d}\theta , \tag{59}$$

$$C_{pnE} = \pi \frac{y_{pE}}{\theta_0} \int_0^{\theta_0} J_{ME} \bar{J}'_{ME} \, \mathrm{e}^{-2s_E \cos\theta} \theta^2 \,\mathrm{d}\theta . \tag{60}$$

If in the integrands of A_{pnH}, B_{pnH}, C_{pnH}, A_{pnE}, B_{pnE}, C_{pnE} the subscripts H and E are interchanged, there result in succession E_{pnE}, D_{pnE}, F_{pnE}, E_{pnH}, D_{pnH}, F_{pnH}. There denote

$$s_H = -j \frac{\beta_{nH} a}{2 \sin\theta_0} , \qquad s_E = -j \frac{\beta_{nE} a}{2 \sin\theta_0} \tag{61), (62}$$

$$J_{MH} = J_M\left(y_{nH} \frac{\theta}{\theta_0} \right) , \qquad \bar{J}_{MH} = J_M\left(y_{pH} \frac{\theta}{\theta_0} \right) , \tag{63), (64}$$

$$J_{ME} = J_M\left(y_{nE} \frac{\theta}{\theta_0} \right) , \qquad \bar{J}_{ME} = J_M\left(y_{pE} \frac{\theta}{\theta_0} \right) . \tag{65), (66}$$

The quantities marked with an asterisk in (49) and (50) are the complex conjugate values. The primed quantities in (55) to (60), e.g. J'_{MH}, are the derivatives with respect to the argument. With $M = 0$ there results from (50) eqn (56) in Piefke, (1961).

With $\cos\theta = 1 - \theta^2/2$ and expansion of the exponential function into a power series the integrals can be reduced to integrals of the form

$$\int x^m Z_p(kx)\bar{Z}_q(lx)\,dx\,, \qquad m = -1,0,1,2,\ldots \tag{67}$$

which can be solved corresponding to Straubel (1942) by (59) in Piefke (1961). In (67) $Z_p(kx)$ and $\bar{Z}_q(1x)$ denote cylinder functions of the orders p and q, respectively.

3.3. *Power*

The power of a mode is

$$P = P_w + jP_{bl} = \frac{1}{2}\int_F [EH^*]\,dF\,, \tag{68}$$

if the amplitudes of the field strengths are inserted. In (68) P_w is the active, P_{bl} the reactive power. H^* is the complex conjugate of H. The quantity dF is the surface element.

After inserting (29), (30), (35), (36) there result the power values listed in the following, corresponding to (43) to(45).

(a) *Waveguide*

(1) H_{Mn}-mode: $P_{HMn} = |r_{nH}|^2 \dfrac{E_0^2a^2}{Z_{nH}}\dfrac{\pi y_{nH}^2}{2\delta_M}[J_M(y_{nH})]^2\left[1-\left(\dfrac{M}{y_{nH}}\right)^2\right]$,

$$M = 0,1,2,\ldots;\quad n = 1,2,3,\ldots, \tag{69}$$

(2) E_{Mn}-mode: $P_{EMn} = |r_{nE}|^2 E_0^2 a^2 \dfrac{Z_{nE}}{Z_0^2}\dfrac{\pi y_{nE}^2}{2\delta_M}[J_{M+1}(y_{nE})]^2$,

$$M = 0,1,2,\ldots;\quad n = 1,2,3,\ldots. \tag{70}$$

(b) *Conical horn*

(1) H_{Mp}-mode: $P_{HMp} = (P_{HMp})_w + j(P_{HMp})_{bl} =$

$$= |t_{pH}|^2 \frac{E_0^2a^2}{Z_{pH}^{(2)*}}\frac{\pi^2 y_{pH}^2}{4\delta_M}\beta_0 r|H_\eta^{(2)}(\beta_0 r)|^2[J_M(y_{pH})]^2\left[1-\left(\frac{M}{y_{pH}}\right)^2\right], \tag{71}$$

$$(P_{HMp})_w = |t_{pH}|^2 \frac{E_0^2a^2}{Z_0}\frac{\pi y_{pH}^2}{2\delta_M}[J_M(y_{pH})]^2\left[1-\left(\frac{M}{y_{pH}}\right)^2\right], \tag{72}$$

$$M = 0,1,2,\ldots;\quad p = 1,2,3,\ldots.$$

(2) E_{Mp}-mode: $P_{EMp} = (P_{EMp})_w + j(P_{EMp})_{bl} =$

$$= |t_{pE}|^2 E_0^2 a^2 \frac{Z_{pE}^{(2)}}{Z_0^2}\frac{\pi^2 y_{pE}^2}{4\delta_M}\beta_0 r|H_\zeta^{(2)}(\beta_0 r)|^2[J_{M+1}(y_{pE})]^2, \tag{73}$$

$$(P_{EMp})_w = |t_{pE}|^2 \frac{E_0^2 a^2}{Z_0} \frac{\pi y_{pE}^2}{2\delta_M} [J_{M+1}(y_{pE})]^2 , \tag{74}$$

$$M = 0,1,2, \ldots ; \quad p = 1,2,3, \ldots .$$

ACKNOWLEDGEMENT

The author thanks Mrs. Strube for the numerical calculation of the curves. She has in particular programmed the electronic computor used for the calculation, supervised the calculating processes, and evaluated the results.

REFERENCES

MAGNUS, W., und OBERHETTINGER, F. (1948) *Formeln und Sätze für die speziellen Funktionen der mathematischen Physik*, 2. Aufl. Springer-Verlag, Berlin.

PIEFKE, G. (1954) Die Ausbreitung elektromagnetischer Wellen in einem Pyramidentrichter, *Z. angew. Phys.* **6**, No. 11, 499–507 (Nov.)

PIEFKE, G. (1961) Reflexion und Transmission beim Einfall einer H_{0n}-Welle auf einen kegelförmigen Übergang zwischen zwei Hohlleitern, *Arch. elektr. Übertrag* **15**, No. 10, 444–454 (Oct.)

SCHNETZLER, K. (1960) Die Reflexion der Grundwelle an den Knickstellen eines Hohlleiters, insbesondere bei einem stetigen Übergang von einem rechteckigen auf einen runden Hohlleiter, *Arch. elektr. Übertrag.* **14**, No. 4, 177–182 (April).

STRAUBEL, R. (1942) Unbestimmte Integrale mit Produkten von Zylinderfunktionen (II. Mitteilung), *Ingenieur-Archiv* **13**, 14–20.

SCATTERING DIAGRAMS IN ELECTROMAGNETIC THEORY

GEORGES A. DESCHAMPS

University of Illinois, Urbana, Illinois, U.S.A.

ABSTRACT

The formalism of quantum mechanics, as described for example by Dirac (1947), suggests a notation convenient also in classical field theories. This notation leads to simple representations by means of diagrams, or graphs, similar to those introduced by Feynman (1949). In this paper both the notation and the associated diagrams are adapted to problems in electromagnetics. They are applied mostly to a review of known results: equivalence principles, formulation of scattering and diffraction problems, Born–Kirchhoff approximations. It will be clear, however, that they provide a systematic method for formulating problems and expressing their solutions. Furthermore, as illustrated by the problem of propagation over a mixed boundary, the method can be used as a tool for finding meaningful approximations.

1. INTRODUCTION

The concept of field was introduced in physics as a means of representing interactions at a distance between objects placed in a given environment. The field plays the role of an intermediary between interacting objects A and B: the object A determines, or "produces", a field F_a independently of B, and the action of A on B is a function of F_a and B independently of A. This scheme describes, for example, the force between two masses in Newtonian mechanics or between two charges in electrostatics. It applies also to a wide variety of situations in physics where the "action" of A on B may be a quantity other than a force.

Cases of most interest are those where the "action" can be described by a scalar function linear with respect to A and B. This implies that the objects under consideration can be represented by elements in a linear vector space $\mathscr{L}$, somewhat as the state vectors in quantum mechanics. The effect of A on B can be denoted by

$$\langle AGB \rangle \tag{1}$$

representing a bilinear function of A and B specified by G.

Expression (1) can be considered as a scalar product of $\langle A$ by $GB\rangle$ or of $\langle AG$ by $B\rangle$. Thus $\langle AG$, element of a linear space $\mathscr{F}$ dual of $\mathscr{L}$, is the field

F_a due to A while G is a linear operator mapping $\mathscr{L}$ into $\mathscr{F}$. The operator G characterizes the environment and will be called its *propagator*.

It may help at this point, as it helps in reading Dirac, to think of $B\rangle$ as a column vector (ket), of $\langle A$ as a row vector (bra), and of the operator G as a matrix. The notation can also be made more explicit by introducing variables of integration x, y, similar to the indices in ordinary vector analysis,

$$\langle A(x)\ G(x, y)\ B(y)\rangle. \tag{2}$$

The variables x, y represent positions and eventually they also include a discrete index to take spin or polarization into account. In expression (2) the integration must be carried out with respect to the repeated variables, somewhat as in Einstein's Summation Convention.

A diagram is associated with the expressions (1) or (2) as shown on Fig. 1. The interacting objects A and B are represented by regions to which the

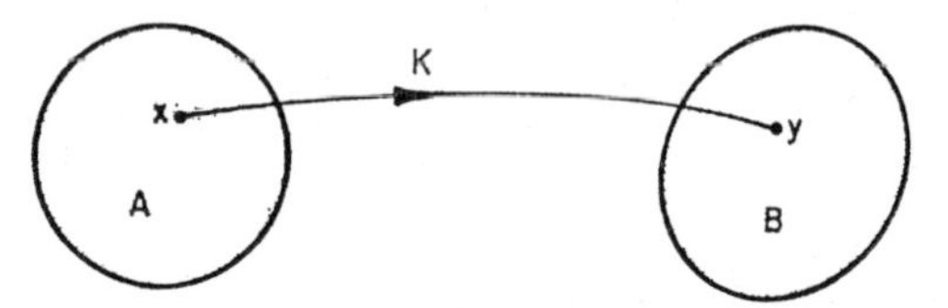

FIG. 1. Diagram representing $\langle AG_kB\rangle$.

points x and y belong. The propagator is represented by a line joining x to y and carrying eventually a label which characterizes the environment or medium. This line is oriented from A toward B to indicate an action of A on B. The action of B on A would be represented by a line oriented in the opposite sense, or by the expression $\langle BGA\rangle$ which in general could be different from $\langle AGB\rangle$. The change of order in (1) or (2) defines the transpose G^T of G by

$$G^T(x, y) = G(y, x) \tag{3}$$

or by the condition

$$\langle AG^TB\rangle = \langle BGA\rangle \tag{4}$$

for all A and B. When $G = G^T$, reciprocity holds, arrows on the branches of the diagrams can be omitted, and the distinction between the objects producing the field and those undergoing its effect is unnecessary.

To give full generality to this theory the elements A, B must not be restricted to ordinary functions. Distributions have to be included as they were in fact in Dirac's work. Integrations indicated by expression (2) are sometimes singular. For example, the support of $A(x)$ (complement of the open set over which A is zero) may be a volume, a surface, a line or a point. The corresponding proper element of integration must be used in each case.

2. ELECTROMAGNETIC FIELDS AND SOURCES

Consideration will now be limited to problems in electromagnetics at a single frequency. The relevant quantities are represented by complex vectors. A field F is a pair of vectors (E, H) functions of position; and a source A is made up of two vector distributions (J, K), the electric and magnetic currents. The field F_a is related to its source A by Maxwell's equations

$$\mathcal{M} F_a = A \tag{5}$$

where $\mathcal{M}$ is the differential operator

$$\begin{bmatrix} -j\omega\epsilon & \text{curl} \\ -\text{curl} & -j\omega\mu \end{bmatrix}.$$

It is known that the other Maxwell's equations are then unessential and that charges can be considered as derived quantities. The environment or medium is defined by ϵ and μ, functions of position, and by side conditions on the field at some material boundaries and at infinity (radiation condition). It shall be assumed that these conditions together with expression (5) are sufficient to insure that the field is uniquely defined. It is then proper to say that the field F_a is produced by the source A and to represent it by

$$\langle F_a = \langle A G_1 \tag{6}$$

where G_1 is the propagator characterizing the environment.

To complete the description according to the introduction one must define the "effect" of the field $F_a = (E_a, H_a)$ on another source $B = (J_b, K_b)$. The appropriate choice is the integral of $J_b \,.\, E_a - K_b \,.\, H_b$ taken over the support of the current distribution B -- a quantity that Rumsey (1954) has called the *reaction*. The result is denoted† by $\langle F_a B\rangle$ or $\langle A G_1 B\rangle$. As will be seen in the next section, in a medium where ϵ and μ are scalars, reciprocity holds and there is no need to distinguish between sources and sinks.

To see how $\langle A G_1$ describes the electromagnetic field consider at some point in space an electric dipole p. Then $\langle A G_1 p\rangle$ gives the scalar product of the E field at that point by the vector p. This in short will be called the "field at p". Similarly taking a magnetic dipole q, $\langle A G_1 q\rangle$ is the "field at q", i.e. the scalar product of the H field at the location of q by the vector q.

The bilinear function G is truly a generalization of the Green's function of the medium. It reduces to it when A and B are dipoles. Conversely the value of G for a pair of arbitrary distributions A and B can be deduced from

† Note the difference between this notation and the original one proposed by Rumsey (1954). Rumsey considers the reaction as a number defined by two sources and their associated fields in a single environment. The environment is not explicitly shown in the notation, at least not in a systematic manner. Here the reaction is a scalar product $\langle$source, field$\rangle$ or if the field is due to some other source in an environment (1), it is expressed by $\langle$source A, propagator G_1, source $B\rangle$.

the Green's function by integration (sometimes in the sense of distributions) over A and B.

3. LORENTZ RELATION

3.1. *Statement of the Relation*

Consider, in a given medium specified by the constants ϵ and μ, two electromagnetic fields F_a and F_b which satisfy Maxwell equations respectively with the currents A and B

$$\begin{aligned} \mathcal{M}F_a &= A \\ \mathcal{M}F_b &= B. \end{aligned} \tag{7}$$

It is well known that these equations imply

$$\operatorname{div}(E_a \times H_b - E_b \times H_a) = (J_a \,.\, E_b - K_a \,.\, H_b) - (J_b \,.\, E_a - K_b \,.\, H_a). \tag{8}$$

By integration over a finite volume V, bounded by the surface $S = \partial V$, one obtains the Lorentz relation

$$\int_S (E_a \times H_b - E_b \times H_a) \,.\, n\mathrm{d}s$$
$$= \int_V (J_a \,.\, E_b - K_a \,.\, H_b)\,\mathrm{d}v - \int_V (J_b \,.\, E_a - K_b \,.\, H_a)\,\mathrm{d}v. \tag{9}$$

The surface S is oriented, the unit normal vector n pointing outward from the volume V.

Equation (9) holds under conditions more general than those stated originally. The two fields F_a and F_b may be produced in different media provided that these media coincide inside of the volume V. The characteristics (ϵ, μ) of these media may vary with position even discontinuously. The fields F_a and F_b do not need to be produced by A and B; they may fail to satisfy the radiation condition or some of the boundary conditions outside of V.

3.2. *Expression in Bra-Ket Notation*

The first task will be to recast eqn (9) in terms of the notation of the preceding sections. The right-hand side of eqn (9) is obviously a reaction. To indicate that the integration should be limited to the region V we introduce the characteristics function of this region. This function, which will be denoted by V is equal to 1 over V and to 0 on its complement. The right-hand side of eqn (9) becomes

$$\langle AVF_b \rangle - \langle BVF_a \rangle.$$

In order to express the left-hand side of eqn (9) we introduce a current distribution, carried by the surface S, and having the surface density

$$\begin{aligned} J_s &= n \times H_a \\ K_s &= -n \times E_a. \end{aligned} \tag{10}$$

This current distribution depends linearly on F_a and will be denoted by $\langle F_a S$, considering eqn (10) as the definition of the linear operator S associated with the oriented surface S. The Lorentz relation becomes

$$\langle F_a S F_b \rangle = \langle A V F_b \rangle - \langle B V F_a \rangle. \tag{11}$$

The symbol $\langle F_a S F_b \rangle$ can be computed for any pair of fields (F_a, F_b) and any surface S provided the flux of the vector $E_a \times H_b - E_b \times H_a$ through the surface S can be integrated. The result will be called the *cross-flux* of the fields F_a and F_b through S.

3.3. *Formal Properties*

The cross-flux depends linearly on the three factors F_a, S, F_b which means that it can be expanded as an ordinary product when some of the factors take the form of a sum.

If the order of the factors is reversed the result is

$$\langle F_b S F_a \rangle = - \langle F_a S F_b \rangle. \tag{12}$$

This may be expressed by saying that the transpose S^T of S is $-S$, i.e. corresponds to the same surface with reversed orientation.

The reaction $\langle A V F_b \rangle$ is also linear with respect to its three factors and V can be considered as an operator multiplying either A or F_b. This operator is obviously equal to its transpose, or symmetric, and we could write $\langle F_b V A \rangle$ instead of $\langle A V F_b \rangle$.

Both the reaction and the cross-flux integrals must be understood with complete generality remembering that A and B are distributions. Some computations lead to singular integrals that have to be evaluated by taking their "principal value".

3.4. *Discontinuities and Surface Impedance*

In proving eqns (9) or (11) for a medium which may be discontinuous it is advantageous to consider all differential operators in the sense of the theory of distributions (Schwartz, 1950/1). Boundary conditions at a transition between two continuous regions are then included automatically in the statement of Maxwell equations and no special proof is needed in that case. From a more elementary point of view the eqn (11) can be extended to piecewise continuous regions by using the boundary conditions deduced from the integral form of Maxwell equations.†

To illustrate this point, and also the use of the bra-ket notation, consider a finite region $V = V_1 + V_2$ made up of the sum of two continuous regions.

† The viewpoint of distributions is nothing more than a systematic use of this integral form which preserves the advantages of the differential notation.

The two regions are separated by a surface Σ oriented from V_1 to V_2 (Fig. 2).

The fields F_a and F_b are discontinuous across Σ but their tangential components are continuous. Let us displace the surface Σ by an infinitesimal

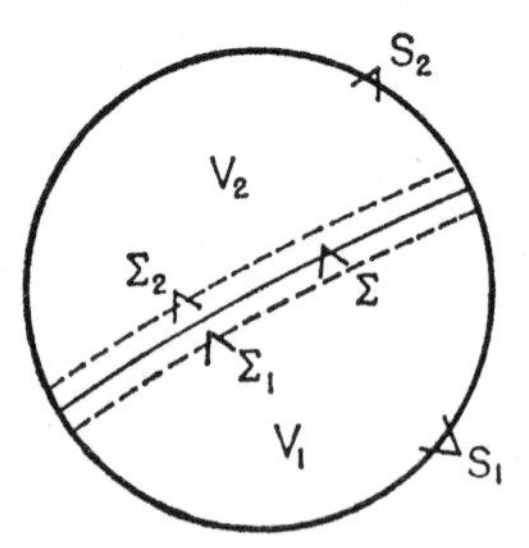

FIG. 2. Lorentz relation for a discontinuous region. The orientation of the surfaces is indicated by arrowheads.

amount in the positive direction to Σ_2 and in the negative direction to Σ_1. The boundaries of V_1, V_2, and V are

$$\begin{aligned} \partial V_1 &= S_1 + \Sigma_1 \\ \partial V_2 &= S_2 - \Sigma_2 \\ \partial V &= S_1 + S_2, \end{aligned}$$

and the Lorentz relation applied to V_1 and V_2 gives

$$\begin{aligned} \langle F_a(S_1 + \Sigma_1)\, F_b \rangle &= \langle A V_1 F_b \rangle - \langle B V_1 F_a \rangle \\ \langle F_a(S_2 - \Sigma_2)\, F_b \rangle &= \langle A V_2 F_b \rangle - \langle B V_2 F_a \rangle . \end{aligned}$$

Adding and using linearity, the following relation obtains

$$\langle F_a(S_1 + S_2 + \Sigma_1 - \Sigma_2)\, F_b \rangle = \langle A V F_b \rangle - \langle B V F_a \rangle .$$

The wanted relation follows from $\langle F_a(\Sigma_1 - \Sigma_2)\, F_b \rangle = 0$ which is satisfied since the tangential components of E and H are continuous across the surface Σ.

Another type of discontinuity occurs at the surface of an object which is not penetrated appreciably by the field. The condition on such a surface is sometimes adequately represented by an impedance relation of the form

$$E_{\text{tan}} = Zn \times H_{\text{tan}}. \tag{13}$$

If both F_a and F_b satisfy this condition on a surface S,

$$\langle F_a S F_b \rangle = 0. \tag{14}$$

This applies in particular at the surface of a conductor ($Z = 0$) or on a magnetic wall ($Z = \infty$).

3.5. *Radiation Condition and Reciprocity*

The field produced by a source of finite extent satisfies at infinity a radiation condition that has been expressed in various ways (Wilcox, 1959). A convenient form of this condition states that on a large sphere S, centered at some point in the source, the field satisfies asymptotically an impedance condition of the form of eqn (13) where Z is the intrinsic impedance of space. It results from this that two radiating fields have a zero cross-flux through a large sphere,

$$\langle F_a S F_b \rangle = 0. \tag{15}$$

When a region V contains no source of either field,

$$\langle F_a \partial V F_b \rangle = 0.$$

Thus when S is distorted continuously to another surface S' without crossing any current distribution, $S - S'$ is the boundary of a source-free region and

$$\langle F_a S F_b \rangle = \langle F_a S' F_b \rangle. \tag{16}$$

As a consequence of this property the surface S in eqn (15) may be replaced by any surface surrounding the sources A and B.

The Lorentz relation becomes

$$\langle A F_b \rangle = \langle B F_a \rangle \tag{17}$$

which is a well-known reciprocity theorem.

Re-introducing the propagator G of the medium this becomes

$$\langle AGB \rangle = \langle BGA \rangle \tag{18}$$

and proves a statement made in Section 2 about the symmetry of G.

If reciprocity did not apply, as in the case of a magnetized ferrite or of magneto plasma, G^T would be different from G, and the corresponding diagrams would have their branches oriented. With this precaution most of the following considerations would apply.

4. EQUIVALENCE PRINCIPLES

An equivalence principle is a rule for replacing some sources by others which produce the same effect in some particular region of space.

In circuit theory, Thevenin and Norton theorems show how sources inside a network can be replaced by a single source connected to its terminals. This source produces the same effect as the original one in any network connected to these terminals. In field theory the network becomes a region V of space and its terminals become the surface $S = \partial V$. Let A be a source inside V and B another one outside. If G_1 is the propagator of the medium, the problem is to express the reaction $\langle AG_1B \rangle$ in terms of some sources carried by S.

Applying Lorentz eqn (11) to the fields $G_1B\rangle$ and $G_1A\rangle$ over the region V gives

$$\langle AG_1SG_1B\rangle = \langle AVG_1B\rangle - \langle BVG_1A\rangle$$

or since $BV = 0$ and $AV = A$,

$$\langle AG_1SG_1B\rangle = \langle AG_1B\rangle \tag{19}$$

This is an expression of Huygen's principle; as far as B is concerned the source $\langle A$ may be replaced by $\langle AG_1S$.

If the two sources A and B are inside S, the Lorentz relation becomes

$$\langle AG_1SG_1B\rangle = 0 \tag{20}$$

which means that the Huygen's source $\langle AG_1S$ produces zero field inside V (zero effect on any B).

Since the fields $G_1A\rangle$ and $G_1B\rangle$ by definition obey the radiation condition, it can be shown that both eqns (19) and (20) apply even when V is not finite. The proof consists of reducing V to a finite region by taking its intersection with the inside of a surface S_∞ containing both sources. Making use of eqn (15) eliminates the contribution of S_∞. The proof of eqn (15) was based on the assumption that the region at large distance from the sources was homogeneous. It can, however, be generalized to other cases.

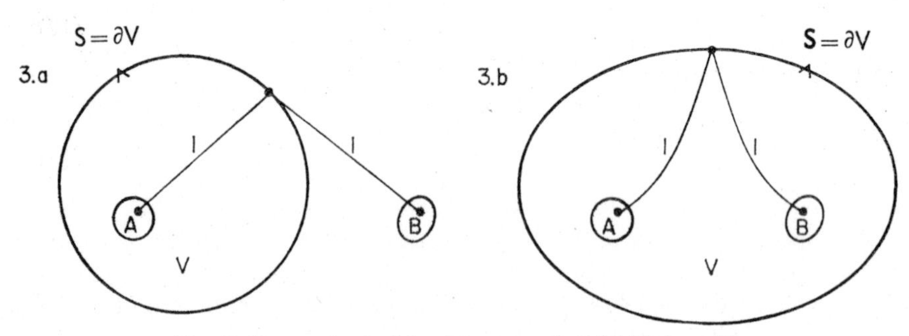

FIG. 3. Huygens' principle. (a) represents $\langle AG_1B\rangle$ (eqn 19); (b) is equivalent to 0 (eqn 20).

The diagrams representing the eqns (19) and (20) are shown in Fig. 3. A line labelled 1 joins a point in A to a point variable on S. Another line also labelled 1 joins this point to a point in B.

A generalization of eqn (19) obtains, if one considers a medium (2) which coincides with medium (1) inside V but may differ outside. The fields $G_2A\rangle$ and $G_1B\rangle$ can be combined in the Lorentz relation

$$\langle AG_2SG_1B\rangle = \langle AVG_1B\rangle - \langle BVG_2A\rangle$$

or since $BV = 0$ and $AV = A$,

$$\langle AG_2SG_1B\rangle = \langle AG_1B\rangle. \tag{21}$$

The diagram in Fig. 4b representing eqn (21) shows a line 2 from A to a point on S and a line 1 from that point to B. The result of the triple integration symbolized by this is equal to the result of the double integration indicated by the single line 1 joining A to B. If the two media were identical outside V and different inside V, the two labels 1 and 2 would be interchanged.

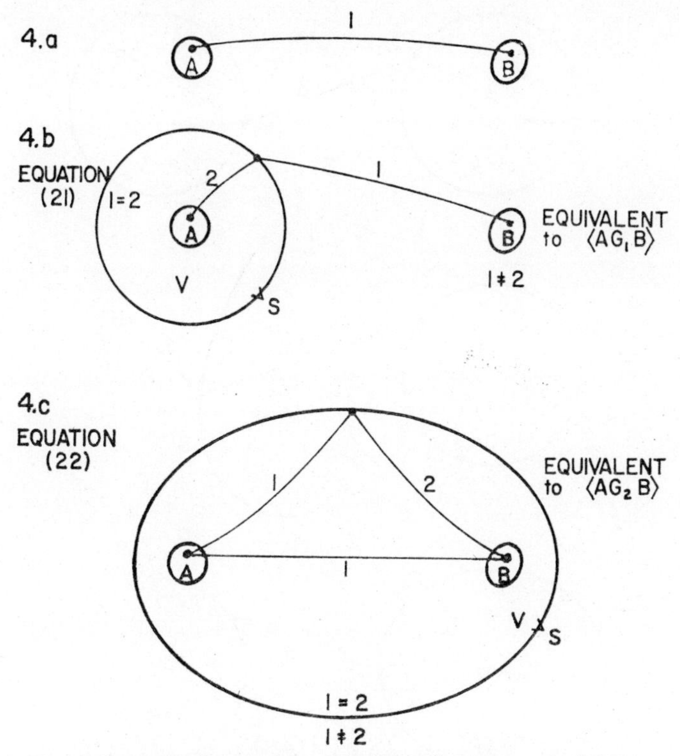

FIG. 4. Generalized equivalence principle. 4b represents $\langle AG_1B\rangle$ (eqn 21); 4c represents $\langle AG_2B\rangle$ (eqn 22).

Equation (21) shows that in medium 1 and as far as B is concerned, the source $\langle AG_2S$ is equivalent to the original source $\langle A$. This generalization of Huygen's principle is due to Schelkunoff (1951).

Equation (20) can also be generalized to a source A producing the field $G_1A\rangle$ in medium 1 and a source B producing the field $G_2B\rangle$ in medium 2. When the two media are identical inside a surface S which surrounds both A and B,

$$\langle AG_1SG_2B\rangle = \langle AG_2B\rangle - \langle AG_1B\rangle. \tag{22}$$

This relation is often used to express the effect of some modification of a medium on the reaction between two sources A and B. If the modification occurs outside of a surface S surrounding A and B, eqn (22) applies. The modified reaction $\langle AG_2B\rangle$ is equal to the original one $\langle AG_1B\rangle$ plus a term $\langle AG_1SG_2B\rangle$ which is the cross-flux through S of the field produced by A in (1) and the field produced by B in (2). The diagram in Fig. 4c shows this equivalence.

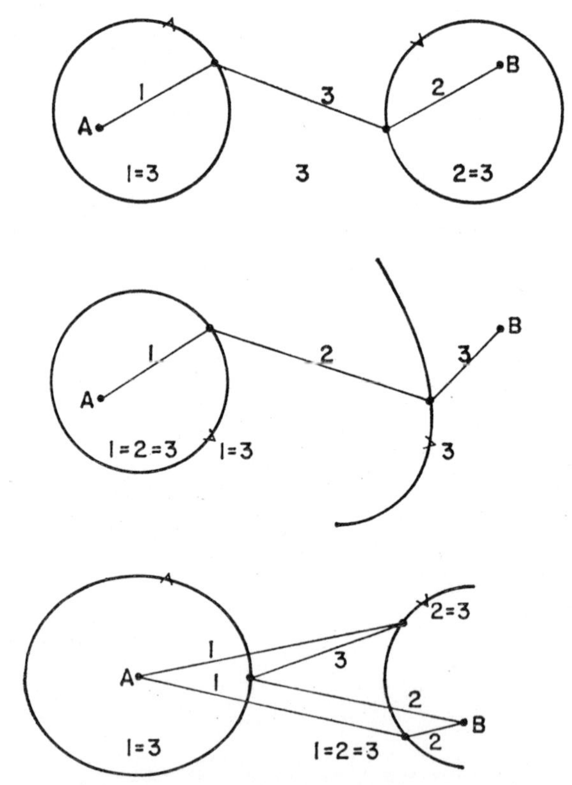

FIG. 5. Three diagrams equivalent to $\langle AG_3B\rangle$. Regions where some of the media are identical are indicated.

Other extensions can be obtained by combining several of the diagrams in Figs. 3 and 4. Several examples are given in Fig. 5 where two intermediary surfaces are used and three different media, or environments, are involved. The regions where these media coincide are shown in the figure.

5. SCATTERING BY A CONDUCTOR

Let G_1 be the propagator when the conductor is absent and G_2 the propagator when the conductor is in place. If a surface S is drawn around the conductor, the conditions of eqn (22) and Fig. 4c are satisfied for a region

V outside of the conductor such that $\partial V = S + S_\infty$. The diagram in Fig. 6a gives the reaction $\langle AG_2B\rangle$. Since $\langle AG_2$ is a field satisfying the condition $E_{\tan} = 0$ on S, knowledge of $H_{\tan}$ is sufficient to solve the problem. This is equivalent to a knowledge of the current induced on the surface of the

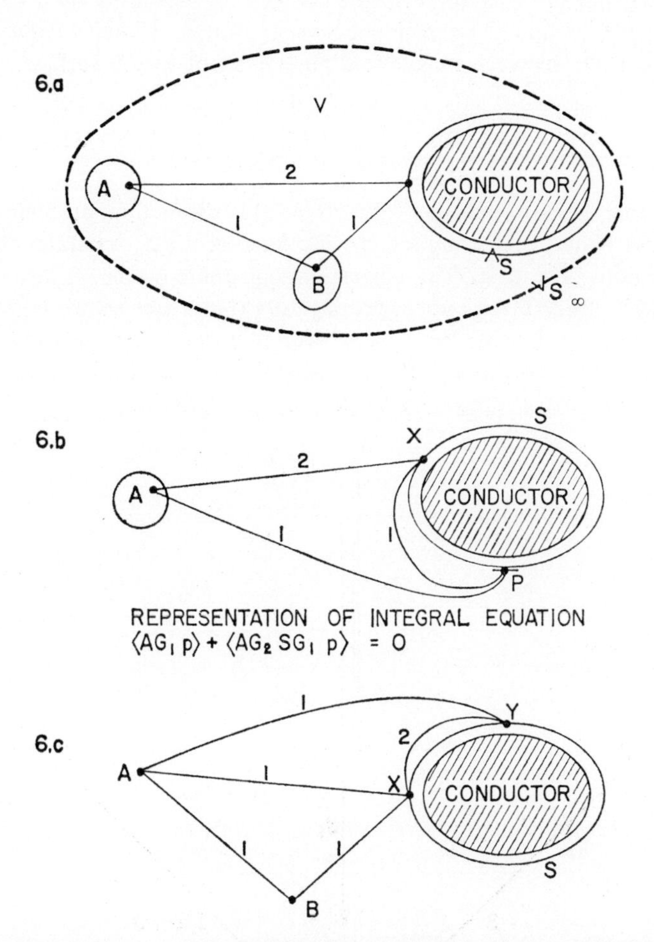

FIG. 6. Scattering by a conductor: (a) shows that the induced currents can be taken as unknown, (b) represents the integral equation for the currents, (c) is a reduction of the problem.

conductor. The usual integral equation for this current may be obtained by taking for B an electric dipole p tangent to S and writing that the reaction is zero. This equation is expressed by the diagram in Fig. 6b.

A transformation of the original diagram (Fig. 6c) results from replacing the path $(A - 2 - x)$ by the sum of a direct path $(A - 1 - x)$ and a path

$(A - 1 - y - 2 - x)$. The term $\langle AG_1SG_1B \rangle$ reduces to zero because of eqn (20) and the result is

$$\langle AG_2B \rangle = \langle AG_1B \rangle + \langle AG_1SG_2SG_1B \rangle. \tag{23}$$

This shows that in order to solve the problem of scattering by a conductor it is sufficient to find the reaction between sources carried by the surface, more specifically between magnetic dipoles tangent to the surface.

6. DIFFRACTION THROUGH AN APERTURE IN A PLANE SCREEN

As another illustration some aspects of this diffraction problem will be reviewed. A plane conducting screen with an aperture S_1 separates the space in two regions, I and II. The corresponding unknown propagator will be designated by G_3 and the known propagators G_0 for free space, G_1 for a full conducting screen, and G_2 for a plane magnetic wall will be introduced. The last two are obtained through the consideration of images. What use can be made of the known propagators to simplify the problem?

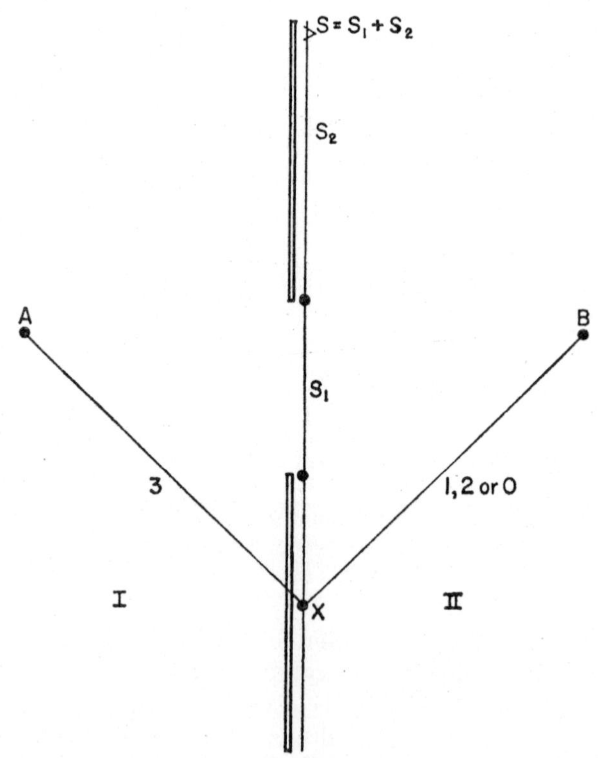

FIG. 7. Diffraction through an aperture.

Consider the surface S fitting the screen and the aperture on side II. Assume A in I and B in II. The equivalence eqn (21) can be applied as shown in Fig. 7. The line from A to a point x on S is labelled 3 and the line from x to B may be labelled either 0, 1, or 2 since all these media coincide with 3 over II. Taking for example G_1,

$$\langle AG_3B\rangle = \langle AG_3SG_1B\rangle. \tag{24}$$

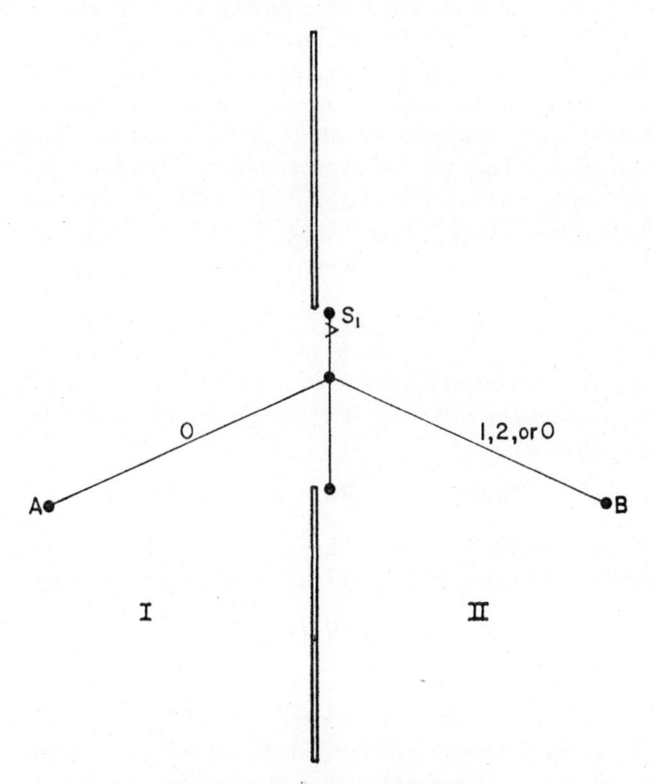

FIG. 8. Kirchhoff approximations.

If we decompose S into $S_1 + S_2$, the aperture S_1 and the screen S_2, we note that $\langle AG_3S_2G_1B\rangle = 0$ since $G_1B\rangle$ and $G_3A\rangle$ satisfy the same impedance condition on S (14). Equation (24) reduces to $\langle AG_3S_1G_1B\rangle$ which would be known if the tangential electric field $\mathbf{e}$ in $G_3A\rangle$ was known. The tangential magnetic field does not enter because the electric field in $G_1B\rangle$ is normal to S_1. A conclusion is that $\mathbf{e}$ can be taken as the unknown function for the problem.

Using the equivalence $\langle AG_3SG_2B\rangle$ shows in a similar manner that the tangential magnetic field $\mathbf{h}$ on the surface S_2 (back of the screen) can be taken as the unknown function. The contribution $\langle AG_3S_1G_2B\rangle$ is not zero but is rigorously equal to $\langle AG_0S_1G_2B\rangle$ and therefore is known. This equality results from symmetry considerations which show that the actual H_{tan}-field in the aperture equals the incident H_{tan}-field.

The Kirchhoff approximation consists of assuming that the field is zero over S_2 and equals the incident field over S_1. For an electromagnetic problem different meanings can be given to the last statement since it can be made to apply to E_{tan}, H_{tan}, or both. The corresponding approximations shown in Fig. 8, are $\langle AG_0S_1G_1B\rangle$, $\langle AG_0S_1G_2B\rangle$, and $\langle AG_0S_1G_0B\rangle$. The last one, often used for large apertures, is sometimes known as the Stratton-Chu formula. If $\langle AG_0S_1$ is a current distribution (J_s, K_s), the first two formulas can be transformed by using images to $2\langle K_sG_0B\rangle$ and $2\langle J_sG_0B\rangle$ while the last one gives $\langle(J_s, K_s)G_0B\rangle$. Thus the last approximation is the average of the first two.

7. SCATTERING BY A DIELECTRIC REGION

A medium described by the propagator G_1 is modified over a volume V by changing its dielectric constant from ϵ_0 to ϵ. The propagator becomes G_2. By using the method which led from eqn (7) to the Lorentz eqn (9), but taking into account the difference between the operators $\mathscr{M}$, the following relation is established

$$\langle AG_2B\rangle = \langle AG_1B\rangle + \langle AG_2MG_1B\rangle. \tag{25}$$

Here M is a diagonal operator which represents the scattering properties of the volume V. Showing the variables of integration the last term becomes

$$\langle A(x)\ G_2(x, y)\ M(y)\ G_1(y, z)\ B(z)\rangle. \tag{26}$$

Explicitly $M(y)$ represents the following linear operation on the field $\langle AG_2$; take the electric vector in this field at point y, multiply it by $j\omega(\epsilon - \epsilon_0)$, and consider the result as the electric current density of a new source $\langle AG_2M$. The scattering diagram corresponding to (25) is shown in Fig. 9a. If $\epsilon - \epsilon_0$ is small, the field $\langle AG_2$ does not differ much from $\langle AG_1$. Substituting in eqn (25) gives the Born approximation†

$$\langle AG_2B\rangle \simeq \langle AG_1B\rangle + \langle AG_1MG_1B\rangle. \tag{27}$$

Another equivalent diagram (9b) results from replacing, according to the initial scheme, the branch $(x - 2 - y)$ by the sum

$$(x - 1 - y) + (x - 2 - u - 1 - y).$$

† It should be noted that the field of $\langle AG_2M$ or of $\langle\ AG_1M$ inside V leads to a singular integral which must be reuglarized (van Bladel, 1961).

Replacing 2 by 1 then gives the next approximation. Continuing this process leads to the Neumann Series for the operator G_2. This series can also be

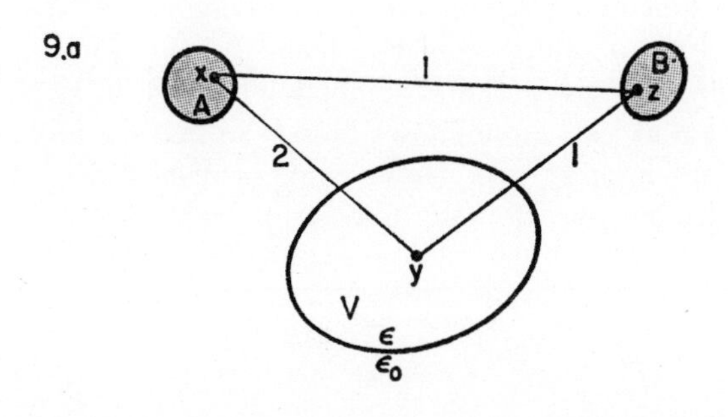

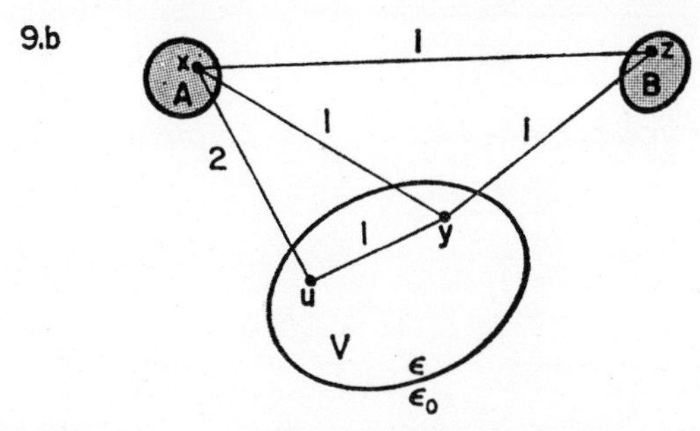

FIG. 9. Scattering diagrams for a dielectric region. Born approximation consists of replacing 2 by 1 on the xy branch in 9a. The next approximation is shown in 9b.

obtained by dropping A and B in eqn (25), thus writing eqn (25) in operator form

$$G_2 = G_1 + G_2MG_1 \tag{28}$$

and expanding formally

$$G_2 = G_1(1 - MG_1)^{-1}. \tag{29}$$

The series converges when $\epsilon - \epsilon_0$ is small.

8. PROPAGATION OVER A MIXED BOUNDARY

As a last example let us consider the problem of propagation from antenna A to antenna B over a flat earth made up of two sections having different surface impedance Z_1 and Z_2. Call G_3 the unknown propagator for this composite region and G_1, G_2 the known propagators for an infinite plane earth with surface impedance respectively Z_1 and Z_2. For the surface $S = S_1 + S_2$ covering the ground with S_1 over Z_1 and S_2 over Z_2, several equivalent diagrams are shown in Fig. 10. The four diagrams 10a, 10b, 10c, and 10d are rigorous expressions of $\langle AG_3B \rangle$. Diagram (a) is obtained by considering that medium (3) results from modifying Z_1 into Z_2 on the surface S_2. Diagram (b) is obtained similarly, 1 and 2 being exchanged. Applying twice the diagram (a) gives (c) and combining (a) and (b) gives (d). All these diagrams contain at least an unknown branch so that the problem is at most transformed into another one. However, approximations can now be obtained by neglecting one of these branches or replacing it by a branch of

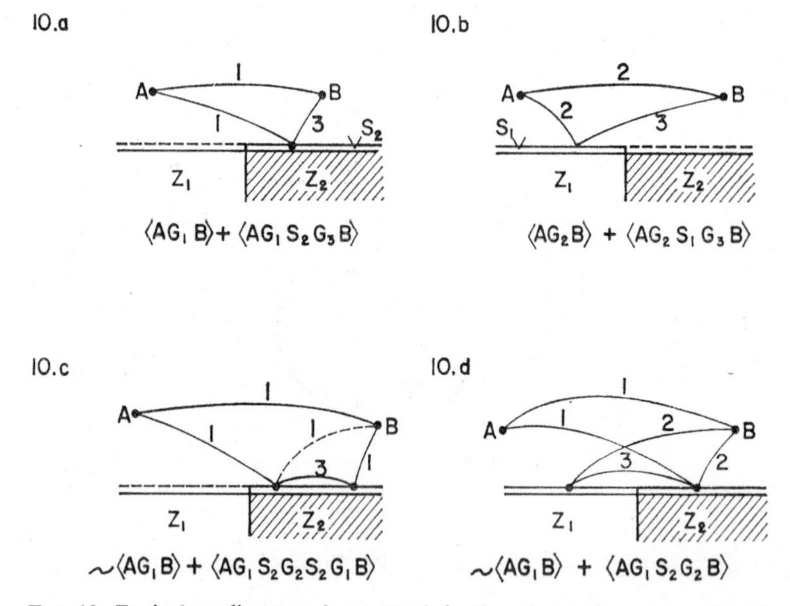

FIG. 10. Equivalent diagrams for $\langle AG_3B \rangle$ in the propagation over a mixed boundary. In 10c the dotted branch can be omitted since $\langle AG_1S_2G_1B \rangle = 0$.

known propagator approximately equivalent to it. Assume that A and B are respectively above each of the two surfaces S_1, S_2 and close to the ground. In diagram (d) the path $\langle AG_1S_2G_3S_1G_2B \rangle$ can be neglected because the phase of the differential contribution to the integral varies rapidly. For the same reason it is sufficient to integrate $\langle AG_1S_2G_2B \rangle$ from the intersection of the two regions up to the projection of B on the ground. In diagram (c) the path $\langle AG_1S_2G_1B \rangle$ is zero and in $\langle AG_1S_2G_3S_2G_1B \rangle$ one can replace G_3 by G_2 because the reaction of two sources on the ground depends mostly on the properties of the ground between them. The reduction of the problem to one dimension, discussed for example by Bremmer (1954) justifies this statement. Furthermore, the scattering diagram applies to the reduced scalar problem and results can be deduced in this manner which were previously obtained by Feinberg (1945), Wait (1956), and others.

9. CONCLUSION

The notations proposed are close relatives of similar ones used successfully in physics. They have been applied here mostly to a re-formulation of known problems in order to facilitate their assimilation by the reader and to simplify the exposition. It will be recognized, however, that they provide a systematic method of formulating many other problems.

This method is abstract in the sense that it could apply without change of form to different situations, scalar waves or vector waves of various dimensions satisfying various partial or ordinary differential equations. It is somewhat related to the abstract point of view advocated by Marcuvitz (1962). Solutions can be translated from one domain to another by a simple re-interpretation of the formalism.

The representation by means of diagrams in a convenient notation. It is an effective shorthand which simplifies the handling of complicated systems of integral equations. Finally it can suggest approximate solutions to specific problems.

REFERENCES

BREMMER, H. (1954) The extension of Sommerfeld's formula for the propagation of radiowave over a flat earth to different conductivities of the soil, *Physica* **20**, 441–460.

DIRAC, P. A. M. (1947) *The Principles of Quantum Mechanics*, 3rd edition. Oxford University Press.

FEINBERG, E. L. (1944) *J. Tech. Phys. USSR* **8**, 317; (1945) ibid. **9**, 1.

FEYNMAN, R. P. (1949) The theory of positrons, *Phys. Rev.* **76**, 749–769.

MARCUVITZ, N. (1962) Abstract operator methods in electromagnetic diffraction, *Electromagnetic Waves*, edited by R. E. Langer, 109–128, The University of Wisconsin Press, Madison.

RUMSEY, V. H. (1954) Reaction concept in electromagnetic theory, *Phys. Rev.* **94**, 1483–1491 (June).

SCHELKUNOFF, S. A. (1951) Field equivalence theorems, *Comm. on Pure and Appl. Math.* **4**, 43–59.

SCHWARTZ, L. S. (1950/1) *Theorie des Distributions*, Paris.

VAN BLADEL, J. (1961) Some remarks on Green's dyadic for infinite space, *I.R.E. Trans.* **AP-9**, 563–566 (November).

WAIT, J. R. (1956) Mixed path ground propagation, *NBS* **57**, 1.

WILCOX, H. (1959) Spherical means and radiation conditions, pp. 133–148, *Archive for Rational Mechanics and Analysis* **3**, No. 2. Springer-Verlag, Berlin.

GENERALIZED VARIATIONAL PRINCIPLES FOR ELECTROMAGNETIC VIBRATIONS; APPLICATION TO THE THEORY OF WAVEGUIDE JUNCTIONS

DAVID M. KERNS

National Bureau of Standards, Boulder, Colorado

ABSTRACT

Variational principles for boundary-value problems of the electromagnetic field in finite regions and with exponential time-dependence are generalized to apply in the case of arbitrary linear media. The generalized variational principles represent generalizations of a type originally given by Sir Joseph Lanmon; they cover both forced- and free-vibration problems, and are stated in the context of a theory of waveguide junctions.

1. INTRODUCTION

We consider the problem of determining the complex vectors **E**, **H** of the electromagnetic field in a finite region V having a boundary S; the tangential components of **E** and **H** are respectively to take on prescribed values on B_1 and on B_2, where B_1 and B_2 are complementary parts of S. Dependence of all electromagnetic field quantities upon the time t is represented by the omitted factor $\exp(j\omega t)$, where j is the imaginary unit and $\omega/(2\pi)$ is the frequency.

Variational principles for such boundary-value problems in the case of homogeneous media characterized by real symmetric dyadic permittivity and permeability have been given by Synge (1957).

In this paper variational principles of the type given by Synge are generalized to apply in the case of arbitrary linear media, for which the underlying differential equations are not self-adjoint, and to apply to mixed integral products between distinct fields (corresponding to distinct boundary conditions) that can exist in the given system or in its adjoint. Both of these generalizations require the construction of bilinear, rather than quadratic, functional forms. The generalized variational principles follow a pattern abstractly indicated by Lippmann (1955) and attributed to Schwinger.

The variational principles given in this paper (as well as those given by Synge) are fairly closely related to a variational principle first given by

Sir Joseph Lanmon (1900), which yields the vacuum Maxwell's equations (including arbitrary time-dependence) as necessary conditions. The variational principles now considered include explicitly the effects of sources as represented by boundary conditions and determine the field, as well as the field equations, for the case of exponential time-dependence.

Although the emphasis here is upon the consideration of forced vibrations, it is noted below that if homogeneous boundary conditions are prescribed, the generalized variational principles become variational statements of the vectorial eigenfunction and eigenvalue problem for electromagnetic fields. As such, they represent generalizations of formulas given by Berk (1956) and by Rumsey.

At some risk of partly obscuring the variational principles themselves, they are stated in the context of a theory of waveguide junctions. In this context the quantities rendered stationary acquire significance as elements of the impedance and the admittance matrices of the junction considered. The needed formulation of the elements of a theory of waveguide junctions has been given as a part of another paper (Kerns, 1951). A brief restatement of what is needed is made here, with certain simplifications and adaptations.

2. ELECTROMAGNETIC FORMULATION

For the present purposes, a waveguide junction is a linear electromagnetic system possessing ideal waveguide leads and is subject to excitation only through nonattenuated modes in these leads. The domain of the electromagnetic field is the finite region V; the surface S, the complete boundary of V, consists of a part S_0 coinciding with a perfectly conducting surface, and the parts $S_1, S_2, \ldots, S_n$ where S_m is the terminal surface in the mth of the n waveguide leads (Fig. 1). Within V the complex vectors $\mathbf{E}$, $\mathbf{H}$ of the time-exponential electromagnetic field satisfy Maxwell's equations, which are written

$$\mathbf{E} = \mathscr{E}(\mathbf{H}), \qquad \mathbf{H} = \mathscr{H}(\mathbf{E})$$

using the operators

$$\mathscr{E} = (j\omega\epsilon)^{-1} \cdot \nabla\times, \qquad \mathscr{H} = -(j\omega\mu)^{-1} \cdot \nabla\times \tag{1}$$

as abbreviations. Here μ, ϵ are in general complex nonsymmetric dyadic point-functions, which reduce to real scalar constants in the ideal portions of the waveguides.

From waveguide theory it is known that the tangential components $\mathbf{E}_t$, $\mathbf{H}_t$, of $\mathbf{E}$, $\mathbf{H}$ on the terminal surface S_m are expressible in the form

$$\mathbf{E}_t = \sum_{\mu=1}^{\nu_m} v_{m\mu}\, \mathbf{e}^o_{m\mu}, \qquad \mathbf{H}_t = \sum_{\mu=1}^{\nu_m} i_{m\mu}\, \mathbf{h}^o_{m\mu}, \tag{2}$$

where $\mathbf{e}^o_{m\mu}$ and $\mathbf{h}^o_{m\mu}$ are real basis-fields for the mode μ in waveguide m, $v_{m\mu}$

and $i_{m\mu}$ are scalar coefficients, and ν_m is the number of propagated modes in waveguide m. These terminal or waveguide basis-fields are subject to (i) the power normalization and orthogonality relations

$$\int_{S_m} \mathbf{e}^o_{m\mu}\, \mathbf{h}^o_{m\lambda}\, \mathbf{n}_m\, \mathrm{d}S = \delta_{\mu\lambda}, \tag{3}$$

where $\delta_{\mu\lambda}$ is a Kronecker delta and $\mathbf{n}_m$ denotes the inward normal on S_m

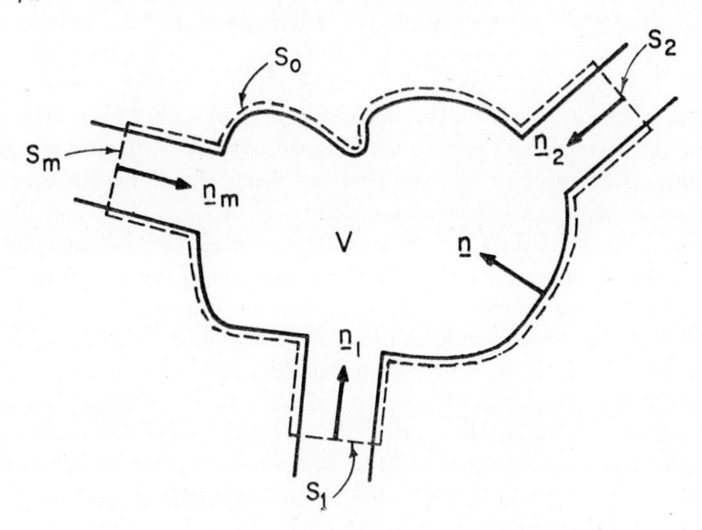

FIG. 1. Schematic illustration of region V and surfaces $S_0, S_1, \ldots\ S_n$.

(here and subsequently integrands in surface integrals are scalar triple products), and to (ii) the impedance normalization

$$\mathbf{h}^o_{m\mu} = \zeta^o_{m\mu}\, \eta_{m\mu}\, \mathbf{n}_m \times \mathbf{e}^o_{m\mu}, \tag{4}$$

where $\eta_{m\mu}$ is the wave-admittance of mode μ in waveguide m and $\zeta^o_{m\mu}$ is the arbitrary characteristic impedance of this mode. These normalizations determine the terminal basis-fields up to the choice of a sign.

On S_0, the homogeneous boundary condition $\mathbf{n} \times \mathbf{E} = 0$ applies. The additional prescription $v_{l\lambda} = \delta_{lm}\delta_{\lambda\mu}$ for given m and μ determines (through (2)) an electromagnetic field in V, which is denoted by $\mathbf{e}_{m\mu}, \mathscr{H}(\mathbf{e}_{m\mu})$. Similarly the prescription $i_{l\lambda} = \delta_{lm}\delta_{\lambda\mu}$ determines an electromagnetic field denoted by $\mathscr{E}(\mathbf{h}_{m\mu}), \mathbf{h}_{m\mu}$. The fields $\mathbf{e}_{m\mu}$ and $\mathbf{h}_{m\mu}$ are appropriately called electric and magnetic junction basis-fields respectively. If we now define the impedance matrix Z and the admittance matrix Y of the junction by writing

$$v_{l\lambda} = \sum_{m\mu} Z_{l\lambda,\, m\mu} i_{m\mu}, \quad i_{l\lambda} = \sum_{m\mu} Y_{l\lambda,\, m\mu} v_{m\mu}, \tag{5}$$

then we have for the matrix elements the basic expressions

$$Z_{l\lambda,\, m\mu} = \int\limits_{S_l} \mathscr{E}(\mathbf{h}_{m\mu})\, \mathbf{h}_{l\lambda}\mathbf{n}_l \,\mathrm{d}S, \quad Y_{l\lambda,\, m\mu} = \int\limits_{S_l} \mathbf{e}_{l\lambda}\mathscr{H}(\mathbf{e}_{m\mu})\, \mathbf{n}_l \,\mathrm{d}S. \tag{6}$$

These may be verified with the aid of (1), (2), (3), (5), and the definitions of the junction basis-fields.

We observe that, in view of the prescribed boundary conditions for the fields involved, the integrals (6) could as well be taken over the entire boundary S instead of over just the terminal surface S_l. This property permits us to obtain volume-integral expressions for the immittance elements with the aid of the divergence theorem. The desired expressions, however, will involve fields of the system "adjoint" to the original one and will exhibit a generalized form of reciprocity relation that has been stated by Rumsey and is attributed to M. H. Cohen (Rumsey, 1954).

By the system "adjoint" to a given system is meant one having constitutive parameters $\hat{\mu}$, $\hat{\epsilon}$ equal respectively to the transposes of the μ, ϵ of the original system, and being otherwise identical to the original system. Terminal basis-fields and boundary conditions are to be the same for the adjoint system as for the original. If μ, ϵ are symmetric, the given system may be said to be "self-adjoint". In terms of operators similar to (1), Maxwell's equations for the adjoint system are $\hat{\mathbf{H}} = \hat{\mathscr{H}}(\hat{\mathbf{E}})$, $\mathbf{E} = \hat{\mathscr{E}}(\hat{\mathbf{H}})$. Quantities associated with the adjoint system are distinguished by the circumflex throughout this paper.

In what follows it will suffice to use single-letter indices (such as $p, q, \ldots$) to indicate both waveguide and mode.

We now rewrite expressions (6) as

$$Z_{pq} = \int\limits_{S} \mathscr{E}(\mathbf{h}_q)\, \hat{\mathbf{h}}_p\mathbf{n} \,\mathrm{d}S, \quad Y_{pq} = \int\limits_{S} \hat{\mathbf{e}}_p\mathscr{H}(\mathbf{e}_q)\, \mathbf{n} \,\mathrm{d}S, \tag{7a, b}$$

where the integrals now go over the entire surface S. The use of the basis fields $\hat{\mathbf{h}}_p$ and $\hat{\mathbf{e}}_p$ (instead of $\mathbf{h}_p$ and $\mathbf{e}_p$ which respectively have identical tangential components on S) indicates the continuation into the region V to be taken in forming the volume-integral expressions below. For the matrix elements of the adjoint system the expressions corresponding to (6) are

$$\hat{Z}_{qp} = \int\limits_{S} \hat{\mathscr{E}}(\hat{\mathbf{h}}_p)\, \mathbf{h}_q\mathbf{n} \,\mathrm{d}S, \quad \hat{Y}_{qp} = \int\limits_{S} \mathbf{e}_q\hat{\mathscr{H}}(\hat{\mathbf{e}}_p)\, \mathbf{n} \,\mathrm{d}S. \tag{8a, b}$$

Here the basis-fields $\mathbf{e}_p$ and $\mathbf{h}_p$ are used advisedly.

The following four equations may be obtained from (7) and (8) with the aid of the divergence theorem:

$$Z_{pq} = j\omega \int\limits_{V} [\hat{\mathbf{h}}_p \,.\, \mu \,.\, \mathbf{h}_q + \hat{\mathscr{E}}(\mathbf{h}_p) \,.\, \epsilon \,.\, \mathscr{E}(\mathbf{h}_q)]\,\mathrm{d}V = \hat{Z}_{qp}; \tag{9}$$

$$Y_{pq} = j\omega \int\limits_{V} [\hat{\mathscr{H}}(\hat{\mathbf{e}}_p) \,.\, \mu \,.\, \mathscr{H}(\mathbf{e}_q) + \hat{\mathbf{e}}_p \,.\, \epsilon \,.\, \mathbf{e}_q]\,\mathrm{d}V = \hat{Y}_{qp}. \tag{10}$$

These equations exhibit the generalized reciprocity relations; in the self-adjoint case the circumflex is without force (e.g. $\hat{Z}_{qp} = Z_{qp}$, $\hat{Y}_{qp} = Y_{qp}$) and the above equations yield the ordinary reciprocity relations.

We now have the expressions and relations needed for the construction of the variational principles.

3. VARIATIONAL PRINCIPLES

3.1. *Forced Vibrations*

Regarded as a bilinear functional of $\mathbf{E}$ and $\hat{\mathbf{F}}$ the expression

$$Z[\mathbf{E}, \hat{\mathbf{F}}] = \int_S (\mathbf{E}\hat{\mathbf{h}}_p + \hat{\mathbf{F}}\mathbf{h}_q)\,\mathbf{n}\,\mathrm{d}S - j\omega \int_V [\hat{\mathscr{H}}(\hat{\mathbf{F}}) \,.\, \mu \,.\, \mathscr{H}(\mathbf{E}) + \hat{\mathbf{F}} \,.\, \epsilon \,.\, \mathbf{E}]\,\mathrm{d}V, \quad (11)$$

in which $\mathbf{E}$ and $\hat{\mathbf{F}}$ are to satisfy $\mathbf{n} \times \mathbf{E} = \mathbf{n} \times \hat{\mathbf{F}} = 0$ on S_0, is a stationary expression for Z_{pq}. More specifically: if $Z[\mathbf{E}, \hat{\mathbf{F}}]$ is stationary with respect to all variations of $\mathbf{E}$ on $S - S_0$ and in V, then $\hat{\mathbf{F}} = \hat{\mathscr{E}}(\hat{\mathbf{h}}_p)$; and if $Z[\mathbf{E}, \hat{\mathbf{F}}]$ is stationary with respect to similar variations of $\hat{\mathbf{F}}$, then $\mathbf{E} = \mathscr{E}(\mathbf{h}_q)$. The converses of both parts of this statement hold. Moreover, the stationary value of $Z[\mathbf{E}, \hat{\mathbf{F}}]$ obtaining when both $\mathbf{E}$ and $\hat{\mathbf{F}}$ have their correct values is Z_{pq}.

The last-stated property becomes evident upon inserting the correct fields in (11) and referring to (7a), (8a), and (9). The other parts of the theorem may be established by using the methods of the variational calculus (see, for example, Courant-Hilbert (1953)). Thus varying $\mathbf{E}$ (and holding $\hat{\mathbf{F}}$ fixed), one obtains the first variation

$$\delta_E Z = \int_{S-S_0} \delta\mathbf{E}\hat{\mathbf{h}}_p\mathbf{n}\,\mathrm{d}S - j\omega \int_V [\hat{\mathscr{H}}(\hat{\mathbf{F}}) \,.\, \mu \,.\, \mathscr{H}(\delta\mathbf{E}) + \hat{\mathbf{F}} \,.\, \epsilon \,.\, \delta\mathbf{E}]\,\mathrm{d}V,$$

remembering that $\mathbf{n} \times \delta\mathbf{E} = 0$ on S_0. The first term in the integrand of the volume integral may be written

$$\hat{\mathscr{H}}(\hat{\mathbf{F}}) \,.\, \mu \,.\, \mathscr{H}(\delta\mathbf{E}) = -\frac{1}{j\omega}\{\delta\mathbf{E} \,.\, \nabla \times \hat{\mathscr{H}}(\hat{\mathbf{F}}) + \nabla \,.\, [\delta\mathbf{E} \times \hat{\mathscr{H}}(\hat{\mathbf{F}})]\};$$

applying the divergence theorem, one finds

$$\delta_E Z = \int_{S-S_0} \delta\mathbf{E}[\hat{\mathbf{h}}_p - \hat{\mathscr{H}}(\hat{\mathbf{F}})]\,\mathbf{n}\,\mathrm{d}S + \int_V [\nabla \times \hat{\mathscr{H}}(\hat{\mathbf{F}}) - j\omega\hat{\mathbf{F}} \,.\, \epsilon] \,.\, \delta\mathbf{E}\,\mathrm{d}V.$$

The requirement that $\delta_E Z$ should vanish for arbitrary $\delta\mathbf{E}$ on $S - S_0$ and in V therefore leads to the Euler differential equation

$$\nabla \times \hat{\mathscr{H}}(\mathbf{F}) - j\omega\hat{\epsilon} \,.\, \hat{\mathbf{F}} = 0 \quad (12)$$

holding in V and to the natural boundary condition

$$[\hat{\mathbf{h}}_p - \hat{\mathscr{H}}(\hat{\mathbf{F}})] \times \mathbf{n} = 0 \tag{13}$$

holding on $S - S_0$. Now, the fulfillment of the Euler equation (which turned out to be one of Maxwell's equations), together with the definition of $\hat{\mathscr{H}}(\hat{\mathbf{F}})$, insures that the pair of functions $\hat{\mathbf{F}}$, $\hat{\mathscr{H}}(\hat{\mathbf{F}})$ constitutes an electromagnetic field satisfying both of Maxwell's equations; furthermore, the natural boundary condition plus the *a priori* boundary condition on $\hat{\mathbf{F}}$ constitute the boundary conditions defining the junction magnetic basis-field $\hat{\mathbf{h}}_p$. Hence $\delta_E Z = 0$ implies $\hat{\mathbf{F}} = \hat{\mathscr{E}}(\hat{\mathbf{h}}_p)$. The converse is easily seen to hold for if $\hat{\mathbf{F}} = \hat{\mathscr{E}}(\hat{\mathbf{h}}_p)$, then (12) and (13) hold and hence $\delta_E Z = 0$.

In a similar fashion, by varying $\hat{\mathbf{F}}$ it may be shown that for $\delta_F Z$ to be zero it is necessary and sufficient that $\mathbf{E} = \mathscr{E}(\mathbf{h}_q)$. Thus the proof of the theorem may be completed.

A similar theorem holds for the admittance elements. Viz., the bilinear functional of $\mathbf{H}$, $\hat{\mathbf{M}}$

$$Y[\mathbf{H}, \hat{\mathbf{M}}] = \int_S (\hat{\mathbf{e}}_p \mathbf{H} + \mathbf{e}_q \hat{\mathbf{M}})\, \mathbf{n}\, \mathrm{d}S - j\omega \int_V [\hat{\mathbf{M}} \cdot \mu \cdot \mathbf{H} + \hat{\mathscr{E}}(\hat{\mathbf{M}}) \cdot \epsilon \cdot \mathscr{E}(\mathbf{H})]\, \mathrm{d}V, \tag{14}$$

with no *a priori* boundary conditions on $\mathbf{H}$, $\hat{\mathbf{M}}$, is a stationary expression for Y_{pq}. For the given expression to be stationary with respect to variations of $\mathbf{H}$ and $\hat{\mathbf{M}}$ it is necessary and sufficient that $\mathbf{H} = \mathscr{H}(\mathbf{e}_q)$ and $\hat{\mathbf{M}} = \hat{\mathscr{H}}(\hat{\mathbf{e}}_p)$; the stationary value is Y_{pq}. The proof, which is very similar to the preceding one, is omitted.

In the special case of diagonal immittance elements of a self-adjoint system, (11) and (14) may be replaced by quadratic variational expressions: for Z_{qq},

$$Z[\mathbf{E}] = 2 \int_S \mathbf{E}\mathbf{h}_q\mathbf{n}\, \mathrm{d}S - j\omega \int_V [\mathscr{H}(\mathbf{E}) \cdot \mu \cdot \mathscr{H}(\mathbf{E}) + \mathbf{E} \cdot \epsilon \cdot \mathbf{E}]\, \mathrm{d}V, \tag{15}$$

with $\mathbf{E}$ subject to $\mathbf{n} \times \mathbf{E} = 0$ on S_0; and for Y_{qq},

$$Y[\mathbf{H}] = 2 \int_S \mathbf{e}_q\mathbf{H}\mathbf{n}\, \mathrm{d}S - j\omega \int_V [\mathbf{H} \cdot \mu \cdot \mathbf{H} + \mathscr{E}(\mathbf{H}) \cdot \epsilon \cdot \mathscr{E}(\mathbf{H})]\, \mathrm{d}V, \tag{16}$$

with no *a priori* boundary condition on $\mathbf{H}$. The proofs differ only slightly from the pattern of the preceding. For example, for the variation of (16) one finds

$$\delta Y = 2 \int_S \mathbf{e}_q\, \delta\mathbf{H}\mathbf{n}\, \mathrm{d}S - 2j\omega \int_V [\mathbf{H} \cdot \mu \cdot \delta\mathbf{H} + \mathscr{E}(\mathbf{H}) \cdot \epsilon \cdot \mathscr{E}(\delta\mathbf{H})]\, \mathrm{d}V$$

using the symmetry of the tensor parameters. This expression may be transformed to

$$\delta Y = 2 \int_S [\mathbf{e}_q - \mathscr{E}(\mathbf{H})]\, \delta\mathbf{H}\mathbf{n}\, \mathrm{d}S - 2 \int_V [j\omega\mu \cdot \mathbf{H} + \nabla \times \mathscr{E}(\mathbf{H})] \cdot \delta\mathbf{H}\mathrm{d}V$$

and the remainder of the argument is evident.

3.2. *Free Vibrations*

We note that variational characterizations of free-vibration problems may be obtained essentially as corollaries of the variational principles given above. Suppose that it is required to determine the eigenfunctions and the eigenfrequencies for the region V containing an arbitrary linear medium, and subject to the simple homogeneous boundary conditions $\mathbf{n} \times \mathbf{E} = \mathbf{n} \times \hat{\mathbf{F}} = 0$ on B_1, $\mathbf{n} \times \mathbf{H} = \mathbf{n} \times \hat{\mathbf{M}} = 0$ on B_2, where (as in the first section of this paper) B_1 and B_2 are complementary parts of the boundary S of V.

To adapt (11) to the homogeneous problem we first replace $\hat{\mathbf{h}}_p$ and $\mathbf{h}_q$ in the surface integral by zeros. (One might identify B_1 with S_0 and B_2 with the aggregate of the terminal surfaces of a waveguide junction, but for the purposes of the present paragraphs the special geometry and the waveguide junction context are quite irrelevant.) Secondly, in order to make the frequency ω explicit, we replace the Maxwell operators $\mathscr{H}, \hat{\mathscr{H}}$ by their definitions. Thus one may obtain the bilinear functional

$$\int_V (\nabla \times \hat{\mathbf{F}} \,.\, \mu^{-1} \,.\, \nabla \times \mathbf{E} - \omega^2 \hat{\mathbf{F}} \,.\, \epsilon \,.\, \mathbf{E})\, \mathrm{d}V \tag{17}$$

in which $\mathbf{E}$ and $\hat{\mathbf{F}}$ are to satisfy the boundary condition on B_1 but are free on B_2. The correct fields render this expression stationary (about the value zero); the electric-field eigenfunctions are the non-trivial solutions of this variational problem, and the eigenfrequencies are the values of ω for which non-trivial solutions exist.

From the variational properties of (14) one may in a similar manner obtain an alternative formulation of the same free-vibration problem, this time in terms of magnetic-field functions. In this case one obtains the functional

$$\int_V (\omega^2 \hat{\mathbf{M}} \,.\, \mu \,.\, \mathbf{H} - \nabla \times \hat{\mathbf{M}} \,.\, \epsilon^{-1} \,.\, \nabla \times \mathbf{H})\, \mathrm{d}V \tag{18}$$

in which $\mathbf{H}$ and $\hat{\mathbf{M}}$ are to satisfy the boundary condition on B_2 but are free on B_1. The correct magnetic fields render this expression stationary (about the value zero); the magnetic-field eigenfunctions are the non-trivial solutions of this variational problem, and the eigenfrequencies are the values of ω for which non-trivial solutions exist.

REFERENCES

BERK, A. D. (1956) Variational principles for electromagnetic resonators and waveguides, *I.R.E. Trans. on Antennas and Propagation*, **AP-4,** 104–111 (April), with referee's comment by V. H. Rumsey.

COURANT, R. and HILBERT, D. (1953) *Methods of Mathematical Physics*. Interscience Publishers, New York, vol. 1, 1st English ed., chap. IV.

KERNS, D. M. (1951) Analysis of symmetrical waveguide junctions, *J. Research N.B.S.* **46,** 267–282 (April).

LANMON, J. (1900) *Aether and Matter*. Cambridge University Press, chap. VI

LIPPMANN, B. A. (1955) Variational formulation of a grating problem, Nuclear Development Associates report No. 18-8.

RUMSEY, V. H. (1954) The reaction concept in electromagnetic theory, *Phys. Rev.* **94,** 1483–1491 (June). Errata, **95,** 1705 (Sept.).

SYNGE, J. L. (1957) *The Hypercircle in Mathematical Physics.* Cambridge University Press, pp. 408–409.

THE SOMMERFELD–RUNGE LAW AND GEOMETRIC OPTICS IN FOUR DIMENSIONS†

H. POEVERLEIN‡

SUMMARY

The Sommerfeld–Runge law, expressing the irrotationality of the propagation vector, leads in four-dimensional generalization to a geometric optics in space-time, which covers wave-fields and media varying with time. The relationship between group and phase velocities $uv = c^2$ is seen to result from Lorentz invariance of the dispersion equation, which corresponds to isotropy in space-time. The four-dimensional Sommerfeld–Runge law is applicable to boundaries (surfaces moving in space or discontinuities in time) and to inhomogeneous media. The laws of group propagation emerge from the equation characterizing the medium together with the Sommerfeld–Runge law.

The Sommerfeld–Runge law in three dimensions states that the propagation vector in a wave-field is irrotational (Sommerfeld and Runge, 1911). The four dimensional-propagation vector§

$$\vec{\mathbf{k}} \equiv (\vec{k}, i\omega/c) \tag{1}$$

is subject to the irrotationality condition

$$\square \times \vec{\mathbf{k}} = 0 \tag{2}$$

as follows from the existence of a uniquely defined wave function. Equation (2) represents the four-dimensional generalization of the Sommerfeld–Runge law.

In application to boundaries between different media (in general moving boundaries), the four-dimensional Sommerfeld–Runge law requires continuity of the $\vec{\mathbf{k}}$ components parallel to the boundary in space-time, while the wave passes through the boundary. This continuity condition is the four-dimensional generalization of Snell's law. In addition to the refracted wave

† More detailed publications on "The Sommerfeld–Runge law in three and four dimensions" and on "Raum-Zeit-Isotropie der Wellenausbreitung" will appear elsewhere.

‡ Air Force Cambridge Research Laboratories, Bedford, Mass.—Presently: Elektrophysikalisches Institut der Technischen Hochschule, München, Germany.

§ Bold type is used to characterize a vector as four-vector. The symbol $\square$ denotes the vector operator

$$\left(\frac{\partial}{\partial x}, \frac{\partial}{\partial y}, \frac{\partial}{\partial z}, -\frac{i}{c}\frac{\partial}{\partial t}\right)$$

there is a reflected wave, which in combination with the incident wave must fulfill the same continuity condition at the boundary.

A medium that may be anisotropic in space-time and inhomogeneous is characterized by the equation

$$F(\vec{\mathbf{r}}, \vec{\mathbf{k}}) = 0 \tag{3}$$

with a given function F. Equations (2) and (3) are usable in dealing with modulated waves and (not too fast) variations of the medium with time. A geometrical-optical approximation of a wave-field, disregarding amplitudes, is derivable from (2) and (3) when $\vec{\mathbf{k}}$ is given along some boundary or some statements on the emission of waves are made. Equation (3) for a given space-time point is the dispersion equation and describes at the same time the variation of the propagation constant k (magnitude of the three-vector $\vec{k}$) with the direction of the wave normal.

A medium is isotropic in space-time when the magnitude of the four-vector $\vec{\mathbf{k}}$ is independent of its direction, or

$$\vec{k} \cdot \vec{k} - \omega^2/c^2 = \text{const} \tag{4}$$

for varying direction of $\vec{k}$ in space and varying ω. The dispersion equation (3) with fixed $\vec{\mathbf{r}}$ is in this case equivalent with (4), and is invariant for rotation of the coordinate system in which $\vec{\mathbf{k}}$ is considered, i.e. Lorentz-invariant. From three dimensions it is known that isotropy leads to coincidence of ray direction and wave normal. In four dimensions this coincidence, being the consequence of four-dimensional isotropy, is expressed by

$$\vec{\mathbf{k}} \times \mathrm{d}\vec{\mathbf{s}} = 0 \tag{5}$$

with $\mathrm{d}\vec{\mathbf{s}}$ denoting the group path element. The two four-vectors $\mathbf{k}$ and $\mathrm{d}\vec{\mathbf{s}}$ according to (5) have the same ratio of space to time components and this has the consequence that the product of phase and group velocities is c^2. This relationship between phase and group velocities, which is encountered in some cases of wave propagation, is seen to result from Lorentz invariance of the dispersion equation.

Synge (1954) in his treatment of matter waves developed a four-dimensional theory of rays and waves by means of Hamiltonian methods in four-dimensional generalization, and discussed in this connection four-dimensional isotropy and its consequences. The present study, largely dealing with the same ideas but being based on different formulations, suggests inclusion of other types of waves, e.g. electromagnetic waves under various conditions.

Some deductions from (2) and (3) lead to a definition of group paths in four-dimensionally anisotropic media (being the analogues to ray paths in three dimensions). The group path element is found to be parallel to $\Box_k F$, the gradient of F with respect to the $\vec{\mathbf{k}}$ components. In traveling with a wave group, the requirement of continuity of the tangential $\vec{\mathbf{k}}$ components, originally stated for boundaries between media, applies to properly defined differential steps of the medium.

At a discontinuous variation in time, the spatial vector $\vec{k}$ stays unvaried, but ω varies in accordance with the sudden variation of the dispersion equation. There are again a refracted and a reflected wave, both being now propagated in the medium that passed through the discontinuous variation. The two waves are characterized by positive and negative ω of the same amount. The dispersion equation with the given $\vec{k}$ admits the two signs of ω. Reflection of an expanding wave-field at a discontinuous variation in time leads to a collapsing wave-field, returning to the source of waves.

Slowly and fast moving boundaries differ from the limiting cases of a stationary boundary and a sudden simultaneous variation of the medium mainly, in that they produce shifts of ω and $\vec{k}$, which in one case are proportional to the velocity of the boundary (Doppler effect of the reflected wave), in the other inversely proportional to the velocity.

Keller's geometrical theory of diffraction (Keller, 1958) is usable in four dimensions. Diffraction at one-dimensional and two-dimensional structures (moving points and lines) and the processes of refraction and reflection at three-dimensional structures require the same type of continuity considerations. The main difference is in the number of continuity conditions, which is determined by the number of dimensions of the structure at which continuity (of tangential $\vec{\mathbf{k}}$ components) is to be observed.

REFERENCES

KELLER, J. B. (1958) A geometrical theory of diffraction, Research Report No. EM-115, Division of Electromagnetic Research, Inst. of Math. Sci., New York University, New York.

SOMMERFELD, A., and RUNGE, J. (1911) Anwendung der Vektorrechnung auf die Grundlagen der geometrischen Optik, *Ann. Physik.* (*Leipzig*), 4th ser. **35**, 277–298.

SYNGE, J. L. (1954) *Geometrical Mechanics and DeBroglie Waves*, pp. 6–59. Cambridge University Press, Cambridge.

ANGULAR MOMENTUM OF ELECTROMAGNETIC RADIATION

G. TORALDO DI FRANCIA

Centro Microonde, Firenze, Italy

SUMMARY

It is well known that an electromagnetic field carries a momentum, which can be assumed to be distributed with the volume density S/c^2, where S denotes the Poynting vector and c the velocity of light. The flux density of this momentum across any surface (or the pressure exerted on a perfectly absorbing surface) can be evaluated by means of the Maxwell stress tensor. It is therefore evident that, in general, the field must also have an angular momentum with respect to a fixed point or axis. However, in this case the evaluation is less simple and gives sometimes rise to paradoxa. For instance, in the case of a circularly polarized plane wave, starting from the formula

$$M = \frac{1}{c^2} \int r \wedge S \, \mathrm{d}V$$

one may be tempted to conclude that the angular momentum M does not have a component parallel to the direction of propagation (say z). Yet the wave can exert a torque on a material plate perpendicular to z, as was predicted many years ago by Sadowsky[1] and by Poynting[2]. The paradox can be solved by considering a wave with only finite extension in the x,y plane[3]. The flux density of the angular momentum about z turns out to be S/ω, which is in agreement with the quantum mechanical result that the spin of a photon of energy $\hbar\omega$ is $\hbar$.

The angular momentum of a circularly polarized wave was first revealed by Beth[4] in the case of light, by means of a doubly refrating plate. Since the torque is inversely proportional to ω, it was pointed out by Carrara[5] that the experiment should become much simpler by using microwaves instead of light waves. Carrara could easily reveal the torque exerted by a circularly polarized wave with $\omega = 2\pi 9.36 \; 10^9 \; Hz$ propagating in a waveguide on a screen which could either absorb that wave or convert the polarization into linear form.

It was soon realized that in the case of microwave measurements one cannot evaluate the torque by simply multiplying S/ω by the area of the screen, for two main reasons. First, the polarization in a waveguide cannot be everywhere circular (it is in general elliptic). Second, since the size of the screen is necessarily of the same order of magnitude as the wavelength, the effective cross-section for angular momentum absorption can be much different from the geometric cross-section. To specify the problem, let us assume that the material obstacle upon which the wave impinges is perfectly conducting in the x-direction and perfectly insulating in the directions perpendicular to x. In practice this type of screen will be composed of parallel thin wires. Let the impinging plane wave be elliptically polarized and specified by

$$E_x^i = E_0 \exp(ikz), \qquad E_y^i = ieE_0 \exp(ikz)$$

where E_0 and e denote complex constants and a time factor $\exp(-i\omega t)$ has been omitted. In particular, the wave is linearly polarized when $\text{Re}\, e = 0$ and circularly polarized when $e = \pm 1$. If we define the flux density of complex angular momentum[6] by

$$m = \frac{E_0 E_0^*}{Z\omega}\, e$$

where Z denotes free space impedance, the quantity $\bar{m} = \text{Re}\, m$ represents the angular momentum about z carried by the wave per unit surface, per unit time. We can also define a complex scattering cross-section by $s = W/S^i$ where W represents the complex power[7] scattered by the obstacle and S^i the Poynting vector of the impinging wave. The ordinary scattering cross-section will therefore be represented by $\bar{s} = \text{Re}\ s$. Finally we will introduce a complex torque M, defined as $\frac{1}{2}$ the torque exerted by the impinging wave upon the complex conjugates of the charges and currents induced in the obstacle, so that the real torque will be given by $\overline{M} = \text{Re}\, M$. The ratio $s_m = M/m$ will be termed the complex cross-section for angular momentum absorption, while $\bar{s}_m = \overline{M}/\bar{m} = \text{Re}\,(e\, s_m)/\text{Re}\, e$ will represent the real cross-section for angular momentum absorption. It is then possible to show that

$$s_m = \frac{1 + ee^*}{2}\, s\,.$$

This formula enables us to evaluate the torque from a knowledge of the complex scattering cross-section. In particular, one can derive the result that in the case of circular polarization the real cross-sections for scattering and for angular momentum absorption are equal.

Particular attention has been given to the case when the uni-directionally conducting obstacle is a plane screen. The complex scattering cross-section of a disc of this type with small radius r was evaluated by the author[8] up to terms in r^{10}. The result is

$$\mathrm{Re}\, s = \frac{2}{1+ee^*}\frac{64}{27\pi}k^4r^6[1+(\tfrac{27}{25}-\tfrac{1}{5}\sin^2\phi)(kr)^2+$$

$$+(\tfrac{4682}{6125}-\tfrac{1144}{7875}\sin^2\phi+\tfrac{3}{175}\sin^4\phi)(kr)^4+\ldots]$$

$$\mathrm{Im}\, s = \frac{2}{1+ee^*}\tfrac{8}{3}kr^3[1+\tfrac{1}{5}(3+\tfrac{1}{3}\sin^2\phi)(kr)^2+$$

$$+\tfrac{1}{525}(139-\tfrac{4}{3}\sin^2\phi-\sin^4\phi)(kr)^4+\ldots]$$

where ϕ denotes the angle of incidence, assumed to be parallel to the plane y,z.

Other results on diffraction by uni-directionally conducting screens are due to Karp[9] and Radlow[10], who solved the half-plane problem for a plane wave and for a dipole wave respectively, and to the author[11] who gave the correct form for Babinet's principle for such screens.

In the case or propagation in a waveguide, elliptic polarization can be obtained by the superposition of two modes, either degenerate or not. If the modes are degenerate and belong to the same eigenvalue, the polarization is the same in any cross-section of the waveguide. If however two non-degenerate modes are superimposed, the polarization varies from right-handed to left-handed and vice versa as the wave propagates. This apparently paradoxical situation can be explained by showing[12] that the field is continually exchanging angular momentum with the walls of the waveguide. Further, angular momentum is found to travel with a velocity equal to the arithmetic mean of the group velocities of both modes.

Electromagnetic angular momentum can have a number of applications. A recent suggestion is that it can be employed to control the attitude of a space vehicle. A research program on this subject is being carried out at the Microwave Center of Florence, under contract with the U.S. Air Force.

REFERENCES

1. SADOWSKY, A. (1899) *Acta et Comment. Imp. Univers. Jurievensis*, **7**, No. 1–3; (1900) **8**, No. 1–2.
2. POYNTING, J. H. (1909) *Proc. Roy. Soc.* **A82**, 560.
3. HEITLER, W. (1957) *Quantum Theory of Radiation*, p. 40. Oxford Univ. Press.
4. BETH, R. A. (1936) *Phys. Rev.* **50**, 115.
5. CARRARA, N. (1949) *Nature*, **164**, 882.
6. TORALDO DI FRANCIA, G. (1956) *Boll. Un. Mat. Ital.* **11**, 332.
7. SCHELKUNOFF, S. A. (1943) *Electromagnetic Waves*, p. 31. van Nostrand.
8. TORALDO DI FRANCIA, G. (1957) *Nuovo Cimento*, **6**, 150.
9. KARP S. N. Institute Math. Sci., New York. Univ., Research Report N. EM-108.
10. RADLOW, J. *Inst. Math. Sci.*, New York Univ. Research Report N. EM-105.
11. TORALDO DI FRANCIA, G. (1958) *Nuovo Cimento*, **9**, 309.
12. TORALDO DI FRANCIA, G. (1960) *Alta Frequenza*, **29**, 148.

PREDICTED LUNAR TEMPERATURE VARIATIONS DURING FUTURE ECLIPSES

K. M. SIEGEL

Conductron Corporation, Ann Arbor, Michigan

SUMMARY

In References 1–3 we have shown that our theoretical results predicted all known experiments of radar scattering by the moon. In two recent reports (Refs. 4 and 5) we have found apparent discrepancies with our previous published theory. Analyses of these experiments yield the fact that the discrepancies were fatuous (Ref. 6). Other published papers agree with the radar cross-sections predicted by us when use is made of the modulation loss, law in the way described by Reference 3, but disagree with the way we obtained the dielectric constant. Some authors assume a dielectric constant similar to their favorite rock on earth and use this with their own measurements at a single frequency and single pulse length, and then postulate a fit to their experimental curve and state they can predict other experiments. These "laws" prove to be wrong by either one of two tests. The first test is to see how the radar cross-section varies as a function of pulse length. Invariably authors who make measurements at pulse lengths above 200 μsec never have a pulse length dependence in their "theories", as above 200 μsec the cross-section is not sensitive to pulse length variations. The second test is to compare the constants used in the theory with constants obtained from radiation data, and here the postulated theories which assume a dielectric constant always fail. In our theory, we derive for scattering centers on the surface of the moon that their surface has a ratio of permittivity ε to the permeability μ and then, assuming that the permeability is close to the permeability of free space, we derive a relative permittivity of $\varepsilon \sim 1.08$. We would not expect the relative permeability to be greater than 1.6, and as a result we would expect that

$$1.75 \geqslant \varepsilon > 1.08\,.$$

By utilizing our theoretical curve and deduced value of relative permittivity (Ref. 2), Giraud (Ref. 7) derived values for the thermal conductivity and the volumetric specific heats of the lunar surface by using optical eclipse and radio emission data. The values which would go with $\varepsilon = 1.1$ yield a thermal conductivity $k = 1.66 \times 10^{-5}$ cal sec^{-1}cm^{-1}deg^{-1} and a volumetric specific heat $c = 0.07$ cal deg^{-1}cm^{-3}.

V. S. Troitski (Ref. 8) reporting on the work of himself and associates such as V. D. Krotikov, A. G. Kislyakov, M. R. Zelinskaya, L. I. Fedoseev, K. M. Strezhneva and others, finds from the radio emission data that he can derive a value of relative permittivity. The value he reports is

$$\varepsilon = 1.5 \pm 0.3 \, .$$

This value, which is obtained for the *average* relative permittivity of the lunar surface, is sufficiently close to our value for the scattering centers (his 1.5 ± 0.3 corresponds to our results at $\epsilon = 1.1$). We consider that a good independent check of both theories, namely, the passive radiation of Troitski *et al.*, agrees for all intents and purposes with our values derived from the radar data. These results indicate the strong compatibility between the optical eclipse, the infrared radiation, the microwave emission, and the radar scattering data (100 Mc to 10,000 Mc). This then allows us to have greater faith in the derived values of volumetric specific heats and thermal conductivity. However, now, utilizing a range in relative permittivity we find that the tendency would be to increase the value of thermal conductivity, and to decrease the value of volumetric specific heats. The range of values found then allows us to obtain a range in $(kc)^{-1/2}$. The value of $(kc)^{-1/2}$ allows us to derive the differential temperature variation of the lunar surface during an eclipse. The derived differential temperature variation agrees with experiment. While the fixed value of temperature also agrees with the constants derived from the above papers. The results allow us to predict the radiated temperature variation as a function of receiver frequency for future eclipses.

REFERENCES

1. SENIOR, T. B. A. and SIEGEL, K. M. (1959) Radar reflection characteristics of the moon, IAU/URSI Symposium on Radio Astronomy, Paris (July–August 1958). Paris Symposium on Radio Astronomy, Paper No. 5, pp. 29–46, Stanford University Press. Edited by R. N. Bracewell.
2. SENIOR, T. B. A. and SIEGEL, K. M. (1960) A theory of radar scattering by the moon, *J. Res. NBS*, **64D**, No. 3, pp. 217–229 (May/June).
3. SIEGEL, K. M. and SENIOR, T. B. A. (1962) A lunar theory re-asserted, *J. Res. NBS*, **66D**, 227–229.
4. A study of ionospheric and lunar characteristics by radar techniques, General Electric Technical Information Series, TIS R61EMH40, Aug. 1961.
5. An experimental investigation of lunar and auroral scattered signals. Part I—Lunar echoes, Cornell Aeronautical Laboratory, Inc., CAL Report No. CM-1393-P-7, Contract No. AF 19(604)-6116, Jan. 1962.
6. Private communication with Millman of General Electric and Flood of Cornell Aeronautical Laboratory has shown that Fig. 14 of the Cornell Aeronautical Lab. Report is not correct and that the data published in Millman's report are not typical. Thus, after correspondence the apparent discrepancies disappeared.
7. GIRAUD, A. (1962) Characteristics of the moon's surface layer: an analysis of its radio emission, *Astrophys. J.* **135**, No. 1, Jan. 1962.
8. TROITSKI, V. S., Some results of the moon exploration by radiophysical methods. Report to the International Space Science Symposium, Washington, D.C

REPRESENTATION OF ELECTRODYNAMIC FIELDS IN CURVILINEAR NON-ORTHOGONAL SPACE COORDINATES BY MEANS OF TWO SCALAR FUNCTIONS AND THE INVERSE VARIATIONAL PROBLEM FOR THE INTRODUCED FUNCTIONS

N. A. KUZMIN

Institute of State's Committee of Radioelectronics, Moscow, U.S.S.R.

ABSTRACT

It is well known to what extent it is possible to facilitate the solution of Maxwell's equations for given boundary conditions with applications of special potentials. In many cases the success to receive a rigorous solution of the boundary-value problems depends upon the fortunate choice of the similar special potential functions. For example, one can find in a rather simple manner the solution of problems connected with waveguides if it is possible to use only one (component) of the electrical or magnetic Hertz functions. More complicated is the construction of the solution when two Hertz functions are introduced at the same time. In addition to that it is known[1] that one can represent the fields by means of two Hertz functions if the rather rigorous conditions are satisfied for metric space factors for the given boundary-value problem. Briefly one can formulate these conditions: one of the metric factors of the quadratic form for orthogonal space is constant and the others do not depend on the coordinate for which the given metric space factor is constant. One of coordinate systems for which this condition is satisfied is the cylindrical one with rectilinear axis. In the lateral plane of this system one can choose any one orthogonal coordinate system.

In practice there are many problems which are not convenient and practically, cannot be worked out in cylindrical coordinates for example the diffraction waves in the case of a sphere, paraboloid, cone, bodies of rotation, the problem of wave propagation along the curved guides (toroidal type) etc. For all these cases it is not convenient to use the simple Hertz functions and one introduces the special potential functions, for sphere—Debye potentials, for paraboloid case—Fock potentials, etc.

In this work we make an attempt to introduce two new scalar functions which can be used for reducing the system of Maxwell's equations to the system of two differential equations of second order in partial derivatives for less rigorous conditions for metric factors of space.[2]

Let us introduce the space metric in the quadratic form

$$ds^2 = \sum_{i=1}^{3} \sum_{j=1}^{3} g_{ij} du^i du^j \tag{1}$$

where g_{ij}—metric factors as functions of two variables u^1 and u^2 that for different indexes do not vanish at zero. The indexes i and j run through all values 1, 2, 3. The geometrical configuration (1) corresponds to non-orthogonal basis with coordinate vectors $\mathbf{e}_1$, $\mathbf{e}^2$, $\mathbf{e}_2$. According to definition $g_{ij} = \mathbf{e}_i$, $\mathbf{e}_j$ and $\mathbf{e}_1$, $\mathbf{e}_2$, $\mathbf{e}_3$—covariant vectors whose modulus do not equal unity.

The case when the metric tensor components g_{ij} are functions of two co-ordinates, for example, corresponds to problems of electromagnetic wave propagation without local reflection, along the coordinate of which the metric factors do not depend. By means of such a metric one can describe the different cases of wave propagation along the curved surface in space, wave guides etc. for which both of the radius of axis curvature (curvature and torsion) are constant.

For $\partial/\partial t = \mathbf{i}\omega$ and in absence of sources, the Maxwell's equations are given by

$$\operatorname{rot} \mathbf{H} + i K \mathbf{E} = 0\,, \tag{2}$$

$$\operatorname{rot} \mathbf{E} - i K \mathbf{H} = 0\,, \tag{3}$$

$$\operatorname{div} \mathbf{E} = 0\,, \tag{4}$$

$$\operatorname{div} \mathbf{H} = 0\,, \tag{5}$$

ε and μ in medium for consideration of simplicity considered equal unity. If necessary they (ε and μ) can be introduced without difficulties in final the equations with "K". Considering the condition that the factors in metric form (1) do not depend on u^3, the relation of the **E** and **H** fields from u^3 can be written in the form

$$\{\mathbf{E}\,;\quad \mathbf{H}\} \sim e^{ihu^3}\,. \tag{6}$$

We can show that by introducing two functions ψ and $\tilde{\psi}$ which depend on u^1 and u^2, and u^3 as in (6), and related with **E** and **H** as given that

$$\mathbf{E} = \mathbf{e}_3 \frac{K^2}{K^2 g_{33} - h^2} \psi + \frac{ih}{K^2 g_{33} - h^2} \nabla\psi - iK \frac{[\mathbf{e}_3 \nabla \tilde{\psi}]}{K^2 g_{33} - h^2} \tag{7}$$

$$\mathbf{H} = \mathbf{e}_3 \frac{K^2}{K^2 g_{33} - h^2} \tilde{\psi} + \frac{ih}{K^2 g_{33} - h^2} \nabla\tilde{\psi} + iK \frac{[\mathbf{e}_3 \nabla \psi]}{K^2 g_{33} - h^2}\,. \tag{8}$$

The expressions (7) and (8) become solutions of the electrodynamic equations (2)–(5) if ψ and $\tilde{\psi}$ satisfy the following system of differential equations:

$$\Delta\psi + K^2\psi - \frac{K^2}{K^2 g_{33} - h^2} \left\{ \nabla g_{33} \nabla \psi + \frac{h}{K g_{33}} [\nabla g_{33} \nabla \tilde{\psi}] \mathbf{e}_3 \right\} -$$
$$- iK \frac{\tilde{\psi}}{\sqrt{g \cdot g_{33}}} \left\{ \frac{\partial}{\partial u^1} (g_{23} g_{33}) - \frac{\partial}{\partial u^2} (g_{13} g_{33}) \right\} = 0\,, \tag{9}$$

$$\Delta\tilde{\psi} + K^2\tilde{\psi} - \frac{K^2}{K^2 g_{33} - h^2}\left\{\nabla g_{33}\nabla\tilde{\psi} - \frac{h}{Kg_{33}}[\nabla g_{33}\nabla\psi]\mathbf{e}_3\right\} +$$

$$+ iK\frac{\psi}{\sqrt{g\cdot g_{33}}}\left\{\frac{\partial}{\partial u^1}(g_{23}g_{33}) - \frac{\partial}{\partial u^2}(g_{13}g_{33})\right\} = 0\,, \quad (10)$$

where ∇ and Δ operators of grad and Laplace in the curvilinear non-orthogonal coordinate system with basis $\mathbf{e}_1$, $\mathbf{e}_2$, $\mathbf{e}_3$ are given in the form:

$$\nabla\ldots = \mathrm{grad}\ldots = \frac{1}{g}\begin{vmatrix} \mathbf{e}_1 & \mathbf{e}_2 & \mathbf{e}_3 & 0 \\ g_{11} & g_{12} & g_{13} & \frac{1}{\sqrt{g}}\frac{\partial\ldots}{\partial u^1} \\ g_{21} & g_{22} & g_{23} & \frac{1}{\sqrt{g}}\frac{\partial\ldots}{\partial u^2} \\ g_{31} & g_{32} & g_{33} & \frac{ih}{\sqrt{g}}\ldots \end{vmatrix} = \sum_{i=1}^{3}\mathbf{e}^i\frac{\partial\ldots}{\partial u^i} \quad (*)$$

$$\Delta\ldots = \frac{1}{\sqrt{g}}\begin{vmatrix} \frac{\partial\ldots}{\partial u^1} & \frac{\partial\ldots}{\partial u^2} & ih & 0 \\ g_{11} & g_{12} & g_{13} & \frac{1}{\sqrt{g}}\frac{\partial\ldots}{\partial u^i} \\ g_{21} & g_{22} & g_{23} & \frac{1}{\sqrt{g}}\frac{\partial\ldots}{\partial u^2} \\ g_{31} & g_{32} & g_{33} & \frac{ih}{\sqrt{g}}\ldots \end{vmatrix}$$

and where g is the Gramm determinant for metric tensor g_{ij} equal to

$$g = \begin{vmatrix} (\mathbf{e}_1\mathbf{e}_1) & (\mathbf{e}_1\mathbf{e}_2) & (\mathbf{e}_1\mathbf{e}_3) \\ (\mathbf{e}_2\mathbf{e}_1) & (\mathbf{e}_2\mathbf{e}_2) & (\mathbf{e}_2\mathbf{e}_3) \\ (\mathbf{e}_3\mathbf{e}_1) & (\mathbf{e}_3\mathbf{e}_2) & (\mathbf{e}_3\mathbf{e}_3) \end{vmatrix} = \begin{vmatrix} g_{11} & g_{12} & g_{13} \\ g_{21} & g_{22} & g_{23} \\ g_{31} & g_{32} & g_{33} \end{vmatrix},$$

In the last representation (*), $\mathbf{e}_i (i = 1,2,3)$ means the coordinate vector for mutual (countervariant) basis as determined by the expressions

$$\mathbf{e}^1 = \frac{[\mathbf{e}_2\mathbf{e}_3]}{\sqrt{g}}, \qquad \mathbf{e}^2 = \frac{[\mathbf{e}_3\mathbf{e}_1]}{\sqrt{g}}, \qquad \mathbf{e}^3 = \frac{[\mathbf{e}_1\mathbf{e}_2]}{\sqrt{g}}.$$

In addition to that the basis vectors are related according to

$$\mathbf{e}_i \mathbf{e}^j = \delta_{ij}$$

where δ_{ij} is Kroeneker's symbol. The given expressions represent all operators in (7)–(10) in expanded form.

Now we can show that **E** and **H** given by means of ψ and $\tilde{\psi}$ taking into account (9) and (10) satisfy all Maxwell's equations. It's enough to prove only the equation (2), because for (3) the proof can be received by means of corresponding substitution ψ to $\tilde{\psi}$. The equations (9) and (10) can be received from the conditions (4) and (5), so ψ and $\tilde{\psi}$ automatically satisfy the second group of Maxwell's equations. Now after corresponding substitutions of the expressions for **E** and **H** to (2) and some vector transformations we receive

$$-\frac{ih}{K^2 g_{33} - h^2}[\nabla g_{33} \nabla \psi] - \frac{K^2 \psi}{K^2 g_{33} - h^2}[\nabla g_{33} \mathbf{e}_3] + \psi \operatorname{rot} \mathbf{e}_3 + \\ + iK \frac{\mathbf{e}_3(\nabla g_{33} \nabla \psi)}{K^2 g_{33} - h^2} - \frac{i}{K}(\nabla \tilde{\psi} \operatorname{grad})\mathbf{e}_3 + \frac{i}{K}(\mathbf{e}_3 \operatorname{grad})\nabla \tilde{\psi} + \\ + \frac{h}{K}\nabla \tilde{\psi} - iK\mathbf{e}_3 \tilde{\psi} = 0 . \quad (11)$$

It is possible to show, using the peculiarities of Christoffel's symbols[3], that

$$(\nabla \tilde{\psi} \operatorname{grad})\mathbf{e}_3 - (\mathbf{e}_3 \operatorname{grad})\nabla \tilde{\psi} - ih\nabla \tilde{\psi} = 0 .$$

Further, taking into account, that

$$\operatorname{rot} \mathbf{e}_3 = \frac{1}{\sqrt{g}}\left\{\mathbf{e}_1 \frac{\partial g_{33}}{\partial u^2} - \mathbf{e}_2 \frac{\partial g_{33}}{\partial u^1} + \mathbf{e}_3\left(\frac{\partial}{\partial u^1} g_{23} - \frac{\partial}{\partial u^2} g_{13}\right)\right\},$$

it is simple to show the direct transition of equation (11) to the basic equation (10) for functions ψ and $\tilde{\psi}$.

For example, we give below in expanding form the initial set of equations (9) and (10) for the space in which the coordinates u^1 and u^2 are cartesian x and y, and u^3 curved axis non-orthogonal to x and y.

For this case the Gramm's determinant is expressed as follows

$$g = \begin{vmatrix} 1 & 0 & g_{13} \\ 0 & 1 & g_{23} \\ g_{31} & g_{32} & g_{33} \end{vmatrix} = g_{33} - (g_{13})^2 - (g_{23})^2 .$$

The set of the equations (9) and (10) after simple transformations can be written in the form

$$\frac{g_{33}-g_{23}^2}{\sqrt{g}}\begin{Bmatrix}\psi_{xx}\\ \tilde{\psi}_{xx}\end{Bmatrix}+\frac{g_{33}-g_{13}^2}{\sqrt{g}}\begin{Bmatrix}\psi_{yy}\\ \tilde{\psi}_{yy}\end{Bmatrix}+2\frac{g_{13}g_{23}}{\sqrt{g}}\begin{Bmatrix}\psi_{xy}\\ \tilde{\psi}_{xy}\end{Bmatrix}+$$

$$+\left[\frac{\partial}{\partial x}\frac{g_{33}-g_{23}^2}{\sqrt{g}}+\frac{\partial}{\partial y}\frac{g_{13}g_{23}}{\sqrt{g}}-2ih\frac{g_{13}}{\sqrt{g}}-\frac{K^2}{\sqrt{g}(K^2g_{33}-h^2)}\times\right.$$

$$\left.\times\left\langle(g_{33}-g_{23}^2)\frac{\partial g_{33}}{\partial x}+g_{13}g_{23}\frac{\partial g_{33}}{\partial y}\right\rangle\right]\times$$

$$\times\begin{Bmatrix}\psi_{x}\\ \tilde{\psi}_{x}\end{Bmatrix}+\left[\frac{\partial}{\partial y}\frac{g_{33}-g_{13}^2}{\sqrt{g}}+\frac{\partial}{\partial x}\frac{g_{13}g_{23}}{\sqrt{g}}-2ih\frac{g_{23}}{\sqrt{g}}-\frac{K^2}{\sqrt{g}(K^2g_{33}-h^2)}\times\right.$$

$$\left.\times\left\langle(g_{33}-g_{13}^2)\frac{\partial g_{33}}{\partial y}+g_{13}g_{23}\frac{\partial g_{33}}{\partial x}\right\rangle\right]\times$$

$$\times\begin{Bmatrix}\psi_{y}\\ \tilde{\psi}_{y}\end{Bmatrix}\mp\frac{Kh}{g_{33}(K^2g_{33}-h^2)}\left\langle g_{33}\left(\frac{\partial g_{33}}{\partial x}\begin{Bmatrix}\tilde{\psi}_{y}\\ \psi_{y}\end{Bmatrix}-\frac{\partial g_{33}}{\partial y}\begin{Bmatrix}\tilde{\psi}_{x}\\ \psi_{x}\end{Bmatrix}\right)-\right.$$

$$\left.-ih\left(g_{23}\frac{\partial g_{33}}{\partial x}-g_{13}\frac{\partial g_{33}}{\partial y}\right)\begin{Bmatrix}\tilde{\psi}\\ \psi\end{Bmatrix}\right\rangle+$$

$$+\left[\frac{K^2g-h^2}{\sqrt{g}}-ih\frac{\partial}{\partial x}\frac{g_{13}}{\sqrt{g}}-ih\frac{\partial}{\partial y}\frac{g_{23}}{\sqrt{g}}+ih\frac{K^2}{\sqrt{g}(K^2g_{33}-h^2)}\times\right.$$

$$\left.\times\left(g_{13}\frac{\partial g_{33}}{\partial x}+g_{23}\frac{\partial g_{33}}{\partial y}\right)\right]\times$$

$$\times\begin{Bmatrix}\psi\\ \tilde{\psi}\end{Bmatrix}-iK\frac{\frac{\partial}{\partial x}(g_{23}g_{33})-\frac{\partial}{\partial y}(g_{13}g_{33})}{g_{33}}\begin{Bmatrix}\tilde{\psi}\\ \psi\end{Bmatrix}=0.$$

From the relations (7) and (8) we can formulate the boundary conditions for the perfectly conducting surface, i.e.

$$\psi\Big|_{(\Gamma)}=0\,,$$

$$\left.\frac{\partial\tilde{\psi}}{\partial n}\right|_{(\Gamma)}=0\,,$$

where Γ is a contour on the coordinate plane (u^1, u^2), $\mathbf{n}$ a normal unit equal to

$$\mathbf{n}=\frac{[\mathbf{e}_2\mathbf{e}_3]}{\sqrt{(g_{22}g_{33}-g_{23}^2)}}=\frac{\mathbf{e}^1}{\sqrt{g^{11}}}\,.$$

It is assumed that the conducting surface coincides with one of the coordinate planes. Evidently only in this case can one use a non-orthogonal coordinate system, when the metric (1) can be written in the form

$$ds^2 = g_{11}(du_1)^2 + g_{22}(du_2)^2 + g_{33}(du_3)^2 . \tag{1'}$$

The fields **E** and **H** and the potentials ψ and $\tilde{\psi}$ with explicitly written exponential factor are determinated from the following equations:

$$\left.\begin{aligned} \mathbf{E} &= \frac{ih\nabla_{\perp}\psi - iK\sqrt{g_{33}}[\mathbf{e}_3\nabla_{\perp}\tilde{\psi}]}{K^2 g_{33} - h^2} e^{ihu_3} + \mathbf{e}_3 \frac{1}{\sqrt{g_{33}}} \psi e^{ihu_3} \\ \mathbf{H} &= \frac{ih\nabla_{\perp}\tilde{\psi} + iK\sqrt{g_{33}}[\mathbf{e}_3\nabla_{\perp}\psi]}{K^2 g_{33} - h^2} e^{ihu_3} + \mathbf{e}_3 \frac{1}{\sqrt{g_{33}}} \tilde{\psi} e^{ihu_3} \end{aligned}\right\}, \tag{12}$$

$$\Delta_{\perp}\psi + \kappa^2 \frac{1}{g_{33}} \psi + K^2 \frac{g_{33} - 1}{g_{33}} \psi - \frac{K^2(g_{33}+1) - \kappa^2}{K^2(g_{33}-1) + \kappa^2} \nabla_{\perp} \ln \sqrt{(g_{33})} \nabla_{\perp}\psi - \\ - \frac{2K\sqrt{(K^2 - \kappa^2)}}{K^2(g_{33}-1) + \kappa^2} [\mathbf{e}_3 \nabla_{\perp} \sqrt{g_{33}}] \nabla_{\perp}\tilde{\psi} = 0 , \tag{13}$$

$$\Delta_{\perp}\tilde{\psi} + \kappa^2 \frac{1}{g_{33}} \tilde{\psi} + K^2 \frac{g_{33} - 1}{g_{33}} \tilde{\psi} - \frac{K^2(g_{33}+1) - \kappa^2}{K^2(g_{33}-1) + \kappa^2} \nabla_{\perp} \ln \sqrt{(g_{33})} \nabla_{\perp}\tilde{\psi} - \\ - \frac{2K\sqrt{(K^2 - \kappa^2)}}{K^2(g_{33}-1) + \kappa^2} [\mathbf{e}_3 \nabla_{\perp} \sqrt{g_{33}}] \nabla_{\perp}\psi = 0, \tag{14}$$

where

$$\kappa^2 = K^2 - h^2, \qquad \|\mathbf{e}_3\| = 1 .$$

$\Delta_{\perp}$ and $\nabla_{\perp}$ are two-dimensional Laplace and Hamilton operators.

The set of eqns (13) and (14) express the physical phenomena which have taken place during the propagation of electromagnetic waves in curved coordinate systems with constant radii of curvature. For this case the waves are mixed E- and H-types.

For the sake of generality we assume that on some surface the boundary conditions in impedance form, expressed by tensors Z_1 and Z_2, are given

$$\left.\begin{aligned} (\mathbf{s}\mathbf{E}_{\perp}) &= Z_2 H_3 \\ E_3 &= -Z_1(\mathbf{s}\mathbf{H}_{\perp}) \end{aligned}\right\}.$$

The functions ψ and $\tilde{\psi}$ on this surface must satisfy the following boundary conditions

$$\frac{\partial\psi}{\partial n} - i \frac{K^2 g_{33} - h^2}{K g_{33} Z_1} \psi + \frac{h}{K\sqrt{g_{33}}} \frac{\partial\tilde{\psi}}{\partial s} = 0 \tag{15}$$

$$\frac{\partial\tilde{\psi}}{\partial n} - iZ_2 \frac{K^2 g_{33} - h^2}{K g_{33}} \tilde{\psi} - \frac{h}{K\sqrt{g_{33}}} \frac{\partial\psi}{\partial s} = 0 . \tag{16}$$

For the case of the perfectly conducting surface the impedances Z_1 and Z_2 become equal to zero and the boundary conditions can be written in simple form

$$\psi = 0, \qquad \frac{\partial \tilde{\psi}}{\partial n} = 0 .$$

In expressions (15) and (16), **n** is the external normal to the surface and **s** the direction of tangential component orthogonal to $\mathbf{e}_3$.

As the coordinate system becomes more complicated the equations for determining the special potentials for the field representation also become more complicated. In the general case it is not possible to find the solutions of these equations. It seems to us very important to reduce the given electrodynamic boundary-value problems to the inverse problem of the theory of variations, i.e. to find for the boundary conditions expressed by given equations such stationary functionals for which the Euler's equations coincide with the given set of equations, and as physical conditions serve the boundary conditions of the problem under consideration (for instance, when the function does not equal zero at the boundary for $\psi = 0$ the last condition is omitted).

It is apparent that such a problem is soluble for all cases and the initial set of equations must satisfy the definite requirements.

In mathematics the inverse problem is considered for a single equation and it is shown that this equation must be of the self-conjugated type. So in this case the stationary functional can be easily constructed. For the systems treated as the inverse variational problem the criteria are not formulated. One can generalize the conclusions for systems related with a single equation when every equation of the system is of the semi-conjugated type. But this condition is too hard and for the problems of electrodynamics very often cannot be satisfied. In connection with this we made an attempt to formulate the conditions of the inverse variational problem for systems and to show the method of the stationary functionals construction. Note that we are interested only the in linear differential operators so the functional must be only bilinear or quadratic. In consideration of simplicity we shall explain the theorem for the sets of two equations with two independent variables. Finally the formulation of theorem can be generalized to sets with many variables.

THEOREM: For the set of differential equations of second order with partial derivatives of second order

$$\left.\begin{aligned} L_{11}(u) + L_{12}(v) = 0,\\ L_{21}(u) + L_{22}(v) = 0 \end{aligned}\right\} \tag{N1}$$

with operators

$$L = L_{ij} = A_{ij}\frac{\partial^2 \ldots}{\partial x^2} + 2B_{ij}\frac{\partial^2 \ldots}{\partial x \partial y} + C_{ij}\frac{\partial^2 \ldots}{\partial y^2} + D_{ij}\frac{\partial \ldots}{\partial x} + E_{ij}\frac{\partial \ldots}{\partial y} + F_{ij} \ldots, \tag{N2}$$

the inverse variational problem exists when the following conditions are satisfied:

(a) The diagonal operators in (N1) must be of the self-conjugated type

(b) The coefficients of system

$$\left.\begin{aligned} A_{ij} &= A_{ji} \\ B_{ij} &= B_{ji} \\ C_{ij} &= C_{ji} \end{aligned}\right\}$$

for the derivates of the highest order must be symmetrical.

(c) The coefficients of the derivatives of first order must satisfy the relation of expanded self-conjugation

$$\left.\begin{aligned} \frac{\partial}{\partial x} A_{ij} + \frac{\partial}{\partial y} B_{ij} &= \tfrac{1}{2}(D_{ij} + D_{ji}) \\ \frac{\partial}{\partial y} C_{ij} + \frac{\partial}{\partial x} B_{ij} &= \tfrac{1}{2}(E_{ij} + E_{ji}) \end{aligned}\right\}. \qquad \text{II}$$

(d) The coefficients of the functions must satisfy to the relation

$$F_{ij} - F_{ji} = \frac{1}{2}\left(\frac{\partial}{\partial x} D_{ij} - \frac{\partial}{\partial x} D_{ji}\right) + \frac{1}{2}\left(\frac{\partial}{\partial y} E_{ij} - \frac{\partial}{\partial y} E_{ji}\right). \qquad \text{III}$$

The proof of this theorem is simple. One constructs the bilinear functional

$$\mathscr{T} = \iint_{(S)} \{uL_{11}(u) + uL_{12}(v) + vL_{21}(u) + vL_{22}(v)\}\, dx\, dy$$

and then it is shown that it is necessary to satisfy the formulated requirements to the operators L_{ij} for coincidence of the Euler's equations (for this functional) with the given initial system. It is necessary to emphasize that non-diagonal operators in the inverse variational problem are admitted non self-conjugated.

Using the first and the second generalized Green's expressions

$$\iint_{(S)} \{Au_x v_x + B(u_x v_y + u_y v_x) + Cu_y v_y + (A_x u_x v + B_y u_x v + Duv_x + D_x uv) +$$
$$+ (B_x u_y v + C_y u_y v + Euv_y + E_y uv) - Fuv\}\, dx\, dy$$
$$= -\iint_{(S)} \{Avu_{xx} + 2Bvu_{xy} + Cvu_{yy} + Dvu_x + Evu_y + Fvu\}\, dx\, dy +$$
$$+ \oint_{(\Gamma)} \{(Avu_x + Bvu_y + Duv)\, dy - (Bvu_x + Cvu_y + Euv)\, dx\},$$

$$\iint_{(S)} \{vL(u) - uL(v)\}\, dx\, dy$$

$$= \iint_{(S)} \{A_x(uv - vu_x) + B_y(uv_x - vu_x) + D(u_x v - uv_x) +$$

$$+ B_x(uv_y - u_y v) + C_y(uv_y - u_y v) + E(u_y v - uv_y)\}\, dx\, dy +$$

$$+ \oint_{(\Gamma)} \{[A(u_x v - uv_x) + B(u_y v - uv_y)]\, dy -$$

$$- [B(u_x v - uv_x) + C(u_y v - uv_y)]\, dx\} ,$$

and after some simple transformations we receive the functional

$$J = \iint_{(S)} \Big\{ A_{11}u_x^2 + 2B_{11}u_x u_y + C_{11}u_y^2 + 2D_{11}uu_x + 2E_{11}uu_y +$$

$$+ \left(\frac{\partial}{\partial x} D_{11} + \frac{\partial}{\partial y} E_{11} - F_{11}\right) u^2 + A_{22}v_x^2 + 2B_{22}v_x v_y + C_{22}v_y^2 +$$

$$+ 2D_{22}vv_x + 2E_{22}vv_y + \left(\frac{\partial}{\partial x} D_{22} + \frac{\partial}{\partial y} E_{22} - F_{22}\right) v^2 + 2A_{12}u_x v_x +$$

$$+ 2B_{12}(u_x v_y + v_x u_y) + 2C_{12}u_y v_y + 2D_{12}vu_x +$$

$$+ \frac{D_{21} - D_{12}}{4} u_x v + 2D_{21}uv_x + \frac{D_{12} - D_{21}}{2} uv_x + 2E_{12}u_y v +$$

$$+ \frac{E_{21} - E_{12}}{2} vu_y + 2E_{22}uv_y + \frac{E_{12} - E_{21}}{2} uv_y +$$

$$+ \left(\frac{\partial}{\partial x} D_{12} + \frac{\partial}{\partial x} D_{21} + \frac{\partial}{\partial y} E_{12} + \frac{\partial}{\partial y} E_{21} - F_{12} - F_{21}\right) uv \Big\}\, dx\, dy$$

and natural conditions of the boundary-value problem

$$\{A_{11}u_x + B_{11}u_y + D_{11}u + A_{12}v_x + B_{12}v_y + \tfrac{1}{4}(D_{21} - D_{12})v + D_{12}v\} \frac{dy}{ds} -$$

$$- \{C_{11}u_y + B_{11}u_x + E_{11}u + C_{12}v_y + B_{12}v_x + \tfrac{1}{4}(E_{21} - E_{12})v + E_{12}v\} \frac{dx}{ds} = 0$$

$$\{A_{22}v_x + B_{22}v_y + D_{22}v + A_{21}u_x + B_{12}u_y + \tfrac{1}{4}(D_{12} - D_{21})u + D_{21}u\} \frac{dy}{ds} -$$

$$- \{C_{22}v_y + B_{22}v_x + E_{22}v + E_{21}u_y + B_{21}u_x + \tfrac{1}{4}(E_{12} - E_{21})u + E_{21}u\} \frac{dx}{ds} = 0. \tag{18}$$

Now we can write the final conditions for the existence of the inverse variational problem for three-dimensional space.

If $\mathscr{L}(u)$ is a three-dimensional operator in the form

$$\mathscr{L}(u) = \mathscr{L}_{ij}(u) = A_{ij}u_{xx} + B_{ij}u_{yy} + C_{ij}u_{zz} + 2D_{ij}u_{xy} + 2E_{ij}u_{xz} + \\ + 2F_{ij}u_{yz} + Gu_x + Hu_y + Ku_z + Qu = 0$$

A, B, C,... and are the functions x, y, z, the condition for common self-conjugation of the operator is

$$\left.\begin{aligned} A_x + D_y + E_z &= G, \\ D_x + B_y + F_z &= H, \\ E_x + F_y + C_z &= K, \end{aligned}\right\} .$$

For this case the theorem of the inverse variational problem can be generalized in such a way. The paragraphs (a) and (b) are formulated without changes conforming to according coefficients of the system of operators. Under condition of the expanded self-conjugation the paragraph (c) must now be understood as follows

$$\left.\begin{aligned} \frac{\partial}{\partial x} A_{ij} + \frac{\partial}{\partial y} D_{ij} + \frac{\partial}{\partial z} E_{ij} &= \frac{G_{ij} + G_{ji}}{2} \\ \frac{\partial}{\partial x} D_{ij} + \frac{\partial}{\partial y} B_{ij} + \frac{\partial}{\partial z} F_{ij} &= \frac{H_{ij} + H_{ji}}{2} \\ \frac{\partial}{\partial x} E_{ij} + \frac{\partial}{\partial y} F_{ij} + \frac{\partial}{\partial z} C_{ij} &= \frac{K_{ij} + K_{ji}}{2} \end{aligned}\right\} . \tag{15}$$

The paragraph (d) is now formulated in form

$$Q_{ij} - Q_{ji} = \frac{1}{2}\left(\frac{\partial}{\partial x} G_{ij} - \frac{\partial}{\partial x} G_{ji}\right) + \frac{1}{2}\left(\frac{\partial}{\partial y} H_{ij} - \frac{\partial}{\partial y} H_{ji}\right) + \frac{1}{2}\left(\frac{\partial}{\partial z} K_{ij} - \frac{\partial}{\partial z} K_{ji}\right) .$$

It is possible to prove that for the given sets of equations the requirements of the theorem are satisfied. Therefore, for the simplest boundary conditions the quadratic stationary functional and functions ψ and $\tilde{\psi}$ can be found by variational methods.

In conclusion we give in final form the stationary functional for the boundary-value problem with metric (1′). The functional has the form

$$J = \iint\limits_{(S\perp)} \left\{ \frac{\sqrt{g_{33}} \cdot (K^2 - h^2)}{K^2 g_{33} - h^2} \left(\nabla_\perp^2 \psi + \nabla_\perp^2 \tilde{\psi} + 2 \frac{h}{K\sqrt{g_{33}}} \mathbf{e}_3 [\nabla_\perp \psi \nabla_\perp \tilde{\psi}] \right) - \right. \\ \left. - \frac{K^2 - h^2}{\sqrt{g_{33}}} (\psi^2 + \tilde{\psi}^2) \right\} dS_\perp - i \frac{K^2 - h^2}{K} \oint\limits_{(\Gamma)} \left\{ \frac{\psi^2}{\sqrt{g_{33}} \cdot Z_1} + Z_2 \frac{\tilde{\psi}^2}{\sqrt{g_{33}}} \right\} dl = 0, \tag{19}$$

where $S_\perp$ is the two-dimensional region at the coordinate surface (u_1, u_2) limited by the contour Γ.

For the region $S_\perp$ external to contour Γ and expanding to infinity (for example, the case of studying the wave phenomenon in the open waveguides) the existence of integrals for region $S_\perp$ is necessary. For example for the surface waveguides these integrals are limited, and the problem of comparing the surface waves which arose owing to different curvatures may be solved completely by the variational method and using the direct method and for similar boundary conditions $\psi|_\Gamma = 0$ and $\frac{\partial\tilde{\psi}}{\partial n}|_\Gamma = 0$. For $g_{33} = 1$ the stationary functional (19) splits into two well-known independent functionals

$$J_1 = \iint\limits_{(S_\perp)} \{\nabla_\perp^2\psi - (K^2 - h^2)\psi^2\}\, dS_\perp = 0,$$

$$J_2 = \iint\limits_{(S_\perp)} \{\nabla_\perp^2\tilde{\psi} - (K^2 - h^2)\tilde{\psi}^2\}\, dS_\perp = 0.$$

1. DE BROGLIE, LOUIS (1948) *Electromagnetic Waves in Waveguides and Cavities*, Techn. edition, U.S.S.R.
WEVEDENSKY, W. and ARENBERG, A. (1946) The method of Hertz-vector in practical problems of electrodynamics, *Proc. of the Academy of Sciences, U.R.S.S.*, Tech. Section, No. 9.
2. KUSMINE, N. (1959) Problemi di trasmissione di energia a microonde in sistemi in guida d'onda, VI Rassegna internazionale elettronica e nucleaze, Roma.
3. LAGALLY, M. (1936) *Vectorial Calculation*, Tech. edition, U.S.S.R.

SECTION B

ANISOTROPIC AND STRATIFIED MEDIA

WAVE PROPAGATION IN ANISOTROPIC PLASMAS†

WILLIAM P. ALLIS‡

SUMMARY

The results of the Appleton–Hartree theory applied to a collisionless plasma will be discussed in terms of a plot whose coordinates are the normalized electron plasma frequency

$$\alpha_-^2 = \omega_{p-}^2/\omega^2 = Ne^2/\varepsilon_0 m_- \omega^2 ,$$

and the normalized electron cyclotron frequency

$$\beta_-^2 = \omega_{b-}^2/\omega^2 = (eB_0/m_-\omega)^2 .$$

This α_-^2–β_-^2 plane is divided into eight regions, which are given numbers from 1 to 8 by the following lines:—

Plasma cutoff:	$\alpha_-^2 = 1$
Cyclotron cutoff:	$\alpha_-^2 \pm \beta_- = 1$
Plasma resonance:	$\alpha_-^2 + \beta_-^2 = 1$
Cyclotron resonance:	$\beta_-^2 = 1$

At each cutoff one of the two electromagnetic waves in the plasma has an infinite phase velocity. At cyclotron resonance the right circularly polarized wave travelling along the magnetic field has zero phase velocity, and at plasma resonance the extraordinary wave travelling across the magnetic field has zero phase velocity. In three regions of the diagram there exists a resonant cone such that waves propagating along this cone have zero velocity.

The phase velocity of an electromagnetic wave is a function of α_-, β_-, and θ, the angle which the propagation vector **k** makes with the magnetic field. At each point of the diagram the nature of electromagnetic propagation is given by a phase velocity surface, and this surface remains topologically the same throughout any one region of the plane but varies from region to region. In four regions, the phase velocity surface consists of two non-intersecting sheets on which the polarizations of the waves are orthogonal.

† This work was supported in part by the U.S. Army Signal Corps, the Air Force Office of Scientific Research, and the Office of Naval Research, and in part by U.S. Atomic Energy Commission and the National Science Foundation.

‡ Department of Physics and Research, Laboratory of Electronics, Massachusetts Institute of Technology, Cambridge, Massachusetts, U.S.A.

In three regions only one polarization propagates and the surface is a single sheet. In one region no electromagnetic wave propagates. The so-called whistler mode appears in the upper right hand region of this diagram.

When the motions of ions are included in the Appleton–Hartree theory the $\alpha_-^2 - \beta_-^2$ plane must be extended in the β-direction to include the normalized ion cyclotron frequency $\beta_+ = \omega\beta_+/\omega$. It is convenient to use

$$\alpha^2 = \frac{Ne^2}{\varepsilon_0\omega^2}\cdot\frac{m_+ + m_-}{m_+ m_-} \quad \text{and} \quad \beta^2 = \beta_+\beta_- \quad \text{as}$$

coordinates and to use a mass ratio $m_+/m_- = 3$ in drawing the resonance and cutoff lines. The extended plane has a new cyclotron resonance line at $\beta_+ = 1$, and the plasma resonance line has a new branch which runs from $\alpha^2 = 0$, $\beta_+^2 = 1$ to $\alpha^2 = \infty$, $\beta_+\beta_- = 1$. The $\alpha^2 - \beta^2$ plane now has thirteen regions each with a distinct phase velocity surface topology, but we now observe a periodicity: the regions and velocity surfaces in the upper half of the plane are similar to those in the lower half but the right and left hand directions of polarization are interchanged. The Alfven velocity appears in the upper right hand region of the $\alpha^2 - \beta^2$ plane as it is here that the approximations of magnetohydrodynamics apply.

The thirteen regions of the diagram are not altered by the introduction of plasma pressure in the theory, provided the thermal motions are non-relativistic. The phase velocity surfaces are altered by the addition of two more sheets, termed plasma-electron and plasma-ion waves, in which the electrons and ions move out of phase and in phase, respectively. These two waves propagate roughly at the electron and ion thermal speeds and therefore, generally, are one inside the other, and are both internal to the electromagnetic sheets of the velocity surface.

These relationships are violated near resonances. The plasma-electron wave has a cutoff at the electromagnetic resonance cone so that these two surfaces join smoothly along this cone.

The plasma-electron wave has a resonance cone at an angle different from the electromagnetic resonance, and this cone intersects the plasma-ion velocity surface. Coupling terms in the equations keep the velocity surfaces from intersecting but the transitions from plasma-electron to plasma-ion wave are surprising. Similar transitions take place in the magnetohydrodynamic region with increasing α^2 as the Alfven speed decreases, and passes first the electron and then the ion thermal speed.

Phenomena whose existence depends on the presence of particles of a particular velocity will not be discussed. Landau damping and resonances at harmonics of the cyclotron frequency are the principal examples. These phenomena are not given by the transport equations but require use of the Boltzmann equation for their description.

THE RELATION BETWEEN HYDROMAGNETIC WAVES AND THE MAGNETO-IONIC THEORY†

C. O. HINES

Dept. Geophysical Sciences, University of Chicago, Illinois

ABSTRACT

The term "hydromagnetic" is often applied to a variety of waves which in fact fall into quite distinct categories. Among these are the waves treated originally by Alfvén and those studied subsequently by Åström. The physical relationship between all of these waves can be understood readily on the basis of a magneto-ionic approach to the problem, and this approach also clarifies their relationship to radiowaves propagating in ionized gases.

In this paper, the physical relationship is illustrated by means of a step-by-step extension of the equations of charge motion that enter conventional magneto-ionic theory. En route, certain potentially misleading views that are common in hydromagnetic discussions are called into question.

1. INTRODUCTION

The study of hydromagnetic waves in a conducting fluid was initiated by Alfvén (1942, 1950), with the use of a scalar conductivity σ such that the current density $\mathbf{J}$ was related to the electric field $\mathbf{E}$, the fluid velocity $\mathbf{V}$, and the magnetic induction $\mathbf{B}$, through the relatively simple equation

$$\mathbf{J} = \sigma(\mathbf{E} + \mathbf{V} \times \mathbf{B})\,; \tag{1}$$

the force density acting on the fluid was taken to be

$$\mathbf{F} = \mathbf{J} \times \mathbf{B} \tag{2}$$

(in m.k.s.a. units). With these and other assumptions, and with σ sufficiently large, it was found that waves would propagate in the direction of a dominating $\mathbf{B}$ at the speed $B\,(\mu\rho)^{-1/2}$, now commonly known as the Alfvén speed, where μ is the magnetic permittivity and ρ the fluid density.

While (1) is normally quite satisfactory for application to conducting liquids, it is often invalid in problems that concern ionized gases, including most of those problems that arise in cosmic, stellar, and ionospheric physics. Often a more appropriate approach to these problems is provided by the study of fully ionized gases, essentially in the manner adopted by Åström (1950). This approach makes use of magneto-ionic theory (cf. Ratcliffe, 1959), which had already been developed for radiowave propagation in an ionized gas

† Contribution from the Defense Research Board, Ottawa, Canada, under project PCC D48-95-10-27.

(e.g. Appleton, 1932), but which required extension by the inclusion of ion as well as electron motions. At sufficiently low frequencies, with ions included, Åström's analysis revealed that the speed of propagation in the direction of a dominant **B** was again $B(\mu\rho_c)^{-1/2}$, though ρ_c here indicates the mass density of the charged species alone, being the only constituents present in his model.

The formal similarity of the speeds deduced by Alfvén and by Åström has often led to the conclusion, either explicitly or implicitly, that the two categories of wave were in some sense equivalent, and both are now commonly termed "hydromagnetic". The fact that ρ included (or was in fact dominated by) the neutral constituents in Alfvén's case and included only ionized constituents in Åström's case was in this view a natural consequence of the difference in their physical models. Such an interpretation is at best misleading, and it certainly obscures the true relationship between the conducting-liquid approach of Alfvén and the magneto-ionic approach of Åström.

This relationship was established by Hines (1953), who extended the magneto-ionic approach by the addition of collisions between the neutral and charged constituents, and at the same time left the neutral constituents free to move as they would under the influence of the collisions. (Electron-neutral collisions were an inherent feature of the early magneto-ionic theory, and ion-neutral collisions were incorporated formally by Bailey (1948). Dungey (1951) abstracted from Bailey's work the hydromagnetic waves derived by Åström, modified by acoustic waves, by ignoring the collisional interaction that Bailey had included by treating in effect the ionized constituents alone. Bailey's analysis did not encompass the original Alfvén waves, for it took the neutral gas to be unaccelerated. The analysis of Hines included as distinct special cases the circumstances treated by Alfvén and others that were equivalent, insofar as the waves were concerned, to those treated by Åström; and it showed further that a speed of the form $B(\mu\rho)^{-1/2}$ was obtained in other circumstances, not treated by either Alfvén or Åström.)

The contrast between a "two-fluid" (ion-electron) system and a "three-fluid" (ion-electron-neutral) system has been discussed also by Piddington (1954), in one of a fairly comprehensive series of papers on hydromagnetic waves. Piddington's analysis employed an anisotropic conductivity tensor $\boldsymbol{\sigma}$, such that

$$\mathbf{J} = \boldsymbol{\sigma}\cdot(\mathbf{E} + \mathbf{V} \times \mathbf{B}) . \tag{3}$$

This has the advantage of generality over Alfvén's formulation and the advantage of formal simplicity over the magneto-ionic approach. However the latter must be adopted in any event when an evaluation of $\boldsymbol{\sigma}$ is to be made, unless consideration is restricted (as by Piddington) to the case where a "d.c." conductivity tensor is applicable. Moreover, the form of (3) is not generally applicable, though it can be justified in suitably restricted circumstances by a rather artificial definition which carries little physical content and sometimes adds unwarranted complications. (See Appendix. The usual justification of (3), as a generalization from the form $\mathbf{J} = \boldsymbol{\sigma} \cdot \mathbf{E}$ that would apply in a coordinate system moving with the fluid, is inapplicable when the fluid itself undergoes accelerated—i.e., non-inertial—motion as in a hydromagnetic wave.) For these reasons, the following discussion will be confined to the magneto-ionic approach.

The purpose of the discussion will be to put into perspective, by means of the magneto-ionic approach, the physical relationship between radiowaves, Åström waves, and Alfvén waves. The discussion may be considered as a physical counterpart to the earlier mathematical development (Hines, 1953), though it will not treat the complete variety of circumstances that were included there nor will it attempt to include all the qualifications that mathematical rigour would demand. The reader is referred to the earlier article for the more general circumstances and qualifications it includes, and of course to the great profusion of papers on hydromagnetic and magneto-acoustic waves that have appeared in the past decade for further generalizations and rigour. The objective here is a physical perspective alone.

2. THE BASIC MAGNETO-IONIC APPROACH

The magneto-ionic approach may be described from an electromagnetically biased point of view. Maxwell's equations are taken then as the primary relationships, with $\mathbf{D} = \varepsilon_0 \mathbf{E}$ and $\mathbf{B} = \mu_0 \mathbf{H}$ in the usual (m.k.s.a.) notation, but the functional form of $\mathbf{J} = \mathbf{J}(\mathbf{E})$ remains to be determined. It is to be found from

$$\mathbf{J} = Ne(\mathbf{V}_i - \mathbf{V}_e) \tag{4}$$

for a simple ionized gas, in which only one species of ion is present, singly charged, with number density N, charge e, and mean local velocity $\mathbf{V}_i$, together with an equal concentration of electrons of mean local velocity $\mathbf{V}_e$. The problem is then converted into one of finding $\mathbf{V}_i$ and $\mathbf{V}_e$ as functions of E, after which substitution into (4) and thence into Maxwell's equations would permit a wave solution to be determined. The velocities $\mathbf{V}_i$ and $\mathbf{V}_e$ are to be determined from appropriate equations of motion.

3. MAGNETO-IONIC WAVES

At sufficiently high frequencies, the pertinent linearized equations of motion may be approximated by

$$i\omega \mathbf{V}_e = -\frac{e}{M_e}\mathbf{E} \tag{5}$$

$$i\omega \mathbf{V}_i = \frac{e}{M_i} \mathbf{E} \tag{6}$$

where a time variation of the form $\exp(i\omega t)$ is assumed, M_i and M_e are the masses of an ion and an electron respectively, and the electric field is taken to be the source of the only important force that acts on the charged constituents. Clearly, in these circumstances

$$V_i/V_e = M_e/M_i \ll 1$$

and the ionic contribution to **J** may be ignored. The solution of (5) for $\mathbf{V}_e$, and its insertion through (4) into Maxwell's equations, will lead to the well-known refractive index $(1 - \omega_e^2/\omega^2)^{1/2}$ for plane wave propagation, ω_e being the "electron plasma frequency". In ionospheric applications, unless Faraday rotation, absorption, or some other special effect is being studied, this approximation is reasonably good at frequencies in the v.h.f.-range and higher.

When a background magnetic induction B is present, (5) and (6) must in principle be extended to

$$i\omega \mathbf{V}_e = -\frac{e}{M_e} \mathbf{E} - \mathbf{V}_e \times \mathbf{\Omega}_e \tag{7}$$

$$i\omega \mathbf{V}_i = \frac{e}{M_i} \mathbf{E} + \mathbf{V}_i \times \mathbf{\Omega}_i \tag{8}$$

where $\mathbf{\Omega}_e = e\mathbf{B}/M_e$ and $\mathbf{\Omega}_i = e\mathbf{B}/M_i$; Ω_e and Ω_i are the electron and ion cyclotron frequencies, respectively, and $\Omega_i/\Omega_e = M_e/M_i \ll 1$. It can now be seen that the validity of (5) depends on the gross inequality $\omega \gg \Omega_e$ being satisfied, and that this condition in turn implies $\omega \gg \Omega_i$ and hence confirms the validity of (6) up to this point.

When $\omega \sim \Omega_e$, all terms in (7) must be retained and the full complexity of standard magneto-ionic theory comes into play. An order-of-magnitude comparison reveals that

$$V_i/V_e \sim M_e/M_i \ll 1\,, \tag{9}$$

and hence that the neglect of $\mathbf{V}_i$ in (4) is still justified at these frequencies.

At yet lower frequencies, when $\omega \ll \Omega_e$, (7) may be approximated by

$$0 = -\frac{e}{M_e} \mathbf{E} - \mathbf{V}_e \times \mathbf{\Omega}_e \tag{10}$$

provided $\mathbf{V}_e$ is not directed too nearly along **B**. In application to waves, this latter requirement will normally be met if propagation is not nearly transverse to B; in particular, it would be satisfied for propagation along **B**, since $\mathbf{V}_e \cdot \mathbf{B} = 0$ then. From (10) it may be seen that $V_e \sim (e/M_e\Omega_e)E = E/B$. If $\omega \gg \Omega_i$, then (8) still yields $V_i \sim (e/M\omega)E$, and so

$$V_i/V_e \sim \Omega_i/\omega \ll 1\,;$$

it is still appropriate to neglect V_i in (4). These circumstances lead to the classical form of "whistler" dispersion in ionospheric propagation (Storey, 1953).

It is worth noting that (10) is sufficient to imply that the electrons are "frozen" to the magnetic field lines, in the conventional hydromagnetic terminology, and yet the speed of propagation along B is not given by $B(\mu\rho_e)^{-1/2}$, with ρ_e the mass density of the electrons. This fact suggests that some danger exists in the analogy that is often employed in "explanation" of the Alfvén speed, that the magnetic field lines behave as vibrating stretched strings, loaded by the mass of fluid that is "frozen" to them. Such an explanation might be expected to carry over to the whistler case, but it certainly does not.

The ion velocities become important at frequencies comparable to Ω_i, for then V_i is comparable to E/B and so to V_e. (It should perhaps be noted that a frequency near $(\Omega_i\Omega_e)^{1/2}$ marks a transition point as well, for waves propagating across B(Åström, 1950); this transition may be important in certain v.l.f.-transmissions (Hines, 1957).) The full equation (8) must be employed, and it introduces a high degree of complexity just as the full equation (7) did at $\omega \sim \Omega_e$. Waves in this frequency range are still considered to be "magneto-ionic" in nature, in contrast to the "hydromagnetic" waves at yet lower frequencies.

4. ÅSTRÖM WAVES

At frequencies well below Ω_i, (8) may be approximated by

$$0 = \frac{e}{M_i}\mathbf{E} + \mathbf{V}_i \times \mathbf{\Omega}_i, \tag{12}$$

which is analogous to (10) for electrons and which shows that $V_i \sim E/B$. The ions now are also "frozen" on to the magnetic field lines, and the speed of propagation along $\mathbf{B}$ is indeed $B(\mu\rho_c)^{-1/2}$ where ρ_c is the mass density of the ions and electrons (Åström, 1950).

It should be noted, however, that the approximations (10) and (12) are inadequate to produce this conclusion. It will be recalled that, in waves propagating along $\mathbf{B}$, $\mathbf{V}_e \cdot \mathbf{B} = 0$, similarly $\mathbf{V}_i \cdot \mathbf{B} = 0$. For such waves, (10) and (12) then imply

$$\mathbf{V}_e = \mathbf{E} \times \mathbf{B}/B^2 = \mathbf{V}_i, \tag{13}$$

and hence $\mathbf{J} = 0$. If such were indeed the case, the waves would be found to have the vacuum speed of light rather than the hydromagnetic speed cited above. The hydromagnetic behaviour of the actual waves results from departures of $\mathbf{V}_e$ and $\mathbf{V}_i$ from the approximate values given in (13), departures that still lie perpendicular to $\mathbf{B}$ and that arise from the residual inertial terms on the left-hand sides of (7) and (8). These terms in turn represent departures from the "frozen-on" condition, so again the use of that condition in explanation of the hydromagnetic speed is in danger of being misleading.

The waves discussed in this section, "Åström waves", are those that arise in most problems of cosmic and stellar physics.

5. ABSORPTION IN MAGNETO-IONIC AND ÅSTRÖM WAVES

The treatment to this point has neglected interaction between the charged particles and any neutral gas that might be present, and indeed Åström's development was initially intended to apply to a totally ionized gas. In the presence of a neutral gas, collisional interactions must in principle be introduced. The classical magneto-ionic theory took them into account through a friction-like term, by modifying (7) to the form

$$i\omega \mathbf{V}_e = -\frac{e}{M_e}\mathbf{E} - \mathbf{V}_e \times \mathbf{\Omega}_e - \nu_e \mathbf{V}_e \tag{14}$$

where ν_e is a "frictional" frequency closely related and nearly equal to the electron-neutral collision frequency. A similar alteration may be anticipated in the equation of ion motion

$$i\omega \mathbf{V}_i = \frac{e}{M_i}\mathbf{E} + \mathbf{V}_i \times \mathbf{\Omega}_i - \nu_i \mathbf{V}_i\,, \tag{15}$$

where ν_i is an ion-neutral "frictional" frequency corresponding to ν_e. (Electron-ion collisions should in principle have been included all along; they have been neglected for simplicity, but can be important in some circumstances in cosmic, stellar and even ionospheric propagation.)

It is now seen that the validity of the magneto-ionic approximations employed previously demands that $\nu_e \ll \omega$ or $\nu_e \ll \Omega_e$, whichever is the less restrictive, and further that $\nu_i \ll \omega$ in the frequency range $\omega \lesssim \Omega_i$. If these gross inequalities are not satisfied, then the more complicated formula (14) must be retained at the higher frequencies and both (14) and (15) at the lower. Even when the gross inequalities are satisfied, the more complicated formulas must be retained if absorption is to be treated (since otherwise the components of $\mathbf{V}_e$ and $\mathbf{V}_i$ are in phase quadrature with any parallel components of $\mathbf{E}$, and no net work is done on the charge in the course of a full cycle), or if singularities are to be removed. Physically, absorption in these circumstances may be thought of as a frictional dissipation, due to the electrons and ions moving backwards and forwards through a milieu of relatively stationary neutral particles with which they are constantly colliding.

In the frequency range $\omega \ll \Omega_i$, Åström's analysis will still apply provided the collisional terms are small in comparison with all the other terms in the equations of motion. Subject to modifications that will be introduced in the next section, this limitation appears to demand that $\nu_e \ll \omega \gg \nu_i$. And within this limitation, Åström's study of fully ionized gases can be carried over to partially ionized gases (unless absorption or the removal of mathematical singularities is to be studied). This case applies in the propagation of micropulsations through the ionosphere above about 500 km; below that level, collisions become more important (Fejer, 1960).

It should be noted in particular that the relevance of Åström's analysis to partially ionized gases requires the collisional terms to be small relative to the inertial terms in (14) and (15), even though the latter have already been assumed small in comparison to the **E** and $\mathbf{V} \times \mathbf{B}$ terms. This is because, as will be recalled, it is the residual *departures* from the "frozen-on" motions (13) that determine the characteristics of wave propagation when $\omega \ll \Omega_i$, and those departures must be inertial in nature if the speed $B(\mu\rho_c)^{-1/2}$ is to be attained. More specifically, when $\Omega_e \gg \nu_e \gg \omega \ll \nu_i \ll \Omega_i$, it will be seen that the ions and electrons are still virtually frozen-on to the magnetic field lines—the approximations (13) are still valid—and yet no propagating wave is obtained. This result is *not* directly analogous to Alfvén's (1950) "finite conductivity" case, for in the latter the departures from "infinite conductivity" conditions become severe only when departures from the "frozen-on" conditions also become severe.

6. LOWEST-FREQUENCY WAVES, MAGNETICALLY DOMINATED

The magneto-ionic approach may now be extended by noting that the collisional force implied by the final terms in (14) and (15) must be matched by a reaction on the neutral gas, and that that reaction may be capable of setting the neutral gas into motion. If the neutral gas is indeed in motion, then the collisional force terms introduced in (14) and (15) must be modified, for these terms cannot depend on absolute velocities but only on relative velocities. Thus, if the neutral constituents have local mean velocity $\mathbf{V}_n$, (14) and (15) must be modified to

$$i\omega\mathbf{V}_e = -\frac{e}{M_e}\mathbf{E} - \mathbf{V}_e \times \mathbf{\Omega}_e + \nu_e(\mathbf{V}_n - \mathbf{V}_e) \tag{16}$$

and

$$i\omega\mathbf{V}_i = \frac{e}{M_i}\mathbf{E} + \mathbf{V}_i \times \mathbf{\Omega}_i + \nu_i(\mathbf{V}_n - \mathbf{V}_i)\,. \tag{17}$$

With this modification, $\mathbf{V}_e$ and $\mathbf{V}_n$ can no longer be determined immediately as functions of **E**, nor then can **J**. Instead, it is necessary to eliminate $\mathbf{V}_n$ from explicit appearance, and this requires further equations characterizing the motion of the neutral gas. The simplest circumstances arise when the collisional force is the only one operating on the neutral gas, and then

$$i\omega\rho_n\mathbf{V}_n = \rho_e\nu_e(\mathbf{V}_e - \mathbf{V}_n) + \rho_i\nu_i(\mathbf{V}_i - \mathbf{V}_n)\,. \tag{18}$$

(Here ρ_n is the mass density of the neutral gas. It enters, as do the mass densities ρ_e of electrons and ρ_i of ions, since the force terms employed previously were forces per unit mass, whereas it is the force per unit volume that must be balanced in action and reaction.)

It is clear that the set (16)–(18) is adequate to permit the suppression of any explicit dependence on $\mathbf{V}_n$, and to provide $\mathbf{V}_e = \mathbf{V}_e(\mathbf{E})$ and $\mathbf{V}_i = \mathbf{V}_i(\mathbf{E})$ as required for the magneto-ionic development. Various approximations are available, depending on the relative magnitudes of Ω_e, Ω_i, ν_e, ν_i, and ω, but the pertinent inequalities may now depend on the ratios ρ_e/ρ_n and ρ_i/ρ_n in addition to ρ_e/ρ_i which has already been employed (in the form M_e/M_i). Only two cases need be discussed for present purposes, one in this section and the other in the next.

The case to be treated here is one in which geomagnetic domination persists to the point that (13) is an adequate approximation for the reduction of (18). (It should be recalled that this approximation depends to some extent on the direction of propagation in the case of waves. For simplicity, attention may be confined to propagation along $\mathbf{B}$ again, when $\mathbf{V}_e \cdot \mathbf{B} = 0 = \mathbf{V}_i \cdot \mathbf{B}$.) The latter then becomes

$$i\omega\rho_n\mathbf{V}_n = (\rho_e\nu_e + \rho_i\nu_i)(\mathbf{V}_e - \mathbf{V}_n)\,, \tag{19}$$

which may be rewritten equally well with $\mathbf{V}_i$ replacing $\mathbf{V}_e$. With this approximation, it will be seen that the $\mathbf{V}_n$-term on the right may be dropped if $\omega \gg (\rho_e\nu_e + \rho_i\nu_i)/\rho_n$ and that $\mathbf{V}_e \gg \mathbf{V}_n \ll \mathbf{V}_i$ then. In these circumstances, the $\mathbf{V}_n$'s may be dropped from (16) and (17) and we return to the damped Åström waves.

On the other hand, when $\omega \ll (\rho_e\nu_e + \rho_i\nu_i)/\rho_n$, the term on the left of (19) becomes smaller than the V_n term on the right, and (19) then leads to the approximation $\mathbf{V}_n = \mathbf{V}_e$. In these conditions the neutral gas is swept up in the motion of the electrons and ions, its inertia being overcome by collisional forces—mainly those associated with the ions, in practice, since $\rho_e\nu_e \ll \rho_i\nu_i$. At the same time, the collisional dissipation of energy is reduced by the reduction in relative velocities that results, and a return is made to nearly freely propagating conditions.

In these new conditions, the collisional terms in (16) and (17) represent only small residual departures from the "frozen-on" motion (13), but we have already seen that residuals are important. They may be derived conveniently from (19) by inserting the approximation $\mathbf{V}_n = \mathbf{V}_e$ into the left of (19), whence

$$\mathbf{V}_e - \mathbf{V}_n = i\omega\rho_n(\rho_e\nu_e + \rho_i\nu_i)^{-1}\mathbf{V}_e \tag{20}$$

and similarly
$$\mathbf{V}_i - \mathbf{V}_n = i\omega\rho_n(\rho_e\nu_e + \rho_i\nu_i)^{-1}\mathbf{V}_i\,. \tag{21}$$

These residual velocity differences may now be inserted in (16) and (17) respectively, to yield

$$i\omega[1+\rho_n\nu_e(\rho_e\nu_e+\rho_i\nu_i)^{-1}]\mathbf{V}_e=-\frac{e}{M_e}\mathbf{E}-\mathbf{V}_e\times\boldsymbol{\Omega}_e \tag{22}$$

and

$$i\omega[1+\rho_n\nu_i(\rho_e\nu_e+\rho_i\nu_i)^{-1}]\mathbf{V}_i=\frac{e}{M_i}\mathbf{E}+\mathbf{V}_i\times\boldsymbol{\Omega}_i\,. \tag{23}$$

These equations have precisely the same form as (7) and (8) respectively, and so lead to solutions of the same form as those obtained by Åström. The extra terms on the left of (22) and (23) in effect just alter the multiplier of the electron and ion masses in Åström's formulas, and do so in such a way that the total effective mass density is $\rho_t \equiv \rho_n + \rho_i + \rho_e$ in place of $\rho_c \equiv \rho_i + \rho_e$. In particular, waves that propagate along **B** now do so with speed $B(\mu\rho_t)^{-1/2}$ (Hines, 1953). The frequency range in which this speed applies suffers a corresponding restriction, in effect to the range where $\omega \ll \Omega_i\rho_i/(\rho_i+\rho_n)$. Other restrictions must also be applied, in order to ensure the legitimacy of (20) and (21), but these restrictions need not be discussed here.

7. ALFVÉN WAVES

Although waves with speed $B(\mu\rho_t)^{-1/2}$ have now been derived, they still are not those treated originally by Alfvén. Alfvén's use of (1) as a constitutive relation implies an isotropic conductivity, and such a conductivity is not obtained in the circumstances treated up to this point. It implies, in fact, that $\nu_e \gg \Omega_e \gg \omega \ll \Omega_i \ll \nu_i$ (Hines, 1953).

These conditions when applied to (16) and (17) suggest the approximations

$$0=-\frac{e}{M_e}\mathbf{E}+\nu_e(\mathbf{V}_n-\mathbf{V}_e) \tag{24}$$

$$0=\frac{e}{M_i}\mathbf{E}+\nu_i(\mathbf{V}_n-\mathbf{V}_i)\,, \tag{25}$$

but these approximations are inadequate for the purposes of (18) since there they would imply $i\omega\rho_n V_n = 0$. Instead, residuals are again important, and a better approximation to (16) and (17) must be adopted. This is achieved by retention of the $\mathbf{V}\times\boldsymbol{\Omega}$ terms in (16) and (17), since they must dominate over the inertial terms in the circumstances treated by Alfvén. With the $\mathbf{V}\times\boldsymbol{\Omega}$ terms included, (18) leads to

$$i\omega\rho_n\mathbf{V}_n=\mathbf{J}\times\mathbf{B}\,, \tag{26}$$

while manipulation of (24) and (25) leads to

$$\mathbf{V}_i-\mathbf{V}_e=e(M_i^{-1}\nu_i^{-1}+M_e^{-1}\nu_e^{-1})\mathbf{E} \tag{27}$$

from which the conventional d.c. conductivity formula may be obtained. These equations lead in straightforward fashion to the hydromagnetic waves of Alfvén, although without the confinement to propagation along **B** which resulted from his further assumption of incompressibility. Waves propagating in the direction of **B** do so with speed $B(\mu\rho_n)^{-1/2}$ when the conductivity is sufficiently large.

The conditions implicit in Alfvén's work may be relaxed somewhat, to $\Omega_e \ll \nu_e \gg \omega \ll \nu_i \gg \Omega_i$. The approximations (24) and (25) are then still valid for the production of (27), and still inadequate for the reduction of (18). The residuals pertinent to that equation may now require the retention of the inertial terms in (16) and (17), in addition to the $\mathbf{V} \times \mathbf{\Omega}$ terms. One consequence of their inclusion is the replacement of ρ_n by ρ_t in the speed of propagation—a replacement that may well be of no significance in specific applications. An observationally more serious consequence can result, however, for when $\Omega_e \ll \omega \gg \Omega_i$ the various vectors that describe the oscillation in full are subject to different polarization groupings than before. This subject will not be pursued here, but it is mentioned to stress further an underlying theme of this paper; that the appearance of a characteristic speed of propagation will often represent only superficial agreement with the original hydromagnetic analysis, and that considerable attention must be paid to the circumstances of a specific problem before the hydromagnetic attributes of that problem can be known.

8. SUMMARY

A broad range of waves and other phenomena is commonly included under the term "hydromagnetic", and this range has certain features in common. One of the more widely quoted of these features is the speed of propagation in the direction of a dominating magnetic induction **B**, given by $B(\mu\rho)^{-1/2}$ where ρ is a mass density. It has been noted, however, that formal similarity of this type overlies quite disparate conditions, and that care ought to be exercised in the application of any one set of hydromagnetic conclusions to conditions that may be fundamentally different. Again, the usual "explanation" of the Alfvén speed has been seen to be potentially misleading, and some caution in its use has been advocated.

But the main theme of the paper has been the physical connections that relate the various hydromagnetic phenomena to one another, and that connect hydromagnetic waves to the radiowaves of conventional magneto-ionic theory. It has been shown in particular that the hydromagnetic waves occur characteristically at low frequencies.

When the collisional frequencies are small in comparison with the cyclotron frequencies, as often occurs in an ionized gas, the hydromagnetic system can include Åström waves, in which $\rho = \rho_c$, as a simple low-frequency extension of conventional magneto-ionic theory with ion motions included. At yet lower frequencies, if the gas is only partially ionized, it can include a further sequence of propagating waves for which $\rho = \rho_t$; but these two sequences are separated by a frequency band near $\omega \sim \rho_i \nu_i/\rho_n$ in which dissipation is severe, and they cannot then be considered to be mere modifications of one another.

When the collisional frequencies are large in comparison with the cyclotron frequencies, then the hydromagnetic behaviour sets in once the wave frequency is well below the collisional frequencies, and the density that enters the propagation speed is ρ_t. This case includes, when the wave frequency is well below the cyclotron frequencies, the conditions treated by Alfvén.

Conditions much more general than those discussed here have been treated by a variety of authors, by the inclusion of additional force terms in one or more of (16)–(18); pressure gradients and ion–electron collisions are the most notable absentees here, though others can on occasion be as important or more so. The physical relationship between all of these can be seen in the magneto-ionic approach.

APPENDIX

When any number of ionized species are present, each subject to an equation of motion of the form (17), then the full current density is given (Hines, 1953, with altered notation) by

$$\left.\begin{aligned}
\mathbf{J}_L &= \sum [N_r e_r^2/M_r \nu_r']\mathbf{E}_L + \sum [N_r e_r \nu_r/\nu_r']\mathbf{V}_{nL} \\
\mathbf{J}_T &= \sum \left[\frac{N_r e_r^2 \nu_r'}{M_r(\nu_r'^2 + \Omega_r^2)}\right]\mathbf{E}_T + \sum \left[\frac{N_r e_r^3}{M_r^2(\nu_r'^2 + \Omega_r^2)}\right]\mathbf{E}_T \times \mathbf{B} \\
&\quad + \sum \left[\frac{N_r e_r \nu_r \nu_r'}{\nu_r'^2 + \Omega_r^2}\right]\mathbf{V}_{nT} + \sum \left[\frac{N_r e_r^2 \nu_r}{M_r(\nu_r'^2 + \Omega_r^2)}\right]\mathbf{V}_n \times \mathbf{B}.
\end{aligned}\right\} \quad (28)$$

Here N_r is the number density, M_r the mass, and ν_r the collision frequency (with neutral particles) of the rth type of charged particle; e_r is its charge and Ω_r its cyclotron frequency, with sign included in both cases (in contrast to the definition given for Ω_e in the main text); $\nu_r' \equiv \nu_r + i\omega$, where variations of a form $\exp(i\omega t)$ are assumed; the subscripts L and T indicate the longitudinal and transverse vector components, as measured with respect to the direction of $\mathbf{B}$; and the summations extend over all charged species. The force density $\mathbf{F}$ that acts on the neutral gas as a collisional reaction is given by

$$\left.\begin{aligned}
\mathbf{F}_L &= \sum [N_r e_r \nu_r/\nu_r']\mathbf{E}_L - i\omega \sum [N_r M_r \nu_r/\nu_r']\mathbf{V}_{nL} \\
\mathbf{F}_T &= \sum \left[\frac{N_r e_r \nu_r \nu_r'}{\nu_r'^2 + \Omega_r^2}\right]\mathbf{E}_T + \sum \left[\frac{N_r e_r^2 \nu_r}{M_r(\nu_r'^2 + \Omega_r^2)}\right]\mathbf{E}_T \times \mathbf{B} \\
&\quad - \sum \left[N_r M_r \nu_r \frac{i\omega \nu_r' + \Omega_r^2}{\nu_r'^2 + \Omega_r^2}\right]\mathbf{V}_{nT} + \sum \left[\frac{N_r e_r \nu_r^2}{\nu_r'^2 + \Omega_r^2}\right]\mathbf{V}_n \times \mathbf{B}.
\end{aligned}\right\} \quad (29)$$

These full expressions include all cases discussed in the present paper. It should be noted that, while a temporal oscillation $\exp(i\omega t)$ has been assumed, no spacial variation has been introduced; the formulas are then applicable to phenomena other than waves.

It will be seen from the existence of an $F_L \neq 0$ that, in general,

$$\mathbf{F} \neq \mathbf{J} \times \mathbf{B}. \tag{30}$$

It is readily shown from the underlying equations, however, that

$$\mathbf{F} = \mathbf{J} \times \mathbf{B} - i\omega \sum N_r M_r \mathbf{V}_r, \tag{31}$$

so that $\mathbf{J} \times \mathbf{B}$ may be treated as a force density acting on the "plasma as a whole", which is defined as a medium of mass density $\rho_t = \rho_n + \Sigma N_r M_r$ moving with velocity $\mathbf{U}$ given by

$$\rho_t \mathbf{U} = \rho_n \mathbf{V}_n + \sum N_r M_r \mathbf{V}_r. \tag{32}$$

Thus, if no additional forces operate on the neutral gas,

$$i\omega\rho_t \mathbf{U} = \mathbf{J} \times \mathbf{B}. \tag{33}$$

This relationship is a useful one, in that it is simple and has wide applicability; it has been used often before now.

No such simple reduction or interpretation applies to the constitutive relations for the current. Specifically, the coefficients of E are not the same as the coefficients of $\mathbf{V}_n \times \mathbf{B}$, except to a certain degree of approximation when $\omega \ll$ all ν_r, and there is no general relationship between $\mathbf{E}$ and $\mathbf{V}_n$ which would permit their separate contributions to be formally redistributed between them. Thus it is impossible to represent $\mathbf{J}$ generally by the form

$$\mathbf{J} = \boldsymbol{\sigma} \cdot (\mathbf{E} + \mathbf{V}_n \times \mathbf{B}) \tag{34}$$

even with $\boldsymbol{\sigma}$ a tensor. This difficulty is not alleviated even if the constitutive relations are rewritten with U as an explicit variable in place of V_n; the form

$$\mathbf{J} = \boldsymbol{\sigma} \cdot (\mathbf{E} + \mathbf{U} \times \mathbf{B}) \tag{35}$$

is equally inapplicable in general. If, however, the collisional reaction is the only force acting on the neutral gas, then it is possible to specify V_n in terms of V_e and V_i, and to find the latter velocities in terms of E alone. In these circumstances, (28) may be rewritten in any one of the forms (34), (35) or

$$\mathbf{J} = \boldsymbol{\sigma} \cdot \mathbf{E}, \tag{36}$$

the $\boldsymbol{\sigma}$ being different in each case. Such a rewrite is at best artificial and of little physical content; at worst it is misleading, for the $\boldsymbol{\sigma}$ s involved are not the parameters commonly employed except in suitable limiting cases. In practice, for a general treatment, it would be far simpler to apply the full equations (28) than to derive artificial alternatives.

While not immediately pertinent to the present analysis, a correction to one of the formulas of Hines (1953) may be noted, the factor K_r in the coefficient of E_T in (39) should be replaced by K_r' (or ν_r by ν_r' in present notation). Equation (41) results only from the correct expression.

REFERENCES

ALFVÉN, H. (1942) On the existence of electromagnetic-hydrodynamic waves, *Nature*, **150**, 405–406 (Oct.).

ALFVÉN, H. (1950) *Cosmical Electrodynamics*, chapter IV. Clarendon Press, Oxford.

APPLETON, E. V. (1932) Wireless studies of the ionosphere, *J. Inst. Elect. Engrs.* **71**, No. 430, 642–650 (Oct.).

ÅSTRÖM, E. (1950) On waves in an ionized gas, *Ark. Fys.*, **2**, Paper 42, 443–457.

BAILEY, V. A. (1948) Plane waves in an ionized gas with static electric and magnetic fields present, *Aust. J. Sci. Res. A*, **1**, No. 4, 351–359.

DUNGEY, J. W. (1951) Derivation of the dispersion equation for Alfvén's magneto-hydrodynamic waves from Bailey's electromagneto-ionic theory, *Nature*, **167**, No. 4260, 1029–1030 (June).

FEJER, J. A. (1960) Hydromagnetic wave propagation in the ionosphere, *J. Atmosph. Terr. Phys.* **18**, Nos. 2/3, 135–146.

HINES, C. O. (1953) Generalized magneto-hydrodynamic formulas, *Proc. Camb. Phil. Soc.* **49**, Part 2, 299–307.

HINES, C. O. (1957) Heavy-ion effects in audio-frequency radio propagation, *J. Atmosph. Terr. Phys.* **11**, No. 1, 36–42.

PIDDINGTON, J. H. (1954) The motion of ionized gas in combined magnetic electric and mechanical fields of force, *Mon. Not. Roy. Astr. Soc.* **114**, No. 6, 651–663.

RATCLIFFE, J. A. (1959) *Magneto-ionic Theory*. University Press, Cambridge.

STOREY, L. R. O. (1953) An investigation of whistling atmospherics, *Phil. Trans. Roy. Soc. A*, **246**, No. 908, 113–141.

NOTES ON WAVES IN PLASMAS

BERTIL AGDUR

Royal Institute of Technology, Stockholm 70, Sweden

SUMMARY

A dispersion anomaly in passive plasma waveguide systems is briefly discussed.

INTRODUCTION

It has been found that the dispersion curve (β,ω-curve) in, for instance, plasma waveguide systems (Trivelpiece, 1958; Sellberg, 1961) or in systems containing an isotropic dielectric rod (Clarricoats, 1961), may have the property that $\partial\omega/\partial\beta$ changes sign at some points along the β,ω-curve. The character of such a dispersion curve around the points at which $\partial\omega/\partial\beta$ is zero is briefly indicated.

DISCUSSION OF THE DISPERSION ANOMALY

The dispersion equation for the plane parallel lossless plasma waveguide system shown in Fig. 1 can be written in the following form (Sellberg, 1961),

$$\frac{k_{II}}{\varepsilon_{II}}\tanh(k_{II}d) = -k_I\tanh[k_I(h-d)]$$

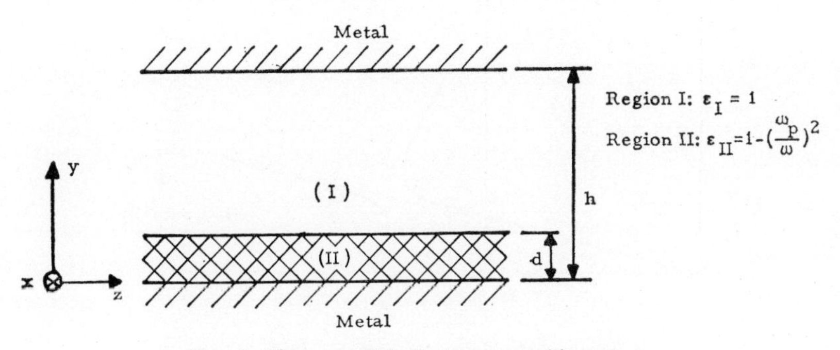

FIG. 1. Plane parallel plasma waveguide system.

where a field variation exp $[j(\omega t\text{-}\beta z)]$ is assumed and

$$\varepsilon_{11} = \left[1 - \left(\frac{\omega_p}{\omega}\right)^2\right]$$

$$k_1 = \sqrt{\left[\beta^2 - \left(\frac{\omega}{c}\right)^2\right]}$$

$$k_{11} = \sqrt{\left[\beta^2 - \varepsilon_{11}\left(\frac{\omega}{c}\right)^2\right]}$$

d = thickness of plasma sheath

h = distance between the metal walls.

Figure 2 shows characteristic features of the β,ω-diagram for real values of β with typically $d/h = 0.05$ and $\frac{\omega_p d}{c} = 0.1$.

If the dispersion equation, in a generalized complex form, is expanded around the points where $\partial\omega/\partial\beta = 0$, one finds that four branches of the dispersion curve are starting from these points; two of which are shown in Fig. 2 and two of which are complex conjugate (not shown in Fig. 2). These

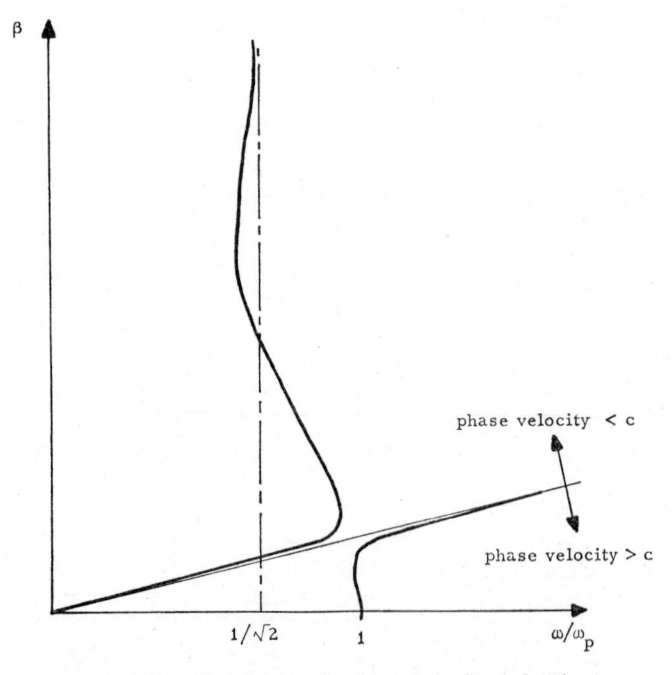

FIG. 2. Dispersion diagram for the system shown in Fig. 1.

latter branches are purely real at the point where $\partial\omega/\partial\beta = 0$. If for instance losses are introduced into the system there will be uniquely determined transitions from the branches shown in Fig. 2 to the complex conjugate branches, and there are no connections between the branches shown in Fig. 2.

The dispersion equation of the system shown in Fig. 1 is chosen as an example because of its simplicity. The properties of the dispersion curve discussed above are of course general.

REFERENCES

CLARRICOATS, P. J. B. (1961) Inst. of E.E. Monograph No. 451E.

SELLBERG, F. (1961) Internal Report, Microwave Department, Royal Institute of Technology, Stockholm.

TRIVELPIECE, A. W., Techn. Rep. No. 7, Cal. Inst. of Techn., El. Tube and Microw. Lab. (May).

DISCONTINUOUS FLOW OF PLASMA

KRYSTYN BOCHENEK

Polish Academy of Science, Warsaw, Poland

SUMMARY

The idea proposed in this paper is to modify the system of equations of magnetohydrodynamics to obtain a smooth approximation of shock waves, consistent with shock conditions (of Rankin–Hugoniot type) in the sense defined below. Two systems of equations, represented by (1) and (7), are considered; the first gives a smooth approximation but does not respect the energetic shock condition, whereas the second respects the total set of shock conditions.

Let us consider the flow of plasma defined by the set of equations†

$$
\begin{aligned}
\dot{\mathbf{E}} &= c^2\, \nabla \times \mathbf{B} + \kappa(\mathbf{B} \times \mathbf{v} - \mathbf{E}) \\
\dot{\mathbf{B}} &= -\nabla \times \mathbf{E} \\
\mathbf{v} &= -(\mathbf{v} \cdot \nabla)\mathbf{v} - a^2 \frac{1}{\rho} \nabla \rho + \frac{1}{\mu\rho} (\nabla \times \mathbf{B}) \times \mathbf{B} \\
\dot{\rho} &= -\nabla(\rho \mathbf{v})
\end{aligned}
\tag{1}
$$

where a and c are respectively the speed of sound and the speed of light, and $\kappa = \sigma/\varepsilon$. When the conductivity σ and its related parameter κ are sufficiently great, we may expect that the flow will be approximately described by the simpler set of equations:

$$
\begin{aligned}
\dot{\mathbf{B}} &= \nabla \times (\mathbf{v} \times \mathbf{B}) \\
\dot{\mathbf{V}} &= -(\mathbf{v} \cdot \nabla)\mathbf{v} - a^2 \frac{1}{\rho} \nabla \rho + \frac{1}{\mu\rho} (\nabla \times \mathbf{B}) \times \mathbf{B} \\
\dot{\rho} &= -\nabla(\rho \mathbf{v})\,.
\end{aligned}
\tag{2}
$$

Both systems are quasi-linear and hyperbolic, but they have different families of characteristics. Bearing in mind the well-known relation of discontinuities

† Let us remark that in this set of equations the displacement current is introduced in the first equation, but not in the third one. A similar remark applies in the set (7). (Note the form of the fifth equation in set (7)). The influence of the displacement current term appears for large κ only in the region of rapid variation of the field, but in that region we are only interested in the consistency with the shock conditions to be proved directly.

of the solutions of hyperbolic systems to their characteristics, we can ask which kind of solutions of the system (1) correspond to the discontinuous solutions of the system (2), or to the solutions of that system with a singular field of characteristics. An asymptotic phenomenon should be expected when $\kappa \to \infty$; namely the continuous solution of (1) converges to the solution of (2) non-uniformly, and for large κ there exists a narrow domain (boundary layer) in which the changes of the solution of system (1) are very rapid.

To be more specific let us consider the case of two dimensions, t (time) and x (space coordinate). This simple case is very important from the point of view of possible applications of numerical methods and electronic computers. In this case the equations for the x-components of $\mathbf{E}$ and $\mathbf{B}$ can be omitted, and for the vector

$$\mathbf{w} = E_y, E_z, B_y, B_z, v_x, v_y, v_z, \rho)$$

we obtain from (1) the equation

$$\dot{\mathbf{w}} = \mathbf{A}\cdot\mathbf{w},_x + \kappa(-\mathbf{u} + \mathbf{f}(\mathbf{l})) \tag{3}$$

where we define

$$\mathbf{u} = (E_y, E_z, 0, \ldots), \qquad \mathbf{l} = (0,0,B_y,B_z,v_x,v_y,v_z,\rho)$$

and

$$\mathbf{f}(\mathbf{l}) = ((\mathbf{B}\times\mathbf{v})_y, (\mathbf{B}\times\mathbf{v})_z, 0, \ldots).$$

The characteristic equation of the matrix $\mathbf{A}$ is

$$(\lambda^2 - c^2)^2(\lambda - v_x)^2[(\lambda - v_x)^2 - a^2] = 0 .$$

The system (2) is now equivalent to the equation

$$\dot{\mathbf{l}} = \mathbf{L}\cdot(\mathbf{f}(\mathbf{l}) + \mathbf{l}),_x = \mathbf{M}\cdot\mathbf{l},_x \tag{4}$$

where $\mathbf{L}$ is the lower part of the matrix $\mathbf{A}$. The characteristic equation of the matrix $\mathbf{M}$ is

$$\left[(\lambda - v_x)^2 - \frac{B_x^2}{\mu\rho}\right]\left[(\lambda - v_x)^2 - \frac{B^2}{\mu\rho} - a^2 + \frac{a^2 B_x^2}{\mu\rho}\right] = 0 .$$

We see that the roots of the two characteristic equations are different. In the first system we have roots corresponding to speed of light and speed of sound, and in the second system the roots correspond to the speed of hydromagnetic waves.

It seems profitable to use the full system (3) in numerical computations of flows in which the simpler system leads to a caustic of characteristic curves (shock wave). In this case the complication introduced by two more dependent variables is not so great as the difficulties of calculation in the neighbourhood of the caustic. This is similar to Richtmyer's[2] introduction of a pseudo-viscosity.

The behavior of the solution in the neighbourhood of a discontinuity of the simpler system can be calculated by means of another kind of asymptotic solution of boundary layer type. The idea is to introduce an orthogonal coordinate system (r, s) in which the line $s = 0$ is the line of discontinuity. Then we replace the variable s by

$$S = \kappa s$$

and retain in (3) only the terms of the order of κ. This results in a system of ordinary differential equations,

$$\left(\frac{\dot{s}}{s_{1x}}\mathbf{I} - \mathbf{A}\right)\cdot \mathbf{w}_{,s} = \frac{1}{s_{1x}}[-\mathbf{u} + \mathbf{f}(\mathbf{l})]\,. \tag{5}$$

This system takes into account the rapid variation of the solution in the direction perpendicular to the curve $s = 0$. The system (5) describes also the solutions of (3) depending on one variable

$$\xi = x - Ut$$

which is a very common approach[2,3] in the calculations of shock waves.

Let us integrate the system (5) by assuming that the solution has a limit for $S \to \pm\infty$ (and $\dot{s}$, $s_{1x} = \text{const}$). This results in the following shock relations for the system (4) at $s = \pm 0$:

$$\{(v_x - U)\rho\} = 0, \qquad \{\mathbf{i} \times \mathbf{E} - U\mathbf{B}_t\} = 0, \qquad \{\mathbf{E} - \mathbf{B} \times \mathbf{v}\} = 0,$$

$$\left\{Mv_x + a^2\rho + \frac{1}{2\mu}B_t^2\right\} = 0, \qquad \left\{M\mathbf{v}_t - \frac{1}{\mu}B_x\mathbf{B}_t\right\} = 0\,, \tag{6}$$

where $\mathbf{i}$ is the unit vector in the direction of the x-axis, t denotes the transverse components, and $U = \frac{\dot{s}}{s_{1x}}$, $M = \text{const}$. These relations do not take into account the dissipation of energy at the shock and probably are suitable for small dissipation. In the general case the equations (3) are to be replaced by the equations

$$\dot{\mathbf{E}} = c^2\mathbf{i} \times \mathbf{B}_{1x} + \kappa(\mathbf{B} \times \mathbf{v} - \mathbf{E})\,; \quad \dot{\mathbf{B}} = -\,\mathbf{i} \times \mathbf{E}_{1x}$$

$$\rho\dot{\mathbf{v}} = -\,\rho v_x\mathbf{v} - \mathbf{i}p_{1x} + \frac{1}{\mu}(\mathbf{i} \times \mathbf{B}_{1x}) \times \mathbf{B} \tag{7}$$

$$\dot{\rho} = -\,(\rho v_x)_{1x}, \quad \rho(\dot{\varepsilon} + v_x\varepsilon_{1x} + p\dot{V} + pv_xV_{1x}) = \frac{1}{\mu}(\mathbf{E} - \mathbf{B} \times \mathbf{v})(\mathbf{i} \times \mathbf{B}_{1x})$$

$$\varepsilon = F(p,\rho)\,.$$

The last equation is the equation of state, $V = 1/\rho$ and ε is the internal energy per unit mass. From this system we obtain by integration the following exact Landau[2] shock relations of Parkin–Hugoniot type.

$$\{(v_x - U)\rho\} = 0\,, \qquad \{\mathbf{i} \times \mathbf{E} - U\mathbf{B}_t\} = 0\,, \qquad \{\mathbf{E} - \mathbf{B} \times \mathbf{v}\} = 0\,,$$

$$\left\{Mv_x + p + \frac{1}{2\mu}\mathbf{B}_t^2\right\} = 0\,, \qquad \left\{M\mathbf{v}_t - \frac{1}{\mu}B_x\mathbf{B}_t\right\} = 0\,, \tag{8}$$

$$\left\{M\varepsilon + M\frac{p}{\rho} + M\frac{\mathbf{v}_t^2}{2} + M^3\frac{1}{2\rho^2} + M\frac{\mathbf{B}_t^2}{\mu\rho} - B_x\frac{\mathbf{v}_t \cdot \mathbf{B}_t}{\mu}\right\} = 0\,.$$

An excellent description of the shock relations and boundary-value problems in magnetohydrodynamics (for the simpler case $v_y = v_z = B_x = B_y = E_x = E_z = 0$) is given by Whitham[(3)].

REFERENCES

1. Landau, L. D. and Lifshic, E. M. (1957) *Elektrodynamika sploshnyh sred.*
2. Richtmyer, R. D. (1957) *Difference Methods for Initial-value Problems.*
3. Whitham, G. B. (1959) *Some Comments on Wave Propagation ... Comm. Pure and Appl. Math.* Vol. XII, No. 1, 113–158, Feb.

UNSTABLE TRANSVERSE MODES OF DRIFTING CHARGED PARTICLES IN A PLASMA IN A MAGNETIC FIELD†

FRED I. SHIMABUKURO

Aerospace Corporation, El Segundo, California.

ABSTRACT

When a fast beam of charged particles, electrons, ions, or a plasma, traverses along magnetic field lines in a plasma, there is a possibility of unstable transverse modes of propagation. The normal modes are a pair of oppositely circularly polarized waves. For drifting electrons or ions the waves exhibit an instability over a very narrow frequency band near the ion and electron cyclotron frequencies. When a plasma drifts through a plasma, in addition to the instabilities noted above, there can be an instability near zero frequency for both polarizations.

1. INTRODUCTION

Whenever there is a beam of charged particles drifting through a plasma in a magnetic field, in addition to the unstable longitudinal mode (for $\omega < \omega_e$), which has been discussed extensively in the literature, there exists the possibility of unstable transverse modes of propagation. The propagation vector, the drift velocity and the magnetic field will be assumed to be in the z-direction. The salient features of the interaction will be examined by considering only a cold, collisionless plasma.

2. GENERAL FORMULATION

The dispersion will be obtained by incorporating the beam-plasma effects into an equivalent dielectric tensor and substituting into the wave equation. The equations needed are Maxwell's equations, the macroscopic equation of motion, and a definition of current density. All equations will be linearized and waves of the form $e^{j(\omega t - k_z z)}$ will be assumed. The subscripts 0 and 1 will be used to denote zero order and first order quantities respectively. Maxwell's equations are

$$\nabla \times \bar{E}_1 = -\frac{\partial \bar{B}_1}{\partial t} \tag{1}$$

† This work was sponsored by the U.S. Army Signal Research and Development Laboratory, Fort Monmouth, New Jersey, under Contract DA36–039 SC–85317, Department of the Army Task No. 3A 99–13–001–05.

$$\nabla \times \bar{H}_1 = \bar{J}_1 + \frac{\partial \bar{D}_1}{\partial t} \tag{2}$$

$$\nabla \cdot \bar{D}_1 = \rho_1 \tag{3}$$

$$\nabla \cdot \bar{B}_1 = 0 \,. \tag{4}$$

The equation of motion for a charged particle with charge to mass ratio η is

$$\frac{\partial \bar{u}_1}{\partial t} + u_0 \cdot \frac{\partial \bar{u}_1}{\partial z} = \eta(\bar{E}_1 + \bar{u}_0 \times \bar{B}_1 + \bar{u}_1 \times \bar{B}_0) \,. \tag{5}$$

The current density is defined

$$\bar{J}_1 = \bar{u}_1 \rho_0 + \bar{u}_0 \rho_1 \,. \tag{6}$$

Taking the divergence of (2) and using (3) one obtains the continuity equation

$$\nabla \cdot \bar{J}_1 = - \frac{\partial \rho_1}{\partial t} \,. \tag{7}$$

One then defines the dielectric tensor

$$\bar{J}_1 + \frac{\partial \bar{D}_1}{\partial t} = j\omega\varepsilon \cdot \bar{E}_1 \,. \tag{8}$$

Using (5), (6), and (7). Taking the curl of (1) and using (2) and (8) one obtains the wave equation

$$\nabla \times \nabla \times \bar{E}_1 - \omega^2 \mu \varepsilon \cdot \bar{E}_1 = \mathbf{L} \cdot \bar{E}_1 = 0 \,. \tag{9}$$

The dispersion is obtained by setting the determinant of the operator **L** equal to zero.

3. TRANSVERSE MODES OF INSTABILITY

3.1. *Electron Beam in a Plasma*

The dispersion for an electron beam in a plasma where k, u_0, and B have the same direction can be written as

$$\left[1 - \frac{c^2k^2}{\omega^2} - \frac{\omega_e^2}{\omega(\omega + \omega_{ce})} - \frac{\omega_i^2}{\omega(\omega - \omega_{ci})} - \frac{\omega_{ed}^2(\omega - u_0 k)}{\omega^2(\omega - u_0 k + \omega_{ce}')}\right] \times$$

$$\times \left[1 - \frac{c^2k^2}{\omega^2} - \frac{\omega_e^2}{\omega(\omega - \omega_{ce})} - \frac{\omega_i^2}{\omega(\omega + \omega_{ci})} - \frac{\omega_{ed}^2(\omega - u_0 k)}{\omega^2(\omega - u_0 k - \omega_{ce}')}\right] \times$$

$$\times \left[1 - \frac{\omega_e^2}{\omega^2} - \frac{\omega_{ed}^2}{(\omega - u_0 k)^2}\right] = 0, \tag{10}$$

where ω_{ce} and ω_{ci} are the electron and ion cyclotron frequencies, and ω_e, ω_i and ω_{ed} are the plasma frequencies of the electrons, ions, and drifting electron beam respectively. Equation (10) is conveniently written in the form

$$D_+(k,\omega)D_-(k,\omega)D_l(k,\omega) = 0\,. \tag{11}$$

The following convention will be used. The + sign denotes the circularly polarized wave in which the field gyrates in the same sense as the ions, and the − sign denotes the wave which gyrates in the same sense as the electrons.

The third factor in (11) is the longitudinal mode and will be ignored. The first dispersion exhibits an instability near $\omega \approx \omega_{ci}$, while the second is stable. The second dispersion can display an instability near $\omega \approx \omega_{ce}$ if a fast beam of ions is present. This case will be examined later. This instability was predicted previously. (Dawson and Bernstein, 1958). In this paper the instability will be examined in a slightly different way and in more detail. The instability will be examined by looking at the undisturbed waves and examining regions where the waves are near synchronism so that there is a possibility of an interaction and a resulting growing wave. Particularly look for the slow electron beam wave interacting with the stationary plasma wave, see Fig. 1 for the plot of the undisturbed waves of $D_+(k,\omega) = 0$. It is seen that the undisturbed waves are close to synchronism near ω_{ci} and there should be an interaction in that frequency vicinity. This will now be examined in some detail.

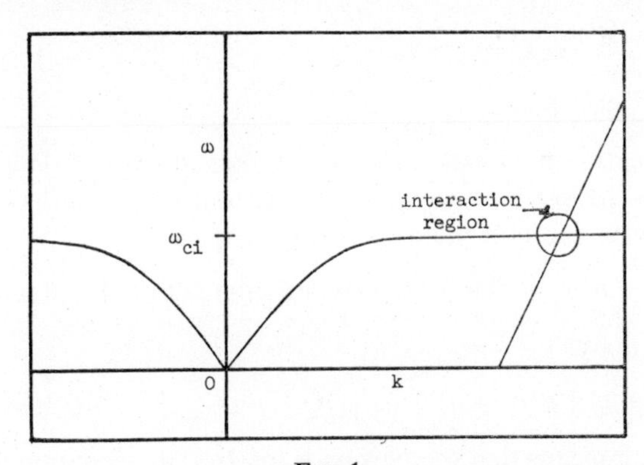

FIG. 1

Instability near ω_{ci}. To examine the instability the dispersion is written in the form

$$\left\{\frac{c^2k^2}{\omega^2} - \left[1 - \frac{\omega_e^2}{\omega(\omega+\omega_{ce})} - \frac{\omega_i^2}{\omega(\omega-\omega_{ci})}\right]\right\}\left(k - \frac{\omega+\omega_{ce}}{u_0}\right) + \\ + \frac{\omega_{ed}^2}{\omega^2}\left(k - \frac{\omega}{u_0}\right) = 0\,. \tag{12}$$

Since interaction is expected near $\omega \approx \omega_{ci}$, k is approximately equal to ω_{ce}/u_0 and the dispersion can be written approximately

$$(k - k_1)(k - \beta_{ce} - \beta_e) + \frac{\omega_{ed}^2}{2c^2} = 0 \tag{13}$$

where

$$k_1^2 \approx -\frac{\omega\omega_i^2}{c^2(\omega - \omega_{ci})}, \qquad \beta_{ce} = \frac{\omega_{ce}}{u_0}, \qquad \beta_e = \frac{\omega}{u_0} \approx \frac{\omega_{ci}}{u_0}.$$

The dispersion can be analyzed in the same manner as done by Pierce (1950). Now a few parameters are defined,

$$A^2 = \frac{u_0^2}{2c^2}\frac{\omega_{ed}^2}{\omega_{ce}^2} \tag{14}$$

$$k = \beta_{ce}(1 + jA\delta)$$

$$k_1 = \beta_{ce}(1 + Ab - jAd)$$

b = differential phase velocity between k and k_1

$$\beta_e = \frac{\omega}{u_0} \approx \frac{\beta_{ce}}{\lambda}$$

$$\lambda = \omega_{ce}/\omega_{ci}$$

$$Q = \frac{1}{\lambda A^2}.$$

$$\gamma = u_0/c$$

A is the interaction parameter and corresponds to the C in Pierce's notation. A depends on the beam velocity and density and the strength of the magnetic field. From (13) and (14) the dispersion becomes

$$\delta^2 + \delta(jQA + jb + d) + QA(jd - b) - 1 = 0. \tag{15}$$

The roots of δ are

$$\delta_{1,2} = -\tfrac{1}{2}(jQA + jb + d) \pm \tfrac{1}{2}\sqrt{[(jQA + jb + d)^2 - 4QA(jd - b)]}. \tag{16}$$

If $d = 0$, which means that the plasma is lossless, the equation for δ becomes

$$\delta_{1,2} = -\frac{j}{2}(QA + b) \pm \tfrac{1}{2}\sqrt{[-(QA + b)^2 + 4(1 + QAb)]} = x \pm jy. \tag{17}$$

The gain condition can be determined from (17) and is

$$2 > |QA - b| \quad \text{gain condition}. \tag{18}$$

Maximum growth occurs at $b = QA$. Plots of δ in the growth region are shown in Figs. 2, 3, 4 and 5. The bandwidth for growth can be determined

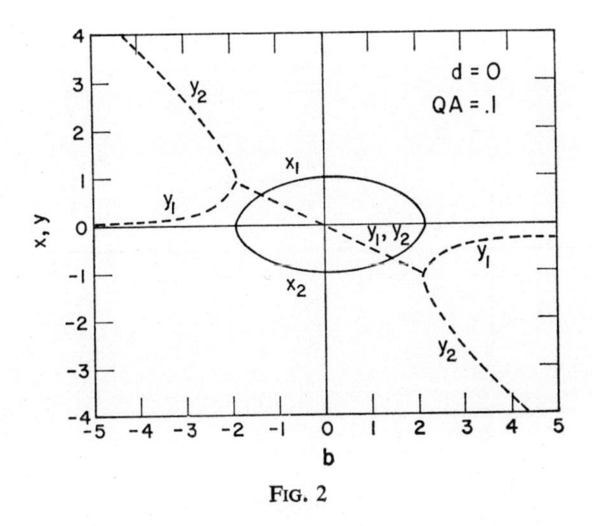
d = 0
QA = .1
y_2
x_1
y_1
y_1, y_2
y_1
x_2
y_2
x, y
4
3
2
1
0
-1
-2
-3
-4
-5
-4
-3
-2
-1
0
1
2
3
4
5
b

FIG. 2

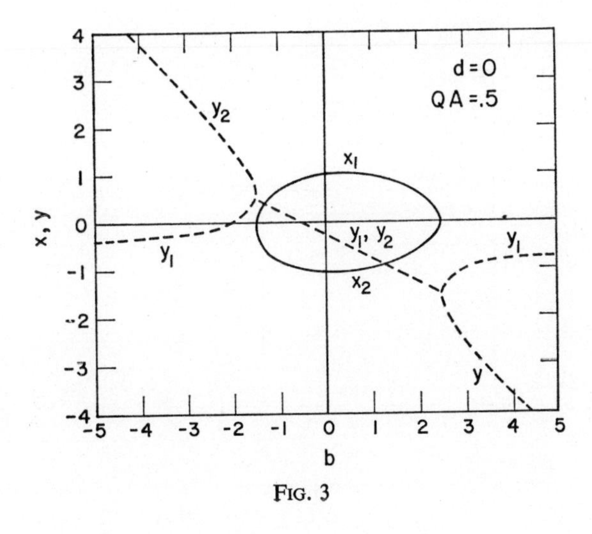
d = 0
QA = .5
y_2
x_1
y_1
y_1, y_2
y_1
x_2
y
x, y
4
3
2
1
0
-1
-2
-3
-4
-5
-4
-3
-2
-1
0
1
2
3
4
5
b

FIG. 3

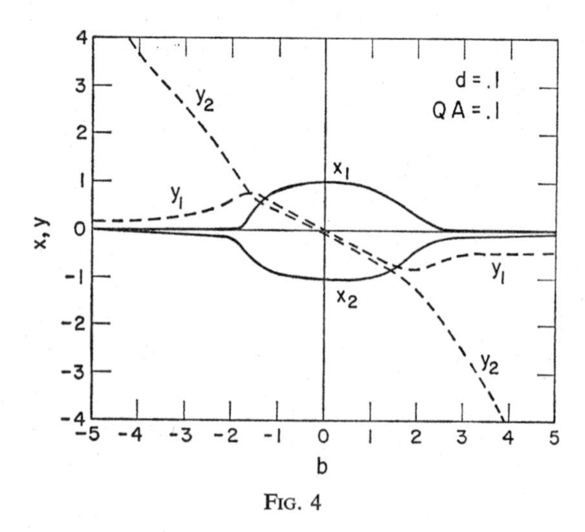
d = .1
QA = .1
y_2
y_1
x_1
x_2
y_1
y_2
x, y
b
4
3
2
1
0
-1
-2
-3
-4
-5
-4
-3
-2
-1
0
1
2
3
4
5

Fig. 4

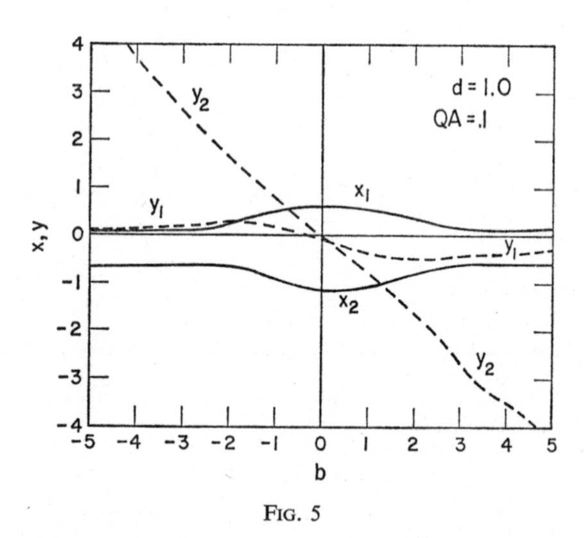
d = 1.0
QA = .1
y_2
y_1
x_1
y_1
x_2
y_2
x, y
b
4
3
2
1
0
-1
-2
-3
-4
-5
-4
-3
-2
-1
0
1
2
3
4
5

Fig. 5

approximately. If $p = \omega_i/\omega_{ci}$ and $z = \omega/\omega_{ci}$, and using the fact $z \approx 1$ and

$$k_1 \approx \beta_{ce} \frac{\gamma p}{\lambda} \frac{z^{1/2}}{(1-z)^{1/2}} \approx \beta_{ce},$$

one obtains

$$\frac{dk_1}{dz} \approx \beta_{ce} \frac{1}{1-z} \tag{19}$$

and using (13) and (17)

$$\Delta k = 4\beta_{ce} A, \tag{20}$$

and the bandwidth over which there is gain is then $\Delta z = 4A(1-z)$ or

$$\Delta\omega = 2\sqrt{2} \cdot \frac{u_0^3}{c^3} \frac{\omega_{ed}\omega_{ci}\omega_i^2}{\omega_{ce}^3}. \tag{21}$$

A sample calculation can be made to see the quantities involved. For $\gamma = 0.02$, $p = \omega_i/\omega_{ci} = 100(\omega_e/\omega_{ce} = 2.32)$, $\lambda = 1837$ (hydrogen)

$$\Delta z \approx 4.93 \times 10^{-9}$$

or

$$\Delta\omega \approx 4.93 \times 10^{-9} \omega_{ci}. \tag{22}$$

It is seen that there is gain over a very limited range.

3.2. *Ion Beam in a Plasma*

If there were a fast beam of ions instead of electrons, the $D_-(k,\omega)$ dispersion would become unstable near $\omega \approx \omega_{ce}$. The dispersion can be solved for in the same manner, making the conversion $\omega_{ed} \to \omega_{id}$, $-\omega'_{ce} \to \omega_{cid}$. Notice that although the interaction takes place at different frequencies the β's are approximately the same.

3.3. *Drifting Plasma in a Plasma*

The ω-k plots of a stationary and a drifting plasma for the + wave are shown in Fig. 6. Notice that for $u_0 > V_a$ ($V_a = c\omega_{ci}/\omega_i$, the Alfvén velocity) there is a wave near zero frequency close to synchronism with the undisturbed stationary plasma wave. The dispersion for the drifting plasma is obtained by making a Lorentz transformation to the stationary dispersion. For the slow waves under discussion and non-relativistic drift velocities, the transformation is a simple one

$$\omega = \omega' + u_0 k'$$
$$k = k' \tag{23}$$

where the primed system is the one in which the plasma is stationary, and the unprimed system is the one in which the plasma is drifting. The instability to be discussed was first recognized by Dokuchaev (1961), and here the

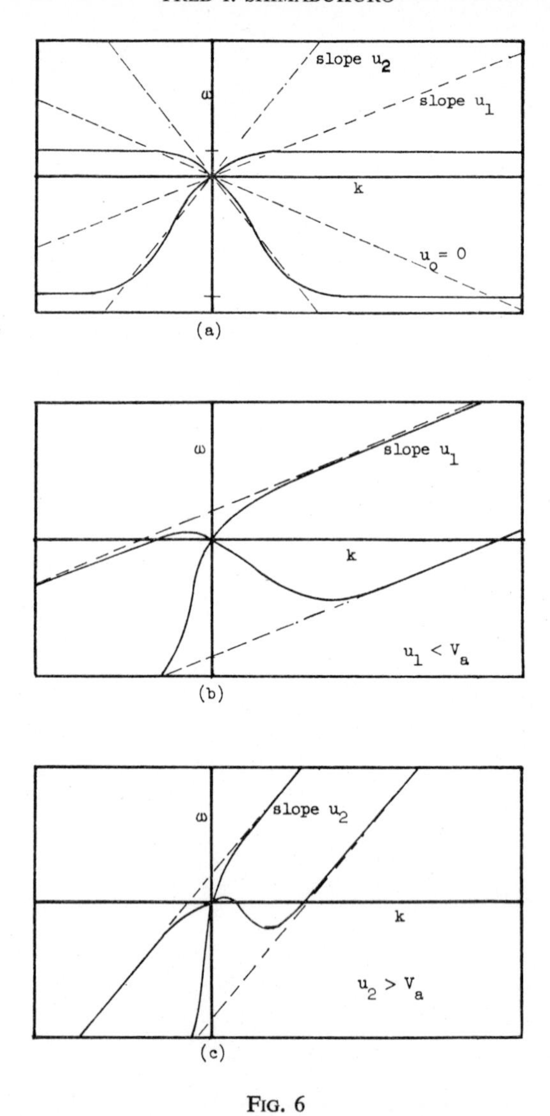

FIG. 6

analysis will be carried out more fully, the instability being investigated as an interaction of the drifting plasma wave and the stationary plasma wave.

For the – wave, the stationary and drifting dispersion is the same as in Fig. 6, if the positive and negative ω-axes are interchanged.

The dispersion for the transverse waves can be written as

$$D_+(k,\omega)D_-(k,\omega) = 0 \tag{24}$$

where

$$D_+(k,\omega) = \frac{c^2k^2}{\omega^2} - \left[1 - \frac{\omega_e^2}{\omega(\omega + \omega_{ce})} - \frac{\omega_i^2}{\omega(\omega - \omega_{ci})} - \right.$$
$$\left. - \frac{\omega_{ed}^2(\omega - u_0 k)}{\omega^2(\omega - u_0 k + \omega_{ce})} - \frac{\omega_{id}^2(\omega - u_0 k)}{\omega^2(\omega - u_0 k - \omega_{cid})}\right] \tag{25}$$

$$D_-(k,\omega) = \frac{c^2k^2}{\omega^2} - \left[1 - \frac{\omega_e^2}{\omega(\omega - \omega_{ce})} - \frac{\omega_i^2}{\omega(\omega + \omega_{ci})} \right.$$
$$\left. - \frac{\omega_{ed}^3(\omega - u_0 k)}{\omega^2(\omega - u_0 k - \omega_{ce})} - \frac{\omega_{id}^2(\omega - u_0 k)}{\omega^2(\omega - u_0 k + \omega_{cid})}\right]. \tag{26}$$

These two cases will now be examined.

$D_+(k,\omega) = 0$: From Fig. 6 it is seen that a necessary condition for instability is that $u_0 > V_{ad}$, where V_{ad} is the Alfvén velocity if the plasma is stationary. The other condition, of course, is on the density of the stationary plasma. An analysis will now be carried out which is valid near the origin. For the case $\omega_{cid} \gg u_0 > \omega/k$ and using the relations $\omega_{ce} \gg \omega_{ci},\omega_{cid}$, (24) can be put in the form by an expansion

$$k^2 = \frac{\omega^2}{c^2} p_s^2 + \frac{p_d^2}{c^2}(\omega - u_0 k)^2 \tag{27}$$

where

$$p_s = \frac{\omega_i^2}{\omega_{ci}^2} = \frac{c^2}{V_{as}^2}, \qquad p_d^2 = \frac{\omega_{id}^2}{\omega_{cid}^2} = \frac{c^2}{V_{ad}^2},$$

and solving for the propagation constant,

$$k = \frac{\omega}{u_0^2 - V_{ad}^2}\left[u_0 \pm \frac{V_{ad}}{V_{as}}\sqrt{\{(V_{ad}^2 + V_{as}^2) - u_0^2\}}\right]. \tag{28}$$

The condition for growth is

$$u_0^2 > V_{as}^2 + V_{ad}^2. \tag{29}$$

This derivation was carried out under the assumption that $\omega \ll \omega_{ci},\omega_{cid}$. To determine the propagation constant over all ranges of frequency, the full dispersion in (25) must be solved. Figure 7 shows the real and imaginary parts of k in the unstable region.

Incidentally, for this case there is still the interaction near $\omega \approx \omega_{ci}$ which was worked out in Section 3.1.

$D_-(k,\omega) = 0$: In the same way, the condition for a growing wave for this case can be determined by examining the dispersion near zero frequency. If the expansion is carried out under the same assumption as before, the dispersion is the same as (27) and the gain condition is the same. In (25) there is no amplification for $\omega > \omega_{ci}$ but (26) can be unstable for $\omega > \omega_{ci}$. A plot of k in the growth region is shown in Fig. 8. In addition there is still the interaction near ω_{ce} between the slow wave of the drifting ions and the plasma wave.

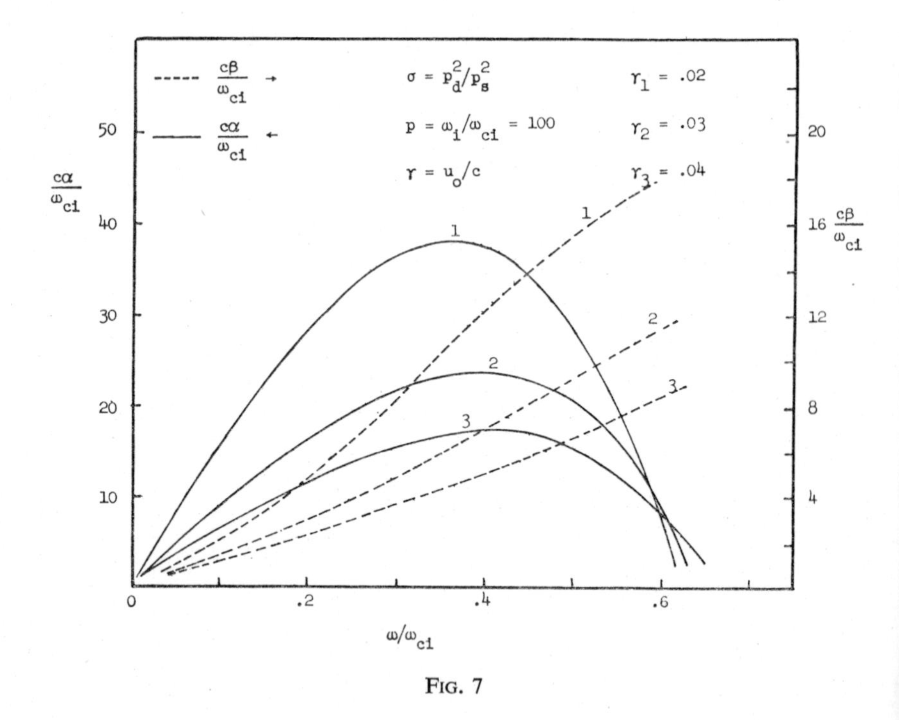

FIG. 7

3.4. *Comparison of the Growth Rates of the Transverse and Longitudinal Modes*

Suppose there is a region where both the transverse and longitudinal waves can exhibit instability. These modes have different rates of growth, and growth occurs over different frequency ranges. For example, the longitudinal mode is unstable from zero frequency to the plasma frequency, with maximum growth at the plasma frequency. For typical parameter values, the + wave of the drifting plasma in a plasma system can be unstable over a frequency band between zero frequency and several tenths of the ion cyclotron frequency with maximum growth occurring near mid-frequency; and the – wave can be unstable from zero frequency to several times the ion cyclotron frequency with maximum growth occurring at a frequency somewhat below mid-frequency. In the system of a drifting electron or ion beam through a plasma, there is an instability only in the immediate vicinity of the ion and electron frequency, respectively.

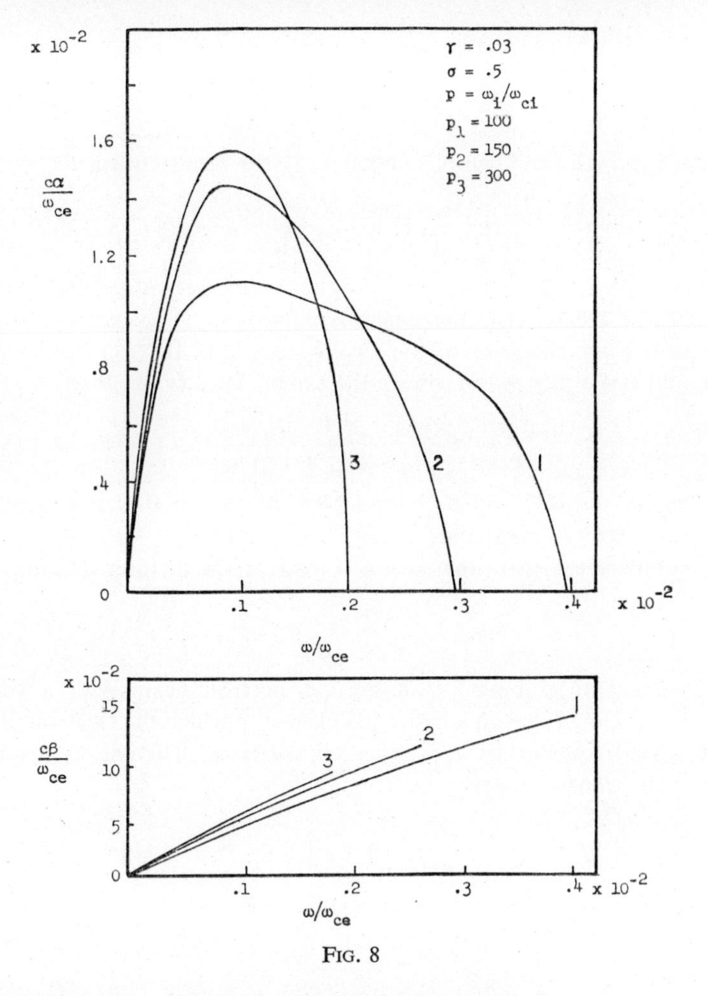

FIG. 8

The propagation constant for the longitudinal mode is

$$k_l = \frac{\omega}{u_0}\left[1 \pm \frac{\omega_{ed}/\omega_e}{\sqrt{(\omega^2/\omega_e^2 - 1)}}\right]. \tag{30}$$

For the + and − waves of the drifting plasma in a plasma system, valid near zero frequency

$$k_{+,-} = \frac{\omega}{u_0(1 - V_{ad}^2/u_0^2)}\left[1 \pm \frac{V_{ad}}{V_{as}}\sqrt{\left(\frac{(V_{as}^2 + V_{ad}^2)}{u_0^2} - 1\right)}\right] \tag{31}$$

and at maximum growth for the drifting electron beam

$$k_e = \frac{\omega_{ce}}{u_0}\left[1 \pm \frac{j}{\sqrt{2}}\frac{u_0}{c}\frac{\omega_{ed}}{\omega_{ce}}\right] \tag{32}$$

and for the drifting ion beam

$$k_i = \frac{\omega_{ce}}{u_0}\left[1 \pm \frac{j}{\sqrt{2}}\frac{u_0}{c}\frac{\omega_{id}}{\omega_{ce}}\right]. \tag{33}$$

Now the growth rates can be compared. Near zero frequency

$$\frac{\alpha_l}{\alpha_{+,-}} \approx \frac{(V_{as}/V_{ad})(\omega_{ed}/\omega_e)(1 - V_{ad}^2/u_0^2)}{\sqrt{[1 - (V_{as}^2 + V_{ad}^2)/u_0^2]}\sqrt{(1 - \omega^2/\omega_e^2)}}. \tag{34}$$

Examining (34), it is seen that depending on parameter values the transverse growth can be greater than the longitudinal growth. For example, suppose the electron beam densities of the two systems and the ion masses of the drifting and stationary plasmas were the same. Then for $\omega_e \gg \omega$

$$\frac{\alpha_l}{\alpha_{+,-}} \approx \frac{(\omega_{ed}/\omega_e)^2(1 - V_{ad}^2/u_0^2)}{\sqrt{[1 - (V_{as}^2 + V_{ad}^2)/u_0^2]}}. \tag{35}$$

For $\omega_{ed}/\omega_e = 1.0$, 0.1, and 0.01, and $(V_{as}^2 + V_{ad}^2)/u_0^2 = 0.8$, $\alpha_l/\alpha_{+,-} = 1.34$, 0.0134, and 0.000134 respectively.

Now compare the two other cases. First, for the drifting electron beam

$$\frac{\alpha_l}{\alpha_e} \approx \sqrt{2}\cdot\frac{\omega_{ci}/\omega_e}{u_0/c} \qquad \omega \approx \omega_{ci}. \tag{36}$$

Take a numerical example. Consider an electron beam with a velocity $u_0 = 0.02c$ traveling through a hydrogen plasma in which the electron plasma and cyclotron frequency are approximately the same. Then at the maximum growth of the transverse wave

$$\frac{\alpha_l}{\alpha_e} \approx 3.8 \times 10^{-2}. \tag{37}$$

For the case of the drifting ion beam in a plasma

$$\frac{\alpha_l}{\alpha_i} \approx \sqrt{2}\cdot\frac{\omega_{ce}/\omega_e \cdot \omega_{ed}/\omega_{id}}{u_0/c}. \tag{38}$$

For an ion and electron beam with equal number density traveling through a hydrogen plasma at $u_0 = 0.02c$ with the electron plasma and cyclotron frequencies equal,

$$\frac{\alpha_l}{\alpha_i} = 303. \tag{39}$$

It is seen that here the longitudinal growth rate is much larger than the transverse rate.

4. CONCLUSIONS

It is seen that whenever fast charged particles move in a plasma along magnetic lines of force, unstable transverse modes can propagate in such a system. These instabilitites are viewed as the interaction of the slow wave of the drifting particles and the wave of the stationary plasma. If the beam is composed of either electrons or ions, growing waves can exist over a very narrow-frequency range in the vicinity of the ion and electron cyclotron frequencies. If the beam consists of a neutral plasma, in addition to the instabilities noted above, there can be an instability near zero frequency. For typical values the + wave is unstable from zero frequency to several tenths of the ion cyclotron frequency, whereas the − wave is unstable from zero frequency to several times the ion cyclotron frequency. These waves are possible in ionospheric, solar and stellar regions.

SUMMARY

When a fast beam of charged particles composed of electrons, ions, or a plasma, traverses along magnetic field lines in a plasma, there is a possibility of unstable transverse modes of propagation. The normal modes are a right and a left circularly polarized wave. These instabilities are viewed as an interaction between the slow wave of the drifting particles and the stationary plasma wave. Then instabilities should occur in regions where these waves are near synchronism. If the drifting particles are either electrons or ions, the circularly polarized waves exhibit an instability over a very narrow range near the ion and electron cyclotron frequencies, respectively. When a plasma drifts through a plasma, in addition to the instabilities noted above, there can be an instability near zero frequency for both polarizations. For typical parameter values, for the wave in which the field gyrates in the same sense as the ions, the instability occurs between zero frequency and several tenths of the ion cyclotron frequency. For the other polarization there can be instability between zero frequency and several times the ion cyclotron frequency.

For parameter values generally encountered in practice, except for the instability in the vicinity of the electron cyclotron frequency, these transverse modes have greater growth constants than the longitudinal mode in the frequency range of application. These unstable transverse modes are possible in ionospheric, solar, and stellar regions.

ACKNOWLEDMENT

The author wishes to express his gratitude to Pro. R. W. Gould for the many enlightening discussions on the subject.

REFERENCES

DAWSON, J. and BERNSTEIN, I. B. (1958) Hydromagnetic Instabilities Caused by Runaway Electrons, Controlled Thermonuclear Conference, 360–365 (February).

DOKUCHAEV, V. P. (1961) On the growth of magnetohydrodynamic waves in a plasma stream moving through an ionized gas, *Soviet Physics JETP* **12**, No. 2, 292–293 (February).

PIERCE, J. R. (1950) *Traveling Wave Tubes*, chapters 7 and 8. Van Nostrand.

PROPAGATION DES ONDES DANS UN GUIDE REMPLI DE PLASMA EN PRESENCE D'UN CHAMP MAGNETIQUE

M. CAMUS et J. LE MEZEC

Ingénieurs des Télécommunications
au Centre National d'Etudes des Télécommunications

SOMMAIRE

On étudie la propagation d'ondes dans un guide circulaire rempli de plasma, en présence d'un champ magnétique dirigé suivant l'axe du guide. On ne tient pas compte des collisions, ni de l'agitation thermique, ni du mouvement des ions. La méthode suivie est celle qui a été exposée par H. Suhl et L. R. Walker (*B.S.T.J.* **33**, n° 3–4–5, 1954). On en déduit les fréquences de coupure des modes de révolution et une équation de dispersion approchée pour les ondes lentes. On donne l'évolution de la forme des courbes de dispersion lorsque le rayon du guide croît.

I. INTRODUCTION

Cette étude a pour but de préciser quelques particularités de la propagation d'ondes dans un guide métallique creux contenant un plasma, en présence d'un champ magnétique H_0 dirigé suivant l'axe du guide. Ce problème a déjà été étudié par plusieurs auteurs. Les uns, en particulier R. W. Gould et A. W. Trivelpiece (1959), ont utilisé une méthode approchée, dite "des états quasi-stationnaires": cette méthode met en évidence l'existence d'ondes lentes, mais ne donne pas toutes les ondes qui peuvent se propager dans le guide à plasma. D'autres auteurs, en particulier H. Suhl et L. R. Walker (1954), ont employé une méthode rigoureuse qui aboutit malheureusement à une équation de dispersion transcendante, dont il est difficile de tirer la forme des courbes de dispersion. Nous verrons qu'il est cependant possible d'en déduire des résultats intéressants dans certains cas, plus particulièrement en ce qui concerne les modes de révolution d'un guide circulaire rempli de plasma.

Dans ce qui suit, nous négligeons le mouvement des ions, les collisions entre électrons et molécules, ainsi que les vitesses thermiques. Comme nous étudions un système de révolution, nous mettons toutes les grandeurs sous la forme:

$$f(r, \Theta, z, t) = F_0 + F_1(r, \Theta, z) \exp(j\omega t)$$
$$= F_0 + F(r, \Theta) \exp(j\omega t - \Gamma z)$$

avec

$$\Gamma = \alpha + j\beta$$

Nous linéarisons les équations en négligeant les termes du second ordre par rapport aux composantes alternatives. Nous suivrons la méthode de Suhl et Walker, dont nous déduirons l'équation de dispersion des modes de révolution du guide circulaire rempli de plasma. Nous étudierons les fréquences de coupure de ces modes et les conditions dans lesquelles la vitesse de phase des ondes devient très petite (ondes lentes). Puis nous donnerons la forme des courbes de dispersion.

II. METHODE DE RESOLUTION ET EXPRESSION DES CHAMPS

Les composantes alternatives des champs électrique et magnétique satisfont les équations de Maxwell, écrites à l'aide du tenseur diélectrique $||\epsilon||$:

$$\begin{cases} \nabla \times E = -j\omega\mu_0 H \\ \nabla \times H = j\omega\epsilon_0 \, ||\epsilon|| \, E \\ \nabla \, . \, ||\epsilon|| \, E = 0 \\ \nabla \, . \, H = 0 \end{cases}$$

Le tenseur diélectrique s'écrit:

$$||\epsilon|| = \begin{Vmatrix} \epsilon_{rr} & j\epsilon_{r\Theta} & 0 \\ -j\epsilon_{\Theta r} & \epsilon_{\Theta\Theta} & 0 \\ 0 & 0 & \epsilon_{zz} \end{Vmatrix} \quad \text{avec} \quad \begin{cases} \epsilon_{rr} = \epsilon_{\Theta\Theta} = 1 + \dfrac{\omega_p^2}{\omega_c^2 - \omega^2} \\ \epsilon_{r\Theta} = \epsilon_{\Theta r} = \dfrac{\omega_c}{\omega} \cdot \dfrac{\omega_p^2}{\omega_c^2 - \omega^2} \\ \epsilon_{zz} = 1 - \dfrac{\omega_p^2}{\omega^2} \end{cases}$$

Dans ces expressions, le plasma est caractérisé par la fréquence de plasma

$$\frac{\omega_p}{2\pi} = \frac{1}{2\pi} \sqrt{\eta \frac{\rho_0}{\epsilon_0}}$$

et le champ magnétostatique par la fréquence cyclotron électronique

$$\frac{\omega_c}{2\pi} = \frac{\eta}{2\pi} B_0.$$

En éliminant entre ces équations les composantes transversales des champs, on aboutit pour les composantes longitudinales E_z et H_z aux deux équations de propagation:

$$\begin{cases} \epsilon_{rr} \nabla_T^2 E_z - \Gamma\omega\mu_0 \, \epsilon_{r\Theta} H_z + \epsilon_{zz}(\Gamma^2 + k^2\epsilon_{rr}) E_z = 0 \\ \epsilon_{rr} \nabla_T^2 H_z + [\epsilon_{rr}(\Gamma^2 + k^2 \epsilon_{rr}) - k^2\epsilon_{r\Theta}^2] H_z + \\ \qquad + \Gamma\omega\epsilon_0 \, \epsilon_{r\Theta} \, \epsilon_{zz} E_z = 0 \end{cases}$$

avec $k^2 = \omega^2 \epsilon_0 \mu_0 = \dfrac{\omega^2}{c^2}$

et

$$\nabla_T^2 = \frac{1}{r}\frac{\partial}{\partial r} + \frac{\partial^2}{\partial r^2} + \frac{1}{r^2}\frac{\partial^2}{\partial \Theta^2}$$

On constate que E_z et H_z sont solutions d'un système de deux équations non indépendantes: il y a en effet un couplage se traduisant par la présence des termes $\Gamma\omega\mu_0\epsilon_{r\Theta}H_z$ et $\Gamma\omega\epsilon_0\epsilon_{r\Theta}\epsilon_z E_z$. Il ne peut donc pas exister de solutions pour lesquelles on ait séparément $E_z = 0$ ou $H_z = 0$. Contrairement au guide vide, des ondes purement TE ou purement TM ne peuvent donc pas se propager dans un guide à plasma en présence d'un champ magnétique qui rend le plasma anisotrope: cette propriété disparaît avec l'anisotropie du plasma quand le champ magnétique H_0 s'annule ($\epsilon_{r\Theta} = 0$). D'autre part, lorsque les termes de couplage deviennent négligeables, on peut parler d'ondes quasi-TE ou quasi-TM: ceci se produit, en particulier, au voisinage des fréquences de coupure où $\Gamma = 0$.

Pour résoudre le système d'équations couplées, nous cherchons deux combinaisons linéaires de E_z et de H_z, sous la forme:

$$\Psi = E_z + j\Lambda \sqrt{\left(\frac{\mu_0}{\epsilon_0\ \epsilon_{zz}}\right)}\ H_z$$

qui satisfassent séparément à deux équations aux dérivées partielles du second ordre. On trouve que si Λ est l'une des solutions Λ_1 ou Λ_2 de l'équation:

$$\Lambda^2 + j\Lambda \frac{(\Gamma^2 + k^2)\ \epsilon_{r\Theta}}{\Gamma k(\epsilon_{rr} - 1)\sqrt{\epsilon_{zz}}} - 1 = 0$$

les fonctions Ψ_1 et Ψ_2 correspondantes satisfont séparément les deux-équations:

$$\nabla_T^2\ \Psi_{1,\,2} + T_{1,\,2}^2\ \Psi_{1,\,2} = 0$$

où l'on a posé:

$$T_{1,\,2}^2 = \frac{1}{\epsilon_{rr}}\ \{j\Lambda_{1,\,2}\ k\Gamma\epsilon_{r\Theta}\sqrt{(\epsilon_{zz})} + \epsilon_{zz}(\Gamma^2 + k^2\epsilon_{rr})\}$$

On retrouve ainsi des équations bien connues de la théorie de la propagation dans les guides, dont les solutions sont de la forme:

$$\Psi_{1,\,2} = [B_{1,\,2}\ J_n(T_{1,\,2}\ r) + C_{1,\,2}\ Y_n(T_{1,\,2}\ r)]\ e^{jn\Theta}$$

où J_n et Y_n sont les fonctions de Bessel de première et de seconde espèce d'ordre n; $B_1\ C_1\ B_2\ C_2$ sont quatre constantes déterminées par les conditions aux limites. A chaque valeur entière, positive ou négative de n correspondent différents modes. Les expressions de E_z et H_z se déduisent facilement de celles de Ψ_1 et Ψ_2:

$$\begin{cases} E_z = \dfrac{\Lambda_2\Psi_1 - \Lambda_1\Psi_2}{\Lambda_2 - \Lambda_1} \\ H_z = j\sqrt{\left(\dfrac{\epsilon_0\,\epsilon_{zz}}{\mu_0}\right)}\,\dfrac{\Psi_1 - \Psi_2}{\Lambda_2 - \Lambda_1} \end{cases}$$

On remarquera que les différentes composantes s'expriment en général à l'aide de quatre fonctions de Bessel alors qu'il suffit de deux telles fonctions pour exprimer les modes *TE* et *TM* dans un milieu isotrope.

III. EQUATION DE DISPERSION D'UN GUIDE COMPLETEMENT REMPLI DE PLASMA

Les conditions aux limites sont, dans ce cas:

(1) Sur l'axe du guide ($r = 0$), les différentes grandeurs doivent être finies. Puisque $Y_n(0)$ est infini, cela entraîne:

$$\Lambda_2 C_1 = \Lambda_1 C_2 \qquad (E_z \text{ fini})$$

et

$$C_1 = C_2$$

Comme en général $\Lambda_1 \neq \Lambda_2$, on en déduit $C_1 = C_2 = 0$.

(2) Sur la paroi du guide supposé parfaitement conducteur ($r = a$), on sait que la seule condition à satisfaire est que la composante tangentielle du champ électrique soit nulle, c'est-à-dire:

$$E_z(a) = 0 \quad \text{et} \quad E_\Theta(a) = 0$$

Ces deux équations constituent un système d'équations linéaires et homogènes en B_1 et B_2. Leur condition de compatibilité—qui impose en définitive une relation entre ω et Γ pour qu'il ait une solution—constitue l'équation de dispersion. En introduisant les deux fonctions:

$$F_n(T_{1,\,2}a) = \frac{1}{T_{1,\,2}a} \cdot \frac{J_n'(T_{1,\,2}a)}{J_n(T_{1,\,2}a)}$$

cette équation de dispersion s'écrit:

$$\Lambda_1 \left\{ \frac{k^2a^2}{\epsilon_{r\Theta}} \left[\left(\frac{\Gamma^2}{k^2} + \epsilon_{rr} \right)^2 - \epsilon_{r\Theta}^2 \right] F_n(T_1a) + n \right\}$$
$$= \Lambda_2 \left\{ \frac{k^2a^2}{\epsilon_{r\Theta}} \left[\left(\frac{\Gamma^2}{k^2} + \epsilon_{rr} \right)^2 - \epsilon_{r\Theta}^2 \right] F_n(T_2a) + n \right\}$$

Lorsque cette relation est satisfaite, on trouve que B_1 et B_2 sont reliés par l'unique équation:

$$B_1 \,.\, \Lambda_2 J_n(T_1a) = B_2 \,.\, \Lambda_1 J_n(T_2a)$$

ce qui permet de déterminer les différentes composantes à un facteur près.

Dans le cas particulier des modes de révolution ($n = 0$), cette équation se simplifie et devient:

$$\Lambda_1 F_0(T_1 a) = \Lambda_2 F_0(T_2 a)$$

En effet l'équation $(\Gamma^2/k^2 + \epsilon_{rr})^2 - \epsilon_{r\Theta}^2 = 0$ est l'équation de dispersion des ondes planes transversales; comme ces ondes ne satisfont pas les conditions aux limites, elles ne constituent pas une solution.

En calculant les différentes composantes des champs, on trouve que les valeurs de E_z et H_z sur l'axe sont liées, dans ce cas particulier ($n = 0$) par la relation:

$$\Lambda_1 \Lambda_2 [J_0(T_2 a) - J_0(T_1 a)] \sqrt{\mu_0} \,.\, H_z(0) = j\sqrt{\epsilon_{zz}}[\Lambda_1 J_0(T_2 a) - \Lambda_2 J_0(T_1 a)] \sqrt{\epsilon_0} \,.\, E_z(0)$$

Cette relation permet de préciser les notions d'ondes quasi-*TE* et quasi-*TM*. Nous définirons les ondes quasi-*TM* par la condition:

$$\sqrt{\mu_0} \,.\, H_z(0) \ll \sqrt{\epsilon_0} \,.\, E_z(0)$$

et les ondes quasi-*TE* par la condition:

$$\sqrt{\epsilon_0} \,.\, E_z(0) \ll \sqrt{\epsilon_0} \,.\, H_z(0)$$

Nous nous limiterons dans la suite à l'étude des modes de révolution du guide complètement rempli de plasma. Dans l'équation de dispersion les trois grandeurs a (rayon du guide), ω_p (fréquence de plasma) et ω_c (fréquence cyclotron) n'interviennent que par l'intermédiaire des deux paramètres:

$$m^2 = \frac{c^2}{a^2 \omega_p^2} \quad \text{et} \quad M^2 = \frac{\omega_c^2}{\omega_p^2}$$

Le paramètre m caractérise à un coefficient près les fréquences de coupure du guide vide par rapport à la fréquence de plasma. On sait, en effet, que les fréquences de coupure du guide vide ont pour valeurs:

—pour les ondes *TE*: $p_{1\nu} \,.\, \dfrac{c}{a} = \omega_{1\nu}$

—pour les ondes *TM*: $p_{0\nu} \,.\, \dfrac{c}{a} = \omega_{0\nu}$

où $p_{0\nu}$ et $p_{1\nu}$ sont respectivement les ν ièmes racines des fonctions de Bessel J_0 et J_1. On a donc par exemple:

$$m^2 p_{0\nu}^2 = \frac{\omega_{0\nu}^2}{\omega_p^2}$$

D'autre part, l'équation de dispersion ne fait intervenir que ω^2 et Γ^2, ce qui conduit à introduire les nouvelles variables réduites X et Y définies par:

$$X = -\frac{\Gamma^2 c^2}{\omega_p^2} \quad \text{et} \quad Y = \left(\frac{\omega}{\omega_p}\right)^2$$

Nous introduisons, d'autre part, comme intermédiaires de calcul, les quatre grandeurs $A_{1,\,2}$ et $Z_{1,\,2}$, plus aisées à manipuler que $\Lambda_{1,\,2}$ et $T_{1,\,2}$ et définies par:

$$\begin{cases} A_{1,\,2} = 2j\dfrac{\Gamma}{k} \cdot \dfrac{(\epsilon_{rr} - 1)\sqrt{\epsilon_{zz}}}{\epsilon_{r\theta}} \cdot \Lambda_{1,\,2} \\ \text{et} \\ Z_{1,\,2} = m^2 a^2 T_{1,\,2}^2 \end{cases}$$

c'est-à-dire:

$$\begin{cases} A_{1,\,2} = 1 - \dfrac{X}{Y} \pm \sqrt{\left[\left(1 - \dfrac{X}{Y}\right)^2 + \dfrac{4}{M^2} \cdot \dfrac{X}{Y}(Y - 1)\right]} \\ Z_{1,\,2} = Y - 1 + \dfrac{1}{1 + M^2 - Y}\left[\dfrac{M^2}{2} A_{1,\,2} - \dfrac{X}{Y}(Y - 1)(M^2 - Y)\right] \end{cases}$$

Les courbes de dispersion sont alors définies par l'équation:

$$A_1 F_0\left(\frac{\sqrt{Z_1}}{m}\right) = A_2 F_0\left(\frac{\sqrt{Z_2}}{m}\right)$$

qui dépend de deux paramètres m et M.

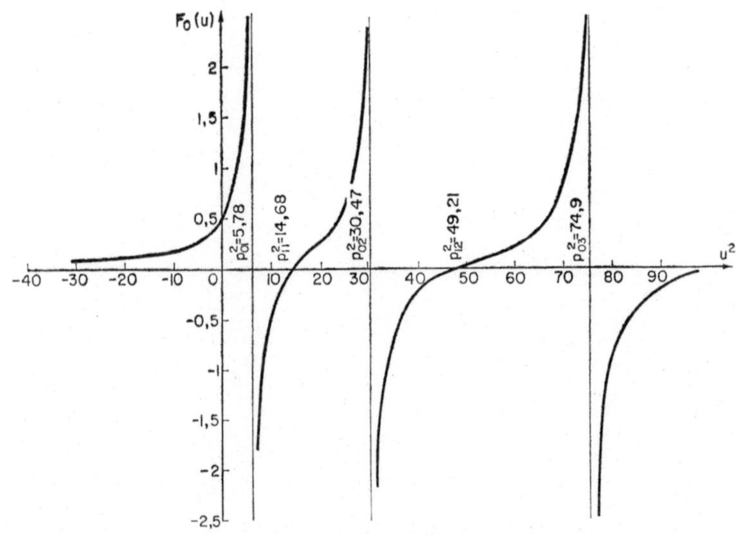

FIG. 1. Courbe $F_0\,(u)$ en fonction de u^2.

La fonction

$$F_0(u) = \frac{1}{u}\frac{J_1(u)}{J_0(u)}$$

est représentée, figure 1, en fonction de u^2. Pour u^2 négatif, elle devient:

$$F_0\,(u \text{ pour } u^2 < 0) = \frac{1}{|u|} \cdot \frac{I_1(|u|)}{I_0(|u|)}$$

IV. FREQUENCES DE COUPURE DES MODES DE REVOLUTION DU GUIDE REMPLI DE PLASMA

Par définition, les fréquences de coupure sont celles pour lesquelles $\Gamma = 0$, c'est-à-dire $X = 0$. Quand on franchit une telle fréquence, on passe d'une onde progressive à amplitude constante ($X > 0$) à une onde évanescente ($X < 0$).

Pour $X = 0$, les valeurs de A_1, A_2, Z_1 et Z_2 sont données par le tableau ci-dessous:

A_1	$A_{10} = 2$
A_2	$A_{20} = 0$
Z_1	$Z_{10} = \dfrac{M^2 Y - (Y-1)^2}{1 + M^2 - Y}$
Z_2	$Z_{20} = Y - 1$

Le fait important est que, pour $X = 0$, on a: $A_2 = 0$.

L'équation de dispersion ne peut donc être vérifiée que si l'on a:

$$F_0\left(\frac{\sqrt{Z_1}}{m}\right) = 0, \text{ c'est-à-dire } \frac{\sqrt{Z_1}}{m} = p_{1\nu} \text{ (ou } T_1 a = p_{1\nu})$$

ou

$$F_0\left(\frac{\sqrt{Z_2}}{m}\right) \text{ infiniment grand, c'est-à-dire:}$$

$$\frac{\sqrt{Z_2}}{m} = p_{0\nu} \text{ (ou } T_2 a = p_{0\nu})$$

IV.1. *Fréquences de coupure telles que $Z_2 = m^2 p_{0\nu}^2$ (ou $Z_2 = \omega_{0\nu}^2/\omega_p^2$)*

En tenant compte de ce que $Z_{20} = Y - 1$, on voit que ces fréquences de coupure sont données par:

$$Y = 1 + m^2 p_{0\nu}^2$$

A chaque fréquence de coupure $\omega_{0\nu}$ due mode $TM_{0\nu}$ du guide vide ($\omega_{0\nu} = p_{0\nu} \,.\, c/a$) correspond par conséquent une fréquence de coupure

$\omega'_{0\nu}$ du guide rempli de plasma, telle que:

$$\omega'^2_{0\nu} = \omega^2_{0\nu} + \omega^2_p$$

Cette fréquence de coupure est indépendante du champ magnétique. Il est facile de montrer qu'au voisinage de cette fréquence, l'onde correspondante est quasi-*TM*. D'autre part, en un tel point la pente $\mathrm{d}Y/\mathrm{d}X$ peut être positive, négative ou nulle suivant les valeurs de M et de m. Si elle est positive, on passe quand la fréquence croît en passant par $\omega_{0\nu}$ d'une onde évanescente à une onde progressive, et inversement dans le cas contraire. Il est possible de vérifier que l'on se trouve toujours dans le premier cas lorsque le rayon du guide est suffisamment petit, ou le nombre ν suffisamment grand pour que $\omega^2_{0\nu} \gg \omega^2_p + \omega^2_c$.

IV.2. *Fréquences de coupure telles que* $Z_1 = m^2 p^2_{1\nu}$ (*ou* $Z_1 = \omega^2_{1\nu}/\omega^2_p$)

En se reportant à la valeur de Z_{10}, on voit que ces fréquences sont déterminées par l'équation:

$$Y^2 - Y(M^2 + m^2 p^2_{1\nu} + 2) + 1 + m^2 p^2_{1\nu}(1 + M^2) = 0$$

A chaque fréquence de coupure $\omega_{1\nu} = p_{1\nu}\, c/a$ d'un mode $TE_{0\nu}$ du guide vide correspondent donc deux fréquences de coupure du guide rempli de plasma, $\omega'_{1\nu}$ et $\omega''_{1\nu}$ données par l'équation:

$$\omega^4 - \omega^2(\omega^2_c + \omega^2_{1\nu} + 2\omega^2_p) + \omega^4_p + \omega^2_{1\nu}(\omega^2_p + \omega^2_c) = 0$$

On peut montrer qu'à leur voisinage, les ondes correspondantes sont quasi-*TE*.

Les expressions de $\omega'_{1\nu}$ et $\omega''_{1\nu}$ sont:

$$\left.\begin{matrix}\omega'^2_{1\nu} \\ \omega''^2_{1\nu}\end{matrix}\right\} = \tfrac{1}{2}(\omega^2_c + \omega^2_{1\nu} + 2\omega^2_p) \pm \tfrac{1}{2}\sqrt{[(\omega^2_{1\nu} - \omega^2_c)^2 + 4\omega^2_c\omega^2_p]}$$

On peut en donner des expressions approchées lorsque:

$$2\omega_c\omega_p \ll |\,\omega^2_{1\nu} - \omega^2_c\,|$$

c'est-à-dire soit lorsque $\omega_{1\nu}$ est très supérieur à la fois à ω_p et à ω_c, soit lorsque ω_c est très supérieur à la fois à $\omega_{1\nu}$ et à ω_p. Ces expressions approchées de $\omega'^2_{1\nu}$ et $\omega''^2_{1\nu}$ sont:

$$\omega^2_p + \omega^2_c \quad \text{et} \quad \omega^2_{1\nu} + \omega^2_p$$

Le premier cas ($\omega_{1\nu}$ grand) se produit, soit pour les modes d'ordre élevé, soit pour tous les modes d'un guide de rayon suffisamment petit: on trouve alors pour chaque valeur de ν une fréquence de coupure voisine de

$$\sqrt{(\omega^2_p + \omega^2_c)}$$

comme dans la méthode des états quasi-stationnaires. Le deuxième cas se produit lorsque le champ magnétique est fort pour les modes d'ordre ν suffisamment petits pour que $\omega_{1\nu} \ll \omega_c$.

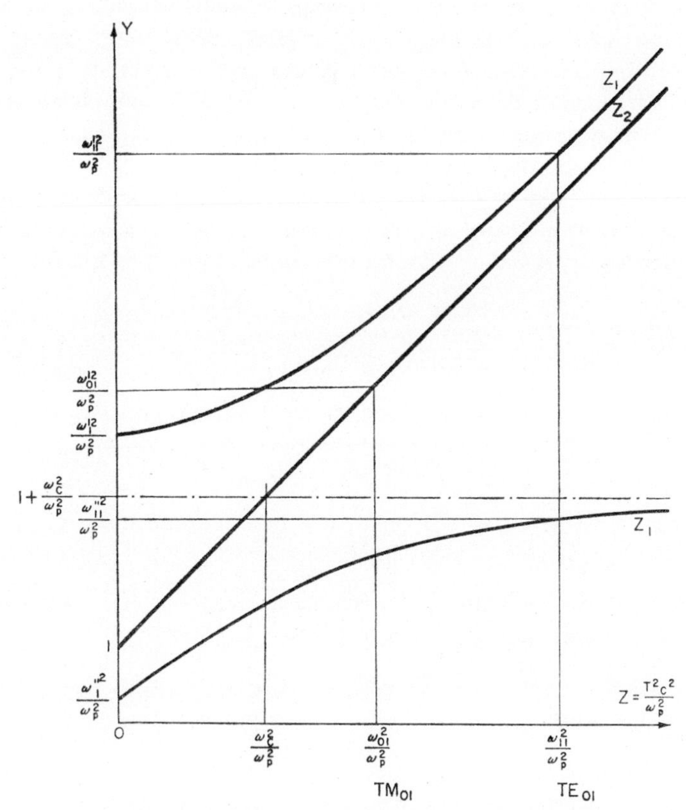

FIG. 2. Frequences de coupure d'un guide rempli de plasma.

IV.3. *Variation des fréquences de coupure avec les dimensions du guide*

Sur la figure 2 on a représenté les variations de $Z_2 = Y - 1$ et de

$$Z_1 = \frac{M^2 Y - (Y-1)^2}{1 + M^2 - Y}$$

en fonction de $Y = \omega^2/\omega_p^2$ pour $X = 0$: ces variations sont représentées respectivement par une droite et par une hyperbole dont les asymptotes sont $Y = 1 + M^2$ et $Z = Y - 1$. Pour avoir les fréquences de coupure $\omega_{0\nu}'$, il suffit de couper la première courbe ($Z_2 = Y - 1$) par les droites d'abscisse $Z = \omega_{0\nu}^2/\omega_p^2$; de même les fréquences de coupure $\omega_{1\nu}'$ et $\omega_{1\nu}''$ sont obtenues en coupant l'hyperbole $Z_1(Y)$ par les droites d'abscisse $Z = \omega_{1\nu}^2/\omega_p^2$. On voit que les fréquences de coupure $\omega_{1\nu}'$ sont toujours supérieures à $\sqrt{(\omega_p^2 + \omega_c^2)}$

tandis que les fréquences de coupure $\omega''_{1\nu}$ sont toujours inférieures à cette valeur.

Sur cette construction graphique, on suit facilement l'évolution des fréquences de coupure lorsque les dimensions du guide varient.

Pour un guide suffisamment petit (a petit, donc $\omega_{01} \gg \sqrt{[\omega_p^2 + \omega_c^2]}$), toutes les fréquences de coupure ω''_{11} sont très voisines de $\sqrt{(\omega_p^2 + \omega_c^2)}$; toutes les fréquences de coupure ω'_{01} et ω'_{11} sont beaucoup plus grandes que $\sqrt{(\omega_p^2 + \omega_c^2)}$ et se rencontrent dans l'ordre $\omega'_{01}, \omega'_{11}, \omega'_{02}, \omega'_{12}, \ldots$

Lorsque le rayon du guide augmente, les fréquences $\omega'_{1\nu}$ décroissent moins vite que les fréquences $\omega'_{0\nu}$. En particulier, lorsque ω_{01} devient inférieure à ω_c, ω'_{01} devient inférieure à $\sqrt{(\omega_p^2 + \omega_c^2)}$ et se trouve donc comprise entre les fréquences $\omega''_{1\nu}$ et $\omega''_{1,\,\nu+1}$. Pour a suffisamment grand tel que

$$\omega'_{0,\,k} < \omega_c < \omega'_{0,\,k+1},$$

on a donc la situation suivante:

—entre les fréquences 0 et ω_p, on trouve des fréquences $\omega''_{1\nu}$ avec

$$\nu \leqslant \zeta \text{ tel que } \omega_{1\zeta} < \omega_p < \omega_{1,\,\zeta+1}.$$

—entre ω_p et $\sqrt{(\omega_p^2 + \omega_c^2)}$ on trouve des fréquences $\omega'_{0\eta}$ ($1 \leqslant \eta$ entier $\leqslant k$) enchevêtrées avec les fréquences $\omega''_{1\nu}$ ($\nu > p$)

—entre $\sqrt{(\omega_p^2 + \omega_c^2)}$ et ω'_{0n}, les fréquences $\omega'_{0\eta}$ ($k < \eta < n$) enchevêtrées avec les fréquences $\omega'_{1,\,\nu}$ ($1 \leqslant \nu \leqslant n$)

—au-dessus de ω'_{0n}, les fréquences $\omega'_{0,\,n+1}$, $\omega'_{1,\,n+1}$, $\omega'_{0,\,n+2}$, etc. ... dans cet ordre.

Cette situation peut se rencontrer dans un guide suffisamment grand si le champ magnétique est élevé. A la limite, lorsque le rayon du guide devient très grand, les fréquences $\omega'_{0\nu}$ tendent vers ω_p tandis que les fréquences $\omega'_{1\nu}$ et $\omega''_{1\nu}$ tendent vers les valeurs limites ω'_1 et ω''_1 données par:

$$\left.\begin{matrix}\omega_1'^2\\ \omega_1''^2\end{matrix}\right\} = \tfrac{1}{2}(\omega_c^2 + 2\omega_p^2) \pm \tfrac{1}{2}\omega_c\sqrt{(\omega_c^2 + 4\omega_p^2)}$$

Ces limites sont les fréquences de coupure des ondes planes qui se propagent dans un plasma, quel que soit l'angle du champ magnétique avec la direction de propagation. En particulier, pour les ondes planes se propageant parallèlement au champ magnétique, ω''_1 est la fréquence de coupure de l'onde extraordinaire et ω'_1 la fréquence de coupure de l'onde ordinaire. On peut retrouver une expression plus connue de ces fréquences de coupure limite en remarquant que ω'_1 et ω''_1 sont les racines positives de l'équation bicarrée:

$$\omega^4 - \omega^2(\omega_c^2 + 2\omega_p^2) + \omega_p^4 = 0$$

qui peut se décomposer en deux équations du second degré:

$$\begin{cases} \omega^2 + \omega_c\,\omega - \omega_p^2 = 0 \\ \omega^2 - \omega_c\,\omega - \omega_p^2 = 0 \end{cases}$$

D'où les expressions:

$$\begin{cases} \omega_1' = \tfrac{1}{2}\omega_c + \tfrac{1}{2}\sqrt{(\omega_c^2 + 4\omega_p^2)} \\ \omega_1'' = -\tfrac{1}{2}\omega_c + \tfrac{1}{2}\sqrt{(\omega_c^2 + 4\omega_p^2)} \end{cases}$$

IV.4. *Relation entre les fréquences de coupure et la propagation d'ondes planes dans le plasma*

Le comportement des ondes dans le guide au voisinage des fréquences de coupure, est lié à celui des ondes planes se propageant perpendiculairement au champ magnétique. Ce fait apparaît simplement si l'on considère les ondes guidées par deux plans parallèles conducteurs: les modes guidés résultent alors de la superposition d'ondes planes se réfléchissant sur les plans conducteurs; lorsqu'on se rapproche d'une fréquence de coupure, ces ondes planes tendent à se propager de plus en plus normalement aux plans conducteurs et les fréquences de coupure sont déterminées par la condition $a = n(\lambda_c/2)$ ou $\beta_c a = n\pi$ (a: écartement des plans; λ_c: longueur d'onde dans le vide; $\beta_c = 2\pi/\lambda_c$).

Lorsque les plans conducteurs sont séparés par un plasma soumis à un champ magnétique H_0 parallèle aux plans et à la direction de propagation des modes guidés, il devient difficile de décrire ces modes par réflexion d'ondes planes parce que les caractéristiques de propagation des ondes planes, et en particulier leurs courbes de dispersion $\beta\,(\omega)$ dépendent de l'angle ϕ du champ magnétique avec la direction de propagation. Néanmoins, au voisinage de la coupure, on pourra se reporter aux caractéristiques $\beta_T(\omega)$ relatives à la propagation d'ondes planes normalement au champ magnétique statique. Pour les ondes guidées entre deux plans, on doit alors s'attendre à une condition de coupure de la forme $a = n(\lambda_T/2)$ ou $\beta_T a = n\pi$. Dans le cas d'un guide circulaire de rayon a, cette condition deviendrait $\beta_T a = p_{0\nu}$ (ou $p_{1\nu}$).

L'équation de dispersion pour les ondes planes se propageant perpendiculairement au champ magnétique (voir Denisse et Delcroix, Théorie des ondes dans les plasmas, 1961, ch. VII, page 92, ou équation générale de dispersion, page 16, en faisant $\omega_L = 0$) se décompose en:

$$\begin{cases} \beta_T^2\, c^2 - \omega^2 - \omega_p^2 = 0 \\ (\omega^2 - \beta_T^2\, c^2)\,(\omega^2 - \omega_p^2 - \omega_c^2) - \omega_p^2(\omega^2 - \omega_p^2) = 0 \end{cases}$$

ou, dans le plan $(Z,\ Y)$, en tenant compte du fait que dans le guide à la coupure T et β_T sont identiques,

$$\begin{cases} Z - Y - 1 = 0 \\ Y^2 - Y(M^2 + Z + 2) + 1 + Z(1 + M^2) = 0 \end{cases}$$

Ce sont les équations des courbes de la figure 2 qui nous ont servi à déterminer les fréquences de coupure du guide complètement rempli de plasma.

La première équation correspond à la propagation d'une onde plane où le champ électrique alternatif et la vitesse alternative des électrons sont parallèles au champ magnétostatique: cette onde est, par conséquent, indépendante du champ magnétique; par réflexion sur des parois parallèles au champ magnétostatique, elle donne naissance à une onde *TM*. Ces propriétés sont bien en accord avec celles des ondes dans le guide plein de plasma, au voisinage des fréquences de coupure des modes quasi-*TM*.

La deuxième équation correspond à la propagation d'une onde plane où le champ magnétique alternatif est parallèle au champ magnétostatique, tandis que le champ électrique alternatif et le mouvement des électrons lui sont perpendiculaires. Ces propriétés permettent de retrouver celles des modes quasi-*TE* au voisinage des fréquences de coupure correspondant aux zéros de J_1.

En dehors des fréquences de coupure, les modes du guide circulaire pourraient s'obtenir par superposition des deux types d'ondes planes qui peuvent se propager dans le plasma à une fréquence donnée. Si l'on cherchait à préciser cette superposition, il ne faudrait pas oublier que les courbes de dispersion des ondes planes dépendent de l'angle du champ magnétostatique avec la direction de propagation.

V. ONDES LENTES DANS LE GUIDE CIRCULAIRE REMPLI DE PLASMA

Sur l'équation de dispersion:

$$A_1 F_0 \left(\frac{\sqrt{Z_1}}{m}\right) = A_2 F_0 \left(\frac{\sqrt{Z_2}}{m}\right)$$

on peut étudier directement les branches des courbes de dispersion pour lesquelles la constante de propagation Γ devient infiniment grande, la fréquence ω restant finie. En fait nous nous restreindrons à la partie du diagramme où $X = -\Gamma^2 c^2/\omega_p^2$ positif, et par suite où $\Gamma = j\beta$, c'est-à-dire celle qui correspond à des ondes progressives lentes.

Les développements des grandeurs A et Z pour X infiniment grand positif, Y fini, sont donnés par le tableau ci-dessous:

$A_1 \sim \frac{2}{M^2}(Y-1)$	$Z_1 \sim \frac{X}{Y} \cdot \frac{(Y-1)(Y-M^2)}{1+M^2-Y} + \frac{(Y-1)(2+M^2-Y)}{1+M^2-Y}$
$A_2 \sim -2\frac{X}{Y}$	$Z_2 \sim -X$

Comme Z_2 est infiniment grand négatif, $F_0(\sqrt{Z_2}/m)$ est un infiniment petit équivalent à $1/(\sqrt{|Z_2|}/m) \sim m/\sqrt{X}$. Le second membre de l'équation de dispersion, qui équivaut donc à $(-2m\sqrt{X}/Y)$ est lui aussi infiniment grand. Comme A_1 est fini, il s'ensuit que $(\sqrt{Z_1}/m)$ est infiniment voisin d'une racine $p_{0\nu}$ de J_0. On obtient ainsi une équation de dispersion approchée pour les ondes lentes:

$$Z_1(X \to +\infty) = m^2 p_{0\nu}^2$$

c'est-à-dire, en tenant compte du développement asymptotique de Z_1:

$$\frac{\beta^2 c^2}{\omega^2} = \frac{\omega_{0\nu}^2(\omega_p^2 + \omega_c^2 - \omega^2)}{(\omega^2 - \omega_p^2)(\omega^2 - \omega_c^2)} - \frac{2\omega_p^2 + \omega_c^2 - \omega^2}{\omega^2 - \omega_c^2}$$

ou encore, en utilisant les composantes du tenseur diélectrique:

$$\beta^2 = -\frac{\omega_{0\nu}^2}{c^2} \cdot \frac{\epsilon_{rr}}{\epsilon_{zz}} + \frac{\omega^2}{c^2}(2\epsilon_{rr} - 1)$$

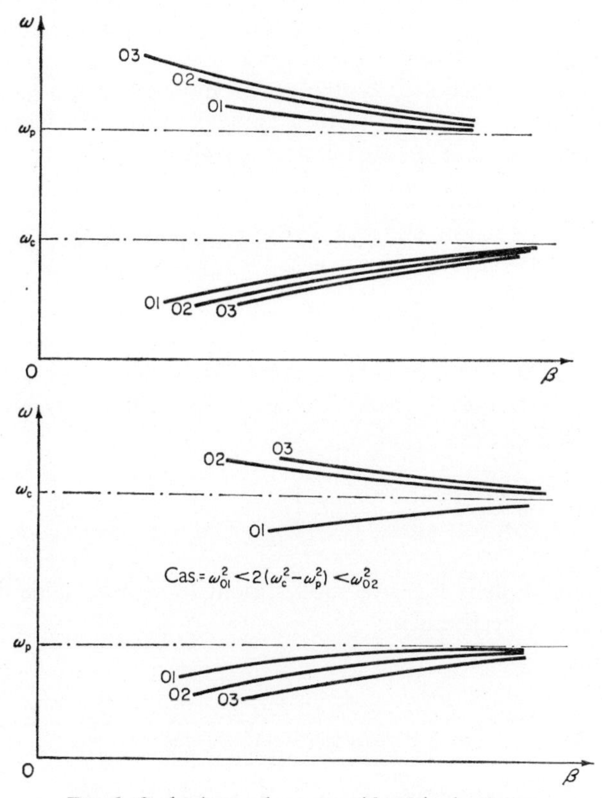

FIG. 3. Ondes lentes dans un guide plein de plasma.

Sur cette formule de dispersion approchée, il est évident que les seules asymptotes horizontales des courbes de dispersion sont $\omega = \omega_p$ ($X = 1$) et $\omega = \omega_c$ ($X = M$). Si l'on néglige le deuxième terme du second membre, l'on retrouve la formule des états quasi-stationnaires: il est facile de voir que ce terme est négligeable au voisinage de $\omega = \omega_p$, mais non de $\omega = \omega_c$. Il en résulte que la disposition des courbes par rapport à leurs asymptotes peut être différente de celle donnée par la théorie des états quasi-stationnaires. En fait, au voisinage de ω_p, les deux théories donnent qualitativement les mêmes résultats: ondes progressives non atténuées pour $\omega > \omega_p$ lorsque ω_p est plus grand que ω_c; pour $\omega < \omega_p$ dans le cas contraire. Il n'en est pas de même au voisinage de ω_c. Les branches infinies des courbes de dispersion sont bien toutes au dessous de l'asymptote $\omega = \omega_c$ lorsque $\omega_c < \omega_p$. Par contre lorsque $\omega_c > \omega_p$, les branches d'indice ν des courbes de dispersion telles que:

$$\omega_{0\nu}^2 < 2(\omega_c^2 - \omega_p^2)$$

sont au-dessous des asymptote ; les autres sont au-dessus. Qualitativement, il n'y a donc de différence avec la théorie des états quasi-stationnaires que pour certains modes d'un guide suffisamment grand pour que la relation ci-dessus puisse être satisfaite.

La position des branches de courbes correspondant à des ondes lentes est représentée, figure 3. On peut vérifier d'après la formule reliant H_z et E_z qu'il s'agit d'ondes quasi-*TM* (H_z infiniment petit par rapport à E_z).

VI. ONDES A VITESSE DE PHASE VOISINE DE CELLE DE LA LUMIERE DANS LE VIDE

L'équation de dispersion permet d'obtenir facilement les solutions pour les fréquences très élevées ($\omega \to \infty$, donc $Y \to +\infty$), la vitesse de phase ω/β restant finie. Dans ces conditions, en prenant des développements limités, on voit que $A_1 \# -A_2$ et $Z_1 \# Z_2 \# -X + Y - 1$. Une forme approchée de l'équation de dispersion est alors:

$$J_0\left[\frac{\sqrt{(Y-X-1)}}{m}\right] \qquad J_1\left[\frac{\sqrt{(Y-X-1)}}{m}\right] = 0$$

Elle conduit donc à deux types de modes, dont les équations de dispersion approchées sont respectivement:

$$\beta^2c^2 = \omega^2 - \omega_p^2 - \omega_{0\nu}^2$$

et

$$\beta^2c^2 = \omega^2 - \omega_p^2 - \omega_{1\nu}^2$$

Les premiers sont quasi-*TM*; les seconds sont quasi-*TE*. Ces équations sont les équations de dispersion exactes pour les modes TE et TM du guide rempli de plasma en l'absence de champ magnétique.

VII. FORME GENERALE DES COURBES DE DISPERSION

Pour résoudre l'équation de dispersion, on se donne une valeur réelle de Y (donc de ω) et on cherche les valeurs de X correspondantes. Du fait de la présence du radical dans les expressions de A et de Z, pour que les solutions $X(Y)$ soient réelles, il faut que:

$$\left(1 - \frac{X}{Y}\right)^2 + \frac{4}{M^2}\frac{X}{Y}(Y-1) > 0$$

On définit ainsi une "zone interdite" du plan (X, Y) en dehors de laquelle doivent se trouver les courbes de dispersion (X et Y réels). Elle est représentée, figure 4.

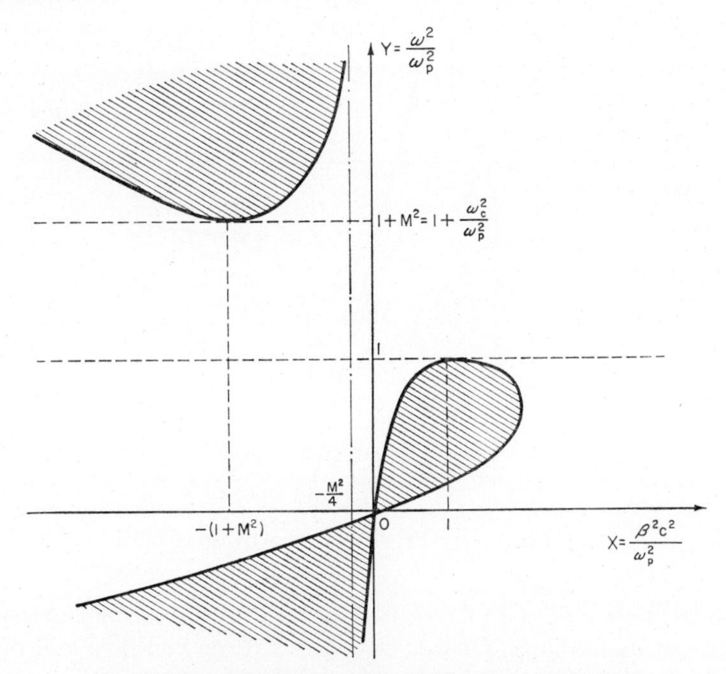

FIG. 4. "Zone interdite" du plan (X, Y): zone où X et Y ne peuvent pas etre simultanément réels.

La résolution peut se faire graphiquement. Prenons par exemple le cas $\omega_c = 2\omega_p$ ($M^2 = 4$); $\omega_{0v} \gg \omega_c$ et $Y < 1$ (par exemple $Y = \omega^2/\omega_p^2 = 0{\cdot}5$). La figure 5 représente l'hyperbole $A(X)$ qui donne les valeurs A_1 et A_2; la figure 6, l'hyperbole $Z(X)$ qui donne Z_1 et Z_2. Sur cette courbe on peut repérer les valeurs de X définies par $Z = m^2 p_{0\nu}^2$ et $Z = m^2 p_{1\nu}^2$: elles donnent respectivement les asymptotes et les zéros de la fonction $F_0(\sqrt{Z}/m)$. Sur la figure 7, on a représenté les variations des deux membres de l'équation $A_1 F_0(\sqrt{Z_1}/m) = A_2 F_0(\sqrt{Z_2}/m)$, en se servant de la fonction $F_0(u^2)$ (figure 1).

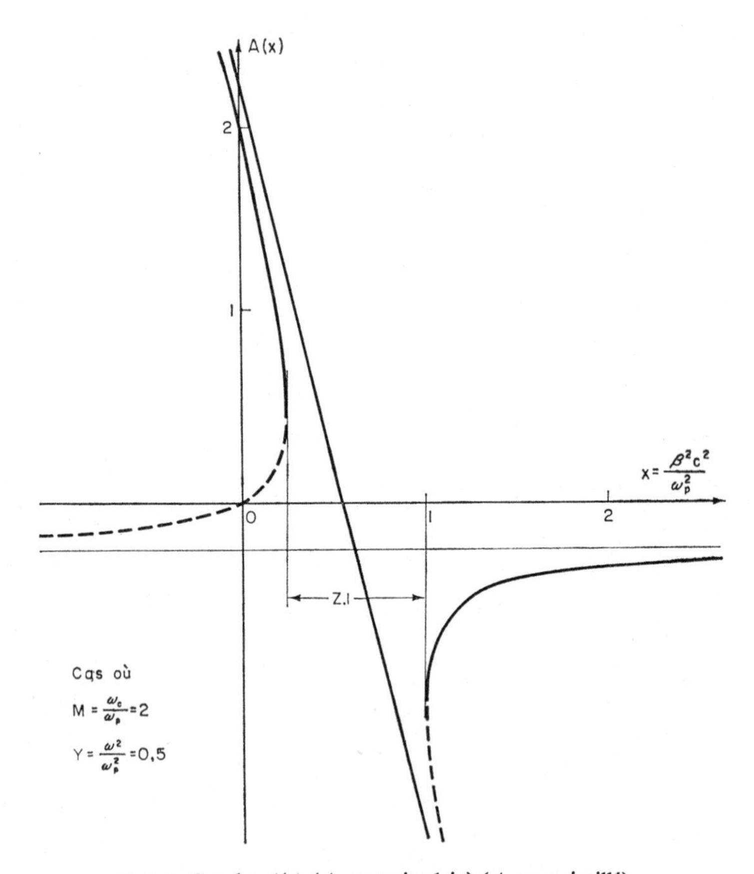

FIG. 5. Courbe $A(x)$ (A_1 en trait plein) (A_2 en pointillé).

On obtient les solutions $X(Y)$ par intersection des courbes donnant les variations des deux membres. Dans le cas où $Y < 1$, on voit qu'il y a, du côté $X > 0$ une infinité de solutions voisines des asymptotes verticales de $A_1 F_0(\sqrt{Z_1}/m)$, données par $Z_1 = m^2 p_{0\nu}^2$: nous avons déjà trouvé cette équation comme équation de dispersion approchée des ondes lentes. Du côté $X < 0$, (ondes évanescentes) on trouve une infinité de solutions voisines des zéros de $A_1 F_0(\sqrt{Z_1}/m)$, donnés par $Z_1 = m^2 p_{1\nu}^2$: en prenant comme valeur approchée de Z_1 dans cette région la valeur $Z_1 = Y - X$ (asymptote de $Z(X)$), on obtient pour relation de dispersion approchée des ondes évanes-

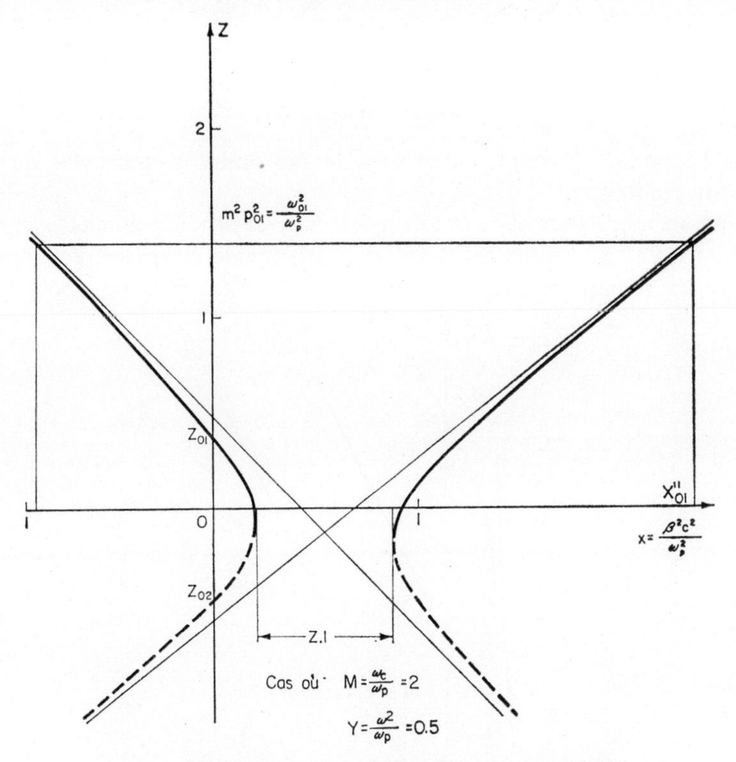

FIG. 6. Courbe $Z(x)$ (Z_1 en trait plein) (Z_2 en pointillé).

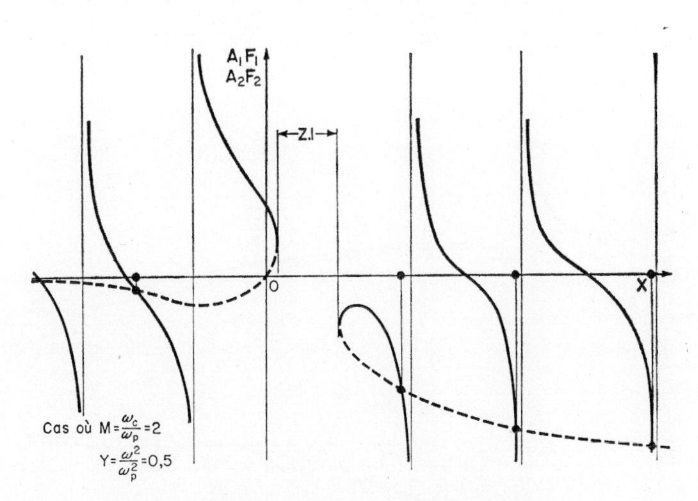

FIG. 7. Resolution graphique de l'equation de dispersion $A_1 F_1 = A_2 F_2$
$A_1 F_1$ en trait plein — $A_2 F_2$ en pointillé).

centes aux basses fréquences:

$$Y - X = m^2 p_{1\nu}^2$$

c'est-à-dire:

$$\alpha^2 c^2 = -\omega^2 + \omega_{1\nu}^2$$

C'est l'équation donnant l'atténuation α des ondes évanescentes du mode $TE_{0\nu}$ du guide vide.

Cette méthode permet de la même facon de déterminer qualitativement les solutions $X(Y)$ lorsque $\omega_{0\nu} \gg \sqrt{(\omega_p^2 + \omega_c^2)}$ (guide de petit rayon), puis de voir ce qui se passe lorsque le rayon du guide croît. Elle se transpose directe-

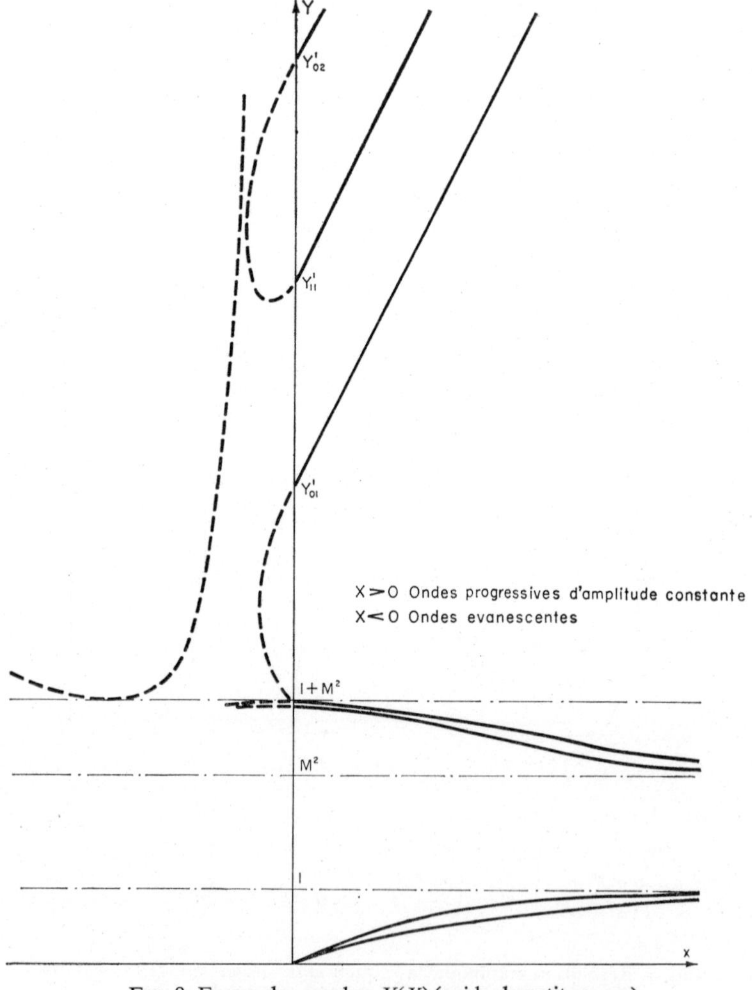

FIG. 8. Forme des courbes $Y(X)$ (guide de petit rayon.)

ment pour résoudre l'équation de dispersion à l'aide d'une machine à calculer.

Une autre méthode permet d'avoir assez rapidement la forme des courbes de dispersion. Nous avons déjà vu que dans de nombreux cas l'on trouvait

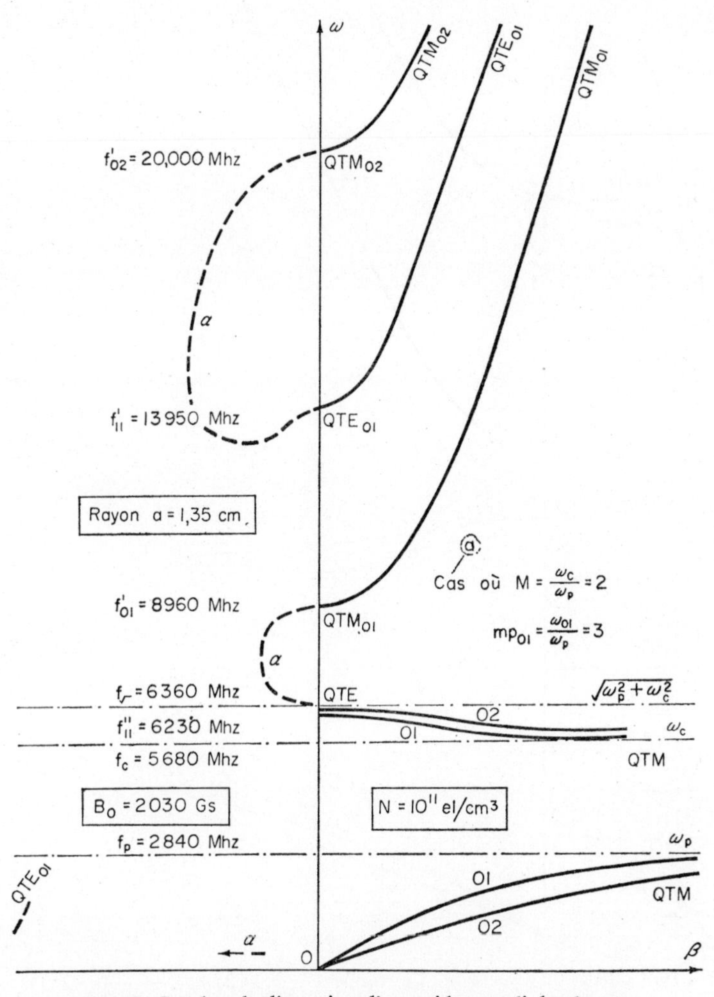

FIG. 9. Courbes de dispersion d'un guide rempli de plasma.

des solutions au voisinage de $Z = m^2p_{0\nu}^2$ ou de $Z = m^2p_{1\nu}^2$, c'est-à-dire au voisinage des couples de valeur (X, Y) qui rendent nul ou infini l'un des membres. La méthode "rapide" consiste alors à tracer dans le plan (X, Y) les courbes $Z =$ constante, plus précisément: $Z = m^2p_{0\nu}^2$ et $Z = m^2p_{1\nu}^2$.

Par un point du plan (X, Y) passent deux courbes $Z =$ constante, correspondant aux deux valeurs Z_1 et Z_2. Un point où se coupent deux courbes $Z_1 = m^2p_{0\nu}^2$ et $Z_2 = m^2p_{0\nu}^2$ se trouve sur une courbe de dispersion, puisque l'équation de dispersion est alors satisfaite, les deux membres étant infinis.

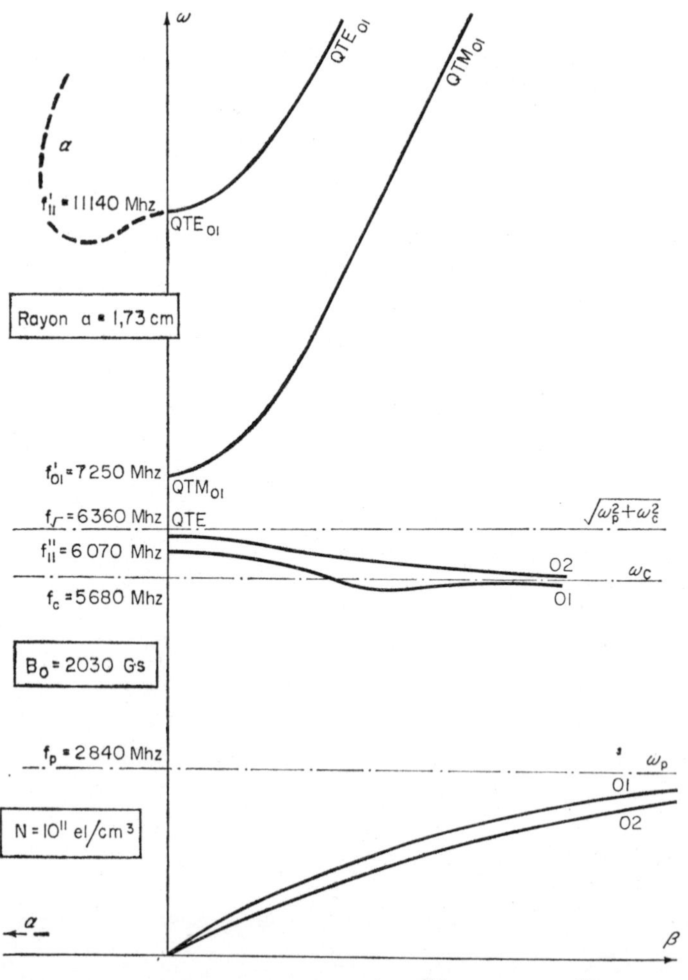

FIG. 10. Courbes de dispersion. Cas où $M = \dfrac{\omega c}{\omega p} = 2$; $mp_{01} = \dfrac{\omega_{01}}{\omega p} = 2{\cdot}34$.

Le réseau $Z = m^2p_{0\nu}^2$ découpe ainsi le plan (X, Y) en quadrilatères curvilignes tels que les courbes de dispersion ne peuvent que passer par leurs sommets (en effet, sur l'une de ces courbes, l'un des membres de la courbe de dispersion étant infini, l'autre doit l'être aussi; le seule exception est constitutée par les fréquences de coupure, où $A_1 = 0$).

Ces méthodes ont été utilisées pour obtenir les figures 8 à 14 qui montrent la forme des courbes de dispersion dans le cas $M = \omega_c/\omega_p = 2$, pour des guides de plus en plus larges. Sur la figure 8 donnant, en partie, la forme des courbes $Y(X)$ pour un guide petit, on remarquera surtout la forme de ces

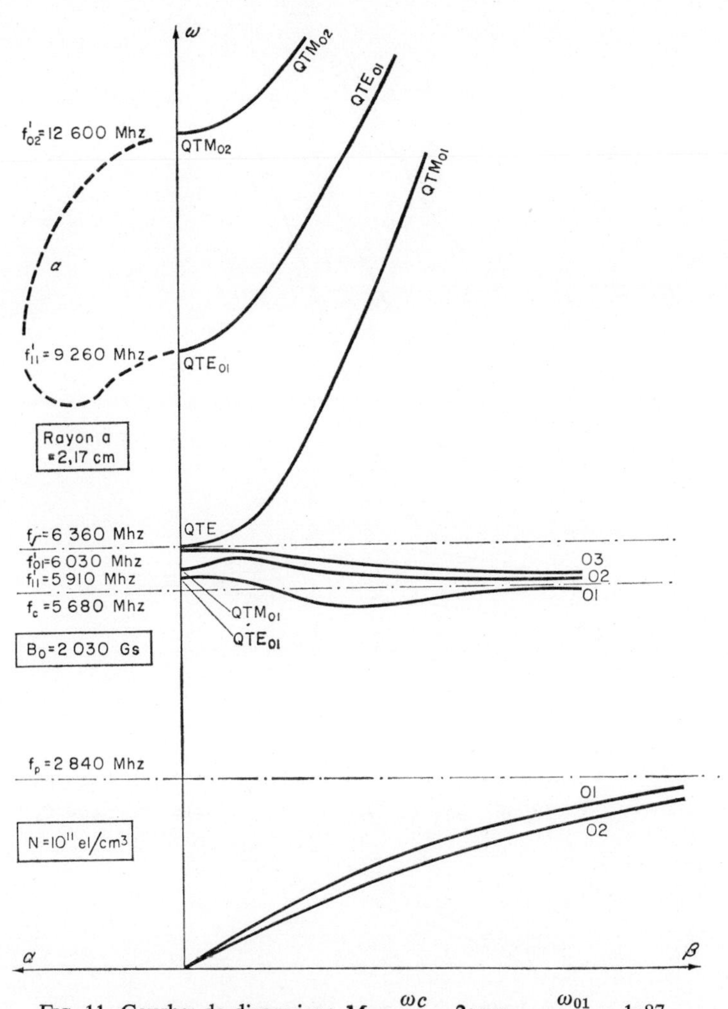

FIG. 11. Courbes de dispersion: $M = \dfrac{\omega c}{\omega p} = 2$; $mp_{01} = \dfrac{\omega_{01}}{\omega p} = 1{\cdot}87$.

courbes pour $Y > 1 + M^2$ et $X < 0$. Sur les figure 9 à 13, donnant les courbes (ω, β), on a indiqué une échelle de fréquences, en supposant une densité d'électrons de $N = 10^{11}$ électrons/cm³ dans le plasma. Figure 10, on voit que dans la bande intermédiaire, voisine de la fréquence cyclotron, il y a un mode dont la courbe de dispersion passe au-dessous de l'asymptote $\omega = \omega_c$, bien que sa fréquence de coupure soit supérieure à ω_c; la courbe $\omega(\beta)$ présente alors un point à tangente horizontale ($\partial\omega/\partial\beta = 0$). La figure 11 correspond au cas où la fréquence de coupure ω'_{01} est inférieure à $\sqrt{(\omega_p^2 + \omega_c^2)}$,

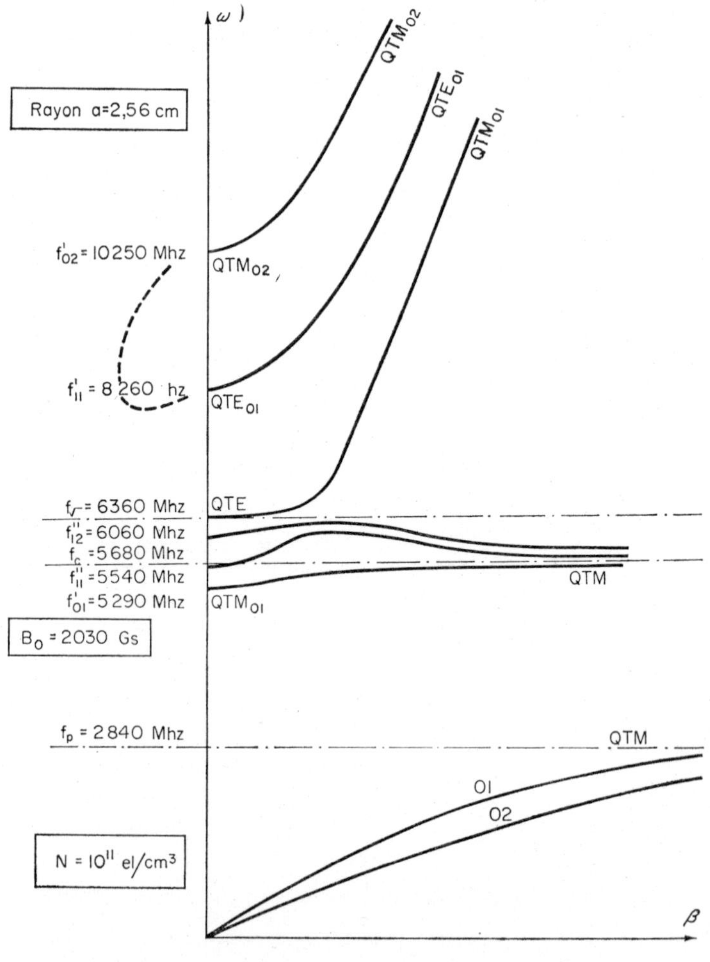

FIG. 12. Courbes de dispersion: $M = \dfrac{\omega c}{\omega p} = 2$; $mp_{01} = \dfrac{\omega_{01}}{\omega p} = 1{\cdot}88$.

(c'est-à-dire $\omega_{01} < \omega_c$): la forme des courbes s'explique alors très bien par la théorie du couplage des modes telle qu'elle a été exposée par B. A. Auld (1962). Sur la figure 12, l'une des branches intermédiaires donne lieu à une onde directe, et non à une onde inverse comme le prévoyait la théorie des états quasi-stationnaires. La figure 14, où l'échelle de fréquence est fixée pour une densité d'électrons de $N = 10^{12}$ électrons/cm³ montre un cas encore plus complexe, correspondant à un guide relativement grand: cette figure a

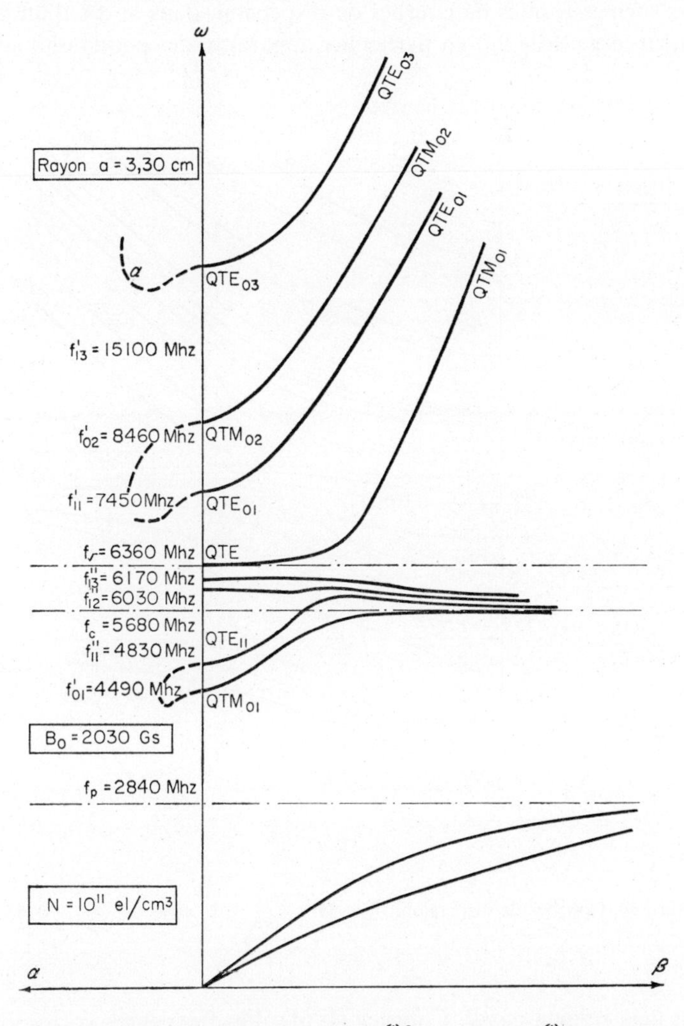

FIG. 13. Courbes de dispersion: $M = \dfrac{\omega c}{\omega p} = 2$; $mp_{01} = \dfrac{\omega_{01}}{\omega p} = 1 \cdot 23$.

été obtenue en grande partie, à l'aide d'une machine à calculer I.B.M. 1620; cependant la figure 15 montre comment on peut obtenir rapidement l'allure des courbes $Y(X)$ dans ce cas, à partir du réseau $Z = m^2 p_{0\nu}^2$.

VIII. CONCLUSION

Cette étude a permis de montrer qu'une théorie rigoureuse de la propagation dans un guide contenant un plasma, tout en confirmant l'existence d'ondes lentes prévue par la théorie des états quasi-stationnaires, n'aboutit pas aux mêmes formes de courbes de dispersion, dans le cas d'un guide relativement large. Elle fait en particulier apparaître des points où $\partial\omega/\partial\beta = 0$,

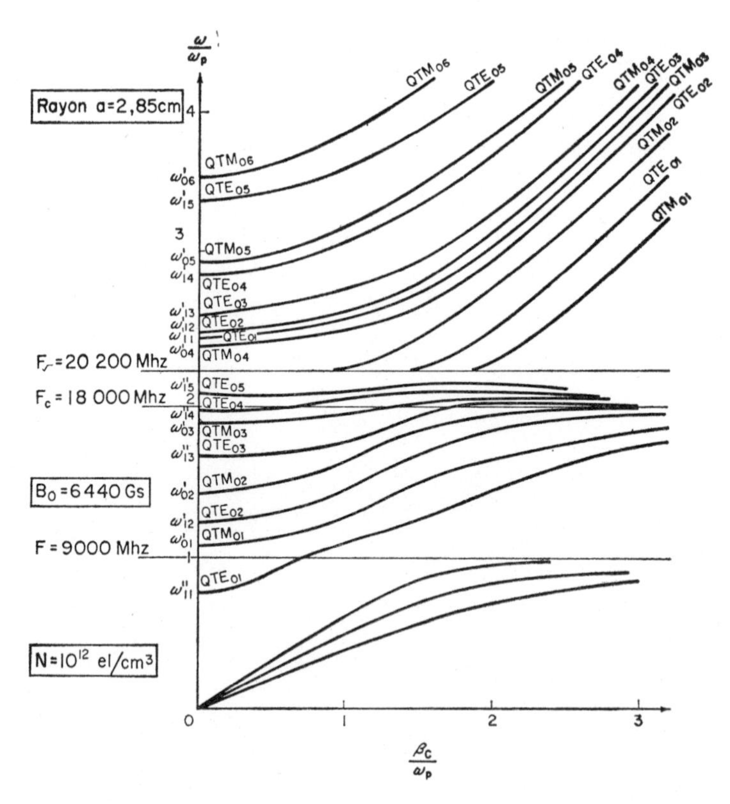

FIG. 14. Courbes de dispersion. Cas: $M = \dfrac{\omega c}{\omega p} = 2$; $mp_{01} = \dfrac{\omega_{01}}{\omega p} = 0{\cdot}45$.

résultat déjà signalé par P. Chorney (1961). Les fréquences correspondantes limitent fort probablement des bandes où la constante de propagation Γ est complexe. Cette étude devrait donc être complétée par une étude des solutions où Γ est complexe, les singularités observées aux points où $\partial\omega/\partial\beta = 0$ disparaîtraient sans doute si l'on tenait compte d'un mécanisme dissipatif tel que les collisions entre électrons et molécules. Elle devrait être étendue aux modes qui n'ont pas la symétrie de révolution.

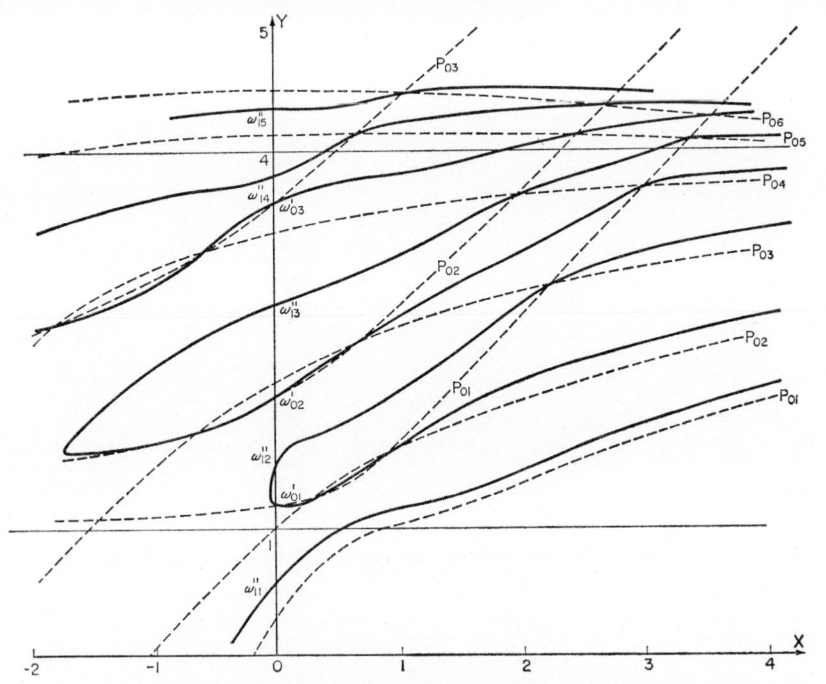

FIG. 15. Position relative des courbes $Y_0(x)$ et des courbes $Z = m^2 p_{0\nu}{}^2)$
Cas: $M = 2$; $m^2 p_{01}{}^2 = 0 \cdot 2$.

RÉFÉRENCES

AULD, B. A. (1962) Coupling of Electromagnetic and Magnetostatic Modes in Axially Magnetized Ferrites Waveguides. Symposium sur la Théorie Electromagnétique et les Antennes. Copenhague, 25–30 juin 1962.

CHORNEY, P. (1961) Power and energy relations in bidirectional waveguides, *Proceedings of the Symposium on Electromagnetics and Fluid Dynamics of Gaseous Plasmas*. Polytechnic Press, New York (April).

DENISSE, J. F. et DELCROIX, J. L. (1961) Théorie des ondes dans les plasmas. Monographie Dunod. Dunod ed. Paris.

SUHL, H. et WALKER, L. R. (1954) Topics in guided-wave propagation through gyromagnetic media, *Bell system Technical Journal* **33**, n° 3–4–5 (mai–juillet–septembre).

TRIVELPIECE, A. W. et GOULD, R. W. (1959) Space charge waves in cylindrical plasma columns, *J.A.P.* **30**, 1784–1793 (Nov.).

SELF-INTERACTION OF LONGITUDINAL PLASMA OSCILLATIONS WITH GENERATION OF ELECTROMAGNETIC RADIATION†

APPLICATION TO THE NARROW-BAND RADIO BURSTS OF THE SUN

RUDOLF-WILHELM LARENZ

Institute for Theoretical Physics, Technical University of Hannover, Germany

ABSTRACT

In a short review of longitudinal and transversal r.f. waves linearily coupled, attention is drawn first to the difficulty of separation of the transversal part to get radiation out of the plasma, and second to the lack of a substantial source of radiation, because these waves represent free plasma wave modes. It is shown that non-linear interaction of space-charge waves yields a radiation source of pure transversal waves, which at the second harmonic of the plasma frequency can run through the plasma and leave it. Formulas for the radiation field at large distances from the source are given and applied to the narrow-band solar bursts. The observed exhibition of the second harmonic as well as the spectral distribution can easily be explained.

I. INTRODUCTION

If one considers the mechanisms of generation of non-thermal radio-frequency radiation in cosmic objects and recently in experimental plasma devices too, one has to distinguish essentially between two groups: On the one hand the non-coherent radiation phenomena of single electrons through the cyclotron acceleration process and the Cerenkov effect, on the other hand emission of radiation from coherent motions of more extended ensembles of electronic charges. Here only the latter domain of plasma waves with their longitudinal and transversal or mixed character will be dealt with. For the irradiation problem the longitudinal waves are of main importance on account of energetic reasons. To emit radiation the plasma has to be supplied first with mechanical energy, this being achieved in astronomical plasmas by powerful gas flows which in the form of longitudinal shock waves can excite longitudinal space charge oscillations only. For the following the existence of such primary longitudinal oscillations will be presumed without considering the energy transfer to them, as here only the problem of electromagnetic radiation will be discussed.

† This research is sponsored by the Air Force Cambridge Research Laboratories, Office of Aerospace Research under Contract AF 61 (052)—161.

2. GENERAL CONSIDERATIONS

It is a well-known difficulty, that in a homogenous and isotropic plasma longitudinal space charge oscillations are not accompanied by a magnetic field and hence not by a transverse radiation field, as the conduction current and the displacement current compensate each other at any place, within the frame of a linear theory. If, however, anisotropy effects or inhomogeneities are present, as for instance plasma boundaries, one obtains a linear coupling between longitudinal and transversal wave parts. Local anisotropies effective in this sense, are represented essentially by the vector of a static magnetic field $\mathbf{H}_0$ (as has been known for decades in ionosphere physics) or of a general field of flow $\mathbf{v}_0$ (Larenz, 1955). In the first case the coupling is effected by the Lorentz force, in the latter case by the convected space charge, which represents a convection current. Both cases need components of the anisotropy vector which are not parallel to the direction of wave propagation.

The introduction of such anisotropies, however, does not solve the radiation problem completely. To emerge from the region of plasma oscillations and to leave the plasma as electromagnetic radiation, the transversal wave part has to separate from the longitudinal one. This can occur to a marked extent only in regions where the generally very different group velocities of the two high-frequency waves are very small and comparable to one another. (The group velocity of the coupled wave with dominant longitudinal part is of the order of the electron sound velocity $a = \sqrt{(5kT/3m)}$ which is very small compared to the velocity of light c). The conditions needed here are encountered if the refractive index, which depends on the plasma density, goes to zero or undergoes rapid changes (points of reflection!). This means that one has to presume density inhomogeneities or especially plasma boundaries to get wave separation (Field, 1956; Tidman, 1960; Kritz and Mintzer, 1960; Bernstein and Trehan, 1960). To obtain an effective radiation, considerable density changes within the dimensions of a wavelength are needed, which give rise to questions of their reality in cosmic objects. Instead of macroscopic density gradients the scattering of longitudinal waves at local random density fluctuations can also be considered as responsible for a transversal radiation (Ginzburg and Zhelezniakov, 1958; Tidman and Weiss, 1961a).

All mechanisms hitherto discussed, with the exception of the last (random fluctuations) exhibit a common feature. They deal with free plasma oscillations or waves, the longitudinal and transversal parts of them being coupled linearily. There is no substantial source of an electromagnetic radiation field, as this would require an emitting an antenna device in the case of radiation into free space. (One has to remember that the current density and charge density terms appearing in the linear Maxwellian equations for a homogeneous ionized medium are inherently connected with the wave propagation within such a medium, and therefore cannot be considered as source terms, in contrast to the case of current carrying surfaces radiating into a non-conducting medium).

3. THE NON-LINEAR INTERACTION

Now it was possible to find such a source of the radiation field in the field of the primary electron plasma oscillations itself, even in the case of a plasma which is homogeneous and isotropic in its rest state. From the facts referred to at the beginning it is clear that this source can only be supplied by the non-linear behavior of the plasma equations. The theoretical derivation was given by Burkhardt, Fahl and Larenz (1961). The way of the derivation of the source consists in a consequent decomposition of all vector quantities which enter in the combined Maxwell and hydrodynamic plasma equations into a divergence type (longitudinal) part and a rotational (transversal) part. The result is represented by the following equations, if one retains only those terms which are essential for the problem and are not of a higher order than of second degree in the amplitude of the electrostatic space-charge potential ϕ,

$$a^2 \Delta\phi - \ddot{\phi} - \omega_p^2 \phi = 0 \tag{1}$$

$$c^2 \Delta\mathbf{H} - \ddot{\mathbf{H}} - \omega_p^2 H = \frac{ec}{m\omega_p^2} [\operatorname{grad} \dot{\phi} \times \operatorname{grad} \Delta\phi]. \tag{2}$$

Instead of the second equation for the magnetic field **H** of the radiation field, an equivalent equation for the transversal electric field or the vector potential **A** may be written down. Furthermore the scalar potential ϕ may be replaced by the longitudinal electric field vector $\mathbf{E}_q = -\operatorname{grad} \phi = \nabla\phi$. If the source term on the right-hand side of (2) vanishes, two independent free wave equations for a conducting medium remain, expressing the non-existence of a coupling between longitudinal and transversal fields, mentioned at the beginning.

Within the region of plasma oscillations the source term cannot be zero if the electrostatic field strength vector $\nabla\phi$, originating from the space charge, is not parallel to the gradient of the space-charge density. This occurs for instance in plane plasma waves of different frequencies and different directions of propagation (wave packet) which are crossing each other. This interaction of longitudinal plasma waves leads to outgoing radiation if a part of the transversal field, excited in this manner, obeys the dispersion relation for free transversal wave propagation in a plasma. The outgoing transversal field fraction can be computed in a relatively simple manner, as both the equations (1) and (2), the homogeneous and the inhomogeneous one, are linear differential equations, where any solution of the first equation may be inserted on the right-hand side of the second. The physical reason for the excitation of a transversal field rests upon the fact that a source-type electrostatic field strength $\nabla\phi$ can produce rotational components of the current density, in a space charge which is not uniformly distributed in space. The rotational current components then constitute the origin of the transversal radiation field. It seems to be possible that this type of mechanism can be related to a wave scattering process suggested by Ginzburg and Zhelezniakov (1958). Furthermore an interaction phenomenon between two longitudinal waves and one transversal wave, described by Sturrock (1961), seems to be related to the mechanism discussed here. A simple interaction case at zero temperature was studied by Tidman and Weiss (1961b).

4. COMPUTATION OF THE FAR FIELD

In most cases one is interested in the radiation field at large distances from the source. For this reason the far-field solution of (2) will be given here.

Describing any plasma wave solution of (1) by a Fourier wave integral

$$\phi = \int_{-\infty}^{+\infty} C(\mathbf{k})\, e^{i\mathbf{kr}-i\alpha(k)t}\, d\mathbf{k} \tag{3}$$

with the frequency $\alpha = \pm\sqrt{(\omega_p^2 + a^2k^2)}$, the general solution of (2) is immediately represented by

$$\mathbf{H} = \frac{iec}{m\omega_p^2} \iint_{-\infty}^{+\infty} \frac{[\mathbf{k}\times\mathbf{k}']\,\alpha(k)\,k'^2\,C(\mathbf{k})\,C(\mathbf{k}')\,e^{i(\mathbf{k}+\mathbf{k}',\mathbf{r})-i(\alpha(k)+\alpha(k'))\,t}}{c^2(\mathbf{k}+\mathbf{k}')^2 - (\alpha(k)+\alpha(k'))^2 + \omega_p^2}\, d\mathbf{k}\, d\mathbf{k}'. \tag{4}$$

Now at a far distant point only the vicinity of the "resonance poles", where the denominator of the integrand (4) vanishes, yields an appreciable contribution. This means, that the outgoing waves have to obey the Eccles dispersion relation for the combined wave vector $\mathbf{k} + \mathbf{k}'$ and the combination frequency $\alpha(k) + \alpha(k')$ of the two interacting plasma waves. By integration by parts over the position angle of $\mathbf{k} + \mathbf{k}'$ one first separates that part of the **H**-field which is inversely proportional to the distance r since this is a known feature of the far field. Then one performs the integration over the magnitude $\mathbf{k} + \mathbf{k}'$, this integration yielding the pole contribution. One obtains the far field (5)

$$\mathbf{H}_{r\to\infty} = \frac{1}{r}\cdot\frac{\pi^2 e}{mc\omega_p^2}\int_{-\infty}^{+\infty}[\mathbf{k}\times\mathbf{l}]\,\alpha(k)\,(\mathbf{l}-\mathbf{k})^2\,C(\mathbf{k})\,C(\mathbf{l}-\mathbf{k})\,e^{ilr-\omega t}\,d\mathbf{k} \tag{5}$$

where $\mathbf{l}$ and ω now represent wave vector and frequency as seen by the far distance observer. The pole integration performed here represents an analogy to the known computation of the response of any oscillatory system subjected to an external force.

The wave vector $\mathbf{l}$ and the frequency ω are somewhat complicated functions of the original wavenumber $\mathbf{k}$ because the waves $\mathbf{k}$ and $\mathbf{k}'$ ($\mathbf{k}'$ contained in $\mathbf{l}$)

whose interaction yields far-field contributions, are selected by the dispersion relation (6)

$$c^2 l^2_{(\mathbf{k})} = \omega^2_{(\mathbf{k})} - \omega^2_p = (\alpha_{(k)} + \alpha_{(\mathbf{l}-\mathbf{k})})^2 - \omega^2_p \,. \tag{6}$$

But with the exception of a very narrow range near $k \simeq 0$, $\alpha \simeq \omega_p$, it is generally true with good approximation that the wave vectors $\mathbf{k}$ and $\mathbf{k}'$ are of about equal magnitude and are directed opposite to each other at an angle which scarcely deviates from π. In fact only waves of approximately equal frequency which are impinging against each other within a very small solid angle of the order $3\pi a^2/c^2$ will show the desired interaction. The electron sound velocity a, being very small compared to the velocity of light c, accounts for the fact that the transversal wavenumber l is generally very small compared to the longitudinal k, and this simplifies the dispersion relation to (6a)

$$c^2 l^2 = \omega^2 - \omega^2_p \simeq 4\omega^2_p + 4a^2k^2 - \omega^2_p = 3\omega^2_p + 4a^2k^2 \ll c^2k^2. \tag{6a}$$

Having obtained the magnetic far-field (5) one passes to the electric vector $\mathbf{E}$ and to the Poynting vector $\mathbf{S}$ of energy flux density by well-known formulas of electrodynamics

$$\mathbf{E} \underset{r\to\infty}{=} -\frac{\omega}{cl^2}\,[\mathbf{l} \times \mathbf{H}] \tag{7}$$

$$\mathbf{S} = \frac{c}{4\pi}\,[\mathbf{E} \times \mathbf{H}]. \tag{8}$$

The next aim must be to derive time-independent quantities. After insertion of (5) and (7) into (8) a time integration is easily performed yielding the total energy per unit area received by the observer. The formula obtained (as a double $\mathbf{k}$—spectrum integral) includes one general $\mathbf{k}$-integration leading to the expression (9), where $\mathbf{r}_0$ means the unit vector in the direction of r,

$$\int\limits_{-\infty}^{+\infty} \mathbf{S}\,\mathrm{d}t \underset{r\to\infty}{\simeq} \frac{\mathbf{r}_0}{r^2}\,\frac{3\pi^5 e^2}{2m^2c^4\omega^4_p} \int\limits_{-\infty}^{+\infty} [\mathbf{k} \times \mathbf{l}]^2\,\frac{\alpha(k)}{l(k)}\,k^5 |\, C(\mathbf{k})\,|^4\,\mathrm{d}\mathbf{k}. \tag{9}$$

Performing this latter integration the results with respect to the interaction solid angle have to be considered. Finally integrating over the whole spherical surface at the distance r one gets the total radiated energy (10) of the plasma oscillation disturbance,

$$r^2 \iint\limits_{-\infty\ \Omega}^{+\infty} S\,\mathrm{d}t\,\mathrm{d}\Omega \simeq \frac{4\sqrt{3}\pi^6 e^2 \omega_p}{m^2c^5} \int\limits_{-\infty}^{+\infty} k^7\,|\,C(\mathbf{k})\,|^4\,\mathrm{d}\mathbf{k} \ll U. \tag{10}$$

This radiated energy must be low compared to the electrostatic energy

$$U = \frac{1}{8\pi} \int (\mathrm{grad}\,\phi)^2\,\mathrm{d}V = \pi^2 \int\limits_{-\infty}^{+\infty} k^2\,|\,C(\mathbf{k})\,|^2\,\mathrm{d}\mathbf{k}$$

of the primary longitudinal plasma wave, so as to avoid violation of the energy principle and to avoid reaction of the transversal field on the longitudinal one. This energy condition may impose certain restrictions to the choice of the original spectral function $C(\mathbf{k})$; otherwise terms of higher order have to be considered, which were neglected in the derivation of the basic equations (1) and (2).

It is easily possible to pass to the frequency spectrum $S(\omega)$ of the radiation by expressing the variable k in terms of the frequency ω in (9) or (10). This can be done because one has from (6a)

$$\omega \simeq 2\omega_p + \frac{a^2k^2}{\omega_p}.$$

This relation will be important for the next section, as it reveals the fact that essentially the doubled plasma frequency will be radiated, or more precisely the sum of the frequencies of two interacting plasma waves. This is simply the consequence of the fact that the potential ϕ or its derivatives enter in the second power, in the source term of (2). The difference of the original frequencies, which should also appear under these circumstances, cannot be radiated, as the dispersion relation (6) does not admit this case.

5. APPLICATION TO THE SOLAR BURSTS

Radio-astronomical observations show that, in connection with solar radio bursts of type II and III, the emission occurs of the fundamental plasma frequency (first harmonic) and simultaneously the emission of the second harmonic with comparable intensity. But higher harmonics are never observed. As the mechanism discussed in this paper yields only the second harmonic, one may conclude that non-thermal radiation of plasma oscillations within cosmical plasmas generally exhibits the second harmonic, which at a refractive index $\sqrt{(3)/2} \simeq 0{\cdot}85$ admits unrestrained propagation through the plasma. In cases where the fundamental is observed too, then the other mechanisms discussed in Section 2 are effective in addition.

Now to draw more conclusions from the average spectral energy distribution of the radiation $\langle S(\omega)\rangle$, one has to assume a suitable spectral function $C(\mathbf{k})$ of the original plasma waves. The following function will be chosen

$$C(\mathbf{k}) = -\frac{2i\beta^5}{\pi^{3/2}}\widehat{\nabla\phi}\,.\,k\cos\vartheta\,\mathrm{e}^{-\beta^2k^2} \tag{11}$$

where β represents a characteristic length of the plasma waves and $\widehat{\nabla\phi}$ a field strength amplitude. This function leads to a dipole-type plasma disturbance

$$\phi = \widehat{\nabla\phi}\,.\,z\,\mathrm{e}^{-(x^2+y^2+z^2)/4\beta^2} \tag{11a}$$

(at time $t = 0$)

which is reproduced qualitatively in Fig. 1. With growing time the disturbance spreads out and decays.

Applying the last part of Section 4, one obtains from (11) a spectral distribution of the received radiation

$$\langle S(\omega)\rangle \propto (\omega - 2\omega_p)^6 \, e^{-4\beta^2\omega_p(\omega-2\omega_p)/a^2} \tag{12}$$

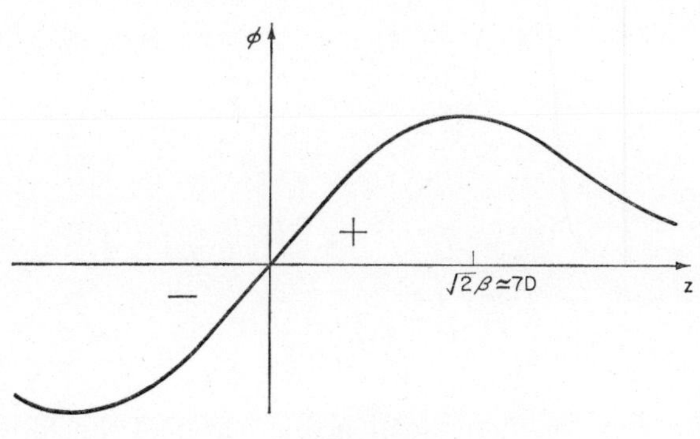

FIG. 1.

with $\omega \geqslant 2\omega_p$. This is reproduced qualitatively in Fig. 2, which shows the narrow frequency band structure actually observed. For the maximum average energy flux per second and per unit area at the distance r there results

$$\langle S(\omega_0)\rangle \simeq \frac{\sqrt{3}}{10}\left(\frac{3}{e}\right)^6 \cdot \frac{e^2\omega_p^2\beta^8}{m^2c^5a^2} \cdot \frac{N}{r^2}(\widehat{\nabla\phi})^4 \propto \frac{T^3}{n^3} \tag{13}$$

where N represents the number of disturbances of the assumed dipole-type per second. For a given value of the product $N(\nabla\phi)^4$, the quantity $\langle S(\omega_0)\rangle$ is proportional to the third power of the ratio of electron temperature T to electron density n.

This latter conclusion follows from a determination of the parameter length β by postulating that the theoretical relative bandwidth

$$\frac{\Delta\omega}{2\omega_p} \simeq \frac{a^2}{3\beta^2\omega_p^2} \tag{14}$$

corresponds to the observed one, which is of the order of 5 per cent. This comparison yields for β the order of magnitude of 5 Debye shielding lengths D ($D^2 = KT/4\pi e^2 n = 3a^2/5\omega_p^2$). This means that the original characteristic plasma wavelength λ will be of the order of 30 D, a very plausible result. One may modify the assumptions about the original spectrum C; but this is not of any essential consequence with respect to the narrow-band structure of $\langle S(\omega)\rangle$ and the conclusion about the characteristic electron wavelength.

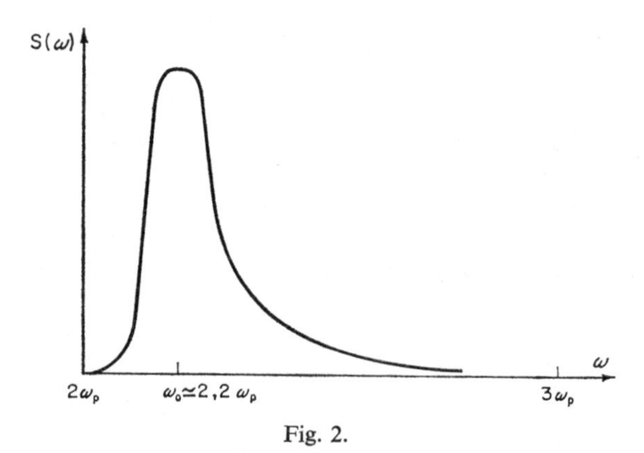

Fig. 2.

Finally the energy condition mentioned in Section 4 should be examined. With the assumed C-function (11) it turns out that the field strength amplitude $\widehat{\nabla\phi}$ of the primary plasma oscillations should be lower than the (very high) value of 100 kV/cm.

REFERENCES

BERNSTEIN, I. B. and TREHAN, S. K. (1960) Plasma oscillations (I), *Nuclear Fusion*, **1**, 3–41.

BURKHARDT, G., FAHL, C. and LARENZ, R. W. (1961) Die Kopplungsmechanismen zwischen longitudinalen und transversalen Wellen in einem Plasma, *Z. Physik*, **161**, 380–387.

FIELD, G. B. (1956) Radiation by plasma oscillations, *Astrophysical J.* **124**, 555–570.

GINZBURG, V. L. and ZHELEZNIAKOV V. V. (1959) On the mechanisms of sporadic solar radio emission, Paper 103 of Paris Symposium on Radio Astronomy 1958, ed. by R. N. Bracewell, Stanford Univ. Press 1959, pp. 574–582. (See further references there.)

KRITZ, A. H. and MINTZER, D. (1960) Propagation of plasma waves across a density discontinuity, *Physical Rev.* **117**, No. 2, 382–386 (January).

LARENZ, R. W. (1955) Zur Entstehung der überthermischen kosmischen Radiofrequenzstrahlung, *Z. Naturforschung* **10a**, 901–913.

STURROCK, P. A. (1961) Non-linear effects in electron plasmas, *J. Nuclear Energy*, Part C: Plasma Physics, **2**, 158–163.

TIDMAN, D. A. (1960) Radio emission by plasma oscillations in non-uniform plasmas, *Physical Rev.* **117**, No. 2, 366–374 (January).

TIDMAN, D. A. and WEISS, G. H. (1961) (a) Radio emission by plasma oscillations in non-uniform plasmas, *Physics of Fluids*, **4**, No. 6, 703–710 (June).

(b) Radiation by a large-amplitude plasma oscillation, *Physics of Fluids*, **4**, No. 7, 866–868 (July).

WAVE PROPAGATION IN ANISOTROPIC MEDIA

PENTTI MATTILA†

SUMMARY

The author uses matrix and tensor analysis in this paper. It is well known that the expressions for wave propagation become complicated in an anisotropic case. It seems that a systematic treatise would be of much value in this field. For these reasons it seems desirable to seek out simplifying methods and particularly those possibilities which matrix and tensor analysis offer.

If we study, for example, expressions for a dielectric tensor and permeability tensor we obtain them in different forms depending on what kind of media are being considered. Wave propagation has been thoroughly studied, especially in crystals and in the ionosphere. It still seems, however, to be worth making a systematic examination of these cases. The method of treatment leads, as is well known, to a case in which it is necessary to use a theory of finite dimensional vector spaces. It turns out that in a medium like the ionosphere, it is necessary to make a transition to a complex space. In this way it is possible, for example, to determine the refractive index of a medium, the principal axis components of the electric intensity and the electric polarisation. The matrix equation in the ionosphere can be put in the following form

$$-\varepsilon_0 X \left\| \begin{matrix} E_1 \\ E_2 \\ E_3 \end{matrix} \right\| = \left\| \begin{matrix} 1+Y & 0 & 0 \\ 0 & 1-Y & 0 \\ 0 & 0 & 1 \end{matrix} \right\| \cdot \left\| \begin{matrix} P_1 \\ P_2 \\ P_3 \end{matrix} \right\| - jZ \left\| \begin{matrix} 1 & 0 & 0 \\ 0 & 1 & 0 \\ 0 & 0 & 1 \end{matrix} \right\| \cdot \left\| \begin{matrix} P_1 \\ P_2 \\ P_3 \end{matrix} \right\|$$

where E_i = the electric intensity, $i = 1, 2, 3$ and P_i = the electric polarization, $i = 1, 2, 3$

$$Z = \frac{\nu}{\omega}, \qquad X = \left(\frac{\omega_N}{\omega}\right)^2 \qquad \text{and} \qquad Y = \frac{\omega_H}{\omega}.$$

It is important to note that the values $1 + Y$, $1 - Y$ and 1 are just eigenvalues of a certain Hermitian matrix. They are important quantities in explaining the triple splitting in the ionosphere. The directions of P_i and E_i, $i = 1, 2, 3$ are called principal axis components. When i is 1 or 2 they are complex and whenever $i = 3$ then the direction is a real one. The constant term in the matrix equation above is due to the dissipation in the medium.

† Finland Institute of Technology, Albertinkatu 40, Helsinki, Finland.

By a method which is well known it is also possible to derive the matrix for an ε-tensor (where the elements are functions of refractive indices) and put it in diagonal form.

As a conclusion we may say that when studying wave propagation in an anisotropic medium we can effectively use a complex space theory. In different media we get different kinds of ε- and μ-tensors. Two types of matrices are especially important, namely a symmetric and a Hermitian form. It is to be noted that in quantum mechanics the Hermitian matrix plus a constant is also important.

Now the basic problem considered in this paper, wave propagation in a homogeneous anisotropic medium, can be studied. The comparison between different media can be made very clearly if we put tensors in matrix form. As in the isotropic case, the wave equation for the anisotropic medium is important. The equations in which transverse components are determined from longitudinal components are of fundamental importance in many ways. It turns out that there is also a transverse electromagnetic wave (TEM) in the anisotropic medium and it is possible to arrive at the propagation constant of the wave in this case. In the isotropic medium there were in addition TE- and TM-waves. In anisotropic media higher order modes can be divided into two groups which are called quasi-TE- and quasi-TM-modes. We pay special attention to the transition from an anisotropic medium to an isotropic one. Therefore the boundary value problems become important as was also the case for isotropic media. In this paper the author especially tries to show how advantageous it is to use tensor and matrix analysis in this connection.

In the last part of the paper guided waves in anisotropic media are studied, especially gyromagnetic and gyroelectric media under different boundary conditions. It turns out that there is no TEM-mode in a wave guide containing anisotropic media. Existing waves may be divided into two groups, which are called quasi-TE- and quasi-TM-modes. It is a difficult problem to find the general solution of Maxwell's equations under the boundary conditions imposed by the wall. From the theoretical point of view the question here is one of a transition from a complex space to a real one. From a physical point of view it is evident that there must be a solution to this problem. The author also studies some special cases, such as a wave guide of circular cross-section, a rectangular waveguide and a waveguide consisting of two parallel planes. It is the author's purpose to show how a tensor and matrix analysis can be applied to the solution of these problems.

VECTOR INTEGRAL EQUATIONS FOR THE ELECTRIC FIELD IN AN INHOMOGENEOUS MAGNETO-IONIC MEDIUM

W. C. HOFFMAN†

SUMMARY

Maxwell's equations for an inhomogeneous magneto-ionic medium can be written as

$$\text{curl}\,\mathbf{E} = -\mu_0 \frac{\partial \mathbf{H}}{\partial t} \qquad \text{curl}\,\mathbf{H} = \frac{\partial \check{\mathbf{D}}}{\partial t} = \mathbf{J} + \varepsilon_0 \frac{\partial \mathbf{E}}{\partial t} \tag{1), (2}$$

$$\text{div}\,\check{\mathbf{D}} = 0 \qquad \text{div}\,\mathbf{B} = 0 \tag{3}$$

where **J** is related to **E** by the conductivity tensor

$$\mathbf{J} = \mathfrak{S}\cdot\mathbf{E} \tag{4}$$

and the effective electric induction $\check{\mathbf{D}}$ is given by

$$\check{\mathbf{D}} = \varepsilon_0\, \mathfrak{K}\cdot\mathbf{E}\,, \tag{5}$$

$\mathfrak{K}$ being the effective dielectric tensor

$$\mathfrak{K} = \mathfrak{J} + \mathfrak{S}/i\omega\varepsilon_0\,. \tag{6}$$

Then, assuming harmonic time dependence $e^{i\omega t}$, one is led in the usual way to the vector wave equation

$$\text{curl}\,\text{curl}\,\mathbf{E} = k_0^2\, \mathfrak{K}\cdot\mathbf{E}\,. \tag{7}$$

If an external source is present an additional source term in **J** must be supplied on the right-hand side.

Previous approaches to the problem of a full wave solution for electromagnetic radiation in the exterior of the earth have for the most part assumed horizontal stratification, and have been based either upon the system (1), (2) or the wave equation (7). Such an assumption is inappropriate to such phenomena as propagation in a dipole geomagnetic field and transmission along field-aligned ionization. In the present work the problem will be formulated as a vector integral equation in order to achieve a more general

† Boeing Scientific Research Laboratories, Seattle 24, Wash., U.S.A.

geometry. This will be carried out in two different ways. The first form applies to a bounded distribution of ionization with an axially symmetric ambient magnetic field. The second formulation as a vector integral equation is based on the dyadic Green's formula and defines the electromagnetic field in the exterior of a perfectly conducting magnetized sphere situated in an inhomogeneous ionized medium. Several different approaches to the solution of this equation are discussed.

We now proceed to a detailed consideration of the first formulation. The ambient magnetic field will be assumed to lie along the z-axis, so that the effective dielectric tensor takes the form

$$\mathfrak{K} = \begin{pmatrix} 1 - \dfrac{UX}{U^2 - Y^2} & i\,\dfrac{XY}{U^2 - Y^2} & 0 \\ -i\,\dfrac{XY}{U^2 - Y^2} & 1 - \dfrac{UX}{U^2 - Y^2} & 0 \\ 0 & 0 & 1 - X/U \end{pmatrix} \tag{8}$$

where U, X, and Y have the usual meanings of magneto-ionic theory. The Stratton-Chu formula

$$\int_V \{\mathbf{E}\cdot\text{curl curl }\mathbf{G} - \mathbf{G}\cdot\text{curl curl }\mathbf{E}\}\,\mathrm{d}V = - \int_{\text{bdry }V} \{\mathbf{E} \wedge \text{curl }\mathbf{G} - \mathbf{G} \wedge \text{curl }\mathbf{E}\}\cdot\mathbf{n}\,\mathrm{d}\Sigma \tag{9}$$

applied to formula (7) and $\mathbf{G} = \mathbf{a}\,\mathrm{e}^{-ik_0R}/R$, where $\mathbf{a}$ is a constant (but otherwise arbitrary) vector, yields the following integral equation for the scattered field $\mathbf{E}^s$:

$$\mathbf{E}^s(P) = \mathbf{F}^i(P) + \frac{1}{4\pi}\int_{V_3} \frac{\mathrm{e}^{-ik_0R}}{R}\left\{k_0^2\mathfrak{M}(Q)\cdot\mathbf{E}^s(Q) + \left(ik_0 + \frac{1}{R}\right)R(\nabla\cdot\mathbf{E}^s)\right\}\mathrm{d}Q \tag{10}$$

where $\mathfrak{M} = \mathfrak{K} - \mathfrak{J}$. The integral in (10) is extended over all of 3-space V_3. We now assume that (i) the ionization is confined to a finite volume of space V_c; (ii) the distance R from any point of that volume to the field point P is such that $R \gg \lambda\!\!\!^{-}_0$; (iii) $k_0 \ll 1$; and (iv) $|\nabla\cdot(\mathfrak{M}\cdot\mathbf{E})| \leq |\mathfrak{M}\cdot\mathbf{E}|$. Then (10) may be approximated by the vector Fredholm equation

$$\mathbf{E}^s(P) = \mathbf{F}^i(P) + \frac{k_0^2}{4\pi}\int_{V_3} \frac{\mathrm{e}^{-ik_0R}}{R}\,\mathfrak{M}\cdot\mathbf{E}^s(Q)\,\mathrm{d}Q\,. \tag{11}$$

Equation (11) may be uncoupled by redefining the field quantities as

$\mathscr{E}_{1,2} = E_1 \pm iE_2$, $\mathscr{F}_{1,2} = F_1 \pm iF_2$, and the elements of the conductivity tensor as $\mathscr{B}_{1,2} = -X\cdot(U \pm Y)/(U^2 - Y^2)$, $\mathscr{A} = A - 1 = -X/U$. The result is the uncoupled system of singular integral equations

$$\begin{aligned} \mathscr{E}^s_{1,2}(P) &= \mathscr{F}^i_{1,2}(P) + \frac{k_0^2}{4\pi}\int_{V_3} \frac{e^{-ik_0R}}{R}\mathscr{B}_{1,2}(Q)\mathscr{E}^s_{1,2}(Q)\,dQ \\ E^s_3(P) &= F^i_3(P) + \frac{k_0^2}{4\pi}\int_{V_3} \frac{e^{-ik_0R}}{R}\mathscr{A}(Q)E^s_3(Q)\,dQ \end{aligned} \tag{12}$$

which under certain general conditions can be solved by mean-square convergent Neumann series.

We now pass to a consideration of the second integral equation. Consider a perfectly conducting sphere of radius a with an antenna located at the point $(a + h, \theta_1, \phi_1)$, $(h \geq 0)$. An ambient dipole magnetostatic field emanates from the center of the sphere. Outside the sphere there is a distribution of ionization N_e which vanishes below some given height. Silver–Müller type radiation conditions are to apply at infinity.

The vector Green's theorem appropriate to a dyadic Green's function $\mathfrak{G}$ is

$$\begin{aligned} &\int_V \{\mathbf{F}\cdot\text{curl curl }\mathfrak{G} - \text{curl curl }\mathbf{F}\cdot\mathfrak{G}\}\,dV \\ &= \int_V \{\mathbf{F}\cdot\text{curl curl }\mathfrak{G} - \mathfrak{G}^T\cdot\text{curl curl }\mathbf{F}\}\,dV \\ &= \int_\Sigma \{\text{curl }\mathbf{F}\cdot\mathbf{n}\wedge\mathfrak{G} - \mathbf{n}\wedge\mathbf{F}\cdot\text{curl }\mathfrak{G}\}\,d\Sigma\,, \end{aligned} \tag{13}$$

where $\mathbf{F}$ is a field vector, $\mathfrak{G}^T$ denotes the transpose of the dyadic $\mathfrak{G}$, $\mathbf{n}$ is the exterior normal, and the order of the inner products is essential. Let $\mathbf{F} = \mathbf{E}$, where $\mathbf{E}$ is governed by (7) plus a term $\mathbf{J}$ due to the presence of an antenna, and let $\mathfrak{G}(P,Q)$ be the dyadic Green's function for the exterior problem for a perfectly conducting sphere in free space:

$$\text{curl curl }\mathfrak{G} = k_0^2\mathfrak{G} + \mathfrak{J}\delta(P - Q)\,, \qquad \mathbf{n}\wedge\mathfrak{G} = \mathfrak{O} \quad \text{at} \quad r = a\,.$$

Insertion of these expressions into (13) yields the integral equation

$$\mathbf{E}(P) = \mathbf{F}(P) - k_0^2\int_{V_a} \mathfrak{G}^T(P,Q)\cdot X(Q)\mathfrak{N}(Q)\cdot\mathbf{E}(Q)\,dQ \tag{15}$$

where V_a is the volume outside the sphere, $X(Q) = \omega_e^2/\omega^2$; $\mathbf{F}(P)$ is defined by

$$\mathbf{F}(P) = -i\omega\mu_0\int_{V_a} \mathfrak{G}^T(P,Q)\cdot\mathbf{J}(Q)\,dQ \tag{16}$$

and the dyadic $\mathfrak{N}$ is, in matrix form,

$$\mathfrak{N}(Q) = \frac{1}{U(U^2 - Y^2)} \begin{pmatrix} U^2 - Y_r^2 & -Y_r Y_\theta & iY_\theta U \\ -Y_r Y_\theta & U^2 - Y_\theta^2 & -iY_r U \\ -iY_\theta U & iY_r U & U^2 \end{pmatrix} . \quad (17)$$

The Green's dyadic for the sphere in free space has been derived by C. T. Tai. Tai's expression (modified for $e^{i\omega t}$ time dependence) reads as follows

$$\mathfrak{G}(P,Q) = \sum_{n=1}^{\infty} \left(-ik_0 \frac{2n+1}{n(n+1)} \right) \sum_{m=0}^{n} \varepsilon_m \frac{(n-m)!}{(n+m)!} \times$$

$$\times \begin{cases} \left[\mathbf{M}^{r(1)}_{\substack{e\\o}mn}(Q) + R_n^b \mathbf{M}^{r(4)}_{\substack{e\\o}mn}(Q)\right] \mathbf{M}^{r(4)}_{\substack{e\\o}mn}(P) + \\ \quad + \left[\mathbf{N}^{r(1)}_{\substack{e\\o}mn}(Q) + R_n^e \mathbf{N}^{r(4)}_{\substack{e\\o}mn}(Q)\right] \mathbf{N}^{r(4)}_{\substack{e\\o}mn}(P) , \qquad r_Q < r_P & (18a) \\ \mathbf{M}^{r(4)}_{\substack{e\\o}mn}(Q) \left[\mathbf{M}^{r(1)}_{\substack{e\\o}mn}(P) + R_n^b \mathbf{M}^{r(4)}_{\substack{e\\o}mn}(P)\right] + \\ \quad + \mathbf{N}^{r(4)}_{\substack{e\\o}mn}(Q) \left[\mathbf{N}^{r(1)}_{\substack{e\\o}mn}(P) + R_n^e \mathbf{N}^{r(4)}_{\substack{e\\o}mn}(P)\right] , \qquad r_Q > r_P & (18b) \end{cases}$$

where $\mathbf{M}^{r(1)}_{\substack{e\\o}mn}(P)$, $\mathbf{M}^{r(4)}_{\substack{e\\o}mn}(P)$, $\mathbf{N}^{r(1)}_{\substack{e\\o}mn}(P)$, $\mathbf{N}^{r(4)}_{\substack{e\\o}mn}(P)$ are spherical wave functions. Denote the form of (18) appropriate to $r_Q < r_P$ by $\mathfrak{G}_<(P,Q)$, that for $r_Q > r_P$ by $\mathfrak{G}_>(P,Q)$, then the integral equation (15) may be written as

$$\mathbf{E}(P) = \mathbf{F}(P) - k_0^2 \left\{ \int_a^r r'^2 \int_{\Omega_Q} X(Q) \mathfrak{S}^T_< (P,Q) \cdot \mathfrak{N}(Q) \cdot \mathbf{E}(Q) \, d\Omega' \, dr' + \right.$$

$$\left. + \int_r^\infty r'^2 \int_{\Omega_Q} X(Q) \mathfrak{S}^T_> (P,Q) \cdot \mathfrak{N}(Q) \cdot \mathbf{E}(Q) \, d\Omega' \, dr' \right\} . \quad (19)$$

Equation (19) is our basic integral equation. When N_e is spherically symmetric $X(Q) = X(r')$; when N_e is field-aligned $X(Q) = X(\alpha')$, where α' is one of a set of "dipolar coordinates" $\alpha' = r'/\sin^2\theta'$, $\beta' = \cos\,\theta'/r'^2$, $\phi' = \phi'$.

Equation (19) may be solved by iteration provided the kernel (that is to say, $X(Q)$) is such as to insure convergence. The Born approximation $\mathbf{E}_1(P)$ is evaluated for vertically and horizontally polarized antennas located at a point near the earth's surface.

For field points P above the base of the ionosphere, so that $\mathfrak{N}(P) \neq \mathfrak{I}$, we may multiply (19) through by $\mathfrak{N}(P)$ to obtain an equivalent integral equation in $\mathbf{И}(P) = \mathfrak{N}(P) \cdot \mathbf{E}(P)$,

$$\mathbf{И}(P) = \mathbf{\Phi}(P) - k_0^2 \left\{ \int_a^r r'^2 \int_{\Omega_Q} X(Q) \cdot \mathfrak{N}(P) \cdot \mathfrak{G}^T_< (P,Q) \cdot \mathbf{И}(Q) \, d\Omega' \, dr' + \right.$$

$$+ \int_r^{r_0} r'^2 \int_{\Omega_Q} X(Q) \mathfrak{N}(P) \cdot \mathfrak{G}^T < (P,Q) \cdot \mathbf{И}(Q) \, d\Omega' \, dr' \quad (20)$$

where r_0 is the value of r such that $X(r,\Omega)$ essentially vanishes for $r > r_0$. If we now attempt to develop $\mathbf{\mathit{И}}(P)$ in a series of spherical vector wave functions

$$\mathbf{\mathit{И}}(P) = \sum_{p,q} \Big\{ e_{pq}^{(M^{(1)})}(r)\mathbf{M}_{\substack{e_{pq}\\0}}^{(1)}(P) + e_{pq}^{(M^{(4)})}(r)\mathbf{M}_{\substack{e_{pq}\\0}}^{(4)}(P) + \\ + e_{pq}^{(N^{(1)})}(r)\mathbf{N}_{\substack{e_{pq}\\0}}^{(1)}(P) + e_{pq}^{(N^{(4)})}(r)\mathbf{N}_{\substack{e_{pq}\\0}}^{(4)}(P) \Big\} \quad (21)$$

the coefficients cannot, on account of the double summation in the kernel, be determined in the usual way from an equivalent infinite system of algebraic equations. However, by properly re-defining the region of integration, (20) can be reduced to a single scalar integral equation over the region $(a, 3r_0 - 2a) \times \Omega$ that appears well suited for numerical solution.

HARMONIC EXCITATION IN AND RERADIATION FROM NON-UNIFORM IONIZED REGIONS

LEWIS WETZEL

Brown University, Providence, Rhode Island

ABSTRACT

The interaction between an electromagnetic wave and a non-uniform ionized region is shown to produce a spectrum of harmonic electric currents localized within the non-uniformity. The magnitudes of the dominant harmonic currents decreased from one order to the next by a factor proportional to $[(e/m)\mathbf{E}\cdot\nabla n_0]/\omega^2 n_0$, where $\mathbf{E}$ is the electric field, ω the primary frequency, and n_0 the stationary electron density. These harmonics are comparable, locally, to those produced by magnetic field interactions. A simple example of coherent re-radiation from such a current distribution is discussed.

1. INTRODUCTION

The several sources of non-linearity in the interaction between a plasma and an electromagnetic field are responsible for a wide variety of interesting phenomena. Some of these non-linearities give rise to a significant change in the electron temperature even for relatively low field strengths. The effects of such changes on the constitutive parameters of the plasma have been discussed at length in the excellent review article by Ginsburg and Gurevich (1960). On the other hand, the existence of non-linear interactions between an electromagnetic field and any physical system always suggests the possibility of exciting harmonic electric currents which could re-radiate detectable, possibly useful, energy at harmonics of the frequency of the primary field. From time to time such currents have been traced to some one of the many sources of non-linearity in the field-plasma interactions. Our purpose here will be to add still another source of harmonic currents to the growing list, and to illustrate by a simple example how such currents might produce coherent re-radiation.

Before proceeding to the source we have in mind, let us review briefly a few of the physical processes that have been suggested as possible sources of harmonic currents. It is helpful to keep in mind that if we ignore ion motion, the induced electric current density can be written as $\mathbf{j} = -en\mathbf{v}$, where n is the electron density (the electron charge is $-e$) and $\mathbf{v}$ is the average electron velocity. Accordingly, harmonics in $\mathbf{j}$ may be traced to processes which produce harmonics in n and $\mathbf{v}$. For example, the usually neglected oscillating magnetic vector of the electromagnetic field perturbs the orbit of an electron in such a way as to produce a harmonic spectrum in **v** (Maiman, 1957; Chen, 1962). Similarly, the anisotropy induced in a plasma by a d.c. magnetic field may, in the presence of a propagating wave, give rise to an oscillating space-charge component ($-en$) which can couple to the fundamental and harmonics of **v** (Feinstein, 1950; Ginsburg, 1958). Ignoring magnetic effects, it has been shown that if the collision frequency of the electrons is velocity-dependent, the average velocity **v** contains a third-harmonic component (Vilenskii, 1953; Rosen, 1961). In the case of an electric field strong enough to produce its own ionization, the fact that the ionization frequency is proportional to the absolute value of **v** results in an even harmonic spectrum in the electron density n (Baird and Coleman, 1961). These authors show that an additional d.c. electric field destroys the symmetry of the n-spectrum, and both odd and even harmonics can be produced.

Two further sources of non-linearity have been identified in the presence of non-uniform electron distributions. One of these sources is located at the point of plasma resonance on an electron density gradient, and has its origin in the non-sinusoidal electron oscillations produced by the sharp spatial variation of the electric field at such a point (Försterling and Wüster, 1951). When, however, a relatively uniform electric field moves the electron distribution parallel to the density gradient, the inhomogeneity of the distribution in the neighborhood of a point produces an oscillating component of electron density at the frequency of the driving field. These electron density perturbations have been proposed as a source of non-linear interaction between two electromagnetic fields of different frequencies (Ginsburg, 1958; Wetzel, 1961a) and recent experiments tend to support the existence of such interactions (Dreicer, 1962). In what follows, we will examine the way in which these particular electron density oscillations can contribute to the production of harmonic currents in regions of plasma inhomogeneity. Since all real plasmas are somewhere non-uniform, harmonic currents excited by this mechanism could be expected whenever an electromagnetic field penetrates a plasma.

2. THE HARMONIC CURRENTS IN A NON-UNIFORM IONIZED REGION

It is clear that once one admits non-linearities, they enter in great profusion and in many guises. However, when the role of a particular non-linearity has been examined, its source may provisionally be ignored in the search for other origins of harmonic currents. In this way, a catalog of non-linear interactions can be drawn up, from which the particular physical mechanisms most likely to contribute harmonic radiation in a given situation can easily be identified (Wetzel, 1962). In the calculation below, then, although we will take the exciting field to be an electromagnetic wave, we will ignore the magnetic field vector of the wave and exclude the presence of a d.c. magnetic field. The plasma may be adequately represented by the Langevin equation for the electron velocity, augmented by a phenomenological force due to electron pressure gradients,

$$m\dot{\mathbf{v}} = -e\mathbf{E}(\mathbf{r}, t) - \nu_c m\mathbf{v} - \nabla p/n, \tag{1}$$

and the continuity equation

$$\frac{\partial n}{\partial t} + \nabla \,.\, (n\mathbf{v}) = Q(\mathbf{r}) - S(\mathbf{r}), \tag{2}$$

where m and $-e$ are the mass and charge of the electron, ν_c is the effective collision frequency for momentum transfer, p is the scalar electron pressure and $\mathbf{E}(\mathbf{r}, t)$ is the local electric field. The stationary functions $Q(\mathbf{r})$ and $S(\mathbf{r})$ describe local electron sources and sinks respectively, and are assumed to be independent of the field $\mathbf{E}$, although each of these functions can be a source of non-linearity in other contexts. For simplicity, let us initially take $\mathbf{E}(\mathbf{r}, t)$ to be the electric vector of a transverse plane wave, and write

$$\mathbf{E} = \mathbf{E}_1 \cos(\boldsymbol{\varkappa} \,.\, \mathbf{r} - \omega t),$$

where $\boldsymbol{\varkappa}$ is the propagation vector and ω the angular frequency of the wave, and $\boldsymbol{\varkappa} \,.\, \mathbf{E}_1 = 0$. We further assume that the plasma is isothermal and the electron velocity distribution isotropic, so that $p = nkT$, where k is Boltzmann's constant and T is the electron temperature. Under these conditions, (1) can be written

$$\dot{\mathbf{v}} + \nu_c\mathbf{v} = -(e/m)\, \mathbf{E}_1 \cos(\boldsymbol{\varkappa} \,.\, \mathbf{r} - \omega t) - (kT\nabla n)/mn. \tag{3}$$

Equations (2) and (3) form the basis from which we will proceed.

2.1. *The Steady-State Electron Density*

The steady-state electron density, denoted by $n_0(\mathbf{r})$, may be given either as an empirical function, or as an analytical solution of (2) and (3) with all time-varying quantities put equal to zero. Solving (3) for the stationary electron drift velocity $\mathbf{v}_0$, we get

$$\mathbf{v}_0 = -(kT\nabla n_0)/mn_0\nu_c. \tag{4}$$

Under steady-state conditions, (2) takes the form

$$\nabla \,.\, (n_0\mathbf{v}_0) = Q(\mathbf{r}) - S(\mathbf{r}),$$

which, upon substitution of (4), yields

$$-D_e\nabla^2 n_0(\mathbf{r}) = Q(\mathbf{r}) - S(\mathbf{r}) \tag{5}$$

which must be solved for $n_0(\mathbf{r})$. In many cases the free electron diffusion coefficient, $D_e = kT/m\nu_c$, must be replaced by the ambipolar diffusion coefficient D_a, which takes into account the space-charge forces arising from the separation of the electrons and ions in the plasma. We need not, however, be concerned with the origin of the electron inhomogeneities in what follows, and will imagine $n_0(\mathbf{r})$ to be some given function of position.

2.2. *The Second-Harmonic Current*

Let us begin by ignoring the ∇n term in (3) and solving for $\mathbf{v}_1^0$, the zeroth-order approximation to the fundamental component of the electron velocity. This is easily found to be

$$\mathbf{v}_1^0 = \frac{(e/m)\,\mathbf{E}_1}{(\omega^2 + \nu_c^2)} [\omega \sin (\varkappa \,.\, \mathbf{r} - \omega t) - \nu_c \cos (\varkappa \,.\, \mathbf{r} - \omega t)]. \quad (6)$$

The corresponding fundamental component of the electron density is determined by substituting (6), together with $n_0(\mathbf{r})$, into (2), and setting the stationary quantities Q and S equal to zero:

$$\frac{\partial n_1^0}{\partial t} = -\nabla \,.\, [n_0(\mathbf{r})\, \mathbf{v}_1^0(\mathbf{r}, t)]. \quad (7)$$

Assuming a transverse wave, $\nabla \,.\, \mathbf{v}_1^0 = 0$, and the solution to (7) becomes

$$n_1^0 = -\frac{(e/m)\,\mathbf{E}_1 \,.\, \nabla n_0(\mathbf{r})}{(\omega^2 + \nu_c^2)} \left[\frac{\nu_c}{\omega} \sin (\varkappa \,.\, \mathbf{r} - \omega t) + \cos (\varkappa \,.\, \mathbf{r} - \omega t)\right]. \quad (8)$$

The assumptions underlying (6) and (8) are no longer valid once the existence of an oscillating space-charge is admitted, and if (6) and (8) are to represent the dominant components of the fundamental electron velocity and density, the introduction of n_1^0 must produce only a negligible effect on these variables. There are two groups of such effects: those due to the departure of the driving field from that of a given transverse plane wave, and those due to the dynamical behavior of the electrons in the plasma. The conditions under which these may be ignored are discussed in detail elsewhere (Wetzel, 1962), and only the results will be given here. The longitudinal part of the field will be negligible if

$$\lambda \omega_{pm}^2 / \Lambda(\omega^2 + \nu_c^2) \ll 1, \quad (9)$$

where λ is the free space wavelength of the wave, ω_{pm} is the maximum plasma frequency encountered in the inhomogeneity, and Λ is a length associated with the inhomogeneity by the relation $|\nabla n_0 / n_0| = 1/\Lambda$. The contributions to $\mathbf{v}_1$ and n_1 arising from electron pressure gradients and stationary electron flow can be ignored, provided that,

$$\nu_c l_m^2 / \omega \Lambda^2 \ll 1 \quad \text{and} \quad \nu_c l_m^2 / c\Lambda \ll 1, \quad (10)$$

where l_m is the mean free path of the electron, and c is the speed of light.

The conditions (10) are virtually always satisfied, while that of (9) is simply a reminder that the electric field structure in a plasma inhomogeneity has a complicated form when the inhomogeneity is abrupt and plasma resonance is approached (Budden, 1961). It will be assumed in the following that the above conditions are satisfied.

The dominant component of the second-harmonic electric current is found from the product $-en_1^0\mathbf{v}_1^0$, the result being

$$\mathbf{j}_2^0(\mathbf{r}, t) = e\,\frac{(e/m)^2\,\mathbf{E}_1(\mathbf{E}_1 \,.\, \nabla n_0(\mathbf{r}))}{2(\omega^2+\nu_c^2)^2}\,\omega\left\{\left(1-\frac{\nu_c^2}{\omega^2}\right)\sin(2\varkappa\,.\,\mathbf{r}-2\omega t) - 2\frac{\nu_c}{\omega}\cos(2\varkappa\,.\,\mathbf{r}-2\omega t)\right\}. \quad (11)$$

When $\nu_c \ll \omega$, this expression simplifies to

$$\mathbf{j}_2^0(\mathbf{r}, t) \approx e\,\frac{(e/m)^2\,\mathbf{E}_1(\mathbf{E}_1 \,.\, \nabla n_0)}{2\omega^3}\sin(2\varkappa\,.\,\mathbf{r}-2\omega t). \quad (12)$$

This is the *only* component of the second-harmonic current that can be found without solving for second-harmonic components of n and $\mathbf{v}$. These, however, contribute terms smaller than (12) by the factors given in (10), so we can be fairly confident that the dominant second-harmonic radiation from the interaction being discussed here will be produced by $\mathbf{j}_2^0$ in most cases.

2.3. *Higher-Order Currents*

In order to proceed to the third-harmonic current, it will be necessary to calculate the second-harmonic component of n. Using the product $(n_1^0\mathbf{v}_1^0)$ for the second-harmonic of the electron flow in the continuity equation, we write in analogy to (7),

$$\frac{\partial n_2^0}{\partial t} = -\nabla\,.\,[n_1^0\mathbf{v}_1^0]. \quad (13)$$

For simplicity, let us assume from now on that $\nu_c \ll \omega$, so that

$$\frac{\partial n_2^0}{\partial t} \approx \nabla\,.\left\{\frac{(e/m)^2\,\mathbf{E}_1(\mathbf{E}_1 \,.\, \nabla n_0(\mathbf{r}))}{2\omega^3}\sin(2\varkappa\,.\,\mathbf{r}-2\omega t)\right\}, \quad (14)$$

which, for a transverse field, yields the solution

$$n_2^0(\mathbf{r}, t) \approx \frac{(e/m)^2\,(\mathbf{E}_1 \,.\, \nabla)^2\,n_0(\mathbf{r})}{4\omega^4}\cos(2\varkappa\,.\,\mathbf{r}-2\omega t), \quad (15)$$

where the operator $(\mathbf{E}_1 \,.\, \nabla)^2$ applies only to the argument of $n_0(\mathbf{r})$. The third-harmonic current is obtained from the product $-en_2^0\mathbf{v}_1^0$, which becomes

$$-en_2^0\mathbf{v}_1^0 \approx -\,e\,\frac{(e/m)^3\,\mathbf{E}_1(\mathbf{E}_1 \,.\, \nabla)^2\,n_0(\mathbf{r})}{4\omega^5}\sin(\varkappa\,.\,\mathbf{r}-\omega t)\cos(2\varkappa\,.\,\mathbf{r}-2\omega t). \quad (16)$$

The trigonometric product separates into two parts:

$$\sin(\varkappa \,.\, \mathbf{r} - \omega t) \cos(2\varkappa \,.\, \mathbf{r} - 2\omega t) = \tfrac{1}{2}\sin(3\varkappa \,.\, \mathbf{r} - 3\omega t) - \tfrac{1}{2}\sin(\varkappa \,.\, \mathbf{r} - \omega t)$$

and (16) is therefore the sum of a third-harmonic current and a perturbation of the fundamental current. This is characteristic of non-linear analysis by Fourier expansion, since the product of two harmonic terms of order m and n respectively will generally contribute mixture terms of order $m + n$ and $m - n$. In this way contributions from higher-order terms are constantly fed back into lower order terms, and it is impossible to terminate the analysis at any stage without ensuring that these additional contributions are small. Assuming that they are small in this case, the third-harmonic component of (16) is

$$j_3^0(\mathbf{r}, t) \approx - e \frac{(e/m)^3 \, \mathbf{E}_1(\mathbf{E}_1 \,.\, \nabla)^2 \, n_0(\mathbf{r})}{8\omega^5} \sin(3\varkappa \,.\, \mathbf{r} - 3\omega t). \tag{17}$$

The procedure for finding the higher-order currents in this main sequence is now clear. The current of order p is obtained by multiplying the $(p - 1)$th harmonic of the oscillating space-charge by the fundamental component of the electron velocity ($\mathbf{v}_1^0$). The continuity equation then provides the pth space-charge harmonic which, in turn, produces the $(p + 1)$th harmonic current. The amplitudes of the currents have the following relationship:

$$|\, j_p \,| = \left[\frac{(e/m) \, \mathbf{E}_1 \,.\, \nabla n_0}{2(p - 1) \, \omega^2 \, n_0}\right] |\, j_{p-1} \,|, \tag{18}$$

so that the higher-order currents contain information about the higher-order derivatives of the electron distribution. Unfortunately, these currents drop off very quickly in most cases. This may be illustrated by assuming that $\mathbf{E}_1$ is parallel to ∇n_0, and replacing $|\, \nabla n_0/n_0 \,|$ by $1/\Lambda$. Approximating $(e/m)\, E_1/\omega$ with v_1^0 and writing $c = \omega\lambda/2\pi$, the factor in (18) may be written

$$\frac{1}{4\pi(p - 1)} \left(\frac{\lambda}{\Lambda}\right) \left(\frac{v_1^0}{c}\right), \tag{19}$$

which is generally very small. However, this also ensures that the correction fed back to the $(p - 2)$nd current from the pth may be ignored, since this correction will be proportional to the square of (19).

3. AN EXAMPLE OF COHERENT HARMONIC RE-RADIATION

An extremely simple, yet not completely useless, example of coherent re-radiation from an induced second-harmonic current distribution is the case in which the primary plane wave illuminates a linear transition region between two ionized half-spaces, as in Fig. 1. We must assume that the electron densities in both half-spaces and the frequency of the primary wave are so chosen that $\omega_p^2/\omega^2 \ll 1$ throughout the transition region, hence everywhere. This means that the refraction of the incident wave can be ignored, and that the propagation constant for the second-harmonic waves is just twice that of the primary wave.

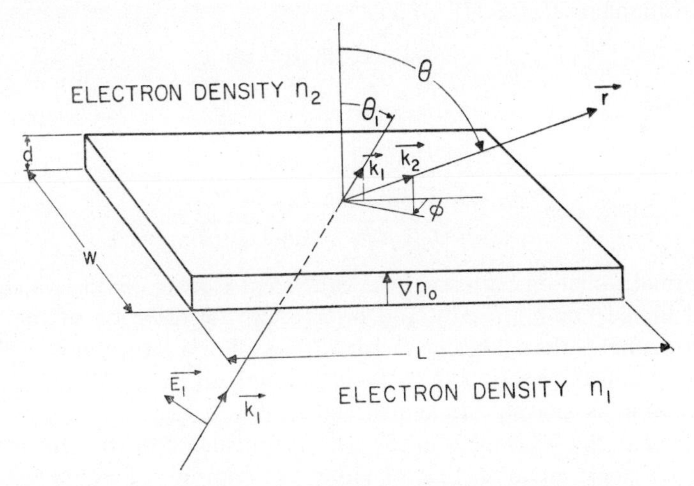

FIG. 1. Illumination of a non-uniformity in an ionized region.

The problem may be solved by finding the Hertz vector produced by the polarization associated with the second-harmonic current distribution. That is, we write

$$\mathbf{\Pi}_2(\mathbf{r}, t) = \frac{1}{4\pi\epsilon} \int_{Vol} \mathbf{P}_2(\mathbf{r}', t) \frac{\exp[ik_2 |\mathbf{r} - \mathbf{r}'|]}{|\mathbf{r} - \mathbf{r}'|} \, dv', \tag{20}$$

where $\mathbf{P}_2$ is the polarization vector derived from the second-harmonic currents, the volume over which the integral is taken is just the volume of these currents, and $k_2 \approx 2k_1$ is the propagation constant for the secondary radiation. If we define a complex current $\mathbf{J}_2$ such that $\mathrm{Re}(\mathbf{J}_2) = \mathbf{j}_2^0$, then $\mathbf{P}_2 = i\mathbf{J}_2/2\omega$ where ω is the frequency of the primary wave. Using the simplified form for $\mathbf{j}_2^0$ given in (12), we find that

$$\mathbf{P}_2 = e \frac{(e/m)^2 \, \mathbf{E}_1(\mathbf{E}_1 \, . \, \nabla' n_0(\mathbf{r}'))}{4\omega^4} \exp[i(2\mathbf{k}_1 \, . \, \mathbf{r}' - 2\omega t)]. \tag{21}$$

Substituting this into (20), and performing the elementary integrations over the rectangular volume shown in Fig. 1, the Hertz vector may be written in the form

$$\mathbf{\Pi}_2(\mathbf{r}) \approx \frac{1}{4\pi\epsilon} \, \mathbf{p}_2 \, \frac{\exp(ik_2 r)}{r}, \quad (k_2 = 2k_1), \quad r \to \infty \tag{22}$$

with an equivalent dipole moment $\mathbf{p}_2$ given by

$$\mathbf{p}_2 = \left[\frac{e^3 E_1^2(n_2 - n_1) \sin \theta_1}{4m^2\omega^4} (wL) F(\theta, \phi)\right] \hat{E}_1. \tag{23}$$

The directionality factor $F(\theta, \phi)$ is

$$F(\theta, \phi) = \frac{\sin \psi_1}{\psi_1} \cdot \frac{\sin \psi_2}{\psi_2} \cdot \frac{\sin \psi_3}{\psi_3}, \tag{24}$$

where

$$\begin{aligned} \psi_1 &= k_1 d (\cos \theta_1 - \cos \theta), \\ \psi_2 &= k_1 w (\sin \theta \sin \phi), \\ \psi_3 &= k_1 L (\sin \theta_1 - \sin \theta \cos \phi). \end{aligned} \tag{25}$$

The normal radiation pattern of the equivalent electric dipole, which is imbedded in the non-uniformity and oriented in the direction of the driving field, is strongly affected by the factor $F^2(\theta, \phi)$. For a large patch of illuminated non-uniformity ($k_1 w, k_1 L \gg 1$), it can be seen from (24) and (25) that the radiation is sharply attenuated unless $\phi \approx 0$ and $\theta = \theta_1$ or $\pi - \theta_1$; that is, unless the observer is in the plane of incidence of the primary wave, at a polar angle equal to that of either the transmitted or the specularly reflected primary wave. If $k_1 d \ll 1$, then $\sin \psi_1/\psi_1 \approx 1$ and the radiation in these two directions is unaffected. As $k_1 d$ increases, however, destructive interference takes place in the direction of the specularly reflected wave ($\theta = \pi - \theta_1$), and when $k_1 d \gg 1$, the re-radiated second-harmonic signal emerges only in the direction of the transmitted primary wave ($\phi = 0, \theta = \theta_1$).

4. CONCLUSIONS

In suggesting another source of harmonic currents, it would be of interest to compare the strength of this source with that of other known sources. This is hard to do in general because of the special conditions underlying the identification of each particular non-linearity. At least one of these, however, has been verified experimentally, and will serve as a useful comparison; namely, the production of second-harmonic currents in a magnetoplasma at cyclotron resonance (Maiman, 1957). With $\mathbf{E}_1$ normal to the d.c. magnetic field, the dominant second-harmonic current found by Naiman (with $\nu_c \ll \omega$) may be written

$$j_{2M} \approx e n_0 \frac{(e/m)^2 E_1^2}{4\nu_c \omega c} = e n_0 \frac{\pi}{2} \frac{(e/m)^2 E_1^2}{\nu_c \omega^2 \lambda}. \tag{26}$$

We have not considered the effect of a d.c. magnetic field on the harmonic currents generated in a non-uniformity, but the effect of such fields on $\mathbf{v}_1^0$ is well known, and the form of n_1^0 in a magnetic field has been discussed in another place (Wetzel, 1961b). Using these results, and assuming the electric field to be parallel to the electron density gradient, we may again form the product $-en_1^0\mathbf{v}_1^0$ and find the second-harmonic current strength at cyclotron resonance to be

$$j_2^0 \approx en_0 \frac{(e/m)^2 E_1^2}{2\nu_c^2 \omega \Lambda}. \tag{27}$$

The ratio of the two current strengths is therefore

$$\frac{j_2^0}{j_{2M}} = \left(\frac{\omega}{\nu_c}\right) \frac{\lambda}{\pi \Lambda}, \tag{28}$$

which can be quite large. Note, however, that j_2^0 exists only over the extent of the inhomogeneity, while j_{2M} is present throughout the plasma. At exciting frequencies well above cyclotron resonance, the factor (ω/ν_c) in (28) is replaced by the number 2.

The harmonic currents produced in a non-uniform ionized region make up a phased source distribution following the geometry of the non-uniformity, and could therefore give rise to interesting and unexpected radiation patterns. However, in common with other sources of induced harmonic radiation in a plasma, the current strengths are small, and decay sharply with decreasing field strength and increasing frequency. For this reason, such effects will be detected most readily under low frequency, high field conditions.

The work described in this paper was supported in part by the Air Force Cambridge Research Laboratories, Office of Aerospace Research, under contract AF 19(604)–4561, and by the Office of Naval Research and the David Taylor Model Basin.

REFERENCES

Baird, J. R. and Coleman, P. D. (1961) Frequency conversion in a microwave discharge, *Proc. I.R.E.* **49**, No. 12, 1890–1900 (December).

Budden, K. G. (1961) *Radio Waves in the Ionosphere*, Cambridge Univ. Press.

Dreicer, H. (1962) Incoherent scattering of microwaves by plasma fluctuations, *Bull. Amer. Phys, Soc.*, Ser. II, 7, No. 2, 151.

Feinstein, J. (1950) Higher order approximations in ionospheric propagation, *J. Geophys. Res.* **55**, 161–170.

Försterling, K. and Wüster, H. (1951) Uber die Entstehung von Oberwellen in der Ionosphäre, *J. Atm. Terr. Phys.* **2**, 22–31.

Ginsburg, V. L. (1958) Non-linear interaction of radio waves propagating in a plasma, *Zhur. Eksptl. i. Teoret. Fiz.* **35**, 1573.

Ginsburg, V. L. and Gurevich, A. V. (1960) Non-linear phenomena in a plasma located in an alternating electromagnetic field, *Usp. Fiz. Nauk*, **70**, 201–246 and 393–428.

Maiman, T. H. (1957) Solid State Millimeter Wave Generation Study, Hughes Aircraft Company Physics Laboratory, Final Report on Contract DA 36–039SC–73063 (June).

Rosen, P. (1961) Generation of the third harmonic by an electromagnetic signal in a plasma, *Phys. Fluids*, **4**, 341–344.

Vilenskii, I. M. (1953) The effect of non-linearities on radio waves propagating in the ionosphere, *Doklady Akad. Nauk SSSR*, **92**, 525–528.

WETZEL, L. (1961) Wave interaction in plasma inhomogeneities, *J. Appl. Phys.* **32**, No. 2, 327–328 (February).

WETZEL, L. (1961) Electric-field-induced anisotropies in an inhomogeneous plasma, *Phys. Rev.* **123**, No. 3, 722–726 (August).

WETZEL, L. (1962) Non-linear Interactions between an Electromagnetic Wave and a Plasma: Excitation of Harmonic Current, Brown University Scientific Report AF4561/19 (December).

IMPEDANCES AND REFLECTION COEFFICIENTS FOR ANISOTROPIC MEDIA†

IRVIN KAY

ABSTRACT

A general definition is given for the impedance of a one-dimensional linear wave propagation system. According to this definition the reflection coefficient is a special case of the impedance in a particular vector coordinate system.

Necessary and sufficient conditions are given for the existence of the impedance in different vector coordinate systems. Conditions for the existence and uniqueness of impedances and reflection coefficients satisfying prescribed boundary conditions are also given. In connection with the concurrent analysis a differential equation for the impedance, originally due to Sternberg and Kaufman, is derived by a method which stems naturally from the point of view adopted in this article.

These results are applied to obtain conditions, equivalent to some obtained by Redheffer, for the conservation of energy in the system.

1. INTRODUCTION

In order to obtain some physical insight, or to simplify a mathematical problem, it is often convenient to replace an electrical structure in a linear system by its input impedance. In making such a replacement one generally assumes that the reflection and transmission properties of the structure can be defined adequately in terms of the impedance. However, except in the case of simple structures which have been analyzed completely, an assumption of this nature may be unjustified, at least from a mathematical point of view. In this article some of the relevant mathematical properties of an input impedance associated with a general one-dimensional wave propagation system (such as an electromagnetic plane wave in an inhomogeneous plane stratified anisotropic medium) will be considered.

From the point of view adopted here the reflection coefficient is itself an impedance matrix relative to a particular vector representation or coordinate system. We shall investigate the relationship between impedance matrices relative to different coordinate systems, and this will include the case where one of the impedances is the reflection coefficient. The question of when an impedance matrix relative to a given coordinate system exists will be raised and settled.

† Sponsored by the Mathematics Research Center, U.S. Army, Madison, Wisconsin, under Contract No. DA-11-022-ORD-2059.

Various authors have dealt, in terms of the impedance concept, with generalizations of the simple linear system characterized by the uniform two-wire transmission line. For example, Rice[(1)] has investigated the properties of the uniform $2n$ wire line, patterning his treatment after the standard analysis of the two-wire line. Pierce[(2)], Walker and Wax[(3)] have derived differential equations for the impedance and reflection coefficient of a non-uniform two-wire line. Similar equations have been obtained for the non-uniform $2n$ wire line by Sternberg and Kaufman[(10)]. Using a different procedure we shall derive equivalent differential equations for the case of a somewhat more general one-dimensional linear system. These differential equations will be used here to connect the impedance matrices relative to different coordinate systems.

In Section 4, as an application of the general analysis of the earlier sections, we shall give necessary and sufficient conditions on the distributed electrical parameters of the system for the conservation of energy. These results are similar to some obtained by Redheffer[(11)], who derived differential equations for the elements of the scattering matrix and used them to discuss the conservation and loss of energy for $2n$ wire transmission lines.

2. IMPEDANCES

We shall be concerned with a $2n$ dimensional vector space of electromagnetic vectors E, which we may regard as column vectors having $2n$ components. These vectors are a sum $I + V$ of two vectors from complementary subspaces of dimension n each. Corresponding to this decomposition we define the projection matrices P and $Q = 1 - P$, where 1 is the identity matrix, such that

$$PE = I, \; QE = V. \tag{1}$$

As projectors the matrices P and Q satisfy

$$P^2 = P, \; Q^2 = Q\,, \tag{2}$$

and hence from (1)

$$PQ = QP = 0 \text{ (the zero matrix)}\,. \tag{3}$$

In an appropriate representation the vectors I, V and the matrices P, Q have the forms

$$I = \begin{pmatrix} 0 \\ i \end{pmatrix}, \; V = \begin{pmatrix} v \\ 0 \end{pmatrix}, \; P = \begin{pmatrix} 0 & 0 \\ 0 & 1 \end{pmatrix}, \; Q = \begin{pmatrix} 1 & 0 \\ 0 & 0 \end{pmatrix}, \tag{4}$$

where single elements in (4) stand for n dimensional column vectors in I and V and $n \times n$ matrices in P and Q. We shall use P and Q for the matrices explicitly indicated in (4) throughout this article regardless of the representation being discussed. Also, if a $2n \times 2n$ matrix M has three distinct $n \times n$ zero submatrices the non-trivial fourth $n \times n$ submatrix of M will be denoted by the symbol ${}_nM_n$. That is, if M has any of the forms

$$M = \begin{pmatrix} X & 0 \\ 0 & 0 \end{pmatrix}, \begin{pmatrix} 0 & X \\ 0 & 0 \end{pmatrix}, \begin{pmatrix} 0 & 0 \\ X & 0 \end{pmatrix}, \begin{pmatrix} 0 & 0 \\ 0 & X \end{pmatrix},$$

then

$${}_nM_n = X.$$

Thus, for example,

$${}_nP_n = 1 \text{ and } {}_nQ_n = 1\ ,$$

where 1 stands for the $n \times n$ identity matrix.

The network or transmission line supporting an electromagnetic vector is a restriction on the vectors which the system can support, in general, so that there will be a proper subspace of the $2n$ dimensional space consisting of those vectors which are admissible in a given physical situation. This subspace is characterized by an impedance which is a matrix Z of the form

$$Z = QZP \tag{5}$$

such that for every admissible E

$$V = ZI\ , \tag{6}$$

or by (1)

$$QE = Z(PE) = ZE\ . \tag{7}$$

From (2), (3) and (7) it follows that $P + Z$ is a projection Π from the full vector space into the space of all admissible vectors; i.e. if E is admissible

$$(P + Z)E = \Pi E = E\ , \tag{8}$$

and since

$$\Pi^2 = \Pi \tag{9}$$

any vector of the form ΠE is admissible.

On occasion it is convenient to transform the electromagnetic vector space into an equivalent vector space, where certain physical properties may become more apparent. This is done usually by applying a one-to-one transformation, that is, by multiplying each vector on the left by a given non-singular $2n \times 2n$ matrix T,

$$\hat{E} = TE\ .$$

We can then define a new subspace of admissible vectors by defining a new impedance $\hat{Z}$ of the form (5),

$$\hat{Z} = Q\hat{Z}P\ . \tag{10}$$

Then the new admissible vectors will all satisfy

$$\hat{\Pi}\hat{E} = (P + \hat{Z})\hat{E} = \hat{E}\ , \tag{11}$$

and $\hat{Z}$ will have properties analogous to those of Z.

Definition: We shall say that $\hat{Z}$ is compatible with Z if

$$E = IIE$$

implies

$$\hat{E} = \hat{I}\hat{I}\hat{E} \, .$$

According to this definition a necessary and sufficient condition that $\hat{Z}$ be compatible with Z is that

$$\hat{I}\hat{I}TII = TII. \tag{12}$$

By applying P and Q separately to (12) we see that (12) is completely equivalent to

$$\hat{Z}TII = QTII \, . \tag{13}$$

On interchanging Z and $\hat{Z}$ and T and T^{-1} we obtain the analogous definition of the compatibility of Z with $\hat{Z}$. It is clear from (12) that T maps the null space of $1 - II$ into the null space of $1 - \hat{I}\hat{I}$. Since these two matrices each have rank n (because of (5)) the mapping is one to one, and hence T^{-1} must map the null space of $1 - \hat{I}\hat{I}$ into the null space of $1 - II$. Thus

$$IIT^{-1}II = T^{-1}\hat{I}\hat{I}$$

and it follows that whenever $\hat{Z}$ is compatible with Z then Z must be compatible with $\hat{Z}$.

A condition for the compatibility of $\hat{Z}$ with Z, depending only on Z and T is easy to find.

THEOREM 1. *A necessary and sufficient condition that there exist a suitably defined $\hat{Z}$ compatible with Z is that the $n \times n$ submatrix ${}_n[(Q - Z)T^{-1}Q]_n$, which acts essentially on the subspace of vectors QE, have an inverse, i.e. that there exist a matrix QKQ such that*

$$QKQ(Q - Z)T^{-1}Q = (Q - Z)T^{-1}Q(QKQ) = Q \, .$$

When this condition is satisfied $\hat{Z}$ is determined uniquely by Z and T. In fact $\hat{Z}$ will be given by

$$\hat{Z} = Q + QAQ(Q - Z)T^{-1} \tag{14}$$

where QAQ is the unique solution of

$$QAQ(Q - Z)T^{-1}Q + Q = 0 \, . \tag{15}$$

Proof. To prove the necessity we start by assuming that there is a $\hat{Z}$ which satisfies (13) and (10). We then add the matrix $\hat{Z}T(Q - Z)$ to both sides of (13) to obtain

$$\hat{Z}T = QTII + \hat{Z}T(Q - Z) \, . \tag{16}$$

We now add and subtract the matrix $QT(Q - Z)$ on the right side of (16) to obtain

$$\hat{Z}T = QT + (\hat{Z}T - QT)(Q - Z)$$

which is equivalent to

$$\hat{Z} = Q + (\hat{Z}T - QT)(Q - Z)T^{-1}. \tag{17}$$

We now define

$$QAQ = (\hat{Z}T - QT)Q \tag{18}$$

so that we can write in place of (17)

$$\hat{Z} = Q + QAQ(Q - Z)T^{-1}. \tag{19}$$

After multiplying (19) on the right by Q we have from (10)

$$QAQ(Q - Z)T^{-1}Q + Q = 0\,. \tag{20}$$

If QAQ satisfying (20) exists it must have a submatrix ${}_n(QAQ)_n$ with an inverse, the matrix

$$-{}_n[(Q - Z)T^{-1}Q]_n\,.$$

The necessity of the condition in the theorem now follows.

To prove the sufficiency we suppose that ${}_n[(Q - Z)T^{-1}Q]_n$ has an inverse and define QAQ as the (unique) solution of (20). Let $\hat{Z}$ be defined by (19). Then, since by (5)

$$ZQ = 0$$

we have

$$(\hat{Z}T - QT)Q = QAQ\,.$$

Thus (17) holds with this definition of $\hat{Z}$, and by reversing the steps in the first part of this proof we can arrive at (13) and hence (12). To see that $\hat{Z}$ has the form (10) we observe from (19) that

$$P\hat{Z} = 0$$

and from (19) and (20) that

$$\hat{Z}Q = 0\,.$$

The uniqueness of Z follows from the uniqueness of QAQ as a solution of (20). Thus the whole theorem is proved.

As an example let us consider the uniform transmission line equations for a two-dimensional vector E,

$$E' = WE \tag{21}$$

where the prime indicates differentiation with respect to distance x along the line. Since the line is uniform W will be a 2×2 matrix with constant elements,

$$W = -i\omega\begin{pmatrix} 0 & L \\ C & 0 \end{pmatrix},$$

in which ω is the frequency, L the distributed inductance and C the distributed capacitance. The impedance has the form

$$Z = \begin{pmatrix} 0 & Z \\ 0 & 0 \end{pmatrix}.$$

If we apply the linear transformation T which diagonalizes W the vector E goes into a vector $\hat{E}$ whose components represent a downgoing wave d and an upgoing wave u.‡ The impedance decomposition in this representation gives d in terms of u, and hence the impedance $\hat{Z}$ must be a reflection coefficient. To emphasize this fact the letter R will be used in place of $\hat{Z}$ whenever we encounter such a representation. The transformation is given by

$$T = \begin{pmatrix} \frac{1}{2} & -\zeta/2 \\ \frac{1}{2} & \zeta/2 \end{pmatrix}, \qquad T^{-1} = \begin{pmatrix} 1 & 1 \\ -\zeta^{-1} & \zeta^{-1} \end{pmatrix},$$

where $\zeta = (L/C)^{1/2}$ is the characteristic impedance of the line. The matrix $(Q - Z)T^{-1}Q$ of Theorem 1 is given by

$$(Q - Z)T^{-1}Q = \begin{pmatrix} (\zeta + z)/\zeta & 0 \\ 0 & 0 \end{pmatrix}.$$

The matrix QAQ can be obtained by solving (15). Thus

$$QAQ = \begin{pmatrix} -\zeta/(\zeta + z) & 0 \\ 0 & 0 \end{pmatrix}.$$

The reflection coefficient $R = \hat{Z}$ now follows from (14),

$$R = \begin{pmatrix} 0 & (z - \zeta)/(z + \zeta) \\ 0 & 0 \end{pmatrix}. \tag{22}$$

The relation (22) is equivalent to the usual expression for the terminal reflection coefficient of a uniform 2-wire transmission line in terms of its characteristic and terminal impedances.

3. THE NON-UNIFORM TRANSMISSION LINE

We now consider the impedance associated with a non-uniform $2n$ wire transmission line; i.e. one whose distributed parameters are functions of distance x along the line. The impedance Z is now a function of x, but the relations (5), (7), (8) and (9) still hold. We suppose that the vector E satisfies the transmission line equation

$$E'(x) = W(x)E(x) \tag{23}$$

where $W(x)$ is a $2n \times 2n$ matrix whose elements are continuous functions of x.

The question of the existence of an impedance $Z(x)$ for such a non-uniform transmission line now arises. This question is answered by the following theorem.

THEOREM 2. *Corresponding to a given constant $2n \times 2n$ matrix $Z(0)$ of the form* (5) *and any solution $E(x)$ of* (23) *for which* (8) *holds at $x = 0$, there exists a*

‡ Cf. Section 4.

unique continuously differentiable matrix $Z(x)$ *of the form* (5) *such that* (8) *holds for all* x.

Proof. We begin by finding a necessary condition for $Z(x)$. If $Z(x)$ exists, by (8) we must have

$$E' = Z'E + IIE'.$$

According to (23) this is equivalent to

$$Z'IIE + IIWIIE = WIIE. \tag{24}$$

Using (24) as a guide, we shall now define a matrix $Z(x)$ having the form (5) and the initial value $Z(0)$ at $x = 0$ as the solution of the differential equation

$$Z' + ZWII = QWII.$$

By adding $PW\,II$ to both sides we see that this is equivalent to

$$Z' + IIWII = WII. \tag{25}$$

We now define a vector U of the form

$$U(x) = PU(x)$$

such that

$$U(0) = PE(0) \tag{26}$$

and

$$U' = PWIIU. \tag{27}$$

Then by (5) and the definition of II,

$$IIU' = IIWIIU. \tag{28}$$

If we substitute from (25) on the right of (28) we obtain

$$Z'U + IIU' = WIIU.$$

This can be written

$$(IIU)' = W(IIU). \tag{29}$$

Now for the chosen matrix $Z(x)$ which satisfies (25) and the given initial condition we can define a class of vectors $E(x)$ such that (8) holds for all x. To do this we let the initial value of the solution U of (27) be given by (26) with $E(0)$ arbitrary. The desired class of vectors $E(x)$ is then defined to be such that each vector in the class is given by

$$E(x) = II(x)U(x) \tag{30}$$

for all possible choices of $U(0)$ under the condition (26). According to (29) the vector $E(x)$ defined by (30) satisfies (23). Because of (9), $E(x)$ satisfies (8). Thus the matrix $Z(x)$ defined as a solution of (25) is the impedance we desire. The existence of $Z(x)$ has therefore been proved.

To prove the uniqueness of $Z(x)$ we write $E(x)$ in the form

$$E = F(x)II(0)E_0, \tag{31}$$

where E_0 is an arbitrary vector and $F(x)$ is a $2n \times 2n$ matrix satisfying the differential equation

$$F' = WF \tag{32}$$

and the initial condition

$$F(0) = 1\,. \tag{33}$$

From the standard theory of linear differential equations we have the fact that $F(x)$ is non-singular for all x. Thus since $\Pi(0)$ has rank n, as we can see from (5) and the definition of P and Q, the matrix $F(x)\Pi(0)$ is a matrix solution of (23) having rank n. Now since by hypothesis

$$\Pi E = E$$

which is the condition (8), we have from (31)

$$\Pi(x)F(x)\Pi(0)E_0 = F(x)\Pi(0)E_0$$

for all E_0. Therefore

$$\Pi(x)F(x)\Pi(0) = F(x)\Pi(0)$$

and hence

$$\Pi(x)F(x)\Pi(0) = \Pi(x)PF(x)\Pi(0)$$

has rank n. Thus $PF(x)\Pi(0)$ has rank n.

Now if there were two matrices $Z_1(x)$ and $Z_2(x)$ such that

$$[P + Z_1(x)]E(x) = E(x)$$

and

$$[P + Z_2(x)]E(x) = E(x)\,,$$

then we should have, using (5) and (31),

$$\{Q[Z_1(x) - Z_2(x)]P\}\{PF(x)[P + Z(0)]P\} = 0\,.$$

But since $PF(x)[P + Z(0)]P = PF(x)\Pi(0)$ has rank n and both $Q[Z_1 - Z_2]P = Z_1 - Z_2$ and $PF(x)\Pi(0)$ are effectively $n \times n$ matrices this implies that

$$P[Z_1(x) - Z_2(x)]Q = Z_1(x) - Z_2(x) = 0\,.$$

Therefore, the impedance is unique, and the proof of the theorem is completed.

Equation (25) can be written in a more symmetrical form

$$\begin{aligned} \Pi' + \Pi W \Pi &= W\Pi \\ \text{or} \qquad \Pi' &= \Gamma W \Pi \end{aligned} \tag{34}$$

where Γ is the complementary projector $1 - \Pi$ having the properties

$$\Gamma^2 = \Gamma,\ \Gamma\Pi = \Pi\Gamma = 0\,.$$

Equation (34) also takes a simple form in terms of Z,

$$Z' = (Q - Z)W(P + Z). \tag{35}$$

It is possible to construct a linear differential equation whose solution leads to a solution of (34). First we recall that for any differentiable non-singular matrix A

$$(A^{-1})' = -A^{-1}A'A^{-1},$$

which follows from a differentiation of the identity

$$AA^{-1} = 1 .$$

We now define a matrix Φ as a non-singular solution of

$$\Phi' = W\Pi\Phi . \tag{36}$$

If W is non-singular we substitute for Π from (36) into (34) and obtain the equation

$$\Phi'' = (W'W^{-1} + W)\Phi',$$

which is a linear first-order differential equation in Φ'. Of course Φ can be obtained from Φ' by a quadrature, and Π can be obtained from Φ and Φ' by solving (36) algebraically. One must select the two arbitrary constant matrices appearing in the general solution for Φ and Φ' so that Φ is non-singular and so that the initial value of Π is correct.§

If W is singular the procedure cannot be carried out directly. However, eqn (23) can be transformed into one of similar form by setting

$$E = e^{-\lambda x}\hat{E} ,$$

λ constant. Equation (23) then becomes

$$\hat{E}' = (W - \lambda 1)\hat{E} ;$$

hence W is effectively transformed into $W + \lambda 1$ while Π does not change. If λ is chosen sufficiently large $W + \lambda 1$ will be non-singular in any desired interval where W has bounded elements. This result could have been obtained directly from (34) which, because Γ and Π are complementary projectors, does not change if we replace W by $W + \lambda 1$.

Suppose we have a non-uniform transmission line with an associated impedance $Z(x)$. Suppose that there is given a non-singular transformation from $E(x)$ to $\hat{E}(x)$ which determines an associated impedance $\hat{Z}(x)$. If $\hat{Z}(0)$ is compatible with $Z(0)$ will $\hat{Z}(x)$ be compatible with $Z(x)$ everywhere along the line? This is a natural question whose answer is provided in the following theorem.

THEOREM 3. *Let $W(x)$ in* (23) *be a continuous* $2n \times 2n$ *matrix. Let $T(x)$ be a non-singular continuously differentiable* $2n \times 2n$ *transformation matrix which takes $E(x)$ into $\hat{E}(x)$. Let $Z(x)$ be the impedance matrix of $E(x)$. Then there exists an impedance matrix $\hat{Z}(x)$ of $\hat{E}(x)$ which is compatible with $Z(x)$ everywhere whenever there is a $\hat{Z}(x_0)$ compatible with $Z(x_0)$ at any point x_0.*

§ The relation of the linear equation (37) to (34) is contained in the work of Levin[4].

Proof. Let $\hat{\Pi}(x)$ be the impedance projection matrix corresponding to $Z(x)$ and $\hat{\Pi}x)$ the projection matrix corresponding to $\hat{Z}(x)$. Then we have an equation which defines $\Pi(x)$ analogous to (34)

$$\hat{\Pi}' = \hat{\Gamma}\,\hat{W}\hat{\Pi}\,, \tag{37}$$

where

$$\hat{W} = TWT^{-1} + T'T^{-1}. \tag{38}$$

We define a matrix K,

$$K = \hat{\Pi}T\Pi - T\Pi. \tag{39}$$

From the trivial identity

$$\hat{\Gamma}T'\Pi + \hat{\Pi}T'\Pi = T'\Pi$$

we obtain

$$\hat{\Gamma}T'T^{-1}\hat{\Pi}T\Pi + \hat{\Pi}T'\Pi - \hat{\Gamma}T'T^{-1}K = T'\Pi\,. \tag{40}$$

From the identity

$$\Gamma TW\Pi + \hat{\Pi}TW\Pi - T\Pi W\Pi = TW\Pi - T\Pi W\Pi$$

we obtain

$$\Gamma TWT^{-1}\hat{\Pi}T\Pi + \hat{\Pi}T\Gamma W\Pi - \Gamma TWT^{-1}K + KW\Pi = T\Gamma W\Pi. \tag{41}$$

If we now add (40) and (41), making use of (34) and (37), we obtain the differential equation

$$K' - \Gamma\,\hat{W}K + KW\Pi = 0\,. \tag{42}$$

If the initial condition

$$K(x_0) = 0 \tag{43}$$

is satisfied for some x_0, (42) implies that

$$K(x) = 0$$

for all x; that is

$$\hat{\Pi}T\Pi = T\Pi$$

for all x. This is the necessary and sufficient condition (12) that $\hat{Z}(x)$ be compatible with $Z(x)$. Since (43) is the condition that $\hat{Z}(x_0)$ be compatible with $Z(x_0)$ the theorem follows.

4. CONSERVATION OF ENERGY

The transmission line can be regarded as an instrument for transferring energy from one point to another. In general, energy may flow in two directions, upward in the direction of increasing x, or downward in the direction of decreasing x. This implies that the vector E decomposes into a sum of two vectors, one representing a wave traveling upward, the other representing a wave traveling downward. It seems reasonable to represent this decomposition of E by means of an impedance in the manner described in Section 2. Thus, if we let U and D be the upgoing and downcoming components of E we have

$$E = U + D \qquad D = RU,$$

where the letter R has been used instead of Z for the impedance in order to emphasize the special physical meaning we are attaching to this decomposition of E. Accordingly we shall refer to R as the reflection coefficient rather than the impedance. Nevertheless, R will have all of the general properties of an impedance described previously. The set of all upgoing vectors U and the set of all downcoming vectors D form disjoint n dimensional subspaces. In some representation they will have the forms

$$U = PE \qquad D = QE$$

and the matrix R will then have the form (5),

$$R = QRP. \|$$

The energy of the electromagnetic vector can be defined as the norm E^*E. The energy of the upgoing and downcoming waves will be U^*U and D^*D, or in the appropriate representation, E^*PE and E^*QE. In terms of the reflection coefficient the energy in the downcoming wave is

$$D^*D = U^*R^*RU = E^*R^*RE$$

since

$$R^* = PR^*Q \, .$$

It is clear from the properties of P and Q that the total energy at a point is just a sum of the upgoing and downcoming contributions

$$E^*E = E^*PE + E^*QE = E^*PE + E^*R^*RE \, .$$

If the independent variable x had the physical meaning of time instead of distance we might expect to have a simple energy conservation law in a passive lossless system, namely, $E^*E =$ constant. Mathematically, this law would follow if the matrix W in (23) were anti-hermitian, i.e. if $W^* = -W$. However, since x has the physical meaning of distance rather than time we have no reason to expect such a law to hold. We must instead formulate a law which refers to the transfer of energy from one point on the line to another.

If there are no losses or energy inputs in the line and all of the upgoing energy is reflected at some point x_0 we expect physically that all of the upgoing energy will be reflected at any other point x along the line. The mathematical

|| Cf. the example of Section 2. Several authors have defined upgoing and downcoming waves in the case of a slowly varying medium (Bremmer[5], Schelkunoff[6], Landauer[7], Keller[8, 9]). The definition of Keller, in which the appropriate representation is obtained by applying a non-singular transformation T which diagonalizes W just as in the case of a uniform line, is the most general of these. Keller's definition includes the $2n$ wire line whereas the definition of the others applies just to the 2-wire line and is therefore a special case of Keller's.

translation of this requirement is that if the non-trivial $n \times n$ submatrix of R is "unitary"¶ at any point x_0, i.e.

$$R^*R = P$$

which implies

$$RR^* = Q({}_nR^*_{nn}R_n = {}_nR_{nn}R^*_n = 1)\,,$$

then it remains unitary at every point x along the line. We shall call this *the principle of total reflection.*

We can state an apparently more inclusive conservation law, however. Consider the section of the transmission line between any two points x and x_0, $x_0 > x$. We can say that energy is conserved in this section if the sum of the upgoing energy at x_0 and the downcoming energy at x is equal to the sum of the downcoming energy at x_0 and the upgoing energy at x.† That is, energy is conserved when

$$E^*(x)PE(x) + E^*(x_0)R^*(x_0)R(x_0)E(x_0) = \\ = E^*(x_0)PE(x_0) + E^*(x)R^*(x)R(x)E(x)\,.$$

The meaning of this requirement is that the energy fed into the section from both ends is equal to the energy flowing out of the section at both ends.

We shall require that energy be conserved in any section of a passive lossless line. Thus we can formulate the law of conservation of energy for the line as a whole simply by requiring

$$E^*(x)PE(x) - E^*(x)R^*(x)R(x)E(x) = \text{constant.} \tag{44}$$

To guarantee (44) there exists a necessary and sufficient condition on the matrix W in the transmission line equation (23). Moreover, the same necessary and sufficient condition holds for the principle of total reflection. That is, the principle of total reflection and the law of conservation of energy are logically equivalent. In demonstrating these facts we may treat the principle of total reflection as a lemma which will help establish the law of conservation of energy.

LEMMA. *A necessary and sufficient condition that the principle of total reflection hold is that the bordering $n \times n$ submatrices of which W is composed satisfy the following relations,*

$$(L)\quad \begin{aligned} &(PWP)^* = \lambda P - PWP,\ (QWQ)^* = \lambda Q - QWQ \text{ (anti-hermitian)},\\ &(PWQ)^* = QWP,\ (QWP)^* = PWQ \text{ (hermitian)}, \end{aligned}$$

where λ is an arbitrary complex scalar.

¶ Henceforth we shall often call such a matrix R, itself, unitary and, in fact treat it as a transformation from a vector space in which P is the identity to one in which Q is the identity.

† If we regard the upgoing wave at x and the downcoming wave at x_0 as the two components of a vector representing an *incoming* wave and the downcoming wave at x and upgoing wave at x_0 as the two components of an outgoing wave, we can define a scattering matrix which transforms an incoming wave into an outgoing wave. Then requiring that energy be conserved is the same as requiring the scattering matrix to be unitary.

Proof. First let us write (35), the impedance differential equation, in terms of R and the submatrices of W. We have

$$R' + RPWP - QWQR + RPWQR - QWP = 0\,. \tag{45}$$

If we take the hermitian adjoint of (45) we obtain a differential equation for R^*,

$$R^* + (PWP)^*R^* - R^*(QWQ)^* + R^*(PWQ)^*R^* - (QWP)^* = 0\,. \tag{46}$$

We now define $R_{-1}(x)$ to be the solution of the differential equation

$$R'_{-1} = -R_{-1}R'R_{-1} \tag{47}$$

subject to the initial condition

$$R_{-1}(x_0) = R^*(x_0)\,. \tag{48}$$

If we make the hypothesis of the lemma that at some value x_0

$$R^*(x_0)R(x_0) = P \tag{49}$$

then (48) implies

$$R_{-1}(x_0)R(x_0) = P. \tag{50}$$

Now by (47) we have

$$(R_{-1}R)' = R_{-1}R'(P - R_{-1}R)\,. \tag{51}$$

Let $S(x)$ be the solution of the differential equation

$$S' = R_{-1}R'(P - S) \tag{52}$$

subject to the initial condition

$$S(x_0) = P. \tag{53}$$

From the uniqueness of the solution of (52) subject to (53) we obtain the result

$$S(x) = P.$$

A comparison of (51), (50), (53) shows that

$$R_{-1}(x)R(x) = P. \tag{54}$$

It follows at once from the form (5) which R must also take that

$$R(x)R_{-1}(x) = Q \tag{55}$$

since ${}_nR_{-1n}$ must be the inverse of ${}_nR_n$.

Now we multiply (45) on the left and on the right by R_{-1} and also by the scalar -1 and use (47) to obtain

$$R'_{-1} - PWPR_{-1} + R_{-1}QWQ - PWQ + R_{-1}QWPR_{-1} = 0\,. \tag{56}$$

If (L) is satisfied a comparison of (56) with (46) shows that R_{-1} and R^* satisfy the same differential equation. If they are also equal at a point x_0 then they

14

must be equal for all x along the line, by the uniqueness of the solution of the initial value problem. Thus the sufficiency part of the lemma is proved.

To prove the necessity, we assume that the principle of total reflection holds and observe that R_{-1} must satisfy equation (56), where now R_{-1} is defined by (54) and (55). On the other hand R^* still satisfies equation (46). We subtract (56) from (46), assuming that $R_{-1} = R^*$, and obtain

$$[(PWP)^* + PWP]R^* - R^*[(QWQ)^* + QWQ] + \\ + R^*[(PWQ)^* - QWP]R^* = (QWP)^* - PWQ. \quad (57)$$

Now (57) must hold for R^* equal to any unitary matrix at any point x since in the principle of total reflection we may select any value of x as the initial point. We can therefore select any unitary R^* and then $-R^*$. If we add the results of these two substitutions in (57) we obtain

$$R^*[(PWQ)^* - QWP]R^* = (QWP)^* - PWQ. \quad (58)$$

If we replace R^* in (58) by iR^* and add the result to (58) we obtain

$$(QWQ)^* = PWQ. \quad (59)$$

We now let R^* be the unitary matrix (see note, p. 386)

$$\begin{pmatrix} 0 & 0 \\ 1 & 0 \end{pmatrix},$$

where the elements are $n \times n$ submatrices. Then we have from (57) for the $n \times n$ non-trivial submatrices

$$\begin{aligned} {}_n(PWP)_n^* + {}_n(PWP)_n = {}_n(QWQ)_n^* + {}_n(QWQ)_n = A \\ A({}_nR_n^*) - ({}_nR_n^*)A = 0 \end{aligned} \quad (60)$$

where R^* is any unitary (see note, p. 386) matrix. By Schur's lemma, then, the matrix A must be proportional to the $n \times n$ identity matrix. Hence we can write, for some complex scalar λ,

$$A = \lambda 1. \quad (61)$$

Our lemma now follows from (61), (60) and (59).

By the use of our lemma we can prove that conservation of energy is equivalent to the principle of total reflection. In fact, we have.

THEOREM 4. *Condition* (L) *is necessary and sufficient for the conservation of energy relation* (44) *to hold.*

Proof. We can write (44) in the form

$$E^*(P - R^*)E = \text{constant}. \quad (62)$$

If (62) holds then, in particular, when R is initially unitary the constant is zero, and R must be unitary for all x. From our lemma the necessity of condition (L) for this theorem follows at once.

To prove sufficiency we differentiate the expression on the left of (62) to obtain

$$E^{*\prime}(P - R^*R)E + E^*(P - R^*R)E' - E^*(R^{*\prime}R + R^*R')E$$

which, after use has been made of (8), (23), and (25), becomes

$$
\begin{aligned}
E^*\Pi^*\{W^*(P-R^*R)+(P-R^*R)W-(\Pi^*W\Gamma^*R+R^*\Gamma W\Pi)\}\Pi E \\
&= E^*\{\Pi^*W^*(P-R)+(P-R^*)W\Pi\}E \\
&= E^*\{(P+R^*)W^*(P-R)+(P-R^*)W(P+R)\}E\,.
\end{aligned}
$$

Now if we apply condition (L) and use the fact that R has the form (5) we see that this expression vanishes, and the theorem is proved.

As an example let us consider the case of a plane wave propagating in an inhomogeneous anisotropic electromagnetic medium. We shall assume that ε and μ are both diagonal tensors and depend only on the distance variable x. We shall also assume that the plane waves propagate in the x-direction and that E and H are polarized in a plane perpendicular to the x-axis. These transverse components satisfy a system of differential equations of the form (23), where

$$
E=\begin{pmatrix}E_2\\E_3\\H_2\\H_3\end{pmatrix} \text{ and } W=-i\omega\begin{pmatrix}0&0&0&\mu_3\\0&0&-\mu_2&0\\0&-\varepsilon_3&0&0\\\varepsilon_2&0&0&0\end{pmatrix}. \tag{63}
$$

The x-component of the average power $\frac{1}{2}$ Re $(E^*\times H)$ (the Poynting vector) can be expressed in the form

$$
E^*\hat{P}E \tag{64}
$$

where

$$
\hat{P}=\begin{pmatrix}0&0&0&1\\0&0&-1&0\\0&-1&0&0\\1&0&0&0\end{pmatrix}.
$$

For convenience of notation we define the "characteristic impedance" elements

$$
\zeta_{ij}=(4\varepsilon_i/\mu_j)^{1/4},\quad i=2,3,j=2,3\,.
$$

We also define the transformation matrices

$$
\begin{aligned}
T&=\begin{pmatrix}\zeta_{23}/2&0&0&\zeta_{23}^{-1}\\0&\zeta_{32}/2&-\zeta_{32}^{-1}&0\\\zeta_{23}/2&0&0&-\zeta_{23}^{-1}\\0&\zeta_{32}/2&\zeta_{32}^{-1}&0\end{pmatrix}\\
T^{-1}&=\begin{pmatrix}\zeta_{23}^{-1}&0&\zeta_{23}^{-1}&0\\0&\zeta_{32}^{-1}&0&\zeta_{32}^{-1}\\0&-\zeta_{32}/2&0&\zeta_{32}/2\\\zeta_{23}/2&0&-\zeta_{23}/2&0\end{pmatrix}.
\end{aligned} \tag{66}
$$

It is easily verified that T diagonalizes the matrix W in (63) and thus provides a representation in which the impedance can be interpreted as a reflection coefficient. At the same time T reduces the quadratic form (64) to the diagonalized form (44) in terms of

$$\hat{E} = TE.$$

According to (38) the transmission line equation is transformed by T to one associated with the distributed parameter matrix $\hat{W}$ replacing W in (23) and given by

$$\hat{W} = \begin{pmatrix} -i\omega\sqrt{\varepsilon_2\mu_3} & 0 & \zeta'_{23}/\zeta_{23} & 0 \\ 0 & -i\omega\sqrt{\varepsilon_3\mu_2} & 0 & \zeta'_{32}/\zeta_{32} \\ \zeta'_{23/23} & 0 & i\omega\sqrt{\varepsilon_2\mu_3} & 0 \\ 0 & \zeta'_{32/32} & 0 & i\omega\sqrt{\varepsilon_3\mu_2} \end{pmatrix}. \quad (67)$$

It is easily seen by inspection of (67) that the submatrices of $\hat{W}$ satisfy the condition (L). Hence conservation of energy is assured for this medium.

REFERENCES

1. RICE, S. O. (1941) Steady state solutions of transmission line equations, *Bell Syst. Tech. J.* **20**, 131–178.
2. PIERCE, J. R. (1943) A note on the transmission line equation in terms of impedance, *Bell Syst. Tech. J.* **22**, 263–265.
3. WALKER, L. R. and WAX, N. (1946) Non-uniform transmission lines and reflection coefficients, *J. Appl. Phys.* **17**, 1043–1045.
4. LEVIN, J. J. (1959) On the matrix Riccati equation, *Proc. Am. Math. Soc.* **10**, 4, 519–524.
5. BREMMER, H. (1951) The W.K.B. approximation as the first term of a geometrical optics series, Symposium on the Theory of Electromagnetic Waves, Interscience, pp. 169–179.
6. SCHELKUNOFF, S. A. (1951) Remarks concerning wave propagation in stratified media, Symposium on the Theory of Electromagnetic Waves, Interscience pp. 181–191.
7. LANDAUER, R. (1951) Reflections in one dimensional wave mechanics, *Phys. Rev.* **82**, 2, 80–83.
8. KELLER, J. B. and KELLER, H. B. (1951) On systems of linear ordinary differential equations, N.Y.U. Inst. Math. Sc., Res. Rept. EM-33 (July).
9. KELLER, H. B. (1953) Ionospheric propagation of plane waves, N.Y.U. Inst. Math. Sc., Res. Rept. EM-56 (August).
10. STERNBERG, R. L. and KAUFMAN, H. (1952) Application of the theory of systems of differential equations to multiple non-uniform transmission lines, *J. Math. and Phys.*, **31**, 244–252.
11. REDHEFFER, R. M. (1959) Inequalities for a matrix Ricatti equation, *J. Math and Mech.* **8**, 349–367; (1957) The Ricatti equation: initial values and inequalities, *Math. Ann.* **133**, 235–250.

THEORY OF RADIATION FROM SOURCES IN ANISOTROPIC MEDIA

PART I: GENERAL SOURCES IN STRATIFIED MEDIA

E. Arbel† and L. B. Felsen‡

ABSTRACT

Part I of this paper deals with the radiation from arbitrary source distributions in plane stratified, anisotropic media. The formal solutions are obtained by an extension of modal procedures familiar from the analysis of isotropic waveguide regions. Special attention is given to the formulation of a radiation condition which requires the flow of energy away from the sources, and to its interpretation utilizing the refractive index surfaces descriptive of plane wave propagation in an anisotropic medium. These concepts are illustrated in detail for an ionized plasma under the influence of an external steady magnetic field.

1. INTRODUCTION

The propagation of plane electromagnetic waves in regions with anisotropic permittivity or permeability has received considerable attention in the literature—the former in connection with the study of the optical properties of crystals(1) and the electromagnetic behavior of the ionosphere(2,3), and the latter in connection with magnetized ferrites(4). In these applications, the sources of the field have been of minor concern, and emphasis has been placed on the determination of the wave spectrum in the region. However, some recent problems—for example, the investigation of discontinuities in ferrite or plasma loaded waveguides, Cerenkov effects due to moving charged particles in a magneto-ionic medium(5), or radio communication from satellites passing through the ionosphere—require a knowledge of the radiation characteristics of localized sources.

This paper deals with the radiation from arbitrary time harmonic source distributions (varying like $\exp(j\omega t)$) in a transversely unbounded region filled with a dielectric medium characterized by a tensor permittivity ε as shown in (1). The permeability μ_0 is assumed to be constant. ε may be a piecewise constant function of the longitudinal variable z, thereby admitting linear stratification. By an extension of the modal procedure described for isotropic

† Dept. of Theoretical Physics, Hebrew University, Jerusalem.
‡ Dept. of Electrophysics, Polytechnic Institute of Brooklyn.

regions by Marcuvitz and Schwinger[6], the transverse (to z) electromagnetic fields $\mathbf{E}_t(x, y, z)$ and $\mathbf{H}_t(x, y, z)$ are represented as a superposition of orthogonal transverse vector mode fields having the form $V_i(z)\bar{\mathbf{e}}_i(x, y)$ and $I_i(z)\bar{\mathbf{h}}_i(x, y)$, respectively; each mode is a solution of the source-free Maxwell field equations. Because of the infinite extent of the cross-sectional (x, y) domain and the non-variability of ε with x and y, the transverse vector mode functions $\bar{\mathbf{e}}_i$ and $\bar{\mathbf{h}}_i$ have an exponential dependence of the form $\exp(j\xi x + j\eta y)$, where the modal index $i \rightarrow (\xi, \eta)$ is continuously variable. As regards their vectorial characteristics, the modes are found to separate into two sets, "ordinary" and "extraordinary" (see Appendix A), in terms of which the electromagnetic fields can be represented as in (7a, b). The modal amplitudes V and I are found to satisfy the first-order differential equations (8) (transmission line equations), thereby permitting the use of network methods in their determination. The continuity requirements on the tangential electric and magnetic fields across an interface between two different anisotropic regions, or the imposition of boundary conditions at a terminal surface of the region, give rise to coupling between the ordinary and extraordinary modal amplitudes. Their systematic calculation via equivalent network procedures is emphasized in Section 2.

The transmission line equations (8) descriptive of the spatial behavior of V and I in each layer contain as parameters the propagation constant κ and the characteristic impedance Z. As shown in Appendix A, κ in an anisotropic medium is generally a complicated multivalued function of the transverse wavenumbers (ξ, η). Its proper definition on the integration path, essential for a unique specification of the fields via the modal representations (7a, b), is accomplished by recourse to a radiation condition which requires that the energy flow due to a localized source distribution is outward from the source region.[7] The associated restrictions on κ, complicated by the fact that the directions of energy and phase propagation generally differ in an anisotropic medium, are discussed in Section 3—both analytically, and through use of the refractive index surfaces for the medium which specify the variation in refractive index as a function of the direction of propagation of a plane wave. Special attention is given to the case of a plasma under the influence of a longitudinal d.c. magnetic field, and to the effect of the plasma parameters on determining the propagation characteristics of electromagnetic waves. The definition of the multivalued function κ for various plasma parameters is examined in detail in Section 4, thereby rendering unique the formal solution of this class of radiation problems. This aspect of the analysis, essential for a complete formulation, has generally been omitted in related studies which can be found in the literature.

The results of the above analysis have been applied elsewhere[8,9] to the detailed study of the radiation field of an electric current element in an infinite, and semi-infinite, homogeneous, anisotropic plasma.

Investigations by other authors have dealt primarily with radiation problems in infinite anisotropic plasma media. In this category, we cite the work of Abraham[10], Bunkin [11], Kogelnik[12], Kuehl[13], and Mittra[14]. Abraham† has employed a two-dimensional Fourier integral representation for the fields similar to that presented here, and has also pointed out the utility of the refractive index surfaces in dealing with the radiation condition; his analysis in connection with the latter does not, however, enter into the function-theoretic questions treated in Sections 3 and 4. The approach of the other authors differs in that they proceed via a three-dimensional Fourier integral formulation which, though imbued with a certain formal elegance for the infinite medium problem, appears less convenient for an asymptotic analysis of the radiation field, and also seems not directly suited to the study of stratified media. Hodara[15] has recently been concerned with the problem of radiation through an anisotropic plasma slab; he employs a two-dimensional Fourier transform procedure but omits any function-theoretic discussion of the type mentioned above. It also seems that the matching of the boundary conditions at the slab interface and at the source could be more easily achieved by the network procedure described here. Several studies of two-dimensional excitation of anisotropic media have likewise been carried out.[16,17] Other contributions have been presented at the Symposium on Electromagnetic Theory and Antennas.[18–20] (See also references 27–35.)

Green's function representations for general anisotropic waveguides have been obtained by Bresler and Marcuvitz[21] via abstract operator methods. For the special class of problems herein, our procedure, though less general, yields the result more directly and by conventional methods.

2. FORMAL SOLUTION

(a) *Reduction of the Field Equations*

Consider an arbitrary (but prescribed) distribution of time-harmonic sources of electric current $\mathbf{J}(\mathbf{r})$ and magnetic current $\mathbf{M}(\mathbf{r})$ in a medium comprising a series of transversely unbounded, parallel anisotropic layers (Fig. 1). A coordinate system is chosen so that the z-axis is perpendicular to the layer interfaces, and each layer is assumed to be characterized by a scalar permeability μ_o and a tensor dielectric constant

$$\hat{\boldsymbol{\epsilon}}^{(\alpha)} = \hat{\boldsymbol{\epsilon}}_t^{(\alpha)} + \mathbf{z}_o\mathbf{z}_o\hat{\epsilon}_z^{(\alpha)}, \quad \hat{\boldsymbol{\epsilon}}_t^{(\alpha)} \rightarrow \begin{pmatrix} \hat{\epsilon}_1^{(\alpha)} & j\hat{\epsilon}_2^{(\alpha)} \\ -j\hat{\epsilon}_2^{(\alpha)} & \hat{\epsilon}_1^{(\alpha)} \end{pmatrix}, \tag{1}$$

where $^{(\alpha)}$ denotes the αth layer, $\hat{\boldsymbol{\epsilon}}_t$ is a transverse (to z) dyadic whose representative in an $x-y$ coordinate system is as shown, and $\mathbf{z}_o$ is a unit vector along the z-direction. These dielectric tensors have the form appropriate to an

† The authors are indebted to G. Meltz of the Air Force Cambridge Research Laboratories for calling this reference to their attention during the preparation of this manuscript.

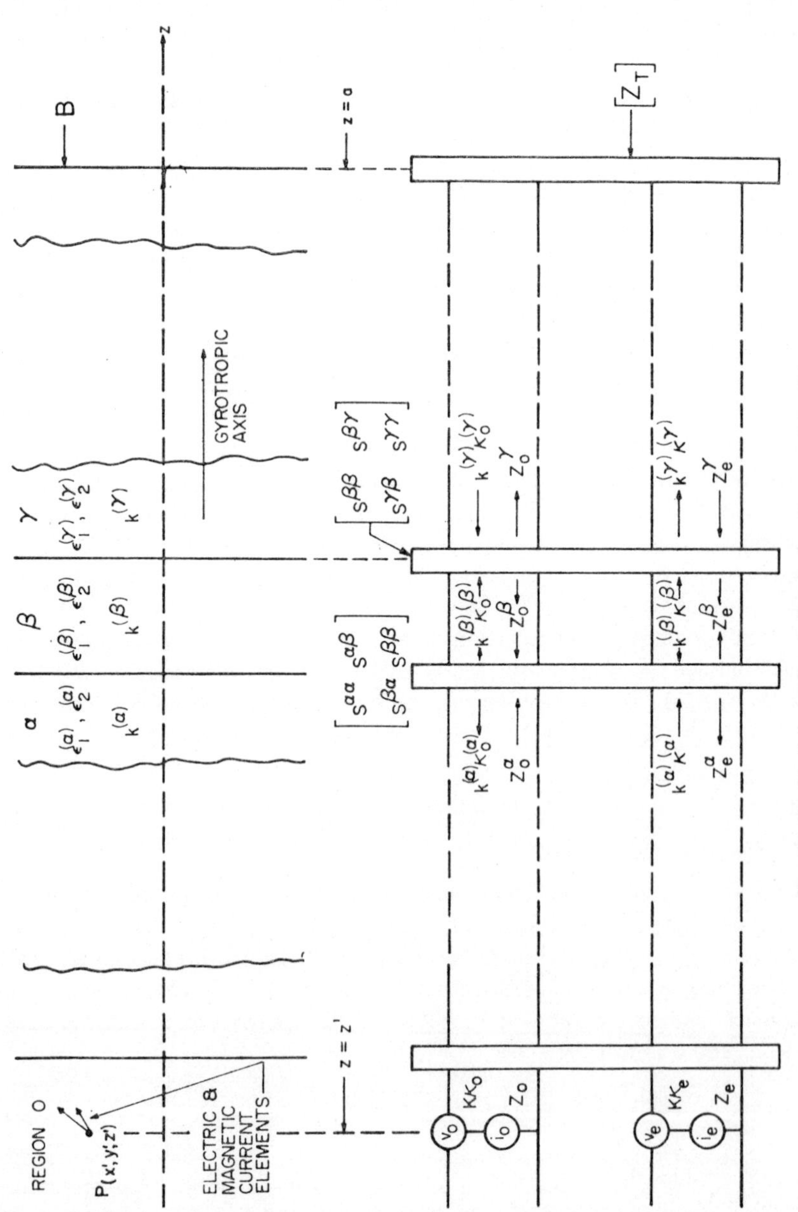

Fig. 1. Physical structure and associated network problem.

ionized plasma medium under the influence of a d.c. magnetic field along the z-axis. We seek a solution of the steady-state Maxwell field equations

$$\nabla \times \mathbf{E}^{(\alpha)} = -j\omega\mu_o \mathbf{H}^{(\alpha)} - \mathbf{M}^{(\alpha)}, \qquad \nabla \times \mathbf{H}^{(\alpha)} = j\omega\hat{\boldsymbol{\epsilon}}^{(\alpha)} \cdot \mathbf{E}^{(\alpha)} + \mathbf{J}^{(\alpha)}, \quad (2)$$

in each layer, subject to the required continuity of the transverse electromagnetic fields $\mathbf{E}_t^{(\alpha)}$ and $\mathbf{H}_t^{(\alpha)}$ at each interface, to prescribed boundary conditions at the z-termini of the region (if any), and to a radiation condition at infinity. A time variation $\exp(j\omega t)$ is understood throughout.

If the configuration in Fig. 1 is viewed as a waveguide with axis along z, the formal solution of the boundary-value problem in (2) can be effected by an application of guided wave techniques developed for isotropic and anisotropic regions.[6,21,22] First, by taking the scalar and vector products of eqns (2) and the longitudinal unit vector z_o, one derives after some rearrangement the equations for the transverse field components (note: $\mathbf{z}_o \times \mathbf{1} \cdot \hat{\boldsymbol{\epsilon}}_t = \hat{\boldsymbol{\epsilon}}_t \cdot \mathbf{z}_o \times \mathbf{1}$),

$$-\frac{\partial \mathbf{E}_t}{\partial z} = j\omega\mu_o \left[\mathbf{1}_t + \frac{\nabla_t \nabla_t}{k^2}\right] \cdot \mathbf{H}_t \times \mathbf{z}_o + \hat{\mathbf{M}}_t \times \mathbf{z}_o, \quad (3a)$$

$$-\frac{\partial \mathbf{H}_t}{\partial z} = j\omega\varepsilon \left[\boldsymbol{\varepsilon}_t + \frac{\nabla_t \nabla_t}{k^2}\right] \cdot \mathbf{z}_o \times \mathbf{E}_t + \mathbf{z}_o \times \hat{\mathbf{J}}_t, \quad (3b)$$

from which the longitudinal components can be obtained via the relation,

$$E_z = \frac{1}{j\omega\varepsilon} [\nabla_t \cdot (\mathbf{H}_t \times \mathbf{z}_o) - J_z], \qquad H_z = \frac{1}{j\omega\mu_o} [\nabla_t \cdot (\mathbf{z}_o \times \mathbf{E}_t) - M_z]. \quad (3c)$$

To simplify the notation, the superscripts$^{(\alpha)}$ identifying the various layers have been omitted, and the following definitions have been introduced:

$$\boldsymbol{\varepsilon}_t = \frac{1}{\varepsilon} \hat{\boldsymbol{\epsilon}}_t \rightarrow \begin{pmatrix} \varepsilon_1 & j\varepsilon_2 \\ -j\varepsilon_2 & \varepsilon_1 \end{pmatrix}, \quad \hat{\epsilon}_z \equiv \varepsilon = \varepsilon_o \varepsilon_z, \quad k^2 = k_o^2 \varepsilon_z, \quad k_o^2 = \omega^2 \mu_o \varepsilon_o, \quad (3d)$$

where ε_0 is the dielectric constant for free space, ε_z is the normalized longitudinal dielectric constant $\hat{\epsilon}_z/\varepsilon_0$, and $\varepsilon_{1,2} = \hat{\epsilon}_{1,2}/\varepsilon$ are normalized to ε.† $\nabla_t = (\nabla - \mathbf{z}_o \, \partial/\partial z)$ represents the transverse gradient operator, $\mathbf{1}_t$ the transverse unit dyadic, and

$$\hat{\mathbf{J}}_t = \mathbf{J}_t - \frac{1}{j\omega\mu} \mathbf{z}_o \times \nabla_t M_z, \qquad \hat{\mathbf{M}}_t = \mathbf{M}_t + \frac{1}{j\omega\varepsilon} \mathbf{z}_o \times \nabla_t J_z, \quad (3e)$$

are equivalent transverse electric and magnetic source current distributions.

Next, we seek a representation of the transverse field components in terms of a set of transverse vector eigenfunctions which are complete, orthogonal,

† For the plasma case, normalization to ϵ implies that $\boldsymbol{\epsilon}_t = \mathbf{1}_t$ in the absence of an external d.c. magnetic field (see Section 4).

and individually satisfy the homogeneous field equations and the required boundary conditions, in the transverse domain. As shown in Appendix A, the electric mode functions $\bar{\mathbf{e}}$ and the magnetic mode functions $\bar{\mathbf{h}}$ comprise a continuous distribution of plane waves whose vectorial characteristics can be grouped into two categories, "ordinary" and "extraordinary", to be denoted by the subscripts o and e, respectively:

$$\bar{\mathbf{e}}_{o,e}(x, y\,;\xi, \eta) = \mathbf{e}_{o,e}(\xi, \eta)\frac{k\,\mathrm{e}^{-jk(\xi x+\eta y)}}{2\pi}, \quad -\infty < (\xi, \eta) < \infty\,, \tag{4a}$$

$$\bar{\mathbf{h}}_{o,e}(x, y\,;\xi, \eta) = \mathbf{h}_{o,e}(\xi, \eta)\frac{k\,\mathrm{e}^{-jk(\xi x+\eta y)}}{2\pi}, \tag{4b}$$

where

$$\mathbf{e}_{\substack{o\\e}} = \frac{\sigma^2 \mp \Delta}{\mp\Delta\sqrt{2}\cdot\sigma}\boldsymbol{\sigma} + \frac{j\delta\sqrt{2}}{\mp\Delta\sigma}(\boldsymbol{\sigma}\times\mathbf{z}_o), \qquad \boldsymbol{\sigma} = \mathbf{x}_o\xi + \mathbf{y}_o\eta\,, \tag{4c}$$

$$\mathbf{h}_{\substack{o\\e}}\times\mathbf{z}_o = \frac{1}{\sqrt{2}\cdot\sigma}\boldsymbol{\sigma} + j\frac{\sigma^2 \pm \Delta}{2\sqrt{2}\cdot\delta\sigma}(\boldsymbol{\sigma}\times\mathbf{z}_o)\,, \qquad \sigma^2 = \xi^2 + \eta^2\,, \tag{4d}$$

and

$$\delta = \frac{\varepsilon_2}{\varepsilon_1 - 1}, \qquad \Delta = \sqrt{[\sigma^4 + 4\delta^2(1-\sigma^2)]}\;\mathrm{sgn}(\varepsilon_1 - 1)\,. \tag{4e}$$

While the eigenvalue problem could have been treated for general dissipative media,[21] the above analysis has been restricted to the lossless case for which ε, ε_1 and ε_2 are real. It is noted that the wavenumbers ξ and η have been normalized to k, thereby implying that $\varepsilon > 0$ and $k > 0$. The modifications introduced when $\varepsilon < 0$ are discussed at the end of Appendix A. The orthogonality properties of the position vectors $\mathbf{e}_{o,e}$ and $\mathbf{h}_{o,e}$ differ according to whether Δ is real or imaginary (the square root in (4e) is defined to be positive when real and negative imaginary otherwise):

$$\begin{aligned} &\mathbf{e}^*_{\substack{o\\e}}\cdot\mathbf{h}_{\substack{o\\e}}\times\mathbf{z}_o = 1, \qquad \mathbf{e}^*_{\substack{o\\e}}\cdot\mathbf{h}_{\substack{e\\o}}\times\mathbf{z}_o = 0, \qquad \Delta \text{ real},\\ &\mathbf{e}^*_{\substack{o\\e}}\cdot\mathbf{h}_{\substack{o\\e}}\times\mathbf{z}_o = 0, \qquad \mathbf{e}^*_{\substack{o\\e}}\cdot\mathbf{h}_{\substack{e\\o}}\times\mathbf{z}_o = 1, \qquad \Delta \text{ imaginary},\end{aligned} \tag{5}$$

where the asterisk denotes the complex conjugate. Thus, the mode functions in (4a, b) satisfy the bi-orthogonality relation,

$$\int_{-\infty}^{\infty}\mathrm{d}x\int_{-\infty}^{\infty}\mathrm{d}y\,\bar{\mathbf{e}}^*(x, y\,;\xi, \eta)\cdot\bar{\mathbf{h}}(x, y\,;\xi', \eta')\times\mathbf{z}_o = \mathbf{e}^*\cdot\mathbf{h}\times\mathbf{z}_o\delta(\xi-\xi')\delta(\eta-\eta')\,, \tag{6}$$

which, upon insertion of the appropriate modal subscripts o or e, is simplified further from (5).

The various vector functions in (3a,b) can now be represented as follows:

$$\mathbf{E}_t(x, y, z) = \frac{k}{2\pi}\int_{-\infty}^{\infty} \mathrm{d}\xi \int_{-\infty}^{\infty} \mathrm{d}\eta$$
$$\times [V_o(z\,;\,\xi, \eta)\mathbf{e}_o(\xi, \eta) + V_e(z\,;\,\xi, \eta)\mathbf{e}_e(\xi, \eta)]\, \mathrm{e}^{-jk(\xi x + \eta y)}, \quad (7a)$$

$$\mathbf{H}_t(x, y, z) = \frac{k}{2\pi}\int_{-\infty}^{\infty} \mathrm{d}\xi \int_{-\infty}^{\infty} \mathrm{d}\eta$$
$$\times [I_o(z\,;\,\xi, \eta)\mathbf{h}_o(\xi, \eta) + I_e(z\,;\,\xi, \eta)\mathbf{h}_e(\xi, \eta)]\, \mathrm{e}^{-jk(\xi x + \eta y)}, \quad (7b)$$

$$\mathbf{z}_o \times \hat{\mathbf{J}}_t(x, y, z) = \frac{k}{2\pi}\int_{-\infty}^{\infty} \mathrm{d}\xi \int_{-\infty}^{\infty} \mathrm{d}\eta$$
$$\times [i_e(z\,;\,\xi, \eta)\mathbf{h}_e(\xi, \eta) + i_o(z\,;\,\xi, \eta)\mathbf{h}_o(\xi, \eta)]\, \mathrm{e}^{-jk(\xi x + \eta y)}, \quad (7c)$$

$$\hat{\mathbf{M}}_t(x, y, z) \times \mathbf{z}_o = \frac{k}{2\pi}\int_{-\infty}^{\infty} \mathrm{d}\xi \int_{-\infty}^{\infty} \mathrm{d}\eta$$
$$\times [v_e(z\,;\,\xi, \eta)\mathbf{e}_e(\xi, \eta) + v_o(z\,;\,\xi, \eta)\mathbf{e}_o(\xi, \eta)]\, \mathrm{e}^{-jk(\xi x + \eta y)}. \quad (7d)$$

The completeness and existence of these representations (for spatially confined source distributions) in the $(x - y)$ function space follows from the theory of the Fourier integral and from the field behavior at infinity, while that in the 2×2 vector space is assured by the linear combination of the *e*- and *o*-eigenvectors. Upon substituting eqns (7a–d) into (3a, b), interchanging the orders of differentiation and integration and noting that the operator ∇_t can then be replaced by $-jk\boldsymbol{\sigma}$, one may equate the resulting Fourier transforms on both sides of these equations. Dot product multiplication of the first and second of these equations with $(\mathbf{h}^*_{o,e} \times \mathbf{z}_o)$ and $(\mathbf{z}_o \times \mathbf{e}^*_{o,e})$, respectively, and use of eqns (A4) in Appendix A and of the orthogonality relations (5) then yields the following expressions to be satisfied by the transforms $V_{o,e}$ and $I_{o,e}$:

$$-\frac{\mathrm{d}V}{\mathrm{d}z} = jk\kappa ZI + v, \qquad -\frac{\mathrm{d}I}{\mathrm{d}z} = jk\kappa YV + i. \quad (8)$$

These equations, valid separately for both the *o*- and *e*-modes (for real or imaginary Δ), can evidently be interpreted as transmission line equations, with V, I, κ, Z, and Y playing the role of voltage, current, normalized propagation constant, characteristic impedance and characteristic admittance, respectively, while v and i represent voltage and current generator distributions (Fig. 1(b)). κ and $Z = 1/Y$ are determined from the eigenvalue problem in Appendix A as

$$\kappa_{\substack{o\\e}} = \sqrt{[U \pm \sqrt{(U^2 - W)}]}, \qquad Y_{\substack{o\\e}} = \kappa_{\substack{o\\e}} \frac{4\delta^2}{\zeta(\Delta^2 \pm \sigma^2\Delta)}, \qquad \zeta = \sqrt{\left(\frac{\mu_o}{\varepsilon}\right)}, \quad (8a)$$

where

$$U = \varepsilon_1 - \frac{\varepsilon_1 + 1}{2}\sigma^2, \qquad W = \varepsilon_1(1 - \sigma^2)(\sigma_2^2 - \sigma^2), \qquad \sigma_2^2 = \frac{\varepsilon_1^2 - \varepsilon_2^2}{\varepsilon_1}. \tag{8b}$$

$\sqrt{(U^2 - W)}$ can alternatively be written as $(\varepsilon_1 - 1)(\Delta/2)$. The voltage and current generator strengths are evaluated by inverting eqns (7c,d) via (6):

$$i_{\substack{o\\e}} = \int_{-\infty}^{\infty} \mathrm{d}x \int_{-\infty}^{\infty} \mathrm{d}y\, \bar{\mathbf{e}}_{\substack{o\\e}}^* \cdot \hat{\mathbf{J}}_t\,, \qquad v_{\substack{o\\e}} = \int_{-\infty}^{\infty} \mathrm{d}x \int_{-\infty}^{\infty} \mathrm{d}y \bar{\mathbf{h}}_{\substack{o\\e}}^* \cdot \hat{\mathbf{M}}_t\,, \qquad \Delta \text{ real}; \tag{9}$$

For imaginary Δ, $\bar{\mathbf{e}}^*_{o,e}$ and $\bar{\mathbf{h}}^*_{o,e}$ are replaced by $\bar{\mathbf{e}}^*_{e,o}$ and $\bar{\mathbf{h}}^*_{e,o}$, respectively.†

From the two-dimensional divergence theorem

$$\int_{-\infty}^{\infty} \mathrm{d}x \int_{-\infty}^{\infty} \mathrm{d}y \mathbf{A} \cdot \nabla_t f = -\int_{-\infty}^{\infty} \mathrm{d}x \int_{-\infty}^{\infty} \mathrm{d}y f \nabla_t \cdot \mathbf{A} + \oint_s \mathrm{d}s f \mathbf{A} \cdot \mathbf{n}\,,$$

where $\mathbf{A}$ and f are suitably continuous vector and scalar functions, s denotes a contour bounding the transverse cross-section at infinity in the x-y plane and $\mathbf{n}$ is a unit vector normal to s, one observes that if $f = 0$ on s, the double integral on the left-hand side is equal to the double integral on the right-hand side. Upon applying this result to eqn (9) (after substituting (3e) and assuming that J_z and M_z are spatially confined sources), one obtains alternatively,

$$i_{\substack{o\\e}} = \int_{-\infty}^{\infty} \int_{-\infty}^{\infty} \mathbf{J}_t \cdot \bar{\mathbf{e}}_{\substack{o\\e}}^* \,\mathrm{d}x\,\mathrm{d}y - \frac{k}{\omega\mu} \int_{-\infty}^{\infty} \int_{-\infty}^{\infty} M_z \bar{\mathbf{e}}_{\substack{o\\e}}^* \cdot \boldsymbol{\sigma} \times \mathbf{z}_0 \,\mathrm{d}x\,\mathrm{d}y, \tag{9a}$$

$$v_{\substack{o\\e}} = \int_{-\infty}^{\infty} \int_{-\infty}^{\infty} \mathbf{M}_t \cdot \bar{\mathbf{h}}_{\substack{o\\e}}^* \,\mathrm{d}x\,\mathrm{d}y - \frac{k}{\omega\varepsilon} \int_{-\infty}^{\infty} \int_{-\infty}^{\infty} J_z \bar{\mathbf{h}}_{\substack{o\\e}}^* \times \mathbf{z}_0 \cdot \boldsymbol{\sigma} \,\mathrm{d}x\,\mathrm{d}y\,, \tag{9b}$$

which result can now be applied also to discontinuous current distributions.

The proper definition of the multivalued function $\kappa(\sigma)$, and the disposition of the integration path in (7a, b) in relation to its singularities, will be discussed further in Section 3. For the present we note only that κ_o and κ_e are associated, respectively, with the + and − signs in (8a), and that $\sqrt{(U^2 - W)}$ is defined to be positive when real. When κ has an imaginary part, the associated wave is non-propagating (along z), while κ real represents a propagating wave. Since i and v are specified in terms of known quantities, the solution of the Maxwell field equations (2) has been reduced to the solution of the transmission line eqns (8) in each layer, subject to the required continuity conditions at each interface and to the specified boundary conditions at the z-termini of the region.

† One observes from (4c,d) that the analytic continuation of the functions $e^*_{o,e}$ and $h^*_{o,e}$ from the domain of real Δ into the domain of imaginary Δ yields the functions $e^*_{e,o}$ and $h^*_{e,o}$, respectively, evaluated for imaginary Δ. Thus, it suffices to evaluate the discontinuously represented inner products in (9) for real Δ and continue these functions into the imaginary–Δ domain.

(b) Solution of the Modal Network Problem

At an interface between two different anisotropic regions, the required continuity of $\mathbf{E}_t$ and $\mathbf{H}_t$ in (7a, b) can be achieved by assuring the continuity of their Fourier transforms. If parameters pertaining to the two regions are distinguished by superscripts α and β, respectively, the boundary conditions are satisfied if

$$k^{(\beta)}[V_o^{(\alpha)}\mathbf{e}_o^{(\alpha)} + V_e^{(\alpha)}\mathbf{e}_e^{(\alpha)}] = k^{(\alpha)}[V_o^{(\beta)}\mathbf{e}_o^{(\beta)} + V_e^{(\beta)}\mathbf{e}_e^{(\beta)}], \tag{10a}$$

$$k^{(\beta)}[I_o^{(\alpha)}\mathbf{h}_o^{(\alpha)} + I_e^{(\alpha)}\mathbf{h}_e^{(\alpha)}] = k^{(\alpha)}[I_o^{(\beta)}\mathbf{h}_o^{(\beta)} + I_e^{(\beta)}\mathbf{h}_e^{(\beta)}], \tag{10b}$$

provided that the normalized transverse wave numbers in the two regions are related via

$$k^{(\alpha)}\xi^{(\alpha)} = k^{(\beta)}\xi^{(\beta)}, \qquad k^{(\alpha)}\eta^{(\alpha)} = k^{(\beta)}\eta^{(\beta)}.\dagger \tag{10c}$$

In view of the orthogonality relation (5), eqns (10a,b) can be reduced to the matrix form:

$$\begin{bmatrix} \mathbf{V}^{(\alpha)} \\ \mathbf{I}^{(\alpha)} \end{bmatrix} = T_z \begin{bmatrix} \mathbf{V}^{(\beta)} \\ \mathbf{I}^{(\beta)} \end{bmatrix}, \qquad T_z \rightarrow \begin{bmatrix} t_{\alpha\alpha} & t_{\alpha\beta} \\ t_{\alpha\beta} & t_{\beta\beta} \end{bmatrix}, \tag{11}$$

where $\mathbf{V}$ and $\mathbf{I}$ are the column vectors

$$\mathbf{V}^{(i)} \rightarrow \begin{bmatrix} V_o^{(i)} \\ V_e^{(i)} \end{bmatrix}, \qquad \mathbf{I}^{(i)} \rightarrow \begin{bmatrix} I_o^{(i)} \\ I_e^{(i)} \end{bmatrix}, \qquad i = \alpha, \beta, \tag{11a}$$

while T_z is the impedance transfer matrix. The matrix elements t_{ij}, $i, j = \alpha, \beta$, are themselves 2×2 matrices whose elements are given by:

$$t_{\alpha\beta} = t_{\beta\alpha} = 0, \tag{11b}$$

$$t_{\alpha\alpha} \rightarrow \frac{k^{(\alpha)}}{k^{(\beta)}} \begin{bmatrix} \mathbf{h}_o^{(\alpha)*} \times \mathbf{z}_o \cdot \mathbf{e}_o^{(\beta)} & \mathbf{h}_o^{(\alpha)*} \times \mathbf{z}_o \cdot \mathbf{e}_e^{(\beta)} \\ \mathbf{h}_e^{(\alpha)*} \times \mathbf{z}_o \cdot \mathbf{e}_o^{(\beta)} & \mathbf{h}_e^{(\alpha)*} \times \mathbf{z}_o \cdot \mathbf{e}_e^{(\beta)} \end{bmatrix}, \tag{11c}$$

$$t_{\beta\beta} \rightarrow \frac{k^{(\alpha)}}{k^{(\beta)}} \begin{bmatrix} \mathbf{e}_o^{(\alpha)*} \cdot \mathbf{h}_o^{(\beta)} \times \mathbf{z}_o & \mathbf{e}_o^{(\alpha)*} \cdot \mathbf{h}_e^{(\beta)} \times \mathbf{z}_o \\ \mathbf{e}_e^{(\alpha)*} \cdot \mathbf{h}_o^{(\beta)} \times \mathbf{z}_o & \mathbf{e}_e^{(\alpha)*} \cdot \mathbf{h}_e^{(\beta)} \times \mathbf{z}_o \end{bmatrix}. \tag{11d}$$

Expressions (11c,d) apply for real or imaginary Δ (see footnote p. 8). Since the submatrices $t_{\alpha\alpha}$ and $t_{\beta\beta}$ are not diagonal, the ordinary and extraordinary modes are coupled at the interface. Because $t_{\alpha\beta} = t_{\beta\alpha} = 0$, the coupling occurs in a particularly simple manner and can be schematized in terms of the transformer network shown in Fig. 2. It is also noted, in view of the

† It is implied that ξ, η in the integral representations (7a, b) are replaced throughout by the appropriate $\xi^{(i)}$, $\eta^{(i)}$, $i = \alpha, \beta$, whence $V_{o,e}{}^{(\alpha)} = V_{o,e}(z;\, \xi^{(\alpha)}, \eta^{(\alpha)})$, $V_{o,e}{}^{(\beta)} = V_{o,e}(z;\, \xi^{(\beta)}, \eta^{(\beta)})$, etc. This is to be borne in mind when interpreting formulas involving parameters with different superscripts.

symmetrical character of eqns (10) as regards α and β, that the elements of the inverse matrix T_z^{-1} are given as in (11c,d) provided that the superscripts α and β are interchanged throughout.

In the interior of each slab region, the equivalent modal network comprises the two transmission lines representative of the o- and e-modes, respectively, as shown in Fig. 1(b). If the slab in question has a length d, the impedance transfer matrix $\hat{T}_z$ for the slab region is obtained from simple transmission line theory as follows:

$$\hat{T}_z \to \begin{bmatrix} t_{11} & t_{12} \\ t_{21} & t_{22} \end{bmatrix}, \tag{12}$$

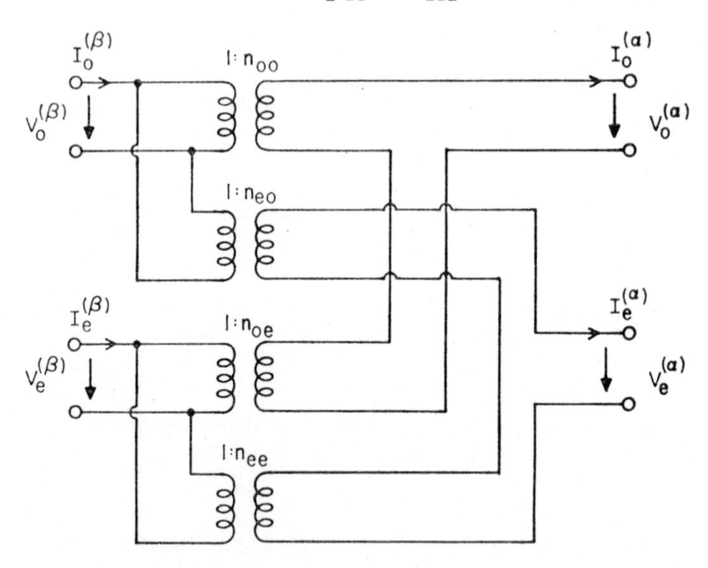

FIG. 2. Equivalent network for interface between two gyrotropic media.

where the 2 × 2 submatrices t_{ij}, $i, j = 1, 2$, are diagonal and are given by

$$t_{11} = t_{22} = \cos(k\hat{K}d), \qquad \hat{Y}t_{12} = \hat{Z}t_{21} = j\sin(k\hat{K}d). \tag{12a}$$

$\hat{K}$ and $\hat{Z}$ are the diagonal propagation constant and characteristic impedance matrices, respectively,

$$\hat{K} \to \begin{bmatrix} \kappa_o & 0 \\ 0 & \kappa_e \end{bmatrix}, \qquad \hat{Z} = \hat{Y}^{-1} \to \begin{bmatrix} Z_o & 0 \\ 0 & Z_e \end{bmatrix}. \tag{12b}$$

The matrices in (12a) are interpreted by recalling that if $\hat{K}$ is a diagonal matrix, with elements κ_i, a matrix $f(\hat{K})$ is also diagonal and has as its elements $f(\kappa_i)\,\delta_{ij}$. The voltage and current vectors $\mathbf{V}^{(1)}$ and $\mathbf{I}^{(1)}$ at the slab face $z = z_1$ are then related to the analogous quantities at $z = z_2$ via

$$\begin{bmatrix} \mathbf{V}^{(1)} \\ \mathbf{I}^{(1)} \end{bmatrix} = \hat{T}_z \begin{bmatrix} \mathbf{V}^{(2)} \\ \mathbf{I}^{(2)} \end{bmatrix}, \qquad z_2 - z_1 = d > 0\,.\dagger \tag{12c}$$

Repeated application of eqns (11) and (12) allows one to express the voltages and currents at any point z_α in the region in terms of the voltages and currents at any other point z_β. The overall transfer matrix descriptive of the network between z_α and z_β is then composed of the ordered product of the transfer matrices of the network constituents in this region.

If the region is terminated at $z = z_o$ in a plane surface on which $\mathbf{E}_t$ and $\mathbf{H}_t$ are related by the boundary condition

$$\mathbf{E}_t(\boldsymbol{\rho}, z_o) = \mathbf{Z}\cdot\mathbf{H}_t(\boldsymbol{\rho}, z_o) \times \mathbf{z}_o\,, \tag{13}$$

where the transverse dyadic $\mathbf{Z}$ denotes a constant anisotropic surface impedance whose representative in an x-y coordinate space is

$$\mathbf{Z} \rightarrow \begin{pmatrix} z_{11} & z_{12} \\ z_{21} & z_{22} \end{pmatrix}, \tag{13a}$$

then the corresponding relation between $V_{o,e}$ and $I_{o,e}$ at $z = z_o$ is given via (7a, b) by

$$V_o\mathbf{e}_o + V_e\mathbf{e}_e = \mathbf{Z}\cdot(I_o\mathbf{h}_o + I_e\mathbf{h}_e) \times \mathbf{z}_o\,. \tag{14}$$

Use of the orthogonality relations (5) allows one to deduce the terminal impedance matrix $\hat{Z}_s$ for this structure as

$$\begin{bmatrix} V_o \\ V_e \end{bmatrix} = \hat{Z}_s \begin{bmatrix} I_o \\ I_e \end{bmatrix}, \tag{15a}$$

where for real or imaginary Δ (see footnote p. 8).

$$\hat{Z}_s \rightarrow \begin{bmatrix} \mathbf{h}_o^* \times \mathbf{z}_o\cdot\mathbf{Z}\cdot\mathbf{h}_o \times \mathbf{z}_o & \mathbf{h}_o^* \times \mathbf{z}_o\cdot\mathbf{Z}\cdot\mathbf{h}_e \times \mathbf{z}_o \\ \mathbf{h}_e^* \times \mathbf{z}_o\cdot\mathbf{Z}\cdot\mathbf{h}_o \times \mathbf{z}_o & \mathbf{h}_e^* \times \mathbf{z}_o\cdot\mathbf{Z}\cdot\mathbf{h}_e \times \mathbf{z}_o \end{bmatrix}, \tag{15b}$$

It is to be noted that even a scalar surface impedance Z_s (for which $\mathbf{Z} = \mathbf{1}_t Z_s$, $Z_s =$ constant) couples the o- and e-modes (see eqns (4c, d). Only when $Z_s = 0, \infty$ (short or open circuit) or when $\sigma = 0$ (normally incident plane wave) does the coupling disappear.

It is frequently more convenient to deal with a traveling wave representation involving incident and reflected waves rather than with the standing wave

† In this equation, the superscripts refer to quantities at different z-locations in the *same* region.

representation employed above. The required transformations, well-known in linear network theory,[23] are summarized below. Let

$$\mathbf{V} = \mathbf{a} + \mathbf{b}, \qquad \mathbf{I} = \hat{\mathbf{Y}}(\mathbf{a} - \mathbf{b}), \tag{16}$$

where the wave vectors **a** and **b** distinguish the amplitudes of waves traveling to the right and left, respectively (see Fig. 3),

$$\mathbf{a} \rightarrow \begin{bmatrix} a_o \\ a_e \end{bmatrix}, \qquad \mathbf{b} \rightarrow \begin{bmatrix} b_o \\ b_e \end{bmatrix}, \tag{16a}$$

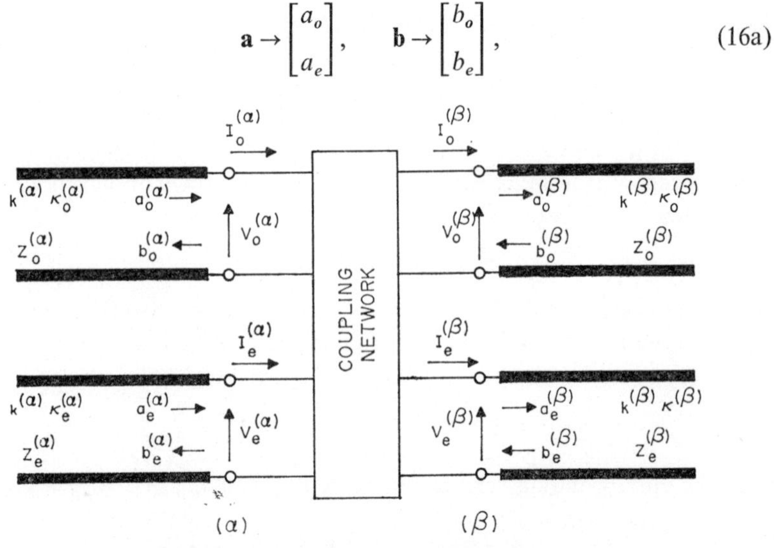

FIG. 3. Definition of traveling and standing wave quantities.

and $\hat{Y}$ is the characteristic admittance matrix (see (12b)). Then the scattering transfer matrix T_s provides the direct connection between the incident and reflected wave amplitudes at terminals α and β as follows

$$\begin{bmatrix} \mathbf{b}^{(\alpha)} \\ \mathbf{a}^{(\alpha)} \end{bmatrix} = T_s \begin{bmatrix} \mathbf{b}^{(\beta)} \\ \mathbf{a}^{(\beta} \end{bmatrix}, \qquad T_s \rightarrow \begin{bmatrix} \tau_{\alpha\alpha} & \tau_{\alpha\beta} \\ \tau_{\beta\alpha} & \tau_{\beta\beta} \end{bmatrix}, \tag{17}$$

where the scattering transfer matrix T_s is related to the impedance transfer matrix T_z in (11) via the linear transformation

$$T_s = \frac{1}{2} \begin{bmatrix} \hat{1} & -\hat{Z}^{(\alpha)} \\ \hat{1} & \hat{Z}^{(\alpha)} \end{bmatrix} T_z \begin{bmatrix} \hat{1} & \hat{1} \\ -\hat{Y}^{(\beta)} & \hat{Y}^{(\beta)} \end{bmatrix}. \tag{18}$$

$\hat{1}$ denotes the 2×2 unit matrix. When applied to an interface between two anisotropic media, $t_{\alpha\beta} = t_{\beta\alpha} = 0$ from (11b), and (18) yields the following expressions for the 2×2 submatrices τ_{ij}:

$$\tau_{\alpha\alpha} = \tfrac{1}{2}[t_{\alpha\alpha} + \hat{Z}^{(\alpha)} t_{\beta\beta} \hat{Y}^{(\beta)}] = \tau_{\beta\beta}, \tag{19a}$$

$$\tau_{\alpha\beta} = \tfrac{1}{2}[t_{\alpha\alpha} - \hat{Z}^{(\alpha)} t_{\beta\beta} \hat{Y}^{(\beta)}] = \tau_{\beta\alpha}. \tag{19b}$$

For a slab of length d, the scattering transfer matrix $\hat{T}_s$ corresponding to $\hat{T}_z$ in (12) has the simple diagonal representation

$$\hat{T}_s \to \begin{bmatrix} \tau_{11} & 0 \\ 0 & \tau_{22} \end{bmatrix}, \qquad \tau_{11} = e^{-j\,\hat{K}d}, \qquad \tau_{22} = e^{+j\,\hat{K}d}, \tag{20}$$

with $\hat{K}$ defined in (12b). Like the impedance transfer representation, the scattering transfer formulation is well suited to the analysis of cascaded networks as encountered in stratified media.

If the sources are located in a semi-infinite medium, the reflection phenomena are analyzed conveniently in terms of a scattering matrix representation which expresses the outgoing waves at terminals (α, β) in terms of the incoming waves via the scattering matrix $\mathscr{S}$:

$$\begin{bmatrix} \mathbf{b}^{(\alpha)} \\ \mathbf{a}^{(\beta)} \end{bmatrix} = \mathscr{S} \begin{bmatrix} \mathbf{a}^{(\alpha)} \\ \mathbf{b}^{(\beta)} \end{bmatrix}, \qquad \mathscr{S} \to \begin{bmatrix} s_{\alpha\alpha} & s_{\alpha\beta} \\ s_{\beta\alpha} & s_{\beta\beta} \end{bmatrix}. \tag{21}$$

The scattering matrix elements are related to those of the T_s matrix via

$$s_{\alpha\alpha} = \tau_{\alpha\beta}\tau_{\beta\beta}^{-1}, \quad s_{\alpha\beta} = \tau_{\alpha\alpha} - \tau_{\alpha\beta}\tau_{\beta\beta}^{-1}\tau_{\beta\alpha}, \quad s_{\beta\alpha} = \tau_{\beta\beta}^{-1}, \quad s_{\beta\beta} = -\tau_{\beta\beta}^{-1}\tau_{\beta\alpha}. \tag{22}$$

The above equations can also be used to describe an interface between an anisotropic and an isotropic medium. In this instance, it is more convenient to employ in the isotropic medium the linearly polarized E and H modes described in Appendix A.†

† Explicit expressions for the various coupling matrix elements, obtained after carrying out the operations and substitutions in the text, are given in reference 8. These rather cumbersome formulas are not listed here because of space limitations.

3. RADIATION CONDITION AND SPECIFICATION OF κ

It was noted in connection with eqn (8a) that $\kappa_{o,e}$ is a multivalued function of $(\xi^2 + \eta^2)^{1/2} = \sigma$, and that its analytic properties must be specified if the integral representations (7a, b) for the fields are to be rendered unique.† The definitions

$$\kappa_o \equiv \sqrt{[U + \sqrt{(U^2 - W)}]} = \sqrt{[U + (\varepsilon_1 - 1)\Delta/2]}, \tag{23a}$$

$$\kappa_e \equiv \sqrt{[U - \sqrt{(U^2 - W)}]} = \sqrt{[U - (\varepsilon_1 - 1)\Delta/2]}, \tag{23b}$$

with U, W, and Δ specified in (8b) and (4e), respectively, imply that κ_o has branch point singularities at $U^2 = W(\Delta = 0)$ and at those values of σ for which $\kappa_o = 0$, while κ_e has branch point singularities at $\Delta = 0$ and at those

† In plane stratified media, $V_{o,e}$ and $I_{o,e}$ may also possess pole singularities on the integration path. For the case of a single interface, these (surface wave) poles have been discussed in reference 8.

values of σ for which $\kappa_e = 0$. Although the branch point at $\Delta = 0$ occurs in both the ordinary and extraordinary integrals in (7a, b)‡, the sum of the ordinary and extraordinary integrands is an even function of Δ and therefore single-valued at $\Delta = 0$. This evenness, resulting from the fact that all ordinary quantities differ from the corresponding extraordinary ones only by the algebraic sign of Δ (see (4c, d), (8a, b) and (9)), implies that a series expansion of the combined integrand about $\Delta = 0$ contains only even powers of Δ, whence $\Delta = 0$ is a regular point. Thus, no special care need be taken in the definition of $\kappa_{o,e}$ at $\Delta = 0$ as regards the total integrand; if the ordinary and extraordinary integrals are treated separately, one may introduce convenient branch cut configurations relative to the branch points at $\Delta = 0$ which render the integrands single-valued on a Riemann surface associated with this singularity.

For a region comprising a series of slabs of finite width along the z-direction, the voltage and current solutions may be even functions of the propagation constants $\kappa_{o,e}^{(\alpha)}$ in each slab. If this is the case, the points where $\kappa_o^{(\alpha)} = 0$ and $\kappa_e^{(\alpha)} = 0$ are regular. The situation is different, however, if one of the regions extends to $z = \pm\infty$. In this instance, the integrands are not even functions of the appropriate propagation constant, and the matter of the definition of $\kappa_{o,e}$ must be studied in detail. The investigation is directly connected with the specification of the boundary condition at $|z| \to \infty$.

Consider a homogeneous anisotropic region which occupies the half space $z_1 < z < \infty$, and assume that all sources are located in the space $z < z_2$ where $z_2 \gtrless z_1$. Then the solution of the transmission line equations (8) at points $z > z_2$ comprises traveling waves characterized by the functions $\exp(-jk\kappa_{o,e}z)$, where k and $\kappa_{o,e}$ have the values appropriate to the region in question, and k is assumed to be positive.† The integration in (7a, b) extends over all real values of σ (see Appendix B). One notes from (23a, b) that κ_o and κ_e may be either real, imaginary, or complex, depending on whether $\kappa_{o,e}^2$ is positive, negative, or complex (the latter case obtains when $U^2 < W$, i.e. Δ is imaginary). For those values of σ for which κ_o or κ_e has a non-vanishing imaginary part, the requirement

$$\operatorname{Im} \kappa_{o,e} < 0 \tag{24}$$

assures that the fields remain finite as $z \to \infty$ and serves to define the multi-valued functions $\kappa_{o,e}$. The associated mode fields decay exponentially with increasing z and represent a non-propagating wave.

If κ_o or κ_e is real, the $\exp(-jk\kappa_{o,e}z)$ solution represents a propagating plane

‡ The two terms in the integrand of (7a) give rise to separate contributions which will be called ordinary (subscript o) and extraordinary (subscript e), respectively. Analogous considerations apply to (7b).

† In view of the remarks at the end of Appendix A, the conclusions derived herein can easily be adapted to imaginary k. In that case, $\exp(-jk\kappa z) \to \exp(-j|k|\kappa z)$, etc.

wave. Since all sources are confined to the region $z < z_2$, there must be a net flow of power toward $z = \infty$ and it appears plausible to suppose that the power flow vector for each constituent propagating plane wave likewise has a component along the $+z$ direction. It will now be shown that the latter condition follows from the former, i.e. that net total power flow toward $z = \infty$ implies that each propagating plane wave carries power in this direction. This radiation condition (see also reference 7) then permits the unique determination of $\kappa_{o,e}$ over the range of σ for which the propagation constants are real. It must be emphasized that the "energy radiation condition" is distinct from a "phase radiation condition" since the directions of phase and energy propagation in an anisotropic medium are generally different.

The requirement of net outward average power flow P_c through a plane at $z = c > z_2$, where c is a real constant, can be phrased as

$$P_c = \mathrm{Re} \int_{-\infty}^{\infty} \mathrm{d}x \int_{-\infty}^{\infty} \mathrm{d}y \, \mathbf{z}_o \cdot \mathbf{S}(x, y, c) \geq 0, \qquad \mathbf{S} = \mathbf{E}_t^* \times \mathbf{H}_t, \tag{25}$$

where $\mathbf{S}$ is the complex Poynting vector. Upon substituting the integral representations (7a, b) into (25), interchanging the orders of integration and recalling the orthogonality relation (6), one obtains:

$$P_c = \mathrm{Re}\left\{ \iint\limits_{\kappa^2 \text{ real}} \mathrm{d}\xi \, \mathrm{d}\eta [V_o^* I_o + V_e^* I_e] + \iint\limits_{\kappa^2 \text{ complex}} \mathrm{d}\xi \, \mathrm{d}\eta [V_o^* I_e + V_e^* I_o] \right\} \geq 0, \tag{26}$$

where the first integral extends over that portion of the $\xi - \eta$ plane for which $\kappa_{o,e}^2$ is real (Δ real), while the second extends over the remaining region wherein $\kappa_{o,e}^2$ is complex (Δ imaginary). Since

$$V_{o,e}(z) = C_{o,e} \, \mathrm{e}^{-jk\kappa_{o,e} z}, \qquad z > z_2, \tag{27a}$$

where C_o and C_e are constants, one notes from the transmission line equations (8) (or from the equivalent network picture in Fig. 1(b)), that

$$I_{o,e}(z) = Y_{o,e} V_{o,e}(z), \qquad z > z_2, \tag{27b}$$

so that (26) can be written as:

$$P_c = \mathrm{Re}\left\{ \iint\limits_{\kappa^2 \text{ real}} \mathrm{d}\xi \, \mathrm{d}\eta [Y_o |V_o|^2 + Y_e |V_e|^2] \right.$$

$$\left. + \iint\limits_{\kappa^2 \text{ complex}} \mathrm{d}\xi \, \mathrm{d}\eta [Y_e V_e V_o^* + Y_o V_o V_e^*] \right\} \geq 0. \tag{28}$$

From (23a, b),

$$\kappa_o^2 = \kappa_e^{*2}, \qquad \text{when } \Delta \text{ is imaginary}, \tag{29a}$$

so that $\kappa_o = \pm\kappa_e^*$. Since κ_o and κ_e must both have negative imaginary parts (from (24)), it follows that

$$\kappa_o = -\kappa_e^*, \qquad \operatorname{Im} \kappa_{o,e} < 0, \qquad \text{when } \Delta \text{ is imaginary,} \tag{29b}$$

whence, from (8a), $Y_o = -Y_e^*$. Hence, the integral over the range of complex κ^2, which arises from a coupling of the ordinary and extraordinary mode energies, is imaginary and does not contribute to P_c. If κ_o^2 and (or) κ_e^2 is negative real, i.e. κ_o and (or) κ_e is imaginary, then Y_o and (or) Y_e is also imaginary (see (8a)) and the resulting integral does not contribute to P_c. Hence, (28) can be simplified to:

$$P_c = \iint\limits_{\kappa_o \text{ real}} Y_o|V_o|^2 \, d\xi \, d\eta + \iint\limits_{\kappa_e \text{ real}} Y_e|V_e|^2 \, d\xi \, d\eta > 0, \tag{30}$$

where the equality has been omitted since we are considering propagating waves which carry energy along the $+z$-direction. The symbol "Re" is superfluous because the integrals are now real. Since the radiation condition (30) must be valid for arbitrary source distributions in the region $z < z_2$, (i.e. for arbitrary V_o and V_e), each of the integrands must satisfy the inequality whence $P_c > 0$ if, and only if,

$$Y_{o,e} > 0 \qquad \text{when } \kappa_{o,e} \text{ is real.} \tag{31}$$

Hence, as stated above, each propagating plane wave must individually satisfy the radiation condition.

(8a) can now be employed to determine the algebraic sign of $\kappa_{o,e}$ in (23a,b) when $\kappa_{o,e}$ is real. Since the characteristic admittances must be positive, the sign of $\kappa_{o,e}$ is identical with that of $(\Delta^2 \pm \sigma^2\Delta)$. This condition evidently depends on the values of the constitutive parameters ε_1, ε_2, and is investigated below in detail for a plasma medium. Equations (24) and (31) specify $\kappa_{o,e}$ uniquely for all real values of ξ and η (or σ), and can be used to determine the analytic continuation of the function $\kappa_{o,e}(\sigma)$ around branch point singularities located on the real σ axis. Some general remarks can be made concerning the determination of the range of propagating modes for which

$$\kappa_{\substack{o\\e}}^2 = U \pm \sqrt{(U^2 - W)} \tag{32}$$

is positive real. One evidently must have $U^2 - W > 0$. If $W > 0$ for some value of σ, then $\sqrt{(U^2 - W)} < |U|$ and $\kappa_{o,e}^2 \gtrless 0$ if $U \gtrless 0$; thus, both the ordinary and extraordinary modes propagate when $U > 0$, and neither propagates when $U < 0$. If $W < 0$, then $\sqrt{(U^2 - W)} > |U|$; in this instance, the ordinary mode propagates for $U \gtrless 0$ while the extraordinary mode does not. The points $\sigma_{3,4}$ at which $U^2 - W = 0$, or $\Delta = 0$, are obtained from (4e) as

$$\sigma_{3,4}^2 = 2\delta[\delta \pm \sqrt{(\delta^2 - 1)}]\,. \tag{33}$$

If $\delta > 1$, these points lie on the real σ axis and give rise to σ-intervals for which $\kappa_{o,e}$ is complex.

While the specification of the function $\kappa_{o,e}(\sigma)$, when real, can be carried out analytically from a study of (31), there also exists a simple graphical procedure utilizing the dispersion curves $\kappa(\sigma)$ vs. σ which locate the real solutions of the dispersion relation in the $\kappa - \sigma$ plane (see (A9) and (A15)). For a plasma medium under the influence of a steady external magnetic field, these dispersion curves fall into various separate categories (distinguished by different ranges of the values of the applied signal, plasma, and cyclotron frequencies) and have been investigated in great detail in connection with plane wave propagation in the ionosphere.[2,3,24,25,26] In particular, it has been shown that a plane wave characterized by the variation

$$e^{-jk[\xi x+\eta y+\kappa(\sigma)z]} = e^{-jk\mathbf{p}\cdot\mathbf{r}}, \tag{34}$$

where $\mathbf{p} = \mathbf{x}_o\xi + \mathbf{y}_o\eta + \mathbf{z}_o\kappa$ is the wave normal, carries real power in a direction perpendicular to the dispersion curve at the point (ξ, η, κ), and that the angle between the real part of the complex Poynting vector $\mathbf{S}$ and the wave normal is less than 90°.[10,26] The direction of energy flow is commonly called the "ray" direction, and the above-mentioned relations are schematized in Fig. 4 for a typical case for which the dispersion curve has a closed and an open branch. The dispersion curves (or refractive index curves†) can be shown to be rotationally symmetric about the d.c. magnetic field direction (z axis) whence a cross-section in the $x - z$ plane suffices. In Fig. 4, the spatial coordinate axes have been superposed upon the $\kappa - \xi$ axes, so that the vector directions of the wave normal and ray are directly those in the $x - z$ plane.

The radiation condition (31) can now be interpreted as corresponding to those values of $\kappa_{o,e}$ and σ for which Re $\mathbf{S}$ has a component in the $+z$ direction. Since $\mathbf{p}\cdot(\text{Re}\,\mathbf{S}) > 0$, the pertinent segments of the dispersion curves are those shown shaded in Fig. 4. One notes that it is possible to have $\mathbf{p}\cdot\mathbf{z}_o = \kappa < 0$ while $(\text{Re}\,\mathbf{S})\cdot\mathbf{z}_o > 0$, corresponding to a "backward" wave in which the directions of energy and phase propagation along z are opposite. While this simple graphical construction allows the identification of those portions of the dispersion curve which contribute propagating waves carrying power in the positive z-direction, it gives no information as to which curve segments correspond to κ_o or κ_e as defined in (23a, b). The assignment of the proper ordinary and extraordinary branches defined herein requires a further study

† Upon writing $\exp[-jk\mathbf{p}\cdot\mathbf{r}] = \exp[-jp\mathbf{k}\cdot\mathbf{r}]$, where k is the wavenumber defined in (3d) and the vector $\mathbf{k}$ is parallel to $\mathbf{p}$, one may interpret p as the refractive index (with respect to a medium having a dielectric constant $\epsilon = \epsilon_0\epsilon_z$) for the wave traveling in the direction $\mathbf{k}$. Since $p^2 - \kappa^2 + \sigma^2$, the distance from the origin to the κ vs. σ surface yields the magnitude of p as a function of the polar angle θ measured from the direction of the applied d.c. magnetic field. Hence, this surface also constitutes the "refractive index surface" for the medium; the latter designation is used extensively in ionospheric propagation theory.

of the behavior of the functions U and W as discussed in connection with (32). These remarks are developed further in the next section where we analyze in detail the nature of the dispersion curves, and the associated disposition of $\kappa_{o,e}$, for a plasma medium.

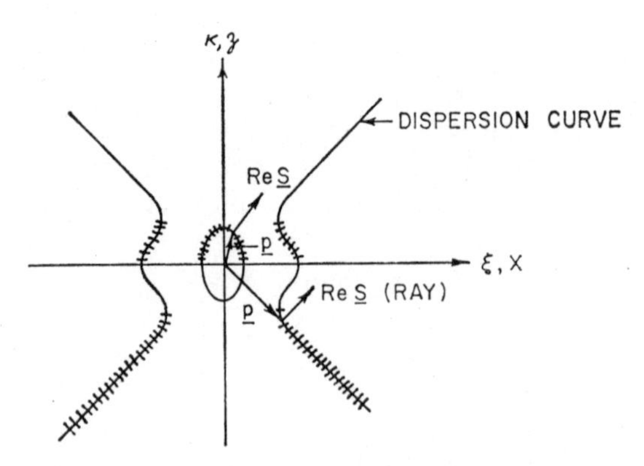

FIG. 4. Dispersion curve, wave normal, and ray.

4. THE FUNCTION $\kappa(\sigma)$ FOR A PLASMA MEDIUM

An ionized plasma medium under the influence of a steady external magnetic field H_o along the z-axis can be characterized by the following constitutive parameters:[2,3]

$$\varepsilon_1 = 1 + \frac{\omega_c^2\omega_p^2}{(\omega_c^2-\omega^2)(\omega^2-\omega_p^2)}, \quad \varepsilon_2 = \frac{\omega\omega_c\omega_p^2}{(\omega_c^2-\omega^2)(\omega^2-\omega_p^2)}, \quad \varepsilon_z = 1 - \frac{\omega_p^2}{\omega^2}, \tag{35}$$

where ω, ω_p and ω_c are the angular applied, plasma, and cyclotron frequencies, respectively. In this simple description of a plasma, only the electrons are considered mobile, and collisions with ions and neutral particles are neglected; also, the impressed a.c. field amplitudes are required to be small. In terms of the electron density N, the electronic charge e, and the electronic mass m, one may express ω_p and ω_c as:

$$\omega_p^2 = \frac{eN}{m\varepsilon_o}, \qquad \omega_c = \frac{\varepsilon\mu_o H_o}{m}. \tag{35a}$$

To assess the properties of κ as a function of σ, it is necessary to investigate the behavior of U and W (see (8b) for $\varepsilon_z > 0$) for various values of ω, ω_c, ω_p. To facilitate this analysis, the parameters ε_1, $2\varepsilon_1/(\varepsilon_1+1)$, σ_2^2, σ_3^2, σ_4^2 have been plotted in Fig. 5 as functions of ω. Figure 5(a) pertains to the case $\omega_p < \omega_c$, while $\omega_p > \omega_c$ in Fig. 5(b). If a scale for σ^2 is superimposed along the positive imaginary axis, the plot may also be used to exhibit the ranges of σ^2 which, at a given frequency ω, correspond to propagating (real κ) or non-propagating

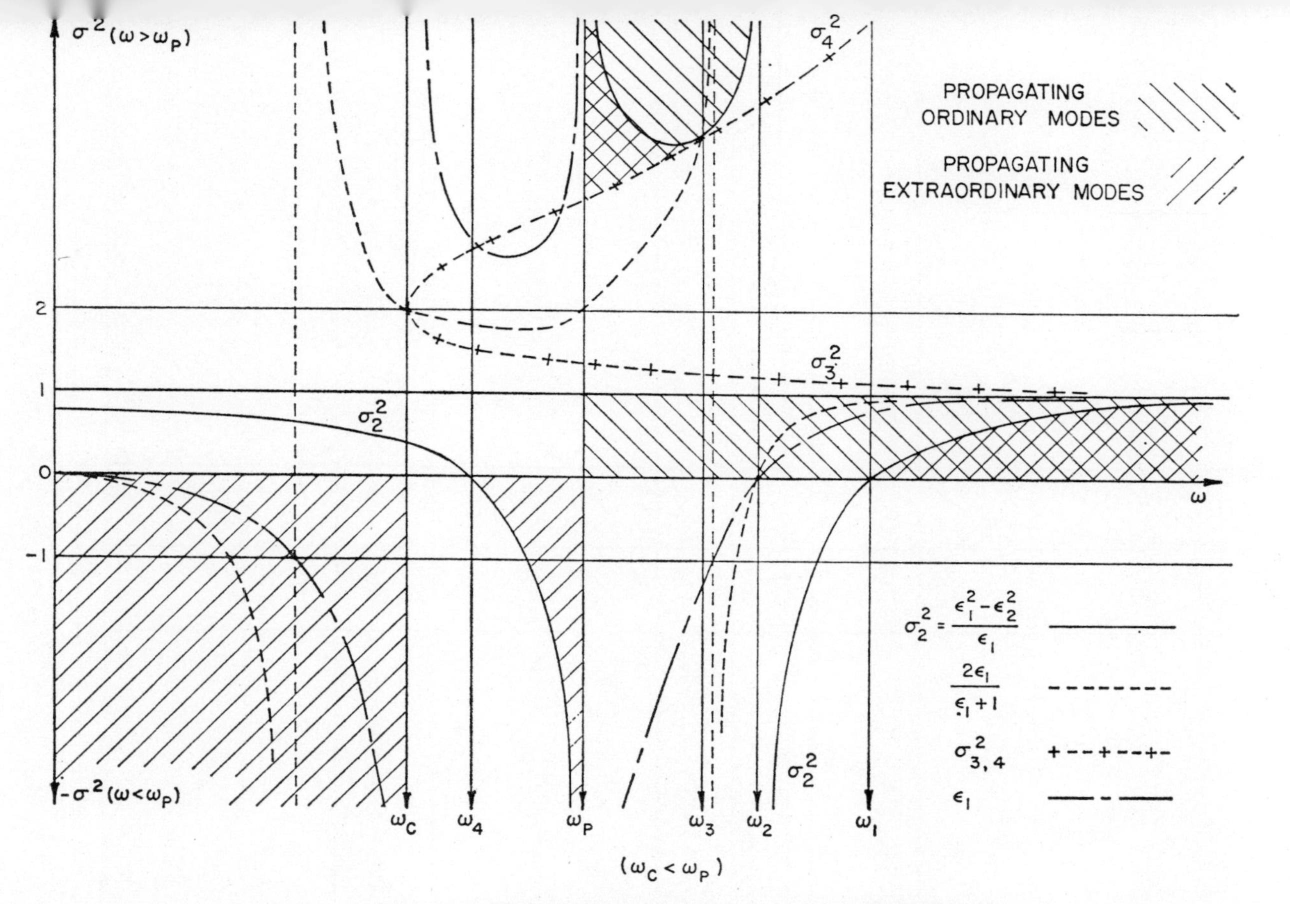

FIG. 5*a*. Frequency dependence of various plasma parameters.

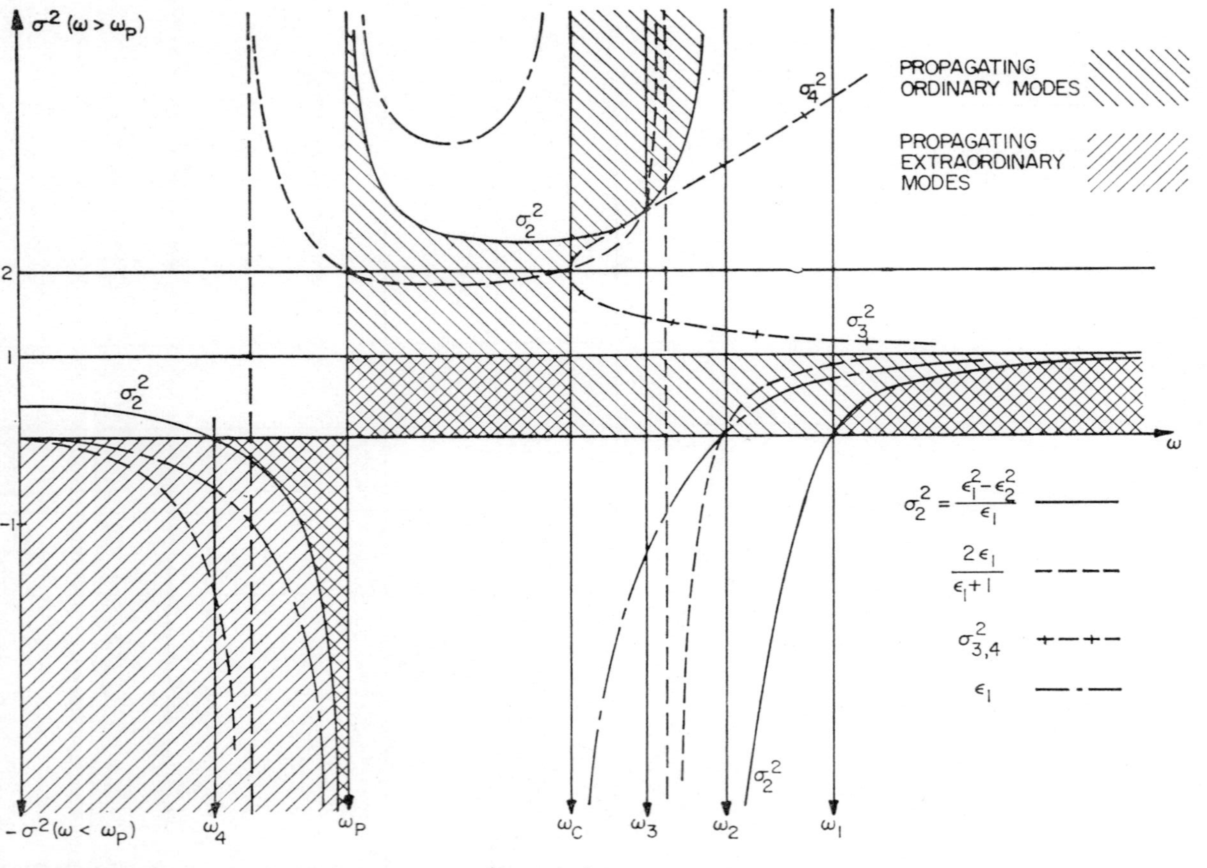
PROPAGATING ORDINARY MODES
PROPAGATING EXTRAORDINARY MODES
$\sigma_2^2 = \frac{\epsilon_1^2 - \epsilon_2^2}{\epsilon_1}$
$\frac{2\epsilon_1}{\epsilon_1 + 1}$
$\sigma_{3,4}^2$
ϵ_1
$\sigma^2 (\omega > \omega_p)$
$-\sigma^2 (\omega < \omega_p)$
ω
ω_4
ω_p
ω_c
ω_3
ω_2
ω_1
$(\omega_p < \omega_c)$
σ_2^2
σ_3^2
σ_4^2
2
1
−1

(complex κ) waves (see (32) et seq.). These propagating wave domains have been shaded in Fig. 5; no propagation obtains outside the shaded regions. Since the defining expressions for $\kappa_{o,e}$ change when $\varepsilon_z < 0$, i.e. when $\omega < \omega_p$ (see (A15)), a separate scale for σ^2 along the negative vertical axis is used in this range. The frequency domain is subdivided naturally into the various intervals exhibited in Fig. 5. ω_1 and ω_4 denote, respectively, the larger and smaller of the frequencies for which $\sigma_2^2 = 0$, ω_2 corresponds to $\varepsilon_1 = 0$, while $2\varepsilon_1/(\varepsilon_1 + 1) = \sigma_2^2$ at ω_3.

The behavior of $\kappa_{o,e}$ for waves satisfying the radiation condition at $z \to +\infty$ in the various frequency ranges is summarized in Table 1. The first column lists the branch point singularities which lie on the positive real σ-axis in the ordinary and extraordinary integrals. The notation a^+ or a^- signifies that a branch point lies at $\sigma = a$, and that the path of integration is indented around it into the upper or lower half of the σ-plane, respectively. To each $a^\pm$ there corresponds a $(-a)^\mp$, i.e. a branch point on the negative real axis, which is not listed explicitly.

Typical dispersion curves in the various frequency ranges are shown in the left half of Fig. 6. These curves represent the real $\kappa - \sigma$ solutions of eqns (A9) or (A15) and the portions corresponding to the ordinary and extraordinary waves, as defined in (23a, b), are labeled "*o*" and "*e*", respectively.† (If no curve is shown, the corresponding wave type does not propagate in this frequency range.) The singularities of $\kappa_o(\sigma)$ and $\kappa_e(\sigma)$ along the real σ-axis are exhibited in the right half of the figure. (The branch points at $\sigma_{3,4}$ in (33) lie on the real axis only when $\delta = \varepsilon_2/(\varepsilon_1 - 1) = \omega/\omega_c > 1$). On the darkened segments, κ is real; elsewhere it is complex. The proper continuation of the functions $\kappa_{o,e}(\sigma)$ around the various branch point singularities is effected in accord with conditions (24) and (31), as summarized in Table 1. To render these functions unique in the multisheeted complex σ-plane, branch cuts have been drawn. This partitioning removes any ambiguities as to the disposition of the integration path with respect to singularities of $\kappa_{o,e}$; the path is distorted around the branch points in a manner illustrated in Fig. 6(a) (a transformation of variables in (7a, b) from $\xi - \eta$ to σ can be accomplished via (Bl)).‡ The integrands in (7a, b) will generally also contain pole singularities on the real σ-axis which must be investigated separately. Such an investigation has been carried out in connection with the problem of radiation from an anisotropic half space (see reference 8). The discussion herein suffices for the complete analysis of radiation in an infinite anisotropic medium where pole singularities

† The asymptotes of the open branches of the dispersion curves (Figs. 6c, d, f, i) can be found by looking for real κ, σ solutions in (A9) and (A15) as $\kappa, \sigma \to \infty$. Solutions exist only when $\epsilon_1 < 0$, in which instance $\kappa \to \pm\sqrt{(-\epsilon_1)}\sigma$ as $\sigma \to \infty$. The open branches arise for the ordinary and extraordinary modes when $\epsilon_z > 0$ and $\epsilon_z < 0$, respectively.

‡ If the range of propagating modes extends to $\sigma \to \pm\infty$ (see Figs. 6c, d, f, i), absolute convergence of the integrals can be secured by deforming the end points of the integration path away from the real σ-axis into a region where the integrands decay exponentially.

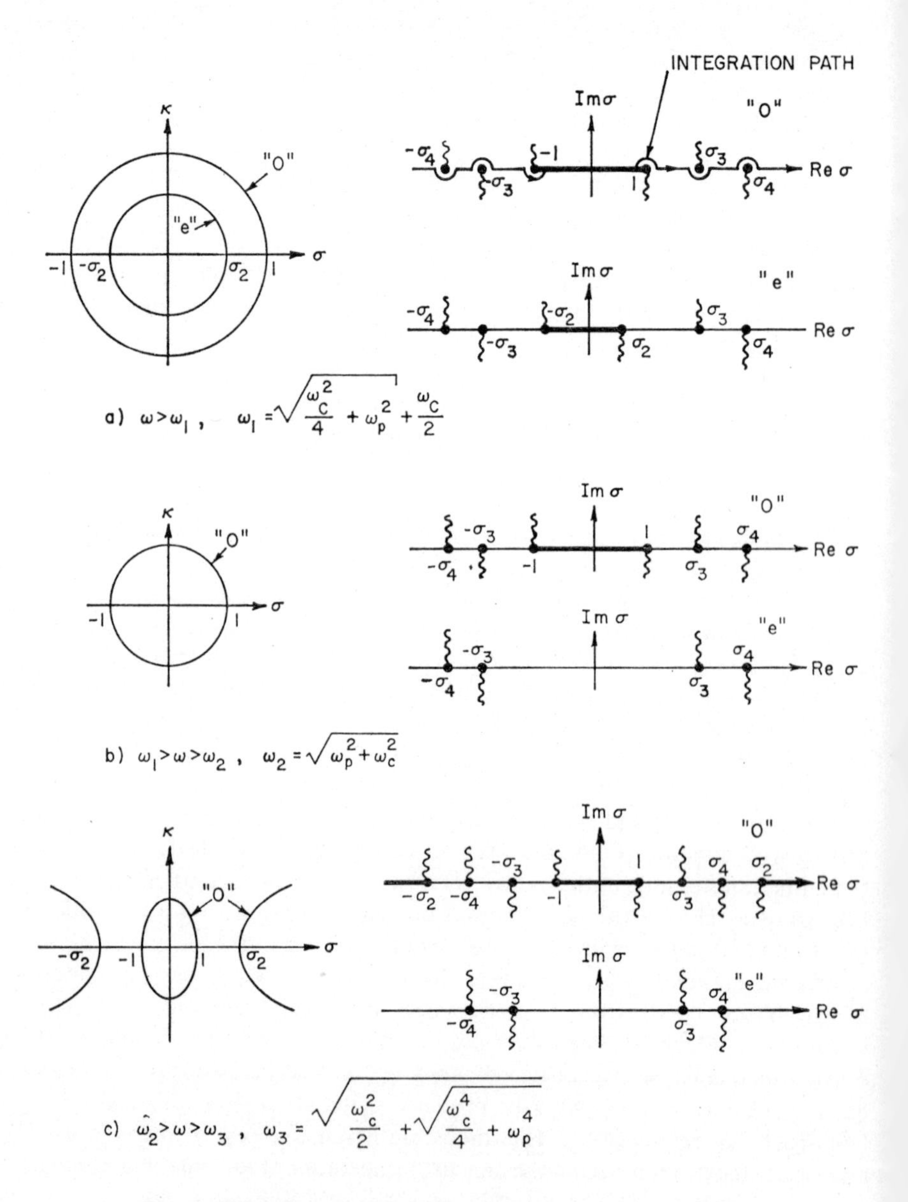

FIG. 6. Dispersion curves and singularities on Re σ-axis: (a)–(c).

The dispersion curves on the left show the behavior of κ vs. σ for propagating waves in various frequency ranges; the designations "o" and "e" distinguish the ordinary and extraordinary bıanches, respectively. Singularities due to $\kappa_{o,e}(\sigma)$ on the real σ-axis are shown on the right. The dark segments denote propagating wave regions.

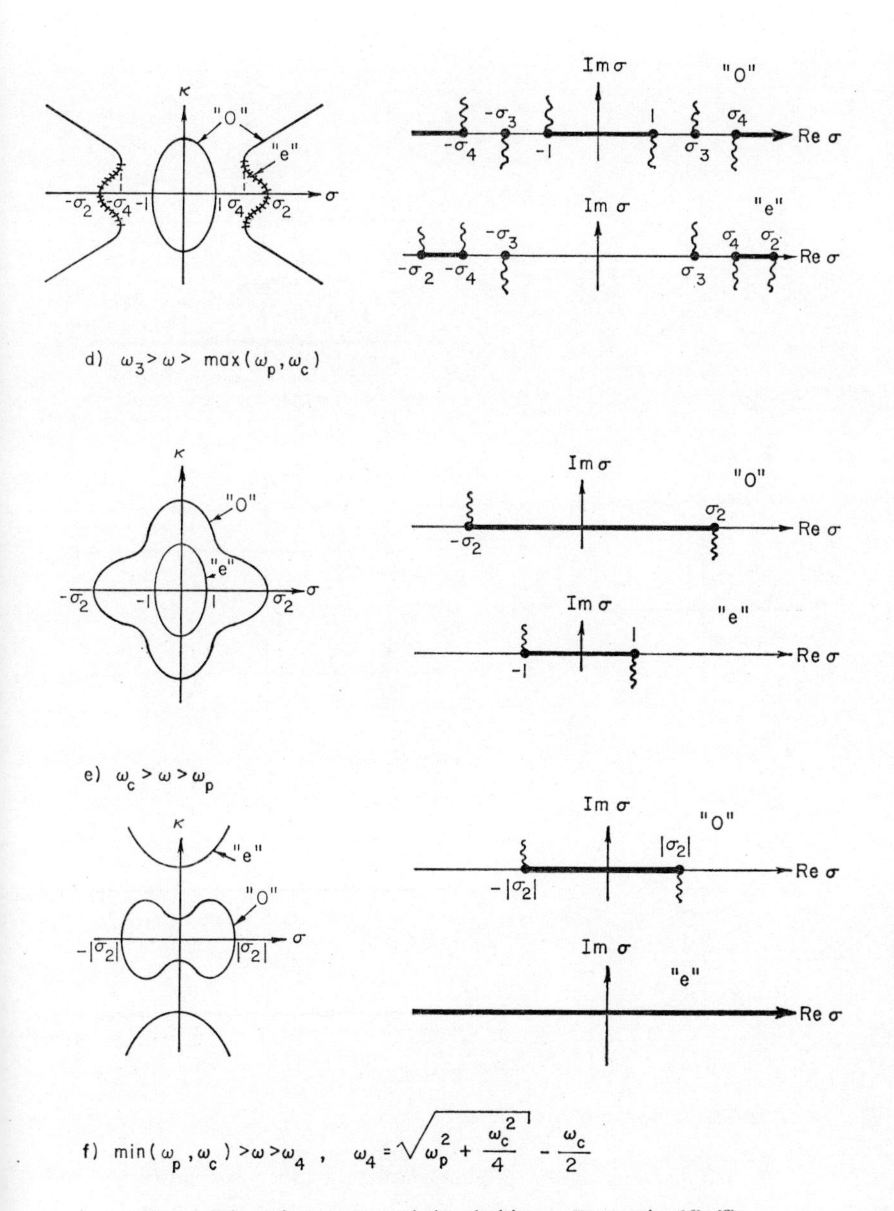

FIG. 6. Dispersion curves and singularities on Re σ axis: (d)–(f).

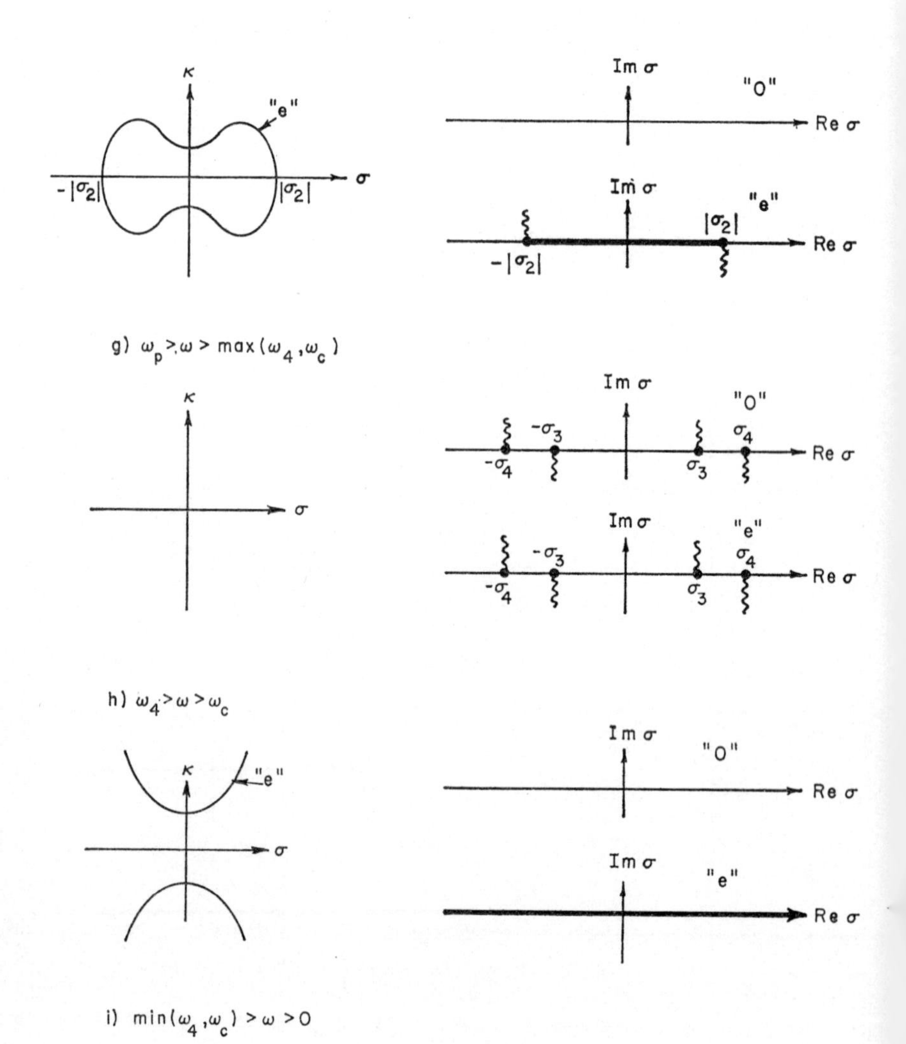

FIG. 6. Dispersion curves and singularities on Re σ axis: (g)–(i).

do not arise. Application of these results to the detailed study of the far field radiated by a dipole source is contained in Part II of this paper.

TABLE 1

Case		Singularities (on real axis)	Range of propagating modes	Sign of κ (when real)
A	o	$1^+, \sigma_3^-, \sigma_4^+$	$\sigma^2 < 1$	$\kappa_o > 0$
$\omega > \omega_1$	e	$\sigma_2^+, \sigma_3^-, \sigma_4^+$	$\sigma^2 < \sigma_2^2$	$\kappa_e > 0$
B	o	$1^+, \sigma_3^-, \sigma_4^+$	$\sigma^2 < 1$	$\kappa_o > 0$
$\omega_1 > \omega > \omega_2$	e	σ_3^-, σ_4^+	none	
C	o	$1^+, \sigma_3^-, \sigma_4^+, \sigma_2^+$	$\sigma^2 < 1,\ \sigma_2^2 < \sigma^2 < \infty$	$\kappa_o > 0,\ \kappa_o < 0$
$\omega_2 > \omega > \omega_3$	e	σ_3^-, σ_4^+	none	
D	o	$1^+, \sigma_3^-, \sigma_4^+$	$\sigma^2 < 1,\ \sigma_4^2 < \sigma^2 < \infty$	$\kappa_o > 0,\ \kappa_o < 0$
$\omega_3 > \omega > \max(\omega_p, \omega_c)$	e	$\sigma_3^-, \sigma_4^+, \sigma_2^+$	$\sigma_4^2 < \sigma^2 < \sigma_2^2$	$\kappa_e > 0$
E	o	σ_2^+	$\sigma^2 < \sigma_2^2$	$\kappa_o > 0$
$\omega_c > \omega > \omega_p$	e	1^+	$\sigma^2 < 1$	$\kappa_e > 0$
F	o	$\lvert\sigma_2\rvert^+$	$\sigma^2 < \lvert\sigma_2^2\rvert$	$\kappa_o > 0$
$\min(\omega_p, \omega_c) > \omega > \omega_4$	e	none	$\sigma^2 < \infty$	$\kappa_e > 0$
G	o	none	none	
$\omega_p > \omega > \max(\omega_4, \omega_c)$	e	$\lvert\sigma_2\rvert^+$	$\sigma^2 < \lvert\sigma_2^2\rvert$	$\kappa_e > 0$
H	o	σ_3^-, σ_4^+	none	
$\omega_4 > \omega > \omega_c$	e	σ_3^-, σ_4^+	none	
I	o	none	none	
$\min(\omega_4, \omega_c) > \omega > 0$	e	none	$\sigma^2 < \infty$	$\kappa_e > 0$

APPENDIX A

Determination of the Vector Eigenfunctions

The transverse electromagnetic fields excited by arbitrary electric or magnetic current distributions in the anisotropic medium are to be represented as a superposition of transverse vector modal solutions which individually satisfy the source-free field equations and the boundary conditions in the transverse domain. To eliminate the z-dependence from (3a, b), characterized by the translation operator $\partial/\partial z$, we assume a separable representation of the form

$$\mathbf{E}_{ti}(\mathbf{r}) = \bar{\mathbf{e}}_i(x, y)\, e^{-jk\kappa_i z}, \qquad k = k_o\sqrt{\varepsilon_z} > 0\,, \tag{A1a}$$

$$\mathbf{H}_{ti}(\mathbf{r}) = Y_i \bar{\mathbf{h}}_i(x, y)\, e^{-jk\kappa_i z}\,, \tag{A1b}$$

where κ_i and the normalization parameter Y_i are constant, and the subscript i

represents the modal index. Equations (3a, b) then define the vector eigenvalue problem for the transverse mode functions $\bar{\mathbf{e}}_i$ and $\bar{\mathbf{h}}_i$:

$$\kappa_i \bar{\mathbf{e}}_i = Y_i \zeta \left[\mathbf{1}_t + \frac{\nabla_t \nabla_t}{k^2} \right] \cdot (\bar{\mathbf{h}}_i \times \mathbf{z}_o), \; \zeta = \sqrt{\left(\frac{\mu_o}{\varepsilon} \right)}, \tag{A2a}$$

$$Y_i \kappa_i \zeta \bar{\mathbf{h}}_i = \left[\boldsymbol{\epsilon}_t + \frac{\nabla_t \nabla_t}{k^2} \right] \cdot (\mathbf{z}_o \times \bar{\mathbf{e}}_i) . \tag{A2b}$$

Since the transverse $(x - y)$ domain is unbounded, the eigenfunctions have a plane wave dependence characterized by

$$\bar{\mathbf{e}}_i = \frac{k}{2\pi} \mathbf{e}(\xi, \eta) \, \mathrm{e}^{-jk(\xi x + \eta y)}, \; \bar{\mathbf{h}}_i = \frac{k}{2\pi} \mathbf{h}(\xi, \eta) \, \mathrm{e}^{-jk(\xi x + \eta y)}, \tag{A3}$$

where the continuous spectrum of the normalized wavenumbers ξ and η runs over the values $-\infty < (\xi, \eta) < \infty$, and the factor $(k/2\pi)$ has been included for convenience in normalization. Since the operator ∇_t in (A2a, b) is now replaceable by $-jk\boldsymbol{\sigma}$, $\boldsymbol{\sigma} = (\mathbf{x}_o \xi + \mathbf{y}_o \eta)$, one obtains the algebraic equations for the transverse position vectors $\mathbf{e}$ and $\mathbf{h}$ (for fixed ξ and η),

$$\kappa \mathbf{e} = Y\zeta \mathbf{R} \cdot (\mathbf{h} \times \mathbf{z}_o), \qquad Y\zeta\kappa(\mathbf{h} \times \mathbf{z}_o) = \mathbf{S} \cdot \mathbf{e}, \tag{A4}$$

where $\mathbf{R}$ and $\mathbf{S}$ are self-adjoint 2×2 dyadics (note: $\boldsymbol{\varepsilon}_t = \boldsymbol{\epsilon}_t^+$),

$$\mathbf{R} = \mathbf{1}_t - \boldsymbol{\sigma\sigma} = \mathbf{R}^+, \qquad \mathbf{S} = \boldsymbol{\epsilon}_t - (\mathbf{z}_o \times \boldsymbol{\sigma})(\mathbf{z}_o \times \boldsymbol{\sigma}) = \mathbf{S}^+. \tag{A4a}$$

The superscript $^+$ denotes the adjoint operator. Equations (A4) can be re-expressed as the eigenvalue problem

$$\mathbf{P} \cdot \mathbf{e} = \kappa^2 \mathbf{e}, \qquad \mathbf{P}^+ \cdot (\mathbf{h} \times \mathbf{z}_o) = \kappa^2 (\mathbf{h} \times \mathbf{z}_o), \qquad \mathbf{P} = \mathbf{R} \cdot \mathbf{S}. \tag{A5}$$

It follows from the theory of linear operators that the two eigenvectors $\mathbf{e}$ and $\mathbf{h} \times \mathbf{z}_o$ belonging to the same eigenvalue κ^2 satisfy a bi-orthogonality relation as in (5). The subscripts o and e are employed to distinguish the two possible eigenvalues, and associated eigenvectors, in (A5).

Equations (A5) are solved readily in a basis comprising the orthogonal position vectors $\boldsymbol{\sigma}$ and $\boldsymbol{\sigma} \times \mathbf{z}_o$, wherein we represent

$$\mathbf{e} = a\boldsymbol{\sigma} + b\boldsymbol{\sigma} \times \mathbf{z}_o, \qquad \mathbf{h} = c\boldsymbol{\sigma} + \mathrm{d}\boldsymbol{\sigma} \times \mathbf{z}_o . \tag{A6}$$

A straight-forward calculation leads to

$$\hat{P} \begin{bmatrix} a \\ b \end{bmatrix} = \kappa^2 \begin{bmatrix} a \\ b \end{bmatrix}, \qquad \hat{P}^+ \begin{bmatrix} c \\ d \end{bmatrix} = \kappa^2 \begin{bmatrix} c \\ d \end{bmatrix}, \tag{A7}$$

$$\hat{P} \rightarrow \begin{bmatrix} \varepsilon_1 (1 - \sigma^2) & -j\varepsilon_2 (1 - \sigma^2) \\ j\varepsilon_2 & (\varepsilon_1 - \sigma^2) \end{bmatrix}, \; \sigma^2 = \xi^2 + \eta^2, \tag{A7a}$$

from which one obtains the normalized eigenvectors in (4). The eigenvalues $\kappa_{o,e}^2$ are the two solutions of the dispersion equation

$$\mathscr{H}(\kappa, \sigma) \equiv \det(\hat{P} - \kappa^2) = 0 \tag{A8}$$

or

$$\kappa^4 + \kappa^2[\sigma^2(\varepsilon_1 + 1) - 2\varepsilon_1] + \varepsilon_1\sigma^4 - (\varepsilon_1^2 - \varepsilon_2^2 + \varepsilon_1)\sigma^2 + \varepsilon_1^2 - \varepsilon_2^2 = 0, \tag{A9}$$

which yields

$$\kappa_{\substack{o\\e}}^2 = U \pm \sqrt{(U^2 - W)} = U \pm (\varepsilon_1 - 1)\Delta/2, \tag{A10}$$

with U, W, and Δ defined in (8b) and (4e), respectively. Evidently,

$$\kappa_{\substack{o\\e}}^2 = \kappa_{\substack{o\\e}}^{*2} \qquad \text{when } U^2 > W(\Delta \text{ real}), \tag{A11a}$$

and

$$\kappa_{\substack{o\\e}}^2 = \kappa_{\substack{e\\o}}^{*2} \qquad \text{when } U^2 < W(\Delta \text{ imaginary}). \tag{A11b}$$

(A9) is the "Booker quartic" for the longitudinal wave number κ (κ is identical with Brooker's q, save for the normalization to k instead of k_o; i.e. $q = \sqrt{\varepsilon} \cdot \kappa$) (see reference 3, eqn (13.13), with $\beta = \delta = 0$).

The subscripts o and e distinguish the two solutions corresponding to the $+$ and $-$ signs in (A10), respectively. The respective definitions "extraordinary" and "ordinary" are commonly applied in plasma theory to those real solutions of the dispersion equation which correspond to waves whose propagation in a direction transverse to the z-axis is, or is not, affected by the presence of the d.c. magnetic field (along z).† Transverse propagation corresponds to $\kappa = 0$; for the ordinary mode one then has $\sigma = 1$, while for the extraordinary mode, $\sigma \neq 1$. It has been customary to label as "ordinary" and "extraordinary" those (real) branches of the κ vs. σ (dispersion) curves which do, or do not, pass through the points $\kappa = 0$, $\sigma = \pm 1$. For our purposes, however, it is more significant to effect a definition on the basis of the analytic properties of the multivalued functions $\kappa_o(\sigma)$ and $\kappa_e(\sigma)$ as defined in (8a). The resulting apportionment, between κ_o and κ_e, of the real branches of the dispersion curves will then not necessarily coincide with that mentioned above (see Section 4). We have nevertheless retained the terminology "ordinary" and "extraordinary" for the o- and e-solutions, respectively, because the real solutions of κ_o do usually include the points $\kappa_o = 0$, $\sigma_o = \pm 1$.

From (A4) and the orthogonality relations (5), one deduces for the characteristic admittance Y:

$$Y_{o,e} = \frac{1}{Z_{o,e}} = \frac{\mathbf{e}_{o,e}^* \cdot \mathbf{S} \cdot \mathbf{e}_{o,e}}{\zeta \kappa_{o,e}} = \frac{\kappa_{o,e}}{\zeta \mathbf{h}_{o,e}^* \times \mathbf{z}_o \cdot \mathbf{R} \cdot \mathbf{h}_{o,e} \times \mathbf{z}_o}. \tag{A12}$$

Substitution of (4d) into the last expression in (A12) yields formula (8a).

If one of the regions is isotropic, $\varepsilon_1 = 1$ and $\varepsilon_2 = 0$ so that $\boldsymbol{\varepsilon}_t = \mathbf{1}_t$. In this instance, $\kappa_o^2 = \kappa_e^2 = 1 - \sigma^2$, and it is convenient to choose the conventional linearly polarized E mode (single primes) and H mode (double primes) eigenfunctions

† When the magnetic field is absent, $\epsilon_2 = 0$ and $\epsilon_1 = 1$.

$$\mathbf{e}' = \mathbf{h}' \times \mathbf{z}_o = \frac{\boldsymbol{\sigma}}{\sigma}, \qquad \mathbf{e}'' = \mathbf{h}'' \times \mathbf{z}_o = \frac{\boldsymbol{\sigma} \times \mathbf{z}_o}{\sigma}. \tag{A13}$$

It can then be shown that the resulting transverse field representation is still given by (7a, b) provided that we replace all subscripts o by single primes and all subscripts e by double primes.† The E and H mode voltages V', V'' and the currents I', I'' satisfy the transmission line eqns (8), with the characteristic impedances defined as

$$Z' = \frac{1}{Y'} = \frac{k\kappa}{\omega\varepsilon}, \qquad Z'' = \frac{1}{Y''} = \frac{\omega\mu}{k\kappa}, \qquad \kappa = \sqrt{(1-\sigma^2)}\,. \tag{A14}$$

The equivalent network for the mode coupling produced by a plane interface between an isotropic and an anisotropic region can then be deduced by proceeding as in (10)–(11).

If $\varepsilon_z < 0$, i.e. $k = -j|k|$, the normalization of the transverse and longitudinal wave numbers ξ, η, and κ is taken with respect to $|k|$. Hence, k in (A1a) and (A3) should be replaced by $|k|$, whence ζ in (A2a) is defined as $\sqrt{(\mu_o/|\varepsilon|)}$. The resulting eigenvalue problem is then found to be the same as in (A5) provided that one replaces σ by $(+j\sigma)$ and $\mathbf{S}$ by $(-\mathbf{S})$. The dispersion relation is now given by

$$\kappa^4 + \kappa^2[\sigma^2(\varepsilon_1+1) + 2\varepsilon_1] + \varepsilon_1\sigma^4 + (\varepsilon_1^2 - \varepsilon_2^2 + \varepsilon_1)\sigma^2 + \varepsilon_1^2 - \varepsilon_2^2 = 0\,, \tag{A15}$$

i.e. (A9) applies provided that κ^2 and σ^2 are replaced by $(-\kappa^2)$ and $(-\sigma^2)$, respectively. Finally, the eigenvectors are given as in eqns (4), if k and σ are replaced by $|k|$ and $(+j\sigma)$, respectively (note: $\boldsymbol{\sigma} \to +j\boldsymbol{\sigma}$). In summary, formulas (7)–(9) appropriate to $\varepsilon_z < 0$ can be obtained from those for $\varepsilon_z > 0$ by letting $\sqrt{\varepsilon_z} \to -j\,|\sqrt{\varepsilon_z}|$, $\sigma \to j\sigma$, $\kappa \to j\kappa$.

APPENDIX B

Cylindrical Wave Representation of the Fields

Instead of the plane wave representation in (7a, b) it is frequently more convenient to employ a cylindrical wave representation based on the well-known transformation

$$\begin{aligned} I &= \int_{-\infty}^{\infty} \mathrm{d}\xi \int_{-\infty}^{\infty} \mathrm{d}\eta f(\sigma)\, \mathrm{e}^{-jk(\xi x + \eta y) + jk(\xi x' + \eta y')} \\ &= 2\pi \sum_{n=-\infty}^{\infty} \mathrm{e}^{-jn(\phi-\phi')} \int_0^{\infty} \sigma f(\sigma) J_n(k\sigma\rho) J_n(k\sigma\rho')\, \mathrm{d}\sigma, \qquad k > 0, \end{aligned} \tag{B1}$$

where

$$\sigma = \sqrt{(\xi^2 + \eta^2)}, \quad x = \rho\cos\phi, \quad y = \rho\sin\phi, \quad x' = \rho'\cos\phi', \quad y' = \rho'\sin\phi'. \tag{B2}$$

† The validity of the formal expressions obtained via this replacement does not necessarily imply that the E and H modes correspond to the limiting values of the o- and e-modes, respectively, as $\varepsilon_t \to \mathbf{1}_t$.

Each of the constituent integrals in (7a, b) resulting upon substitution of (4c, d) can be reduced to the form I upon observing that the vector $\boldsymbol{\sigma}$ can be replaced by the vector operator $(j/k)\nabla_t$, and that the differentiation and integration operations can then be commuted. The scalar constituents of $\mathbf{e}_{o,e}$ and $\mathbf{h}_{o,e}$ remaining in the integrands are functions of σ only. The voltage and current $V_{o,e}$ and $I_{o,e}$ are solutions of the differential equations (8). Since the parameters κ, Y, Z, i, v in this equation depend on (ξ, η) only through σ, as do the coupling matrices (11) and (15) descriptive of interface or terminal effects, $V_{o,e}$ and $I_{o,e}$ can be expressed as functions of σ only, and a cylindrical wave representation is effected as in (B1). The explicit appearance of the factor $\exp[jk(\xi x' + \eta y')]$ in the integrand of (B1) results from consideration of a point current element located at $\mathbf{r}' = (x', y', z') \to (\rho', \phi', z')$ (cf. (9a, b)). The expression for a distributed source is obtained by integration over the source point coordinate $\mathbf{r}'$.

If $f(\sigma) = f(-\sigma)$, i.e. f is an even function of σ, the σ integral can be converted to run over the entire real σ-axis:

$$2\int_0^\infty \sigma f(\sigma) J_n(k\sigma\rho) J_n(k\sigma\rho')\, d\sigma = \int_{\infty e^{-j\pi}}^{\infty} \sigma f(\sigma) J_n(\sigma\rho_<) H_n^{(2)}(\sigma\rho_>)\, d\sigma. \tag{B3}$$

$\rho_>$ and $\rho_<$ denote the greater and lesser of the variables ρ and ρ', respectively. The second representation in (B3), involving an infinite integration contour, is particularly convenient for an asymptotic evaluation of the integral.

ACKNOWLEDGEMENT

The work described herein was sponsored by the U.S. Air Force Cambridge Research Laboratories under Contract No. AF-19(604)-4143. It is based on a dissertation prepared by one of the authors (E.A.) in partial fulfilment of the requirements for the D.E.E. degree at the Polytechnic Institute of Brooklyn.

REFERENCES

1. Born, M. and Wolf, E. (1959) *Principles of Optics*, Ch. XIV. Pergamon Press.
2. Ratcliffe, J. A. (1960) *The Magneto-Ionic Theory and Its Application to the Ionosphere*. Cambridge Univ. Press.
3. Budden, K. G. (1961) *Radio Waves in the Ionosphere*. Cambridge Univ. Press.
4. Clarricoats, P. J. B. (1961) *Microwave Ferrites*. Wiley.
5. Jelley, J. (1958) *Cerenkov Radiation and Its Applications*, Pergamon Press.
6. Marcuvitz, N. and Schwinger, J. (1951) *J.A.P.*
7. Bresler, A. D. (1959) *I.R.E. Trans. PGMTT*, MTT–7, 282.
8. Arbel, E. (1960) Radiation from a Point Source in an Anisotropic Medium, Rep. PIBMRI-861-60, Microwave Res. Inst., Polytech. Inst. of Brooklyn, Nov. 1960. See also doctoral dissertation June 1961.
9. Arbel, E. and Felsen, L. B. Theory of radiation from sources in anisotropic media. Part II. Point source in infinite, homogeneous medium.
10. Abraham, L. G., Jr. (1953) Extensions of the magneto-ionic theory to radio wave propagation in the ionosphere, including antenna radiation and plane wave scattering, Rep. No. 13, School of El. Eng., Cornell U., Ithaca, N.Y., Aug. 1953.
11. Bunkin, F. V. (1957) *J. Exp. Theor. Phys.* (*USSR*), **32**.

12. KOGELNIK, H. (1960) *J. Research, Nat. Bureau Stds.* **64D**.
13. KUEHL, H. (1960) Radiation from an electric dipole in an anisotropic cold plasma, Antenna Lab. Rep. 24, Calif. Inst. of Technology, Oct. 1960; Electromagnetic radiation from a dipole in an anisotropic plasma, USCEC Rep. 79–203, Electrical Engin. Dept. U. of South. Calif., Sept. 1961. See also (1962), *J. Phys. Fluids*, Vol. 5, p. 1095.
14. MITTRA, R. (1962) Solution of Maxwell's equations in a magneto-ionic medium with sources, Sci. Rep. No. 4, Antenna Lab., U. of Illinois, Jan. 1962.
15. HODARA, H. (1962) The radiation characteristics of a slot antenna covered with a plasma slab in the presence of a magnetic field perpendicular to the slot, presented at the I.R.E. International Convention, New York, March 1962.
16. SHORE, R. and MELTZ, G. (1962) Anisotropic plasma-covered magnetic line source, *IRE Trans. PGAP*, **AP–10**, Jan. 1962.
17. ISHIMARU, A. (1962) The effect of a unidirectional surface wave along a perfectly conducting plane on the radiation from a plasma sheath, presented at the Second Symposium on the Plasma Sheath, Boston, April 1962.
18. CLEMMOW, P. C. (1962) On the theory of radiation from a source in a magneto-ionic medium, presented at the Symposium on Electromagnetic Theory and Antennas, Copenhagen, June 1962.
19. MOTZ, H. and KOGELNIK, H. (1962) Electromagnetic radiation from sources embedded in an infinite anisotropic medium, and the significance of that Poynting vector, presented at the Symposium on Electromagnetic Theory and Antennas, Copenhagen, June 1962.
20. MITTRA, R. and DESCHAMPS, G. A. (1962) Field solution for a dipole in an anisotropic medium, presented at the Symposium on Electromagnetic Theory and Antennas, Copenhagen, June 1962.
21. BRESLER, A. D. and MARCUVITZ, N. (1956) Operator methods in electromagnetic theory, Report MRI-R-495-56, Microwave Res. Inst., Polytech. Inst. of Brooklyn.
22. FELSEN, L. B. and MARCUVITZ, N. Modal analysis and synthesis of electromagnetic fields, Reports R-446-55, R-726, R-776, R-841, Microwave Res. Inst., Polytech. Inst. of Brooklyn.
23. FELSEN, L. B. and KAHN, W. K. (1959) Transfer characteristics of 2 N-port networks, Proc. Symp. on Millimeter Waves, Polytech. Inst. of Brooklyn.
24. CLEMMOW, P. C. and MULALLY, F. (1955) *The Physics of the Ionosphere*, The Physical Society (London).
25. ALLIS, W. P. (1961) *I.R.E. Trans. PGMTT*, **MTT–9**.
26. HINES, C. O. (1951) *J. Geophysical Res.* **56**, 63, 197, 207, 535.
27. CHOW, Y. (1962) A note on radiation in a gyro-electric-magnetic medium—an extension of Bunkin's calculation, *IRE Trans.* **AP–10**, 464–469.
28. MEECHAM, W. C. (1961) Source and reflection problems in magneto-ionic media, *J. Phys. of Fluids*, **4**, 1517–1524.
29. WU, C. P. (1962) A study on the radiation from elementary dipoles in a magneto-ionic medium, Report 1021–20, Dept. of Elect. Eng., Ohio State Univ., August 1962.
30. SESHADRI, S. (1962) Excitation of surface waves on a perfectly conducting screen covered with anisotropic plasma, Tech. Rep. No. 366, Cruft Lab., Harvard Univ., May 1962.
31. TUAN, H. S. and SESHADRI, S. R. (1962) Radiation from a line source in a uniaxially anisotropic plasma, Tech. Rep. No. 375, Cruft Lab., Harvard Univ., August 1962.
32. ARBEL, E. and FELSEN, L. B. (1962) On electromagnetic Green's functions for uniaxially anisotropic regions, Report PIBMRI–985–61, Microwave Res. Inst. Polytech. Inst. of Brooklyn, January 1962.
33. FELSEN, L. B. (1962) Lateral waves on an anisotropic plasma interface, *IRE Trans. PGAP*, **AP–10**, 347–349; Radiation from a uniaxially anisotropic half-space, Report PIBMRI–1058–62, Microwave Res. Inst., Polytech. Inst. of Brooklyn, August 1962.
34. HONDARA, H. and COHN, G. I. (1962) Radiation from a gyro-plasma coated magnetic line source, *PGAP*, **AP–10**, p. 581.
35. FORD, G. W. (1961) Electromagnetic radiation from a source in a plasma, *Ann. of Physics*, **16**, No. 2, 185–200.

THEORY OF RADIATION FROM SOURCES IN ANISOTROPIC MEDIA

PART II: POINT SOURCE IN INFINITE, HOMOGENEOUS MEDIUM

E. ARBEL† and L. B. FELSEN‡

ABSTRACT

The previously derived solution is specialized here to the study of the radiation field due to an electric current dipole in an infinitely extended, homogeneous anisotropic plasma. The asymptotic field solution is interpreted in terms of rays along which energy propagates from the source to the observation point. The number of contributing rays, varying from zero to four, depends on the constitutive parameters and is determined by the refractive index curves for the medium. Special attention is given to transition phenomena associated with refractive index curves which have points of inflection or unbounded branches. A set of representative radiation patterns has been calculated.

1. INTRODUCTION

In the previous part of this paper (to be referred to as I)[1], general formal solutions were derived for the electromagnetic fields radiated by arbitrary source distributions in a plane stratified medium with gyrotropic properties, with the gyrotropic axis oriented perpendicular to the planes of stratification. These solutions comprise double Fourier integrals or their equivalents, and cannot generally be evaluated in closed form. To gain an insight into the structure of the radiation field it is therefore necessary to employ asymptotic techniques which permit the reduction of the integral representations. The asymptotic results are then to be interpreted in a manner which provides a physical basis for the explanation of radiation and diffraction phenomena in anisotropic media. Basic to the interpretation of effects arising from stratification is the thorough understanding of radiation from a localized source in an infinite homogeneous anistropic region. This latter problem is considered herein. To gain some simplification of the formal expressions for the electromagnetic fields we treat the case of an oscillating electric current element oriented along the gyrotropic axis. However, the asymptotic procedures and the subsequent interpretation of the results can be applied as well to arbitrarily oriented sources.

† Dept. of Theoretical Physics, Hebrew University, Jerusalem.
‡ Dept. of Electrophysics, Polytechnic Institute of Brooklyn.

In the asymptotic evaluation we make full use of the "spatial dispersion" or refractive index surfaces[2, 3] descriptive of plane wave propagation in the medium. These surfaces (or the curves obtained by intersection with a plane passing through their axis of symmetry, the direction of the applied d.c. magnetic field) aid not only in the proper identification of the multivated propagation constants $\kappa_{o,e}(\sigma)$ which occur in the integrands of the integral representations of the fields (see I); it will be shown that their shape also bears an intimate relation both to the structure and to the amplitude of the radiation field. For the case of a plasma under the influence of a d.c. magnetic field, it is found that the far field generally consists of a superposition of several ordinary and (or) extra-ordinary ray contributions which propagate with different velocities and may produce interference along the radial direction. While each ray is found to carry power radially away from the source, the total average power flow need not be radial due to this interference.

The above-mentioned ray contributions to the far field arise from certain points on the dispersion curves determined by the real saddle points σ_s, $\kappa(\sigma_s) \equiv \kappa_s$ of the integrands. For a specified angle of observation, the normal to the curve at the various (κ_s, σ_s) is parallel to the radius vector r connecting the source and observation points. Interesting phenomena arise when a dispersion curve has a point of inflection or when it contains an open (unbounded) branch. As r approaches the direction perpendicular to the curve at an inflection point, two pertinent saddle points approach one another, and the corresponding ray contributions cannot be treated individually but must be investigated as a unit. The radial dependence of the radiation field for these rays then departs from the conventional $(1/r)$ variation and approaches instead $(1/r^{5/6})$; the detailed transition through this focusing region is discussed in Section 4. As r approaches the direction perpendicular to the asymptote of an unbounded branch of the dispersion curve, the associated saddle point moves to infinity and the asymptotic evaluation becomes invalid. The presence of an unbounded branch, like that of a turning point, implies that the associated ray contribution is confined to a certain limited region of space—for example, to a cone about the dipole axis; the region outside the cone, wherein the ray amplitude becomes exponentially small, is a "shadow" region for this ray.

The transition behavior of the electromagnetic fields through a shadow boundary associated with a turning point of the dispersion curve, as discussed in Section 4, is regular and exhibits focusing effects, but no singularities. Near a shadow boundary arising from an unbounded branch of the dispersion curve, on the other hand, the fields and the average power flow density generated by a constant current dipole take on arbitrarily large values. This behavior is linked with the "hyperbolic" nature of the field equations in this region. It is shown, however, in Section 5 that these singularities occur essentially only for idealized excitations which drop from a finite value abruptly to zero, thereby creating a sharply defined shadow. If the spatial function descriptive of the source distribution is sufficiently smooth, all quantities become regular and actually vanish along the previously singular direction; by confining the spatial extent of the source distribution, the field strengths near this shadow boundary can, however, be made as large as desired. It should be emphasized that the assumption of a lossless medium plays an important role in the derivation of these results. When losses are included the above-mentioned singularities disappear. Moreover, the dielectric tensor descriptive of the plasma is valid only in the "linear approximation" wherein the r.f. field strengths are assumed to be small. This condition is evidently violated near this shadow boundary so that the mathematical model no longer adequately represents the physical medium. Nevertheless, the associated focusing effects most likely occur also in an actual plasma wherein losses are small. It is of interest to note that the field behavior in the vicinity of a shadow boundary associated with an open branch of the dispersion curve for a plasma in a finite d.c. magnetic field is essentially the same as when the d.c. magnetic field is assumed to be infinite. The latter case, corresponding to a uniaxially anisotropic medium, permits an evaluation in a simple closed form as shown in Appendix B.

Concerning other studies of the problem of radiation from a dipole source in an infinite gyrotropic, homogeneous plasma, we cite the work of Abraham[(4)], Bunkin[(5)], Kogelnik[(6)], Kuehl[(7a,b)].† The relation between the formal aspects of these and our investigations has been discussed in I. As regards the asymptotic evaluation, none of these authors has dealt with the transition phenomena arising from the presence of points of inflection, or of open branches, of the dispersion curve. Moreover, Bunkin did not point out the possible existence of several ray contributions to the radiation field. Kuehl, following Bunkin, initially[(7a)] arrived at a similar conclusion but modified his observations after a more detailed analysis[(7b)] of the dispersive characteristics of the plasma medium; he does not utilize the refractive index surfaces which are fundamental to our approach. Abraham made use of the refractive index surfaces, and our study, though much more detailed, resembles his in this regard. An important feature of our asymptotic analysis is the use of a steepest descent, rather than a stationary phase, technique. In the steepest descent analysis, the integration path is deformed away from the real axis into steepest descent contours through the saddle points of the integrand, whence the contribution from integration path segments away from the saddle points is exponentially small and, therefore, negligible. In the stationary phase procedure, on the other hand, the integration path remains on the real axis and the integrands away from the vicinity of the saddle points do not decay

† See also contributions by Clemmow[(8)], Motz and Kogelnik[(9)], and Mittra and Deschamps[(10)], at the Symposium on Electromagnetic Theory and Antennas, held in Copenhagen, June 1962.

exponentially. While the steepest descent and stationary phase procedures generally yield the same first-order result when the integrand has no contributing singularities on the real axis, the stationary phase evaluation may be ambiguous when such singularities are present. This latter circumstance does not arise in the problem of radiation in an infinite medium, but will occur in the presence of interfaces where such additional contributions as surface waves or lateral waves are more easily picked up via the steepest descent analysis.

Attention should also be called to a paper by Lighthill[11] on radiation problems in unbounded anisotropic media, wherein the use of wave normal and ray surfaces is emphasized.

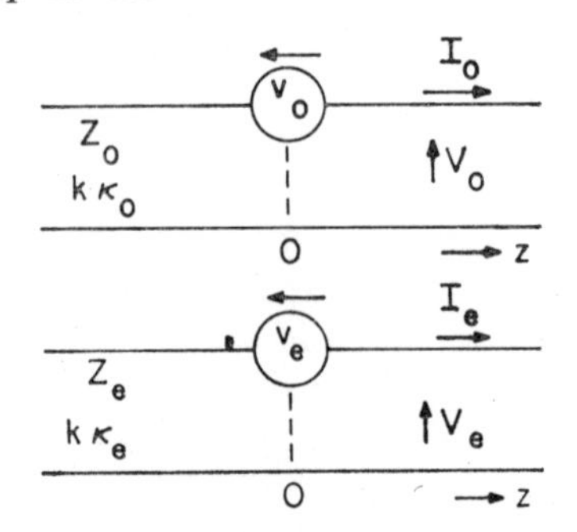

FIG. 1. Equivalent network problem.

2. SOLUTION

Consider an infinite, homogeneous, lossless plasma medium subjected to a d.c. magnetic field along the z-axis of a chosen coordinate system. Subject to well-known restrictions, the electromagnetic properties of the medium can be characterized by a dielectric tensor[12]

$$\boldsymbol{\varepsilon} \rightarrow \varepsilon \begin{pmatrix} \varepsilon_1 & j\varepsilon_2 & 0 \\ -j\varepsilon_2 & \varepsilon_1 & 0 \\ 0 & 0 & 1 \end{pmatrix}, \qquad \varepsilon = \varepsilon_o \varepsilon_z, \tag{1}$$

where

$$\varepsilon_1 = 1 + \frac{\omega_c^2 \omega_p^2}{(\omega_c^2 - \omega^2)(\omega^2 - \omega_p^2)}, \qquad \varepsilon_2 = \frac{\omega \omega_c \omega_p^2}{(\omega_c^2 - \omega^2)(\omega^2 - \omega_p^2)},$$

$$\varepsilon_z = 1 - \frac{\omega_p^2}{\omega^2}. \tag{1a}$$

ε_0 is the dielectric constant of vacuum, and ω, ω_c, ω_p are the (angular) applied electromagnetic, electron cyclotron, and electron plasma frequencies, respectively. This medium is assumed to be excited by an electric current element

$$J(\mathbf{r}) = \mathbf{z}_o J\delta(\mathbf{r}) = \mathbf{z}_o J \frac{\delta(\rho)\delta(z)}{\rho}, \tag{2}$$

oscillating with a suppressed time dependence $\exp(j\omega t)$.

The solution for the electromagnetic fields due to this dipole source can be obtained as a special case of the results in I. Since the medium is unbounded, the equivalent network problem for the ordinary (subscript o) and extra-ordinary (subscript e) modes is as shown in Fig. 1. In view of eqns (2) and (I.9),† only the voltage source terms $v_{o,e}$ are present and have a strength given by

$$v_o = v_e = -\frac{\sigma\zeta kJ}{2\sqrt{2\cdot\pi}}, \qquad \sigma = \sqrt{(\xi^2+\eta^2)}, \qquad \zeta = \sqrt{\left(\frac{\mu_o}{\varepsilon}\right)}, \qquad k = k_o\sqrt{\varepsilon_z}, \tag{3}$$

where k_o is the free-space wavenumber $k_o = \omega\sqrt{(\mu_o\varepsilon_o)}$. From Fig. 1 or from the transmission line equations (I.8), one has

$$V_{o,e}(z) = -(\operatorname{sgn} z)\frac{v_{o,e}}{2}\,e^{-jk\kappa_{o,e}|z|}, \qquad I_{o,e}(z) = \operatorname{sgn}(z)Y_{o,e}V_{o,e}(z), \tag{4}$$

where, from (I.8) and (I.A12),

$$Y_{o,e} = -\frac{\kappa_{o,e}(\sigma^2 \mp \Delta)}{\pm\Delta(1-\sigma^2)}, \qquad \Delta = \operatorname{sgn}(\varepsilon_1 - 1)\sqrt{[\sigma^4 + 4\delta^2(1-\sigma^2)]},$$

$$\delta = \frac{\varepsilon_2}{\varepsilon_1 - 1} = \frac{\omega}{\omega_c}, \tag{4a}$$

$$\kappa_{o,e} = \sqrt{[U \pm \sqrt{(U^2 - W)}]}, \qquad U = \varepsilon_1 - \frac{\varepsilon_1 + 1}{2}\sigma^2,$$

$$W = \varepsilon_1(1-\sigma^2)(\sigma_2^2 - \sigma^2), \qquad \sigma_2^2 = \frac{\varepsilon_1^2 - \varepsilon_2^2}{\varepsilon_1}. \tag{4b}$$

Equations (3) and (4) are valid for real or imaginary Δ. Upon substituting these expressions into (I.7), we obtain the formal solution for the transverse electromagnetic fields; the longitudinal field components are obtained via (I.3c). The multivalued function $\kappa_{o,e}(\sigma)$ is defined in accord with the considerations in Section 3 of I.

For an asymptotic evaluation, it will be convenient to transform the field expressions into the Fourier–Bessel representation discussed in Appendix B of I. The resulting formulas are readily found to be

$$\mathbf{E} = \mathbf{E}_o + \mathbf{E}_e, \qquad \mathbf{H} = \mathbf{H}_o + \mathbf{H}_e, \tag{5}$$

† This denotes eqn (9) of reference 1.

where the "ordinary" and "extraordinary" contributions are given by:†

$$\mathbf{E}_{t,o,e} = Aj(\operatorname{sgn} z)\int_{-\infty}^{\infty} \sigma^2 \left[\boldsymbol{\rho}_o \frac{\sigma^2 \mp \Delta}{\pm\Delta} + \boldsymbol{\phi}_o \frac{j2\delta}{\pm\Delta}\right] H_1^{(2)}(k\sigma\rho)\, \mathrm{e}^{-jk\kappa_{o,e}|z|}\, d\sigma, \tag{5a}$$

$$E_{z,o,e} = A\int_{-\infty}^{\infty} \frac{\sigma^2 \mp \Delta}{\pm\Delta} \frac{\kappa_{o,e}\sigma^3}{1-\sigma^2} H_o^{(2)}(k\sigma\rho)\, \mathrm{e}^{-jk\kappa_{o,e}|z|}\, d\sigma, \tag{5b}$$

$$\mathbf{H}_{t,o,e} = \frac{A}{\zeta}\int_{-\infty}^{\infty} \sigma^2 \left[\boldsymbol{\rho}_o \frac{2\delta\kappa_{o,e}}{\pm\Delta} + \boldsymbol{\phi}_o j \frac{\sigma^2 \mp \Delta}{\pm\Delta} \frac{\kappa_{o,e}}{1-\sigma^2}\right] H_1^{(2)}(k\sigma\rho)\, \mathrm{e}^{-jk\kappa_{o,e}|z|}\, d\sigma, \tag{5c}$$

$$H_{z,o,e} = -(\operatorname{sgn} z)\frac{2jA\delta}{\zeta}\int_{-\infty}^{\infty} \frac{\sigma^3}{\pm\Delta} H_o^{(2)}(k\sigma\rho)\, \mathrm{e}^{-jk\kappa_{o,e}|z|}\, d\sigma, \tag{5d}$$

where $A = k^2\zeta J/16\pi$, and $\boldsymbol{\rho}_o$, $\boldsymbol{\varphi}_o$, $\mathbf{z}_o$ are unit vectors in a cylindrical (ρ, ϕ, z) coordinate system. Due to the presence of the Hankel functions in the integrands, a branch point singularity exists at $\sigma = 0$; the integration path is indented around the branch point into the lower half of the complex σ-plane. The disposition of the integration path in relation to the branch point singularities arising from the functions Δ and $\kappa_{o,e}$ has been discussed in I. (The path must proceed so that Im $\kappa_{o,e} < 0$ on those portions the real σ-axis where κ is complex; when $\kappa_{o,e}$ is real, its algebraic sign must be chosen in accord with the (radiation) condition $Y_{o,e} > 0$).‡ Typical integration paths corresponding to certain choices of the parameter values ω, ω_p, ω_c are obtainable from Fig. 6 of I.

In eqns (3)–(5) it has been assumed that k is positive real, i.e. $\varepsilon > 0$. If $\varepsilon < 0$, the appropriate equations are obtained from the above upon making the following replacements: $\sqrt{\varepsilon} \to -j\sqrt{|\varepsilon|}$, $\sigma \to +j\sigma$, $\kappa \to +j\kappa$ (see end of Appendix A of I).

3. ASYMPTOTIC EVALUATION

If $r = \sqrt{(\rho^2 + z^2)} \gg 1$, the integrals in eqns (5) can be evaluated asymptotically by the method of saddle points.[13] It will be convenient to introduce spherical coordinates (r, θ) via

$$\rho = r \sin\theta, \qquad z = r\cos\theta. \tag{6}$$

If $kr \sin\theta$ is likewise large, i.e. k and $\sin\theta$ are not approximately zero, then we may represent the Hankel functions by their asymptotic values

† The upper and lower signs distinguish the ordinary and extra-ordinary contributions, respectively.

‡ The factor $(1 - \sigma^2)^{-1}$ in the integrand of (5b, c) does not give rise to a pole singularity at $\sigma = \pm 1$ since it is compensated for by other factors.

$$H_n^{(2)}(k\sigma\rho) \sim \sqrt{\left(\frac{2}{\pi k\sigma r \sin\theta}\right)}\, e^{-jk\sigma r \sin\theta + jn\pi/2 + j\pi/4}, \qquad |k\sigma\rho| \gg n, \tag{7}$$

where $k = \omega\sqrt{(\mu_o\varepsilon)}$ is assumed positive. This representation can be substituted into eqns (5) provided that the integration path is distorted away from $\sigma = 0$ so that the inequality $|k\sigma\rho| \gg n$ can be satisfied along the entire path. Each of the integrals is then of the form

$$B = \int_c L(\sigma)\, e^{-jkrM(\sigma)}\, d\sigma, \qquad M(\sigma) = \kappa(\sigma)|\cos\theta| + \sigma\sin\theta, \tag{8}$$

where the large parameter kr appears only in the exponential. The principal contribution to the integral arises from the vicinity of the saddle points σ_i determined by

$$M'(\sigma_i) = 0, \quad \text{or} \quad \kappa'(\sigma_i) = -|\tan\theta|. \tag{9}$$

Only those saddle points for which $\kappa(\sigma_i)$ and σ_i are real, yield propagating wave contributions to the radiation field; the fields associated with complex κ and (or) σ are exponentially damped and will not concern us further.

In the vicinity of a real saddle point σ_i, the factor $\exp[-jkrM(\sigma)]$ decays most rapidly along steepest descent contours inclined at angles of $\pm 45°$ with the real axis. Let us suppose that the original path can be deformed to follow a steepest descent path (SDP) near each saddle point, and that the remaining portions C' of the path proceed in (valley) regions of the complex σ-plane where $\text{Im}\, M(\sigma) < 0$ (see Fig. 2). Then the major contribution to the integral accrues from the SDP segments through each saddle point ($\text{Im}\, M(\sigma_i) = 0$) since the contribution to B from the remainder of the path is proportional

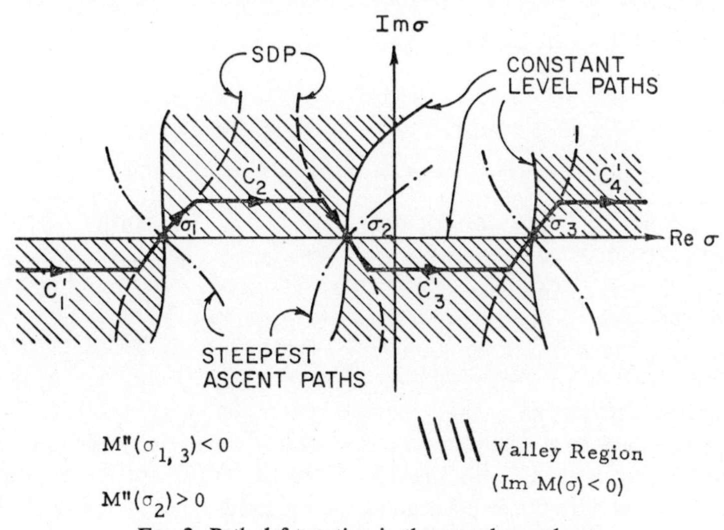

FIG. 2. Path deformation in the complex σ-plane.

to $\exp(-\alpha kr)$, where $\alpha > 0$ is the smallest value of $|\text{Im } M(\sigma)|$ along C' (i.e. $\text{Im } M(\sigma) \leq -\alpha < 0$ along C'). The direction along the original integration path C runs from $-\infty$ to $+\infty$, whence a first-order steepest descent evaluation along a typical SDP segment through a saddle point σ_i yields

$$B_i \sim L(\sigma_i)\, e^{-jkrM(\sigma_i)} \sqrt{\left(\frac{2\pi}{kr|M''(\sigma_i)|}\right)}\, e^{\mp j\pi/4}, \qquad M''(\sigma_i) \gtrless 0. \tag{10}$$

The argument of the vector element $(\sigma - \sigma_i)$ leading away from the saddle point along the SDP is $\mp \pi/4$ for $M''(\sigma_i) \gtrless 0$.

The above-described path deformation can be carried out only if the disposition of the steepest descent paths relative to successive saddle points σ_1 and σ_2 is as shown in Fig. 2; i.e. if $M(\sigma)$ is real on some continuous segment of the Re σ axis which includes the saddle point $\sigma_{1,2,3}$, then the values of $M''(\sigma_{1,2,3})$ must have alternating algebraic signs. That the required condition is satisfied by the analytic function $M(\sigma)$ can be demonstrated from a study of the dispersion curves $\kappa(\sigma)$ vs. σ which have been investigated in I.

We note first that the saddle point condition (9) has a very simple graphical interpretation: it states that the saddle points σ_i locate those points $\kappa(\sigma_i)$, σ_i on the dispersion curve κ vs. σ at which the normal to the curve makes an angle θ with the positive κ-axis (see Fig. 3). If the spatial z, ρ coordinates

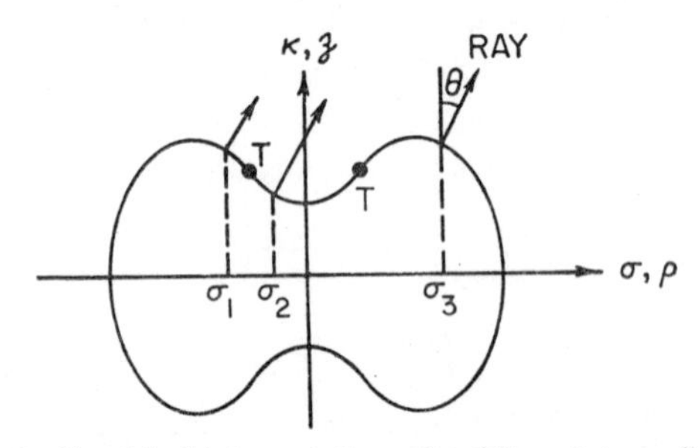

FIG. 3. Graphical interpretation of saddle point condition.

are superimposed upon the κ, σ axes, respectively, the normal to the dispersion curve at the point $\kappa(\sigma_i)$, σ_i, locates the average plane wave power flow direction (ray direction). Hence, the saddle point condition implies that energy reaches the distant observation point (r, θ) via rays which far from the source individually travel along the radial direction from the source point to the observation point. Concerning the algebraic sign of $M''(\sigma_i) = \kappa''(\sigma_i)|\cos\theta|$, one observes from Fig. 3 that, in view of the smoothness of the κ vs. σ curves, the curvature alternates as one moves from one saddle point to the next along a continuous branch. Since the curvature is proportional to $\kappa''(\sigma_i)$, one verifies the disposition of the integration paths in Fig. 2. Thus, we have seen that the entire integration path can be deformed into regions of the complex σ-plane where Im $M(\sigma) < 0$. If $M(\sigma)$† has real saddle points, the integration path must traverse them to maintain Im $M(\sigma) < 0$ along C. To within an accuracy of $0(1/\sqrt{kr})$, the integral B in (8) is then approximated by contributions of the form (10) which locally represent propagating plane waves carrying energy in the radial direction away from the source. It is noted from Fig. 3 that the number of propagating ray contributions at a given observation point can be immediately inferred from the refractive index surface for the medium. The surface also provides information about the ray amplitude since the $M''(\sigma_i)$ term in (10) is directly related to the curvature of the surface at the point $\kappa(\sigma_i)$, σ_i.

From eqns (5), (7), (8), and (10) one obtains the following asymptotic expressions for the fields observed far from the dipole source in the range $0 < \theta < \pi/2$:

$$\mathbf{\rho}_o E_\rho + \mathbf{z}_o E_z \sim 2jA \sum_i \beta_i \sqrt{(R_i)} G_i \frac{e^{-jkrN_i(\theta)}}{kr} (\mathbf{q}_i x \phi_o), \tag{11a}$$

$$E_\phi \sim 2A \sum_i \beta_i \sqrt{(R_i)} F_i \frac{e^{-jkrN_i(\theta)}}{kr}, \tag{11b}$$

$$\mathbf{\rho}_o H_\rho + \mathbf{z}_o H_z \sim -\frac{2A}{\zeta} \sum_i \beta_i \sqrt{(R_i)} F_i \frac{e^{-jkrN_i(\theta)}}{kr} (\mathbf{p}_i x \phi_o), \tag{11c}$$

$$H_\phi \sim -\frac{2jA}{\zeta} \sum_i \beta_i \sqrt{(R_i)} G_i \frac{e^{-jkrN_i(\theta)}}{kr}, \tag{11d}$$

where the summation extends over all contributing saddle points. To emphasize the θ-dependence in the exponential term, the ray refractive index $N_i(\theta)$ has been introduced via the definition

$$M(\sigma_i) \equiv N_i(\theta). \tag{12a}$$

The radius of curvature, R_i, of the dispersion curve at the saddle point $\kappa(\sigma_i) \equiv \kappa_i$, σ_i, is related to $M''(\sigma_i)$ via [16]

$$\frac{1}{R_i} = \frac{|\kappa''(\sigma_i)|}{[1 + \kappa'(\sigma_i)^2]^{3/2}} = \frac{|\kappa''(\sigma_i)|}{[1 + \tan^2 \theta]^{3/2}} = |\kappa_i''(\sigma_i)| \cos^3 \theta = |M_i''(\sigma_i)| \cos^2 \theta. \tag{12b}$$

In addition, the amplitude functions F_i and G_i are given by

$$F_i(\theta) = \sqrt{\left(\frac{\sigma_i}{\sin \theta}\right)} \frac{2\sigma_i \delta}{\gamma_i \Delta_i} \cos \theta, \quad G_i(\theta) = \sqrt{\left(\frac{\sigma_i}{\sin \theta}\right)} \frac{\kappa_i \sigma_i}{1 - \sigma_i^2} \frac{\sigma_i^2 - \gamma_i \Delta_i}{\gamma_i \Delta_i} \cos \theta, ‡ \tag{12c}$$

where $\Delta_i \equiv \Delta(\sigma_i)$, and $\gamma_i = +1$ if the saddle point arises in the ordinary integral, while $\gamma_i = -1$ if the saddle point arises in the extraordinary integral.

† The branch point at $\sigma = 0$ is not crossed during the path deformation.

‡ From (4*b*), one has alternatively: $\gamma_i \Delta(\varepsilon_1 - 1) = 2\left(\kappa^2 - \varepsilon_1 + \frac{\varepsilon_1 + 1}{2} \sigma^2\right)$.

$\beta_i = 1$ if $M''(\sigma_i) < 0$ and $\beta_i = -j$ if $M''(\sigma_i) > 0$. Finally, the polarization vectors $\mathbf{p}_i$ and $\mathbf{q}_i$ are

$$\mathbf{p}_i = \boldsymbol{\rho}_o \sigma_i + \mathbf{z}_o \kappa_i, \qquad \mathbf{q}_i = \boldsymbol{\rho}_o \sigma_i + \mathbf{z}_o \frac{1 - \sigma_i^2}{\kappa_i}. \tag{12d}$$

The relation between the wave normal vector $\mathbf{p}_i$ and the ray refractive index N_i is given via: $k\mathbf{p}_i \cdot \mathbf{r} = krN_i$, i.e. $N_i = \mathbf{p}_i \cdot \mathbf{r}_o$, where the radial unit vector $\mathbf{r}_o$ is parallel to the ray direction and points from the source to the observation point. The (generally elliptic) polarization of the fields in each ray is readily determined from eqns (11).

In these results, it has been assumed that $\varepsilon_z > 0$. The pertinent equations for $\varepsilon_z < 0$ are obtained from (11) and (12) by making the substitutions

$$\sqrt{\varepsilon_z} \to -j\sqrt{|\varepsilon_z|},\ \kappa \to j\kappa,\ \sigma \to j\sigma\ .$$

It is evident from eqns. (11) and the preceding discussion that the structure of the radiation field of a lumped source in an anisotropic plasma is directly dependent upon the shape of the dispersion (or refractive index) curve for the medium. An inspection of the curve yields not only the number of rays which reach a given observation point, but also information pertaining to the ray amplitudes. If the curve has points of inflection, or open branches, then certain rays will contribute only over a limited range of observation angles. In the vicinity of the boundaries of the domains of existence of these rays, formulas (11) become invalid and G_i, F_i, and (or) R_i grow without limit; the correct behavior of the fields in these transition regions is studied in Sections 4 and 5. Although each ray carries power in the radial direction outward from the source, the power flow resulting from several ray contributions need not be radial. If two rays have almost identical refractive indices N_i, the resulting field will be characterized by an interference pattern along the radial direction. Such distinctive features may be of value in the experimental measurement of fields in a plasma, and can motivate the proper choice of any of the parameters ω, ω_c, ω_p.

The types of dispersion curves for a lossless anistotropic plasma medium characterized by different ω, ω_c, ω_p were discussed in I (see also other references listed in I). To gain a quantitative insight into the magnitudes of the various parameters appearing in (11), we have carried out numerical evaluations for a typical case in each of the above-mentioned categories. The results are shown in graphical form in Fig. 4,† which exhibits the dispersion curves, the ray refractive index surfaces (N_i vs. θ), and the amplitude functions

$$\hat{F}_i = F_i\sqrt{R_i} \quad \text{and} \quad \hat{G}_i = G_i\sqrt{R_i}\,.$$

In case A, appropriate to the range of high frequencies, the dispersion curve has two closed branches; the outer (subscript $_1$) corresponds to the

† Four graphs are shown for each of the cases A to I listed in I (except case H which does not yield propagating fields): graph (*a*) represents the dispersion curves, graph (*b*) the ray refractive index curves, and the two graphs (*c*) depict the ray amplitude functions.

ordinary and the inner (subscript $_2$) to the extra-ordinary modes (see also Fig. 6 of I). While the fields are generally elliptically polarized, one notes from Fig. 4(a) that $\hat{F}_{1,2}$ and $\hat{G}_2$ vanish at $\theta = 90°$ so that linear polarization prevails in the equatorial plane. One may also show that $\hat{F}_{1,2}$ and $\hat{G}_{1,2}$ converge to the function $\sin\theta$ as $\omega \to \infty$ (for $\theta = 90°$) so that one recovers the radiation pattern for a current element in an isotropic medium. In case B, only the ordinary modes propagate and their characteristics are similar to those in case A.

The dispersion curve for case C has a closed (subscript $_1$) and open (subscript $_2$) branch, both belonging to the ordinary modes. The rays arising from the open branch exist only in the angular domain $\theta > \theta_A$; for the numerical example in Fig. 4(c), $\theta_A = 43°$.† $N_2(\theta) \to 0$ while $\hat{F}_2$ and $\hat{G}_2$ diverge, as $\theta \to \theta_A$. A detailed study of the fields near the "ray shadow boundary" $\theta = \theta_A$ is carried out in Section 5. Case D is particularly interesting since the dispersion curve has—in addition to a single closed branch as in case A—an open branch which also possesses turning points. Two limiting angles are associated with this branch: $\theta_A = 63°$, arising from the asymptote as in case C, and $\theta_B = 74°$, corresponding to the direction of the normal to the dispersion curve at the turning point T. In the region $\theta < \theta_A$, there exists only a single ray arising from the closed branch of the curve (subscript $_1$). A second ray (subscript $_2$) emerges when $\theta_A < \theta < \theta_B$, corresponding to the portion $M_2 X$ of the unbounded branch as shown in part (a) of Fig. 4(d). Two additional rays exist when $\theta > \theta_B$; they belong to the segments CT (subscript $_3$) and $M_1 T$ (subscript $_4$) of the dispersion curve. The turning points in the dispersion curve give rise to cusps in the ray refractive index curves as shown in part (b) of Fig. 4(d). $\hat{G}_2$ and $\hat{F}_2$ diverge near $\theta = \theta_A$, while $\hat{G}_{3,4}$ and $\hat{F}_{3,4}$ diverge near $\theta = \theta_B$. The field structure near $\theta = \theta_B$ is discussed in Section 4.

Case E is characterized by a dispersion curve with two closed branches: a simple inner branch, and an outer branch which may have two turning points T_1 and T_2 in the first quadrant. For the numerical example in Fig. 4(e), the corresponding limit angles are $\theta_{B1} = 36°$ and $\theta_{B2} = 66°$, respectively. Four rays are possible here—from the inner branch (subscript $_1$), from segment $A T_2$ (subscript $_2$), from segment $T_1 B$ (subscript $_3$) and from segment T_1 T_2 (subscript $_4$). Their behavior is similar to that for case D. The interpretation of the remaining cases is similar to the preceding. It is to be noted that $\omega \approx \omega_p$ in the numerical examples for cases D, E, F, G; only under these circumstances do the points of inflection appear in the curves.

4. MODIFICATION WHEN $\kappa''(\sigma_i) \approx 0$

If the dispersion curve has points of inflection T where the curvature vanishes (see Fig. 3), the simple formula (10) becomes invalid. A difficulty

† The domains of existence of the various rays are observed most easily from the N vs. θ plotes.

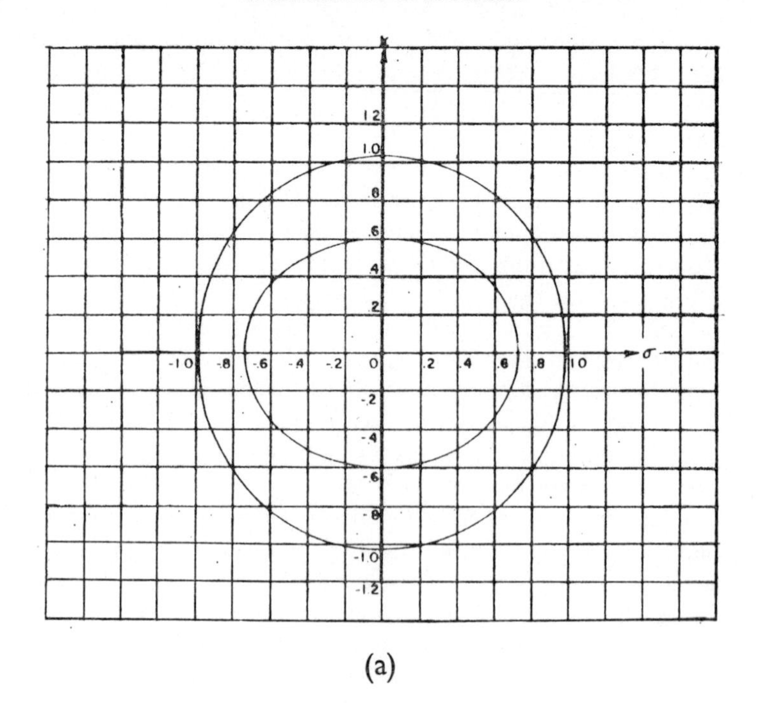
1.2
1.0
.8
.6
.4
.2
-1.0
-.8
-.6
-.4
-.2
0
.2
.4
.6
.8
1.0
σ
-.2
-.4
-.6
-.8
-1.0
-1.2

(a)

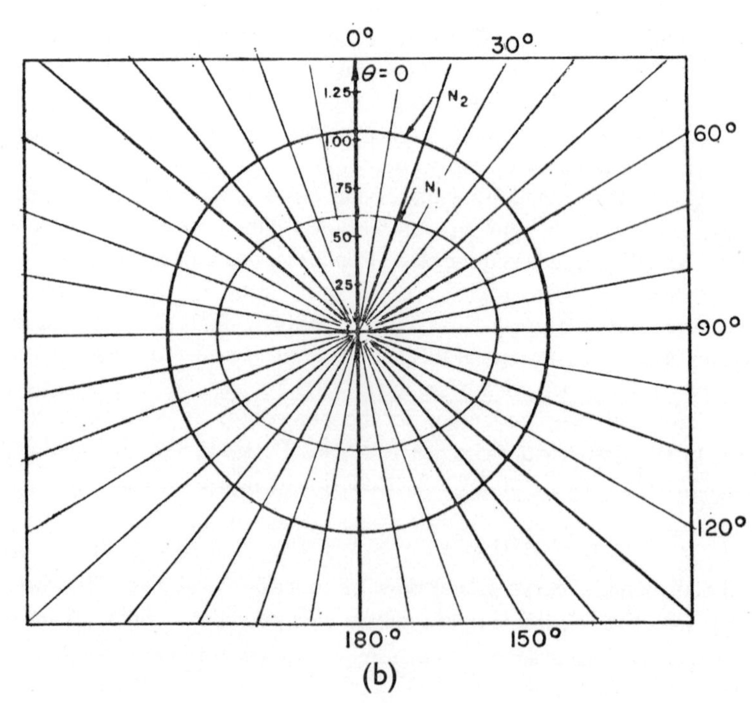
0°
30°
θ = 0
1.25
1.00
.75
.50
.25
N_2
N_1
60°
90°
120°
180°
150°

(b)

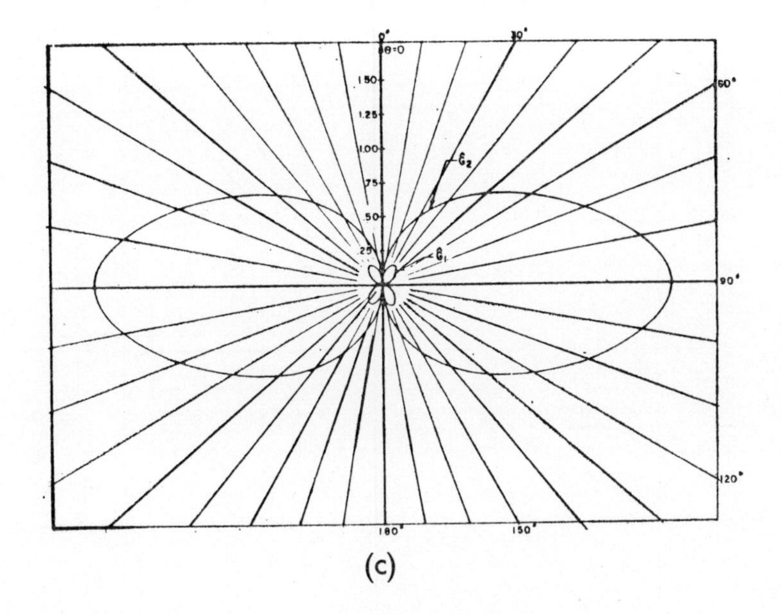

(c)

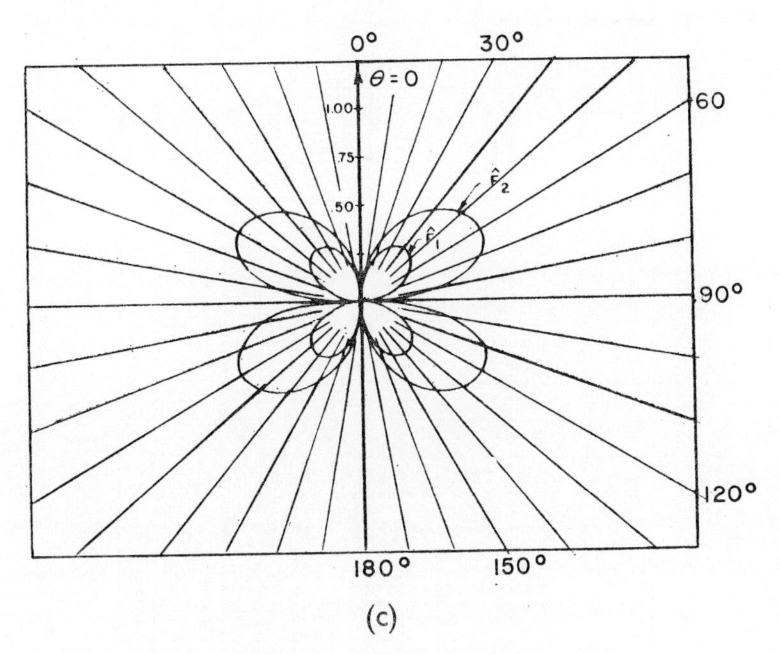

(c)

FIG. 4(a) Case A: $\omega_c/\omega_p = 1.8$, $\omega/\omega_p = 2.4$

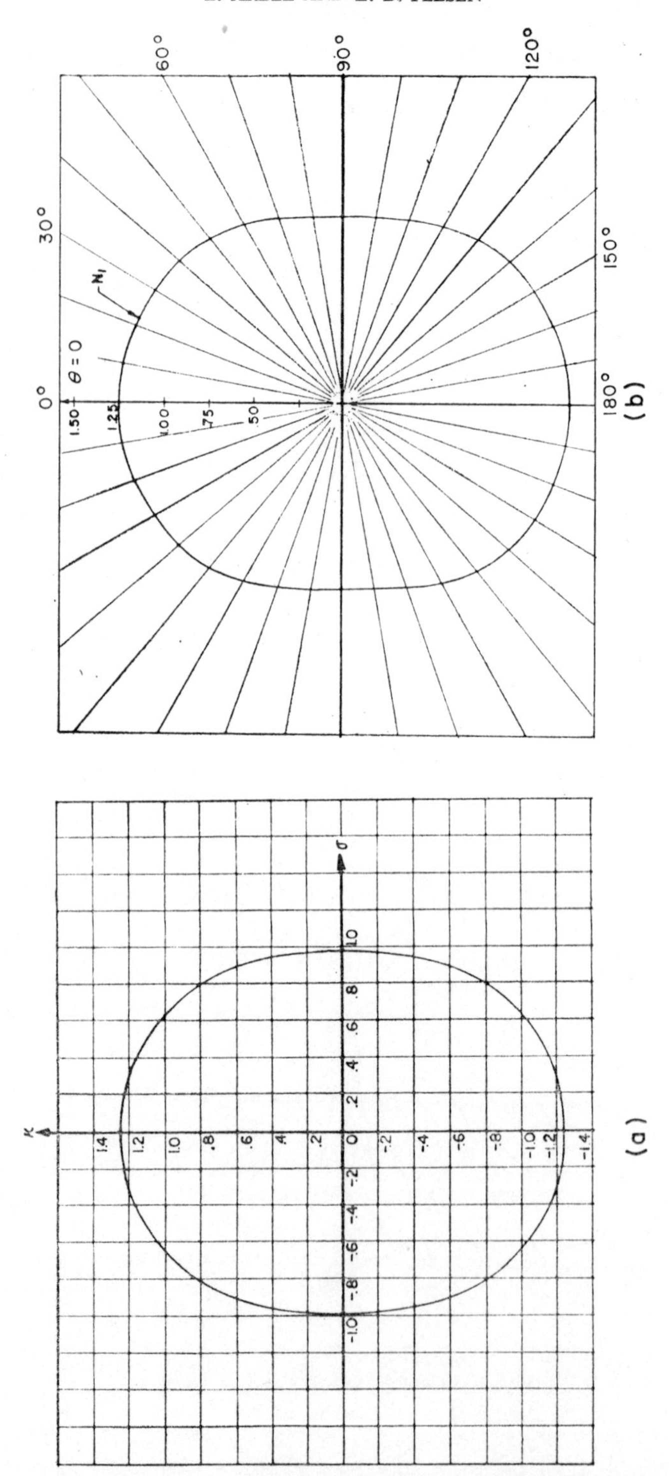
0°
30°
60°
90°
120°
150°
180°
θ = 0
N_1
1.50
1.25
1.00
.75
.50
(b)
κ
σ
1.4
1.2
1.0
.8
.6
.4
.2
0
-.2
-.4
-.6
-.8
-1.0
-1.2
-1.4
(a)

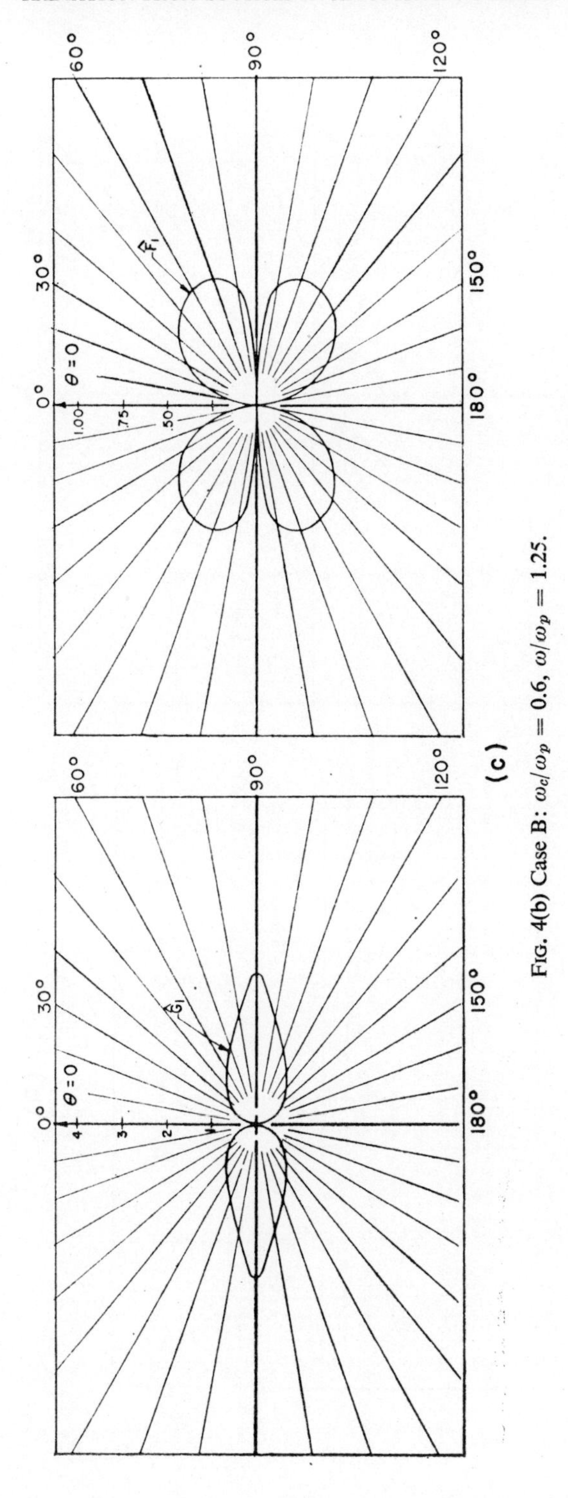

FIG. 4(b) Case B: $\omega_c/\omega_p = 0.6$, $\omega/\omega_p = 1.25$.

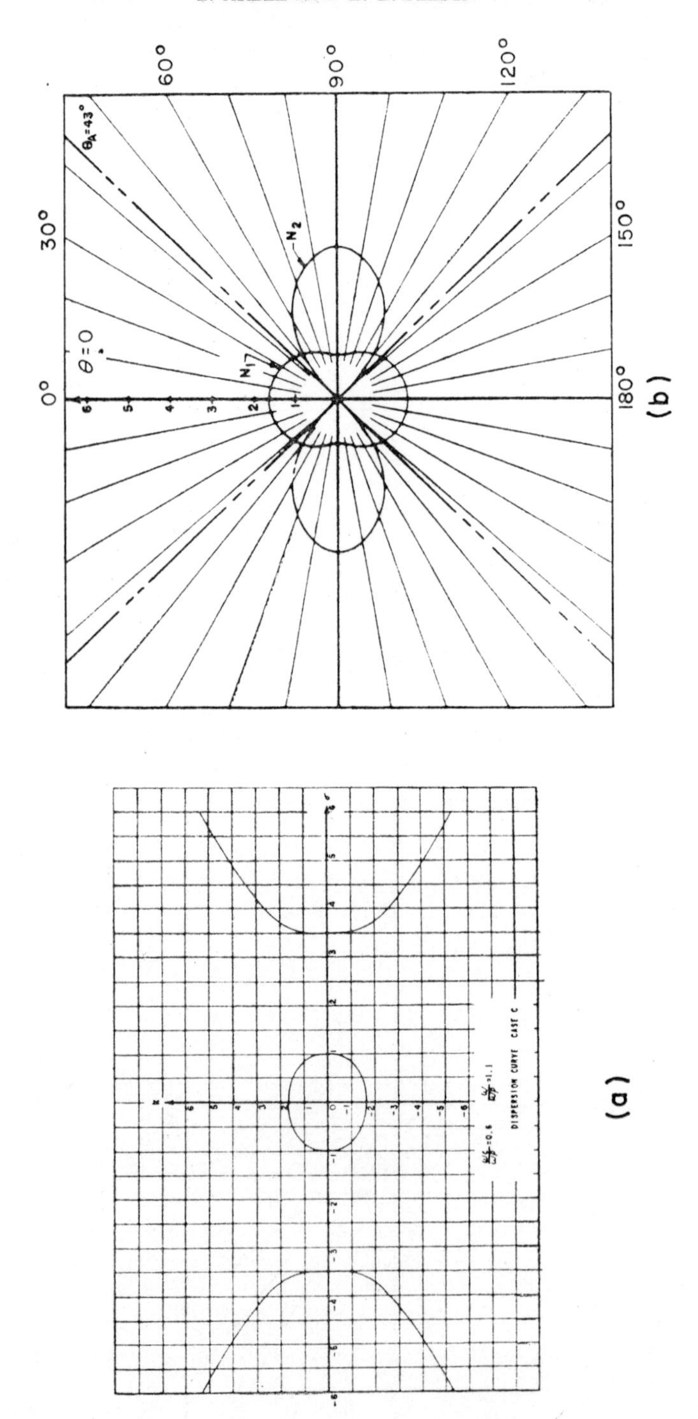
0°
30°
60°
90°
120°
150°
180°
θ_A=43°
θ=0
N_1
N_2
(b)
DISPERSION CURVE CASE C
(a)

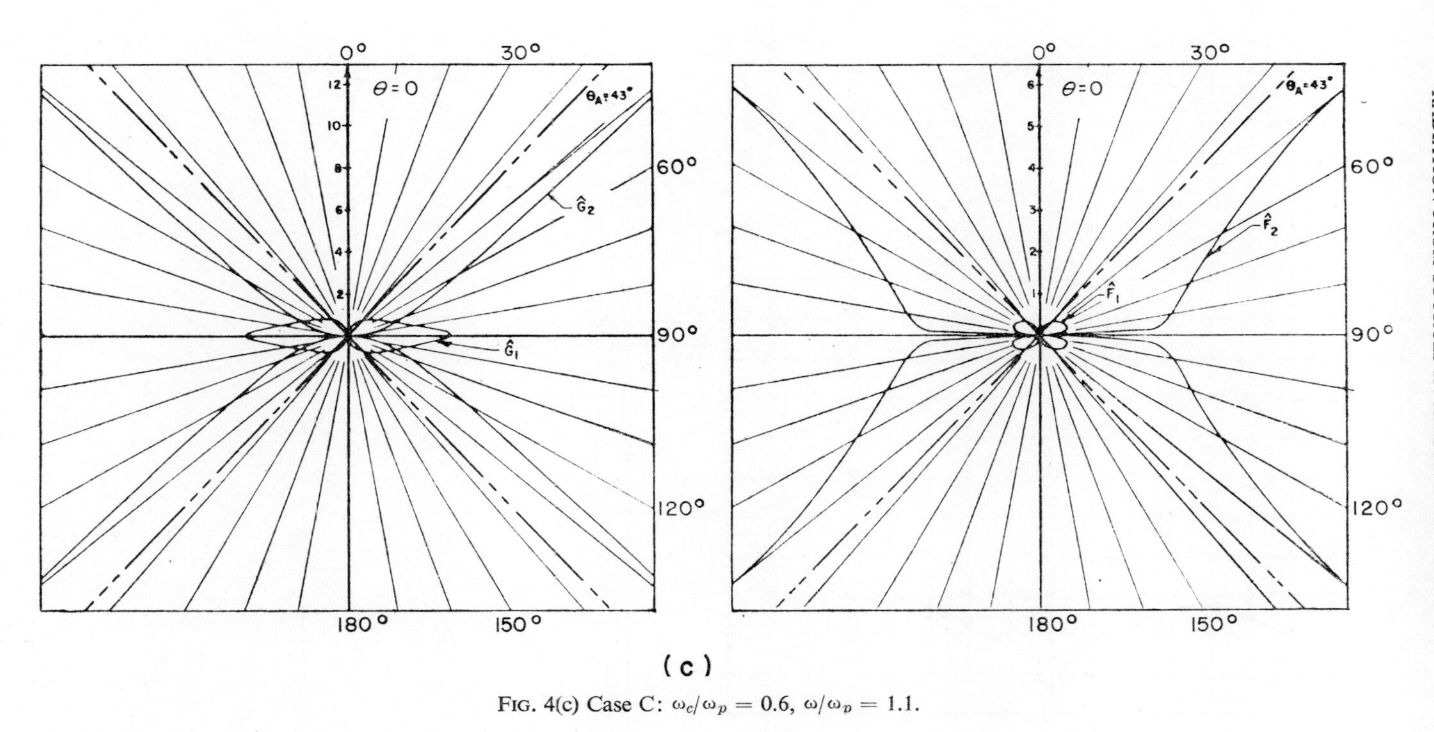

FIG. 4(c) Case C: $\omega_c/\omega_p = 0.6$, $\omega/\omega_p = 1.1$.

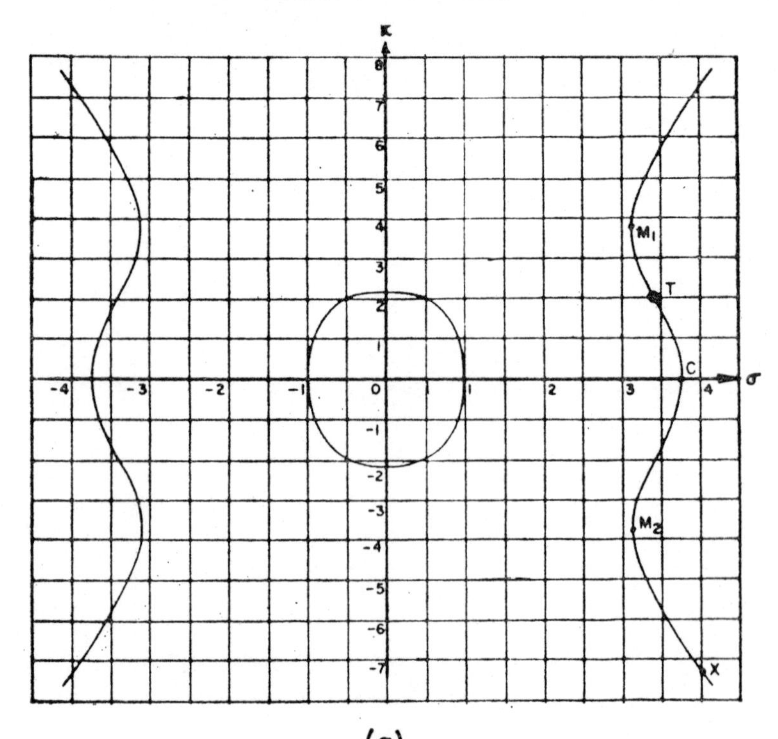
κ
σ
8
7
6
5
4
3
2
1
0
-1
-2
-3
-4
-5
-6
-7
-4
-3
-2
-1
1
2
3
4
M1
T
C
M2
X

(a)

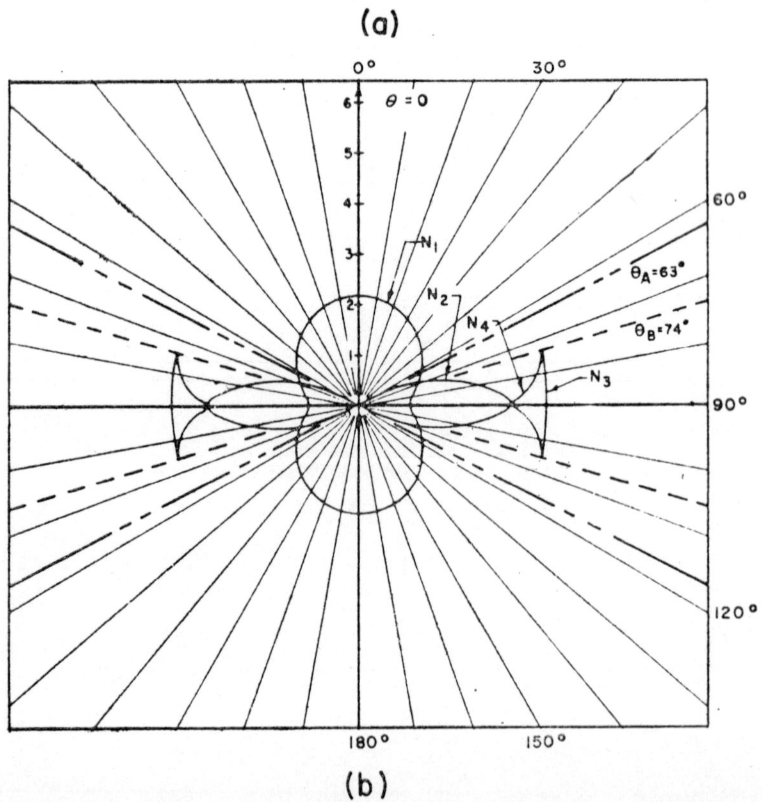
0°
30°
60°
90°
120°
150°
180°
θ = 0
6
5
4
3
2
1
N1
N2
N4
N3
θA=63°
θB=74°

(b)

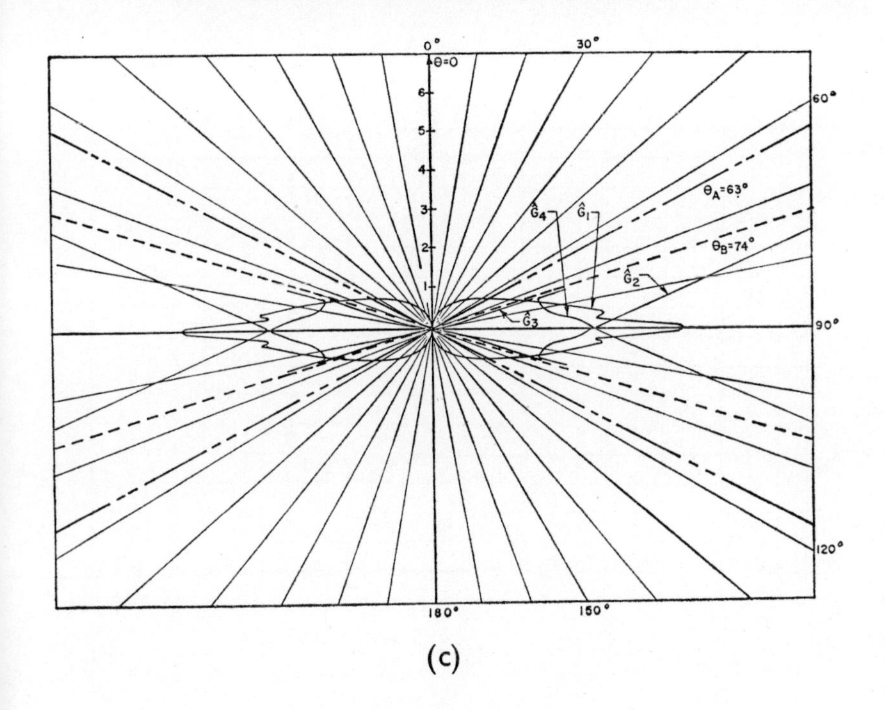

(c)

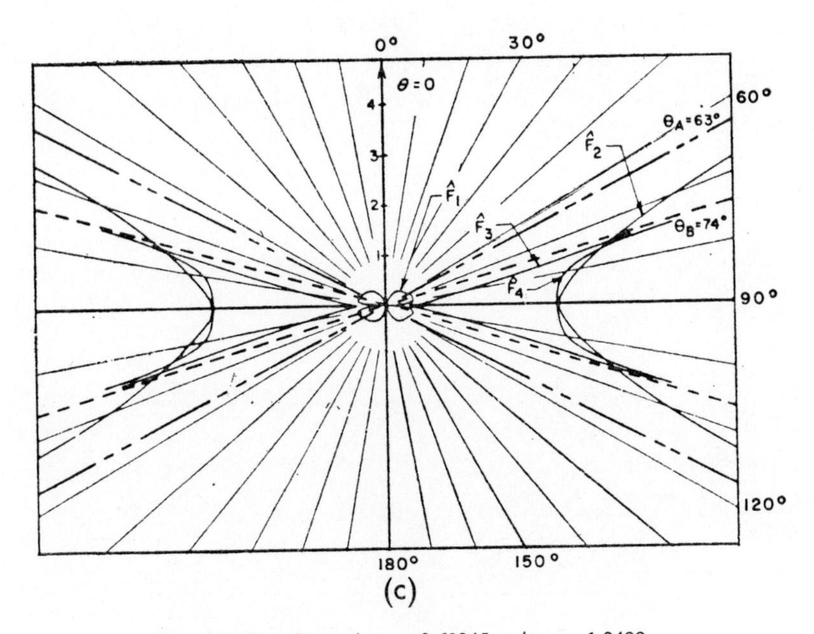

(c)

FIG. 4(d) Case D: $\omega_c/\omega_p = 0.63245$, $\omega/\omega_p = 1.0488$.

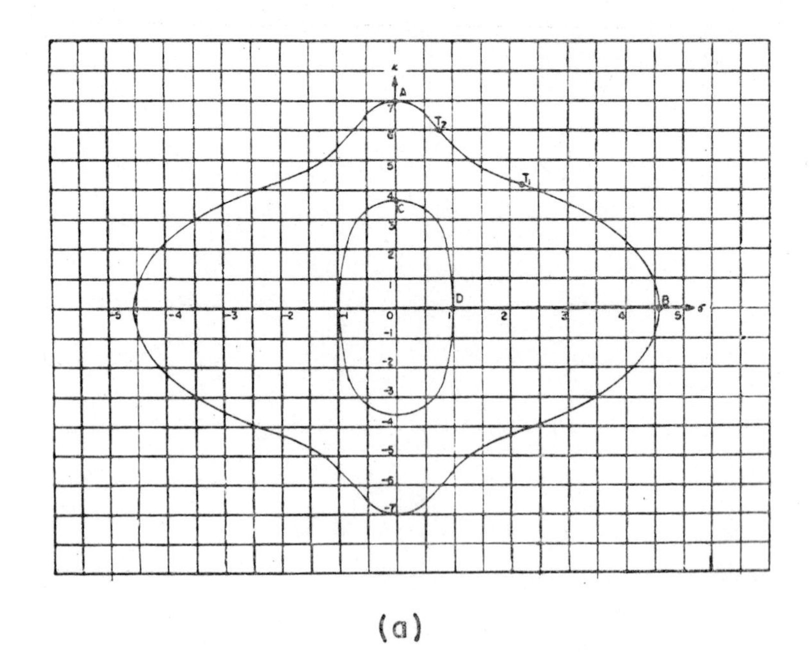

(a)

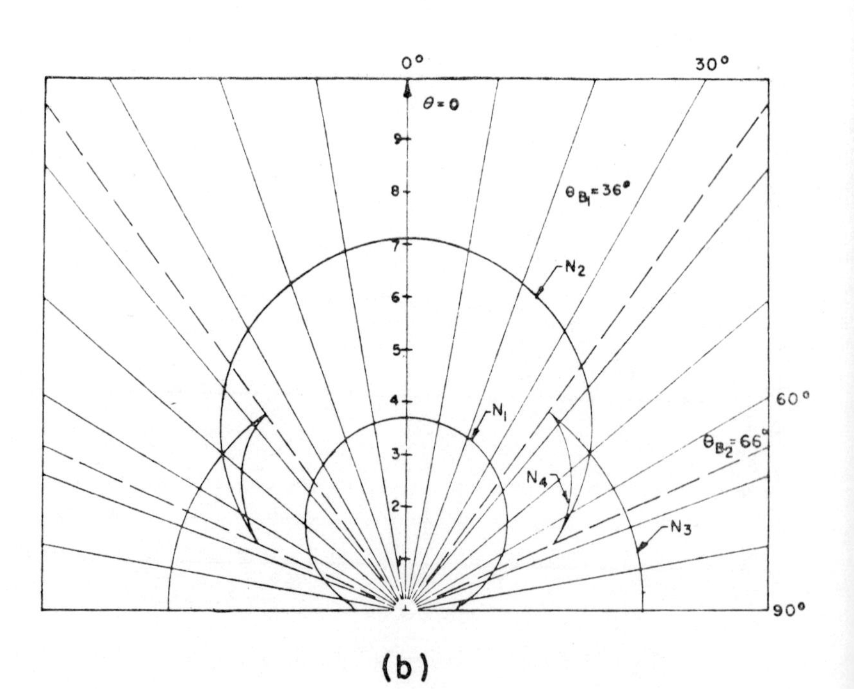

(b)

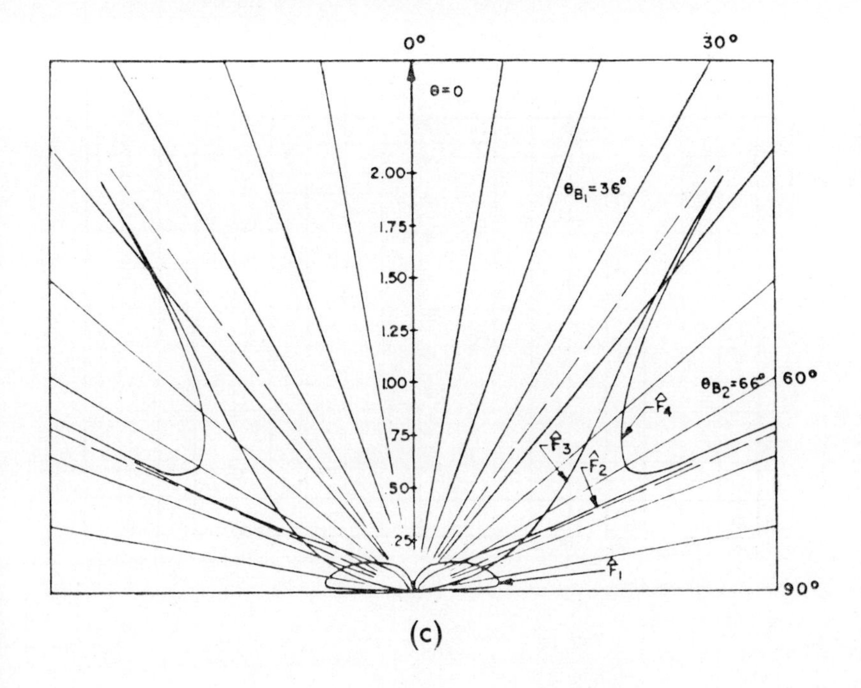

(c)

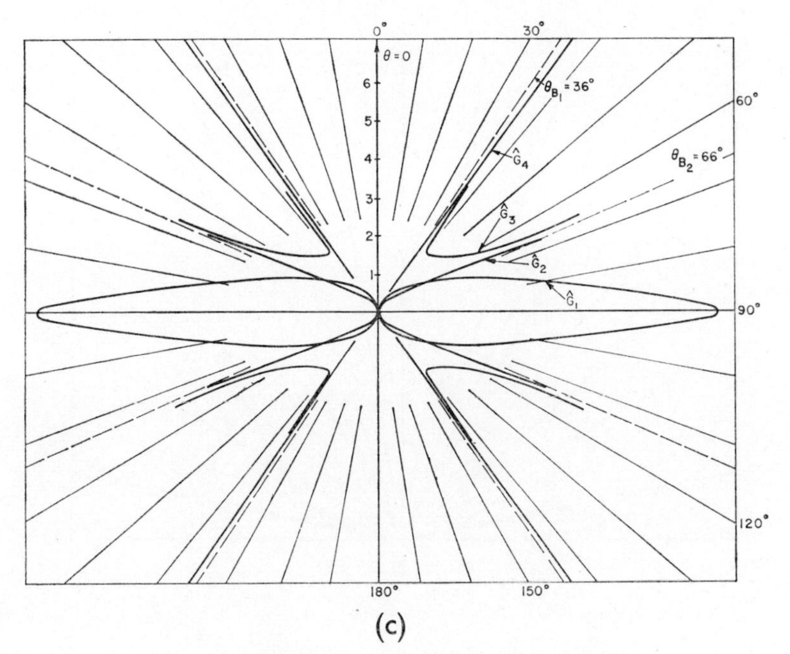

(c)

FIG. 4(e) Case E: $\omega_c/\omega_p = 1.732$, $\omega/\omega_p = 1.0247$.

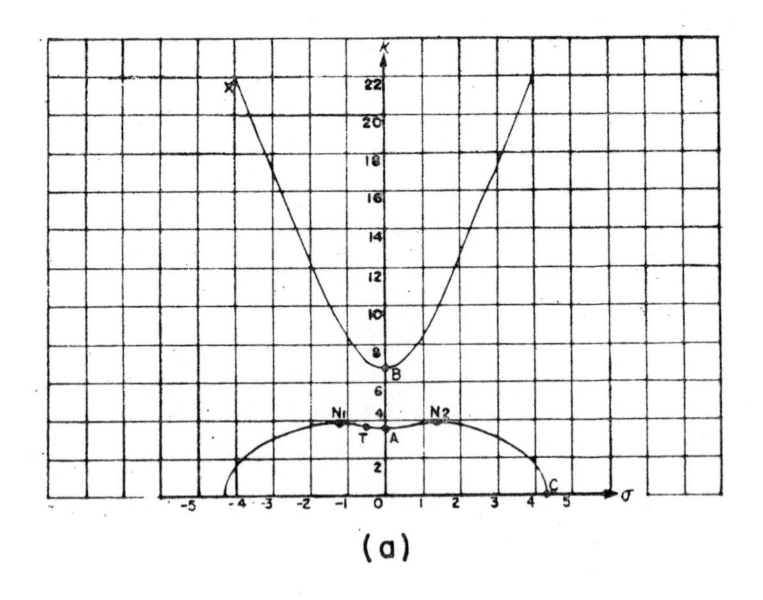
κ
22
20
18
16
14
12
10
8
B
6
N1
4
N2
T
A
2
C
σ
-5
-4
-3
-2
-1
0
1
2
3
4
5
(a)

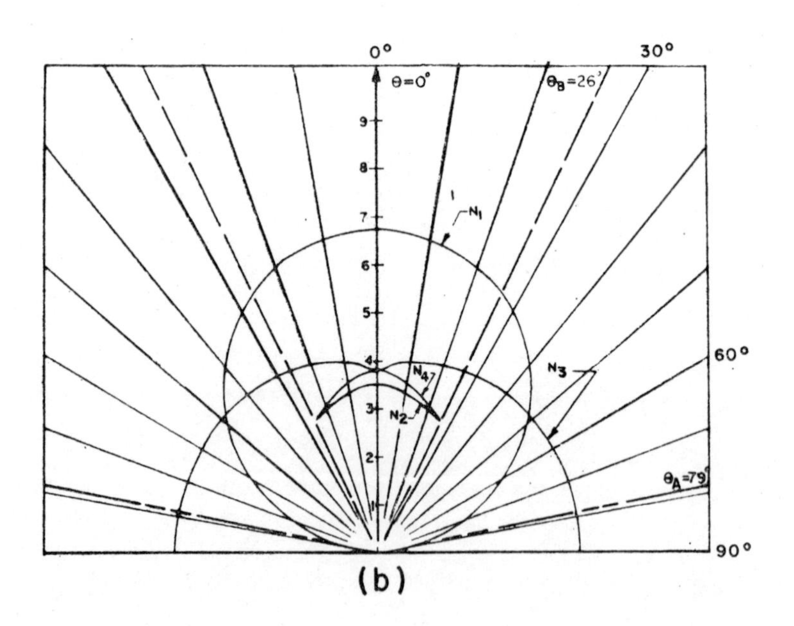
0°
30°
θ=0°
θB=26°
9
8
7
N1
6
5
4
N4
N3
60°
3
N2
2
θA=79°
1
90°
(b)

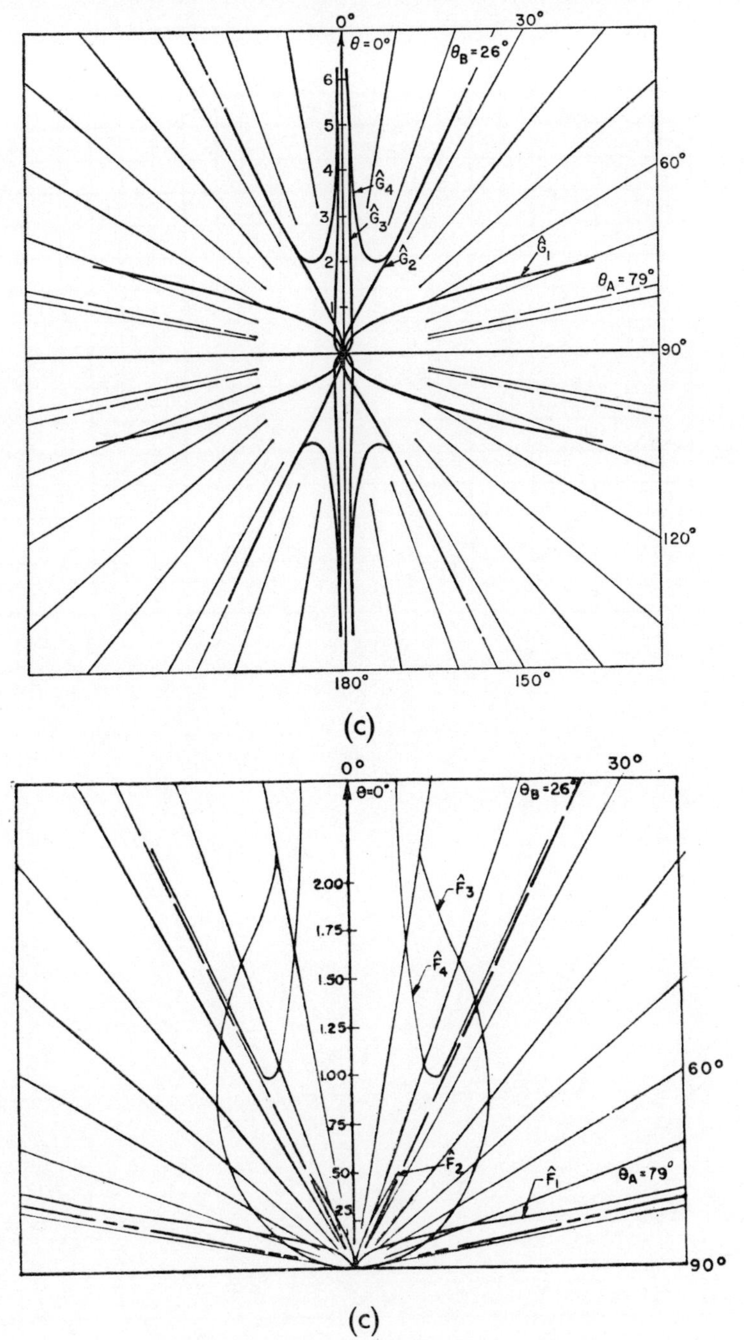

FIG. 4(f) Case F: $\omega_c/\omega_p = 1.732$, $\omega/\omega_p = 0.9747$.

Note: While the amplitude, of rays 3 and 4 diverge as $\theta \to 0$, their sum may be shown to tend to zero.

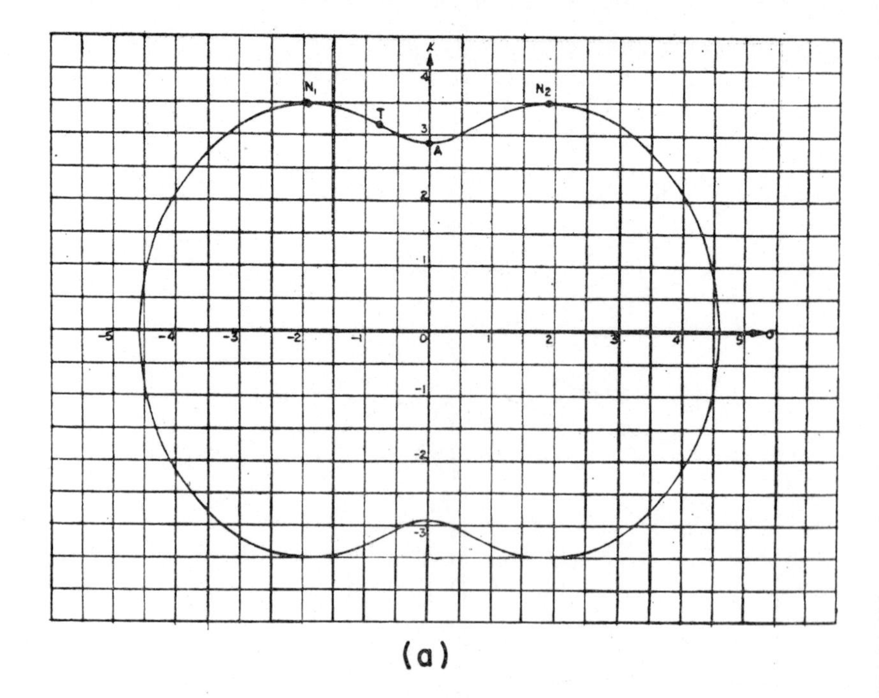
κ
N₁
N₂
T
A
4
3
2
1
-1
-2
-3
-5
-4
-3
-2
-1
0
1
2
3
4
5
σ

(a)

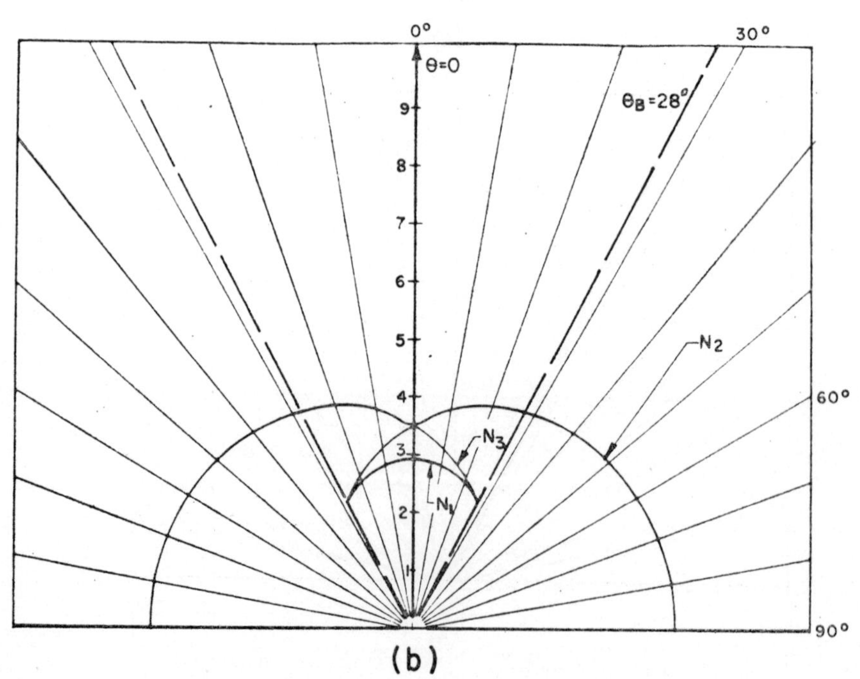
0°
30°
60°
90°
θ=0
θB=28°
N2
N3
N1
9
8
7
6
5
4
3
2
1

(b)

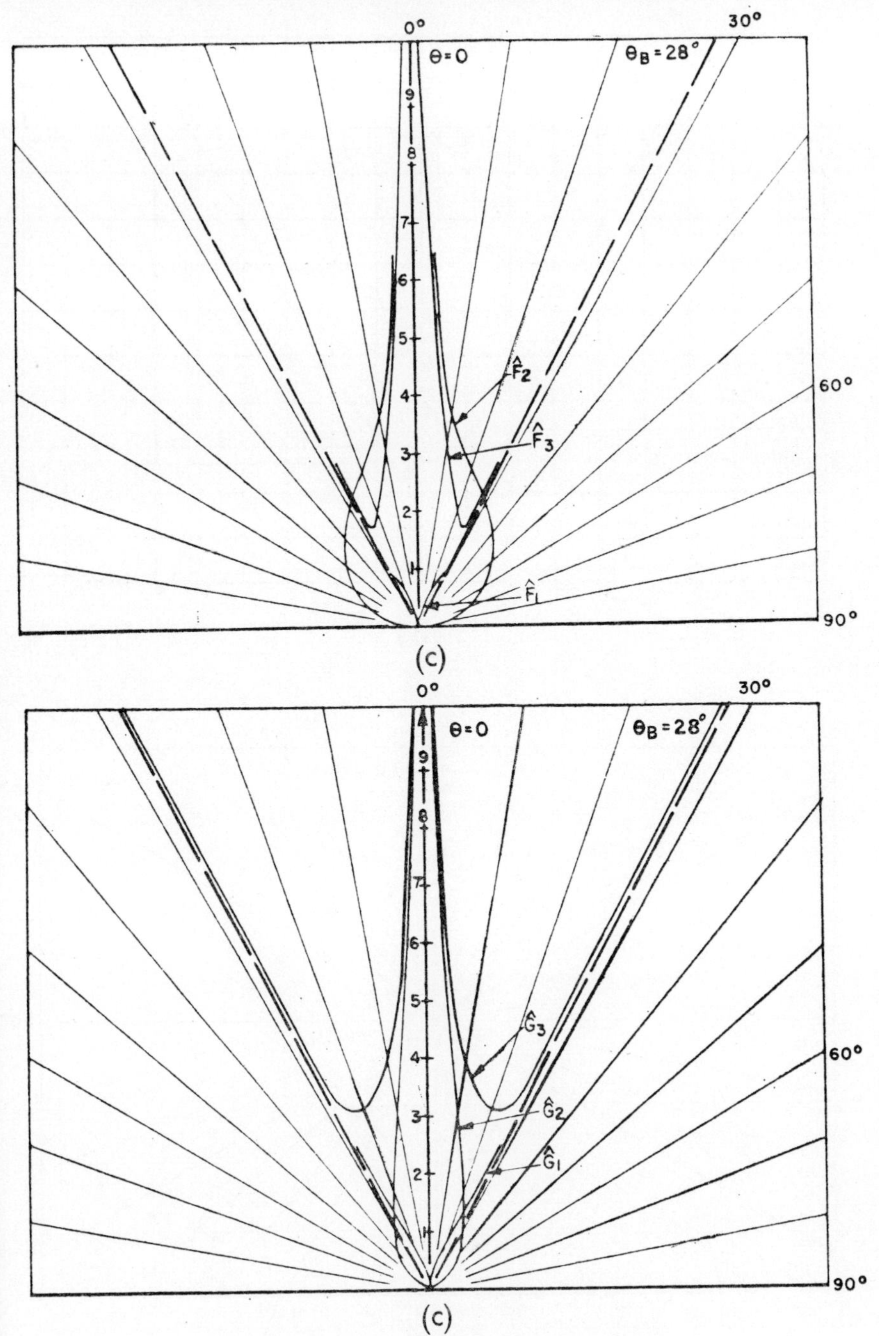

FIG. 4(g) Case G: $\omega_c/\omega_p = 0.6$, $\omega/\omega_p = 0.98$.

Note: The three rays here behave similarly to rays **2**. 3, 4 in case F.

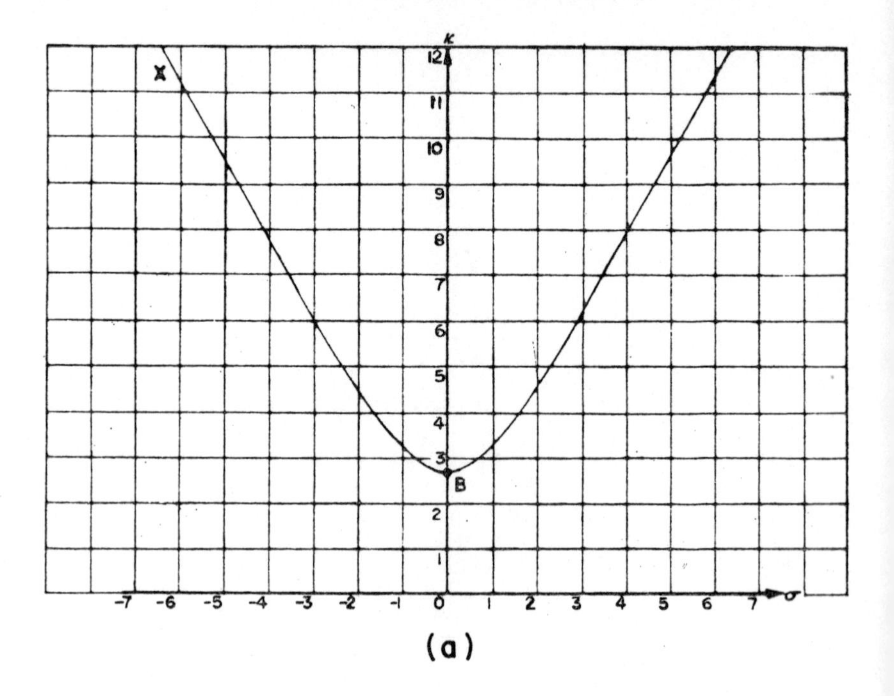
k
12
11
10
9
8
7
6
5
4
3
2
1
X
B
-7 -6 -5 -4 -3 -2 -1 0 1 2 3 4 5 6 7
σ

(a)

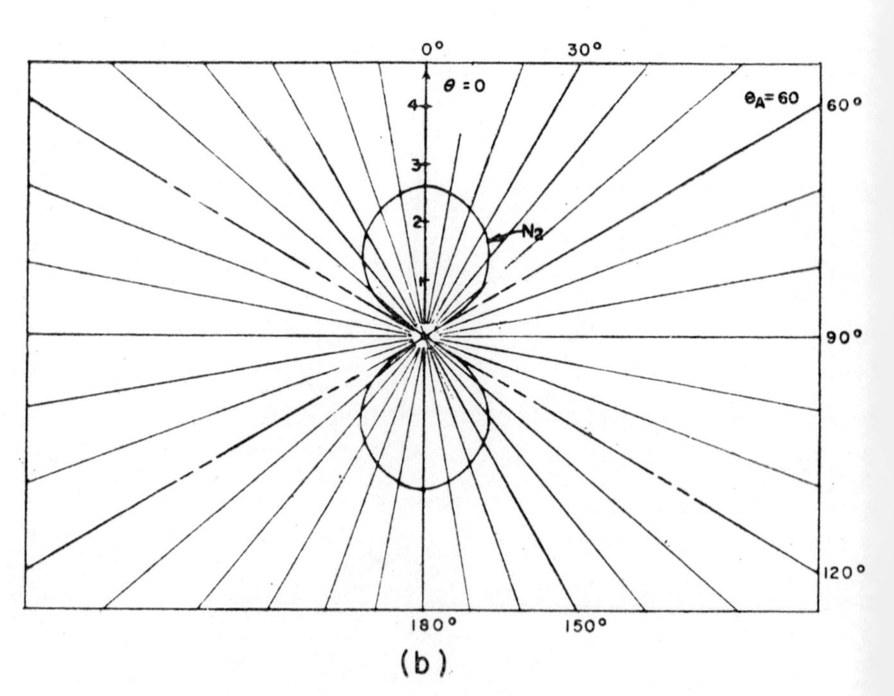
0°
30°
θ = 0
θA = 60
60°
90°
120°
150°
180°
N2
4
3
2
1

(b)

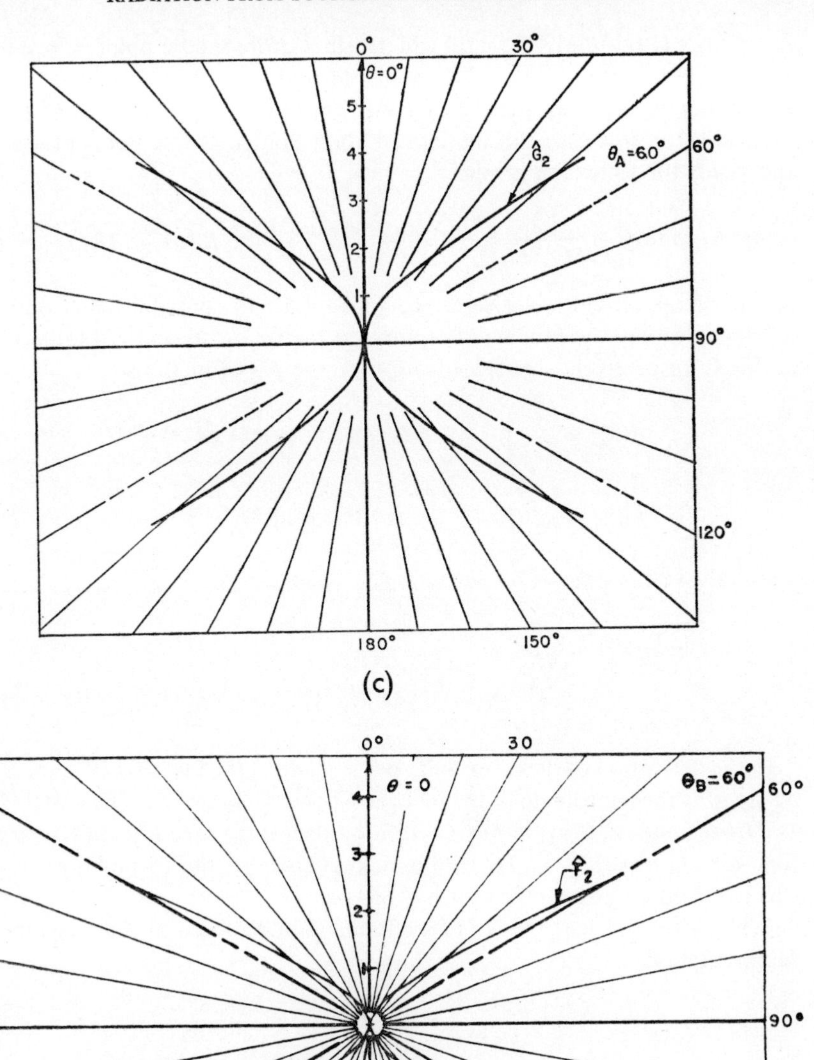

FIG. 4(h) Case I: $\omega_c/\omega_p = 0.6$, $\omega/\omega_p = 0.5$.

arises since (10) is derived for an isolated first-order saddle point ($\kappa''(\sigma_i) \neq 0$), whereas the range $\kappa''(\sigma_i) \to 0$ is characterized by the coalescence of two saddle points ($\sigma_1 \to \sigma_2$ in Fig. 3). An asymptotic approximation for the integral (8) in this case can be derived via a double saddle point procedure, and yields the following result:[14, 15]

$$B_i \sim \pm L(\sigma_i)\left[\frac{2}{krM^{(3)}(\sigma_i)}\right]^{1/3} \mathrm{e}^{-jkr[M(\sigma_1)+M(\sigma_2)]/2} j\pi[Bi(\xi) \mp jAi(\xi)]\,, \tag{13}$$

where the upper and lower signs are appropriate to contributions from path segments leading over the saddle points $\sigma_1[M''(\sigma_1) < 0]$ and $\sigma_2[M''(\sigma_2) > 0]$ in Fig. 2, respectively. $Ai(\xi)$ and $Bi(\xi)$ are the Airy functions

$$Ai(\xi) = \frac{1}{2\pi j}\int_{\infty \mathrm{e}^{-j2\pi/3}}^{\infty \mathrm{e}^{j2\pi/3}} w(t)\,\mathrm{d}t, \qquad w(t) = \mathrm{e}^{\xi t-(1/3)t^3}, \tag{14a}$$

$$Bi(\xi) = jAi(\xi) + \frac{1}{\pi}\int_{\infty \mathrm{e}^{j2\pi/3}}^{\infty} w(t)\,\mathrm{d}t\,, \tag{14b}$$

and

$$\xi = \chi(kr)^{2/3}, \qquad \chi < 0\,, \tag{14c}$$

$$\tfrac{4}{3}\chi^{3/2} = -j[M(\sigma_1) - M(\sigma_2)],$$

$$\text{or} \quad \chi^{1/2} \approx -\frac{j}{2}[\tfrac{1}{2}M^{(3)}(\sigma_1)]^{1/3}(\sigma_1 - \sigma_2) = j|\chi^{1/2}|. \tag{14d}$$

In the derivation of these formulas use is made of the fact that $\sigma_1 \approx \sigma_2 \approx \sigma_0$, where σ_0 is the coordinate of the inflection point $M''(\sigma_0) = 0$. Thus, $M''(\sigma_1) \approx -M''(\sigma_2)$, and $M^{(3)}(\sigma_1) \approx M^{(3)}(\sigma_2) > 0$†; also, since $L(\sigma)$ is a slowly varying function, $L(\sigma_1) \approx L(\sigma_2)$. To within the accuracy of the formula, σ_i in (13) can be taken as either σ_1, σ_2, or σ_0.

When $|\xi| \gg 1$, $\xi < 0$, the Airy functions can be replaced by their asymptotic representations

$$Ai(\xi) \sim \frac{1}{\sqrt{\pi}(-\xi)^{1/4}} \sin\left[\frac{2}{3}(-\xi)^{3/2} + \frac{\pi}{4}\right], \tag{15a}$$

$$Bi(\xi) \sim \frac{1}{\sqrt{\pi}(-\xi)^{1/4}} \cos\left[\frac{2}{3}(-\xi)^{3/2} + \frac{\pi}{4}\right]. \tag{15b}$$

It is noted that $(-\xi) \gg 1$ implies

$$-\chi = [\tfrac{1}{2}M^{(3)}(\sigma_i)]^{2/3}\left(\frac{\sigma_1 - \sigma_2}{2}\right)^2 \gg (kr)^{-2/3}\,, \tag{16}$$

† Since $M''(\sigma_1) \approx -M''(\sigma_2)$ and $M''(\sigma_0) = 0$, one can easily show that $\sigma_0 \approx (\sigma_1 + \sigma_2)/2$, and that $M^{(3)}(\sigma_1) \approx M^{(3)}(\sigma_2) \approx M^{(3)}(\sigma_0)$. Moreover, $M''(\sigma_2) \approx M''(\sigma_0) + M^{(3)}(\sigma_0)(\sigma_2 - \sigma_0) \approx M^{(3)}(\sigma_0)(\sigma_2 - \sigma_0)$, so that $M(\sigma_1) - M(\sigma_2) \approx M'(\sigma_2)(\sigma_1 - \sigma_2) + M''(\sigma_2)(\sigma_1 - \sigma_2)^2/2 + M^{(3)}(\sigma_2)(\sigma_1 - \sigma_2)^3/6 + \ldots \approx -(1/12)M^{(3)}(\sigma_0)(\sigma_1 - \sigma_2)^3$.

which condition can be satisfied for very small values of $(\sigma_1 - \sigma_2)$ since $(kr) \gg 1$ and $M^{(3)}(\sigma_1) \neq 0$. Hence, the approximation $\sigma_i \approx \sigma_1 \approx \sigma_2$ in (13) is justified over the entire range from $\xi = 0$ to $|\xi| \gg 1$. Substitution of (15a, b) into (13) and use of the relation (derivable from (14d)),

$$j\left[\frac{M^{(3)}(\sigma_i)}{2}\right]^{-1/3} \approx \sqrt{\left(\frac{-2\chi^{1/2}}{j|M''(\sigma_{1,2})|}\right)}, \tag{17}$$

reduces eqn. (13) to eqn (10). Thus, (13) constitutes the modification required of (10) when $M''(\sigma_{1,2}) \rightarrow 0$, whence (10) and (13) together permit the smooth calculation of B_i for arbitrary values of $(\sigma_1 - \sigma_2)$, with the understanding that (13) is employed only when $\chi = O[(kr)^{-2/3}]$.

When $\xi = 0$, a second-order saddle point exists at $\sigma_1 = \sigma_2 = \sigma_0$ and the two path segments which pass through the saddle points σ_1 and σ_2 in Fig. 2 coalesce into the single segment shown in Fig. 5. Hence, the contribution to

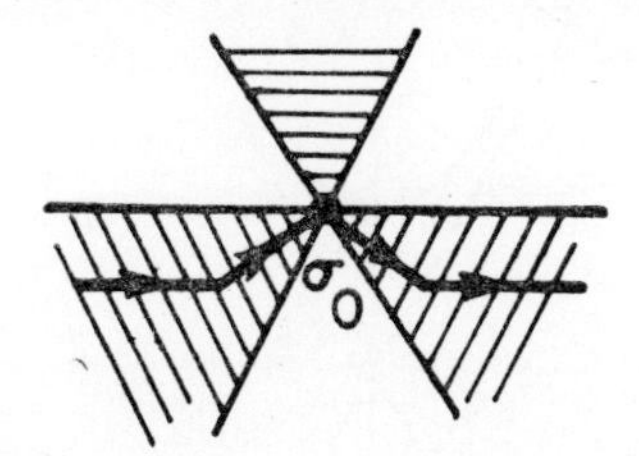

FIG. 5. Integration path for 2nd order saddle point.

B in (8) from the path segment through σ_0 comprises the sum of the two parts of (13) (when $\xi = 0$) and yields:

$$B_0 \sim 2\pi L(\sigma_0)\left[\frac{2}{krM^{(3)}(\sigma_0)}\right]^{1/3} e^{-jkrM(\sigma_0)}Ai(0)\,, \tag{18}$$

where $2\pi\, Ai(0) = \Gamma\left(\tfrac{1}{3}\right) 3^{-1/6}$. The asymptotic behavior of B in (8) therefore varies from $O[(kr)^{-1/2}]$ to $O[(kr)^{-1/3}]$ as two saddle points approach one another, or, equivalently, as two ray contributions in Fig. 3 tend to merge. Since the expressions for the field components contain the additional factor $(kr)^{-1/2}$, the far field behavior passes through a transition from the conventional $(1/kr)$ variation to a variation like $(kr)^{-5/6}$. In a plot of the angular radiation pattern based on a far field behavior of $(1/kr)$, the focusing effects in this transition region lead to infinities in the pattern function whose resolution requires use of eqn (13).

The observation angle θ_B, for which the two saddle points coalesce, constitutes the shadow boundary for the associated ray system. For a dispersion curve as in Fig. 3, the two saddle points separate and remain on the real σ-axis when $\theta < \theta_B$, but they move into the complex plane when $\theta > \theta_B$. This complete transition behavior is contained in eqn (13).

5. FIELD BEHAVIOR NEAR A SHADOW BOUNDARY

It was noted previously (Section 4 of I) that the dispersion curve may have an open branch† when $\varepsilon_1 < 0$, and that the equation for its asymptotes is given by

$$\tan \theta_A = \pm\sqrt{-\varepsilon_1}, \tag{19}$$

where the angle θ_A is measured from the positive σ-axis. The same equation also serves to define the direction of the limiting ray perpendicular to the asymptotes (see Fig. 6) which arises from a saddle point $|\sigma_s| \to \infty$. The cone making an angle θ_A with the z-axis therefore defines the "shadow boundary" for the ray system associated with an unbounded branch of the dispersion

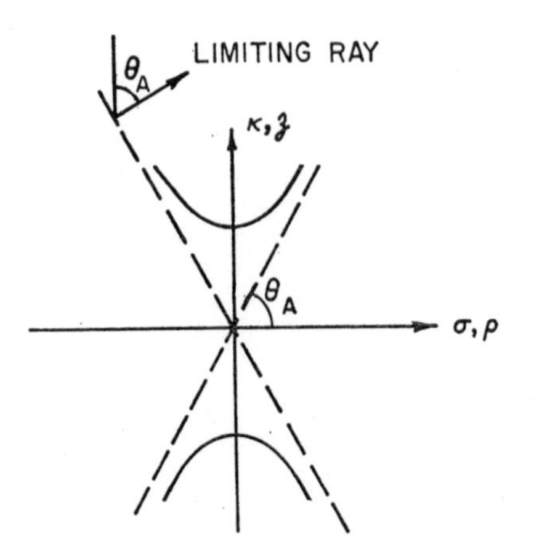

FIG. 6. Dispersion curve with open branch.

curve; for a curve shaped as in Fig. 6 (appropriate to $\varepsilon_z < 0$, $\varepsilon_1 < 0$), no rays exist in the region $\theta > \theta_A$, where the angle θ is measured from the positive z-axis.

Since the range of propagating wave solutions now extends to $\sigma \to \pm\infty$ some modifications of the analysis are in order. Because $\kappa(\sigma)$ is real as $\sigma \to \pm\infty$, convergence of the integrals (5) along the real σ-axis is no longer assured. However, in view of Fig. 2, the integration path can always be deformed so that it terminates in regions of the complex σ-plane where the integrands in (5) are exponentially small. Henceforth, such a path is assumed, and the resulting integrals are absolutely convergent—provided that all saddle points σ_s are confined to finite portions of the real σ-axis. If $\sigma_s \to \infty$, i.e. $\theta \to \theta_A$, the path terminates on the real axis and the integrals for the field components may diverge (see eqns (11)). The behavior

† For $\varepsilon_z > 0$ and $\varepsilon_z < 0$, this branch belongs to the ordinary and extra-ordinary modes, espectively.

of the fields in the vicinity of this shadow boundary will now be studied in some detail.

As $\theta \to \theta_A$, the important contribution to the integrals in eqns (5) arises from the range $|\sigma| \gg 1$ whence the functions in the integrands can be replaced by their approximate values for large σ. It was observed previously that the phenomena in question arise in the ordinary and extra-ordinary integrals when $\varepsilon_z > 0$ and $\varepsilon_z < 0$, respectively. In particular, from (4b) and its counterpart for $\varepsilon_z < 0$, one finds for $|\sigma| \gg 1$,

$$\kappa_o \approx -\sqrt{\left(\varepsilon_1 + \frac{\varepsilon_2^2}{\varepsilon_1 - 1} - \varepsilon_1\sigma^2\right)}, \qquad \varepsilon_z > 0, \quad \varepsilon_1 < 0, \tag{20a}$$

$$\kappa_e \approx \sqrt{\left(-\varepsilon_1 + \frac{\varepsilon_2^2}{1 - \varepsilon_1} - \varepsilon_1\sigma^2\right)}, \qquad \varepsilon_z < 0, \quad \varepsilon_1 < 0, \tag{20b}$$

where $\kappa_o < 0$ while $\kappa_e > 0$ (see Section 3 in Part I) for determination of the algebraic sign of $\kappa_{o,e}$). Also, $\Delta \approx -\sigma^2$, and $\kappa \propto \sigma$, whence the functions in the integrands of (5) multiplying $H_n^{(2)}(k\sigma\rho) \exp[-jk\kappa|z|]$ are approximated by simple powers of σ when $|\sigma| \gg 1$. Since $k\sigma H_1^{(2)}(k\sigma\rho) = -(\partial/\partial\rho)H_0^{(2)}(k\sigma\rho)$, and $(\partial/\partial z) \to -jk\kappa \propto \sigma$, the resulting integrals can be expressed as spatial derivatives of those in Appendix A which are evaluated in closed form. For $\varepsilon_z > 0$, the pertinent closed form expression is proportional to

$$I_1 = \frac{e^{-j|\beta|kr\hat{N}(\theta)}}{r\hat{N}(\theta)}, \qquad \hat{N}(\theta) = \sqrt{\left(\frac{1}{|\varepsilon_1|}\sin^2\theta - \cos^2\theta\right)}, \qquad \beta = \varepsilon_1 + \frac{\varepsilon_2^2}{\varepsilon_1 - 1}, \tag{21a}$$

while for $\epsilon_z < 0$, the appropriate expression is

$$I_2 = \frac{e^{-j|\beta|krN(\theta)}}{rN(\theta)}, \qquad N(\theta) = \sqrt{\left(\cos^2\theta - \frac{1}{|\varepsilon_1|}\sin^2\theta\right)} = \frac{\sqrt{[\sin(\theta_A - \theta)\sin(\theta_A + \theta)]}}{\sin\theta_A}. \tag{21b}$$

The ray refraction index $\hat{N}(\theta)$ is positive when $\theta > \theta_A = \tan^{-1}\sqrt{|\varepsilon_1|}$ and negative imaginary when $\theta < \theta_A$, while $N(\theta) > 0$ when $\theta < \theta_A$ and $N(\theta) = -j|N(\theta)|$ when $\theta > \theta_A$. As $\theta \to \theta_A$, the field components are well approximated by the appropriate ρ and (or) z derivatives of the results (21a, b). (For the case $\varepsilon_2 = 0$, corresponding to an applied infinite d.c. magnetic field, the results in Appendix A are exact for all values of θ) (see Appendix B).

A physical interpretation of these formulas is given in Appendix A where it is pointed out that near $\theta \approx \theta_A$, i.e. $N(\theta)$ or $\hat{N}(\theta) \approx 0$, the near field of the source extends to large distances so that the condition $kr \gg 1$ no longer suffices to specify the far zone. Instead, one must have $kr\hat{N}(\theta)$ or $krN(\theta) \gg 1$. In essence, these latter restrictions, or the equivalent $kr/|\sigma_s| \gg 1$ since

$\sigma_s \propto 1/\hat{N}$ or $1/N$, define the range of validity of the asymptotic results in eqns (11) when $|\sigma_s|$ becomes large. Under these conditions, the dominant contribution arising from spatial derivatives of I_1 or I_2 in (21) is still $O[1/krN(\theta)]$. This situation no longer obtains when $(\theta - \theta_A)$ becomes so small that $kr\hat{N}(\theta)$ or $krN(\theta)$ is likewise small, although $kr \gg 1$. In this instance, previously negligible spatial derivative terms involving higher inverse powers of $krN(\theta)$ or $kr\hat{N}(\theta)$ become dominant and as a result the fields become highly singular as $\theta \to \theta_A$. Thus, we observe that the divergence in the asymptotic formulas (11) in the vicinity of the shadow boundary $\theta = \theta_A$ is not removed by a more careful analysis of the pertinent integrals in this transition region, but that strong field singularities actually do exist there. This is in contrast to transition phenomena of the type discussed in Section 4, where the divergence of the asymptotic formulas (11) does not imply a similar behavior of the true field solution. These remarks are illustrated in detail in Appendix B in connection with the study of the exact closed form solution of radiation from a dipole source in a plasma subjected to an infinite d.c. magnetic field.

The singularities of the dipole fields near the shadow boundary $\theta = \theta_A$ are so strong that the real power flow in that vicinity is also infinite. (Since the dipole current is assumed to be maintained constant, this implies an infinite radiation resistance of the dipole antenna).† The unboundedness of the total power flow P can be verified either from the behavior of the far fields in the neighborhood of $\theta = \theta_A$, or from an examination of the expression given in (I.30):

$$P = 2\pi \int_{\kappa_o \text{real}} d\sigma \sigma Y_o |V_o|^2 + 2\pi \int_{\kappa_e \text{real}} d\sigma \sigma Y_e |V_e|^2. \tag{22}$$

Since $Y_{o,e} \propto 1/|\sigma|$ as $|\sigma| \to \infty$ (see (4a)), and $V_{o,e}$ is given by (4), P will be finite only if

$$\int^{\infty} |v_{o,e}|^2 \, d\sigma < \infty. \tag{23}$$

The integration runs over those values of σ which correspond to propagating waves (real κ); in the case of interest here, this interval extends to infinity. Since $v_{o,e} \propto \sigma$ for a dipole source (see (3)), the integral in (23), and therefore P, diverges.

To limit the spectrum of the source to finite values of σ and thereby secure

† The field singularities described here are of a purely mathematical nature and follow from the assumption of a lossless medium characterized by a tensor of the form (1). This tensor is not adequate to represent a real plasma in regions where strong fields exist since its derivation is based on small signal theory; moreover, the presence of losses would limit the growth of the fields even for media characterizable by such a tensor. Nevertheless, the analysis shows that the fields tend to become large near certain shadow boundaries, and this conclusion most likely holds as well for realistic anisotropic media excited by strongly localized sources.

convergence of the integral in (23), it is necessary to consider a current excitation whose functional dependence on the cross-sectional coordinate ρ is more regular than that in (2). Let us assume, for example, that

$$\mathbf{J}(\mathbf{r}) = \mathbf{z}_o f(\rho)\delta(z), \qquad f(\rho) = \frac{1}{\pi a^2} e^{-\rho^2/a^2}, \tag{24}$$

which is normalized so that $\int_o^\infty f(\rho)\, dS = 1$, $dS = 2\pi\rho\, d\rho$. From (I.9b) and Appendix I.B,

$$v_{o,e}(\sigma) = -\frac{k\sigma\zeta}{\sqrt{2}} \int_o^\infty \rho J_o(k\sigma\rho) f(\rho)\, d\rho, \qquad \zeta = \sqrt{\left(\frac{\mu}{\varepsilon}\right)}, \qquad k = \omega\sqrt{(\mu\varepsilon)}, \tag{25}$$

which can be integrated to yield[17]

$$v_{o,e}(\sigma) = -\frac{k\sigma\zeta}{2\pi\sqrt{2}} e^{-(ka\sigma/2)^2}. \tag{26}$$

For the dipole current in (2), $f(\rho) = J\delta(\rho)/2\pi\rho$, and $v_{o,e}(\sigma)$ is then given by (3). The radial extent of the source distribution (24) is therefore of the order of the parameter "a" and reduces to the delta function case in the limit $a = 0$. For any non-vanishing "a", the integral in (23) converges, and P in (22) is finite though its value increases with decreasing "a".

The expressions for the fields radiated by the source distribution (24) are still given by eqns (5) provided that the factor $\exp[-(ka\sigma/2)^2]$ is included in the integrand. The effect of this weighting factor is to render insignificant the contributions from the large σ range of integration, whence the difficulty associated with $\sigma_s \to \infty$ does not arise. In fact, the algebraic growth (as $\sigma \to \infty$) of the integrands appropriate to the dipole excitation is easily overcome by the exponential decay factor associated with the smooth current distribution (24) so that the fields actually vanish on the shadow boundary $\theta = \theta_A$. By choosing "a" small enough, the radiation fields can be made as large as desired near $\theta = \theta_A$ (see eqns (11), with the factor $\exp[-(ka\sigma_i/2)^2]$ included) but will always vanish at $\theta = \theta_A$; in a similar manner, the radial power flow density $S_r = \mathrm{Re}\ (\mathbf{E} \times \mathbf{H}^* \cdot \mathbf{r}_o)$ at any given distance r can be made as large as desired if "a" is sufficiently small, but will remain finite as long as $a \neq 0$. The dependence on "a" of the radiation pattern function $r^2 S_r$ is plotted in Fig. 7 for a typical case in which $\varepsilon_2 = 0$ (infinite d.c. magnetic field) and $\varepsilon_z < 0$.

The regularity of the fields radiated by the distributed source (24) is due to the fact that such an excitation does not create a sharply defined shadow boundary on which the field behavior is singular. If the intensity of the current distribution (as a function of the spatial coordinates) decays to zero gradually, there are no well-defined illuminated and shadow regions and the transition phenomena are less pronounced. This requirement of a gradually decreasing

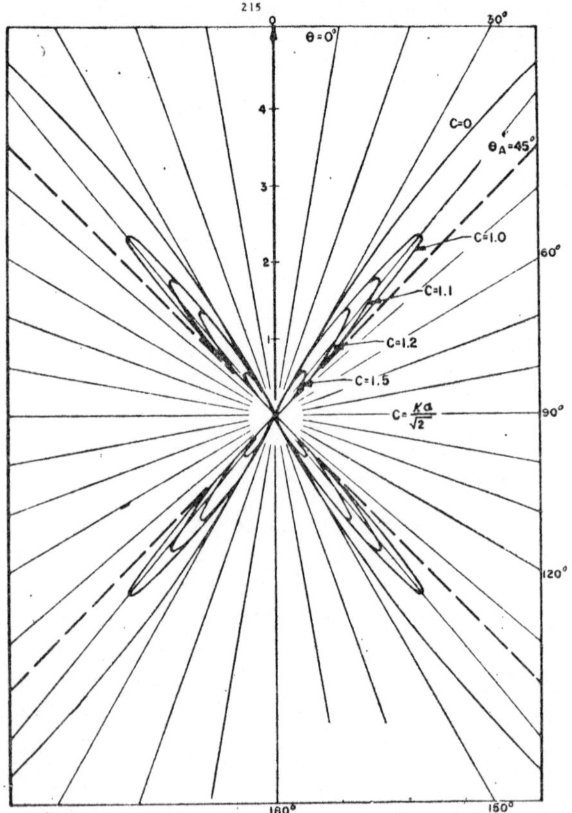

FIG. 7. Radiation pattern for a distributed source, S_r vs. θ. *Note:* $\omega/\omega_p = 0.707$, $C = ka/\sqrt{2}$.

excitation function is illustrated by an examination of a source distribution of the form

$$f(\rho) = C = \frac{1}{\pi b^2}, \qquad 0 \le \rho \le b$$
$$= 0, \qquad \rho > b, \tag{27}$$

which is constant in the vicinity of the origin and abruptly drops to zero at $\rho = b$. Evaluation of the integral (25) shows that $v_{o,e} \propto J_1(k\sigma b)$, i.e. $|v_{o,e}| \sim |\sigma|^{-1/2}$ as $|\sigma| \to \infty$. While this behavior of v as $|\sigma| \to \infty$ is much less violent than that for the delta function distribution in (3), it is not sufficient to assure convergence of the integral (23).

If v in (25) is bounded, i.e. if the Fourier–Bessel transform of $f(\rho)$ exists, we can show that the additional condition

$$\int_o^\infty \rho \left| \frac{\mathrm{d}f}{\mathrm{d}\rho} \right|^2 \mathrm{d}\rho < \infty \tag{28}$$

is sufficient to assure the validity of (23) and therefore the boundedness of the power flow integral (22). Consider the integral

$$A = \int_o^\infty \sigma |v|^2 \, \mathrm{d}\sigma , \tag{29}$$

with v specified in (25). Since $(\mathrm{d}/\mathrm{d}x)\,[xJ_1(x)] = xJ_o(x)$, one has

$$\int_o^\infty \rho f(\rho) J_o(k\sigma\rho)\,\mathrm{d}\rho = \frac{f(\rho)\rho J_1(k\sigma\rho)}{k\sigma}\bigg|_o^\infty - \frac{1}{k\sigma}\int_o^\infty \rho J_1(k\sigma\rho)\frac{\mathrm{d}f}{\mathrm{d}\rho}\,\mathrm{d}\rho\,. \tag{30}$$

If $\rho^2 f(\rho) \to 0$ as $\rho \to 0$ and $\sqrt{\rho} f(\rho)|_\infty = 0$, the first term on the right-hand side of (30) disappears. Substitution of (25), with (30), into (29) (separate integrals are required for v and v^*), interchange of the orders of integration, and use of the formula

$$\int_o^\infty \sigma J_1(k\sigma\rho) J_1(k\sigma\rho')\,\mathrm{d}\sigma = \frac{\delta(k\rho - k\rho')}{k\rho'}, \tag{31}$$

leads to the result that the integral in (29) is proportional to that in (28). Since the convergence difficulties arise from the range $\sigma \to \infty$, condition (23) is certainly satisfied if $A < \infty$. Hence, the restriction (28) on the derivative of the source distribution function $f(\rho)$ is sufficient to assure that the power flow, and hence the radiation resistance, remains finite. The restriction, violated in (2) and (27), is not severe and is certainly satisfied by practical current distributions.

APPENDIX A

Evaluation of Certain Integrals

Consider the integral representation[18]

$$\int_{-\infty}^\infty \sigma H_o^{(2)}(\sigma\rho)\frac{e^{-j\sqrt{(\beta^2-\alpha\sigma^2)}|z|}}{\sqrt{(\beta^2-\alpha\sigma^2)}}\,\mathrm{d}\sigma = +\frac{2j}{\alpha}\frac{e^{-j\beta\sqrt{[z^2+(\rho^2/\alpha)]}}}{\sqrt{[z^2+(\rho^2/\alpha)]}}, \quad \text{(A1}$$

where it is assumed that α and β are positive, the square root in the integrand is positive if real and has a negative imaginary part if complex, and $\sqrt{[z^2+(\rho^2/\alpha)]}$ is positive. The integrand has branch points at $\sigma_{b1} = \beta\alpha^{-1/2} > 0$ and at $\sigma_{b2} = -\beta\alpha^{-1/2}$, and the integration path is indented around them into the first and third quadrants, respectively.† Suppose now that α is allowed to take on complex values in accord with $0 \le \arg\alpha \le \pi$; the branch points then move toward the imaginary σ-axis without crossing the

† $\sigma = 0$ is also a branch point and the path of integration is indented around it into the lower half of the σ-plane.

integration path and both sides of (20) can be continued into this range of α. Thus, for negative real α,

$$\int_{-\infty}^{\infty} \sigma H_o^{(2)}(\sigma\rho) \frac{e^{-j\sqrt{(\beta^2+|\alpha|\sigma^2)}|z|}}{\sqrt{(\beta^2+|\alpha|\sigma^2)}} d\sigma = -\frac{2j}{|\alpha|} \frac{e^{-j\beta\sqrt{[z^2-(\rho^2/|\alpha|)]}}}{\sqrt{[z^2-(\rho^2/|\alpha|)]}}, \qquad \beta > 0. \quad \text{(A2)}$$

To assure absolute convergence of the integrand in (A2), the endpoints of the integration path can be distorted away from the real σ-axis into a region where Im $[\sigma\rho + \sqrt{(\beta^2+|\alpha|\sigma^2)}|z|] < 0$. The continuation implies that $\sqrt{[z^2-(\rho^2/|\alpha|)]}$ is positive when $z^2 > |\rho^2|/\alpha$ and negative imaginary when $z^2 < \rho^2/|\alpha|$, and that Re $\sqrt{(\beta^2+|\alpha|\sigma^2)} > 0$ as $\sigma \to \pm\infty$ along the path.

Alternatively, we can let β vary over a range $0 \geq \arg\beta \geq -\pi/2$ to deduce for negative imaginary β,

$$\int_{-\infty}^{\infty} \sigma H_o^{(2)}(\sigma\rho) \frac{e^{-\sqrt{(|\beta|^2+\alpha\sigma^2)}|z|}}{-j\sqrt{(|\beta|^2+\alpha\sigma^2)}} d\sigma = \frac{2j}{\alpha} \frac{e^{-|\beta|\sqrt{(z^2+(\rho^2/\alpha))}}}{\sqrt{[z^2+(\rho^2/\alpha)]}}, \qquad \alpha > 0, \qquad \text{(A3)}$$

where Re $\sqrt{(|\beta|^2+\alpha\sigma^2)} > 0$. Now let $0 \geq \arg\alpha \geq -\pi$ so that the branch points σ_{b1} and σ_{b2} move back toward the real σ-axis from the fourth and second quadrants, respectively. Correspondingly,

$$\int_{-\infty}^{\infty} \sigma H_o^{(2)}(\sigma\rho) \frac{e^{-j\sqrt{(-|\beta|^2+|\alpha|\sigma^2)}|z|}}{\sqrt{(-|\beta|^2+|\alpha|\sigma^2)}} d\sigma = \frac{2j}{-|\alpha|} \frac{e^{-|\beta|\sqrt{[z^2-(\rho^2/|\alpha|)]}}}{\sqrt{[z^2-(\rho^2/|\alpha|)]}}, \qquad \text{(A4)}$$

where Im $\sqrt{(-|\beta|^2+|\alpha|\sigma^2)} \leq 0$, and $\sqrt{[z^2-(\rho^2/|\alpha|)]}$ is now positive imaginary when $z^2 < \rho^2/|\alpha|$. One also finds that Re$\sqrt{(-|\beta|^2+|\alpha|\sigma^2)} < 0$ as $\sigma \to \pm\infty$ along the path. Absolute convergence of the integrand in (A4) can be secured by considerations analogous to those carried out in connection with (A2).

It is noted that the integral in (A2) represents a propagating wave inside the cone $\rho < \sqrt{(|\alpha|)}z$ and an exponentially attenuated wave in the exterior region $\rho > \sqrt{(|\alpha|)}z$; the converse holds for (A4). In spherical coordinates $z = r\cos\theta$, $\rho = r\sin\theta$,

$$\sqrt{[z^2-(\rho^2/|\alpha|)]} = rN(\theta), \qquad N(\theta) = \sqrt{(\cos^2\theta - |\alpha|^{-1}\sin^2\theta)}, \quad \text{(A5)}$$

so that $\beta N(\theta)/k$ plays the role of an effective refractive index. A singularity exists on the shadow boundary cone $\tan\theta_A = \sqrt{|\alpha|}$ on which $N(\theta) = 0$. Since the wavelength of propagation in a medium increases with a decrease in refractive index, the wavelength becomes infinite as $N \to 0$ so that the near field of the source extends to greater and greater distances. Along the cone $\theta = \theta_A$, the wavelength is infinite and it is impossible to get into the radiation zone of the source.

If one evaluates the integrals in (A2) and (A4) asymptotically for large values of r via (10), he obtains the same results as those given by the closed form expressions. However, while the asymptotic calculation requires

$M''(\sigma_s) \propto N^3(\theta) \not\approx 0$ or, more precisely, $rN(\theta) \gg 1$, where the saddle point $\sigma_s \propto 1/N(\theta)$, the exact evaluation shows that the same formula retains its validity as $N(\theta) \to 0$. Hence, the divergence in the fields exhibited by the asymptotic result as $\sigma_s \to \infty$ $(\theta \to \theta_A)$ is a real one in this case, and not attributable to the limited range of validity of the asymptotic approximation. This fact is highlighted in Appendix B where we examine the exact closed form result for the special case of an infinite applied d.c. magnetic field.

APPENDIX B

Plasma in an Infinite d.c. Magnetic Field

If $\omega_c \to \infty$, corresponding to an infinite external magnetic field, $\varepsilon_2 \to 0$ and $\varepsilon_1 \to 1/\varepsilon_z$. Hence, the dielectric tensor in (1) has the form

$$\boldsymbol{\epsilon} = (\mathbf{x}_0\mathbf{x}_0 + \mathbf{y}_0\mathbf{y}_0)\varepsilon_o + \mathbf{z}_0\mathbf{z}_0\varepsilon_z\varepsilon_o, \qquad \varepsilon_z = 1 - \frac{\omega_p^2}{\omega^2} = 1/\varepsilon_1, \tag{B1}$$

characteristic of uniaxial anisotropy, with the optic axis directed along z. The appropriate expressions for the field components due to an electric current element oriented along the d.c. magnetic field direction are obtained readily from eqns (5) upon letting $\delta = \omega/\omega_c \to 0$, whence $\Delta \to |\sigma^2|$ sgn $(\varepsilon_1 - 1)$. Moreover, from (4b) and its counterpart for $\varepsilon_z < 0$:

$$\kappa_o^2 = \varepsilon_1 - \sigma^2, \qquad \kappa_e^2 = \varepsilon_1(1 - \sigma^2), \qquad \text{when } \varepsilon_z > 0, \tag{B2}$$

$$\kappa_o^2 = -\varepsilon_1 - \sigma^2, \qquad \kappa_e^2 = -\varepsilon_1(1 + \sigma^2), \qquad \text{when } \varepsilon_z < 0. \tag{B3}$$

Thus, the dispersion curve for the ordinary modes is a circle, and these waves propagate as in an isotropic medium with wavenumber k_o (it is recalled that κ and σ are normalized to $k = k_o\sqrt{\varepsilon_z}$). The dispersion curve for the extraordinary waves, on the other hand, is an ellipse when $\varepsilon_z > 0$ and a hyperbola when $\varepsilon_z < 0$. It is noted that E_ϕ, H_ρ and H_z in (5) vanish in the limit so that the radiated field comprises only the components E_ρ, E_z, H_ϕ, which arise in the radiation from a longitudinal electric current element in an isotropic medium. Also, the ordinary integrals contain a factor $(\sigma^2 - \Delta)$ when $\varepsilon_z > 0\,(\varepsilon_1 > 1)$ and a factor $(-\sigma^2 - \Delta)$ when $\varepsilon_z < 0\,(\varepsilon_1 < 0)$, both of which vanish as $\delta \to 0$. Thus, the longitudinal electric current element excites only extraordinary modes which, in this case, degenerate into the E modes familiar from the study of wave propagation in isotropic media.

Instead of dealing with eqns. (5) in the limit $\delta \to 0$, one may show [19] directly from the Maxwell equations that the fields radiated by a longitudinal electric current element of strength J in a uniaxial medium as in (B1) can be derived from a scalar Green's function G via the relation:

$$E_\rho = \frac{-jJ}{\omega\varepsilon_o}\frac{\partial^2 G}{\partial\rho\,\partial z}, \qquad E_z = \frac{jJ}{\omega\varepsilon_o\varepsilon_z}\nabla_t^2 G, \qquad H_\phi = -J\frac{\partial G}{\partial\rho}, \tag{B4}$$

where G satisfies the inhomogeneous differential equation

$$\left(\nabla_t^2 + \varepsilon_z \frac{\partial^2}{\partial z^2} + k_o^2 \varepsilon_z\right) G(\mathbf{r}, \mathbf{r}') = -\delta(\mathbf{r} - \mathbf{r}'), \quad \nabla_t^2 = \frac{1}{\rho} \frac{\partial}{\partial \rho} \rho \frac{\partial}{\partial \rho}, \tag{B5}$$

subject to a radiation condition at $r \to \infty$. Upon introducing a change of variable $z = \sqrt{(\varepsilon_z)} w$, the differential operator in the $\rho - w$ space becomes the ordinary Laplacian for which the solution is known in closed form. Transforming back to the $\rho - z$ space, one obtains

$$G(\mathbf{r}, \mathbf{r}') = \frac{e^{-jk_o\sqrt{(z^2 + \varepsilon_z \rho^2)}}}{4\pi\sqrt{(z^2 + \varepsilon_z \rho^2)}} = \frac{e^{-jk_o r N(\theta)}}{4\pi r N(\theta)}, \qquad N(\theta) = \sqrt{(\cos^2\theta + \varepsilon_z \sin^2\theta)}, \tag{B6}$$

where $k_o = \omega\sqrt{(\mu_o \varepsilon_o)}$. While (B6) has been obtained with the assumption that $\varepsilon_z > 0$, the result can also be continued into the range $\varepsilon_z < 0$, with the proviso that $N(\theta) = -j|N(\theta)|$ when N is imaginary. Substitution of (B6) into (B4) yields *exact* closed form expressions for the fields valid at *arbitrary* observation points (we have transformed to the more convenient spherical coordinates via $E_r = E_\rho \sin\theta + E_z \cos\theta$, $E_\theta = E_\rho \cos\theta - E_z \sin\theta$):

$$H_\phi = \frac{JG\varepsilon_z \sin\theta}{N(\theta)}\left[jk_o + \frac{1}{rN(\theta)}\right], \tag{B7a}$$

$$E_r = -\frac{2jJG\cos\theta}{\omega\varepsilon_o r N(\theta)}\left[jk_o + \frac{1}{rN(\theta)}\right], \tag{B7b}$$

$$E_\theta = \frac{jJ\omega\mu_o G\epsilon_z \sin\theta}{N^2(\theta)}\left[1 + \left(1 + 2\frac{\varepsilon_z - 1}{\varepsilon_z}\cos^2\theta\right)\left(\frac{1}{jk_o r N(\theta)} - \frac{1}{k_o^2 r^2 N^2(\theta)}\right)\right]. \tag{B7c}$$

If one employs instead of the closed form result for G the integral representations (A1) or (A2) for $\varepsilon_z > 0$ and $\varepsilon_z < 0$, respectively (note that $\beta = k_o$, $\alpha = 1/\varepsilon_z$), one obtains via (B4) the limiting form of eqns. (5) as $\delta \to 0$. The observations made in Section 5 and Appendix A concerning the far field behavior in the vicinity of the shadow boundary $\theta = \theta_A$ when $\varepsilon_z < 0$, are borne out clearly by eqns (B7). It is noted that the partial differential equation (B5) is elliptic when $\varepsilon_z > 0$ and hyperbolic when $\varepsilon_z < 0$. The latter circumstance accounts for the presence of "characteristic" surfaces on which singularities can occur.

ACKNOWLEDGEMENT

The work described herein was sponsored by the U.S. Air Force Cambridge Research Laboratories under Contract No. AF–19(604)—4143.

REFERENCES

1. ARBEL, E. and FELSEN, L. B. (1963) Theory of radiation from sources in anisotropic media; Part I: General sources in stratified media.
2. CLEMMOW, P. C. and MULLALY, F. (1954) The dependence of the refractive index in the magneto-ionic theory on the direction of the wave normal, p. 340. *The Physics of the Ionosphere*, The Physical Society, London.
3. BUDDEN, K. G. (1961) *Radio Waves in the Ionosphere*, Sec. 13.17. Cambridge Univ. Press.
4. ABRAHAM, L. G. (1953) Extensions of the magneto-ionic theory to radio wave propagation in the ionosphere, including antenna radiation and plane wave scattering, Rep. No. 13, School of El. Eng., Cornell U., Ithaca, N.Y., Aug. 1953.
5. BUNKIN, F. V. (1957) *J. Exp. Theor. Phys. (USSR)*, **32.**
6. KOGELNIK, H. (1960) *J. Research, Nat. Bureau Stds.* **64D.**
7. KUEHL, H. (a) Radiation from an electric dipole in an anisotropic cold plasma, Antenna Lab. Rep. 24, Calif. Inst. of Technology, Oct. 1960; (b) Electromagnetic radiation from a dipole in an anisotropic plasma, USCEC Rep. 79-203, Electrical Engin. Dept., U. of South Calif., Sept. 1961.
8. CLEMMOW, P. C. (1962) On the theory of radiation from a source in a magneto-ionic medium, presented at the Symposium on Electromagnetic Theory and Antennas, Copenhagen, June 1962.
9. MOTZ, H. and KOGELNIK, H. (1962) Electromagnetic radiation from sources embedded in an infinite anisotropic medium, and the significance of the Poynting vector, presented at the Symposium on Electromagnetic Theory and Antennas, Copenhagen, June 1962.
10. MITTRA, R. and DESCHAMPS, G. A. (1962) Field solution for a dipole in an anisotropic medium, presented at the Symposium on Electromagnetic Theory and Antennas, Copenhagen, June 1962.
11. LIGHTHILL, M. J. (1960) Studies on the magneto-hydrodynamic waves and other anisotropic wave motions, *Phil. Trans. Roy. Soc. London* **252**, 397.
12. Reference 3, Chapter 3.
13. DEBRUIJN, N. G. (1958) *Asymptotic Methods in Analysis*, Ch. 4–6, Interscience, New York.
14. FELSEN, L. B. and MARCUVITZ, N. (1959) Modal analysis and synthesis of electromagnetic fields, Report R-776-59, PIB-705, Microwave Res. Inst., Polytech. Inst. of Brooklyn, Oct. 1959.
15. CHESTER, C., FRIEDMAN, B. and URSELL, F. (1957) An extension of the method of steepest descents, *Proc. Camb. Phil. Soc.* **53**, 599.
16. COURANT, R. (1934) *Differential and Integral Calculus*, Vol. I, p. 281. Interscience, New York.
17. MAGNUS, W. and OBERHETTINGER, F. (1954) *Formulas and Theorems for the Functions of Mathematical Physics*, p. 35. Chelsea, New York.
18. *Ibid.*, p. 34.
19. ARBEL, E. and FELSEN, L. B. (1962) On electromagnetic Green's functions for uniaxially anisotropic regions, Dept. of Electrophysics, Polytech. Inst. of Brooklyn, Report PIBMRI-985-61, Feb. 1962.

ON THE THEORY OF RADIATION FROM A SOURCE IN A MAGNETO-IONIC MEDIUM

P. C. CLEMMOW

Cavendish Laboratory, Cambridge

ABSTRACT

A method is given for writing down, as an angular spectrum of plane waves, the field in a homogeneous magneto-ionic medium of a plane distribution of surface current. In the case when the magnetostatic field is infinite, so that the dielectric tensor is diagonal with two terms equal, two particular problems are considered. The first is that of an electric dipole parallel to the magnetostatic field; it is shown that the complete field is readily obtained, and is closely related to that of the corresponding dipole in a vacuum. The second problem is that of a charged particle moving with uniform velocity parallel to the magnetostatic field; it is shown that the particle radiates by the Cerenkov process, and a simple formula is obtained for the radiated power.

1. INTRODUCTION

The transmission and reception of radio waves by aerials carried on rockets or satellites when they are traversing the ionosphere, has focused attention on the problems associated with the existence of a localized source immersed in an electrically anisotropic medium. If the source is time-harmonic, and if the medium is regarded for theoretical purposes as homogeneous,† two questions, in particular, present themselves: what is the pattern of the field of a given current distribution, and what is the effect of the medium on the impedance of a given aerial? The two questions are related in the sense that a rigorous theory would, in effect, aim to find the field which satisfied the boundary conditions on the surface of the aerial; but any general treatment on these lines is likely to prove complicated, and it would seem worthwhile to attack on more restricted fronts. The main content of this paper is the presentation of one possible method of deriving the field of a given current distribution, and its application to some special cases which are simple enough to be worked out in detail.

The traditional theory of the electromagnetic field in an anisotropic medium is developed in terms of the refractive index and polarization characteristics of time-harmonic plane waves (see, for example, Born and Wolf 1959).

† It should be pointed out that as well as ignoring large scale inhomogeneities the discussion does not include any consideration of such complications as the formation round the aerial of plasma sheaths.

This suggests that a profitable approach might be to express other time-harmonic fields as a superposition of plane waves travelling in different directions. Use of the representation as an angular spectrum of plane waves is a standard technique in discussing the vacuum field of an aperture distribution, and in brief it can be said that what is done here is to introduce the modifications required by the fact that the ambient medium is anisotropic.

In this context the most obvious distinction between anisotropy and isotropy arises from the existence in the former case, in general, of two refractive indices associated with any one given direction. This means that the field representation involves two plane wave spectra more explicitly than in the vacuum case, where the two polarizations associated with each direction have a common refractive index. The other notable distinction is that whereas in an isotropic medium the Poynting vector of a plane wave is in the direction of phase propagation, in an anisotropic medium these directions are in general different. Satisfaction of the radiation condition demands that each plane wave of the spectrum representing the field in a half-space, $z > 0$, say, must have the z component of its Poynting vector positive. In an isotropic medium this is automatically ensured by choosing only directions of phase propagation with positive z components. In an anisotropic medium, however, it may well be that a plane wave with a positive z component of direction of phase propagation has a negative z component of Poynting vector, and the question needs to be examined explicitly.

In Section 2, for the sake of subsequent comparison, the plane wave spectrum representation is stated for the vacuum field of an arbitrary time-harmonic plane surface current distribution, and applications to a dipole source and a charged particle in uniform rectilinear motion are briefly indicated. In Section 3, the plane wave spectrum representation is considered for the field of an arbitrary time-harmonic plane surface current distribution located in a magneto-ionic medium, in which the motions of the electrons only are taken into account, and collisions are neglected. The formulas are set out for the case in which the plane of the current distribution is parallel to the magnetostatic field. Because of its comparative simplicity the problem is then considered when the magnetostatic field is very large. The dielectric tensor characterizing the magneto-ionic medium is then diagonal with two of its diagonal elements equal. For this situation it is shown in Section 4 that the complete field of a dipole lying parallel to the magnetostatic field can be expressed in terms of elementary functions of the type which describe a vacuum dipole field, and in Section 5 that a simple formula exists for the power radiated through the Cerenkov mechanism by a charged particle moving with uniform speed along a magnetostatic field line.

At this stage the general method has been indicated and illustrated by the simplest non-trivial cases. The paper concludes with a section, Section 6, in which further developments and related work are briefly described.

2. A PLANE WAVE SPECTRUM REPRESENTATION OF A VACUUM FIELD

The field associated with a time-harmonic surface current density in the plane $z = 0$ is represented in the form

$$\mathbf{H} = \int_{-\infty}^{\infty} \int_{-\infty}^{\infty} \left(\pm P, \pm Q, \frac{lP + mQ}{n} \right) e^{ik(lx+my\mp nz)} \, dl \, dm, \tag{1}$$

$$\mathbf{E} = Z \int_{-\infty}^{\infty} \int_{-\infty}^{\infty} \left\{ \frac{lmP + (1 - l^2) Q}{n}, - \frac{(1 - m^2) P + mlQ}{n}, \mp (mP - lQ) \right\} e^{ik(lx+my\mp nz)} \, dl \, dm, \tag{2}$$

where

$$n = \sqrt{(1 - l^2 - m^2)}, \tag{3}$$

and the upper sign applies for $z > 0$, the lower for $z < 0$. The integrands in (1) and (2) represent various plane waves (Z is the vacuum impedance, and $k = \omega/c$, where ω is the angular frequency and c the vacuum speed of light). If $l^2 + m^2 \leqslant 1$, then n is real and the plane waves are homogeneous and travel in the directions $(-l, -m, \pm n)$. If $l^2 + m^2 > 1$, then n is purely imaginary, and the waves are inhomogeneous; in fact, in compliance with the requirement on the behaviour of the complete field at infinity, the waves become evanescent away from the plane $z = 0$, so that n is negative imaginary. The polarizations of the plane waves are determined by the spectrum functions $P(l, m)$, $Q(l, m)$, and the inclusion of two such functions gives the representation full generality.

The nature of the symmetry of the complete field about the plane $z = 0$ is made explicit in (1) and (2). Since the surface current density is given by the discontinuity in the tangential component of H, its components appear in the form

$$(J_x, J_y) = 2 \int_{-\infty}^{\infty} \int_{-\infty}^{\infty} (-Q, P) \, e^{ik(lx+my)} \, dl \, dm. \tag{4}$$

The inverse Fourier relations, giving the spectrum functions in terms of the surface current density, are

$$(P, Q) = \frac{k^2}{8\pi^2} \int_{-\infty}^{\infty} \int_{-\infty}^{\infty} (J_y, -J_x) \, e^{-ik(lx+my)} \, dx \, dy. \tag{5}$$

2.1. *The Dipole*

To illustrate the formulas set out above, the case is first considered of a time-harmonic dipole of strength I amp × metre situated at the origin and directed along the x-axis. Then the current density can be expressed by means of delta functions in the form

$$J_x = I\,\delta(x)\,\delta(y), \qquad J_y = 0, \tag{6}$$

so that, from (5),

$$P = 0, \qquad Q = -\frac{Ik^2}{8\,\pi^2}. \tag{7}$$

The substitution for P and Q from (7) into (1) and (2) gives the plane wave spectrum representation of the field. Identification with the standard expression of the dipole field gives the important result

$$\int_{-\infty}^{\infty}\int_{-\infty}^{\infty} \frac{1}{n}\,e^{ik(lx+my\mp nz)}\,dl\,dm = 2\,\pi i\,\frac{e^{-ikr}}{kr}, \tag{8}$$

where $r^2 = x^2 + y^2 + z^2$.

2.2. *The point charge in uniform rectilinear motion*

The case is now considered when the surface current density is that associated with a point charge e travelling along the x-axis with uniform speed v. Here J_x, J_y are the frequency components of the actual surface current density

$$K_x = e\,\delta(x - vt)\,\delta(y), \qquad K_y = 0; \tag{9}$$

that is

$$(K_x, K_y) = \int_{-\infty}^{\infty} (J_x, J_y)\,e^{i\omega t}\,d\omega, \tag{10}$$

which gives

$$J_x = \frac{e}{2\pi}\,\delta(y)\,e^{-i\frac{\omega}{v}x}, \qquad J_y = 0. \tag{11}$$

Then, from (5),

$$P = 0, \quad Q = -\frac{ek}{8\pi^2}\,\delta\left(l + \frac{c}{v}\right). \tag{12}$$

Substitution of (12) into (1) gives for the frequency component of H

$$\mathbf{H}^{(\omega)} = -\frac{e\omega}{8\pi^2 c}\,e^{-i\frac{\omega}{v}x}\int_{-\infty}^{\infty}\left(0, 1, \frac{m}{n'}\right) e^{i\frac{\omega}{c}(my\mp n'z)}\,dm, \tag{13}$$

where

$$n' = \sqrt{\left(1 - \frac{c^2}{v^2} - m^2\right)}. \tag{14}$$

Since v must be less than c, (14) shows that n' is purely imaginary for all (real) values of m, so that all the plane waves in the spectrum are evanescent and there is no radiation. The well-known simple explicit expressions for the field can be recovered by substituting from (13) into†

$$\mathbf{H} = \text{Re}\, 2 \int_0^\infty \mathbf{H}^{(\omega)}\, e^{i\omega t}\, d\omega, \tag{15}$$

carrying out the ω integration and noting that the real part of the resulting integral over m is easily found.

However, a more interesting situation is covered by essentially the same basic formulation. Suppose that, rather than in a vacuum, the charge moves in a homogeneous isotropic medium characterized by a refractive index $\mu(\omega)$. Then the representation (1) can stand unaltered, provided (3) is replaced by

$$n^2 = \sqrt{(\mu^2 - l^2 - m^2)}. \tag{16}$$

This in turn implies that (13) holds with (14) replaced by

$$n' = \sqrt{\left(\mu^2 - \frac{c^2}{v^2} - m^2\right)}. \tag{17}$$

If the parameters are such that $\mu^2 > c^2/v^2$ at some frequencies, then (17) is vitally different from (14) in that it yields some real values of n'. This means that the spectrum of the complete field contains some homogeneous plane waves, and there is radiation.

In this Cerenkov radiation process the total energy radiated by the particle per unit length of path can be written

$$W = 2 \int_{-\infty}^{\infty} \int_{-\infty}^{\infty} E_x H_y\, dy\, dt, \tag{18}$$

evaluated for any x and any $z \geqslant 0$.

If H_y is taken from (15), and E_x from the corresponding prescription for $\mathbf{E}$, the integrations in appropriate order are easily carried out without introducing any approximations, and lead to the familiar formula

$$W = \frac{e^2 \mu_0}{4\pi} \int \left(1 - \frac{c^2}{v^2 \mu^2}\right) \omega\, d\omega. \tag{19}$$

In (19), μ_0 is the vacuum permeability, and the integration ranges over all values of ω for which the integrand is positive.

The result (19) can indeed be derived by a method which is more direct, although it throws no light on the mechanism of the radiation. This method is to calculate the work done against the field in maintaining the motion of the point charge, which per unit length of path is simply eE_x evaluated at $x = vt$, $y = z = 0$. From the foregoing prescription for $\mathbf{E}$ this is readily seen to be

† This particular form for the actual vector $\mathbf{H}$ is convenient because it is assumed in (1) and (2) that ω is positive.

(19). It is interesting to note that the formulation manages to avoid any difficulty associated with the fact that the magnitude of **E** becomes infinite at the point charge. The Fourier transform technique leads to appropriate mean values of E_x and E_y at the point charge itself (that of E_y being zero). The infinity shows up in E_z because, of course, the discontinuity in E_z across $z = 0$ matches the charge density in the plane and therefore has a delta function behavior.

3. A PLANE WAVE SPECTRUM REPRESENTATION OF A FIELD IN A MAGNETO-IONIC MEDIUM

The case is now considered when the whole of space is occupied by a conventional, homogeneous, magneto-ionic medium. It is again supposed that the source of the field is a surface current distribution in the plane $z = 0$ and the object is to represent the field in the respective half-spaces $z > 0$, $z < 0$ as a superposition of various plane waves travelling in different directions. When the magnetostatic field is either parallel or perpendicular to the plane $z = 0$, the field associated with the current has the same symmetry about the plane as the corresponding vacuum field, and for simplicity only these cases are considered.

As a preliminary to writing down the representation analogous to (1), (2) it is noted that the magneto-ionic medium can be characterized by a dielectric tensor (see, for example, Ratcliffe, 1959), and that the time-harmonic plane wave specified by

$$\mathbf{H} = \left(P,\ Q,\ \frac{lP + mQ}{n}\right) e^{ik(lx+my-nz)} \tag{20}$$

is found to give a solution of Maxwell's equations for arbitrarily specified values of l and m provided P, Q and n satisfy certain relations, which are now described. In the first place, n must have one of the two values

$$n_i = \sqrt{(\mu_i^2 - l^2 - m^2)} \qquad (i = 1, 2), \tag{21}$$

where μ_1, μ_2 are the two refractive indices, determinate functions of l and m. And then, correspondingly, P, Q must be one of two pairs P_i, Q_i $(i = 1, 2)$ for which the ratios P_i/Q_i are again determinate functions of l and m. These statements comprise, of course, nothing other than the familiar features of magneto-ionic theory; their form is unconventional only in that (of necessity for the purpose to which they are to be put) l and m are regarded as specified rather than the direction cosines l/μ, m/μ.

It should now be clear that the representation analogous to (1) can be written

$$\mathbf{H} = \mathbf{H}_1 + \mathbf{H}_2, \tag{22}$$

where

$$\mathbf{H}_i = \int_{-\infty}^{\infty}\int_{-\infty}^{\infty}\left(\pm P_i, \pm Q_i, \frac{lP_i + mQ_i}{n_i}\right) e^{ik(lx+my\mp n_i z)}\, dl\, dm \qquad (i = 1, 2), \tag{23}$$

with the upper sign for $z > 0$, the lower for $z < 0$.

The associated surface current density in the plane $z = 0$ is

$$(J_x, J_y) = 2 \int_{-\infty}^{\infty}\int_{-\infty}^{\infty} (-Q_1 - Q_2, P_1 + P_2)\, e^{ik(lx+my)}\, dl\, dm, \tag{24}$$

which corresponds to (4); and the inverse relations are

$$(P_1 + P_2, Q_1 + Q_2) = \frac{k^2}{8\pi^2}\int_{-\infty}^{\infty}\int_{-\infty}^{\infty} (J_y, -J_x)\, e^{-ik(lx+my)}\, dx\, dy. \tag{25}$$

Thus if J_x and J_y are given, P_i, Q_i are determined as functions of l and m by (25) together with the known ratios P_i/Q_i.

As regards the general formulation, it is left to remark that the representation for **E** follows from (22) and (23) through Maxwell's equations, and to recall that the branch of the square root in the expression (21) for n_i has to be determined by the requirement that the Poynting vector of the associated plane wave is directed away from the plane $z = 0$.

The details of the two cases (a) magnetostatic field parallel to $z = 0$, (b) magnetostatic field perpendicular to $z = 0$ are now given in turn.

3.1. *The magnetostatic field parallel to the plane of the surface currents*

If the magnetostatic field is in the x-direction the dielectric tensor of the magneto-ionic medium in the absence of collisions is

$$\mathbf{K} = \begin{pmatrix} 1 - X & 0 & 0 \\ 0 & \dfrac{1 - X - Y^2}{1 - Y^2} & -i\dfrac{XY}{1 - Y^2} \\ 0 & i\dfrac{XY}{1 - Y^2} & \dfrac{1 - X - Y^2}{1 - Y^2} \end{pmatrix}, \tag{26}$$

where X is the square of the ratio of plasma to wave frequency, and Y is the ratio of the gyro to wave frequency. The inverse of (26) is

$$\mathbf{K}^{-1} = \begin{pmatrix} \beta & 0 & 0 \\ 0 & a & \gamma \\ 0 & -\gamma & a \end{pmatrix}, \tag{27}$$

where

$$\alpha = \frac{1 - X - Y^2}{(1 - X)^2 - Y^2}, \qquad \beta = \frac{1}{1 - X}, \qquad \gamma = i\,\frac{XY}{(1 - X)^2 - Y^2}. \tag{28}$$

Then from (23), Maxwell's equations give

$$\mathbf{E} = \mathbf{E}_1 + \mathbf{E}_2 \tag{29}$$

where

$$\begin{aligned}\mathbf{E}_i = Z \int_{-\infty}^{\infty}\int_{-\infty}^{\infty} \Big\{ & \frac{\beta}{n_i}\,[lmP_i + (\mu_i^2 - l^2)\,Q_i], \\ & -\left[\frac{\alpha}{n_i}(\mu_i^2 - m^2) + \gamma m\right] P_i - l\left(\frac{\alpha m}{n_i} - \gamma\right) Q_i, \\ & \mp\left[\frac{\gamma}{n_i}(\mu_i^2 - m^2) - \alpha m\right] P_i \mp l\left(\frac{\gamma m}{n_i} + \alpha\right) Q_i \Big\}\, e^{ik(lx + my \mp n_i z)} \\ & \mathrm{d}l\,\mathrm{d}m \qquad (i = 1, 2),\end{aligned} \tag{30}$$

with the upper sign for $z > 0$ the lower for $z < 0$;

$$\left(1 + \gamma\frac{l^2 m}{n_i} - \alpha\mu_i^2\right) P_i + \gamma\frac{l}{n_i}(\mu_i^2 - l^2)\,Q_i = 0 \qquad (i = 1, 2); \tag{31}$$

and

$$\mu_i^2 = 1 - \frac{X}{1 - \frac{1}{2}\epsilon(1 - l^2) \pm \sqrt{[\frac{1}{4}\epsilon^2(1 - l^2)^2 + \epsilon l^2]}}, \tag{32}$$

with one branch of the radical for $i = 1$, and the other for $i = 2$, and with

$$\epsilon = \frac{Y^2}{1 - X}. \tag{33}$$

3.2. *The magnetostatic field perpendicular to the plane of the surface currents*

With the magnetostatic field in the z-direction the inverse of the dielectric tensor is

$$\mathbf{K}^{-1} = \begin{pmatrix} \alpha & \gamma & 0 \\ -\gamma & \alpha & 0 \\ 0 & 0 & \beta \end{pmatrix} \tag{34}$$

and the equations corresponding to (30), (31) and (32) are

$$\begin{aligned}\mathbf{E}_i = Z \int_{-\infty}^{\infty}\int_{-\infty}^{\infty} \Big\{ & \frac{\beta lmP_i + (1 - \beta l^2)\,Q_i}{n_i}, \; -\frac{(1 - \beta m^2)\,P_i + \beta lmQ_i}{n_i}, \\ & \mp \beta(mP_i - lQ_i)\Big\}\, e^{ik(lx + my \mp n_i z)}\,\mathrm{d}l\,\mathrm{d}m \qquad (i = 1, 2),\end{aligned} \tag{35}$$

with the upper sign for $z > 0$ the lower for $z < 0$,

$$[l(\alpha\mu_i^2 - 1) - \gamma mn_i^2]\,P_i + [m(\alpha\mu_i^2 - 1) + \gamma ln_i^2]\,Q_i = 0 \quad (i = 1, 2) \tag{36}$$

and

$$\mu^2 = 1 - \frac{X}{1 - \frac{1}{2}\epsilon(l^2 + m^2) \pm \sqrt{[\frac{1}{4}\epsilon^2(l^2 + m^2)^2 + Y^2 - \epsilon(l^2 + m^2)]}}, \tag{37}$$

with one branch of the radical for $i = 1$, the other for $i = 2$.

4. DIPOLE PARALLEL TO INFINITE MAGNETOSTATIC FIELD

The double integral expressions for the field given in the previous section are complicated, and even when the source is a dipole it is perhaps hardly to be expected that they can be evaluated in terms of elementary functions in the general case. The derivation of the *far* field is relatively straightforward because it is given by the method of stationary phase, the analytical details of which are closely related to those of ray theory. But this aspect is not pursued here. What is done is to show that there is a special case which, without being trivial, is sufficiently simple to be readily tractable.

The situation envisaged is that in which the magnetostatic field is in the x-direction, as in Section 3.1, and for given wave and plasma frequencies, and hence given X, the gyro frequency, and hence Y, is so great that (28) can be replaced by

$$\alpha = 1, \qquad \beta = \frac{1}{1 - X}, \qquad \gamma = 0. \tag{38}$$

The corresponding dielectric tensor is

$$\mathbf{K} = \begin{pmatrix} 1 - X & 0 & 0 \\ 0 & 1 & 0 \\ 0 & 0 & 1 \end{pmatrix}, \tag{39}$$

as could be deduced directly from the fact that electrons cannot be moved across the infinite magnetostatic field, and when moved parallel to it are unaffected by its presence. The dielectric tensor (30) is of the same form as that of a uniaxial crystal, in that it is diagonal with two of its diagonal elements equal; but there is the interesting possibility of the leading element being negative, which occurs when the medium is overdense in the sense that the plasma frequency exceeds the wave frequency.

The details of the characteristic plane waves supported by the medium are easily worked out from first principles, or may be obtained from the formulas of Section 3.1 together with (38). Thus from (32), letting ϵ tend to infinity,

$$\mu_1^2 = 1, \qquad \mu_2^2 = 1 - X(1 - l^2). \tag{40}$$

Then (31) gives

$$lmP_1 + (1 - l^2) Q_1 = 0, \qquad P_2 = 0, \tag{41}$$

where the first relation is arrived at by noting that μ_1^2 differs from unity by a term of order $1/Y^2$, whereas γ is of order $1/Y$.

If now the source is assumed to be a dipole of strength I amp $\times$ metre, situated at the origin and directed along the axis, the components of the current density are given by (6), and (25) yields (cf. (7))

$$P_1 + P_2 = 0, \qquad Q_1 + Q_2 = -\frac{k^2 I}{8\pi^2}. \tag{42}$$

From (41) and (42)

$$P_1 = P_2 = Q_1 = 0, \qquad Q_2 = -\frac{k^2 I}{8\pi^2}. \tag{43}$$

The field components are obtained by substituting the values (43) into (23) and (30). It is perhaps simplest just to note in the first instance that $\mathbf{H}_1 = 0$ and

$$H_{2x} = 0, \qquad H_{2y} = -\frac{1}{ik}\frac{\partial \Pi}{\partial z}, \qquad H_{2z} = \frac{1}{ik}\frac{\partial \Pi}{\partial y}, \tag{44}$$

where

$$\Pi = -\frac{k^2 I}{8\pi^2} \int_{-\infty}^{\infty}\int_{-\infty}^{\infty} \frac{1}{n_2} \mathrm{e}^{ik(lx+my\mp n_2 z)}\, \mathrm{d}l\, \mathrm{d}m, \tag{45}$$

and, from (21) and (40),

$$n_2 = \sqrt{[(1 - X)(1 - l^2) - m^2]}. \tag{46}$$

It is readily seen that the double integral of (45) for Π is simply that in (8) with y, z replaced by $y\sqrt{(1 - X)}$, $z\sqrt{(1 - X)}$ respectively. Therefore

$$\Pi = -\frac{Ik^2}{4\pi}\, i\, \frac{\mathrm{e}^{-ikR}}{kR}, \tag{47}$$

where

$$R = \sqrt{[x^2 + (1 - X)(y^2 + z^2)]}. \tag{48}$$

To get all the field components is now just a matter of appropriate differentiation of the function (47). In terms of spherical polar coordinates r, θ, ϕ referred to the dipole as origin and direction of dipole as axis ($x = r\cos\theta$), it is found that the non-zero components are

$$H_\phi = \frac{Ik^2}{4\pi}\, i \left(1 - \frac{p^2}{\omega^2}\right) \sin\theta\, \frac{r}{R}\left(1 - \frac{i}{kR}\right)\frac{\mathrm{e}^{-ikR}}{kR}, \tag{49}$$

$$E_\theta = Z\,\frac{Ik^2}{4\pi}\, i\left(1 - \frac{p^2}{\omega^2}\right)\sin\theta \left(\frac{r}{R}\right)^2 \times$$
$$\times \left\{1 - \frac{i}{kR}\left(1 - \frac{i}{kR}\right)\left(1 - \frac{2p^2}{\omega^2 - p^2}\cos 2\theta\right)\right\}\frac{\mathrm{e}^{-ikR}}{kR}, \tag{50}$$

$$E_\gamma = Z \frac{Ik^2}{4\pi} 2 \cos \theta \frac{1}{kR}\left(1 - \frac{i}{kR}\right)\frac{e^{-ikR}}{kR}, \tag{51}$$

where p is the plasma frequency ($X = p^2/\omega^2$), and, from (48),

$$R = r\sqrt{\left[1 - \left(\frac{p}{\omega} \sin \theta\right)^2\right]}. \tag{52}$$

It should be remarked that the integral in (45) has been evaluated by comparing it with that in (8), and the behavior of the field (49), (50), (51) at great distances from the dipole is clearly physically acceptable both for $\omega > p$ $(X < 1)$, and also for $\omega < p$ $(X > 1)$ provided that in the latter case R is taken to be negative pure imaginary for those values of θ which make (52) negative. Thus far, then, there has been no need to find the directions of the Poynting vectors associated with the plane waves of the spectrum (45). In the next section, however, where the dipole source is replaced by a point charge in uniform motion along a magnetostatic field line, it is necessary.

5. POINT CHARGE IN UNIFORM RECTILINEAR MOTION IN THE DIRECTION OF INFINITE MAGNETOSTATIC FIELD

The case is now considered of a point charge e travelling along the x-axis with uniform speed v in the magneto-ionic medium described in Section 4. The procedure is basically the same as that given in Section 2.2. That is, in the first place, the frequency components of the field are obtained.

The frequency component of the surface current density is (11) and substitution in (25) gives (cf. (12))

$$P_1 + P_2 = 0, \qquad Q_1 + Q_2 = -\frac{ek}{8\pi^2}\,\delta\left(l + \frac{c}{v}\right). \tag{53}$$

Together with (41) these equations yield

$$P_1 = P_2 = Q_1 = 0, \qquad Q_2 = -\frac{ek}{8\pi^2}\,\delta\left(l + \frac{c}{v}\right). \tag{54}$$

The frequency components of the field vectors are obtained by substituting the values (54) into (23) and (30), recalling that (38) and (40) hold. If this is done, and the trivial integration carried out, it is found that

$$\mathbf{H}^{(\omega)} = -\frac{ek}{8\pi^2} e^{-i\frac{\omega}{v}x} \int_{-\infty}^{\infty} \left(0, \pm 1, \frac{m}{n'}\right) e^{i\frac{\omega}{c}(my \mp n'z)}\, dm, \tag{55}$$

$$\mathbf{E}^{(\omega)} = -Z\frac{ek}{8\pi^2} e^{-i\frac{\omega}{v}x} \int_{-\infty}^{\infty} \left(\frac{1 - c^2/v^2}{n'}, \frac{c}{v}\frac{m}{n'}, \mp\frac{c}{v}\right) e^{i\frac{\omega}{c}(my \mp n'z)}\, dm, \tag{56}$$

with the upper sign for $z > 0$, the lower for $z < 0$, where

$$n' = \sqrt{\left[(X-1)\left(\frac{c^2}{v^2}-1\right) - m^2\right]}. \tag{57}$$

It is clear from (57) that two quite distinct situations arise according as to whether X is greater or less than unity. If $X < 1$, n' is purely imaginary for all (real) values of m, the plane waves of the spectrum are evanescent and there is no radiation. But if $X > 1$, n' is real for some real values of m, some of the waves in the spectrum are homogeneous and there is radiation.

This conclusion, that there is some Cerenkov radiation at all frequencies less than the plasma frequency, with none at any frequency greater than the plasma frequency, is perhaps clarified by the following observations. Since $\mu_2^2 = l^2 + m^2 + n_2^2$, (40) can be written

$$l^2 + \frac{m^2 + n_2^2}{1 - X} = 1. \tag{58}$$

If θ is the angle the direction of phase propagation makes with the magnetostatic field, then $l^2 = \mu_2^2 \cos^2 \theta$, $m^2 + n_2^2 = \mu_2^2 \sin^2 \theta$, and (58) states that the (plane) radial plot of μ_2 as a function of θ is an ellipse for $X < 1$ and a hyperbola for $X > 1$. Moreover, for any $v(< c)$ there are two points on the hyperbola for which the "coherence" condition $l = -c/v$ is satisfied, but no such points on the ellipse. The situation is depicted in Fig. 1, which also establishes the important fact that at any point on the hyperbola the corresponding ray direction (along the normal to the hyperbola) has its

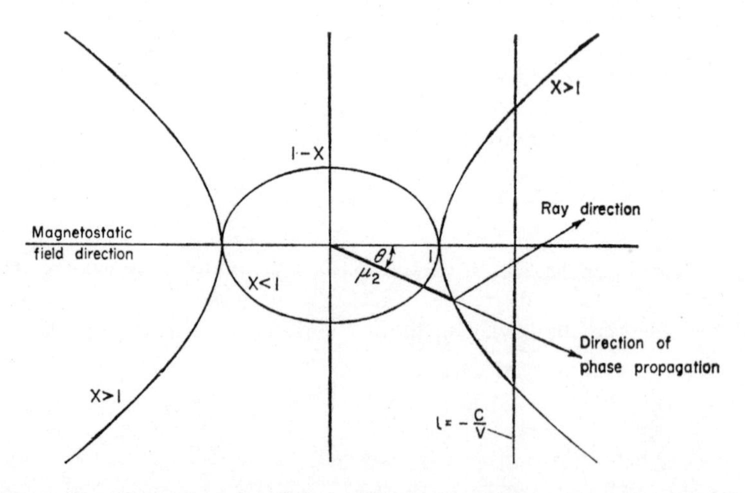

FIG. 1. Showing the (plane) radial plot of refractive index μ_2 as a function of the angle θ which the direction of phase propagation makes with that of the magnetostatic field.

z-component in opposition to that of the direction of phase propagation. This means that the real values of n' given by (57) must be taken as *negative*.

The total energy radiated by the particle per unit length of path is now calculated. It is given by the formula (18), where **H** and **E** are determined by integrating (55) and (56) with respect to ω, as in (15); but the range of integration is just 0 to p, since radiation appears only at frequencies less than the plasma frequency. Likewise the integration variable m in (55) and (56) is restricted to those values for which (57) is real; namely, values between $-m_0(\omega)$ and $m_0(\omega)$, where

$$m_0(\omega) = \sqrt{\left[\left(\frac{c^2}{v^2} - 1\right)\left(\frac{p^2}{\omega^2} - 1\right)\right]}. \tag{59}$$

With these provisos

$$W = \frac{Ze^2}{8\pi^4 c^2}\left(\frac{c^2}{v^2} - 1\right)\int \cdots \int \frac{\omega\omega'}{\sqrt{(m_0^2 - m^2)}} \times$$
$$\cos\left[\omega\left(t + \frac{my}{c}\right)\right]\cos\left[\omega'\left(t + \frac{m'y}{c}\right)\right] \mathrm{d}m\mathrm{d}\omega\mathrm{d}m'\mathrm{d}\omega'\mathrm{d}y\mathrm{d}t. \tag{60}$$

From standard delta function formulas, the integrations with respect to t, ω', y and m' in turn are all elementary. When they are carried out the result is

$$W = \frac{e^2\mu_0}{4\pi^2}\int_0^p \omega \int_{-m_0}^{m_0} \frac{\mathrm{d}m}{\sqrt{(m_0^2 - m^2)}}\,\mathrm{d}\omega, \tag{61}$$

where μ_0 is the vacuum permeability. As explained at the end of Section 2.2, W can also be obtained from eE_x evaluated at $x = vt$, $y = z = 0$; and it is clear from (56) that this method leads directly to the expression (61).

Both integrations in (61) are trivial. Finally, therefore,

$$W = \frac{e^2\mu_0}{8\pi}\left(\frac{c^2}{v^2} - 1\right)p^2. \tag{62}$$

An intriguing implication of this result is that the more slowly the particle travels the greater is the energy radiated, per unit time as well as per unit length of path.

6. FURTHER DEVELOPMENTS

The object of this paper is to draw attention to one possible method for the analytical examination of problems associated with the presence of a localized source in a magneto-ionic medium, and to show what is involved in the mathematically simplest special cases. In this concluding section mention is made of further work on the same or related lines.

The Cerenkov radiation problem as given in Section 5 has been similarly worked out by J. F. McKenzie for the case of the general collisionless magneto-ionic medium characterized by the dielectric tensor (26), and details have been obtained of the dependence of the radiated power on medium parameters.

The angular spectrum technique has been used by J. R. Thompson to consider the rigorous theory of the diffraction of a plane wave by a perfectly conducting half-plane immersed in a magneto-ionic medium. Again it has been found possible, for one non-trivial case, to carry through the analysis when the medium is described by (26).

The result in Section 4, the dipole field, is only the simplest of a class implicit in the following statement, which can be established by a trivial extension of the analysis: if $\mathbf{E}^0(x, y, z)$, $\mathbf{H}^0(x, y, z)$ is the vacuum field generated by a volume current density $\mathbf{J}^0(x, y, z)$ confined to the x-direction, then

$$E_x(x, y, z) = \sqrt{K}\, E_x^0(\sqrt{K}x, \sqrt{K_1}y, \sqrt{K_1}z), \tag{63}$$

$$E_y(x, y, z) = \sqrt{K_1}\, E_y^0(\sqrt{K}x, \sqrt{K_1}y, \sqrt{K_1}z), \tag{64}$$

$$E_z(x, y, z) = \sqrt{K_1}\, E_z^0(\sqrt{K}x, \sqrt{K_1}y, \sqrt{K_1}z), \tag{65}$$

$$H_x(x, y, z) = 0, \tag{66}$$

$$H_y(x, y, z) = \sqrt{(KK_1)}\, H_y^0(\sqrt{K}x, \sqrt{K_1}y, \sqrt{K_1}z), \tag{67}$$

$$H_z(x, y, z) = \sqrt{(KK_1)}\, H_z^0(\sqrt{K}x, \sqrt{K_1}y, \sqrt{K_1}z), \tag{68}$$

is the field generated in the infinite, homogeneous anisotropic medium characterized by the dielectric tensor

$$\begin{pmatrix} K_1 & 0 & 0 \\ 0 & K & 0 \\ 0 & 0 & K \end{pmatrix}, \tag{69}$$

by the current density

$$J_y = J_z = 0, \qquad J_x(x, y, z) = K_1\sqrt{K}\, J_x^0(\sqrt{K}x, \sqrt{K_1}y, \sqrt{K_1}z). \tag{70}$$

A direct examination of the possibility of obtaining, by simple scaling, fields in the anisotropic medium characterized by (69) from arbitrary vacuum fields, has revealed that this can be done provided the vaccum field is first expressed as the super-position of a transverse magnetic field and a coplanar transverse electric field. In this way, for example, the field in the anisotropic medium due to an electric dipole perpendicular to the axis of symmetry of the medium has been found. The expressions for the field components are somewhat lengthier than (49), (50), (51), but are in terms of the same types of elementary function.

In conclusion it should be stressed that the approach set out in this paper is just one way of looking at the problem. There is published work on an alternative method by Bunkin (1957) and Kognelnik (1960), and both rather similar and rather different methods described by other authors in the present proceedings.

REFERENCES

BORN, M. and WOLF, E. (1959) *Principles of Optics*. Pergamon Press.

BUNKIN, F. V. (1957) *J. exp. Theor. Phys. U.S.S.R.* **32,** 338.

KOGELNIK, H. (1960) *J. Res. N.B.S.* **64D,** 515.

RATCLIFFE, J. A. (1959) *The Magneto-ionic Theory and Its Application to the Ionosphere*. Cambridge.

ELECTROMAGNETIC RADIATION FROM SOURCES EMBEDDED IN AN INFINITE ANISOTROPIC MEDIUM AND THE SIGNIFICANCE OF THE POYNTING VECTOR

H. KOGELNIK and H. MOTZ

ABSTRACT

The paper is concerned with the excitation of waves in an infinite and uniform magneto-ionic medium. In particular the power radiated by a small magnetic dipole embedded in such a medium and the power radiated by a modulated ion beam are computed in detail. The work on the magnetic dipole presents the counterpart of work on the electric dipole published in an earlier paper.

One aim of the paper is to present a general method which is capable of reducing similar problems to elementary matrix operations and integrations. This is done by computing dyadic Green's functions of the inhomogeneous wave equation in a uniform anisotropic medium. In the case of the usual magneto-ionic medium they are explicitly computed.

By means of these dyadic Green's functions the electromagnetic field can readily be expressed in the form of a multiple integral. This is evaluated at large distances from the source. Moving away in a given direction from the source, one finds that the wave normal, in general, makes an angle with this given direction: the wave is side-slipping. The evaluation of the fields at large distances leads to a simple geometrical construction for finding the magnitude and direction of the propagation vector.

It is clear that the electromagnetic energy travels in the direction of the group velocity. It is interesting to find out whether the direction of the average Poynting vector coincides with that of the group velocity. The question is answered in the affirmative after a fairly complicated explicit computation of the two directions, which are found to be identical.

1. INTRODUCTION

An ionized gas in a permanent magnetic field is an anisotropic dielectric medium. The dielectric properties of such a medium have been studied by many authors under various assumptions (Nicholls and Schelleng, 1925; Allis, 1956; Ratcliffe, 1959; Drummond, 1958). Our analysis is confined to a medium characterized by a permittivity tensor

$$\hat{\epsilon} = \begin{Vmatrix} \epsilon_1 & -j\epsilon_2 & 0 \\ i\epsilon_2 & \epsilon_1 & 0 \\ 0 & 0 & \epsilon_3 \end{Vmatrix} \tag{1}$$

which is obtained, for example, for a "cold" plasma where the thermal velocities of the electrons and ions are small compared to the phase velocity of the waves. Here the z or x_3-axis of the coordinate system is orientated in the direction of the applied permanent magnetic field. This orientation will be used throughout the paper.

The propagation of electromagnetic waves in such a medium which is sometimes called magnetoplasma or magneto-ionic medium has been studied extensively by many authors (e.g. Ratcliffe, 1959; Bunkin, 1957; Gershmann *et al.*, 1957). In an earlier paper (Kogelnik, 1960) the power radiated by a known distribution of oscillating electric currents was computed by means of a method which reduces the problem to elementary matrix operations and the evaluation of integrals. The example of an electric dipole radiating in an infinite medium was treated in detail (Kogelnik, 1960) and numerical results for the elements of the radiation resistance matrix were presented. In the present paper the computation of the dyadic Green's functions on which the method of Kogelnik is based will be recapitulated and immediately applied so as to find the four Fourier transforms needed when oscillating magnetic currents are included among the sources.

The examples of the power radiated by a modulated ion beam and that of an elementary magnetic dipole will be treated in detail. Any problem of radiation from sources in the infinite medium can be readily reduced by means of our method to elementary matrix operations and integrations. Without further special assumptions the imaginary part of the power radiated diverges in most cases.

The rest of the paper is concerned with the computation of the radiation fields at large distances from the sources. It is shown that the wave normal need not be parallel to the direction of propagation, i.e. to the group velocity. It is shown, that, although the polarization vectors may carry out a complicated motion, the direction of the average Poynting vector is nevertheless the same as that of the group velocity.

2. DEFINITION AND STARTING POINT

We shall only treat homogeneous, infinite, non-magnetic ($\mu_{rel} = 1$) media. A–c quantities shall be described by their complex amplitudes. Column vectors $\mathbf{a}$ when transposed, become row vectors $\tilde{\mathbf{a}}$. The hermitian conjugate vectors are denoted by $\mathbf{a}^+ = \tilde{\mathbf{a}}^*$. Dyades are written as $\mathbf{a}\tilde{\mathbf{a}}$, $\mathbf{k}\tilde{\mathbf{k}}$, etc., the dyade $\nabla\tilde{\nabla}$ stands for a matrix with elements $\partial^2/\partial x_i \partial x_k$. Tensor quantities are indicated by the symbols $\hat{\mathbf{a}}$, $\hat{\mathbf{K}}$, etc. In m.k.s. units Maxwell's equation takes the form

$$\nabla \times \mathbf{E} = -j\omega\mu_0 \mathbf{H} - \mathbf{M} \tag{2a}$$

$$\nabla \times \mathbf{H} = j\omega\epsilon_0 \hat{\epsilon} \mathbf{E} + \mathbf{J} \tag{2b}$$

where we have dropped the factors $e^{j\omega t}$ and $\hat{\epsilon}$ is given by (1) and **M** is the magnetic current density.

Rewriting (2a) as

$$\mathbf{H} = (j/\omega\mu_0)/[\nabla \times \mathbf{E} + \mathbf{M}]$$

and eliminating **H** from eqn (2b) we obtain the wave equation

$$\left(\nabla\tilde{\nabla} - \Delta\hat{1} - \frac{\omega^2}{c^2}\hat{\epsilon}\right)\mathbf{E} = -j\omega\mu_0 J - \nabla \times \mathbf{M} \tag{3}$$

We shall first briefly discuss plane wave solutions of the homogeneous wave equation ($\mathbf{J} = 0$, $\mathbf{M} = 0$). The wave normal defined by the unit vector

$$\mathbf{n} = (n_1, n_2, n_3)$$

with components

$$\begin{aligned} n_1 &= \sin\theta\cos\phi \\ n_2 &= \sin\theta\sin\phi \\ n_3 &= \cos\theta \end{aligned} \tag{4}$$

determines the plane wave

$$\mathbf{E}(\mathbf{r}) = \mathbf{E}_0\, e^{-jk(\mathbf{n}.\mathbf{r})} \tag{5}$$

with propagation vector

$$\mathbf{k} = k\mathbf{n} \tag{6}$$

The wave equation, applied to plane waves (5) yields

$$\hat{W}_E(\mathbf{k}, \mathbf{k}_0)\,\mathbf{E} = 0 \tag{7}$$

where

$$\hat{W}_E(\mathbf{k}, \mathbf{k}_0) = \mathbf{k}\tilde{\mathbf{k}} - (\mathbf{k}.\mathbf{k})\,\hat{1} + k_0^2\hat{\epsilon} \tag{8}$$

$$k_0 = \frac{\omega}{c} \tag{9}$$

For non-vanishing fields we must have

$$\det W_E(\mathbf{k}, \mathbf{k}_0) = 0 \tag{10}$$

One can easily show that, with an $\hat{\epsilon}$ of the form (1)

$$\det \hat{W}(\mathbf{k}, \mathbf{k}_0) = k_0^2(\epsilon_1\sin^2\theta + \epsilon_3\cos^2\theta)\,(k^2 - k_{\mathrm{I}}^2)\,(k^2 - k_{\mathrm{II}}^2) \tag{11}$$

With the refractive indices $k_{\mathrm{I}}^2/k,_0^2\; k_{\mathrm{II}}^2/k_0^2$ given by

$$\frac{k_{\mathrm{I,II}}^2}{k_0^2} = \frac{(\epsilon_1^2-\epsilon_2^2)\sin^2\theta+\epsilon_1\epsilon_3(1+\cos^2\theta)\pm\sqrt{[(\epsilon_1^2-\epsilon_2^2-\epsilon_1\epsilon_3)^2\sin^4\theta+4\epsilon_1^2\epsilon_2^2\cos^2\theta]}}{2(\epsilon_1\sin^2\theta+\epsilon_3\cos^2\theta)} \tag{12}$$

These relations show that there exist two types of plane waves, the "ordinary" and the "extraordinary" wave with complex propagation constants k_{I} and k_{II} whose values depend on the angle between the wave normal and the permanent magnetic field. Equation (12) can be shown to be equivalent to the Appleton equation (e.g. Ratcliffe, 1959, p. 19, eqn 2.6.11).

One also finds from (7) that the electric field vector with propagation constant $k_{\text{I, II}}$ is given by

$$\mathbf{E}_{\text{I, II}} = \frac{n_3 \epsilon_3 E z}{(n_1^2 + n_2^2)(\epsilon_2^2 - \epsilon_1^2 + \epsilon_1 \, k_{\text{I, II}}^2/k_0^2)} \, e_{\text{I, II}} \tag{13}$$

where

$$\mathbf{e}_{\text{I, II}} = \left\{ \begin{array}{c} n_1(\epsilon_1 - k_{\text{I, II}}^2/k_0^2) + jn_2\epsilon_2 \\ n_2(\epsilon_1 - k_{\text{I, II}}^2/k_0^2) - jn_1\epsilon_2 \\ \dfrac{n_1^2 + n_2^2}{n_3\epsilon_3} (\epsilon_2^2 - \epsilon_1^2 + \epsilon_1 k_{\text{I, II}}^2/k_0^2) \end{array} \right\} \tag{14}$$

3. DYADIC GREEN'S FUNCTIONS AND THEIR FOURIER TRANSFORMS

We want to compute the radiation field when the electric and magnetic current distributions are known. Appealing to the linearity of Maxwell's equations we can write

$$\mathbf{E}(\mathbf{r}) = \int \mathrm{d}\mathbf{r}' \hat{G}_{EJ}(\mathbf{r}, \mathbf{r}') \, \mathbf{J}(\mathbf{r}') + \int \mathrm{d}\mathbf{r}' \hat{G}_{EM}(\mathbf{r}, \mathbf{r}') \, M(\mathbf{r}') \tag{15a}$$

$$\mathbf{H}(\mathbf{r}) = \int \mathrm{d}\mathbf{r}' \, \hat{G}_{HJ}(\mathbf{r}, \mathbf{r}') \, \mathbf{J}(\mathbf{r}') + \int \mathrm{d}\mathbf{r}' \, \hat{G}_{HM}(\mathbf{r}, \mathbf{r}') \, M(\mathbf{r}'). \tag{15b}$$

If we could compute the matrices $\hat{G}_{HJ}$, $\hat{G}_{EM}$, $\hat{G}_{HJ}$, $\hat{G}_{HM}$, the dyadic Green's functions (e.g. see Kogelnik, 1960, for the terminology), the problem would be solved.

To do this we define Fourier transforms

$$\mathbf{E}(\mathbf{r}) = \int \mathrm{d}\mathbf{k}\mathbf{E}_{\mathbf{k}} \, \mathrm{e}^{-j\mathbf{k}\cdot\mathbf{r}} \tag{16a}$$

$$\mathbf{E}_{\mathbf{k}} = \frac{1}{8\pi^3} \int \mathrm{d}\mathbf{r}\mathbf{E}(\mathbf{r}) \, \mathrm{e}^{j\mathbf{k}\cdot\mathbf{r}} \tag{16b}$$

and matrices $\hat{g}_{EJ}$, etc., which are the Fourier transforms of the dyadic Green's functions $\hat{G}_{EJ}$, etc.

$$\hat{G}_{EJ}(\mathbf{r}, \mathbf{r}') = \frac{1}{8\pi^3} \int \mathrm{d}\mathbf{k}\hat{g}_{EJ}(\mathbf{k}) \, \mathrm{e}^{-j\mathbf{k}(\mathbf{r}-\mathbf{r}')}. \tag{17}$$

Introducing (16a, b), (17), into (15a, b) we obtain

$$\mathbf{E}_{\mathbf{k}} = \hat{g}_{EJ} \, \mathbf{J}_{\mathbf{k}} + \hat{g}_{EM}\mathbf{M}_{\mathbf{k}} \tag{18a}$$

$$\mathbf{H}_{\mathbf{k}} = \hat{g}_{HJ} \, \mathbf{J}_{\mathbf{k}} + \hat{g}_{HM}\mathbf{M}_{\mathbf{k}}. \tag{18b}$$

4. COMPUTATION OF THE DYADIC GREEN'S FUNCTION

For the computation of the dyadic Green's function it is useful to define a matrix

$$\hat{K} = \begin{Vmatrix} 0 & -k_3 & k_2 \\ k_3 & 0 & -k_1 \\ -k_2 & k_1 & 0 \end{Vmatrix} \tag{19}$$

such that the vector products may be written as a matrix product, e.g.

$$\mathbf{k} \times \mathbf{E} = \hat{K}\mathbf{E}. \tag{20}$$

With this notation the Fourier-transform of Maxwell's equations (2a, b) may be written in the form

$$-j\hat{K}\mathbf{E}_\mathbf{k} = -j\omega\mu_0\mathbf{H}_\mathbf{k} - \mathbf{M}_\mathbf{k} \tag{21a}$$

$$-j\hat{K}\mathbf{H}_\mathbf{k} = j\omega\epsilon_0\hat{\epsilon}\mathbf{E}_\mathbf{k} + \mathbf{J}_\mathbf{k}. \tag{21b}$$

Since det $\hat{K} = 0$, $\hat{K}$ has no inverse. In most cases, however, ϵ^{-1} exists. We can thus solve for $\mathbf{E}_\mathbf{K}$ and $\mathbf{H}_\mathbf{K}$. First we multiply eqns (21) by $1/(\omega j\mu_0)$ and $\epsilon^{-1}/(j\omega\epsilon_0)$ and obtain

$$\mathbf{H}_\mathbf{k} = \frac{1}{j\omega\mu_0}(j\hat{K}\mathbf{E}_\mathbf{k} - \mathbf{M}_\mathbf{k}) \tag{22a}$$

$$\mathbf{E}_\mathbf{k} = -\epsilon^{-1}/j\omega\epsilon_0\,(j\hat{K}\mathbf{H}_\mathbf{k} + \mathbf{J}_\mathbf{k}) \tag{22b}$$

Substituting in (22a, b) we obtain

$$(\hat{K}\hat{K} + k_0^2\hat{\epsilon})\,\mathbf{E}_\mathbf{k} = j\omega\mu_0\,\mathbf{J}_\mathbf{k} - j\hat{K}\mathbf{M}_\mathbf{k} \tag{23a}$$

$$(\hat{K}\epsilon^{-1}\hat{K} + k_0^2)\,\mathbf{H}_\mathbf{k} = j\hat{K}\epsilon^{-1}\,\mathbf{J}_\mathbf{k} + j\omega\epsilon_0\mathbf{M}_\mathbf{k} \tag{23b}$$

It is easily seen that

$$\hat{K}\hat{K} = \mathbf{k}\tilde{\mathbf{k}} - (\mathbf{k}.\mathbf{k})\,1$$

so that (23a) agrees with the Fourier transform of (3). Comparing (23a, b) with (18a, b) we find the Fourier transforms of the dyadic Green's functions

$$\hat{g}_{EJ} = j\omega\mu_0\hat{W}_E^{-1} \tag{24a}$$

$$\hat{g}_{EM} = -j\hat{W}_E^{-1}\hat{K} \tag{24b}$$

$$\hat{g}_{HJ} = j\hat{W}_H^{-1}\hat{K}\hat{\epsilon}^{-1} \tag{24c}$$

$$\hat{g}_{HM} = j\omega\epsilon_0\hat{W}_H^{-1} \tag{24d}$$

where $\hat{W}_E$ is defined by (9) and

$$\hat{W}_H = \hat{K}\epsilon^{-1}\hat{K} + k_0^2. \tag{25}$$

We note that the inverses w_E^{-1} and W_H^{-1} can be evaluated as follows

$$\hat{W}_E^{-1} = \hat{w}_E/(\det \hat{W}_E), \quad \hat{W}_H^{-1} = \hat{w}_H/(\det \hat{W}_H) \tag{26}$$

where $\hat{w}_E$ and $\hat{w}_H$ are the adjoint of $\hat{W}_E$ and $\hat{W}_H$ and $\hat{w}_E$ is given by

$$\hat{w}_E(\mathbf{k}, \mathbf{k}_0) = k^4 \mathbf{n}\tilde{\mathbf{n}} + k^2 k_0^2 \hat{L} + k_0^2 \hat{\mathscr{E}} \tag{27}$$

where $\hat{\mathscr{E}}$, the adjoint of $\hat{\epsilon}$ is given by

$$\hat{\mathscr{E}} = \begin{Vmatrix} \epsilon_1\epsilon_3 & j\epsilon_2\epsilon_3 & 0 \\ -j\epsilon_2\epsilon_3 & \epsilon_1\epsilon_3 & 0 \\ 0 & 0 & \epsilon_1^2 - \epsilon_2^2 \end{Vmatrix} \tag{28}$$

and

$$\begin{aligned} L_{11} &= \epsilon_1(n_1^2 + n_2^2) + \epsilon_3(n_1^2 + n_3^2) & L_{21} &= -j\epsilon_2(n_1^2 + n_2^2) + \epsilon_3 n_1 n_2 \\ L_{12} &= j\epsilon_2(n_1^2 + n_2^2) + \epsilon_3 n_1 n_2 & L_{22} &= \epsilon_1(n_1^2 + n_2^2) + \epsilon_3(n_2^2 + n_3^2) \\ L_{13} &= \epsilon_1 n_1 n_3 + j\epsilon_2 n_2 n_3 & L_{23} &= \epsilon_1 n_2 n_3 - j\epsilon_2 n_1 n_3 \end{aligned} \tag{29}$$

$$L_{31} = \epsilon_1 n_1 n_3 - j\epsilon_2 n_2 n_3$$
$$L_{32} = \epsilon_1 n_2 n_3 + j\epsilon_2 n_1 n_3$$
$$L_{33} = \epsilon_1(1 + n_3^2)$$

while

$$W_H^{-1} = \frac{\tilde{\mathbf{n}}\hat{\epsilon}\mathbf{n} k^4 \mathbf{n}\tilde{\mathbf{n}} - k^2 k_0^2 \hat{S} + k_0^4 \hat{1} \det \hat{\epsilon}}{\det \hat{W}_E} \tag{30}$$

and

$$\begin{aligned} S_{11} &= \epsilon_1\epsilon_3 + n_1^2(\epsilon_1^2 - \epsilon_2^2) & S_{21} &= jn_3^2\epsilon_2\epsilon_3 + n_1 n_2(\epsilon_1^2 - \epsilon_2^2) \\ S_{12} &= -jn_3^2\epsilon_2\epsilon_3 + n_1 n_2(\epsilon_1^2 - \epsilon_2^2) & S_{22} &= \epsilon_1\epsilon_3 + n_2^2(\epsilon_1^2 - \epsilon_2^2) \\ S_{13} &= jn_2 n_3\epsilon_2\epsilon_3 + n_1 n_3\epsilon_1\epsilon_3 & S_{23} &= n_2 n_3\epsilon_1\epsilon_3 - jn_1 n_3\epsilon_2\epsilon_3 \end{aligned} \tag{31}$$

$$S_{31} = -jn_2 n_3\epsilon_2\epsilon_3 + n_1 n_3\epsilon_1\epsilon_3$$
$$S_{32} = n_2 n_3\epsilon_1\epsilon_3 + jn_1 n_3\epsilon_2\epsilon_3$$
$$S_{33} = 2n_3^2\epsilon_1\epsilon_3 + (n_1^2 + n_2^2)(\epsilon_1^2 - \epsilon_2^2).$$

5. POWER RADIATED BY DISTRIBUTIONSOF CURRENT

Let us now consider the power radiated by a distribution of alternating currents $\mathbf{J}(\mathbf{r})$ and $\mathbf{M}(\mathbf{r})$. The mean complex power radiated by a volume element dx dy dz of the distribution is given by

$$\mathrm{d}P = -\tfrac{1}{2}(\mathbf{J}^*\mathbf{E} + \mathbf{H}^*\mathbf{M})\, \mathrm{d}x\mathrm{d}y\mathrm{d}z. \tag{32}$$

Hence the power radiated by this complete distribution is obtained by integration and by the use of the expressions (15a, b) for the fields, in the form

$$P = -\tfrac{1}{2} \int \mathrm{d}\mathbf{r} \int \mathrm{d}\mathbf{r}' \, \{\mathbf{J}^{+}(\mathbf{r}) \, \hat{G}_{EJ} \, \mathbf{J}(\mathbf{r}') + \mathbf{J}^{+}(\mathbf{r}) \, \hat{G}_{EM} \, \mathbf{M}(\mathbf{r}') + \mathbf{M}^{+}(\mathbf{r}') \, \hat{G}_{HM} \, \mathbf{M}(\mathbf{r}) + \mathbf{J}^{+}(\mathbf{r}') \, \hat{G}_{HJ} \, \mathbf{M}(\mathbf{r})\}. \quad (33)$$

In terms of the Fourier transforms of the dyadic Green's functions, the expression (33) for the power reduces by application of well known δ-function identities to

$$P = -4\pi^3 \int \mathrm{d}\mathbf{k} \, \{\mathbf{J}_{\mathbf{k}}^{+} \hat{g}_{EJ} \mathbf{J}_{\mathbf{k}} + \mathbf{J}_{\mathbf{k}}^{+} \hat{g}_{EM} \mathbf{M}_{\mathbf{k}} + (\mathbf{M}_{\mathbf{k}}^{+} \hat{g}_{HM} \, \mathbf{M}_{\mathbf{k}})^{*} + (\mathbf{M}_{\mathbf{k}}^{+} g_{HJ} \mathbf{J}_{\mathbf{k}})^{*}\}. \quad (34)$$

It will be useful for the purpose of the computation of the radiation from a modulated plasma sheet beam to calculate the power per unit area radiated by a plane current distribution. Assuming for this case $\mathbf{M} = 0$ let us take the element of area $\mathrm{d}x \, \mathrm{d}y$. The power per unit area radiated into a cylinder erected vertically over this area is given by

$$-\frac{\mathrm{d}P}{\mathrm{d}x \, \mathrm{d}y} = \frac{1}{2} \int_{-\infty}^{\infty} \mathbf{J}^{*} \mathbf{E} \, \mathrm{d}z. \quad (35)$$

In this case we introduce partial Fourier transforms of the form

$$\mathbf{J}(\mathbf{r}) = \int_{-\infty}^{\infty} \mathrm{d}k_3 \, \mathbf{J}(x, y, k_3) \, \mathrm{e}^{-jk_3 z} \quad (36a)$$

$$\mathbf{J}(x, y, k_3) = \frac{1}{2\pi} \int_{-\infty}^{\infty} \mathrm{d}z \, \mathbf{J}(\mathbf{r}) \, \mathrm{e}^{jk_3 z} \quad (36b)$$

and obtain

$$\frac{\mathrm{d}P}{\mathrm{d}x \, \mathrm{d}y} = -\frac{1}{4\pi} \int \mathrm{d}\mathbf{k} \int \int \mathrm{d}x' \, \mathrm{d}y' \, \{\mathbf{J}^{*}(x, y) \, \hat{g}_{EJ} \, \mathbf{J}(x', y', k_3) \, \mathrm{e}^{-j[k_1(x-x')+k_2(y-y')]}\} \quad (37)$$

6. POWER RADIATED BY A MODULATED ION BEAM IN PLASMA

It has been suggested that a thermonuclear plasma might be heated by injecting a flat, density modulated ion (or electron) beam. We assume that the ions are injected in a plane perpendicular to the permanent magnetic field, and that the distribution of the beam's charge $\rho(\mathbf{r}, t)$ can be described with good approximation by

$$\rho(\mathbf{r}, t) = \left[\rho_0 + \rho \cos \omega \left(t - \frac{x}{v}\right)\right] \delta(z) \quad (38)$$

where ρ_0 and ρ are measured in coulomb/meter, ω is the radian frequency of the density modulation, and v is the velocity of the ions, if the magnetic field is low enough so that the circular motion of the ions can be neglected. If the described model can be applied, the alternating current density produced by the ion beam has a non-zero component in x-direction only, whose amplitude is given by

$$J_x = \rho v \, \mathrm{e}^{-j\frac{\omega x}{v}} \, \delta(z) \equiv I \, \mathrm{e}^{-j\frac{\omega x}{v}} \, \delta(z). \tag{39}$$

The quantity I is measured in amperes per meter of the beam's breadth in y-direction.

We are interested in the power delivered by the beam to the (lossy) plasma. As we propose to employ formula (37) ,we need the partial Fourier transform of the current density, which is obtained by means of eqns (36a, b) as

$$J_x(x, y, k_3) = \frac{I}{2\pi} \, \mathrm{e}^{-j\frac{\omega x}{v}} \tag{40}$$

We can now apply eqn (37) to find the complex power radiated per area unit of the beam

$$\frac{\mathrm{d}P}{\mathrm{d}x \, \mathrm{d}y} = \frac{I^2}{16\pi^3} \int \mathrm{d}k \int\!\!\int_{-\infty}^{\infty} \mathrm{d}x' \, \mathrm{d}y' \, g_{11}(k) \, \mathrm{e}^{-j\left[(x-x')\left(k_1 - \frac{\omega}{v}\right) + k_2(y-y')\right]} \tag{41}$$

where g_{11} is the element in the first row and the first column of the matrix $\hat{g}_{EJ}$. We can carry out the integrations

$$\int_{-\infty}^{+\infty} \mathrm{d}x' \, \mathrm{d}y',$$

with the result

$$\frac{\mathrm{d}P}{\mathrm{d}x \, \mathrm{d}y} = -\frac{I^2}{4\pi} \int \mathrm{d}k g_{11}(k) \, \delta\left(k_1 - \frac{\omega}{v}\right) \, \delta(k_2) \, \mathrm{e}^{-j\left[x\left(k_1 - \frac{\omega}{v}\right) + k_2 y\right]}. \tag{42}$$

From eqn (27) we have

$$\hat{w}_{E11}\left(\frac{\omega}{v}, 0, k_3\right) = \left(\frac{\omega^2}{v^2} - \epsilon_3 k_0^2\right)\left(k_3^2 + \frac{\omega^2}{v^2} - k_0^2 \epsilon_1\right) \tag{43}$$

and the determinant of the wave matrix can be rewritten after some computation in the form

$$\det \hat{W}_E\left(\frac{\omega}{v}, 0, k_3\right) = k_0^2 \epsilon_3 (k_3^2 - K_{\mathrm{I}}^2)(k_3^2 - K_{\mathrm{II}}^2) \tag{44}$$

where

$$K_{\mathrm{I,\,II}}^2 / k_0^2 = \epsilon_1 - \frac{c^2}{2v^2}[1 + \epsilon_1/\epsilon_3] \pm \sqrt{\left\{\epsilon_2^2(1 - c^2/v^2\epsilon_3) + \frac{c^4}{4v^4}[(1 - \epsilon_1/\epsilon_3)]^2\right\}} \tag{45}$$

and c is the velocity of light in vacuum. If we combine eqns (24a), (26), (43) and (44) to substitute in eqn (42) there results

$$\frac{\mathrm{d}P}{\mathrm{d}x\,\mathrm{d}y} = j\omega\mu_0 \frac{I^2}{4\pi}\left(1-\frac{c^2}{v^2\epsilon_3}\right)\int_{-\infty}^{\infty} \mathrm{d}k_3 \frac{k_3^2+\frac{\omega^2}{v^2}-k_0^2\epsilon_1}{(k_3^2-K_{\mathrm{I}}^2)\,(k_3^2-K_{\mathrm{II}}^2)} \tag{46}$$

Applying Cauchy's residue formula we find

$$\int_{-\infty}^{\infty} \frac{\mathrm{d}k_3}{(k_3^2-K_{\mathrm{I}}^2)} = -\frac{\pi j}{K_{\mathrm{I}}} \tag{47}$$

where the sign of the square root of K_{I}^2 has been chosen such that Im $K_{\mathrm{I}}<0$. From this equation follows

$$\int_{-\infty}^{\infty} \mathrm{d}k_3 \frac{k_3^2+\frac{\omega^2}{v^2}-k_0^2\epsilon_1}{(k_3^2-K_{\mathrm{I}}^2)\,(k_3^2-K_{\mathrm{II}}^2)} = \frac{-\pi j}{K_{\mathrm{I}}^2-K_{\mathrm{II}}^2}\left\{K_{\mathrm{I}}-K_{\mathrm{II}}+\left(\frac{\omega^2}{v^2}-k_0^2\epsilon_1\right)\left(\frac{1}{K_{\mathrm{I}}}-\frac{1}{K_{\mathrm{II}}}\right)\right\} \tag{48}$$

by expanding the integrand into partial fractions. If we insert this into eqn (46) we can write our final result in the form

$$\frac{\mathrm{d}P}{\mathrm{d}x\,\mathrm{d}y} = I^2\sqrt{\left(\frac{\mu_0}{\epsilon_0}\right)}\,\frac{\left(1-\frac{c^2}{v^2}\epsilon_3\right)}{4(K_{\mathrm{I}}+K_{\mathrm{II}})/k_0}\left\{1+\frac{\epsilon_1-c^2/v^2}{K_{\mathrm{I}}K_{\mathrm{II}}/k_0^2}\right\}. \tag{49}$$

The type of radiation produced is similar to Čerenkov radiation. This becomes obvious if we apply formula (49) in the case of a lossless isotropic medium, where $K_{\mathrm{I,\,II}}^2/k_0^2 = \epsilon_1-(c^2/v^2)$, and

$$\frac{\mathrm{d}P}{\mathrm{d}x\,\mathrm{d}y} = I^2\sqrt{\left(\frac{\mu_0}{\epsilon_0}\right)}\,\frac{\epsilon_1-c^2/v^2}{4\epsilon_1}.$$

From this can be seen that real power is radiated only if

$$v > c/\sqrt{\epsilon_1}.$$

7. THE MAGNETIC DIPOLE

To show the power of the general method we present as another example the calculation of the real power radiated by a wire-loop immersed in a plasma. A small loop is equivalent to a magnetic dipole which may be represented by a distribution of magnetic current

$$\mathbf{M}(\mathbf{r}) = j\omega\mathbf{p}_m\,\delta(\mathbf{r}) \tag{50}$$

where $\delta(\mathbf{r})$ is Dirac's δ-function. The spatial Fourier transform of the current is

$$\mathbf{M}(k) = j\omega\rho_m/8\pi^3 \tag{51}$$

By application of eqns (34) we find that the complex conjugate of the power can be written as a bilinear form

$$P^* = \frac{\omega^2}{2}\,\mathbf{p}_m^+\,\hat{z}_m\,\mathbf{p}_m \tag{52}$$

with

$$\hat{z}_m = -\frac{1}{8\pi^3}\int \mathrm{d}\mathbf{k}\hat{g}_{HM}. \tag{53}$$

The elements of $\hat{z}_m$ are measured in $\mathrm{ohm}^{-1}\ \mathrm{m}^{-2}$ (admittance-units per unit area). The matrix $\hat{z}_m$ is split into its Hermitian and its anti-Hermitian part

$$\hat{z}_m = \hat{r}_m + j\hat{x}_m \tag{54}$$

and the real power radiated is

$$P_r = \frac{\omega^2}{2}\,\mathbf{p}_m^+\hat{r}_m\mathbf{p}_m. \tag{55}$$

We can substitute in eqn (53) for $\hat{g}_{HM}$ from eqns (24, 30) with the result

$$\hat{z}_m = \frac{-j\omega\epsilon_0}{8\pi^3 k_0}\int_0^{\pi/2}\sin\theta\,\mathrm{d}\theta\int_{-\infty}^{\infty}\mathrm{d}k\int_0^{2\pi}\mathrm{d}\phi\left[\frac{k^2\{\tilde{\mathbf{n}}\hat{\epsilon}\mathbf{n}k^4\mathbf{n}\tilde{\mathbf{n}} - k^2k_0^2 S + k_0^4\hat{1}(\det\hat{\epsilon})\}}{\tilde{\mathbf{n}}\hat{\epsilon}\mathbf{n}\,(k^2-k_{\mathrm{I}}^2)\,(k^2-k_{\mathrm{II}}^2)}\right] \tag{56}$$

The matrices $\hat{M}_m$, $\hat{N}$ are defined by

$$\hat{N} = \frac{1}{\pi}\int_0^{2\pi}\mathbf{n}\tilde{\mathbf{n}}\,\mathrm{d}\phi = \left\|\begin{matrix}\sin^2\theta & 0 & 0\\ 0 & \sin^2\theta & 0\\ 0 & 0 & 2\cos^2\theta\end{matrix}\right\| \tag{57a}$$

$$\hat{M}_m = \hat{M}_m^+ = \tfrac{1}{2}\int_0^{2\pi}\hat{S}\,\mathrm{d}\phi =$$

$$\left\|\begin{matrix}2\epsilon_1\epsilon_3+(\epsilon_1^2-\epsilon_2^2)\sin^2\theta & -2j\epsilon_2\epsilon_3\cos^2\theta & 0\\ 2j\epsilon_2\epsilon_3\cos^2\theta & 2\epsilon_1\epsilon_3+(\epsilon_1^2-\epsilon_2^2)\sin^2\theta & 0\\ 0 & 0 & 4\epsilon_1\epsilon_3\cos^2\theta+2(\epsilon_1^2-\epsilon_2^2)\sin^2\theta\end{matrix}\right\|. \tag{57b}$$

Having performed the integrations with respect to ϕ and k, the final result can be written in the form

$$\hat{r}_m = -\frac{\pi}{2z_0\lambda_0^2}\int_0^1 \frac{du\,\hat{F}_m}{k_0^3[\epsilon_1+(\epsilon_3-\epsilon_1)u^2][k_{\mathrm{I}}^2-k_{\mathrm{II}}^2]} \tag{58}$$

with $z_0 = \sqrt{(\mu_0/\epsilon_0)}$, $\lambda_0 = 2\pi/k_0$ and $\hat{F}_m$ being the Hermitian part

$$\hat{F}_m = \mathrm{Herm}\{\tilde{\mathbf{n}}\hat{\epsilon}\mathbf{n}(k_{\mathrm{I}}^5-k_{\mathrm{II}}^5)\,\hat{N}-k_0^2(k_{\mathrm{I}}^2-k_{\mathrm{II}}^2)\,\hat{M}_m+2k_0^4(k_{\mathrm{I}}^2-k_{\mathrm{II}}^2)\,1(\det\hat{\epsilon})\}.$$

Thus the matrix $\hat{r}_m$ has the form

$$\hat{r}_m = \begin{Vmatrix} r_{m1} & -jr_{m2} & 0 \\ jr_{m2} & r_{m1} & 0 \\ 0 & 0 & r_{m3} \end{Vmatrix} \tag{59}$$

with the elements

$$r_{mi} = -\frac{\pi}{2z_0\lambda_0^2}\int_0^1 \frac{du}{k_0^3[\epsilon_1+(\epsilon_3-\epsilon_1)u^2][k_{\mathrm{I}}^2-k_{\mathrm{II}}^2]} f_{mi}(u) \tag{58a}$$

where

$$f_{m1} = \mathrm{Re}\{\tilde{\mathbf{n}}\hat{\epsilon}\mathbf{n}(k_{\mathrm{I}}^5-k_{\mathrm{II}}^5)\sin^2\theta - k_0^2(k_{\mathrm{I}}^3-k_{\mathrm{II}}^3)\,[2\epsilon_1\epsilon_3+(\epsilon_1^2-\epsilon_2^2)\sin^2\theta] + 2k_0^4(k_{\mathrm{I}}-k_{\mathrm{II}})\det\hat{\epsilon}\}$$

$$f_{m2} = \mathrm{Re}\{-2k_0^2(k_{\mathrm{I}}^3-k_{\mathrm{II}}^3)\,\epsilon_2\epsilon_3\cos^2\theta\}$$

$$f_{m3} = \mathrm{Re}\{2\tilde{\mathbf{n}}\hat{\epsilon}\mathbf{n}(k_{\mathrm{I}}^5-k_{\mathrm{II}}^5)\cos^2\theta - 2k_0^2(k_{\mathrm{I}}^3-k_{\mathrm{II}}^3)\,[2\epsilon_1\epsilon_3\cos^2\theta+(\epsilon_1^2-\epsilon_2^2)\sin^2\theta] + 2k_0^4(k_{\mathrm{I}}-k_{\mathrm{II}})\det\hat{\epsilon}\}$$

8. FIELDS AT LARGE DISTANCE FRQM SOURCES

It is seen from eqn (15) that the dyadic Green's functions determine the field and that, by expanding them at large distance $\mathbf{r}-\mathbf{r}'$ we can find expansions for the fields at large distance from the sources. We take a typical Green's function dyade $\hat{G}_{EJ}$ which can, by means of eqns (17) and (24a), (26) and (27) be expressed in the form

$$\hat{G}_{EJ} = (j\omega\mu_0/8\pi^3)\int d\mathbf{k}[\hat{W}_E/(\det\hat{W}_E)]\,e^{-j\mathbf{k}(\mathbf{r}-\mathbf{r}')} \tag{60}$$

Hence for the fields we obtain expressions like

$$\mathbf{E} = j/8\pi^3\epsilon_0\int d\mathbf{r}'\int d\mathbf{k}[\hat{W}_E/(\det\hat{W}_E)]\,e^{-j\mathbf{k}(\mathbf{r}-\mathbf{r}')} \tag{61}$$

where $\hat{w}_E$ and det $\hat{W}_E$ are fourth degree polynomials in k_1, k_2, k_3, k_0. According to a method of Lighthill (1960) the nine integrals of (60) can be evaluated for large values of $\mathbf{r}-\mathbf{r}'$ with the following result: Contributions to the integrals occur only for values of $\mathbf{k} = \mathbf{k}^{(m)}$ which are roots of

$$D(\mathbf{k}, \mathbf{k}_0) = \det \hat{W}_E = 0$$

and where, moreover, the normal to the surface $D(\mathbf{k}) = 0$ in k-space is parallel to $\mathbf{r}$.

Let the Gaussian curvature of the surface $D(\mathbf{k}, \mathbf{k}_0)$ for a given k_0 be given near $\mathbf{k} = \mathbf{k}^{(m)}$ by K. Then the Green's function $\hat{G}_{EJ}$ can be approximated by a sum of dyadic Green's functions $\hat{G}_{EJ}^{(m)}$

$$\hat{G}_{EJ} = \sum_m \hat{G}_{EJ}^{(m)} \tag{62}$$

$$\hat{G}_{EJ}^{(m)} = (C/2\pi\omega\epsilon_0 r) \left\{ \frac{\hat{W}_E \, e^{-j\mathbf{k}^{(m)}(\mathbf{r}-\mathbf{r}')}}{\nabla_\mathbf{k} D(\mathbf{k}) \sqrt{K}} \right\}_{\mathbf{k}=\mathbf{k}^{(m)}} \tag{63}$$

where $\nabla_\mathbf{k}$ is the gradient in k-space, C is a constant and the expressions (63) have to be evaluated at $\mathbf{k} = \mathbf{k}^{(m)}$.

We illustrate this result by Fig. 1.

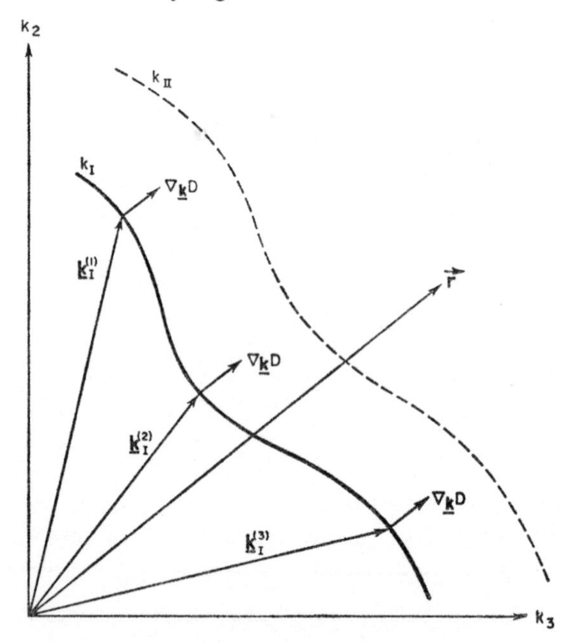

FIG. 1. Two leaves of schematic dispersion surface with normals and wave normals.

The plane of the figure contains the direction of observation $\mathbf{r}$ and the direction of k_3 which is parallel to that of the permanent magnetic field. The figure shows a section of the surface $D(\mathbf{k}) = 0$ which has rotational symmetry with respect to the axis k_3. We know from (12) that there may be two leaves corresponding to $\mathbf{k}_{\mathrm{I}}$ and $\mathbf{k}_{\mathrm{II}}$ and we have chosen a surface which has on leaf I, three positions $\mathbf{k}_{\mathrm{I}}^{(1)}$, $\mathbf{k}_{\mathrm{I}}^{(2)}$, $\mathbf{k}_{\mathrm{I}}^{(3)}$, where the normal to $D(k)$ is parallel to $\mathbf{r}$. Three positions on the second leaf similarly characterized can also be seen on the figure. The radiation propagating in the direction $\mathbf{r}$ may thus have a wave normal $\mathbf{k}^{(m)}$ making an angle with $\mathbf{r}$.

This result may be proved by choosing a coordinate system such that the axis k_1' is parallel to the direction of observation $\mathbf{r}$. In evaluating the triple integral (60) one starts by evaluating

$$\int \mathrm{d}k_1' \, [\hat{w}_E/(\det \hat{W}_E)] \, \mathrm{e}^{-jk_1'(r-r')}.$$

There will be a contribution to the integral only when the integrand has a pole. For real solutions of $D(\mathbf{k}) = \det \hat{W}_E = 0$, a contour for integration in the complex k_3'-plane is not determined. A consideration of the radiation condition which will be mentioned below can be used to show that the poles are situated slightly below the real axis. The integral then reduces to a sum of residues:

$$\hat{G}_{EJ} = (2\pi j/8\pi^3\omega\epsilon_0) \sum_n \int\int \left[\hat{w}_E \Big/ \frac{\partial D(\mathbf{k})}{\partial k_1'}\right]_n \mathrm{e}^{-jk_1'(r-r')} \, \mathrm{d}k_2' \, \mathrm{d}k_3' \qquad (64)$$

where $\mathbf{k}_2' \perp \mathbf{k}_3' \perp \mathbf{k}_1'$. The remaining integral is worked out according to the principle of stationary phase. This implies that contributions arise only when k_1' is stationary, hence our statement that the normal to $D(\mathbf{k}) = 0$ has to be parallel to $\mathbf{r}$.

The radiation condition may be used in the following way to obtain a unique contour determination for the k_1'-integral. Solutions with a time dependence $\mathrm{e}^{\epsilon t}$ are forced by giving the exciting currents such a time dependence. Incoming waves will then be absent in a solution even when $\mathrm{Lim}\ \epsilon \to 0$.

Using this technique it is easy to show that non-vanishing contributions are only obtained when

$$\frac{\mathbf{r} \cdot \nabla_{\mathbf{k}} D(\mathbf{k})}{r \partial D/\partial\omega} < 0 \qquad (65)$$

Now it can be shown that the group velocity, in a medium where frequency and wave number are related by a dispersion equation $D(k_1, k_2, k_3, k_0) = 0$, is given by

$$r_{gr} = -\frac{\nabla_{\mathbf{k}} D(\mathbf{k})}{\partial D/\partial\omega}. \qquad (66)$$

The restriction (65) may thus be written

$$\mathbf{r} \cdot v_{gr} > 0 \qquad (67)$$

and this means that only such waves can occur for which the group velocity has an outward (positive) component along the observation direction.

9. THE FIELD AT LARGE DISTANCE AS A PLANE WAVE

The electric field at large distance is given by

$$E = \sum_{m^+} \mathbf{E}^{(m)} = \sum_{m^+} \int d\mathbf{r}' \, \hat{G}^{(m)} \, \mathbf{J}(\mathbf{r}') \, e^{-j\mathbf{k}^{(m)} \cdot (\mathbf{r}-\mathbf{r}')} \tag{68}$$

where $\hat{G}^{(m)}$ is given by (63)*. The numerator of (63) has the factor $\hat{w}_E{}^{(m)}$ which in the case of a lossless medium may be written as a dyade

$$\hat{w}_E^{(m)} = C_1 \, \mathbf{e}_{\mathrm{I,\,II}}^{(m)} \, \tilde{\mathbf{e}}_{\mathrm{I,\,II}}^{(m)*} \tag{69}$$

where $\mathbf{e}^{(m)}$ is the polarization vector (14) with components

$$\begin{array}{c} n_1^{(m)} \left(\epsilon_1 - \dfrac{k_{\mathrm{I,\,II}}^2}{k_0^2} \right) + jn_2^{(m)} \, \epsilon_2 \\ n_2^{(m)} \left(\epsilon_1 - \dfrac{k_{\mathrm{I,\,II}}^2}{k_0^2} \right) - jn_1^{(m)} \, \epsilon_2 \\ \dfrac{n_1^{(m)} + n_2^{(m)}}{n_3^{(m)} \, \epsilon_3} \left(\epsilon_2^2 - \epsilon_1^2 + \epsilon_1 \dfrac{k_{\mathrm{I,\,II}}^2}{k_0^2} \right) \end{array} \tag{70}$$

and C_1 is a constant.

This follows from the fact that $\hat{w}_E^{(m)}$ may be written as a sum of three dyades

$$\hat{w}_E^{(m)} = \mathbf{w}_1 \tilde{\mathbf{i}}_1 + \mathbf{w}_2 \tilde{\mathbf{i}}_2 + \mathbf{w}_3 \tilde{\mathbf{i}}_3 \tag{71}$$

in terms of its column vectors $\mathbf{w}_1$, $\mathbf{w}_2$, $\mathbf{w}_3$ and the unit vectors $\tilde{\mathbf{i}}_1$, $\tilde{\mathbf{i}}_2$, $\tilde{\mathbf{i}}_3$ of the coordinate system.

$\hat{w}_E^{(m)}$ is defined by

$$\det \hat{W}_E = \hat{W}_E^{(m)} \, \hat{w}_E^{(m)} = \hat{w}_E^{(m)*} \, \hat{W}_E^{(m)*}. \tag{72}$$

Hence

$$\hat{W}_E^{(m)} \, \mathbf{w}_1 = \hat{W}_E^{(m)} \, \mathbf{w}_2 = \hat{W}_E^{(m)} \, \mathbf{w}_3 = 0 \tag{73}$$

and

$$\mathbf{w}_1, \mathbf{w}_2, \mathbf{w}_3 \text{ are proportional to } \mathbf{e}^{(m)} \tag{74}$$

since the polarization vector is proportional to the unique solution of

$$\hat{W}_E^{(m)} \, \mathbf{e}^{(m)} = 0. \tag{7}$$

This shows that $\hat{w}_E^{(m)}$ is indeed a dyade and it is easy to show from

$$\hat{\mathbf{w}}_E^{(m)*} \, \hat{W}^{(m)*} = 0$$

that (69) is in fact true.

Since $\mathbf{e}^{(m)} \, \tilde{\mathbf{e}}^{(m)*} \, \mathbf{J} = \mathbf{e}^{(m)} \, (\mathbf{e}^{(m)*} \, . \, \mathbf{J})$ it follows that

$$\mathbf{E}^{(m)} = C_1 C_2 \int dr' \, \mathbf{e}^{(m)} \, (\mathbf{e}^{(m)*} \, . \, \mathbf{J}(r')) \, e^{j\mathbf{k}^{(m)} \cdot (\mathbf{r}-\mathbf{r}')} \tag{75}$$

(C_2 is another constant) so that each contribution to E is indeed a plane wave with the appropriate polarization vector as calculated before.

* *Note:* The index m^+ indicates that we sum only over terms satisfying (67).

10. THE DIRECTION OF THE AVERAGE POYNTING VECTOR

It seems interesting to consider the significance of the average Poynting vector

$$\bar{\mathbf{S}} = \mathrm{Re}\, \tfrac{1}{2}(\mathbf{E} \times \mathbf{H}^*) \tag{76}$$

in anisotropic wave motion. We shall show in the case of a lossless anisotropic medium characterized by a permittivity tensor of the type (1) that $\bar{\mathbf{S}}$ is parallel to the direction of the group velocity.

We assume that $k = k_{\mathrm{I}}$, or $k = k_{\mathrm{II}}$ is real and introduce the refractive index R, where

$$R^2 = k^2_{\mathrm{I,\,II}}/k_0^2 > 0. \tag{77}$$

Because of the rotational symmetry around k_3 we can put $n_2 = 0$ without loss of generality. We have $\mathbf{E} = C\mathbf{e}$ where C is a constant and

$$\mathbf{e} = \begin{pmatrix} e_1 \\ e_2 \\ e_3 \end{pmatrix} = \begin{pmatrix} n_1(\epsilon_1 - R^2) \\ -j\epsilon_2 n_1 \\ (n_1/n_3\epsilon_3)\,(\epsilon_2^2 - \epsilon_1^2 + \epsilon_1 R^2) \end{pmatrix}. \tag{78}$$

From Maxwell's equation

$$\mathbf{H} = (k/\mu_0\omega)\,(\mathbf{n} \times \mathbf{E}) \tag{79}$$

Hence

$$\bar{\mathbf{S}} = \frac{CC^*}{2\omega\mu_0}\,\mathrm{Re}\,[\mathbf{e} \times (\mathbf{n} \times \mathbf{e}^*)] = \frac{CC^*k}{2\omega\mu_0}\,\mathrm{Re}\,\{\mathbf{n}(\mathbf{e}\,.\,\mathbf{e}^*) - \mathbf{e}^*(\mathbf{e}\,.\,\mathbf{n})\} \tag{80}$$

and since, in view of $n_2 = 0$, $(\mathbf{e}\,.\,\mathbf{n})$ as well as $(\mathbf{e}\,.\,\mathbf{e})$ are real

$$\bar{\mathbf{S}} = \frac{kCC^*}{2\omega\mu_0}\,\{\mathbf{n}(\mathbf{e}\,.\,\mathbf{e}^*) - \mathbf{e}'(\mathbf{e}\,.\,\mathbf{n})\} \tag{81}$$

where

$$\mathbf{e}' = \begin{pmatrix} e_1 \\ 0 \\ e_3 \end{pmatrix}. \tag{82}$$

Calculating the unit vector $\mathbf{s}$

$$\mathbf{s} = \mathbf{S}/(\mathbf{S}^2)^{1/2} \tag{83}$$

and defining the angle χ as that between the wave normal $\mathbf{n}$ and the vector $\mathbf{s}$ we have

$$\mathbf{s}\,.\,\mathbf{n} = \cos\chi \tag{84}$$

and

$$\tan\chi = \sqrt{\frac{1}{(\mathbf{s}\,.\,\mathbf{n})^2 - 1}}. \tag{85}$$

Straightforward, but somewhat tedious, computation leads to

$$\tan \chi = \pm n_1 n_3 \frac{\{R^2(\epsilon_1 - \epsilon_3) - \epsilon_1^2 + \epsilon_2^2 + \epsilon_1\epsilon_3\} \{(R^2 - \epsilon_1)(\epsilon_1 n_1^2 + \epsilon_2 n_2^2) + n_2 n_1^2\}}{\{(R^2 - \epsilon_1)(\epsilon_1 n_1^2 + \epsilon_3 n_3^2) + n_1^2 \epsilon_2^2\} + n_3^2 \epsilon_2^2 \epsilon_3^2}. \tag{86}$$

On the other hand, the equation det $\hat{W}_E = 0$ for the R- or $\mathbf{k}$-surface may be written

$$\epsilon_3(\epsilon_1^2 - \epsilon_2^2) - R^2\{(\epsilon_1^2 - \epsilon_2^2)\sin^2\theta + \epsilon_1\epsilon_3(1 + \cos^2\theta)\} + R^4(\epsilon_1 \sin^2\theta + \epsilon_3 \cos^2\theta) = 0 \tag{87}$$

It is clear from Fig. 2 that

$$\tan \chi' = \frac{dR}{R d\theta} \tag{88}$$

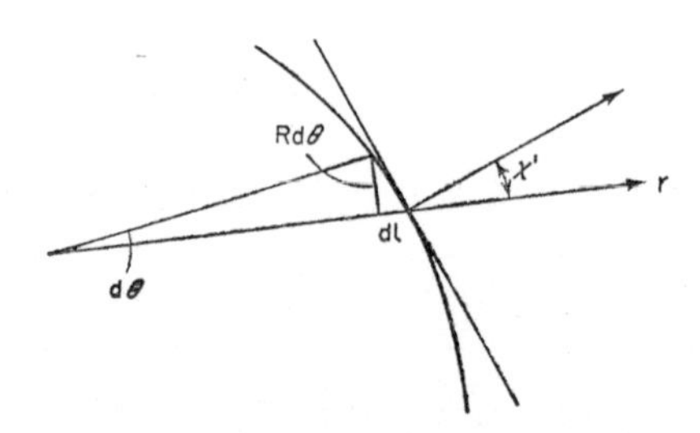

FIG. 2. Dispersion surface and angle between wave normal and normal to the surfece.

where χ' is the angle between n and the normal to the R-surface. The result of the differentiation is

$$\frac{dR}{R d\theta} = \frac{\sin\theta\cos\theta\{R^2 - (\epsilon_1 - \epsilon_3) - \epsilon_1^2 + \epsilon_2^2 + \epsilon_1\epsilon_2\}}{n_1^2(\epsilon_1^2 - \epsilon_2^2) + \epsilon_1\epsilon_3(1 - n_3^2) - 2R^2(\epsilon_1 n_1^2 + \epsilon_3 n_3^2)}. \tag{89}$$

We can show that

$$\tan \chi' = \tan \chi \tag{90}$$

from (86) and (89) by using the identity

$$\epsilon_2^2\epsilon_3^2 n_3^2 = [(R^2 - \epsilon_1)(\epsilon_1 n_1^2 + \epsilon_3 n_3^2) + \epsilon_2^2 n_1^2]\,[(\epsilon_1 n_1^2 + \epsilon_3 n_3^2) R^2 - \epsilon_1\epsilon_3] \tag{91}$$

which can be derived from det $\hat{W}_E = 0$ by multiplication by the factor $(\epsilon_1 n_1^2 + \epsilon_3 n_3^2)$.

It thus appears that considerable computation is necessary to establish our result, i.e. that the direction of the average Poynting vector is that of the normal to the dispersion surface and therefore according to Section 8 of this paper parallel to the group velocity. The Poynting vector itself moves in a complicated fashion.

Acknowledgement—The research reported in this document has been sponsored in part by the Air Force Office of Scientific Research, O.A.R., through the European Office, Aerospace Research, United States Air Force.

REFERENCES

ALLIS, W. P. (1956) Motions of electrons and ions, *Hanb. Phys.* **21,** 383.

BARSUKOV, K. A. (1959) On the Doppler effect in an anisotropic and gyrotropic medium, *J. Exp. Theoret. Phys.* (*U.S.S.R.*) **36,** 1485.

BEGIASHVILE, G. A. and GEDALIN, E. V. (1959) Cerenkov radiation of a magnetic dipole in an anisotropic medium, *J. Expt. Theoret. Phys.* (*U.S.S.R.*) **36,** 1939.

BORN, M. (1933) *Optik*, pp. 413–420. Berlin.

BRESLER, A. D. (1959) The far fields excited by a point source in a passive dissipationless anisotropic uniform waveguide, *Trans. I.R.E.* MTT-7, 282.

BUNKIN, F. V. (1957) On radiation in anisotropic media, *J. Expt. Theoret. Phys.* (*U.S.S.R.*) **32,** 338.

DRUMMOND, J. E. (1958) Basic microwave properties of hot magneto-plasmas, *Phys. Rev.* **110,** 293 (with further references).

GERSHMANN, B. N., GINZBURG, V. L. and DENISOV, N. G. (1957) The propagation of electromagnetic waves in plasma (ionosphere) *Uspekhi fiz Nauk* (*U.S.S.R.*) **61,** 561–612 (with further references).

GINZBURG, V. L. and EIDMAN, V. YA (1959) The radiation reaction in the motion of a charge in a medium, *J. Expt. Theoret. Phys.* (*U.S.S.R.*) **36,** 1823.

HINES, C. O. Wave packets, the Poynting Vector, and Energy flow. *J. Geophysics Research* (1951) **56,** 63, 197, 207, 535.

KOGELNIK, H. (1960) The radiation resistance of an elementary dipole in anisotropic plasmas, *Fourth Int. Conf. Ionis. Phenomena in Gases*, Upsala, III C, 721–25. North Holland Pub. Co.

KOGELNIK, H. (1960) Electromagnetic radiation in anisotropic media. Thesis, Oxford University.

KOGELNIK, H. (1960) On electromagnetic radiation in magneto-ionic media, *J. Res. Nat. Bur. Stand.* **64D,** No. 5, 515–523 (Sept.–Oct.).

KOLOMENSKII, A. A. (1956) Radiation from a plasma electron in uniform motion in a magnetic field, *Doklady Akad. Nauk S.S.S.R.* **106,** 982 (with further references).

KUEHL, H. (1960) California Tech. Antenna Tech. Rep. 24, pp. 1–54 (October).

LIGHTHILL, M. J. (1960) Studies on magnetohydrodynamic waves and other anisotropic wave motions, *Phil. Trans. Roy. Soc. Lond*, Ser. A, **252,** 397–430 (March).

NICHOLS, H. W. and SCHELLENG, J. C. (1925) *B.S.T.J.* **4,** 215.

RATCLIFFE, J. A. (1959) *The Magneto-ionic Theory and Its Application to the Ionosphere.* Cambridge Univ. Press, Cambridge (with further references).

SEIDEL, H. (1957) The character of waveguide modes in gyromagnetic media, *B.S.T.J.* **26,** 409–426.

SITENKO, A. G. and KOLOMENSKII, A. A. (1956) Motion of a charged particle in an optically active anisotropic medium, *J. Expt. Theoret. Phys.* (*U.S.S.R.*) **30,** 311.

FIELD SOLUTION FOR A DIPOLE IN AN ANISOTROPIC MEDIUM

R. MITTRA† and G. A. DESCHAMPS‡

ABSTRACT

In this paper, Maxwell's equations are solved for an anisotropic medium in the presence of an infinitesimally small electric current source with an arbitrary orientation. Three-dimensional Fourier transforms technique is used to obtain the solution of the field equations, and the inversion of the transforms is discussed in detail. The singular terms representing the very near fields are obtained in a closed form and the remainder of the solution is expressed in a finite range integral whose integrand is finite everywhere, making the form of the solution convenient for numerical calculations. Unlike the solutions obtained by the previous workers, the present one is not restricted either to the far field evaluation or to a lossless medium.

1. INTRODUCTION

The knowledge of field solutions of Maxwell's equations with source terms is of considerable importance for many applications. In this paper we have derived a complete solution of the field equations in a form which may be used for the calculation of the fields at arbitrary distances from the source. For an infinitesimal dipole, the very near fields are found in a closed form involving trigonometric functions. The remainder of the solution is expressed in terms of a finite range integral, the integrand of which is finite everywhere. Hence, the solution is very suitable for numerical calculations. It is also pertinent to point out that the solutions derived hold for a very general nature of the elements of the $\bar{\bar{\epsilon}}$ tensor; for instance, the solutions are valid for complex elements and hence are useful for the treatment of a medium with finite losses. Some other previous contributions on this topic, e.g. by Bunkin (1957) and Kogelnik (1960), Kuehl (1960), and Meecham (1961), do not give a general solution such as the one derived in this paper.

2. DERIVATION OF MATRIX EQUATIONS IN A MAGNETO-IONIC MEDIUM WITH SOURCES

In this section we shall derive the matrix equations for the fields in a homogeneous medium with tensor dielectric properties, for the case of im-

† Antenna Laboratory, University of Illinois, Urbana, Illinois, and Radar Division, Ground Systems Group, Hughes Aircraft Company, Fullerton, California.

‡ Antenna Laboratory, University of Illinois, Urbana, Illinois.

pressed electric current sources. The equations will subsequently be solved using the three-dimensional Fourier transforms. Maxwell's equations for the $e^{j\omega t}$ time convention are:

$$\nabla \times \bar{E} = -j\omega\mu\bar{H} \tag{1}$$

$$\nabla \times \bar{H} = j\omega\epsilon_0\bar{\bar{\epsilon}}\bar{E} + \bar{J} \tag{2}$$

$$\nabla \, . \, \bar{H} = 0 \tag{3}$$

$$\nabla \, . \, \epsilon\bar{E} = \frac{\rho}{\epsilon_0} \tag{4}$$

where $\bar{J}$ represents the impressed electric current source term. It will be assumed that the coordinate axes have been so oriented that $\bar{\bar{\epsilon}}$ has the form

$$\bar{\bar{\epsilon}} = \hat{x}\hat{x}\epsilon - j\hat{x}\hat{y}\epsilon' + j\hat{y}\hat{x}\epsilon' + \hat{y}\hat{y}\epsilon + \hat{z}\hat{z}\epsilon_z \tag{5}$$

where $\hat{x}$, $\hat{y}$, $\hat{z}$ are unit vectors in the cartesian system and ϵ_0 is the free space dielectric constant.

It is well known that in a magneto-ionic medium such a form for ϵ results when the z-direction is oriented along the d.c. magnetic field.

Elimination of $\bar{H}$ from (1) and (2) gives

$$\nabla \times \nabla \times \bar{E} = k_0^2\bar{\bar{\epsilon}}\bar{E} - j\omega\mu\bar{J}, \qquad k_0^2 = \omega^2\mu\epsilon_0. \tag{6}$$

Because of the particular form of $\bar{\bar{\epsilon}}$ we shall find it convenient to work with a different coordinate system in which the $\bar{\bar{\epsilon}}$ tensor diagonalizes. Such a system of coordinates is defined below. Let the unit vectors $\hat{U}_1$, $\hat{U}_2$, $\hat{U}_3$ in the new system, hereafter called the U-system be defined by

$$\hat{U}_1 = 2^{-1/2}(\hat{x} + j\hat{y}), \qquad \hat{U}_2 = 2^{-1/2}(\hat{x} - j\hat{y}), \qquad \hat{U}_3 = \hat{z}. \tag{7}$$

An arbitrary vector $\bar{F}$ can be written in the U-system as

$$\bar{F} = \hat{U}_1F_1 + \hat{U}_2F_2 + \hat{U}_3F_3 \tag{8}$$

where

$$F_1 = \bar{F} \, . \, \hat{U}_2, \; F_2 = \bar{F} \, . \, \hat{U}_1 \text{ and } F_3 = \bar{F} \, . \, \hat{U}_3.$$

It may be easily verified that

$$F_1 = 2^{-1/2}(F_x - jF_y), \qquad F_2 = 2^{-1/2}(F_x + jF_y), \qquad F_3 = F_z. \tag{9}$$

Now the representation for $\bar{\bar{\epsilon}}$ in the U-system is

$$\bar{\bar{\epsilon}} = \hat{U}_1\hat{U}_1\epsilon_1 + \hat{U}_2\hat{U}_2\epsilon_2 + \hat{U}_3\hat{U}_3\epsilon_3 \tag{10}$$

where

$$\epsilon_1 = \epsilon + \epsilon' \qquad \epsilon_2 = \epsilon - \epsilon' \qquad \epsilon_3 = \epsilon_z. \tag{11}$$

The expression for the operator ∇ is as follows

$$\nabla = \hat{U}_1d_1 + \hat{U}_2d_2 + \hat{U}_3d_3 \tag{12}$$

where

$$d_1 = 2^{-1/2}\left(\frac{\partial}{\partial x} - j\frac{\partial}{\partial y}\right), \quad d_2 = 2^{-1/2}\left(\frac{\partial}{\partial x} + j\frac{\partial}{\partial y}\right), \quad d_3 = \frac{\partial}{\partial z}. \tag{13}$$

After going through some algebra and the use of (10) and (12), the inhomogeneous field equation for $\bar{E}$ given in (6) may be written in the matrix form as

$$\left\{ -k_0^2 \begin{bmatrix} \epsilon_1 & 0 & 0 \\ 0 & \epsilon_2 & 0 \\ 0 & 0 & \epsilon_3 \end{bmatrix} + \begin{bmatrix} -(d_1d_2+d_3^2), & d_1^2, & d_1d_3 \\ d_2^2, & -(d_1d_2+d_3^2), & d_2d_3 \\ d_3d_2, & d_1d_3, & -2d_1d_2 \end{bmatrix} \right\} \begin{bmatrix} E_1 \\ E_2 \\ E_3 \end{bmatrix} = -j\omega\mu \begin{bmatrix} J_1 \\ J_2 \\ J_3 \end{bmatrix}. \tag{14}$$

The above are the desired matrix equations for the fields. The solution for (14) will be obtained by taking the three-dimensional Fourier transform of the equations and by subsequent inversion of the transforms. This is discussed in the next section.

3. TRANSFORM METHOD OF SOLUTION

Let the three-dimensional Fourier transform $\mathscr{E}$ of E be defined by

$$\mathscr{E}_{1,2,3} = \int\int\int_{-\infty}^{\infty} E_{1,2,3}(x, y, z)\, e^{+j(k_x x + k_y y + k_z z)}\, dx dy dz \tag{15}$$

and let the transforms $\mathscr{J}_{1,2,3}$ of $J_{1,2,3}$ be similarly defined. Using the property of the transforms, the differential operators d_1, d_2, d_3 may be equivalently replaced by

$$d_{1,2} \to 2^{-1/2} j (k_x \mp jk_y) \qquad d_3 \to jk_z. \tag{16}$$

For convenience, we shall introduce the polar form through the definitions:

$$k_x = \Gamma \sin\psi \cos\alpha \qquad k_y = \Gamma \sin\psi \sin\alpha \qquad k_z = \Gamma \cos\psi \tag{17}$$

and rewrite (20) as

$$d_{1,2} = j\, 2^{-1/2}\, \Gamma \sin\psi\, e^{\mp j\alpha}, \qquad d_3 = j\Gamma \cos\psi. \tag{18}$$

Taking the transform of the matrix equation (14) we may then derive the desired equation:

$$\left\{ -k_0^2 \begin{bmatrix} \epsilon_1 & 0 & 0 \\ 0 & \epsilon_2 & 0 \\ 0 & 0 & \epsilon_3 \end{bmatrix} - \Gamma^2 \begin{bmatrix} -\left(\frac{\sin^2\psi}{2}+\cos^2\psi\right), & \frac{1}{2}\sin^2\psi e^{-2j\alpha}, & 2^{-1/2}\sin\psi\cos\psi e^{-j\alpha} \\ \frac{1}{2}\sin^2\psi e^{2j\alpha}, & -\left(\frac{\sin^2\psi}{2}+\cos^2\psi\right), & 2^{-1/2}\sin\psi\cos\psi e^{j\alpha} \\ 2^{-1/2}\sin\psi\cos\psi\, e^{j\alpha}, & 2^{-1/2}\sin\psi\cos\psi e^{-j\alpha}, & -\sin^2\psi \end{bmatrix} \right\} \begin{bmatrix} \mathscr{E}_1 \\ \mathscr{E}_2 \\ \mathscr{E}_3 \end{bmatrix} = -j\omega\mu \begin{bmatrix} \mathscr{J}_1 \\ \mathscr{J}_2 \\ \mathscr{J}_3 \end{bmatrix}. \tag{19}$$

The determinant Δ of the matrix inside the curly brackets of eqn (19), obtained after working out some algebra is expressed as

$$\left.\begin{aligned} \Delta = & -k_0^2\,\Gamma^4(\epsilon\sin^2\psi+\epsilon_3\cos^2\psi)+ \\ & +k_0^4\,\Gamma^2\{\epsilon_3\,\epsilon(1+\cos^2\psi)+(\epsilon^2-\epsilon'^2)\sin^2\psi\} \\ & \qquad -k_0^6(\epsilon^2-\epsilon'^2)\,\epsilon_3 \\ = & -k_0^2(\epsilon\sin^2\psi+\epsilon_3\cos^2\psi)\{\Gamma^2-n_1^2(\psi)\}\{\Gamma^2-n_2^2(\psi)\} \end{aligned}\right\} \tag{20}$$

where n_1^2, n_2^2, are the roots of Δ. The roots are given by

$$n_{1,2}^2 = k_0^2\,\frac{B(\psi)\pm\sqrt{[B^2(\psi)-4A(\psi)\,C(\psi)]}}{2A(\psi)} \tag{21}$$

where

$$\left.\begin{aligned} A(\psi) &= (\epsilon\sin^2\psi+\epsilon_3\cos^2\psi) \\ B(\psi) &= \{\epsilon_3\epsilon(1+\cos^2\psi)+(\epsilon^2-\epsilon'^2)\sin^2\psi\} \\ C(\psi) &= (\epsilon^2-\epsilon'^2)\,\epsilon_3. \end{aligned}\right\} \tag{22}$$

It may be verified that n_1 and n_2 are the indices of refraction for plane extraordinary and ordinary waves, respectively. This is to be expected since the condition that a finite solution of (19) exist even when $\Delta = 0$ is that $\mathscr{J}_1 = \mathscr{J}_2 = \mathscr{J}_3 = 0$, which obviously corresponds to the plane wave case. It should also be pointed out that the coefficient of Γ^6 in the expansion for Δ is identically zero and hence the third order determinant in Γ^2 has only two roots. The coefficient of Γ^6 is zero because of the property of the operator $(\nabla\, x\, \nabla\, x)$ which makes the matrix corresponding to this operator a singular one. Another observation about Δ is that it is independent of α.

The solution of the fields in the transformed domain is now a straightforward step involving the inversion of the matrix operator in (19). In the matrix form the solution is written as

$$[\mathscr{E}] = -j\omega\mu\, [A][\mathscr{J}]. \tag{23}$$

The elements of the inverted matrix are given by $A_{mn} = a_{mn}/\Delta$ where

$$a_{11} = 2^{-1}\Gamma^4 \sin^2 \psi - \Gamma^2 k_0^2\{\epsilon_2 \sin^2 \psi + \epsilon_3(\tfrac{1}{2} \sin^2 \psi + \cos^2 \psi)\} + k_0^4 \epsilon_3 \epsilon_2 \tag{24}$$

$$a_{12} = \mathrm{e}^{-2j\alpha}\, 2^{-1}\Gamma^2 \sin^2 \psi(\Gamma^2 - k_0^2\epsilon_3) \tag{25}$$

$$a_{13} = \mathrm{e}^{-j\alpha}\, 2^{-1/2}\Gamma^2 \sin \psi \cos \psi(\Gamma^2 - k_0^2\epsilon_2) \tag{26}$$

$$a_{21} = \mathrm{e}^{2j\alpha}\, 2^{-1}\Gamma^2 \sin^2 \psi(\Gamma^2 - k_0^2\epsilon_3) \tag{27}$$

$$a_{22} = 2^{-1}\Gamma^4 \sin^2 \psi - \Gamma^2 k_0^2\{\epsilon_1 \sin^2 \psi + \epsilon_3(\tfrac{1}{2} \sin^2 \psi + \cos^2 \psi)\} + k_0^4 \epsilon_1 \epsilon_3 \tag{28}$$

$$a_{23} = \mathrm{e}^{j\alpha}\, 2^{-1/2}\Gamma^2 \sin \psi \cos \psi(\Gamma^2 - k_0^2\epsilon_1) \tag{29}$$

$$a_{31} = \mathrm{e}^{j\alpha}\, 2^{-1/2}\Gamma^2 \sin \psi \cos \psi(\Gamma^2 - k_0^2\epsilon_2) \tag{30}$$

$$a_{32} = 2^{-1/2}\, \mathrm{e}^{-j\alpha}\, \Gamma^2 \sin \psi \cos \psi(\Gamma^2 - k_0^2\epsilon_1) \tag{31}$$

$$a_{33} = \Gamma^4 \cos^2 \psi - k_0^2\Gamma^2(\epsilon_2 + \epsilon_1)(\tfrac{1}{2} \sin^2 \psi + \cos^2\psi) + k_0^4 \epsilon_1 \epsilon_2 \tag{32}$$

and

$$\epsilon_1 = \epsilon + \epsilon', \qquad \epsilon_2 = \epsilon - \epsilon', \qquad \epsilon_3 = \epsilon_z.$$

The main problem of evaluating the inverse transforms still remains and we discuss this in the following.

4. EVALUATION OF INVERSE TRANSFORMS

We are now ready to get into the core of the problem, namely that of evaluating the inverse transforms of the fields. It should be pointed out that the asymptotic evaluation of the inverses, at distances far away from the sources, is a relatively simple task and has been discussed by Bunkin (1957) and others.

In this paper we shall emphasize on the derivation of the field representations which are valid for arbitrary distances from the source and are very useful, in particular, for evaluating the near and intermediate fields. Only a few brief comments will be made on the problem of far field evaluation so as to avoid any repetition of Bunkin's work. Let us assume an infinitesimal source located at the origin, i.e. let

$$\left.\begin{aligned} \bar{J} &= (\hat{x}C_x + \hat{y}C_y + \hat{z}C_z)\,\delta(x)\,\delta(y)\,\delta(z) \\ &= (\hat{U}_1C_1 + \hat{U}_2C_2 + \hat{U}_3C_3)\,\delta(x)\,\delta(y)\,\delta(z) \end{aligned}\right\} \tag{33}$$

where δ is the Dirac δ and C_x, C_y, C_z and hence C_1, C_2 and C_3 are constants, which may be readily obtained from the orientation and moment of the dipole. For an arbitrary distribution of sources in space, an integration over the distribution would yield the desired result. For an infinitesimal source at the origin, the transform components are readily obtained by the use of the property of δ and are given by

$$\mathscr{J}_{1,\,2,\,3} = C_{1,\,2,\,3}. \tag{34}$$

For simplicity, we shall carry out the steps for evaluating the inverse transforms for the case, $C_1 = 1$ and $C_2 = C_3 = 0$. It is obvious that through the use of superposition we can readily derive the complete solution when the current coefficients C_2, C_3 are non-zero, by following through similar routines.

From (24), (27), and (30) we have

$$\mathscr{E}_n = -\,j\omega\mu\,\frac{a_{n1}(\Gamma,\,\psi,\,\alpha)}{\Delta(\Gamma,\,\psi)} \qquad n = 1, 2, 3. \tag{35}$$

The inverse transform of a function $\mathscr{F}(r,\,\psi,\,\alpha)$ is given by

$$F = \frac{1}{(2\pi)^3}\iiint\limits_{-\infty}^{\infty} \mathscr{F}(k_x,\,k_y,\,k_z)\,\mathrm{e}^{-j(k_x x+k_y y+k_z z)}\,\mathrm{d}k_x\,\mathrm{d}k_y\,\mathrm{d}k_z, \tag{36}$$

or in the polar form as

$$\left.\begin{aligned} F = \frac{1}{(2\pi)^3}\int_0^\infty\int_0^\pi\int_0^{2\pi} \mathscr{F}(\Gamma,\,\psi,\,\alpha)\,\mathrm{e}^{-j\Gamma R(\sin\theta\sin\psi\cos(\alpha-\phi)+\cos\theta\cos\psi)} \\ \Gamma^2\sin\psi\,\mathrm{d}\alpha\,\mathrm{d}\psi\,\mathrm{d}\Gamma \end{aligned}\right\} \tag{37}$$

where $x = R\sin\theta\cos\phi$, $y = R\sin\theta\sin\phi$ and $z = R\cos\theta$.

Hence, the inverse of (35) may be written as

$$\left.\begin{aligned} E_n = -\,\frac{j\omega\mu}{(2\pi)^3}\int_0^\infty\int_0^\pi\int_0^{2\pi} \frac{a_{n1}}{\Delta}\,(\Gamma,\,\psi,\,\alpha)\,\mathrm{e}^{-j\Gamma R(\sin\theta\sin\psi\cos(\alpha-\phi)+\cos\theta\cos\psi)} \\ \Gamma^2\sin\psi\,\mathrm{d}\alpha\,\mathrm{d}\psi\,\mathrm{d}\Gamma. \end{aligned}\right\} \tag{38}$$

Let us consider the case $n = 1$. From (24) we see that a_{11} is independent of α hence we can carry out the integration in (38) with respect to α

giving

$$E_1 = -\frac{j\omega\mu}{(2\pi)^2}\int_0^\infty\int_0^\pi \Gamma^2 \frac{a_{11}(\Gamma,\psi)}{\Delta}\, e^{-j\Gamma R\cos\theta\cos\psi} J_0(\Gamma R\sin\theta\sin\psi)\sin\psi\, d\psi\, d\Gamma \tag{39}$$

where we have made use of one of the well-known (1954) integral representations for the Bessel function.

We rewrite (39) in the form

$$E_1 = \frac{j\omega\mu}{k_0^2(2\pi)^2}\nabla^2 I(R,\theta) \tag{40}$$

where

$$\left.\begin{aligned} I(R,\theta) &= k_0^2\int_0^\infty\int_0^\pi \frac{a_{11}}{\Delta}\, e^{-j\Gamma R p(\theta,\psi)} J_0(\Gamma R q(\theta,\psi))\sin\psi\, d\psi\, d\Gamma \\ p(\theta,\psi) &= \cos\theta\cos\psi, \qquad q(\theta,\psi) = \sin\theta\sin\psi \end{aligned}\right\} \tag{41}$$

and ∇^2 is the Laplacian operator. It was necessary to reorganize the integral representation so as to enable us to change the order of integration, which we intend to do shortly. It should be noted that since a_{11} and Δ are both of fourth order in Γ, $a_{11}/\Delta \to$ constant for large Γ and the convergence of the integral representation for $I(R, \theta)$ in (40) is assured for all $R \neq 0$. We anticipate the type of behavior $I(R, \theta) \to 0(1/R)$ as $R \to 0$, so we set out to separate the singular part. When this is accomplished, $I(R, \theta)$ may be written as

$$I(R,\theta) = I_s(R,\theta) + I_f(R,\theta) \tag{42}$$

where $I_s(R, \theta)$ is $0(1/R)$ for small R and I_f is finite for all R and θ. The advantage of doing this is twofold. First, we shall find that it is possible to evaluate I_s (the singular part of the integral) exactly and hence to estimate the very near field behavior, which is dominated by the singular part. The second reason is that this procedure will make it convenient for us as we shall see shortly, to perform the integration in the expression for $\nabla^2 I_f$.

Separation and Evaluation of the Singular Part $I_s(R, \theta)$

Let us consider the ratio a_{11}/Δ. From (20) and (24) we have, after replacing ϵ_1 by $\epsilon + \epsilon'$ and ϵ_2 by $\epsilon - \epsilon'$

$$k_0^2\frac{a_{11}}{\Delta} = -\frac{(1/2)\Gamma^4\sin^2\psi - \Gamma^2 k_0^2\{(\epsilon-\epsilon')\sin^2\psi + \epsilon_3(1/2\sin^2\psi + \cos^2\psi)\} + k_0^4\epsilon_3(\epsilon-\epsilon')}{(\epsilon_3\cos^2\psi + \epsilon\sin^2\psi)(\Gamma^2 - n_1^2)(\Gamma^2 - n_2^2)}$$

$$k_0^2 \frac{a_{11}}{\Delta} = -\frac{1}{2} \frac{\sin^2 \psi}{(\epsilon_3 \cos^2 \psi + \epsilon \sin^2 \psi)} + \frac{\Gamma^2\left[-\frac{\sin^2\psi}{2}(n_1^2 + n_2^2) + k_0^2(\epsilon - \epsilon')\sin^2\psi + k_0^2\epsilon_3(\frac{1}{2}\sin^2\psi + \cos^2\psi)\right] - k_0^4\epsilon_3(\epsilon - \epsilon') + n_1^2 n_2^2 \frac{\sin^2\psi}{2}}{(\epsilon_3 \cos^2\psi + \epsilon\sin^2\psi)(\Gamma^2 - n_1^2)(\Gamma^2 - n_2^2)}. \tag{43}$$

Now let us define I_s as

$$I_s = -\frac{1}{2}\int_0^\infty \int_0^\pi \frac{\sin^2\psi \,.\, \sin\psi}{\epsilon_3\cos^2\psi + \epsilon\sin^2\psi} e^{-j\Gamma R p(\theta,\,\psi)} J_0(\Gamma R q(\theta, \psi))\, d\psi\, d\Gamma \tag{44}$$

then

$$\nabla^2 I_s = \frac{1}{2}\int_0^\infty \int_0^\pi \frac{\Gamma^2 \sin^3\psi}{\epsilon_3\cos^2\psi + \epsilon\sin^2\psi} e^{-j\Gamma R p} J_0(\Gamma R q)\, d\psi\, d\Gamma$$

$$= -\frac{1}{2}\frac{1}{\rho}\frac{\partial}{\partial\rho}\rho\frac{\partial}{\partial\rho}\int_0^\infty \int_0^\pi \sin\psi \times \frac{e^{-j\Gamma R p}}{\epsilon_3\cos^2\psi + \epsilon\sin^2\psi} J_0(\Gamma R q)\, d\psi\, d\Gamma \tag{45}$$

where $\rho = R\sin\theta$.

The integral in (45) permits an exact evaluation as shown in the appendix. Using (A.4) we then have

$$\nabla^2 I_s = -\frac{\pi}{2} \cdot \frac{1}{\rho}\frac{\partial}{\partial\rho}\rho\frac{\partial}{\partial\rho}\frac{\epsilon^{-1/2}}{(\epsilon z^2 + \epsilon_3\rho^2)^{1/2}}. \tag{46a}$$

It is easy to verify, say by solving the static problem, that $\nabla^2 I_s$ has the correct order of singularity at the origin. We also note that the expression for I_s is independent of ϵ', the off-diagonal term of the $\bar{\bar{\epsilon}}$ tensor and that $\nabla^2 I_s$ may be rewritten as

$$\nabla^2 I_s = -\frac{\pi\epsilon^{-1/2}}{2}\frac{1}{\rho}\frac{\partial}{\partial\rho}\rho\frac{\partial}{\partial\rho}\frac{1}{R'}\,, \quad (\epsilon z^2 + \epsilon_3\rho^2)^{1/2} = R' \tag{46b}$$

where R' may be interpreted as a radial distance which depends on the observation angle θ. It is also pertinent to point out that if the applied d.c. magnetic field in a magneto-ionic medium is made indefinitely large, the off-diagonal terms in the $\bar{\bar{\epsilon}}$ tensor reduce to zero. Hence, I_s is seen to have the same behavior as would be obtained by letting $H_{dc} \to \infty$.

Continuing with our solution, we now go on to evaluate $\nabla^2 I_f$ where I_f, the finite part of I is given by $I_f = I - I_s$.

Evaluation of $\nabla^2 I_f$

Although I_f itself is finite at the origin, we shall find that $\nabla^2 I_f$ is not and it has a singular part of the type 0 $(1/R)$. We shall now attempt to separate this singularity from the expression of $\nabla^2 I_f$ by the same technique used in the previous section. We shall obtain an exact representation for thc singular part and a reduced integral for the regular part in $\nabla^2 I_f$.

Using (41), (43) and (44) we have the representation for I_f,

$$I_f(R, \theta) = \int_0^\pi \int_0^\infty \frac{S(\Gamma, \psi)\, \mathrm{e}^{-j\Gamma R p(\theta, \psi)}}{(\Gamma^2 - n_1^2)(\Gamma^2 - n_2^2)} J_0(\Gamma R q(\theta, \psi)) \sin \psi \, \mathrm{d}\Gamma \, \mathrm{d}\psi \tag{47}$$

where

$$S(\Gamma, \psi) = \frac{\Gamma^2\left[-\frac{\sin^2 \psi}{2}(n_1^2 + n_2^2) + k_0^2(\epsilon - \epsilon') \sin^2 \psi + k_0^2 \epsilon_3\left(\frac{1}{2} \sin^2 \psi + \cos^2 \psi\right)\right] - k_0^4 \epsilon_3(\epsilon - \epsilon') + n_1^2 n_2^2 \frac{\sin^2 \psi}{2}}{\epsilon_3 \cos^2 \psi + \epsilon \sin^2 \psi}. \tag{48}$$

Differentiating under the integral we obtain, remembering $p = \cos \theta \cos \psi$, $q = \sin \theta \sin \psi$

$$\nabla^2 I_f = -\int_0^\pi \int_0^\infty \frac{\Gamma^2 S(\Gamma, \psi)}{(\Gamma^2 - n_1^2)(\Gamma^2 - n_2^2)} \mathrm{e}^{-j\Gamma R p} J_0(\Gamma R q) \sin \psi \, \mathrm{d}\Gamma \, \mathrm{d}\psi. \tag{49}$$

Using the expression for $S(\Gamma, \psi)$ from (48), break the integral in (49) into two parts as follows:

$$\nabla^2 I_f = \int_0^\pi \int_0^\infty \frac{M(\psi)\, \mathrm{e}^{-j\Gamma R p}}{\epsilon_3 \cos^2 \psi + \epsilon \sin^2 \psi} J_0(\Gamma R q) \sin \psi \, \mathrm{d}\Gamma \, \mathrm{d}\psi - \int_0^\pi \int_0^\infty \frac{N(\Gamma, \psi)}{(\epsilon_3 \cos^2 \psi + \epsilon \sin^2 \psi)(\Gamma^2 - n_1^2)(\Gamma^2 - n_2^2)} \times \mathrm{e}^{-j\Gamma R p} J_0(\Gamma R q) \sin \psi \, \mathrm{d}\Gamma \, \mathrm{d}\psi \tag{50}$$

where

$$M(\psi) = [-\sin^2 \psi\, (n_1^2 + n_2^2)/2 + k_0^2(\epsilon - \epsilon') \sin^2 \psi + + k_0^2 \epsilon_3 (\tfrac{1}{2} \sin^2 \psi + \cos^2 \psi)] \tag{51}$$

$$N(\Gamma, \psi) = [k_0^4 \epsilon_3(\epsilon - \epsilon') - \tfrac{1}{2} n_1^2 n_2^2 \sin^2 \psi - (n_1^2 + n_2^2) M(\psi)] \Gamma^2 + M(\psi) n_1^2 n_2^2 \quad (52)$$

The first integral which has the behavior of the type 0 $(1/R)$ at the origin permits an exact evaluation. The integration with respect to Γ in the second integral may be carried out using the techniques of contour integration, and an integral over a finite range may be derived. The resulting finite range integral is very suitable for numerical computations because the integrand is bounded in the entire range. The detailed calculation of the first integral and the reduction of the second integral are given in Appendixes B and C, respectively. We shall merely present the final result, which is

$$\nabla^2 I_f = \frac{\pi k_0^2}{2} \left\{ \frac{C_1}{R} + \frac{C_2}{R'} + \frac{C_3 R^2 \cos^2 \theta}{R'^3} \right\}$$

$$- 2\pi j \int_0^{\pi/2} \frac{N(n_1(\psi), \psi)}{A(\psi)} \frac{e^{-jn_1 Rp} J_0(n_1 qR)}{n_1(n_1^2 - n_2^2)} \sin \psi \, d\psi$$

$$+ 4 \int_{\pi/2-\theta}^{\pi/2} \int_0^{\cos^{-1}(p/q)} \frac{N(n_1(\psi), \psi)}{A(\psi)} \frac{\sin\{n_1 R(p - q \cos \alpha)\}}{n_1(n_1^2 - n_2^2)} \sin \psi \, d\alpha \, d\psi$$

$$+ P_{(n_2)}(R, \theta) \quad (53)$$

where $A(\psi) = \epsilon \sin^2 \psi + \epsilon_3 \cos^2 \psi$, and $P_{(n_2)}(R, \theta)$ is obtained by interchanging n_2 and n_1 everywhere in the two integrals appearing in (53). The coefficients C_1, C_2 and C_3 in the first curly bracket are dependent on the medium parameters and may be obtained from the equation (B—6b). It is interesting to observe that whereas I_s obtained before contains terms of the type $1/R'$ with coefficient independent of ϵ', the singular terms in $\nabla^2 I_f$ have $1/R$, $1/R'$ and $R^2 \cos^2 \theta / R'^3$ type of variation and furthermore the coefficients C_1, C_2 and C_3 do depend on all the elements of $\bar{\bar{\epsilon}}$ tensor including ϵ'.

No further reduction of the integrals seems possible for a general choice of elements in the $\bar{\bar{\epsilon}}$ tensor, and numerical integration must be performed to evaluate the regular part of $\nabla^2 I_f$. The desired final form for E_1 is obtained by substituting (46) and (53) in (40). The other components of the electric field may also be evaluated by following through a similar procedure.

It is also pertinent to point out here that if one is merely interested in the far fields, these may be obtained by working directly with (39) and making asymptotic approximations in it. The asymptotic expressions thus obtained, are no different in their form from the ones derived by Bunkin. The only comments we have on his solution are the following:

Referring to eqn (3.13) in his paper (1957) one finds that the saddle point is determined by an equation of the type

$$\frac{d}{d\psi}[n(\psi)\cos(\psi-\theta)]=0 \tag{54}$$

in our notation.

A geometric solution of this equation is simply derived as follows: Plot the $n(\psi)$ surface as shown in Fig. 1, with OQ as the reference direction for the angles ψ.

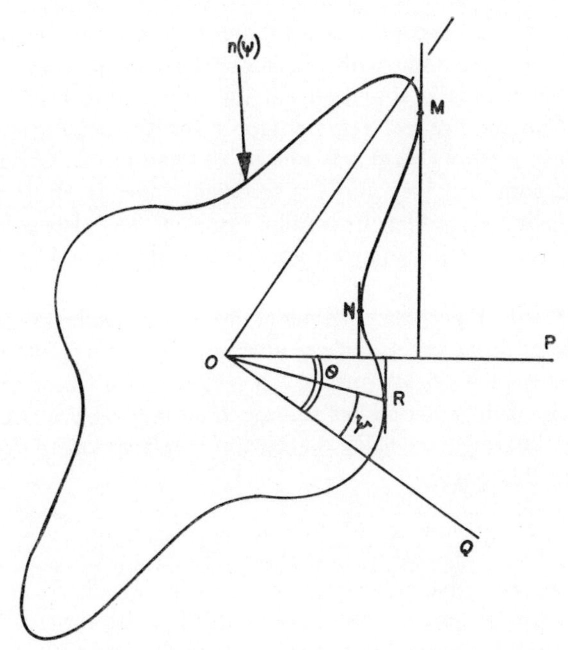

Fig. 1. Geometric solution for saddle point for far field evaluation.

Draw OP such that $\angle POQ = \theta$. Now locate the points such as M, N and R such that the tangents to the $n(\psi)$ curve at these points are perpendicular to the line OP. It is straightforward to verify that the desired solutions of (54) are the angles $\angle MOQ$, $\angle NOQ$ and $\angle ROQ$. The number of such points depend in general on the observation angle θ and the shape of the $n(\psi)$ curve and this number varies between one and three. It should be noted that if ψ is a solution, $\pi+\psi$ is also a solution. The proper choice between this pair has to be made on the basis of the application of the radiation condition. Detailed work along this direction has been carried out by Felsen, his co-workers and also by Motz.

The turning point of the $n(\psi)$ curve where $\{n(\psi)\cos(\psi-\theta)\}'$ and $\{n(\psi)\cos(\psi-\theta)\}''$ are both zero, is a singular point. This point has to be treated in a special way, but there is no analytical difficulty in doing this.

Neither the possible existence of the multiple solutions of (54) nor the special case of $\{n(\psi) \cos (\psi - \theta)\}'' = 0$ were discussed in the paper by Bunkin.

Since our main emphasis in this paper is not on the derivation of far fields, we shall not elaborate on these points any further.

5. COMMENTS AND DISCUSSIONS

In this paper we have obtained solutions of Maxwell's equations with source terms, in a medium with linear dielectric properties. The very near field terms which have a singularity at the origin have been derived in an exact form and an integral representation for the regular part has been obtained. Since no restriction was put on the nature of the elements of the $\bar{\bar{\epsilon}}$ tensor, the solutions are valid for complex elements and hence they are useful for treating a medium with finite losses. It was also pointed out that the integral representations may also be asymptotically evaluated for far field calculations.

In many physical problems, however, it is unrealistic to talk about far fields because of finite losses that are always present and a knowledge of the near and intermediate fields are very desirable. Also, when one is solving the impedance problems, one has to have the knowledge of the behavior of the near and intermediate fields, hence the emphasis on the calculation of these fields in this paper.

6. FUTURE WORK

Although we are able to determine the fields produced at an arbitrary distance by a distribution of sources, there still remains the problem of determining the current distribution in a given radiator when it is immersed in a magneto-ionic medium and excited by a voltage source at its input terminals. It has been found that the usual procedure of calculating the input impedance of a short linear antenna gives unrealistic answers, when for instance, the simple assumption of a constant current distribution is made. It is felt that a satisfactory solution of this problem will be very useful.

Also, it has been found that discontinuous types of solutions result when ϵ and ϵ_z are real and ϵ_z is negative. The significance of these solutions deserves a thorough investigation.

The properties of the solutions for a magnetic current distribution should also be studied. Although no analytical difficulty is expected in tackling this problem, the magnetic dipole case should nevertheless be studied and compared with the electric dipole.

Some of these problems are currently under investigation at the Antenna Laboratory of the University of Illinois as a follow-up of the present work.

Acknowledgement—The research reported in this paper was carried out under Air Force Contracts AF 19(604)–5565 and AF 33(616)–6079. It is also a pleasure to thank the Radar Division of the Ground Systems Group, Hughes Aircraft Company, Fullerton, California, where one of the authors held a summer employment during the course of the preparation of the paper. Thanks are also due to Mr. Keith Balmain of the Antenna Laboratory of the University of Illinois for assistance in the preparation of the paper.

REFERENCES

ARBEL, E. and FELSEN, L. B. On radiation from sources in anisotropic media (to be published).

BUNKIN, F. V. (1957) On radiation in anisotropic media, *J. Exp. Theoret. Phys.* (*U.S.S.R.*) **32,** 338–346 (Feb.).

KOGELNIK, H. (1960) On electromagnetic radiation in magneto-ionic media, *J. Res. N.B.S.* **64D,** No. 5 (Sept.–Oct.), pp. 515–522.

KUEHL, H. (1960) Radiation from an electric dipole in an anisotropic cold plasma, Calif. Inst. of Technology, Antenna Lab., Report No. 24.

MAGNUS, W. and OBERHETTINGER, F. (1954) *Functions of Mathematical Physics*, p. 26. Chelsea Publishing Co.

MEECHAM, W. C. (1961) Source and radiation problems in magneto-ionic media, *Phys. of Fluids*, **4,** No. 12 (Dec.)

MOTZ, H. and KOGELNIK, H. Electromagnetic radiation from sources imbedded in infinite anisotropic medium and the significance of Poynting vector (to be published).

APPENDIX A

Evaluation of the integral appearing in (45):

The integral to be evaluated, say $L(R, \theta)$ is

$$\left.\begin{aligned} L(R,\theta) &= \int_0^\infty \int_0^\pi \frac{e^{-j\Gamma R\cos\theta\cos\psi}}{\epsilon_3\cos^2\psi + \epsilon\sin^2\psi} \\ &\qquad J_0(\Gamma R\sin\theta\sin\psi)\sin\psi\,d\psi\,d\Gamma \\ &= \int_0^\infty \int_0^\pi \frac{e^{-j\Gamma R\cos\theta\cos\psi}}{\epsilon_3\Gamma^2\cos^2\psi + \epsilon\Gamma^2\sin^2\psi} \\ &\qquad J_0(\Gamma R\sin\theta\sin\psi)\,\Gamma^2\sin\psi\,d\Gamma\,d\psi\,. \end{aligned}\right\} \quad \text{(A–1)}$$

Introducing a change in variables through the relations

$$\gamma = \Gamma\cos\psi, \qquad u = \Gamma\sin\psi, \qquad z = R\cos\theta, \qquad \rho = R\sin\theta$$

we derive

$$L(R,\theta) = \int_0^\infty \int_{-\infty}^\infty \frac{e^{-j\gamma z}J_0(u\rho)}{\epsilon_3\gamma^2 + \epsilon u^2}\,u\,d\gamma\,du. \quad \text{(A–2)}$$

Performing the integration with respect to γ we obtain through the use of contour integration methods

$$L(R, \theta) = \pi \int_0^\infty \frac{e^{-(\epsilon/\epsilon_3)^{1/2} uz} J_0(u\rho)}{(\epsilon\, \epsilon_3)^{1/2}} du\,. \tag{A-3}$$

(A–3) is a standard integral and is discussed in books on Bessel Functions. The final form for $L(R, \theta)$ reads

$$\left.\begin{aligned} L(R, \theta) &= \frac{\pi}{(\epsilon\, \epsilon_3)^{1/2}} \frac{1}{\{(\epsilon/\epsilon_3)\, z^2 + \rho^2\}^{1/2}} \\ &= \frac{1}{\epsilon^{1/2}} \frac{\pi}{R(\epsilon \cos^2 \theta + \epsilon_3 \sin^2 \theta)^{1/2}}, \qquad \mathrm{Re}\, (\epsilon/\epsilon_3)^{1/2} > 0\,. \end{aligned}\right\} \tag{A-4}$$

APPENDIX B

Evaluation of the singular part of $\nabla^2 I_f$:

As pointed out in connection with (50), the singular part of $\nabla^2 I_f$ can be evaluated exactly. Let the singular part be represented by $K(R, \theta)$ where

$$K(R, \theta) = \int_0^\pi \int_0^\infty \frac{M(\psi)\, e^{-j\Gamma R p}}{\epsilon_3 \cos^2 \psi + \epsilon \sin^2 \psi} J_0(\Gamma R q) \sin \psi \, d\Gamma \, d\psi \tag{B-1}$$

and $p = \cos\theta \cos\psi$, $q = \sin\theta \sin\psi$.

As a first step toward evaluating the integral, substitute the expressions for n_1 and n_2 from (21) into (51) and rewrite $M(\psi)$ as

$$\left.\begin{aligned} M(\psi) = \frac{k_0^2}{2} &(\epsilon_3 \cos^2 \psi + \epsilon \sin^2 \psi)^{-1} \\ &.\ \{\sin^4 \psi(\epsilon - \epsilon')^2 \\ &\quad + \sin^2 \psi \cos^2 \psi\, \epsilon_3(\epsilon_3 + 2\epsilon - 2\epsilon') + 2\epsilon_3^2 \cos^4 \psi\}\,. \end{aligned}\right\} \tag{B-2}$$

Now multiply the denominator and the numerator of the integrand in (A–5) by Γ^6 and introduce the transformations

$$\gamma = \Gamma \cos \psi \qquad u = \Gamma \sin \psi$$

$$z = R \cos \theta \qquad \rho = R \sin \theta$$

obtaining

$$\left.\begin{aligned} 2K(R, \theta) = k_0^2 \int_0^\infty \int_{-\infty}^\infty &\frac{\{u^4(\epsilon - \epsilon')^2 + u^2\gamma^2\epsilon_3(\epsilon_3 + 2\epsilon - 2\epsilon') + 2\epsilon_3^2\gamma^4\}}{(\epsilon_3\gamma^2 + \epsilon u^2)^2\, (\gamma^2 + u^2)} \\ &.\ e^{-j\gamma z} J_0(u\rho)\, u \, d\gamma \, du\,. \end{aligned}\right\} \tag{B-3}$$

Evaluate the integral with respect to γ using the method of contour integration and obtain

$$2k_0^{-2} K(R, \theta) = \pi \frac{\epsilon^2 + \epsilon'^2 + \epsilon_3^2 - 2(\epsilon\epsilon' + \epsilon_3\epsilon - \epsilon_3\epsilon')}{(\epsilon - \epsilon_3)^2} \int_0^\infty e^{-uz} J_0(u\rho)\, du$$

$$+ \pi \left\{ \frac{(\epsilon_3 - 2\epsilon - 2\epsilon')}{(\epsilon/\epsilon_3)^{1/2} (\epsilon_3 - \epsilon)} + \frac{(\epsilon^2 + \epsilon'^2 - \epsilon\epsilon_3)(\epsilon_3 - 3\epsilon)}{2\epsilon_3(\epsilon/\epsilon_3)^{3/2} (\epsilon - \epsilon_3)^2} \right\}$$

$$\cdot \int_0^\infty e^{-u(\epsilon/\epsilon_3)^{1/2} z} J_0(u\rho)\, du$$

$$- \left(\frac{\pi z}{2}\right) \frac{\epsilon^2 + \epsilon'^2 - \epsilon\epsilon_3}{\epsilon(\epsilon - \epsilon_3)}$$

$$\cdot \int_0^\infty e^{-u(\epsilon/\epsilon_3)^{1/2} z} J_0(u\rho)\, u\, du \, . \qquad \text{(B–4)}$$

The first term in (B–4) is the contribution of the pole at $\gamma = -ju$ whereas the second and third terms are due to the residue of the double pole at $\gamma = -j(\epsilon/\epsilon_3)^{1/2} u$.

Now rewrite the third term as

$$\int_0^\infty e^{-zu(\epsilon/\epsilon_3)^{1/2}} J_0(u\rho)\, u\, du = -\left(\frac{\epsilon_3}{\epsilon}\right)^{1/2} \frac{\partial}{\partial z} \int_0^\infty e^{-zu(\epsilon/\epsilon_3)^{1/2}} J_0(u\rho)\, du \qquad \text{(B–5)}$$

and evaluate the integrals involving the Bessel functions in the same manner as in Appendix A, obtaining after some simplification

$$2\pi^{-1} k_0^{-2} K(R, \theta) = \frac{1}{R} \left\{ \frac{\epsilon^2 + \epsilon'^2 + \epsilon_3^2 - 2(\epsilon\epsilon' + \epsilon_3\epsilon - \epsilon_3\epsilon')}{(\epsilon - \epsilon_3)^2} \right\}$$

$$+ \left\{ \frac{\epsilon_3(\epsilon_3 - 2\epsilon - 2\epsilon')}{\epsilon^{1/2}(\epsilon_3 - \epsilon)} + \frac{\epsilon_3(\epsilon^2 + \epsilon'^2 - \epsilon\epsilon_3)(\epsilon_3 - 3\epsilon)}{2\epsilon^{3/2} (\epsilon - \epsilon_3)^2} \right\}$$

$$\cdot \frac{1}{R(\epsilon \cos^2 \theta + \epsilon_3 \sin^2 \theta)^{1/2}}$$

$$+ \left\{ \frac{1}{2} \frac{\epsilon_3}{\epsilon^{1/2}} \frac{\epsilon^2 + \epsilon'^2 - \epsilon\epsilon_3}{(\epsilon_3 - \epsilon)} \right\}$$

$$\cdot \frac{\cos^2 \theta}{R(\epsilon \cos^2 \theta + \epsilon_3 \sin^2 \theta)^{3/2}}$$

$$= \frac{C_1}{R} + \frac{C_2}{R'} + \frac{C_3 z^2}{R'^3} \qquad \text{(B–6)}$$

where $R' = R(\epsilon \cos^2 \theta + \epsilon_3 \sin^2 \theta)^{1/2}$ and C_1, C_2, and C_3 are given by the expressions inside the first, second and third curly brackets, respectively, in the eqn (B–6).

APPENDIX C

Reduction of regular part of $\nabla^2 I_f$:

We shall discuss the reduction of the integral say $P(R, \theta)$ which is the regular part of $\nabla^2 I_f$ given in (50). Let

$$\left.\begin{aligned} P(R, \theta) &= \int_0^{\pi}\int_0^{\infty} \frac{N(\Gamma, \psi)\sin\psi\, e^{-j\Gamma Rp}}{A(\psi)(\Gamma^2 - n_1^2)(\Gamma^2 - n_1^2)} J_0(\Gamma Rq)\, d\Gamma\, d\psi \\ A(\psi) &= \epsilon \sin^2\psi + \epsilon_3 \cos^2\psi\,. \end{aligned}\right\} \qquad \text{(C–1)}$$

We note that the highest power of Γ in $N(\Gamma, \psi)$ is 2, hence the integral $P(R, \theta)$ is finite even when $R \to 0$. To evaluate the integral, we recast it in the following form after a couple of simple changes of variables:

$$P(R, \theta) = 2\int_0^{\pi/2}\int_{-\infty}^{\infty} \frac{N(\Gamma, \psi)\sin\psi\, e^{-j\Gamma Rp}}{A(\psi)(\Gamma^2 - n_1^2)(\Gamma^2 - n_2^2)} J_0(\Gamma Rq)\, d\Gamma\, d\psi\,. \qquad \text{(C–2)}$$

The advantage of the form of (C–2) over (C–1) is that (C–2) is suitable for integration with respect to Γ using the technique of contour integration because of infinite limits. Before this integration is performed, however, it is convenient to replace the Bessel function by its integral representation and arrive at the form

$$\left.\begin{aligned} P(R, \theta) = &\frac{2}{\pi}\int_0^{\pi/2}\int_{-\infty}^{\infty}\int_0^{\pi/2} \frac{N(\Gamma, \psi)\sin\psi}{A(\psi)} \\ &\qquad\cdot \frac{e^{-j\Gamma R(p+q\cos\alpha)}}{(\Gamma^2 - n_1^2)(\Gamma^2 - n_2^2)}\, d\alpha\, d\Gamma\, d\psi \\ + &\frac{2}{\pi}\int_0^{\pi/2}\int_{-\infty}^{\infty}\int_0^{\pi/2} \frac{N(\Gamma, \psi)\sin\psi}{A(\psi)} \\ &\qquad\cdot \frac{e^{-j\Gamma R(p-q\cos\alpha)}}{(\Gamma^2 - n_1^2)(\Gamma^2 - n_2^2)}\, d\alpha\, d\Gamma\, d\psi\,. \end{aligned}\right\} \qquad \text{(C–3)}$$

Now interchange the order of integration so that the integration is first performed with respect to Γ. Following the usual procedure, extend the integral to an infinite semicircular contour over a complex Γ plane, closing the contour in the upper half plane when the coefficient of Γ in the exponent is positive and in the lower half plane when otherwise. Assuming n_1 and n_2 have finite negative imaginary parts, however small, we see that only the poles at $\Gamma = n_1$ and n_2 are enclosed in the lower half plane and their negative counterparts in the upper half plane. Note that $p + q \cos\alpha$ is always positive for $0 < \psi < \pi/2$ and $0 < \alpha < \pi/2$, hence the first integral in (C–3) is always evaluated by closing the contour in the lower half plane.

The evaluation of the integrals gives

$$P(R, \theta) = P_{(n_1)}(R, \theta) + P_{(n_2)}(R, \theta)$$

where $P_{(n_1)}$ and $P_{(n_2)}$ are contributions due to the poles n_1 and n_2 respectively.

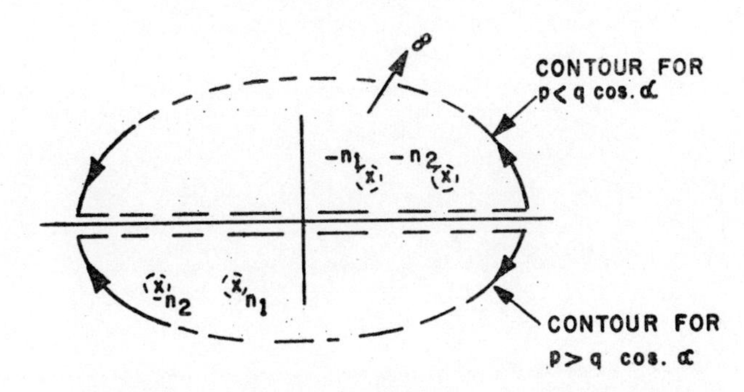

FIG. 2. Location of poles and choice of contours for the second integral in C-3.

The expression for $P_{(n_1)}$ is

$$\left.\begin{aligned}
P_{(n_1)}(R, \theta) = &-2j \int_0^{\pi/2}\int_0^{\pi/2} \frac{N(n_1(\psi), \psi)\sin\psi}{A(\psi)} \cdot \frac{e^{-jn_1R(p+q\cos\alpha)}}{n_1(n_1^2-n_2^2)}\, d\alpha\, d\psi \\
&-2j \int_0^{\pi/2}\int_{\cos^{-1}(p/q)}^{\pi/2} \frac{N(n_1(\psi), \psi)\sin\psi}{A(\psi)} \cdot \frac{e^{-jn_1R(p-q\cos\alpha)}}{n_1(n_1^2-n_2^2)}\, d\alpha\, d\psi \\
&-2j \int_{\pi/2-\theta}^{\pi/2}\int_0^{\cos^{-1}(p/q)} \frac{N(n_1(\psi), \psi)\sin\psi}{A(\psi)} \cdot \frac{e^{jn_1R(p-q\cos\alpha)}}{n_1(n_1^2-n_2^2)}\, d\alpha\, d\psi
\end{aligned}\right\} \quad \text{(C–4)}$$

and $P_{(n_2)}(R, \theta) = P_{(n_1)}(R, \theta)$ with n_2 and n_1 interchanged everywhere. The integrals in (C–4) may be further reduced. To this end, rearrange (C–4) and derive

$$\left.\begin{aligned} P_{(n_1)}(R, \theta) = & - 4j \int_0^{\pi/2} \int_0^{\pi/2} \frac{N(n_1(\psi), \psi)}{A(\psi)} \\ & \cdot \frac{e^{-jn_1Rp} \cos(n_1Rq \cos \alpha)}{n_1(n_1^2 - n_2^2)} \sin \psi \, d\alpha \, d\psi \\ & + 4 \int_{\pi/2-\theta}^{\pi/2} \int_0^{\cos^{-1}(p/q)} \frac{N(n_1(\psi), \psi)}{A(\psi)} \\ & \cdot \frac{\sin \{n_1R(p - q \cos \alpha)\}}{n_1(n_1^2 - n_2^2)} \sin \psi \, d\alpha \, d\psi . \end{aligned}\right\} \quad \text{(C-5)}$$

Note that in writing (C–5) we have made use of the fact that

$$\int_0^{\pi/2-\theta} \int_0^{\cos^{-1} p/q} F(\psi, \theta, \alpha) \, d\alpha \, d\psi = 0$$

since

$$p/q = \frac{\cos \theta \cos \psi}{\sin \theta \sin \psi} > 1, \qquad \text{for } \psi < \pi/2 - \theta$$

and the integral with respect to α is over a real range only.

The first integral with respect to α may be reduced to a Bessel function, yielding the final representation for $P_{(n_1)}$ as given below

$$\left.\begin{aligned} P_{(n_1)}(R, \theta) = & - 2\pi j \int_0^{\pi/2} \frac{N(n_1(\psi), \psi)}{A(\psi)} \frac{e^{-jn_1Rp} J_0(n_1qR)}{n_1(n_1^2 - n_2^2)} \sin \psi \, d\psi \\ & + 4 \int_{\pi/2-\theta}^{\pi/2} \int_0^{\cos^{-1}(p/q)} \frac{N(n_1(\psi), \psi)}{A(\psi)} \\ & \cdot \frac{\sin \{n_1R(p - q \cos \alpha)\}}{n_1(n_1^2 - n_2^2)} \sin \psi \, d\alpha \, d\psi . \end{aligned}\right\} \quad \text{(C-6)}$$

THE IMPEDANCE OF AN AERIAL IMMERSED IN AN ANISOTROPIC MEDIUM

K. G. BUDDEN

Cavendish Laboratory, University of Cambridge

SUMMARY

1. INTRODUCTION

This paper describes a method of calculating the impedance of a radio aerial of simple shape when it is immersed in a magneto-ionic medium. Much previous work on aerial impedances applies to an isotropic medium and it is then possible to reduce the problem to an integral equation using the vector potential or the Hertz potential. For an anisotropic medium these potentials cannot be used and a more elaborate approach is necessary. In the present paper the field of the aerial is expressed as an angular spectrum of waves. A suitable complex amplitude function ($f(z)$ in eqns (2), (3) below) is to be found such that the field fits the boundary conditions at the surface of the aerial. The impedance is then calculated by using the field to find the current flow at the feed point.

The calculations are being made for a cylindrical dipole aerial fed at its center. The boundary conditions are that the electric field at the surface shall be zero except near its center where power is fed in, and the current at the ends shall be zero. The work has been planned in four stages. The method is first being tested for an isotropic medium. It will then be applied to a dipole immersed in a magneto-ionic medium and aligned along the superimposed magnetic field. In the third stage it will be applied to a dipole transverse to the field, and finally to a dipole at a general angle to the field. At the time of the Symposium only the first stage, for an isotropic medium, had been reached. Work on the second stage is proceeding.

2. FORMULATION OF THE PROBLEM FOR AN ISOTROPIC MEDIUM

A cylindrical aerial of length l and radius R is placed symmetrically about the origin of Cartesian coordinates x, y, z, with its axis parallel to the z-axis. Let E_x, E_y, E_z denote the Cartesian components of the electric intensity E.

The aerial is fed with a voltage V applied over a short length s symmetrically about its center. Thus on its curved surface

$$\int_0^z E_z \, dz = \begin{cases} -Vz/s & \text{for } 0 \leqslant z \leqslant \tfrac{1}{2}s \\ -\tfrac{1}{2}V & \text{for } \tfrac{1}{2}s \leqslant z \leqslant \tfrac{1}{2}l. \end{cases} \tag{1}$$

Consider now a distribution of current along the z-axis given by

$$I = f(z)$$

in an isotropic medium of refractive index $\mathfrak{n}$, and suppose that the aerial is temporarily removed. Then it can be shown that the fields, at the position previously occupied by the curved surface of the aerial, satisfy:

$$4\pi \int_0^z E_z \, dz = -kZ_0 \int_0^\infty H_0^{(2)} \{k\mathfrak{n}RC\} \int_{-\frac{1}{2}l}^{\frac{1}{2}l} f(\zeta) \frac{\sin \{Sk\mathfrak{n}(z-\zeta)\}}{Sk\mathfrak{n}} \, d\zeta \, C^2 dS \tag{2}$$

$$4\pi H_\phi(z) = -ik\mathfrak{n} \int_0^\infty H_1^{(2)} \{k\mathfrak{n}RC\} \int_{-\frac{1}{2}l}^{\frac{1}{2}l} f(\zeta) \cos \{Sk\mathfrak{n}(z-\zeta)\} \, d\zeta \, C dS \tag{3}$$

where $k = 2\pi/\lambda$ and λ is the wavelength in free space, $C^2 = 1 - S^2$, Z_0 is the characteristic impedance of free space and H_ϕ is the azimuthal component of the magnetic field.

The integrand in each of these expressions represents a wave with conical wave fronts, with the generators of the cone at a complex angle arc sin S to the z-axis. These waves could be further resolved into plane waves, if the Hankel functions were replaced by Sommerfeld's integral expressions. This resolution is unnecessary in problems which have symmetry about the z-axis, but it will be necessary in later work when the aerial is inclined to the magnetic field.

The problem now is to find a function $f(z)$ such that (2) satisfies the boundary condition (1), and (3) is zero at $z = \pm\frac{1}{2}l$. Then the fields are those that would be radiated by the cylindrical aerial when fed with a voltage V at its center. The current at $z = 0$ can then be found from (3) and hence the input impedance is calculated.

Some results for an aerial in free space are shown in Figs. 1 and 2. In obtaining these the function $f(z)$ was taken to be

$$f(z) = \sin \{k(\tfrac{1}{2}l - |z|)\} [A_0 + A_1 \cos az + \dots + A_n \cos naz] \tag{4}$$

where the unknown coefficients A_r are to be found, and a is a suitably chosen constant. If (4) is substituted in (2), and (1) is used, we obtain

$$kZ_0 \sum_{r=0}^{n} A_r F(z, r) = \begin{cases} 4\pi Vz/s & \text{for } 0 \leqslant z \leqslant \frac{1}{2}s \\ 2\pi V & \text{for } \frac{1}{2}s \leqslant z \leqslant \frac{1}{2}l \end{cases} \tag{5}$$

where the functions $F(z, r)$ involve the S integration in (2). These functions have been computed using EDSAC II, the digital computer in the University Mathematical Laboratory, Cambridge. A series similar to (5) is obtained from (3), and for the end of the aerial, $z = \frac{1}{2}l$, it becomes

$$\sum_{r=0}^{n} A_r G(\tfrac{1}{2}l, r) = 0 \tag{6}$$

where the functions G are also computed.

The complex functions $F(z, r)$ are computed for a number of discrete values of z. Equations (5) and (6) are then solved for the $n+1$ unknown

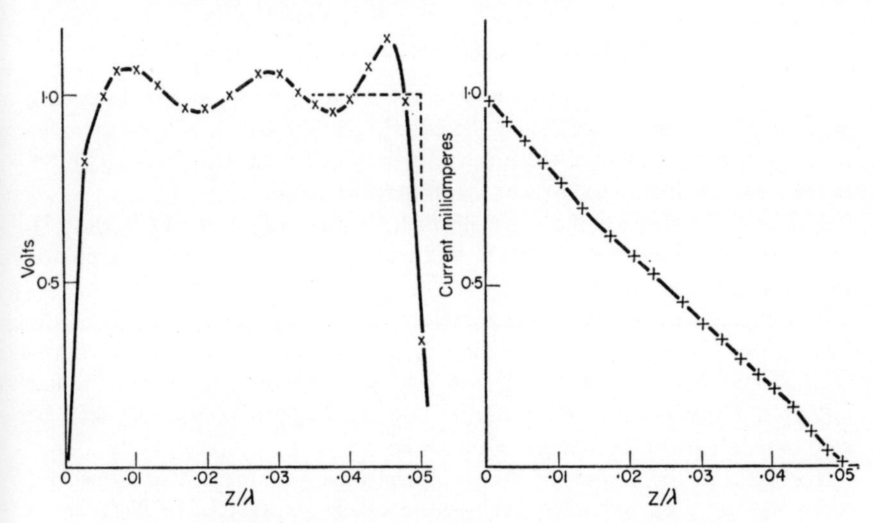

FIG. 1. The computed voltage and current on one half of a cylindrical aerial of length 0·1 λ and diameter 0·0001 λ fed with 2 volts over a length of 1/10 l at its center.

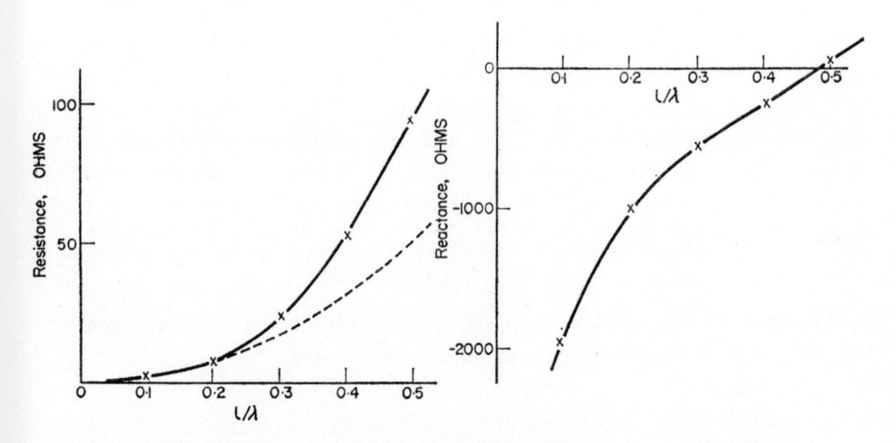

FIG. 2. The impedance of a cylindrical aerial in free space, fed over a length 1/10 l at its center, for various lengths, l. The dotted curve in the left diagram is 197 $(l/\lambda)^2$ ohms, which is the theoretical radiation resistance when $l \ll \lambda$.

complex coefficients A_r by a matrix inversion process. This is done either by using n discrete values of z and obtaining an exact fit at these points, or by using more than n discrete values of z and obtaining a set of A_r's which make the sum of the squares of the moduli of the errors as small as possible. Both methods have worked well for an isotropic medium with $n = 8$.

Other forms for the function $f(z)$ in (4) are being tried. A useful alternative form was suggested by a speaker in the discussion at the Symposium.

3. EXTENSION TO AN ANISOTROPIC MEDIUM

In this section we indicate briefly the modifications to the foregoing method which are needed when the aerial is immersed in a homogeneous magneto-ionic medium, in which the superimposed magnetic field is parallel to the z-axis, that is to the axis of the cylindrical aerial.

For any direction of the wave normal, determined by S in (2) and (3), there are now two waves, ordinary and extraordinary, with different polarizations and refractive indices.

The expressions for the fields therefore now contain two integrals of the form (2) or (3), and in addition there is at the surface of the aerial an azimuthal component E_ϕ of the electric field given by the sum of two similar integrals. There is now an additional boundary condition that E_ϕ shall be zero everywhere on the surface of the aerial.

The fields at the surface of the aerial can thus be expressed in terms of a series like (5) with unknown coefficients, which are then to be found by a matrix inversion process.

4. AERIAL INCLINED TO THE MAGNETIC FIELD

When the aerial is inclined to the magnetic field, the radial dependence of the fields is no longer given by the Hankel functions in (2) and (3) and it will be necessary to replace the single integration with respect to S by a double integration. The fields E_z, E_ϕ will vary with the azimuth angle ϕ, and it will be necessary to impose the boundary conditions at a number of different values of ϕ, for each of the selected values of z used in the simpler problem.

5. CONCLUSIONS

The method described has worked successfully for an isotropic medium. Its extension to an anisotropic medium is proceeding.

LATERAL WAVES ON AIR–MAGNETOPLASMA INTERFACES

G. TYRAS,† A. ISHIMARU,‡ and H. M. SWARM‡

ABSTRACT

We consider the problem of electric and magnetic current line sources situated in a magnetoplasma with a separation boundary. The line sources and the magnetostatic field are parallel to the interface while the sources and magnetostatic field are either parallel or perpendicular to each other. It is shown that the field in air will consist of the radiation and the lateral fields contributed by the saddle-point and the branch-cut integration respectively. The leading terms of the radiation and the lateral fields are found explicitly and without restriction on any of the pertinent parameters.

1. INTRODUCTION

The problem of radiation in free space from sources immersed in a magnetoplasma half-space was first considered by Barsukov (1959). He obtained expressions for the Poynting vector in the air in the case when the magnetostatic field was normal to the boundary and an electric dipole was either normal or parallel to the boundary. A similar problem was also considered by Arbel (1960).

In this paper we treat the problems of electromagnetic radiation from line sources in magnetoplasma when the magnetostatic field is parallel to the separation boundary. In particular, we consider (a) electric current line source and H_{DC} perpendicular, (b) electric current line source and H_{DC} parallel, (c) magnetic current line source and H_{DC} perpendicular and (d) magnetic current line source and H_{DC} parallel.

In what follows we find the integral representation of the fields everywhere, but we evaluate these integrals only for the air-region. Here we find that in addition to the radiation field, given by the saddle-point contribution, one often encounters the lateral field resulting from branch-point singularities. Depending on the direction of the magnetostatic field with relation to the line source, one or two branch points will be found in the complex β-plane resulting in a lateral field consisting of one or two waves respectively.

† Present address: Department of Electrical Engineering, University of Arizona, Tucson, Arizona. On leave from The Boeing Company, Seattle, Washington. This work was done while this author was at Boeing.

‡ Department of Electrical Engineering, University of Washington, Seattle, Washington.

2. ELECTRIC CURRENT LINE SOURCE PERPENDICULAR TO H_{DC}

The geometry of the problem is shown in Fig. 1. The horizontal plane $z = 0$ coincides with the interface between the anisotropic homogeneous magnetoplasma and air. For convenience we shall call the magnetoplasma medium (1) and the air medium (0) and assume that both media have the same magnetic inductive capacity of free space, μ_0.

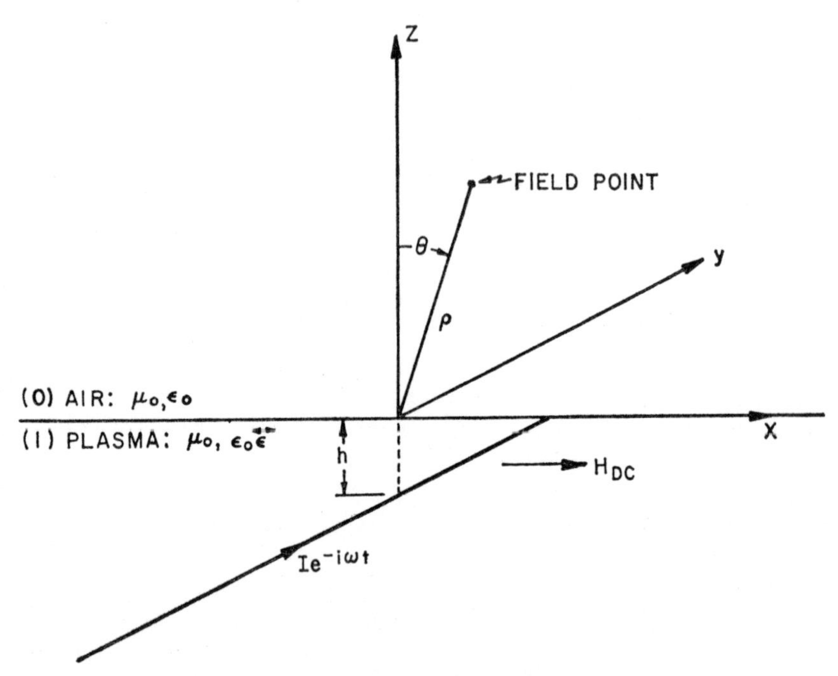

FIG. 1. Geometry of the problem of the electric current line source, H_{DC} perpendicular.

The definition of the present boundary value problem implies solution to Maxwell's equations subject to the usual boundary conditions at the interface and a proper behavior at infinity. The use of the auxiliary vector potentials does not seem to simplify the problem. We shall, therefore, work with the Cartesian components of the field vectors directly.

For the purpose of this problem it will be assumed that the source of the electromagnetic waves consists of a very thin and straight wire of infinite extent carrying an alternating current $Ie^{-i\omega t}$. In the magnetoplasma region, the electric field will then satisfy

$$(-\nabla x \nabla x + k_0^2 \epsilon)\, \mathbf{E} = -\, i\omega\mu_0 I \delta(x)\, \delta(z + h)\, \mathbf{I}_y \tag{1}$$

where $\boldsymbol{\epsilon}$ is the magnetoplasma permittivity tensor given by

$$\boldsymbol{\epsilon} = \begin{bmatrix} \zeta & 0 & 0 \\ 0 & \epsilon & i\eta \\ 0 & -i\eta & \epsilon \end{bmatrix} \tag{2}$$

where

$$\left.\begin{aligned} \zeta &= 1 - \frac{\omega_P^2}{\omega^2 + \nu^2} + \frac{i\nu\omega_P^2}{\omega(\omega^2 + \nu^2)} \\ \epsilon &= 1 - \frac{\omega_P^2(\omega^2 - \omega_H^2 + \nu^2)}{\Delta} + \frac{i\nu\omega_P^2(\omega^2 + \omega_H^2 + \nu^2)}{\omega\Delta} \\ \eta &= \frac{\omega_H\omega_P^2(\omega^2 - \omega_H^2 - \nu^2)}{\omega\Delta} - \frac{i2\omega_H\nu\omega_P^2}{\Delta} \\ \Delta &= [(\omega + \omega_H)^2 + \nu^2][(\omega - \omega_H)^2 + \nu^2] \\ \omega_P &= \sqrt{\left(\frac{e^2N}{m\epsilon_0}\right)};\ \omega_H = \frac{|\,e\,|\,\mu_0 H_{DC}}{m}. \end{aligned}\right\} \tag{3}$$

In the air region the electric field vector satisfies

$$(\nabla^2 + k_0^2)\,\mathbf{E} = 0. \tag{4}$$

The formulation of the problem can be simplified by expressing the field components in the magnetoplasma and the air regions in terms of their double Fourier integral representation in Cartesian coordinates in the transform space as well as in the configuration space. To this end we introduce a double Fourier transform pair defined by

$$\left.\begin{aligned} \tilde{\tilde{F}}(\alpha_1, \alpha_3) &= \frac{1}{2\pi}\iint_{-\infty}^{\infty} F(x, z)\,\mathrm{e}^{-i(\alpha_1 x + \alpha_3 z)}\,\mathrm{d}x\,\mathrm{d}z \\ F(x, z) &= \frac{1}{2\pi}\iint_{-\infty}^{\infty} \tilde{\tilde{F}}(\alpha_1, \alpha_3)\,\mathrm{e}^{i(\alpha_1 x + \alpha_3 z)}\,\mathrm{d}\alpha_1\,\mathrm{d}\alpha_3. \end{aligned}\right\} \tag{5}$$

In what follows we shall also need the transforms of the derivatives. These can be obtained by integrating by parts where the vanishing of the integrated part is assured providing that the fields behave properly at infinity.

Transformation of (1) leads to a system of three simultaneous algebraic equations in three unknowns. These can be reduced to the form

$$\begin{bmatrix} \zeta k_0^2-\alpha_3^2 & 0 & \alpha_1\alpha_3 \\ 0 & \epsilon(\alpha_3^2-s_1^2)(\alpha_3^2-s_2^2) & 0 \\ 0 & -i\eta(\zeta k_0^2-\alpha_3^2) & \zeta(\epsilon k_0^2-\alpha_1^2)-\epsilon\alpha_3^2 \end{bmatrix} \begin{bmatrix} \tilde{\tilde{E}}_{x1} \\ \tilde{\tilde{E}}_{y1} \\ \tilde{\tilde{E}}_{z1} \end{bmatrix} = \frac{-i\omega\mu_0 I[\zeta(\epsilon k_0^2-\alpha_1^2)-\epsilon\alpha_3^2]}{2\pi} \begin{bmatrix} 0 \\ e^{i\alpha_3 h} \\ 0 \end{bmatrix} \tag{6}$$

where

$$s_{1,\,2}^2 = \frac{1}{2}\left\{(\chi+\zeta)\,k_0^2 - \frac{\epsilon+\zeta}{\epsilon}\,\alpha_1^2 \pm \pm\sqrt{\left\langle\left[(\chi-\zeta)\,k_0^2 - \frac{\epsilon-\zeta}{\epsilon}\,\alpha_1^2\right]^2 + 4\left(\frac{\eta}{\epsilon}\right)^2 \zeta k_0^2\alpha_1^2\right\rangle}\right\} \tag{7}$$

and

$$\chi = (\epsilon^2 - \eta^2)\,\epsilon^{-1}.$$

The inversion with respect to the α_3 transform variable can be carried out at once by integrating in the complex α_3-plane along the real axis and then along the semicircular arc in the upper half-plane for $z + h > 0$ and in the lower half-plane for $z + h < 0$. It can be shown that the contribution from the semicircular arc vanishes if $\mathrm{Im}\,\{s_{1,\,2}\} > 0$ and the integral is equal to the sum of residues at the poles $\alpha_3 = \pm\, s_1$ and $\alpha_3 = \pm\, s_2$. Thus, we obtain for the particular integral of the system of equations in (6)

$$E_{y1}^{(P)} = \frac{\omega\mu_0 I}{4\pi\epsilon}\int_{-\infty}^{\infty}\left[\frac{P_E(s_1)}{s_1(s_1^2-s_2^2)}\,e^{is_1|z+h|} - \frac{P_E(s_2)}{s_2(s_1^2-s_2^2)}\,e^{is_2|z+h|}\right] e^{i\alpha_1 x}\,d\alpha_1 \tag{8}$$

$$E_{z1}^{(P)} = \frac{i\eta\omega\mu_0 I}{4\pi\epsilon}\int_{-\infty}^{\infty}\left[\frac{\zeta k_0^2-s_1^2}{s_1(s_1^2-s_2^2)}\,e^{is_1|z+h|} - \frac{\zeta k_0^2-s_2^2}{s_2(s_1^2-s_2^2)}\,e^{is_2|z+h|}\right] e^{i\alpha_1 x}\,d\alpha_1 \tag{9}$$

$$\underset{z+h\lessgtr 0}{E_{x1}^{(P)}} = \mp\,\frac{\eta\omega\mu_0 I}{4\pi\epsilon}\,\partial_x\int_{-\infty}^{\infty}\frac{e^{is_1|z+h|}-e^{is_2|z+h|}}{s_1^2-s_2^2}\,e^{i\alpha_1 x}\,d\alpha_1 \tag{10}$$

$$P_E(s_{1,\,2}) = \zeta(\epsilon k_0^2-\alpha_1^2) - s_{1,\,2}^2. \tag{11}$$

To satisfy the boundary conditions we shall also need the solution to the homogeneous system of (6). This can be obtained in a straightforward manner by substituting $\alpha_3 = i\partial_z$ in (6) which gives

$$(\partial_z^2 + s_1^2)(\partial_z^2 + s_2^2)\,\tilde{\mathbf{E}}^{(c)} = 0 \tag{12}$$

of which the solution appropriate to our problems can be written

$$\begin{bmatrix} \tilde{E}_{x1}^{(c)} \\ \tilde{E}_{y1}^{(c)} \\ \tilde{E}_{z1}^{(c)} \end{bmatrix} = \frac{\omega\mu_0 I}{2\sqrt{(2\pi)}\,\epsilon} \begin{bmatrix} C_{11} & C_{12} \\ C_{21} & C_{22} \\ C_{31} & C_{32} \end{bmatrix} \begin{bmatrix} e^{-is_1 z} \\ e^{-is_2 z} \end{bmatrix}. \tag{13}$$

The coefficients C_{ij} are not independent. For, since $\tilde{\mathbf{E}}^{(c)}$ must satisfy the homogeneous system of (6), the coefficients C_{11}, C_{21}, and C_{31} must satisfy the same system with $\alpha_3 = -s_1$ and C_{12}, C_{22}, and C_{32} with $\alpha_3 = -s_2$ (Ince, 1956, p. 145). As a result one can reduce the number of the arbitrary coefficients to just two and write the desired field representation in the form

$$E_{y1}^{(c)} = \frac{\omega\mu_0 I}{4\pi\epsilon} \int_{-\infty}^{\infty} (A_1\, e^{-is_1 z} + A_2\, e^{-is_2 z})\,.\, e^{i\alpha_1 x}\, d\alpha_1 \tag{14}$$

$$E_{z1}^{(c)} = \frac{i\eta\omega\mu_0 I}{4\pi\epsilon} \int_{-\infty}^{\infty} \left[\frac{(\zeta k_0^2 - s_1^2)\, A_1 e^{-is_1 z}}{P_E(s_1)} + \frac{(\zeta k_0^2 - s_2^2)\, A_2\, e^{-is_2 z}}{P_E(s_2)}\right] e^{i\alpha_1 x}\, d\alpha_1 \tag{15}$$

$$E_{x1}^{(c)} = \frac{\eta\omega\mu_0 I}{4\pi\epsilon}\, \partial_x \int_{-\infty}^{\infty} \left[\frac{s_1 A_1\, e^{-is_1 z}}{P_E(s_1)} + \frac{s_2 A_2\, e^{-is_2 z}}{P_E(s_2)}\right] e^{i\alpha_1 x}\, d\alpha_1. \tag{16}$$

For the air region we choose the following solution to (4)

$$E_{y0} = \frac{-\,\omega\mu_0 I}{2\pi} \int_{-\infty}^{\infty} B_1\, e^{i(\alpha_1 x + s_0 z)}\, d\alpha_1 \tag{17}$$

$$E_{z0} = \frac{i\eta\omega\mu_0 I}{2\pi\epsilon}\, \partial_x^2 \int_{-\infty}^{\infty} \frac{B_2}{s_0}\, e^{i(\alpha_1 x + s_0 z)}\, d\alpha_1 \tag{18}$$

$$E_{x0} = \frac{\eta\omega\mu_0 I}{2\pi\epsilon}\, \partial_x \int_{-\infty}^{\infty} B_2\, e^{i(\alpha_1 x + s_0 z)}\, d\alpha_1 \tag{19}$$

where

$$s_0 = (k_0^2 - \alpha_1^2)^{1/2} \text{ and } \operatorname{Im}\{s_0\} \geqslant 0.$$

The requirement that the tangential components of the fields be continuous across the interface, provides four simultaneous equations from which the coefficients A_1, A_2, B_1 and B_2 can be determined. We obtain

$$\left.\begin{aligned} A_1 = {} & \frac{P_E(s_1)[(s_1 - s_0)(s_2 + \zeta s_0) P_E(s_1) - (s_2 + s_0)(s_1 - \zeta s_0) P_E(s_2)] \, e^{is_1 h}}{s_1(s_1^2 - s_2^2) N_E} - \\ & - \frac{P_E(s_1) P_E(s_2)[(s_2 - s_0)(s_2 + \zeta s_0) - (s_2 + s_0)(s_2 - \zeta s_0)] \, e^{is_2 h}}{s_2(s_1^2 - s_2^2) N_E} \end{aligned}\right\} \tag{20}$$

$$\left.\begin{aligned} A_2 = {} & \frac{P_E(s_2)[(s_2 - s_0)(s_1 + \zeta s_0) P_E(s_2) - (s_1 + s_0)(s_2 - \zeta s_0) P_E(s_1)] \, e^{is_2 h}}{s_2(s_1^2 - s_2^2) N_E} - \\ & - \frac{P_E(s_1) P_E(s_2)[(s_1 - s_0)(s_1 + \zeta s_0) - (s_1 + s_0)(s_1 - \zeta s_0)] \, e^{is_1 h}}{s_1(s_1^2 - s_2^2) N_E} \end{aligned}\right\} \tag{21}$$

$$B_1 = \frac{P_E(s_1)(s_2 + \zeta s_0) \, e^{is_1 h} - P_E(s_2)(s_1 + \zeta s_0) \, e^{is_2 h}}{N_E} \tag{22}$$

$$B_2 = \zeta s_0 \frac{(s_2 + s_0) \, e^{is_1 h} - (s_1 + s_0) \, e^{is_2 h}}{N_E} \tag{23}$$

$$N_E = (s_1 + s_0)(s_2 + \zeta s_0) P_E(s_1) - (s_2 + s_0)(s_1 + \zeta s_0) P_E(s_2). \tag{24}$$

To evaluate the field integrals for the air region we first transform to cylindrical coordinates in both configuration and transform spaces using

$$\left.\begin{aligned} \alpha_1 &= k_0 \sin \beta \\ z &= \rho \cos \theta \\ x &= \rho \sin \theta \end{aligned}\right\} \tag{25}$$

which enables one to write the components of the electric field as follows:

$$E_{y0} = \frac{-\omega \mu_0 I}{2\pi} \int_{\Gamma_1} F_{y0}(\beta) \cos \beta \, e^{ik_0 \rho \cos(\beta - \theta)} \, d\beta \tag{26}$$

$$E_{\theta 0} = \frac{\eta \zeta \omega \mu_0 I}{2\pi k_0 \epsilon} \partial_x \int_{\Gamma_1} F_{\theta 0}(\beta) \cos \beta \cos(\beta - \theta) \, e^{ik_0 \rho \cos(\beta - \theta)} \, d\beta \tag{27}$$

$$E_{\rho 0} = \frac{-\eta \zeta \omega \mu_0 I}{2\pi k_0 \epsilon} \partial_x \int_{\Gamma_1} F_{\theta 0}(\beta) \cos \beta \sin(\beta - \theta) \, e^{ik_0 \rho \cos(\beta - \theta)} \, d\beta \tag{28}$$

where

$$F_{y0} = \frac{(\sigma_2 + \zeta \cos \beta) P_1(\beta) e^{ik_0h\sigma_1} - (\sigma_1 + \zeta \cos \beta) P_2(\beta) e^{ik_0h\sigma_2}}{\mathscr{H}(\beta)} \tag{29}$$

$$F_{\theta 0} = \frac{(\sigma_2 + \cos \beta) e^{ik_0h\sigma_1} - (\sigma_1 + \cos \beta) e^{ik_0h\sigma_2}}{\mathscr{H}(\beta)} \tag{30}$$

$$\mathscr{H}(\beta) = (\sigma_1 + \cos \beta)(\sigma_2 + \zeta \cos \beta) P_1(\beta) - (\sigma_2 + \cos \beta)(\sigma_1 + \zeta \cos \beta) P_2(\beta) \tag{31}$$

$$P_{1,2}(\beta) = R(\beta) \mp \sqrt{[R^2(\beta) + \eta^2\zeta \sin^2 \beta]} \tag{32}$$

$$\sigma_{1,2} = \sqrt{\frac{[\zeta(\epsilon - \sin^2 \beta) - P_{1,2}(\beta)]}{\epsilon}} \tag{33}$$

$$R(\beta) = \tfrac{1}{2} [\epsilon\zeta + (\epsilon - \zeta) \sin^2 \beta - (\epsilon^2 - \eta^2)] \tag{34}$$

and Γ_1 is the appropriate integration path in the complex β-plane.

In what follows we shall evaluate the integrals (26), (27) and (28) by the method of steepest descents. To this end we shall be interested in the singularities of these integrands that may lie between the original path of integration Γ_1 and the path of steepest descent Γ.

The poles of the integrands are the zeros of the denominator $\mathscr{H}(\beta)$ given by $\mathscr{H}(\beta_P) = 0$. Unfortunately, the analytical solution to this equation is extremely difficult to find. On physical grounds, however, one could not expect poles in the range $-\pi/2 \leqslant \beta_P \leqslant \pi/2$ since no resonance phenomenon is likely in a semi-infinite region. Thus, we shall discard the possibility of poles within the region of interest.

The integrands under consideration contain radicals associated with σ_1 and σ_2 as a result of which the points $\beta = \theta_{B1}$ and $\beta = \theta_{B2}$ at which σ_1 and σ_2 respectively vanish will be branch points. At each point β these integrands can take on four different values depending on which sign we choose for the radicals. It will be convenient here to talk about four sheets of the β-plane (formed by a four-sheeted Riemann surface) on which each integrand is single valued. We shall call the "proper sheet" one on which $\text{Im}\,\{\sigma_{1,2}\} \geqslant 0$.

Now it can be readily found that the branch points are located at

$$\pm\, \theta_{B1,2} = \arcsin \sqrt{(\epsilon \pm \eta)} \tag{35}$$

It can also be recalled that $\sqrt{(\epsilon \pm \eta)}$ are, respectively, the indices of refraction of right- and left-hand circularly polarized plane waves propagating in the direction of the steady magnetic field in the plasma.

To evaluate the field integral by the method of steepest descent we note first that each of the integrals is of the form

$$W = \int_{\Gamma_1} F(\beta)\, e^{ik_0\rho \cos (\beta - \theta)}\, d\beta \tag{36}$$

and we wish to evaluate these integrals approximately when the distance ρ is large. The saddle-point of the integrands occurs when $\beta = \theta$ and the path of the steepest descent intersects the real axis at the angle of $\pi/4$.

Now we focus our attention on the following. As we remarked earlier, the integrals under consideration are four valued since they contain the radicals

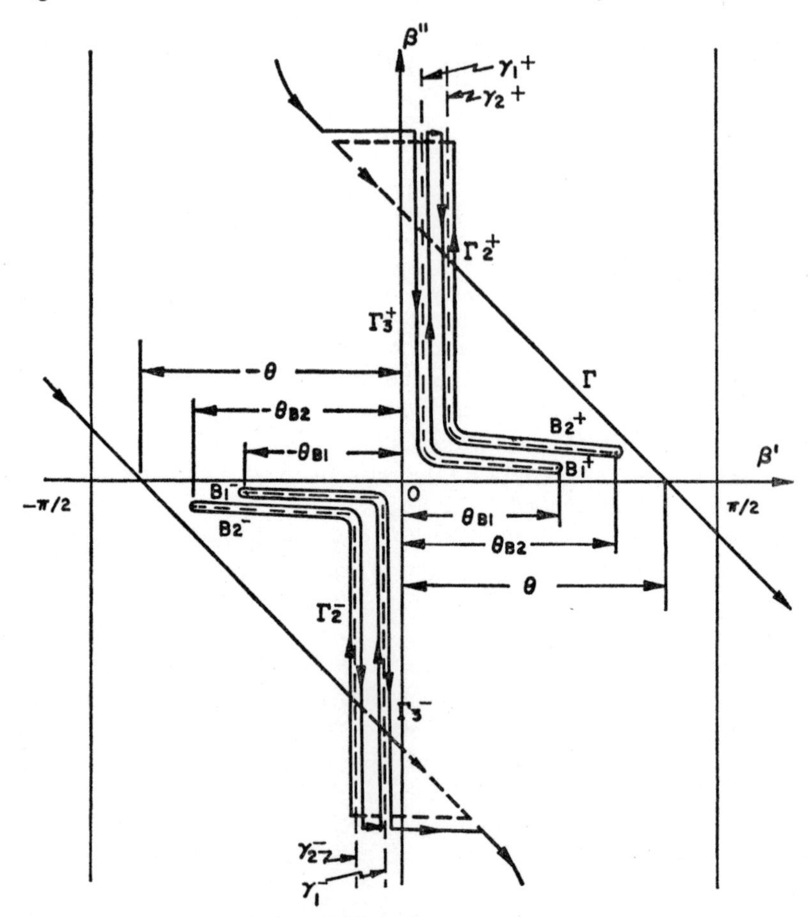

FIG. 2. Path of integration when $\theta > \theta_{B1,\,2}$

of σ_1 and σ_2. The original path of integration passes over the proper sheet of the four-sheeted Riemann surface and can be deformed into the path of the steepest descent only when at least the beginning and the end of it lie on this sheet. In the case when the angle θ does not exceed one of the critical angles θ_{B1} or θ_{B2}, the transition from the path Γ_1 to the path Γ is accomplished without complications and the only contribution to the field integrals will be from the saddle point.

The situation is quite different when the angle θ exceeds one of the critical angles. A more complicated path of integration must be devised supplementing the path of the steepest descent Γ by a contour encompassing each cut in such a way that the beginning and the end of the more complicated path will lie on the proper sheet. The new contour shown in Fig. 2 is similar to that originally proposed by Ott (1942) with the difference that in our case two branch cuts are involved.

As a consequence the complete expression for any of the field integrals will consist of three parts and it can be written as follows:

$$W = W_S + W_{B1}\, u(\theta - \theta_{B1}) + W_{B2}\, u(\theta - \theta_{B2}) \tag{37}$$

where W_S denotes the contribution from the saddle point and W_{B1} and W_{B2} denote the contributions from the borders of the cuts associated with the branch points B_1 and B_2 respectively and $u(\ )$ is a unit step function.

While the saddle-point contribution W_S represents the radiation field, the branch-cut contributions W_{B1} and W_{B2} are associated with the phenomenon of lateral waves which for the case of isotropic interfaces has been discussed in literature (Brekhovskikh, 1960, p. 270). If the losses in the plasma are negligible the lateral field may contribute significantly to the total field especially near the interface. If the plasma is sufficiently lossy (high collision frequency), the critical angles θ_{B1} and θ_{B2} will be complex and a high attenuation of the lateral field will result.

2.1. *The Radiation Field*

The radiation field is given by the saddle-point contribution. This is easy to find (Collin, 1960, p. 495) and we can write the leading term at once

$$E_{y0}^{(R)} \sim -\frac{\omega\mu_0 I}{\sqrt{(2\pi)}}\, F_{y0}(\theta) \cos\theta\, \frac{e^{i(k_0\rho - \pi/4)}}{\sqrt{(k_0\rho)}} \tag{38}$$

$$E_{\theta 0}^{(R)} \sim \frac{i\eta\zeta\omega\mu_0 I}{\epsilon\sqrt{(2\pi)}}\, F_{\theta 0}(\theta) \sin\theta \cos\theta\, \frac{e^{i(k_0\rho - \pi/4)}}{\sqrt{(k_0\rho)}} \tag{39}$$

$$E_{\rho 0}^{(R)} \sim 0. \tag{40}$$

We note that the electric field is still transverse to the radial direction but in general it is not linearly polarized. The time average Poynting vector has only a radial component given by

$$S_{\rho 0}^{(R)} \sim \frac{\omega\mu_0 I^2 \cos^2\theta}{4\pi\rho} \left\{ |\, F_{y0} \,|^2 + \sin^2\theta \,\left|\, \frac{\eta\zeta}{\epsilon} F_{\theta 0} \,\right|^2 \right\} \tag{41}$$

It is worthwhile to note that our solution goes over to the proper limit as the magnetostatic field approaches zero. In fact, we can show that as $H_{DC} \to 0$

or $\omega \to \infty$ then

$$E_{y0}^{(R)} \to -\frac{\omega\mu_0 I}{\sqrt{(2\pi)}} \cdot \frac{\cos\theta\, e^{ik_0 h\sqrt{\zeta-\sin^2\theta}}}{\cos\theta + \sqrt{(\zeta - \sin^2\theta)}} \cdot \frac{e^{i(k_0\rho-\pi/4)}}{\sqrt{(k_0\rho)}} \tag{42}$$

$$E_{\theta 0}^{(R)} \to \frac{i\eta\omega\mu_0 I}{4\zeta\sqrt{(2\pi)}} \cdot \frac{\sin 2\theta\{1 - ik_0 h[\cos\theta + \sqrt{(\zeta - \sin^2\theta)}]\}\, e^{ik_0 h\sqrt{(\zeta-\sin^2\theta)}}}{\sqrt{(\zeta - \sin^2\theta)}[\sqrt{(\zeta - \sin^2\theta)} + \cos\theta] \cdot [\sqrt{(\zeta - \sin^2\theta)} + \zeta\cos\theta]} \cdot \frac{e^{i(k_0\rho-\pi/4)}}{\sqrt{(k_0\rho)}} \tag{43}$$

The component $E_{y0}^{(R)}$ in (42) can be recognized as the proper expression for the electric field in the air due to an electric current line source in an isotropic dielectric having a relative dielectric constant ζ. The component $E_{\theta 0}^{(R)}$ in (43) is then the first order correction for the presence of a small magnetostatic field.

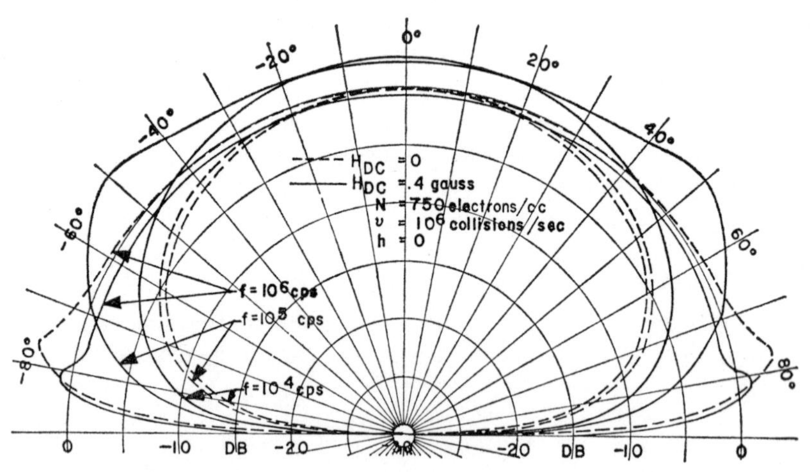

FIG. 3. Radiation patterns of an electric current line source, H_{DC} perpendicular.

The radial component of the Poynting vector (41) is shown in Fig. 3 for a plasma approximately equivalent to that found in the lower edge of the ionosphere and the steady magnetic field equivalent to that of the earth. To show the effect of the steady magnetic field more clearly, the radiation patterns with $H_{DC} = 0$ are superposed and all patterns are normalized at each one of the frequencies shown in such a manner that $S_{\rho 0}^{(R)} = 1$ when H_{DC} and θ are zero. It will be observed that the transmission is generally improved when H_{DC} is present. The peaks in the radiation patterns occur when either one of the radicals σ_1 or σ_2 vanishes corresponding to the critical angles $\theta = \pm\,\theta_{B1,\,2}$.

2.2. *The Lateral Field*

As remarked earlier the integration along the borders of the branch cuts $\gamma_1\pm$ and $\gamma_2\pm$ is associated with the phenomenon of lateral waves (Ott, 1942). To evaluate these contributions we integrate formally on both sides of the cuts and then deform the path of integration in such a way that it goes from the branch point along the line on which the exponent decreases most rapidly (Brekhovskikh, 1960, p. 273). This is equivalent to changing the variable of integration to

$$\beta = \theta_{B1,\,2} + \frac{ix^2}{2} \tag{44}$$

which recasts the field integrals in the form suitable for approximate evaluation as follows:

$$W_B = i e^{ik_0\rho \cos (|\theta| - \theta_B)} \int_0^\infty x\, [F_-(\beta) - F_+(\beta)]\, e^{-k_0\rho \sin (|\theta| - \theta_B) x^2/2}\, dx \tag{45}$$

where $F_-(\beta)$ denotes the value of the function $F(\beta)$ on the left side of the branch cut and $F_+(\beta)$ is the value of the same function on the right side of the cut. Evaluation of the field integrals (26), (27) and (28) using the formulation (45) and when "h" is moderate, gives for the leading terms

$$E_{y0}^{(L1,\,2)} \sim -\frac{i\omega\mu_0 I}{\sqrt{(2\pi)}}\, C_{y0}^{(1,\,2)}\, \frac{e^{i[k_0\rho \cos (|\theta| - \theta_{B1,\,2}) - \pi/4]}}{[k_0\rho \sin (|\theta| - \theta_{B1,\,2})]^{3/2}} \tag{46}$$

$$E_{\theta 0}^{(L1,\,2)} \sim \pm\, \mathrm{sgn}\,(\theta)\, \frac{\zeta\omega\mu_0 I}{\epsilon\sqrt{(2\pi)}}\, C_{\theta 0}^{(1,\,2)} \cos (|\theta| - \theta_{B1,\,2})\, \cdot \frac{e^{i[k_0\rho \cos (|\theta| - \theta_{B1,\,2}) - \pi/4]}}{[k_0\rho \sin (|\theta| - \theta_{B1,\,2})]^{3/2}} \tag{47}$$

$$E_{\rho 0}^{(L1,\,2)} \sim \mathrm{sgn}\,(\theta)\, E_{\theta 0}^{(L1,\,2)} \tan (|\theta| - \theta_{B1,\,2}) \tag{48}$$

where

$$C_{y0}^{(1,\,2)} = \frac{\zeta\sqrt{\{2\zeta[1-(\epsilon\pm\eta)]\sqrt{\langle(\epsilon\pm\eta)[1-(\epsilon\pm\eta)]\rangle}\}}}{b_{E1,\,2}^2\sqrt{\zeta+(\epsilon\pm\eta)}} \cdot \{\zeta\sqrt{[1-(\epsilon\pm\eta)]} + \sqrt{[\mp(\eta/\epsilon)(\zeta+\epsilon\pm\eta)]}\} \cdot [ik_0 h b_{E1,\,2} - a_{E1,\,2} + (1-\zeta)(\epsilon\mp\eta)] \cdot \sqrt{\langle 1-(\epsilon\pm\eta)\rangle}\, \exp\{ik_0 h\sqrt{\langle\mp(\eta/\epsilon)(\zeta+\epsilon\pm\eta)\rangle}\} \tag{49}$$

$$C_{\theta 0}^{(1,\,2)} = \frac{\sqrt{(\epsilon\pm\eta)}[\sqrt{\langle 1-(\epsilon\pm\eta)\rangle} + \sqrt{\langle\pm(\eta/\epsilon)(\zeta+\epsilon\pm\eta)\rangle}]}{\zeta[\zeta\sqrt{\langle 1-(\epsilon\pm\eta)\rangle} + \sqrt{\langle\mp(\eta/\epsilon)(\zeta+\epsilon\pm\eta)\rangle}]} \cdot C_{y0}^{(1,\,2)} \tag{50}$$

$$a_{E1,\,2} = (\zeta+\epsilon\pm\eta)\sqrt{[\mp(\eta/\epsilon)(\zeta+\epsilon\pm\eta)]} + (\zeta^2+\epsilon\pm\eta)\sqrt{[1-(\epsilon\pm\eta)]} \tag{51}$$

$$b_{E1,\,2} = \zeta\sqrt{[1-(\epsilon \pm \eta)]}[(1+\epsilon \pm \eta)\sqrt{\langle\mp(\eta/\epsilon)(\zeta+\epsilon\pm\eta)\rangle} + \\ + (\zeta+\epsilon\pm\eta)\sqrt{\langle 1-(\epsilon\pm\eta)\rangle}] \quad (52)$$

The general form of the field components (46), (47) and (48) is similar to that found by Paul (1959) for an equivalent problem in acoustics and in variance with results obtained by Brekhovskikh (1960, p. 298).

It will be observed that the leading terms of the lateral field components lose their meaning when the plasma is lossless so that the critical angles $\theta_{B1,\,2}$ are real and $\theta = \theta_{B1,\,2}$. In this case the field expressions (46), (47) and (48) are not valid since in their derivations we used the method of steepest descents assuming that the function $F(\beta)$ in (36) was slowly varying which, evidently, is not true in the vicinity of $\beta \rightarrow \theta_{B1,\,2}$. The way of dealing with this situation was proposed by Brekhovskikh (1960, p. 281) but we shall not follow it here because of the unavoidable algebraic complexity. The point we want to stress, however, is the fact that the lateral field does not have a physical singularity at $\theta = \theta_{B1,\,2}$ and the singularities in the field expressions (46), (47) and (48) are artificial ones induced by the approximate evaluation of the field integrals.

The geometrical interpretation of the lateral field becomes more apparent if we make use of the following transformations:

$$\left.\begin{aligned} x &= d_{1,\,2} + g_{1,\,2}\sin\theta_{B1,\,2} \\ z &= g_{1,\,2}\cos\theta_{B1,\,2}. \end{aligned}\right\} \quad (53)$$

As a consequence of which we obtain

$$\left.\begin{aligned} k_0\rho\cos(\theta-\theta_{B1,\,2}) &= k_0\sqrt{(\epsilon\pm\eta)}\cdot d_{1,\,2} + k_0 g_{1,\,2} \\ \rho\sin(\theta-\theta_{B1,\,2}) &= \sqrt{[1-(\epsilon\pm\eta)]}\,d_{1,\,2} \end{aligned}\right\} \quad (54)$$

Using the above results one can recast the components of the lateral field as follows:

$$E_0^{(L1,\,2)} \sim d_{1,\,2}^{-3/2}\,e^{i[k_0\sqrt{(\varepsilon\pm\eta)}\,d_{1,\,2} + k_0 g_{1,\,2}]} \quad (55)$$

In (55) we recognize $k_0(\epsilon\pm\eta)^{1/2}$ as the propagation constants of the plane extraordinary and ordinary rays respectively in a magnetoplasma when the direction of propagation coincides with the direction of the steady magnetic field. The total phase of each one of the lateral waves consists of two parts. The first part, $k_0(\epsilon\pm\eta)^{1/2}$, tells us that the wave travels along the interface in the magnetoplasma through a distance $d_{1,\,2}$ with the phase velocity $c(\epsilon\pm\eta)^{-1/2}$ and the second part, $k_0 g_{1,\,2}$, stands for the phase change due to the travel in free space with phase velocity c through a distance $g_{1,\,2}$.

The geometry of both lateral waves is shown in Fig. 4 where, for convenience, we assumed that the source is located at the interface. The ray OB is a direct ray from the source representing the radiation field as found from the saddle-point contribution. The lateral waves represented by the rays OA_1 and OA_2 propagate along the interface with the phase velocity, respectively, of the extraordinary and ordinary rays in magnetoplasma. Since these velocities are different from the one "allowed" by free space, there is a disturbance created along the interface in the air as a result of which new waves are produced represented by the rays A_1B and A_2B. Since the spatial period of the disturbance along the interface due to, say, the extraordinary wave is equal to λ_1, the wavelength in the air can "fit" this period only if the angle θ_{B1} between the normal to the boundary and the direction of propagation is

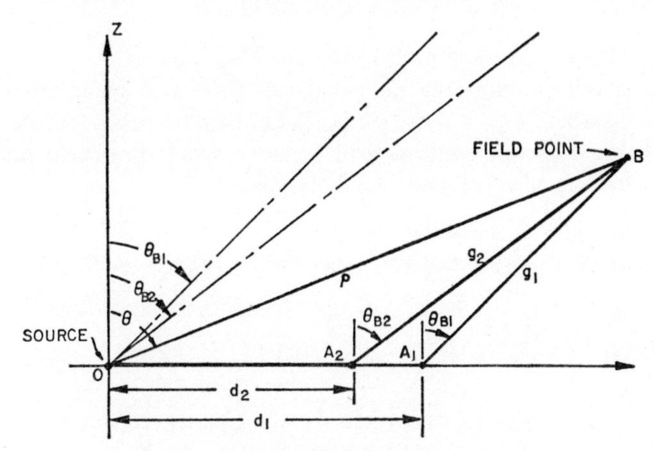

FIG. 4. Geometry of two lateral waves when $\omega_p^2 < \omega(\omega \pm \omega_H)$.

such that the wavelength in the air $\lambda_0 = \lambda_1 \sin \theta_{B1}$. This is just the direction of the lateral wave. The same argument holds for the second lateral wave represented by the ray OA_2B. The resultant field at the point B is a superposition of three waves as concluded earlier in (37).

Figure 4 depicts the situation when the plasma is lossless and

$$\omega_P^2 < \omega(\omega \pm \omega_H).$$

If this inequality is not satisfied, one or both lateral waves will be either highly attenuated or disappear completely.

3. ELECTRIC CURRENT LINE SOURCE —H_{DC} PARALLEL

When the magnetostatic field in the plasma is parallel to an electric line source, then the alternating electric field is always parallel to the magnetostatic field. In this case the only force acting on the free electrons in the plasma is that due to the alternating electric field since the cross-product $\mathbf{u} \times \mathbf{H}_{DC}$ vanishes and consequently the plasma behaves as an ordinary single-refracting medium. Moreover, the boundary conditions are not affected by the presence of the magnetostatic field either. Thus, the solution in this case is identical to the case of an electric current line source situated in an isotropic dielectric with relative dielectric constant ζ. For a line source parallel to the x-axis and buried a distance "h" from the interface in the magnetoplasma the only component of the electric field in the air is the one in the x-direction and is given by

$$E_{x0} = -\frac{\omega\mu_0 I}{2\pi}\int_{\Gamma_1}\frac{\cos\beta e^{ik_0 h\sqrt{(\zeta-\sin^2\beta)}}}{\cos\beta+\sqrt{(\zeta-\sin^2\beta)}}e^{ik_0\rho\cos(\beta-\theta)}\,d\beta \tag{56}$$

where Γ_1 is the appropriate path in the complex β-plane.

The integrand contains the radical $(\zeta\text{-}\sin^2\beta)^{1/2}$ as a result of which the point $\beta = \theta_B$ where $\theta_B = \pm\arcsin(\sqrt{\zeta})$ will be a branch point. As a consequence the field component E_{x0} will consist of radiation field and lateral field. The leading term of these fields is given by

$$E_{x0}^{(R)} \sim -\frac{\omega\mu_0 I}{\sqrt{(2\pi)}}\cdot\frac{\cos\theta e^{ik_0 h\sqrt{(\zeta-\sin^2\theta)}}}{\cos\theta+\sqrt{(\zeta-\sin^2\theta)}}\cdot\frac{e^{i(k_0\rho-\pi/4)}}{\sqrt{(k_0\rho)}} \tag{57}$$

$$E_{x0}^{(L)} \sim \frac{\omega\mu_0 I}{\sqrt{(2\pi)}}\cdot\sqrt[4]{\left(\frac{\zeta}{1-\zeta}\right)}\cdot\frac{(1-ik_0 h)\,e^{i[k_0\rho\cos(|\theta|-\theta_B)+\pi/4]}}{[k_0\rho\sin(|\theta|-\theta_B)]^{3/2}}. \tag{58}$$

4. MAGNETIC CURRENT LINE SOURCE —H_{DC} PERPENDICULAR

The geometry of this problem is the same as in the corresponding case of the electric current line source with $Ke^{-i\omega t}$ replacing $Ie^{-i\omega t}$ in Fig. 1. The corresponding wave equation in the plasma is

$$(-\nabla\times\epsilon^{-1}\nabla\times + k_0^2)\,\mathbf{H}_1 = -i\omega\epsilon_0 K\delta(x)\,\delta(z+h)\,\mathbf{1}_y. \tag{59}$$

Using the Fourier transform (5) one can write (59) as a system of three algebraic equations as follows:

$$\begin{bmatrix} \alpha_1 & 0 & \alpha_3 \\ 0 & (\alpha_3^2-s_1^2)(\alpha_3^2-s_2^2) & 0 \\ 0 & -i\eta\alpha_1^2 & \epsilon(\chi k_0^2-\alpha_1^2-\alpha_2^2) \end{bmatrix}\begin{bmatrix} \tilde{\tilde{H}}_{x1} \\ \tilde{\tilde{H}}_{y1} \\ \tilde{\tilde{H}}_{z1} \end{bmatrix} = \frac{-i\omega\epsilon_0\zeta K(\chi k_0^2-\alpha_1^2-\alpha_3^2)}{2\pi}\begin{bmatrix} 0 \\ e^{i\alpha_3 h} \\ 0 \end{bmatrix} \tag{60}$$

where s_1 and s_2 are given by (7). As before, the inversion with respect to the α_3-transform variable can be carried out by integrating in the complex α_3-plane using the theorem of residues. Moreover, inverting with respect to the α_1-transform variable one obtains the desired integral representation of the primary excitation

$$H_{y1}^{(P)} = \frac{\omega\epsilon_0\zeta K}{4\pi} \int_{-S}^{S} \left[\frac{P_H(s_1)\, e^{is_1|z+h|}}{s_1(s_1^2 - s_2^2)} - \frac{P_H(s_2)\, e^{is_2|z+h|}}{s_2(s_1^2 - s_2^2)}\right] . e^{i\alpha_1 x} . d\alpha_1 \tag{61}$$

$$H_{z1}^{(P)} = \frac{-i\omega\epsilon_0\zeta\eta K}{4\pi\epsilon} \partial_x^2 \int_{-S}^{S} \left[\frac{e^{is_1|z+h|}}{s_1(s_1^2 - s_2^2)} - \frac{e^{is_2|z+h|}}{s_2(s_1^2 - s_2^2)}\right] . e^{i\alpha_1 x}\, d\alpha_1 \tag{62}$$

$$\underset{z+h\gtrless 0}{H_{x1}^{(P)}} = \mp \frac{\omega\epsilon_0\zeta\eta K}{4\pi\epsilon} \partial_x \int_{-S}^{S} \frac{e^{is_1|z+h|} - e^{is_2|z+h|}}{s_1^2 - s_2^2} . e^{i\alpha_1 x} . d\alpha_1 \tag{63}$$

$$P_H(s_{1,\,2}) = \chi k_0^2 - \alpha_1^2 - s_{1,\,2}^2. \tag{64}$$

To find an appropriate solution to the homogeneous system of (60) we use the same method as before and obtain

$$H_{y1}^{(C)} = \frac{\omega\epsilon_0\zeta K}{4\pi} \int_{-S}^{S} (A_1 e^{-is_1 z} + A_2 e^{-is_2 z})\, e^{i\alpha_1 x}\, d\alpha_1 \tag{65}$$

$$H_{z1}^{(C)} = \frac{-i\omega\epsilon_0\zeta\eta K}{4\pi\epsilon} \partial_x^2 \int_{-S}^{S} \left[\frac{A_1 e^{-is_1 z}}{P_H(s_1)} + \frac{A_2 e^{-is_2 z}}{P_H(s_2)}\right] . e^{i\alpha_1 x} . d\alpha_1 \tag{66}$$

$$H_{x1}^{(C)} = \frac{\omega\epsilon_0\zeta\eta K}{4\pi\epsilon} \partial_x \int_{-S}^{S} \left[\frac{s_1 A_1 e^{-is_1 z}}{P_H(s_1)} + \frac{s_2 A_2 e^{-is_2 z}}{P_H(s_2)}\right] e^{i\alpha_1 x} . d\alpha_1 \tag{67}$$

The integral representation of the magnetic field components in the air can be written

$$H_{y0} = \frac{-\omega\epsilon_0\zeta K}{2\pi} \int_{-S}^{S} B_1 e^{i(\alpha_1 x + s_0 z)}\, d\alpha_1 \tag{68}$$

$$H_{z0} = \frac{i\omega\epsilon_0\zeta\eta K}{2\pi\epsilon} \partial_x^2 \int_{-S}^{S} \frac{B_2}{s_0} e^{i(\alpha_1 x + s_0 z)}\, d\alpha_1 \tag{69}$$

$$H_{x0} = \frac{\omega\epsilon_0\zeta\eta K}{2\pi\epsilon} \partial_x \int_{-S}^{S} B_2 e^{i(\alpha_1 x + s_0 z)}\, d\alpha_1 \tag{70}$$

The requirement that the tangential components of the fields be continuous at the boundary $z = 0$ supplies four simultaneous equations from which the coefficients $A_{1,\,2}$ and $B_{1,\,2}$ can be determined. This gives the following

$$\left.\begin{aligned} A_{1,\,2} = {} & \frac{P_H(s_{1,\,2})[(s_{1,\,2} - \zeta s_0)(s_{2,\,1} + s_0)\, P_H(s_{1,\,2}) - (s_{2,\,1} + \zeta s_0)(s_{1,\,2} - s_0)\, P_H(s_{2,\,1})]}{s_{1,\,2}(s_1^2 - s_2^2)\, N_H}\, e^{is_{1,\,2}h} - \\ & - \frac{P_H(s_1)\, P_H(s_2)[(s_{2,\,1} - \zeta s_0)(s_{2,\,1} + s_0) - (s_{2,\,1} + \zeta s_0)(s_{2,\,1} - s_0)]}{s_{2,\,1}(s_1^2 - s_2^2)\, N_H}\, e^{is_{2,\,1}h} \end{aligned}\right\} \tag{71}$$

$$B_1 = \frac{(s_2 + s_0)\, P_H(s_1)\, e^{is_1 h} - (s_1 + s_0)\, P_H(s_2)\, e^{is_2 h}}{N_H} \tag{72}$$

$$B_2 = s_0 \cdot \frac{(s_2 + \zeta s_0)\, e^{is_1 h} - (s_1 + \zeta s_0)\, e^{is_2 h}}{N_H} \tag{73}$$

$$N_H = (s_1 + \zeta s_0)(s_2 + s_0)\, P_H(s_1) - (s_2 + \zeta s_0)(s_1 + s_0)\, P_H(s_2) \tag{74}$$

4.1. *The Radiation Field*

To evaluate the integrals for the air region we first transform using (25) and obtain

$$H_{y0} = \frac{\omega\epsilon_0\zeta K}{2\pi} \int\limits_{\Gamma_1} G_{y0}(\beta) \cos \beta e^{ik_0\rho \cos (\beta-\theta)}\, d\beta \tag{75}$$

$$H_{\theta 0} = \frac{\omega\epsilon_0\zeta\eta K}{2\pi k_0}\, \partial_x \int\limits_{\Gamma_1} G_{\theta 0}(\beta) \cos \beta \cos (\beta - \theta)\, e^{ik_0\rho \cos (\beta-\theta)}\, d\beta \tag{76}$$

$$H_{\rho 0} = \frac{-\,\omega\epsilon_0\zeta\eta K}{2\pi k_0}\, \partial_x \int\limits_{\Gamma_1} G_{\theta 0}(\beta) \cos \beta \sin (\beta - \theta)\, e^{ik_0\rho \cos (\beta-\theta)}\, d\beta \tag{77}$$

$$G_{y0}(\beta) = \frac{(\sigma_2 + \cos \beta)\, P_2(\beta)\, e^{ik_0 h\sigma_1} - (\sigma_1 + \cos \beta)\, P_1(\beta)\, e^{ik_0 h\sigma_2}}{\mathscr{H}(\beta)} \tag{78}$$

$$G_{\theta 0}(\beta) = \frac{(\sigma_2 + \zeta \cos \beta)\, e^{ik_0 h\sigma_1} - (\sigma_1 + \zeta \cos \beta)\, e^{ik_0 h\sigma_2}}{\mathscr{H}(\beta)} \tag{79}$$

and $\sigma_{1,\,2}$, $P_{1,\,2}$ and $\mathscr{H}$ were defined previously.

As in the corresponding case of an electric current line source we note that the integrands contain branch points at $\sigma_1 = 0$ and $\sigma_2 = 0$. Thus, again the resulting field in the air region will consist of the saddle-point contribution to the radiation field and the branch-cut contribution to the lateral field.

We find that the radiation field is given by

$$H_{y0}^{(R)} \sim \frac{\omega\epsilon_0\zeta K}{\sqrt{(2\pi)}} G_{y0}(\theta) \cos\theta \frac{e^{i(k_0\rho-\pi/4)}}{\sqrt{(k_0\rho)}} \tag{80}$$

$$H_{\theta 0}^{(R)} \sim \frac{i\omega\epsilon_0\zeta\eta K}{\sqrt{(2\pi)}} G_{\theta 0}(\theta) \sin\theta \cos\theta \frac{e^{i(k_0\rho-\pi/4)}}{\sqrt{(k_0\rho)}} \tag{81}$$

$$H_{\rho 0}^{(R)} \sim 0. \tag{82}$$

We note that the magnetic field is purely transverse to the radial direction but in general it is not linearly polarized. The time average Poynting vector has a single component in the radial direction given by

$$S_{\rho 0}^{(R)} \sim \frac{\omega\epsilon_0\zeta^2 K^2 \cos^2\theta}{4\pi\rho} [\,|G_{y0}(\theta)|^2 + |\eta \sin\theta \,.\, G_{\theta 0}(\theta)|^2] \tag{83}$$

As before, we can show that our solution goes over to the proper limit as the magnetostatic field approaches zero or the wave frequency becomes high.

4.2. *The Lateral Field*

As in the corresponding case of the electric current line source, we evaluate the lateral field components using the formulation (45) and obtain

$$H_{y0}^{(L1,\,2)} \sim \frac{i\omega\epsilon_0\zeta K}{\sqrt{(2\pi)}} M_{y0}^{(1,\,2)} \,.\, \frac{e^{i[k_0\rho\cos(|\theta|-\theta_{B1,\,2})-\pi/4]}}{[k_0\rho \sin(|\theta| - \theta_{B1,\,2})]^{3/2}} \tag{84}$$

$$\left.\begin{aligned} H_{\theta 0}^{(L1,\,2)} &\sim \pm\,\mathrm{sgn}\,(\theta) \frac{\omega\epsilon_0\zeta K}{\sqrt{(2\pi)}} M_{\theta 0}^{(1,\,2)} \,. \\ &\quad .\cos(|\theta| - \theta_{B1,\,2}) \frac{e^{i[k_0\rho\cos(|\theta|-\theta_{B1,\,2})-\pi/4]}}{[k_0\rho\sin(|\theta| - \theta_{B1,\,2})]^{3/2}} \end{aligned}\right\} \tag{85}$$

$$H_{\rho 0}^{(L1,\,2)} \sim \mathrm{sgn}\,(\theta)\, H_{\theta 0}^{(L1,\,2)} \,.\, \tan(|\theta| - \theta_{B1,\,2}) \tag{86}$$

$$\left.\begin{aligned} M_{y0}^{(1,\,2)} &= \frac{(\epsilon\pm\eta)\sqrt{\{2\zeta[1-(\epsilon\pm\eta)]\sqrt{\langle(\epsilon\pm\eta)[1-(\epsilon\pm\eta)]\rangle}\}}}{b_{H1,\,2}^2\sqrt{\zeta} + (\epsilon\pm\eta)} \,. \\ &\quad .\,[\sqrt{\langle 1-(\epsilon\pm\eta)\rangle} + \sqrt{\langle\mp(\eta/\epsilon)(\zeta+\epsilon\pm\eta)\rangle}] \,. \\ &\quad .\,[ik_0 h b_{H1,\,2} - a_{H1,\,2} + (1-\zeta)^2 \,. \\ &\quad .\,\sqrt{\langle 1-(\epsilon\pm\eta)\rangle}\exp(ik_0 h\sqrt{\langle\mp(\eta/\epsilon)(\zeta+\epsilon\pm\eta)\rangle})] \end{aligned}\right\} \tag{87}$$

$$\left.\begin{aligned} M_{\theta 0}^{(1,\,2)} &= \frac{\zeta\sqrt{[1-(\epsilon\pm\eta)]} + \sqrt{[\mp(\eta/\epsilon)(\zeta+\epsilon\pm\eta)]}}{\sqrt{(\epsilon\pm\eta)}[\sqrt{\langle 1-(\epsilon\pm\eta)\rangle} + \sqrt{\langle\mp(\eta/\epsilon)(\zeta+\epsilon\pm\eta)\rangle}]} M_{y0}^{(1,\,2)} \end{aligned}\right\} \tag{88}$$

$$\begin{aligned} a_{H1,\,2} &= \sqrt{[\mp(\eta/\epsilon)(\zeta+\epsilon\pm\eta)]}[1-(\zeta+\epsilon\pm\eta)] - \\ &\quad - \sqrt{[1-(\epsilon\pm\eta)]}[\zeta^2 + (\epsilon\pm\eta-\zeta)] \end{aligned} \tag{89}$$

$$\left.\begin{aligned} b_{H1,\,2} = \sqrt{[1-(\epsilon \pm \eta)]}\,\cdot \\ \cdot\,\{[1-\zeta(1+\epsilon\pm\eta)]\sqrt{[\mp(\eta/\epsilon)(\zeta+\epsilon\pm\eta)]}\,+ \\ +\,\zeta[1-\zeta-(\epsilon\pm\eta)]\sqrt{[1-(\epsilon\pm\eta)]}\} \end{aligned}\right\} \tag{90}$$

5. MAGNETIC CURRENT LINE SOURCE —H_{DC} PARALLEL

In this problem the magnetic current line source is parallel to the x-axis as well as to the steady magnetic field H_{DC}. One finds that in the air region the magnetic field has a single component as follows:

$$H_{x0} = -\frac{\omega\epsilon_0\chi K}{2\pi}\int_{\Gamma_1}\frac{\cos\beta\, e^{ik_0h\sqrt{(\chi-\sin^2\beta)}}\,.\,e^{ik_0\rho\cos(\beta-\theta)}}{\sqrt{(\chi-\sin^2\beta)}+\chi\cos\beta-i\dfrac{\eta}{\epsilon}\sin\beta}\,.\,d\beta \tag{91}$$

where Γ_1 is the appropriate integration path in the complex β-plane. The integrand contains the radical $(\chi\text{-}\sin^2\beta)^{1/2}$ as a result of which the point $\beta=\theta_B$ where $\theta_B = \pm\,\text{arc}\sin\sqrt{\chi}$ will be a branch point. As a consequence the field component H_{x0} will consist of radiation field and lateral field as before. The leading term of each of these fields is given by

$$\left.\begin{aligned} H_{x0}^{(R)} \sim -\frac{\omega\epsilon_0\chi K}{\sqrt{(2\pi)}}\cdot\frac{\cos\theta\, e^{ik_0h\sqrt{(\chi-\sin^2\theta)}}}{\sqrt{(\chi-\sin^2\theta)}+\chi\cos\theta-i\dfrac{\eta}{\epsilon}\sin\theta}\,\cdot \\ \cdot\,\frac{e^{i(k_0\rho-\pi/4)}}{\sqrt{(k_0\rho)}} \end{aligned}\right\} \tag{92}$$

$$\left.\begin{aligned} H_{x0}^{(L)} \sim \frac{\begin{aligned}\omega\epsilon_0 K\sqrt[4]{[\chi(1-\chi)^3]}[1-ik_0\sqrt{\chi}\,\cdot \\ \cdot\,h(\sqrt{\langle\chi(1-\chi)\rangle}\mp i\eta/\epsilon)]\end{aligned}}{\begin{aligned}\sqrt{(2\pi)}[\sqrt{\langle\chi(1-\chi)\rangle}\mp i\eta/\epsilon]^2\,\cdot \\ \cdot\,[k_0\rho\sin(|\theta|-\theta_B)]^{3/2}\end{aligned}}\,\cdot \\ .\,e^{i[k_0\rho\cos(|\theta|-\theta_B)-\pi/4]} \end{aligned}\right\} \tag{93}$$

We note that the radiation field and the lateral field in the air will not, in general, be symmetric about the vertical axis $\theta = 0$. Moreover, we observe that the reversal of the orientation of the magnetostatic field has the same effect as the reversal of the sign of θ.

The time average Poynting vector corresponding to the radiation field (92) is shown in Fig. 5 for a plasma identical to that in Fig. 3. As before, the corresponding radiation patterns with $H_{DC} = 0$ are superposed for comparison and all patterns are normalized so that $S_{\rho 0}^{(R)} = 1$ when H_{DC} and θ are zero. As concluded from (92) the radiation pattern is not symmetric about the vertical axis and the degree of non-symmetry varies with the wave frequency.

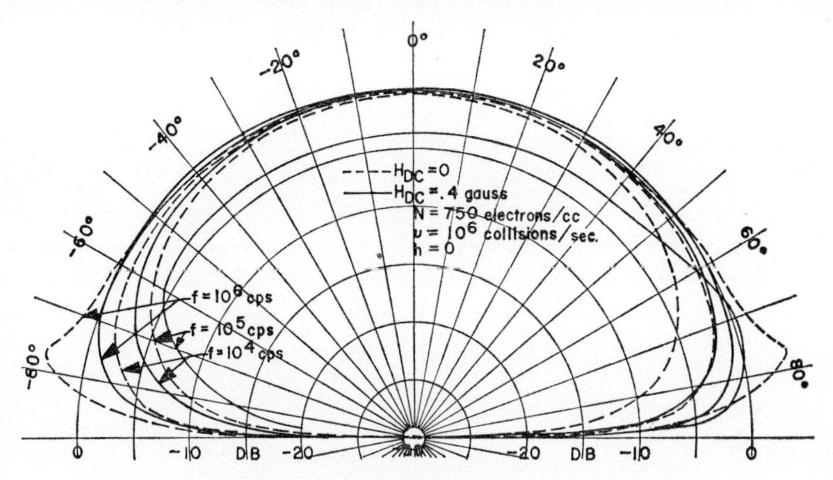

FIG. 5. Radiation patterns of a magnetic current line source, H_{DC} parallel.

REFERENCES

ARBEL, E. (1960) Radiation from a point source in an anisotropic medium, Polytechnic Institute of Brooklyn, Microwave Research Institute, Res. Rep. PIBMRI-861-60 (November).

BARSUKOV, K. A. (1959) Radiation of electromagnetic waves from a point source in a gyrotropic medium with separation boundary, *Radio Engineering and Electronics*, **4**, 1–9.

BREKHOVSKIKH, L. M. (1960) *Waves in Layered Media*, Academic Press, New York-London.

COLLIN, R. E. (1960) *Field Theory of Guided Waves*, McGraw-Hill, New York.

INCE, E. L. (1956) *Ordinary Differential Equations*, Dover Publications.

OTT, H. (1942) Reflexion and Brechnung von Kugelwellen, Effekte Z. Ordnung, *Annalen d. Physik*, **41**, 443–466.

PAUL, D. I. (1959) Wave propagation in acoustics using the saddle point method, *J. Math. and Phys.* **38**, 1–15.

ANTENNA CHARACTERISTICS IN THE PRESENCE OF A PLASMA SHEATH†

G. G. CLOUTIER and M. P. BACHYNSKI
R.C.A. Victor Research Laboratories, Montreal, Canada

ABSTRACT

An experimental investigation of the behavior of a microwave horn antenna in the presence of a plasma sheath generated in helium, has been conducted at frequencies of 9·7 Gc (x-band) and 25 Gc (k-band). The effect of the plasma sheath on the far-field radiation pattern of the horn antenna is to decrease the signal level at normal incidence (by as much as 25 db at "cut-off" conditions) and to redistribute the energy of the radiation pattern over angle so that at large scanning angles the sidelobe level may exceed those measured when the plasma is absent. The plasma does not, however, present a serious mismatch to the antenna, indicating that the plasma is not highly reflecting.

A simplified theoretical model which considers the plasma as a uniform, infinite slab and takes into account diffraction around the edge of the experimental plasma container can be used to account for many of the significant features which are observed.

1. INTRODUCTION

A great deal of attention is being paid at present to the performance of antennas in the presence of ionized media and in particular to the effect of a plasma sheath on communications to and from radio systems engulfed by the plasma sheath. This paper describes experiments designed to examine the influence of a plasma sheath located in front of the aperture of a microwave horn antenna on the far-field radiation pattern and impedance of that antenna.

In the ideal case, it is desired to generate a thin, uniform plasma of "infinite" extent and measure its influence on a microwave radiator located near the plasma as function of the plasma properties. Such a model could readily be compared to theoretical predictions. However, practical limitations dictate a "finite" plasma, which may be non-uniform, contained within a dielectric container. The theoretical interpretation in this latter situation becomes exceedingly more difficult.

An experimental programme has been conducted in which a plasma sheath was generated by a 60 c/s discharge in helium contained in a cylindrical con-

† Performed under Contract AF 19(604)-7334 with Electronics Research Directorate, Air Force Cambridge Research Laboratories and Contract AF 33(616)-7868 with Aeronautical Systems Division, Wright Patterson Air Force Base.

tainer. The effect of this plasma sheath on the radiation characteristics and impedance of microwave horn antennas was then examined in detail at frequencies of 9·7 Gc and 25 Gc, using a fast-acting microwave phase and amplitude measuring system. A simplified theoretical model which treats the plasmas as a uniform infinite slab, and which considers diffraction around the edge of the experimental plasma container, is found to explain many of the important features of the measurements.

2. EXPERIMENTAL TECHNIQUES

2.1. *Containment and Generation of the Plasma*

To approximate a slab of plasma, the plasma was confined in a cylindrical glass container terminated at both ends by parallel low-loss dielectric (polystyrene) plates. The size of the container was limited by the requirements for a uniform discharge throughout the entire volume and by the amount of power required to generate a sufficiently intense plasma (for the x- and k-band microwave region) in this large volume. The final size of the container was 8 in. in diameter and $4\frac{3}{4}$ in. long corresponding to a thickness of the order of 3·5 wavelengths at x-band and 9 wavelengths at k-band.

In order to apply power to the gas in the container, two flat electrodes made of non-magnetic stainless steel were located within the container at a distance of $6\frac{1}{4}$ in. apart. Outlet and inlet connections for evacuating the plasma container and admitting the gas into it were also provided. The plasma was generated by applying a 60 c/s voltage through a 10 : 1 step-up transformer to the electrodes. Typical operating conditions were a voltage of 900 V r.m.s. across the discharge and a current of one ampere, both suitably monitored by an ammeter and voltmeter located in the circuit. Due to the relatively high power generation in the plasma ($\sim$1 kW), it was necessary to operate the discharge in short bursts of not more than one second to prevent overheating the container. Synchronization of the measurements with the firing of the plasma was accomplished from a main control switch.

Helium gas was employed for the discharge and introduced into the plasma container through a fine control needle valve, in series with a pressure regulator and a high-pressure helium reservoir. In order to prevent accumulation of impurities in the plasma container and to insure reproducibility of the plasma conditions, a continuous gas feed system was used in which a small helium flow rate maintained the gas pressure in the plasma container at a constant value. The arrangement of the plasma container, the discharge electrical supply unit and the gas handling system are shown in Fig. 1.

2.2. *Microwave Arrangement and Techniques*

The microwave transmitting horn (15 db gain) is also shown in Fig. 1, together with the plasma container positioned on a rotary turntable. The receiver horn and auxiliary microwave system is located in the far field of the transmitter. The turntable is motor driven and the angle of rotation is determined by a synchro-generator. The entire microwave arrangement is located within an anechoic room in such a way as to eliminate reflection from the walls.

The microwave measurements are carried out using the "multiple probe" technique developed by Osborne (1962). The multiprobe system allows simultaneous measurements of phase and amplitude to be made over a wide dynamic range and at a very rapid sampling rate. The resulting display on an

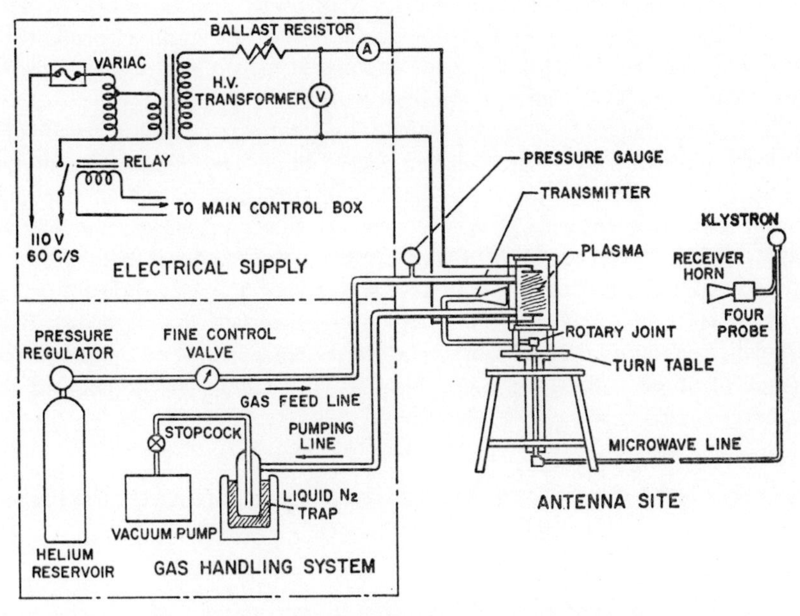

FIG. 1. Schematic diagram of the experimental set-up.

oscilloscope trace is a polar plot of amplitude and phase, i.e. the magnitude of the radius vector corresponds to the signal amplitude, and the polar angle to the phase of the microwave signal. The system can be used for measurement of either the transmitted or reflected signal.

In this experiment a system was developed to display on an oscilloscope, the changes of phase and amplitude of the microwave signal transmitted through or reflected by the plasma as the plasma characteristics changed, during one-quarter of a cycle of the 60 c/s gas discharge. By applying to the oscilloscope a series of sharp brightening pulses at a repetition rate of 10 kc/s, the measurements were displayed as a series of bright spots as shown in Fig. 2. In this manner 40 points or 40 values of the plasma properties could be sampled in one-quarter cycle of the discharge. Figure 2 illustrates a typical display of microwave transmission measurement across the bottle. The points appearing near the edge of the screen correspond to the zero voltage condition across the discharge and it is seen that as the discharge intensity is increased, a large attenuation and phase shift is observed for an x-band (9·7 Gc) signal passing through the plasma. For k-band (25 Gc) frequencies, the same plasma has less effect (plasma frequency/radio frequency and collision frequency/radio frequency less by a factor of 2·5) and hence the attenuation is much less. The phase shift is however considerable, since the effective width of the plasma is increased in inverse proportion to the wavelength. Note the continuous unambiguous display of phase change in excess of an angle of 360°. This is another of the advantages of the multiple probe technique and display. In addition, scale expansion for greater accuracy at the high attenuation levels can also readily be done.

The optimum conditions under which a uniform plasma, in helium of sufficient electron density to markedly affect x-band and k-band signal frequencies, could be produced was experimentally determined to occur at pressures ranging from 0·30 to 1·0 torr. At lower pressures, the electron density is insufficient, while at higher pressures the discharge becomes non-uniform constricting itself to the region between the electrodes. For the measurements presented here, the helium pressure was maintained at 0·85 torr which, assuming a uniform plasma slab, corresponds to a plasma frequency of about 0·8 of the signal frequency at x-band at the peak of the discharge cycle.

3. ANTENNA PATTERN AND IMPEDANCE MEASUREMENTS IN THE PRESENCE OF PLASMA

3.1. *Far-field Antenna Patterns*

The far-field antenna radiation patterns in the presence of a plasma were measured in the following way, since the discharge could only be run for a short interval of time due to the large amount of power dissipated in the plasma container. A record of the phase and amplitude of the received signal for a fixed scanning angle was obtained over a quarter cycle of the discharge by means of the multiple probe sampling system. Consequently, it was only necessary to run the discharge a fraction of a second for the measurements at a given scanning angle. The same procedure was then repeated at two degree intervals of scanning angle. The far-field radiation patterns for a given plasma intensity could then be obtained by plotting the attenuation of a given dot on the multiple probe display as a function of the scanning angle. The phase variation with scanning angle for a specific plasma could similarly be obtained from a plot of the phase angle variation for a given dot. The dots on the displays are synchronized with the gas discharge and can thus be used as time markers corresponding to different plasma properties. The antenna radiation

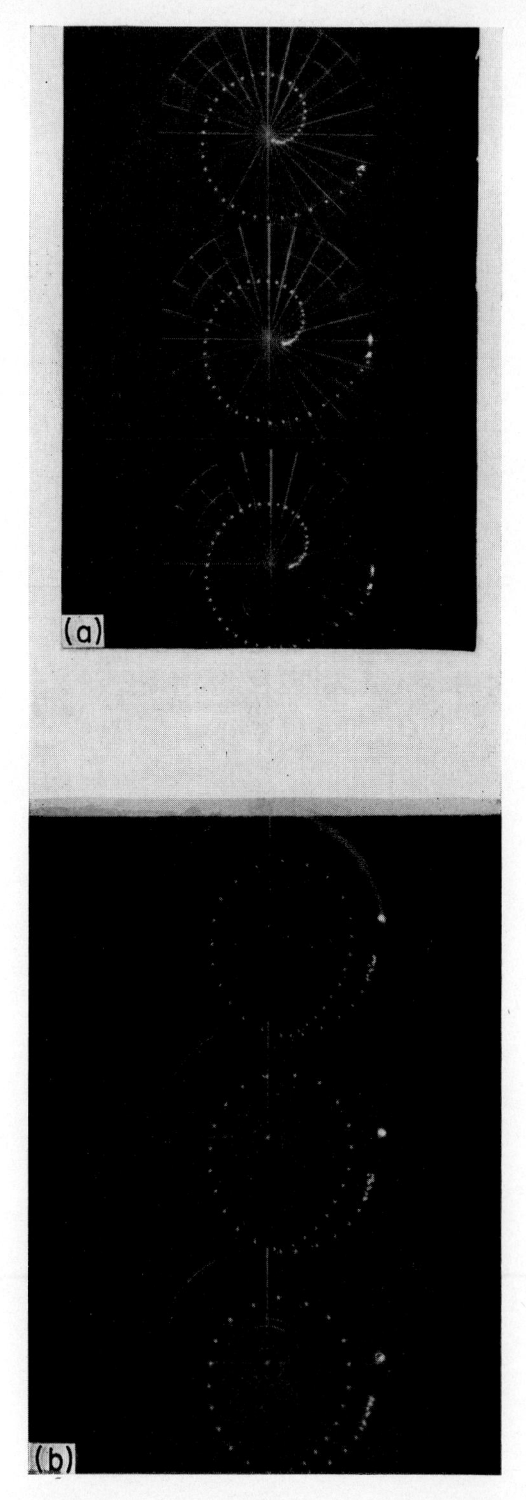

FIG. 2. Typical measurements of transmission of electromagnetic waves through a time-varying plasma using the multiple-probe technique at frequencies of 9·7 Gc and 25 Gc respectively. The radius vector of the polar display corresponds to amplitude while the polar angle is a direct measure of the phase of the microwave signals.

[*facing p. 540*

patterns for other plasma intensities are, of course, constructed from the positions of other dots on the multiple probe displays.

For the above procedure to be valid, the electronic microwave sampling system must be kept stable during a complete set of measurements and the discharge must be accurately reproducible for each angle of the complete scanning range. The tests and procedures to achieve this are described in detail by Cloutier and Bachynski (1962).

Typical measurements of the far-field radiation pattern (both amplitude and phase) of a microwave horn in the presence of a sheath of plasma are shown in Figs. 3 and 4. The variations shown in Fig. 3 were obtained at x-band (9·7 Gc) with the electric vector of the transmitting horn set in the horizontal direction. (A similar result is obtained with the other polarization.) The variation of antenna pattern with plasma properties at k-band (25 Gc) is illustrated in Fig. 4. In all cases measurements were done using a helium plasma and were repeated several times and found to be reproducible.

The general characteristics of the x-band radiation patterns of the horn antenna in the presence of a plasma are the following. A pronounced minimum is observed at normal incidence and this minimum becomes predominant with increasing plasma density. The side lobes of the antenna pattern tend to increase relative to the main lobe power level, and at large scanning angles (30–40°) may be greater than their no plasma value. The phase front emanating from the plasma covered antenna tend to become more plane with increasing electron density. Due to the limited size of the plasma container it is evident that the effects at large scanning angles are seriously affected by diffraction around the edges of the container. At greater electron densities ($\omega_p/\omega > 0 \cdot 75$ where $\omega_p =$ plasma frequency, $\omega =$ signal frequency), the received signal is generally more than 15 db below the signal received at normal incidence in the absence of the plasma. These patterns are not shown in Fig. 3, but their power levels lie well below those illustrated.

The measurements at k-band (Fig. 4) tend to substantiate most of the x-band observations. Due to the higher frequency, the plasma does not affect the r-f signal as strongly and hence the attenuation of the pattern is much less than at x-band. The undulation of the radiation pattern with scanning angle when there is no plasma, tends to be damped out in the presence of the plasma. As at lower frequencies, a marked minimum occurs in the forward direction and the phase fronts become more plane with increasing plasma density.

Auxiliary measurements of phase shift of a microwave signal transmitted directly across the plasma were used to evaluate the electron concentration and hence the plasma frequency. The evaluation of the electron concentration was based on the theory for a uniform slab of plasma of thickness equal to that of the interior of the plasma container. This gave average electron concentrations of 10^{12} electrons/cm^3 at the peak intensity of the gas discharge.

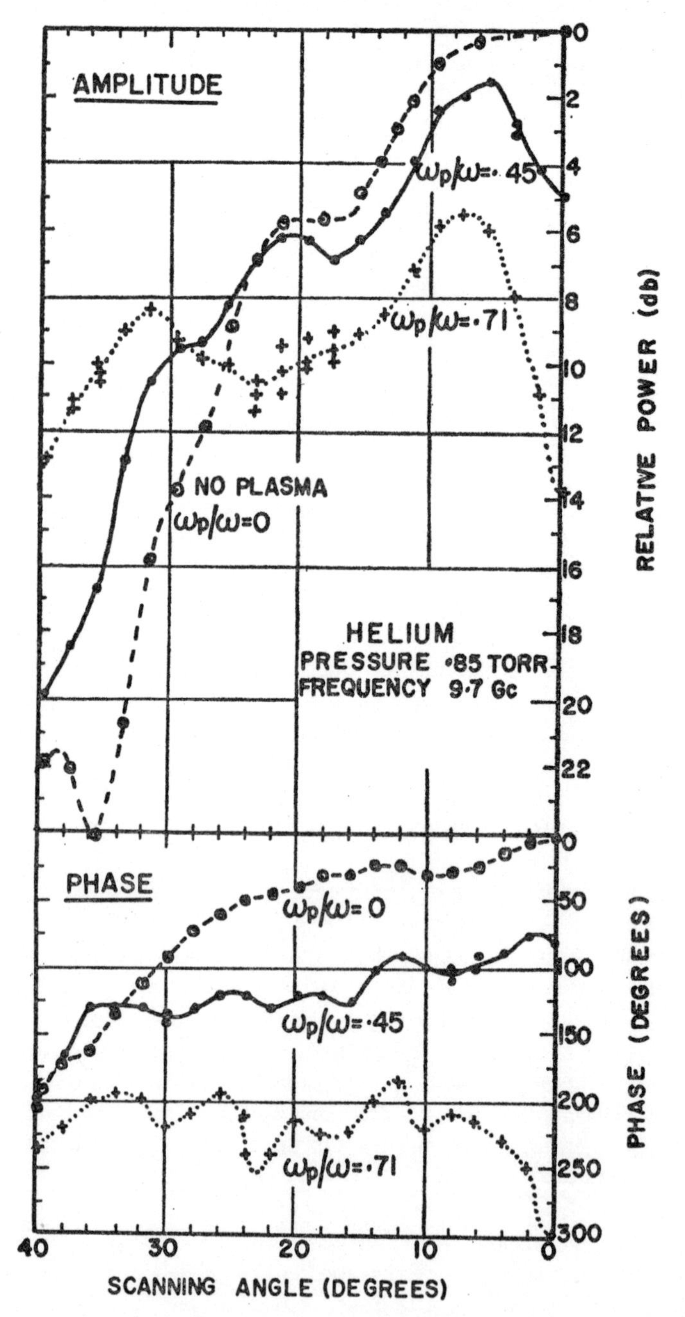

FIG. 3. Measured far-field radiation pattern (intensity and phase) in the presence of a helium plasma at x-band (9·7 Gc).

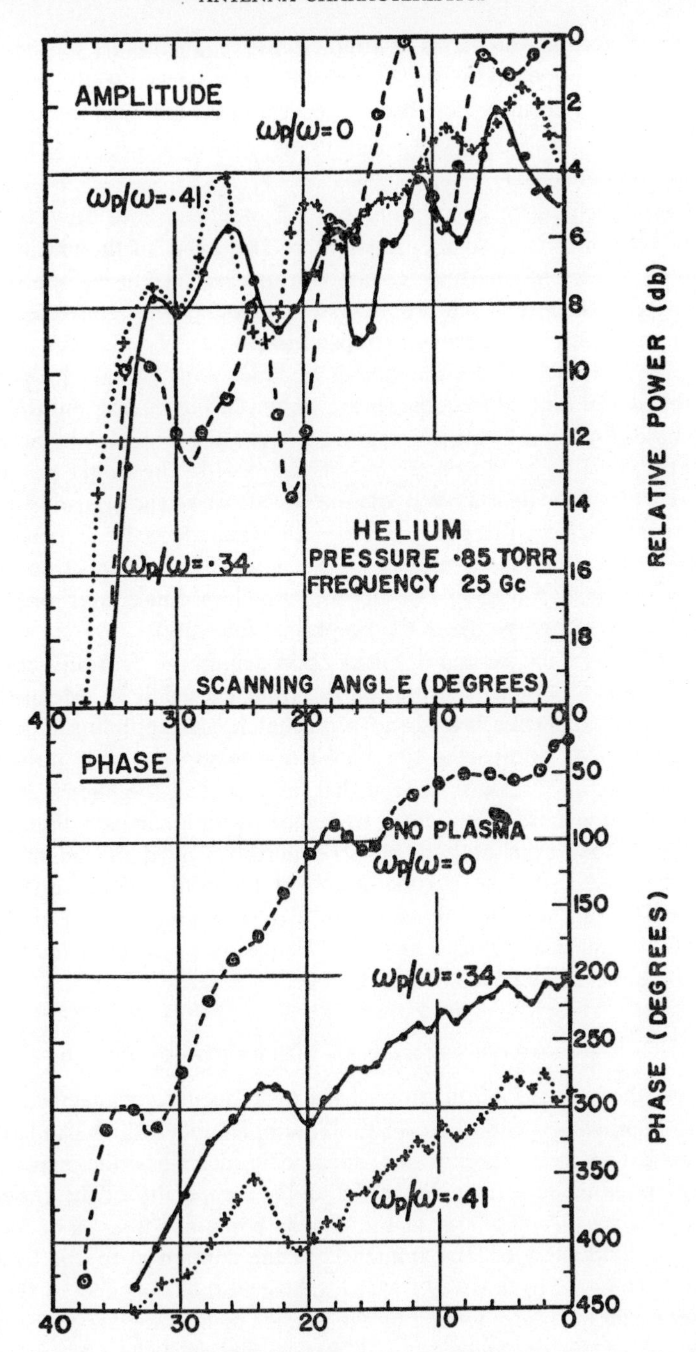

FIG. 4. Measured far-field radiation pattern (intensity and phase) in the presence of a helium plasma at *k*-band (25 Gc).

Since microwave measurements of attenuation are much more sensitive to the geometrical arrangement of the experiment, it is consequently much more difficult to obtain accurate conclusions regarding the collision frequency.

3.2. *Antenna Impedance*

The signal incident on the plasma and reflected back into the transmitting horn creates a mismatch at the transmitter. This effect of the plasma on the impedance of the transmitting horn was also examined using the multiple-probe system. The multiple-probe system gives a very convenient direct Smith chart display of the antenna impedance.

Typical variations of the antenna impedance with plasma properties at both x-band and k-band frequencies are shown in Fig. 5. The multiple-probe display has been expanded to correspond to a preset nominal voltage standing wave ratio (VSWR) at the outer ring of the record. The center point of the display corresponds to a VSWR of one. In all cases the transmitting horn has been matched to the plasma container when there is no plasma. The variation of the antenna impedance is then given by the spiral trace starting from the origin and tracing out the antenna impedance over one-quarter cycle of the gas discharge (from the no plasma intensity).

At x-band, even for plasma densities approaching the "cut-off" condition, the VSWR introduced by the plasma does not exceed $1 \cdot 70$. Meanwhile, at k-band, the same plasma introduces a mismatch corresponding to a VSWR of less than $1 \cdot 25$. Comparing the VSWR-measurements with transmission measurements (Fig. 2) it can be seen that, as expected, the highest mismatch corresponds to conditions of least transmission or greatest attenuation. It thus appears that even at high electron densities most of the microwave energy penetrates into the plasma where at the peak plasma intensities a great fraction of the incident energy is absorbed and/or scattered by the plasma. The results are consistent with calculation based on a lossy slab of plasma (Cloutier, Bachynski, Graf, 1962).

4. THEORETICAL INTERPRETATION

A direct theoretical investigation of the experimental arrangement is prohibitively complicated and hence a suitable simple model which can lend itself to theoretical analysis—and which adequately describes the experimental arrangement, must be sought. Since the major complexity of the experiment is introduced by the finite size of the plasma container, a series of measurements was conducted on the influence of the container on the radiation pattern of the horn. In this way it was established that the effect of the entire container could be approximated by the outside dielectric plate of the plasma container alone. Furthermore using diffraction theory (Bachynski and Bekefi, 1955) and taking into account the directivity of the microwave horn antenna,

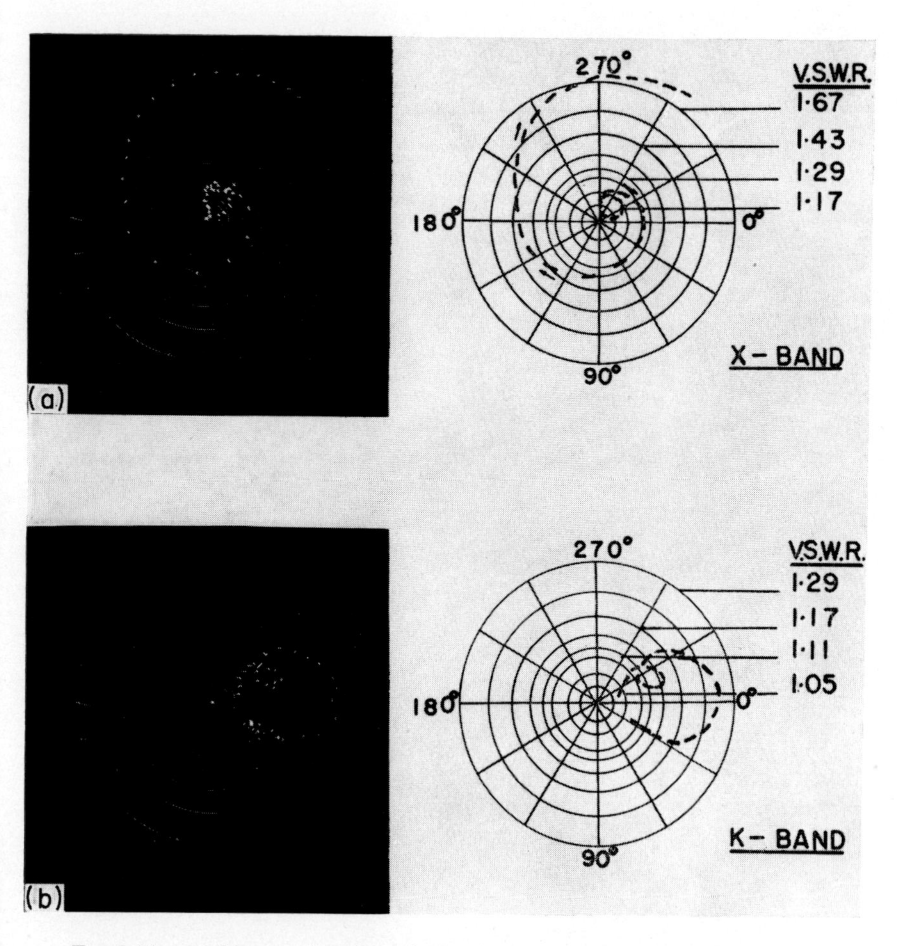

FIG. 5. Measured antenna impedance in the presence of plasma sheaths of different properties (a) *x*-band (9·7 Gc); (b) *k*-band (25 Gc).

the effects predicted analytically agree well with experiment. Consequently the theoretical model adopted for comparison with the experimental determinations of the effect of a plasma on the radiation characteristics of a microwave horn, was that of a horn radiator located directly behind an infinite uniform slab of plasma with a dielectric plate on the opposite side of the plasma representing the plasma container.

With this theoretical model it is possible to apply diffraction theory taking into account the presence of the plasma. The derivation of this solution can now be carried out in a manner analogous to the derivation for diffraction by a dielectric disc conducted previously. The field intensity in the far-field region when a plasma slab is located between the source and the dielectric plate, as given by the scalar Kirchhoff diffraction theory, can be written (Cloutier and Bachynski, 1962)

$$I(q) = [(ka)^2 - q^2]\,[(I_1 + I_2)^2 + (I_3 + I_4)^2], \tag{1}$$

where $q = ka \sin \theta$

$k = 2\pi/\lambda$, λ is the free-space wavelength

a = radius of the dielectric plate

θ = diffraction or scanning angle measured from the normal to the plate

$$I_1 = \int_0^1 \frac{e^{-ka\gamma_1 \sqrt{(p^2+r^2)}}\, e^{-\beta r^2} \cos [ka\alpha_1 \sqrt{(p^2 + r^2)}]\, J_0(qr)}{\alpha_1 \sqrt{(p^2 + r^2)}}\, rdr \tag{2a}$$

$$I_2 = \int_1^{r_0} \frac{e^{-ka\gamma_2 \sqrt{(p^2+r^2)}}\, e^{-\beta r^2} \cos [ka\alpha_2 \sqrt{(p^2 + r^2)}]\, J_0(qr)}{\alpha_2 \sqrt{(p^2 + r^2)}}\, rdr \tag{2b}$$

$$I_3 = \int_0^1 \frac{e^{-ka\gamma_1 \sqrt{(p^2+r^2)}}\, e^{-\beta r^2} \sin [ka\alpha_1 \sqrt{(p^2 + r^2)}]\, J_0(qr)}{\alpha_1 \sqrt{(p^2 + r^2)}}\, rdr \tag{2c}$$

$$I_4 = \int_1^{r_0} \frac{e^{-ka\gamma_2 \sqrt{(p^2+r^2)}}\, e^{-\beta r^2} \sin [ka\alpha_2 \sqrt{(p^2 + r^2)}]\, J_0(qr)}{\alpha_2 \sqrt{(p^2 + r^2)}}\, rdr, \tag{2d}$$

and

$$\alpha_1 = n_{0r} + (n + n_{0r}) \frac{d}{d_1} \qquad k_r = 1 - \left(\frac{\omega_p}{\omega}\right)^2 \frac{1}{1 + (\nu/\omega)^2}$$

$$\alpha_2 = n_{0r} = \left(\frac{|K| + k_r}{2}\right)^{1/2} \qquad k_i = \left(\frac{\omega_p}{\omega}\right)^2 \frac{\nu/\omega}{1 + (\nu/\omega)^2}$$

$$\gamma_1 = n_{0i} \left(1 - \frac{d}{d_1}\right) \qquad |K| = [k_r^2 + k_i^2]^{1/2}$$

$$\gamma_2 = n_{0i} = \left(\frac{|K| - k_r}{2}\right)^{1/2}$$

$d =$ thickness of dielectric plate
$\nu =$ effective electron collision frequency
$n =$ dielectric constant of the plate
$p = d_1/a$ is the normalized distance between the source and the dielectric plate
$d_1 =$ distance from source to dielectric plate, which is the thickness of the plasma;
J_0 is the Bessel function of the first kind,
$r = \rho/a$ is a normalized radius vector
$\rho =$ radius vector in polar coordinates;
$e^{-\beta\rho^2}$ is the illumination function approximating the directivity of the horn pattern; β is selected to fit the main lobe of the radiation pattern;
r_0 is the upper limit of integration.

The analytic solution (1) is sufficiently involved to require computation on an electronic computer. Figure 6 shows such theoretical curves obtained for the following parameters

$$a = 10 \text{ cm} \qquad \lambda = 3 \cdot 1 \text{ cm}$$
$$d = 1 \cdot 9 \text{ cm} \qquad \beta = 1 \cdot 33$$
$$n = 1 \cdot 58 \qquad r_0 = 2.$$
$$d_1 = 14 \text{ cm}$$

These values correspond to the experimental conditions for x-band measurements except that the actual plasma thickness in the experiment was $3 \cdot 5\lambda$. Due to the directivity of the source, the upper limit of integration could be taken as $r_0 = 2$. The computed patterns display only the main lobe of the radiation pattern since the approximations in the theory are no longer valid for scanning angles larger than 20 deg.

In agreement with the experimental observations (Fig. 3) the power received at normal incidence decreases rapidly as the plasma intensity increases, resulting in a deep minimum in the radiation pattern at normal incidence. The effect of increasing collision frequency is also shown and results in an effective lowering of the overall power level of the far-field radiation pattern. The predicted power level is however greater than that measured experimentally. This may be due to the fact that reflection at the plasma boundaries, refraction at the first dielectric plate and non-uniformity in the plasma are not considered in the analysis.

5. CONCLUSIONS

A laboratory experimental investigation has been conducted on the behavior of a horn antenna in the presence of a plasma sheath. The effects of the plasma sheath on the radiation pattern and impedance of the antenna

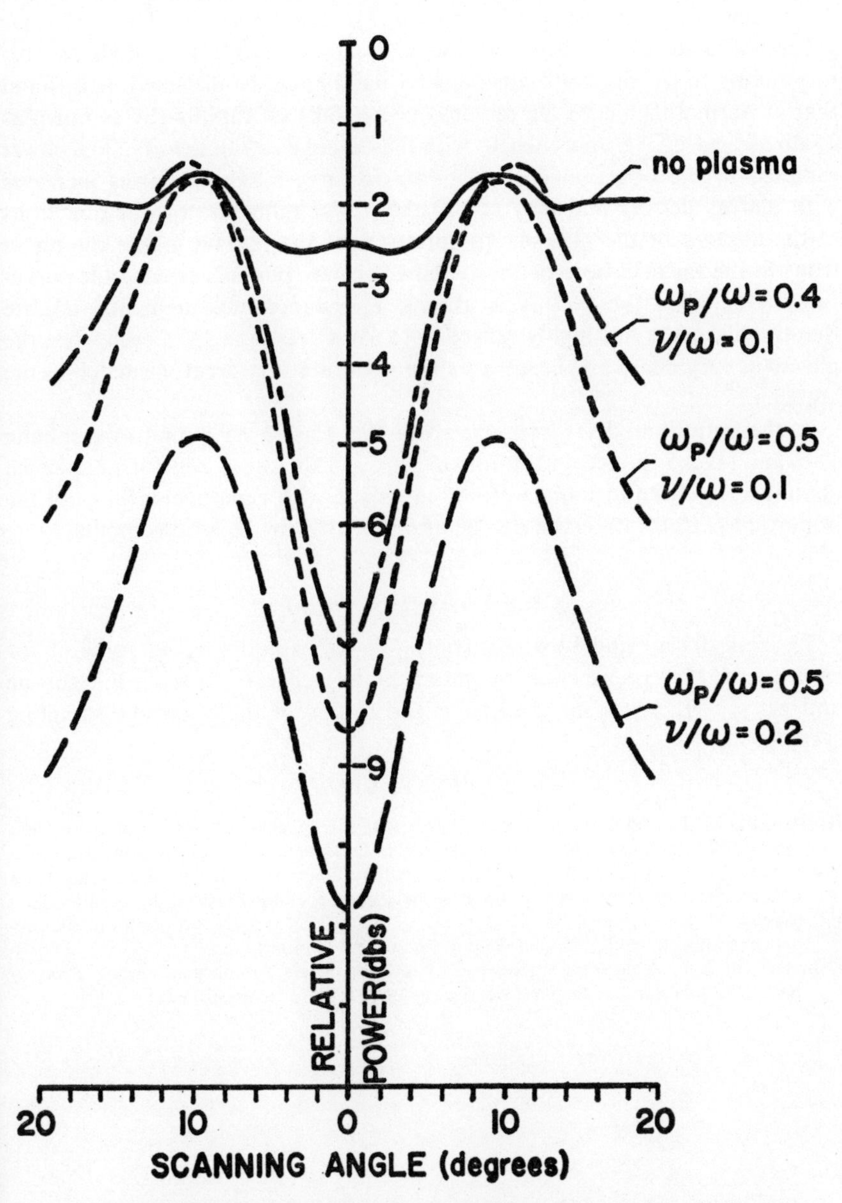

FIG. 6. Theoretical far-field radiation patterns for a microwave horn in the presence of an infinite plasma slab and including diffraction effects due to the edge of the plasma container.

were determined at x-band (9·7 Gc) and k-band (25 Gc) frequencies, using a fast acting microwave phase and amplitude measuring system. Using plasmas generated under controlled conditions, electron densities as high as corresponding to the plasma frequency at x-band could be obtained. It is found that at normal incidence the radiated power falls off rapidly (by as much as 25 db as "cut-off" is approached) with increasing electron density. The power radiated in the direction corresponding to large scanning angles increases with plasma density and may exceed the power radiated in these directions in the absence of the plasma. The presence of the plasma makes the phase front in the far field become more plane (or may possibly reverse the curvature) than when the plasma is absent. Impedance measurements indicate that the plasma is not highly reflecting (VSWR less than 1·7), even when the plasma is very dense and hence does not present a very great mismatch to the antenna.

A theoretical model which considers the plasma as a uniform, infinite slab, and takes into account diffraction around the outer edge of the plasma container, has been formulated for comparison with experiment. Some of the important features observed can be accounted for by using this model.

ACKNOWLEDGEMENTS

The authors are indebted to Mr. B. W. Gibbs and Mr. K. A. Graf for carrying out the experimental measurements as well as to Dr. F. J. F. Osborne and Mr. F. H. Smith for the design and testing of the electronic sampling system.

REFERENCES

BACHYNSKI, M. P. and BEKEFI, G. (1955) Investigation of aberrations in microwave lens systems, Eaton Electronics Laboratory, McGill University Technical Rep. No. 35.

BACHYNSKI, M. P. & CLOUTIER, G. G. Antenna radiation patterns in the presence of a plasma sheath, *Proc. of Symposium on the Plasma Sheath*, Plenum Press (to be published).

BACHYNSKI, M. P., CLOUTIER, G. G. & GRAF, K. A. (1962) Antenna properties in the presence of ionized media, AFCRL Report No. 62-191 (March).

OSBORNE, F. J. F. A versatile microwave diagnostic system for plasma studies, *Proc. of the 3rd Symposium on Engineering Aspects of MHD* (to be published).

COUPLING OF ELECTROMAGNETIC AND MAGNETOSTATIC WAVES IN FERRITE WAVEGUIDES†

B. A. AULD

Microwave Laboratory, W. W. Hansen Laboratories of Physics,
Stanford University, Stanford, California

ABSTRACT

A coupled mode formalism has been developed for treating propagation in ferrite waveguides in terms of coupled electromagnetic and magnetostatic modes. The procedure is described for both axially- and transversely-magnetized structures, and is developed in detail for the axially-magnetized problem. An approximate method is given for deducing by inspection the qualitative nature of the dispersion curves; and some numerical results are given for the case of an axially-magnetized ferrite-filled parallel plane waveguide.

1. INTRODUCTION

A number of detailed treatments of propagation on specific gyromagnetic waveguide structures have appeared in the literature (Gamo, 1953; Kales, 1953; Van Trier, 1953; Mikaelyan, 1954; Suhl and Walker, 1954; Lax and Button, 1955). These have demonstrated the existence of certain anomalous propagation characteristics; for example, propagation in arbitrarily small waveguides, unusual types of cutoff (Suhl and Walker, 1954; Tai, 1960), and backward wave propagation in uniform waveguides (Suhl and Walker, 1954; Trivelpiece, 1958). In these problems the characteristic equations are of complicated transcendental form and specific results are obtained only by resorting to numerical computation. The results are therefore not readily amenable to physical interpretation. A similar situation arises in the case of isotropic composite waveguides. For such guides it has been shown by Bahiana and Smullin (1960) that the principle of mode coupling may be used to formulate solutions which may be evaluated approximately without cumbersome numerical computation and have, at the same time, a simple physical interpretation. The coupled modes in this case are the normal modes for the homogeneously-filled regions of the waveguide cross-section. Application of the same principle to gyromagnetic waveguides, where the coupling is between electromagnetic and static modes, permits a clarification of the

† The research reported in this paper was supported jointly by the U.S. Army Signal Corps, the U.S. Air Force, and the U.S. Navy (Office of Naval Research).

physical origin of the anomalous properties of these waveguides and illustrates clearly the interrelation of these anomalous characteristics. Propagation of uniform plane waves in an unbounded ferrite has recently been discussed from this point of view (Pincus, 1962) (see Fig. 1). In this case the bending upward of the magnetostatic dispersion curves at very short wavelengths is due to the influence of exchange coupling. This effect, which will also be

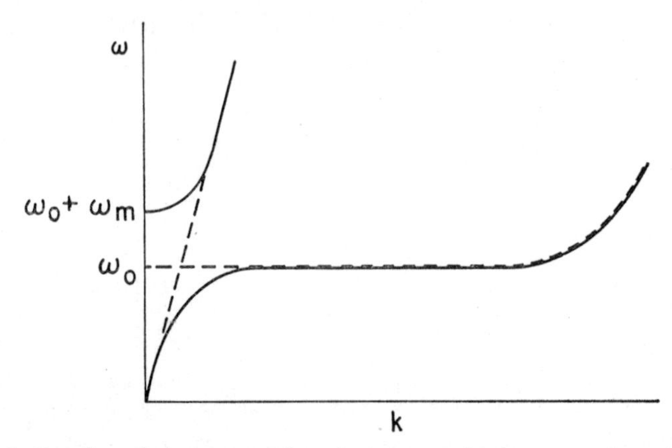

FIG. 1. Coupling of electromagnetic and magnetostatic plane waves in an infinite ferrite medium (Pincus, 1962).

important in the case of guided magnetostatic waves of very high order, is not considered here. Although the theory will be applied only to ferrite waveguides in this paper, it is also useful in plasma waveguide problems.

2. MODE COUPLING THEORY

The theory will be developed for a closed uniform waveguide of arbitrary cross-section, loaded with a uniform ferrite rod. A bias field H_0 is applied in a direction **Z** which is either transverse or parallel to the waveguide axis **z** (see Fig. 2). In the case of transverse bias it is assumed that the ferrite cross-section is ellipsoidal with **Z** along a major or minor axis. Otherwise, the ferrite cross-section is assumed to be arbitrary.

In the small signal approximation the fields satisfy the following equations

$$\begin{aligned} \nabla \times \mathbf{E} &= -j\omega\mu_0(\mathbf{H} + \mathbf{M}) \\ \nabla \times \mathbf{H} &= j\omega\epsilon\,\mathbf{E}, \end{aligned} \tag{1}$$

with the r.f. magnetization specified in terms of the r.f. magnetic field by the linearized equation of motion

$$j\omega\mathbf{M} + \omega_H\mathbf{Z} \times \mathbf{M} = \omega_M\mathbf{Z} \times \mathbf{H}, \tag{2}$$

where

$$\omega_M = \gamma\mu_0 M_s$$

and

$$\omega_H = \gamma\mu_0 H_0.$$

Continuity of the tangential E and H fields is required at all surfaces of discontinuity.

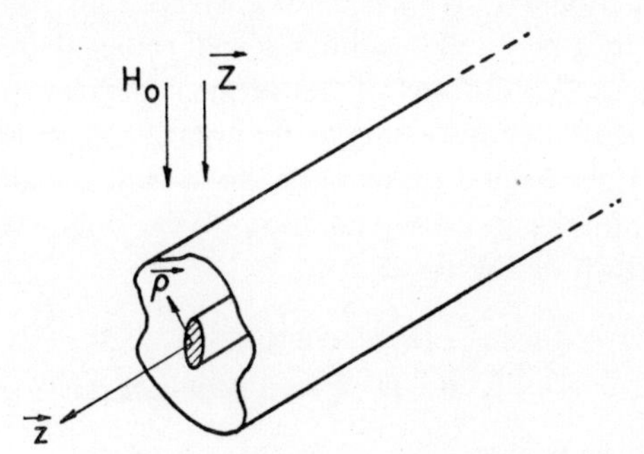

FIG. 2. General ferrite waveguide.

A coupled mode formulation of the problem is obtained by exprcssing the magnetic field in terms of rotational and irrotational parts

$$\mathbf{H}^{(r)} = \mathbf{h}^{(r)}\,(\rho)\,e^{-\gamma z}$$

and

$$\mathbf{H}^{(i)} = \nabla\Phi = \nabla\phi(\rho)\,e^{-\gamma z}.$$

If a Coulomb gauge, $\nabla \cdot \mathbf{H}^{(r)} = 0$, is chosen (Suhl, 1957) (1) and (2) reduce to

$$\nabla \times \mathbf{E} + j\omega\mu_0\mathbf{H}^{(r)} = -j\omega\mu_0(\nabla\Phi + \mathbf{M}) \tag{3a}$$

$$\nabla \times \mathbf{H}^{(r)} - j\omega\epsilon\,\mathbf{E} = 0 \tag{3b}$$

and

$$\begin{aligned} j\omega\mathbf{M} + \omega_H\mathbf{Z} \times \mathbf{M} - \omega_M\mathbf{Z} \times \nabla\Phi &= \omega_M\mathbf{Z} \times \mathbf{H}^{(r)} \quad \text{(Inside ferrite)} \\ \mathbf{M} &= 0 \quad \text{(Outside ferrite)} \end{aligned} \tag{4a}$$

$$\nabla \cdot (\nabla\Phi + \mathbf{M}) = 0. \tag{4b}$$

These equations represent a coupling between the electromagnetic equations

$$\nabla \times \mathbf{E} + j\omega\mu_0\mathbf{H}^{(r)} = 0 \tag{5a}$$

$$\nabla \times \mathbf{H}^{(r)} - j\omega\epsilon\,\mathbf{E} = 0 \tag{5b}$$

obtained by "turning off" the magnetization in the original problem, and the magnetostatic equations

$$\begin{aligned} j\omega\mathbf{M} + \omega_H\mathbf{Z} \times \mathbf{M} - \omega_M\mathbf{Z} \times \nabla\Phi &= 0 \quad \text{(Inside ferrite)} \\ \mathbf{M} &= 0 \quad \text{(Outside ferrite)} \end{aligned} \tag{6a}$$

$$\nabla \cdot (\nabla\Phi + \mathbf{M}) = 0 \tag{6b}$$

obtained by "turning off" the dielectric constant in the original problem. In terms of the permeability tensor, (6a) and (6b) may be expressed alternatively as

$$\nabla \cdot \boldsymbol{\mu} \nabla \Phi = 0, \tag{7}$$

which is Walker's problem (Walker, 1957), except that solutions are sought in a region closed by a perfectly conducting wall rather than in an unbounded region. The boundary conditions of the original problem required the continuity of the tangential components of the total fields at all surfaces of discontinuity. For convenience, these same boundary conditions are applied individually to both the electromagnetic and the magnetostatic problems; that is, at the surface of the ferrite,

$$\begin{aligned} &E_{\tan},\ H_{\tan}^{(r)}\text{—continuous} \\ &\Phi,\ \mathbf{n} \cdot \boldsymbol{\mu} \nabla \Phi\text{—continuous,} \end{aligned} \tag{8}$$

and at the waveguide wall,

$$\begin{aligned} E_{\tan} &= 0 \\ \partial \Phi / \partial n &= 0. \end{aligned} \tag{9}$$

This insures that the electromagnetic and magnetostatic equations are coupled only through terms appearing explicitly in (3) and (4) and not through the boundary conditions.

Coupled mode equations are now derived by expanding (3) in terms of the normal modes of the electromagnetic problem (5) and expanding (4) in terms of the normal modes of the magnetostatic problem (6). The electromagnetic expansion is straightforward, since, for closed waveguides, the electromagnetic modes

$$\mathbf{E}_N = \mathbf{e}_N(\boldsymbol{\rho})\, e^{-\gamma_N z}, \qquad \mathbf{H}_N = \mathbf{h}_N(\boldsymbol{\rho})\, e^{-\gamma_N z}$$

satisfy the orthogonality relations

$$\int_{\substack{\text{cross}\\ \text{section}}} \mathbf{e}_N \times \mathbf{h}_N^* \cdot \mathbf{ds} = \delta_{MN}$$

even in the case of composite structures (Adler, 1952). For simplicity, however, the discussion here is restricted to homogeneously filled waveguides. That is, the region outside the ferrite is assumed to be filled with a lossless dielectric having the same dielectric constant as the ferrite. By following standard procedures (Bahiana and Smullin, 1960), the electric field and the rotational part of the magnetic field are expanded in terms of the electromagnetic modes

$$\begin{aligned} \mathbf{E} &= \sum_N \{V_N \mathbf{e}_{Nt}(\boldsymbol{\rho}) + I_N Z_N \mathbf{e}_{Nz}(\boldsymbol{\rho})\}\, e^{-\gamma z} \\ \mathbf{H}^{(r)} &= \sum_N \{I_N \mathbf{h}_{Nt}(\boldsymbol{\rho}) + V_N Y_N \mathbf{h}_{Nz}(\boldsymbol{\rho}) + \mathbf{m}_z(\boldsymbol{\rho}) - \gamma \phi(\boldsymbol{\rho})\}\, e^{-\gamma z}. \end{aligned} \tag{10}$$

The transverse part of (3a) then reduces to a set of equation pairs:

$$\begin{aligned}\gamma V_N &= \gamma_N Z_N I_N + j\omega\mu_0 \int_{\text{waveguide}} (\nabla\Phi + \mathbf{M})\, e^{\gamma z} \,.\, \mathbf{h}^*_{Nt}\, ds \\ \gamma I_N &= \gamma_N Y_N V_N - \gamma^*_N \int_{\text{waveguide}} (\nabla\Phi + \mathbf{M})\, e^{\gamma z} \,.\, \mathbf{h}^*_{Nz}\, ds,\end{aligned} \tag{11}$$

where $Z_N = 1/Y_N$ is the wave impedance of the Nth mode.

Similarly, the potential function and magnetization in (4a) are expanded in terms of solutions of Walker's equation (7), subject to boundary conditions (8) and (9)

$$\Phi_\nu(\omega', \omega_H') = \Phi_\nu(\omega', \omega_H', \rho)\, e^{-\gamma_\nu(\omega', \omega_H')\, z} \tag{12a}$$

$$\mathbf{M}_\nu = \mathbf{m}_\nu(\rho)\, e^{-\gamma_\nu z} = \chi(\omega', \omega_H')\, \nabla\Phi_\nu, \tag{12b}$$

where χ is the susceptibility tensor of the medium,

$$\begin{aligned}\Phi &= \sum_m A_\nu \Phi_\nu \\ \mathbf{M} &= \sum_m A_\nu \mathbf{M}_\nu.\end{aligned} \tag{13}$$

From (4), the total magnetization $\mathbf{M}$ comprises a part due to the rotational field $\mathbf{H}^{(r)}$ in addition to the part due to the irrotational field $\nabla\Phi$. In making the expansion it is assumed, in the absence of detailed knowledge about the completeness of the $\mathbf{M}_\nu$'s, that both portions of $\mathbf{M}$ may be expanded in terms of the $\mathbf{M}_\nu$'s (Suhl, 1957). The right-hand side of (3a) then contains only magnetostatic mode amplitudes and there is, therefore, no coupling between electromagnetic modes. This magnetostatic expansion cannot be carried out in the most obvious way, by expanding $\phi(\rho)$ and $\mathbf{m}(\rho)$ in terms of $\phi_\nu(\rho)$'s and $\mathbf{m}_\nu(\rho)$'s which are obtained from solutions to Walker's equation for the *same* frequency ω and bias field $\omega_H/\gamma\mu_0$ specified in the original problem. The reason for this is that the left-hand side of (4a) would then reduce identically to zero. This difficulty is avoided by retaining the z-dependence of the magnetostatic modes in the expansion, as in (13), and selecting the modes so that they all have the same specified propagation factor γ. This may be arranged either by selecting a different frequency ω_ν for each mode, keeping the bias field $\omega_H/\gamma\mu_0$ fixed at the value specified in the original problem,

$$\gamma_\nu(\omega_\nu, \omega_H) = \gamma, \tag{12a}$$

or by choosing a different bias field $\omega_{H\nu}/\gamma\mu_0$ for each mode, keeping the frequency fixed at the value specified in the original problem,

$$\gamma_\nu(\omega, \omega_{H\nu}) = \gamma. \tag{12b}$$

The procedure to be used is determined by the choice of dependent variable (ω or ω_H) in the calculation, and entirely distinct orthogonality relations are obtained for the two cases (see Appendix).

In the present analysis ω will be taken as the dependent variable and the orthogonality relations to be used are therefore (A4) and (A8). For the case of axial magnetization the bias axis **Z** coincides with the waveguide axis **z** and it is found, by assuming normalization and applying (13) and (A4) to (4a), that

$$(\omega - \omega_\nu)\, A_\nu = -\omega_M \int_{\text{ferrite}} \mathbf{z} \,.\, [\mathbf{z} \times \mathbf{H}^{(r)}] \times \mathbf{m}_\nu^* \, \mathrm{d}s, \tag{14}$$

where use has been made of

$$-j\omega_\nu \mathbf{M}_\nu = \omega_H \mathbf{z} \times \mathbf{M}_\nu - \omega_M \mathbf{z} \times \nabla \Phi_\nu$$

from (6a). Expressing $\mathbf{H}^{(r)}$ in (14) in terms of (10) then yields the set of equations

$$(\omega - \omega_\nu)\, A_\nu = \sum_N \overline{K}_{\nu N}(\gamma, \omega)\, I_N \tag{15a}$$

where

$$\overline{K}_{\nu N} = \omega_M \int_{\text{ferrite}} \mathbf{m}_\nu^* \,.\, \mathbf{h}_{N_t} \, \mathrm{d}s \,. \tag{16}$$

From (9), after eliminating V_N and substituting (13), there follows the set of equations

$$[\gamma^2 - \gamma_N^2(\omega)]\, I_N = \sum_\nu K_{N\nu}(\gamma, \omega)\, A_\nu \tag{15b}$$

where

$$K_{N\nu} = \int_{\text{ferrite}} \{j\omega\mu_0\, \gamma_N Y_N \mathbf{h}_{Nt}^* \,.\, (\nabla_t \phi_\nu + \mathbf{m}_\nu) + \gamma^2 \gamma_N^* \phi_\nu h_{Nz}^*\} \, \mathrm{d}s. \tag{17}$$

Similarly, coupled-mode equations may be found for the case of transverse magnetization. These have the somewhat more complicated form

$$\begin{aligned} \gamma V_N &= \gamma_N Z_N I_N + \sum_\nu K'_{N\nu} A_\nu \\ \gamma I_N &= \gamma_N Y_N V_N + \sum_\nu K''_{\nu N} A_\nu \\ (\omega - \omega_\nu)\, A_\nu &= \sum_N (\overline{K}'_{\nu N} I_N + \overline{K}''_{\nu N} V_N) + \sum_{\nu'} K_{\nu_{\nu'}}\, A_{\nu'}. \end{aligned} \tag{18}$$

Dispersion curves for the coupled electromagnetic-magnetostatic modes are then found by setting equal to zero the characteristic determinant of the set of coupled eqns (15) or (18). Since it was specified that all the magnetostatic modes should have a particular value of γ, the characteristic equation must be solved by assuming γ and finding ω.

3. DISPERSION CURVES FOR THE UNCOUPLED MODES

3.1. *Magnetostatic Modes*

The propagation of magnetostatic waves on an axially-magnetized circular ferrite rod has been studied by Trivelpiece (1958) for the case of the shielded rod, and by Joseph and Schlömann (1961) for the case of the unshielded rod. For either structure there exists an infinite manifold of backward waves, which propagate only within a finite frequency band in the vicinity of ferromagnetic resonance. Solutions exhibiting the same properties are easily found for an axially-magnetized parallel plane waveguide completely filled with ferrite,

$$\Phi_\nu = \mathrm{e}^{-\gamma_\nu z} \cos(\nu\pi/b)\, y \tag{19a}$$

$$\gamma_\nu^2 = (\nu\pi/b)^2 \,\mu, \tag{19b}$$

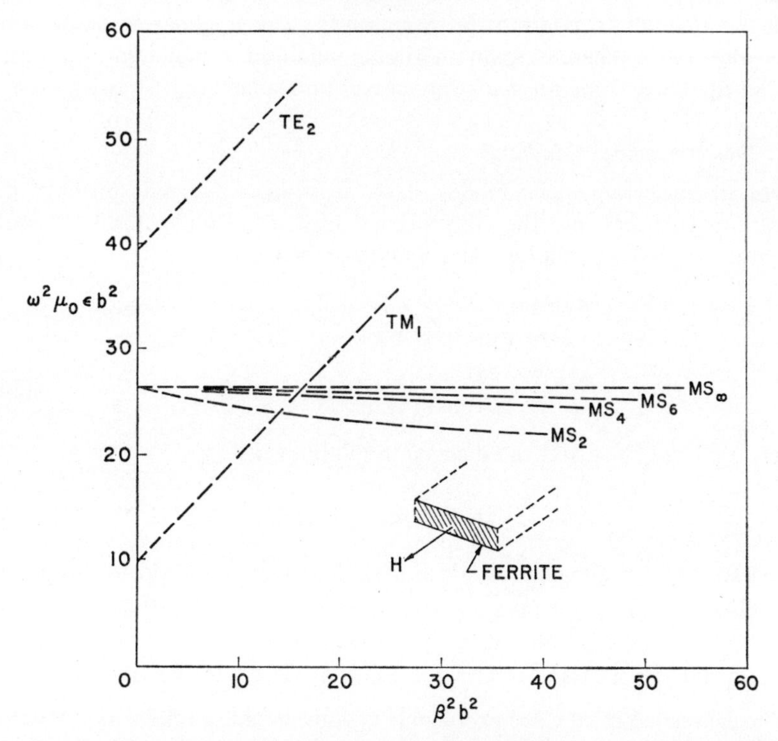

FIG. 3. Dispersion curves for uncoupled electromagnetic and magnetostatic modes on parallel plane waveguide.

where μ is the diagonal element of the permeability tensor. From (19b), propagation occurs only within the region of negative μ, that is, within the frequency band $\omega_H \rightarrow [\omega_H(\omega_H + \omega_M)]^{1/2}$; and the modes are all backward waves because of the frequency dependence of μ in this region (see Fig. 3).

In the specific examples discussed above, all of the magnetostatic modes are backward waves. A simple physical argument shows that this is always the case for an axially magnetized structure (Joseph and Schlömann, 1961). This may be understood by considering the demagnetizing field induced by the surface poles at the surface of the ferrite. At low values of γ in the propagating region, the surface charge produces an additional restoring force and raises the frequency. Since the sign of the surface poles alternates along the axis of the cylinder, the demagnetizing field tends to cancel. The additional restoring force, and change in frequency, therefore decreases as γ increases and the cancellation becomes more complete. It should not be concluded from this, however, that magnetostatic modes are always backward waves. For the completely filled *transversely* biased rectangular guide it is again a simple matter to find the magnetostatic modes, which are identical with the anomalous modes of propagation for a rectangular guide described by Seidel (1957). There is again an infinite manifold of magnetostatic waves; but, in this case, some are backward waves and some are forward waves.

3.2. *Electromagnetic Modes*

For the electromagnetic modes of the isotropic waveguide problem it is most convenient to use the conventional representation in terms of scalar functions θ_N and ψ_N (Bahiana and Smullin, 1960),

$$\left.\begin{aligned}(\nabla_t^2 + p_N^2)\,\theta_N &= 0\\ \theta_N &= 0 \text{ over wall}\end{aligned}\right\}\ \text{E-modes} \tag{20a}$$

$$\left.\begin{aligned}(\nabla_t^2 + p_N^2)\,\psi_N &= 0\\ \mathbf{n}\,.\,\nabla_t\psi_N &= 0 \text{ over wall}\end{aligned}\right\}\ \text{H-modes} \tag{20b}$$

where $p_N^2 = \omega_{cN}^2\,\mu_0\epsilon$ and the dispersion relations are

$$\gamma_N^2(\omega) = p_N^2 - \omega^2\mu_0\epsilon. \tag{21}$$

By contrast with the magnetostatic modes, these modes are spread throughout the entire $\omega^2 - \gamma^2$ plane (Fig. 3) and are always forward waves.

4. DISPERSION CURVES FOR THE COUPLED MODES

The general coupled wave problem is of considerable complexity. However, certain simplifications are possible. In the first place there is no coupling between pairs of electromagnetic waves. For the case of axial magnetization there is, in addition, no coupling between pairs of magnetostatic modes. Furthermore, since waveguide structures usually possess some degree of symmetry, the electromagnetic and magnetostatic modes may be classified in symmetry species according to their transformation properties with respect to the symmetry operators of the structure. It may then be stated immediately that there is no coupling between modes of different species.

For most waveguide structures electromagnetic modes of the same symmetry species are either degenerate or widely spaced in the ω^2, γ^2 plane (Fig. 3), and for many common structures, such as the parallel plane waveguide, there is no degeneracy among modes of the same symmetry type. In general, the dispersion curves of a pair of waves are distorted by coupling only in the vicinity of the cross-over. When the electromagnetic modes are widely spaced, one may therefore, by way of approximation, treat separately the interaction of each electromagnetic mode with the magnetostatic modes. The same device cannot be used in the case of the magnetostatic modes, since they are closely spaced in the ω^2, γ^2 plane.

For an axially-magnetized structure the approximate characteristic equation then takes the form

$$\gamma^2 - (\omega_{cN}^2 - \omega^2)\,\mu_0\epsilon = \sum_\nu \frac{K_{N\nu}\bar{K}_{\nu N}}{\omega - \omega_\nu} \tag{22}$$

where, assuming normalized modes,

$$K^e_{N\nu}\bar{K}^e_{\nu N} = -\frac{\omega_M(\omega^2\mu_0\epsilon)\,\omega_\nu^2\gamma^4}{[\mu(\omega_\nu)]^2}\left|\int_{\text{ferrite}} \phi_\nu\theta_N^*\,ds\right|^2 \tag{23a}$$

for an E-mode interaction, and

$$K^h_{N\nu}\bar{K}^h_{\nu N} = \frac{\gamma^2(\omega_{cN}^2\mu_0\epsilon - \gamma^2)\,(\omega_H^2 - \omega_\nu^2)^2\,\omega^2\mu_0\epsilon}{\omega_M}\left|\int_{\text{ferrite}} \phi_\nu\psi_N^*\,ds\right|^2 \tag{23b}$$

for an H-mode interaction. With transverse magnetization, the approximate secular equation has a more complicated form because of the coupling between the magnetostatic modes.

For the case of axial magnetization the shape of the dispersion curves in the vicinity of each electromagnetic cross-over may then be determined by assuming values of γ in (23) and calculating ω by means of a simple graphical construction (Fig. 4). It is clear from the form of the equations that there will be single solutions, lying respectively above and below the magnetostatic manifold, and a solution lying between each pair of magnetostatic dispersion curves within the manifold. From considerations such as these the general shape of the dispersion curves in the vicinity of each interaction region may be sketched without actual calculation, and the curves for the different interaction regions joined together as smoothly as possible. In this way the qualitative effects on the propagation characteristics of changes in parameters may be estimated in a simple fashion.

Figure 5 illustrates the results of a graphical calculation of the dispersion curves for the modes of even symmetry on a parallel plane transmission line. The electromagnetic interactions have been treated separately and only three magnetostatic modes have been retained in the calculation. Since the curves for the two interaction regions do not join smoothly, it is clear that the approximation involved in treating the ineractions separately is not particularly good in this case. For more accurate results a calculation including both interactions would be required, and the graphical method outlined above would not be applicable. The curves shown in Fig. 5 do, however, illustrate the general character of the dispersion curves.

In the vicinity of cross-over, the uncoupled dispersion curves tend to repel each other, as is usual in coupled systems. Some branches of the coupled curves exhibit regions of opposite slope, with forward wave propagation for small values of β and backward wave propagation for large values of β. This effect has been described by Suhl and Walker (1954) for the case of the cir-

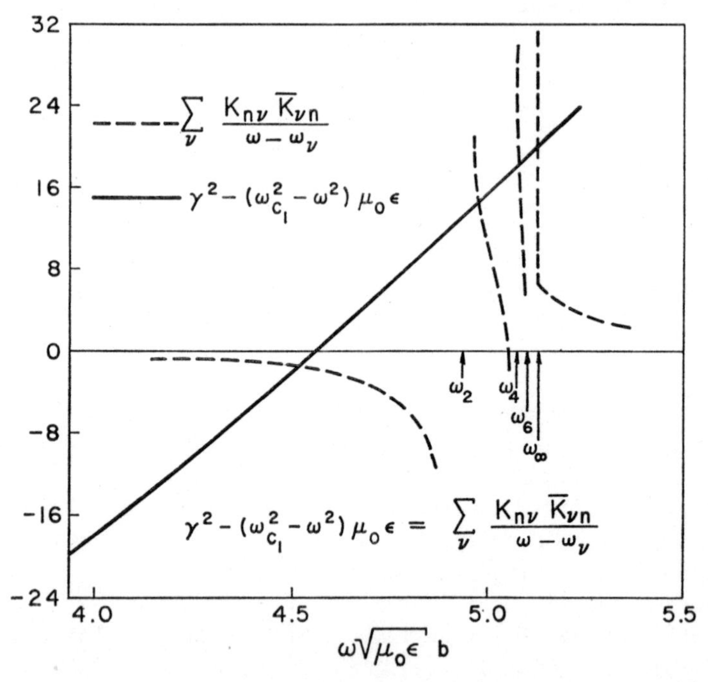

FIG. 4. Graphical solution of the characteristic equation for parallel plane waveguide [$(\gamma b)^2=10\pi^2/9$].

cular waveguide. The points of zero slope may be regarded as common points of cutoff, with finite β, for pairs of forward and backward waves. At frequencies above such cutoff points the cutoff waves should have complex propagation factors, as has been discussed by Tai (1960). A point of apparent cutoff at a finite value of β appears at the top of the magnetostatic manifold, where $\omega = \omega_H(\omega_H + \omega_M)$ and $\mu = 0$. At this point the dispersion curve of the coupled wave appears to terminate abruptly. Actually, this is the point at which coupling takes place with the magnetostatic modes of infinitely high order. Since the coupling to these waves is vanishingly small, the effect of such coupling is not apparent until the point at which cross-over actually occurs.

For this structure the TE_2-mode couples directly only to the MS_2-mode. This coupling accounts for the pushing down of the MS_2-curve and the raising of the TE_2-curve for small values of β. At the point $\beta = 0$ the coupling vanishes and the curves return to their unperturbed positions. As the TE_2 cutoff frequency is lowered by increasing the waveguide dimensions, the MS_2-curve "sticks" at $\omega = \omega_H(\omega_H + \omega_M)^{1/2}$ until the TE_2-cutoff passes

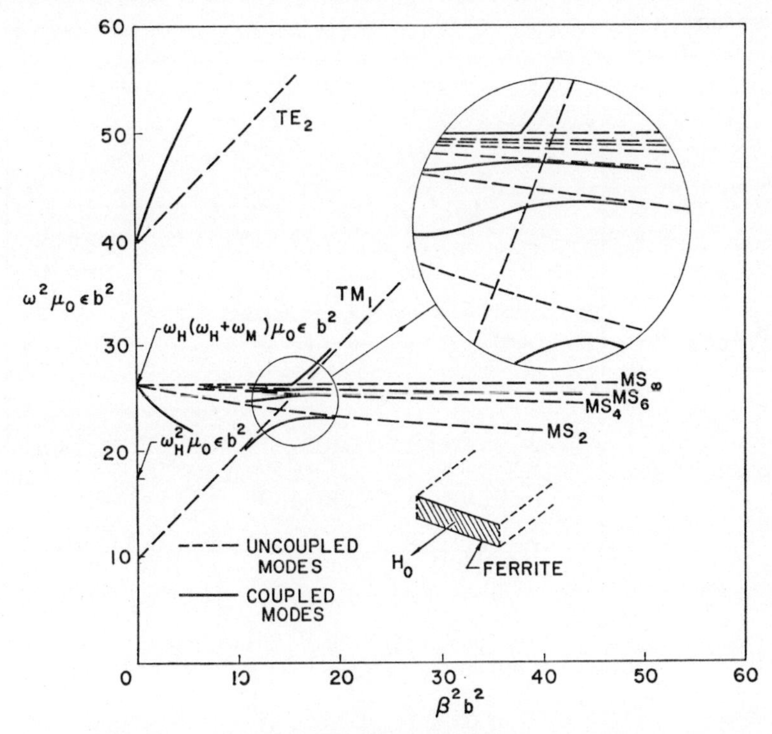

FIG. 5. Dispersion curves for ferrite-filled parallel plane waveguide, calculated by coupled mode theory.

through this frequency. At this point the MS_2-curve begins to move down with the TE_2-cutoff, and the upper branch "sticks" at $\omega = \omega_M(\omega_H + \omega_M)^{1/2}$. Further increase in the waveguide dimensions brings the TM_3-cutoff down toward the manifold. Since this mode couples to all static modes, the static dispersion curves are all depressed in the vicinity of $\beta = 0$. As the TM_3-cutoff passes through the manifold, the lowest static mode is "peeled" off the manifold and the remaining curves in the manifold shift over one step at large values of β, as in the case of the TM_1-interaction illustrated in Fig. 5.

In the Suhl and Walker (1954) treatment of propagation in gyromagnetic waveguides the concept is developed of a "reservoir" of *incipient* modes, from which the higher modes are drawn as the waveguide dimensions are increased. A mode is said to be *incipient* at a particular frequency if propagation occurs at that frequency only in the presence of gyromagnetic anisotropy, and becomes *fully developed* when the electromagnetic cutoff frequency passes through the particular frequency in question. From the above description it is clear that the modes become fully developed as they are "peeled off" the manifold and that the "reservoir" of incipient modes is just the magnetostatic manifold.

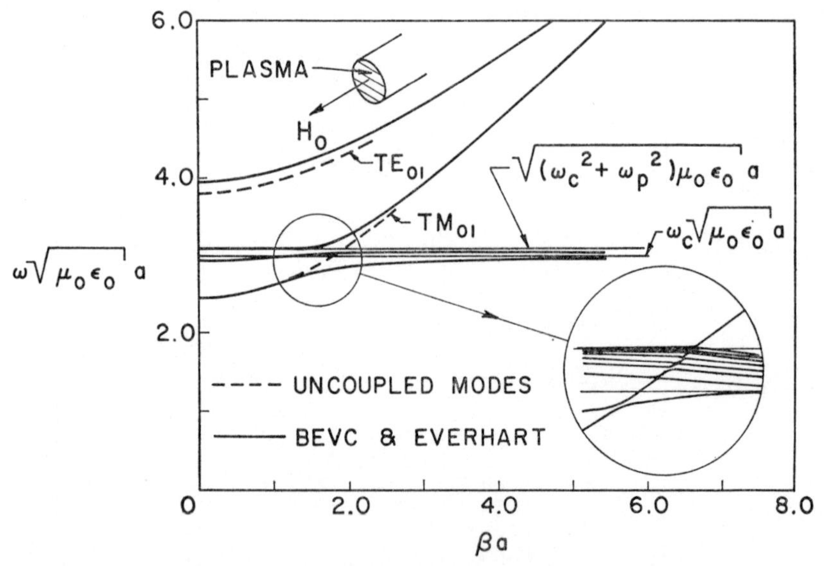

FIG. 6. Dispersion curves for completely-filled circular plasma waveguide, calculated by the direct method (Bevc and Everhart, 1961).

The coupled-mode formalism developed here is also applicable, with some modification, to the problem of axially-magnetized plasma waveguides. In Fig. 6 are shown the dispersion curves for a completely-filled circular plasma waveguide, calculated by the conventional direct method (Bevc and Everhart, 1961). In this problem the coupled mode theory predicts direct coupling of the TM_{01}-mode to the lowest order static mode only; and the multiple splittings in the cross-over region (Fig. 6) will therefore not be predicted when only the TM_{01}-interaction is included (Eidson, 1962). It has been pointed out by Eidson, however, that the coupled-mode theory will predict multiple splittings in the TM_{01}-cross-over region if the TE_{01}-interaction is also included in the analysis. The TM_{01}-mode is then coupled to the higher order static modes by way of the TE_{01}-mode, which couples directly with *all* of the static modes. A similar remark applies to the TE_2-interaction in the ferrite problem discussed previously, where the TE_2-mode interacts indirectly with *all* of the static modes by way of the TM_1-and TM_3-modes.

5. CONCLUSIONS

A formalism has been developed for treating propagation in axially- or transversely-magnetized ferrite waveguides in terms of coupling between electromagnetic and magnetostatic modes. This approach to the problem permits one to develop a physical insight into the origins and interrelation of a variety of anomalous effects which have been reported in the literature. It is also possible to deduce without elaborate computations the qualitative effect on the propagation characteristics of changes in the waveguide parameters, a feature which should be useful in practical device design. In some cases it is necessary to take into account the joint interaction with the magnetostatic manifold of two or more electromagnetic modes. The extent to which this impairs the usefulness of the principle of mode coupling remains yet to be determined.

ACKNOWLEDGEMENT

The author would like to express his appreciation to E. Schlömann, H. J. Shaw, and J. C. Eidson for many helpful discussions.

APPENDIX

Derivation of Walker-type Orthogonality Relations for Magnetostatic Modes in Closed Ferrite Waveguides

For two modes of an axially-magnetized waveguide having the same *real* γ^2, we have

$$\left.\begin{aligned} \nabla_t^2\Phi_\lambda + \gamma^2\Phi_\lambda = 0 \\ \nabla_t^2\Phi_\nu^* + \gamma^2\Phi_\nu^* = 0 \end{aligned}\right\} \text{outside the ferrite}$$
$$\left.\begin{aligned} \nabla_t^2\Phi_\lambda + \gamma^2\Phi_\lambda + \nabla_t \,.\, \mathbf{M}_\lambda = 0 \\ \nabla_t^2\Phi_\nu^* + \gamma^2\Phi_\nu^* + \nabla_t \,.\, \mathbf{M}_\nu^* = 0 \end{aligned}\right\} \text{inside the ferrite} \qquad \text{(A1)}$$

where the subscripts t denote transverse operations. From this we obtain

$$\int_{\substack{\text{waveguide}\\ \text{section}}} (\Phi_\lambda \nabla_t^2 \,\Phi_\nu^* - \Phi_\nu^* \nabla_t^2 \Phi_\lambda)\, ds + \int_{\substack{\text{ferrite}\\ \text{section}}} (\Phi_\lambda \nabla_t \,.\, \mathbf{M}_\nu^* - \Phi_\nu^* \nabla_t \,.\, \mathbf{M}_\lambda)\, ds = 0.$$

By means of Green's identities and (8) and (9), this is reduced to

$$\int_{\substack{\text{ferrite}\\ \text{section}}} (\mathbf{M}_\nu^* \,.\, \nabla_t \Phi_\lambda - \mathbf{M}_\lambda \,.\, \nabla_t \Phi_\nu^*)\, ds = 0. \qquad \text{(A2)}$$

Consider now the case of fixed bias field. From (6), we have

$$\begin{aligned} \nabla_t \Phi_\lambda &= (\omega_H/\omega_M)\,\mathbf{M}_\lambda - j\omega_\lambda/\omega_M \mathbf{z} \times \mathbf{M}_\lambda \\ \nabla_t \Phi_\nu^* &= (\omega_H/\omega_M)\,\mathbf{M}_\nu^* + j\omega_\nu/\omega_M \mathbf{z} \times \mathbf{M}_\nu^* \end{aligned} \tag{A3}$$

since $\mathbf{Z} = \mathbf{z}$. Substituting in (A2)

$$-j(\omega_\lambda - \omega_\mu)/\omega_M \int_{\substack{\text{ferrite}\\ \text{section}}} \mathbf{z} \cdot \mathbf{M}_\lambda \times \mathbf{M}_\nu^*\, ds = 0.$$

This leads to

$$\int_{\substack{\text{ferrite}\\ \text{section}}} \mathbf{m}_\lambda \times \mathbf{m}_\nu^*\, ds = \mathbf{z} D_\nu \delta_{\lambda\nu} \qquad \begin{pmatrix} \omega_H = \text{constant} \\ \gamma^2 \text{ real} \end{pmatrix} \tag{A4}$$

where the z-dependence has been suppressed, and $\mathbf{m}_\lambda$, $\mathbf{m}_\nu$ are defined by (10b).

For the case of fixed frequency, we have

$$\begin{aligned} \nabla_t \Phi_\lambda &= (\omega_{H\lambda}/\omega_M)\,\mathbf{M}_\lambda - j\omega/\omega_M \mathbf{z} \times \mathbf{M}_\nu \\ \nabla_t \Phi_\nu^* &= (\omega_{H\nu}/\omega_M)\,\mathbf{M}_\nu^* + j\omega/\omega_M \mathbf{z} \times \mathbf{M}_\nu. \end{aligned} \tag{A5}$$

Substituting in (A2) yields

$$\int_{\substack{\text{ferrite}\\ \text{section}}} \mathbf{m}_\lambda \cdot \mathbf{m}_\nu^*\, ds = \mathbf{z} C_\nu \delta_{\lambda\nu} \qquad \begin{pmatrix} \omega = \text{constant} \\ \gamma^2 \text{ real} \end{pmatrix}. \tag{A6}$$

The derivation is somewhat different for the transversely magnetized problem. In this case we have

$$\int_{\substack{\text{waveguide}\\ \text{section}}} (\Phi_\lambda \nabla_t \Phi_\nu^* - \Phi_\nu^* \nabla_t^2 \Phi_\lambda)\, ds + \int_{\substack{\text{ferrite}\\ \text{section}}} \{(\Phi_\lambda \nabla_t \cdot \mathbf{M}_\nu^* - \Phi_\nu^* \nabla_t \cdot \mathbf{M}_\lambda) - (\gamma^* M_{\nu z}^* \phi_\lambda - \gamma M_{\lambda z} \phi_\nu^*)\}\, ds = 0.$$

For *propagating* modes ($\gamma = -\gamma^*$) this reduces to

$$\int_{\substack{\text{ferrite}\\ \text{section}}} (\mathbf{M}_\nu^* \cdot \nabla_T \Phi_\lambda - \mathbf{M}_\lambda \cdot \nabla_T \Phi_\nu^*)\, ds = 0, \tag{A7}$$

where the subscripts T denote operations transverse to the bias field. Substituting $\nabla_T \Phi$ from (6) results in

$$\int_{\substack{\text{ferrite}\\\text{section}}} (\mathbf{m}_\lambda \times \mathbf{m}_\nu^*)\, \mathrm{d}s = \mathbf{Z} D\nu\, \delta\lambda\nu \qquad \text{(A8)}$$

$$\begin{pmatrix} \omega_H = \text{constant} \\ \gamma^2 \text{ real and } \textit{negative} \end{pmatrix}$$

or

$$\int_{\substack{\text{ferrite}\\\text{section}}} (\mathbf{m}_\lambda \,.\, \mathbf{m}_\nu^*)\, \mathrm{d}s = \mathbf{Z} C\nu\, \delta\lambda\nu \qquad \text{(A9)}$$

$$\begin{pmatrix} \omega = \text{constant} \\ \gamma^2 \text{ real and } \textit{negative} \end{pmatrix}.$$

REFERENCES

ADLER, R. B. (1952) Waves on inhomogeneous structures, *Proc. I.R.E.* **40,** 339–348 (March).

BAHIANA, L. C. and SMULLIN, L. D. (1960) Coupling of modes in uniform, composite waveguides, *I.R.E. Trans.* **MTT–8,** 454–458 (July).

BEVC, V. and EVERHART, T. E. (1961) Fast waves in plasma-filled waveguides, Electronics Res. Lab., Univ. of Calif., Berkeley, Tech. Rep. No. 61–14 (July).

EIDSON, J. C. (1962) Unpublished memorandum, Microwave Lab., W. W. Hansen Laboratories of Physics, Stanford Univ., Stanford, Calif. (June).

GAMO, H. (1953) The Faraday rotation of waves in a circular waveguide, *J. Phys. Soc. Japan* **8,** 176–182 (March–April).

JOSEPH, R. and SCHLÖMANN, E. (1961) Theory of magnetostatic modes in long, axially-magnetized cylinders, *J. Appl. Phys.* **32,** 1001–105 (June).

KALES, M. L. (1953) Modes in waveguides containing ferrites, *J. Appl. Phys.* **24,** 604–608 (May).

LAX, B. and BUTTON, K. J. (1955) Theory of new ferrite modes in rectangular waveguide, *J. Appl. Phys.* **26,** 1184–1185 (September).

MIKAELYAN, A. L. (1954) Electromagnetic waves in a rectangular waveguide filled with a magnetized ferrite, *Doklady Akad Nauk SSSR* **98,** 941–944 .

PINCUS, P. (1962) Propagation effects on ferromagnetic resonance in dielectric slabs, *J. Appl. Phys.* **33,** 553–556 (February).

SEIDEL, H. (1957) The character of waveguide modes in gyromagnetic media, *BSTJ* **36,** 409–426 (March).

SUHL, H. and WALKER, L. R. (1954) Topics in guided wave propagation through gyromagnetic media, Part I, *BSTJ* **33,** 579–639 (May).

SUHL, H. (1957) Theory of the ferromagnetic microwave amplifier, *J. Appl. Phys.* **28,** 1225–1236 (November).

TAI, C. T. (1960) Evanescent modes in a partially filled gyromagnetic rectangular waveguide, *J. Appl. Phys.* **31,** 220–221 (January).

TRIVELPIECE, A. W. (1958) Slow wave propagation in plasma waveguides, Calif. Inst. Tech. Elec. Tube Microwave Lab., Tech. Rep. No. 7 (May).

VAN TRIER, A. A. TH. M. (1954) Guided electromagnetic waves in anisotropic media, *Appl. Sci. Res.* Sec. B, **3,** 305–371 (September).

WALKER, L. R. (1957) Magnetostatic modes in ferromagnetic resonance, *Phys. Rev.* **105,** 390–399 (January).

GYROMAGNETIC RESONANCES OF THICK FERRITE SLABS EXCITED IN A TRANSVERSE ELECTRIC MODE

HAROLD SEIDEL

Bell Telephone Labs., Murray Hill, N.J.

SUMMARY

In an earlier paper[1] the author considered the observation of resonance of thick ferrite slabs in a parallel plane waveguide to be primarily identified with the appearance of singularities of the propagation constant. It is the purpose of the present paper to review some of the earlier results but with extensions and new results not published previously.

The analyses of gyromagnetic resonance phenomena have generally concerned themselves with perturbation viewpoints[2, 3] in which the rf magnetization is assumed constant within the cross-section of the material. This approach is compatible with that employed in the study of paramagnetic resonance in which the material is very weakly magnetic and reacts weakly, in turn, back on the exciting field. The incident rf magnetic field is then substantially unaffected by the resonance of the medium and the observation of maximum absorption coincides with the intrinsic resonance of the paramagnet.

Interaction of the electromagnetic field with a ferrimagnetic material, on the other hand, produces an extremely strong reaction field. If the rf magnetic field were to penetrate into an ideally zero line width material at an intrinsic resonance condition, an infinite magnetic moment would result, which clearly does not meet the boundary conditions of the guide. A magnetic skin depth therefore results which is related to the line width, with relatively little of the incident energy being permitted to enter into the gyromagnetic medium.

Observation of resonant absorption in a waveguide system is to be identified either with resonant input scattering or with the appearance of resonant imaginary terms in the propagation constant. Since experimental observation yields resonant attenuation substantially proportional to the length of the medium, the resonance must then be identified with a real propagation constant singularity for the ideally zero line width medium. The actual ferrimagnetic line width, phrased in terms of a small additive imaginary frequency, provides the proper picture of an absorption rate inverse to the line width together with an appropriately related observed absorption width for the guided structure.

The peak in absorption is to be distinguished from the intrinsic resonance of the medium, for the reasons indicated above and, therefore, occurs for finite wavenumbers within the medium. Under such circumstance, a singularity of the longitudinal component of the wavenumber vector carries with it an equal magnitude transverse component in phase quadrature. Waveguide resonance for gyromagnetic media therefore corresponds to steeply bound surface waves, or boundary layers, decaying away from transverse interfaces.

Thick slab resonance is then characterized by the features of the interface and is unrelated to the bulk properties of the medium, if these properties deviate from those of the surface. Further, if the gyromagnetic interface is separated by the order of a few skin depths from other interfaces in the guide, the appearance of the resonance is independent of the general features of the guide, and becomes an intrinsic property of the interface alone. The need for a physical interface to bind the surface waves holds true even in the event of non-uniform magnetization of the medium. It is found that a continuous field variation, including that which passes through a susceptibility resonance internal to the medium, is inadequate to form a boundary layer.

A sensitive means for displaying the critical nature of the interface in the observation of waveguide resonance is found in double slab experiments. An interface is formed between uniformly magnetized slabs of unequal magnetizations, and resonant absorption properties are studied. The results are then contrasted to the absorption properties when the slabs are separated. It is found, in reasonable accordance with prediction, that the resonances are non-reciprocal, being drawn closer together in one direction of propagation than the other. Had the resonant absorption process been a bulk phenomenon, this would not have been true since the medium susceptibility is not a function of the direction of propagation.

The non-reciprocal features of resonance at an interface are seen simply in examining the condition for the propagation constant singularity at the double slab interface. This is found to be, in a given propagation direction,

$$\mu_1 + \mu_2 + (k_1 - k_2) = 0 \tag{1}$$

where the two subscripts relate to the two slabs and where μ and k are the Polder tensor components. Since eqn (1) is sensitive to the sign of either k, a reversal of the magnetic field, which is equivalent to a reversal of the direction of propagation, changes the equation. Equation (1) is implicitly quadratic in frequency, providing two solutions for resonance. These solutions differ with the sign of k and provide for the shift of separation indicated above.

Reinforced by the experimental confirmation of the boundary-layer nature of the resonant distribution, the characteristics of the singularity may be examined more critically. It is found, in the case of uniform magnetization that the singularity has a branch point with respect to frequency and is thus onesided. This leads not only to an asymmetry of the line shape of the absorption but leads also to ambiguity in the interpretation of line width. A more interesting result stems from the investigation of propagation constant singularities in transversely non-uniformly magnetized slabs. It is found there that singularities appear at the surface condition of $\mu^2 = k^2$, a result in no way anticipated from analysis of the uniformly magnetized sample. This follows from the one dimensional transverse wave equation which is

$$\left\{\frac{d^2}{\mathrm{d}x^2} - P\,\frac{\mathrm{d}}{\mathrm{d}x} + [Q\beta - \beta^2 + R]\right\} E_z = 0\,, \tag{2}$$

where

E_z is the transverse electric field

$$P = \frac{\mathrm{d}}{\mathrm{d}x}\ln\frac{\mu^2 - k^2}{\mu}$$

$$Q = -\frac{1}{\mu}\left[\frac{\mathrm{d}k}{\mathrm{d}x} - \frac{\mathrm{d}}{\mathrm{d}x}\ln\,(\mu^2 - k^2)\right]$$

$$R = \left[\frac{2\pi}{\lambda}\right]^2 \frac{\mu^2 - k^2}{\mu}$$

λ = free space wavelength in a medium of the gyromagnetic dielectric constant.

Zeros in $\mu^2 - k^2$ or in the μ produce poles in the coefficients of (2) and lead to a hypergeomagnetic equation. Analysis demonstrates, however, that only the zero in $\mu^2 - k^2$, and not in μ, leads to a singularity condition in β, the propagation constant, and forms a second-order pole. It is of interest that the poles in P and Q appear as logarithmic derivates of $\mu^2 - k^2$ and therefore have residues independent of the degree of non-uniformity, or the rate at which $\mu^2 - k^2$ passes through zero. Experimental observation has shown large absorptions at this condition in double slab configurations. Since the slab composite differs markedly from a uniform ellipsoid, field non-uniformity exists at the interface and produces this new resonance situation.

The analysis of the inhomogeneously magnetized slab tends to emphasize the nature of dipolar narrowing in a low line width gyromagnetic medium. Since singularities are isolated in the zero line width case, there can be no broadening of the line. It is only as the line width increases that discrete lines broaden and merge in inhomogeneous fields.

The analysis in this paper is primarily concerned with TE-modes and shape independent resonances. There are many other mechanisms producing resonances related to interference effects and higher mode coupling. Since there exists sets of very slowly propagating modes, many possible resonant situations can be suggested. Under these circumstances, this paper cannot be considered a complete phenomenological description of resonance, but it is hoped that it does form some inroads to its understanding.

REFERENCES

1 Seidel, H. (1957) *I.R.E. WESCON Convention Record*, pp. 58–69 (August).
2 Kittel, C. (1948) *Phys. Rev.* **73,** 155 (January).
3 Sensiper, S. (1956) *Proc. I.R.E.* **44,** 1323 (October).

ON THE POSSIBILITY OF INTRINSIC LOSS OCCURRING AT THE EDGES OF FERRITES

R. A. HURD

Radio and Electrical Engineering Division, National Research Council, Ottawa, Canada

SUMMARY

Diffraction problems involving ferrites are usually solved under two simplifying assumptions:

(1) the a.c. magnetic field components in the ferrite are much smaller than the static magnetic field H;
(2) the ferrite damping constant α, which is introduced in an empirical way into the equation of motion of the electrons in the ferrite, is assumed to be small.

Approximations based on (1) and (2) lead to a permeability matrix of the form

$$(\mu) = \begin{pmatrix} \mu & -j\kappa & 0 \\ j\kappa & \mu & 0 \\ 0 & 0 & \mu_0 \end{pmatrix} \tag{1}$$

with

$$\mu \pm \kappa = 1 + \frac{\gamma M}{\gamma H \mp \omega + j\alpha} \tag{2}$$

where γ is a constant and M is the saturation magnetization of the ferrite.

Under these assumptions we show that an intrinsic loss may occur in some diffraction problems. By intrinsic loss is meant an energy loss which, in a finite region, tends to a non-zero value as the damping constant tends to zero.

To investigate this phenomenon we consider the behavior of the electromagnetic field in the neighborhood of the common edge where three sectors meet. Sector I, from $\phi = 0$ to $\phi = \phi_1$, is composed of a ferrite characterized by constants ϵ, μ, κ, α. Sector II, occupying $\phi = \phi_1$ to $\phi = \phi_2$, comprises an isotropic region with constants ϵ_0, μ_0. The remaining sector from $\phi = \phi_2$ to $\phi = 2\pi$ is a perfect conductor. It is assumed that the static magnetic field H applied to the ferrite lies in a direction parallel to the edge (the z-axis).

Only fields with no z-variation are considered. Not dealt with are fields derivable from H_z since these are unaffected by the anisotropy. The problem then reduces to considering fields with E_z, H_ρ, H_ϕ only.

The electric field in the ith region is expanded in the form

$$E_z^{(i)} = \rho^t[A_0^{(i)} + \rho A_1^{(i)} + \rho^2 A_2^{(i)} + \dots] \tag{3}$$

ρ being the distance from the edge. The leading term of this expression dominates near the edge. From Maxwell's equations this term is found to be

$$A_0^{(i)} = a_0^{(i)} \cos t\phi + b_0^{(i)} \sin t\phi. \tag{4}$$

The imposition of the boundary conditions yields a transcendental equation for the determination of t,

$$\frac{1}{\mu_0} \cot t(\phi_1 - \phi_2) - \frac{\mu}{\Delta} \cot t\phi_1 = -j\kappa/\Delta \tag{5}$$

where $\Delta = \mu^2 - \kappa^2$.

An adequate discussion of the general solution of eqn (5) would be quite difficult. Instead we choose the particular case which occurs when $\phi_2 = 2\phi_1$. For this special case,

$$t = \pi p/\phi_1 + n\pi/\phi_1; \qquad n = 0, \pm 1, \pm 2, \dots \tag{6}$$

where

$$p = -\frac{j}{2\pi} \log \frac{(\kappa + \mu)(\mu_0 + \mu - \kappa)}{(\kappa - \mu)(\mu_0 + \mu + \kappa)}. \tag{7}$$

From eqns (2) and (7) we find that

$$0 < Re(p) < 1 \tag{8}$$

as H or ω varies, providing that $\alpha \neq 0$. The value $\mathrm{Re}(p) = 0$ is approached as $\alpha \to 0$ for $\gamma(H + \frac{1}{2}M) < \omega < \gamma(H + M)$, assuming that H and M are positive. This is called the critical range of ω. Outside this range of ω, for H, $M > 0$, $\mathrm{Re}(p)$ is approximately 1/2. At $\omega = 0$ we find that $p = 1/2$. However the ferrite becomes isotropic at $\omega = 0$; and in this case the lowest permissible eigenvalue has been given by Meixner[1] as $t = \pi/2\phi_1$. Using these facts, we establish that the correct lowest permissible eigenvalue in the ferrite case is $t_0 = \pi p/\phi_1$.

In the limit $\alpha = 0$, and for ω in the critical range, we then have $E_z = 0(1)$ and $H_\phi = 0(\rho^{-1})$ near the edge. This opens up the possibility that intrinsic loss (or gain) may occur.

This loss (or gain) is calculated by integrating the real part of the radial complex Poynting vector around a circle enclosing the edge and letting the circle shrink to zero. When $\mathrm{Re}(p) = 0$ this gives a non-zero result. It is found that the answer always represents power flow *towards* the edge, and hence there is power loss but not gain.

The loss at the edge appears to be intimately connected with the existence of a Bresler wave[2] at the ferrite-metal boundary; in that it occurs for the same range of parameters. Indeed the edge-loss can be regarded as the energy of excitation of the Bresler wave which, because of its infinite phase constant, must be regarded as non-propagating. Thus loss at the edge provides an alternate resolution of the thermodynamic paradox[2, 3] in ferrite-loaded waveguide.

Another energy balance paradox occurs in the problem of input to a ferrite-loaded waveguide, as first noted by Lewin.[4, 5] It has been shown that intrinsic loss at an edge of the ferrite-air interface is just sufficient to restore the power balance.

Finally, we note that the infinite magnetic field at the edge violates assumption (1) so that intrinsic loss probably does not occur in a real ferrite. An accurate analysis would include non-linear effects, and would be difficult to carry out. One should note that the non-linearity would give rise to frequency multiplication and conversion to non T.E.-modes, thus affording mechanisms for resolving the above-mentioned paradoxes.

REFERENCES

1 Meixner, J. (1954) The behavior of electromagnetic fields at edges, New York University, Institute of Mathematical Sciences, Res. Report No. EM-72, (December).

2 Bresler, A. D. (1960) Modes of a ferrite-loaded waveguide, *I.R.E. Trans. on Microwave Theory and Techniques*, **MTT–8**, 81 (January).

3 Button, K. J. and Lax, B. (1956) Theory of ferrites in rectangular waveguide, *I.R.E. Trans. on Antennas and Propagation*, **AP-4,** 531 (July).

4 Lewin, L. (1959) A ferrite boundary-value problem in a rectangular waveguide, *Proc. I.R.E.* **106** B, 559 (November).

5 Lewin, L. (1961) The part played by surface waves on the reflection at a ferrite boundary, *Proc. I.E.E. Monographs* 108C, 359.

A MODAL SOLUTION FOR A RECTANGULAR GUIDE LOADED WITH LONGITUDINALLY MAGNETIZED FERRITE†

GIORGIO BARZILAI and GIORGIO GEROSA

Istituto di Elettronica dell'Università di Roma, Roma, Italy

ABSTRACT

A modal solution for a rectangular guide partially or completely filled with a slab of longitudinally magnetized ferrite is given as a superposition of modes of the parallel plate guide loaded with ferrite magnetized in a direction parallel to the walls. A numerical case relative to a guide completely filled with longitudinally magnetized ferrite is solved. Diagrams showing the field distribution are given.

1. INTRODUCTION

The problem of the electromagnetic wave propagation in a rectangular guide partially or completely filled with a slab of ferrite longitudinally magnetized has not been, to our knowledge, solved exactly up to date.

The difficulties encountered in the solution of problems of this type arise from the lack of reflection symmetry possessed, with respect to certain directions, by an electromagnetic wave traveling in a gyromagnetic medium. To appreciate this let us consider a parallel plate guide partially or completely filled with ferrite magnetized in a direction parallel to the walls. Exact modal solutions for such a structure can be found[1–3]. Such solutions possess reflection symmetry along the direction of the d.c. magnetic field; therefore the corresponding solutions for the rectangular guide, obtained by closing the structure with two metallic plates normal to the axis of the d.c. magnetic field, can be immediately found[1–3]. If however we try to close the structure with two metallic plates parallel to the d.c. magnetic field, the lack of reflection symmetry in the direction normal to the d.c. magnetic field makes the problem difficult since now it is necessary to solve a discontinuity problem.

The purpose of this work is to solve such a discontinuity problem by superimposing an infinite number of modes of the parallel plate guide.

† The research reported in this document has been sponsored by Cambridge Research Laboratories, OAR, through the European Office, Aerospace Research, United States Air Force, under Contract AF61(052)–101.

2. THE DISCONTINUITY PROBLEM

Let us consider a parallel plate guide partially filled with a slab of lossless ferrite magnetized in a direction parallel to the walls, as represented in Fig. 1. Perfectly conducting walls are assumed.

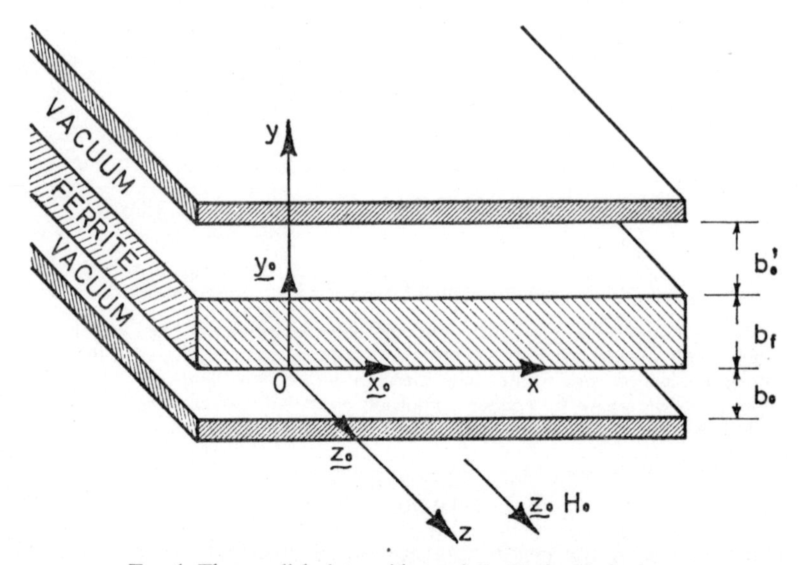

FIG. 1. The parallel plate guide partially filled with ferrite.

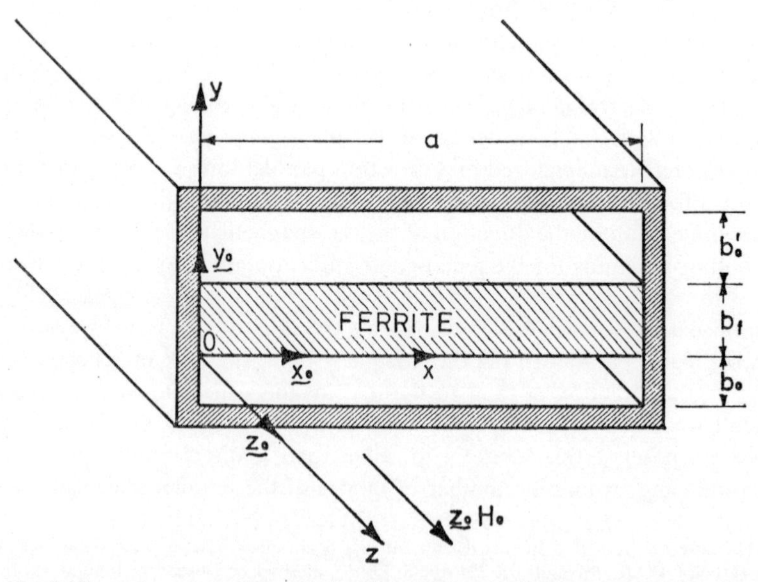

FIG. 2. The rectangular guide partially filled with longitudinally magnetized ferrite.

The spectrum of modes for this structure and for the particular cases obtained by letting $b_0' = 0$, or $b_0' = b_0 = 0$ has been discussed in previous works[1–3]. A typical mode has the following expression (time dependence $\exp[j\omega t]$)

$$\left\{\begin{matrix} \mathbf{E} \\ \mathbf{H} \end{matrix}\right\} = P \left\{\begin{matrix} \mathbf{E}(y) \\ \mathbf{H}(y) \end{matrix}\right\} \exp[j(k_x x + k_z z)]. \tag{1}$$

By closing the structure of Fig. 1 with two perfectly conducting planes normal to the x-axis at a distance a apart we shall obtain the rectangular guide partially filled with longitudinally magnetized ferrite represented in Fig. 2.

Our problem is to find for the structure of Fig. 2 an unattenuated propagating modal solution, i.e. an electromagnetic field having $\exp[jk_z z]$ as spatial dependence along the z-axis (k_z real).

We shall try to express such a modal solution as a superposition of modes relative to the structure of Fig. 1 corresponding to a given k_z.

We shall write therefore

$$\mathbf{E}(x, y) \exp[jk_z z] = \sum_{s=1}^{\infty} P_s \mathbf{E}_s(y) \exp[j(k_{xs} x + k_z z)] \tag{2}$$

$$\mathbf{H}(x, y) \exp[ik_z z] = \sum_{s=1}^{\infty} P_s \mathbf{H}_s(y) \exp[j(k_{xs} x + k_z z)]. \tag{3}$$

It can be proved that the following orthogonality relations are satisfied by the modes of the expansion (2)–(3),

$$\int_{-b_0}^{b_f+b_0'} (\mathbf{x}_0 \times \mathbf{E}_s \,.\, \mathbf{H}_r^* + \mathbf{H}_s \times \mathbf{x}_0 \,.\, \mathbf{E}_r^*)\, dy = 0 \qquad \text{if } k_{xs} \neq k_{xr}^* \tag{4}$$

where the asterisk indicates the complex conjugate.

The boundary conditions to which the field must satisfy are the following,

$$\mathbf{x}_0 \times \mathbf{E}(0, y) = 0 \tag{5}$$

$$\mathbf{x}_0 \times \mathbf{E}(a, y) = 0. \tag{6}$$

We shall now transform conditions (5)–(6) by using the expansion (2)–(3) and the orthogonality relations (4).

From (2)–(3), by virtue of (4), we obtain

$$\int_{-b_0}^{b_f+b_0'} (\mathbf{x}_0 \times \mathbf{E} \,.\, \mathbf{H}_r^* + \mathbf{H} \times \mathbf{x}_0 \,.\, \mathbf{E}_r^*)\, dy \exp[-jk_{xr}^* x] = P_{r*} \psi_{rr*} \tag{7}$$

where

$$\psi_{rr*} = \int_{-b_0}^{b_f+b_0'} (\mathbf{x}_0 \times \mathbf{E}_{r*} \,.\, \mathbf{H}_r^* + {}_r\mathbf{H}_* \times \mathbf{x}_0 \,.\, \mathbf{E}_r^*)\, dy\,,$$

and the subscript r^* refers to the mode such that

$$k_{xr^*} = k_{xr}^*.$$

In particular, for $x = 0$, taking into account (5) and (3), (7) becomes

$$\sum_{s=1}^{\infty} \phi_{rs} P_s = 0 \qquad r = 1, 2, \ldots \tag{8}$$

where

$$\left.\begin{aligned} \phi_{rs} &= \int_{-b_0}^{b_f+b_0'} \mathbf{H}_s \times \mathbf{x}_0 \,.\, \mathbf{E}_r^* \mathrm{d}y \qquad \text{for } s \neq r^* \\ \phi_{rr^*} &= \int_{-b_0}^{b_f+b_0'} \mathbf{H}_{r^*} \times \mathbf{x}_0 \,.\, \mathbf{E}_r^* \mathrm{d}y - \psi_{rr^*} = \int_{-b_0}^{b_f+b_0'} E_{r^*} \times \mathbf{x}_0 \,.\, \mathbf{H}_r^* \,\mathrm{d}y. \end{aligned}\right\} \tag{9}$$

By virtue of the orthogonality relations (4) it can be immediately seen that

$$\phi_{sr} = -\phi_{rs}^* \qquad \text{if } s \neq r^*. \tag{10}$$

Similarly, for $x = a$, taking into account (6) and (3), (7) becomes

$$\sum_{s=1}^{\infty} \phi_{rs} \chi_{rs} P_s = 0 \qquad r = 1, 2, \ldots \tag{11}$$

where

$$\chi_{rs} = \exp\,[j(k_{xs} - k_{xr}^*)\,a]. \tag{12}$$

It should be noted that

$$\chi_{sr} = \chi_{rs}^*. \tag{13}$$

To solve our problem it is necessary to determine the amplitudes P in such a way as to satisfy the eqns (8)–(11). We note that both (8) and (11) are systems of an infinite number of equations in an infinite number of unknowns. In each of these systems the unknowns are the amplitudes P of our expansion (2)–(3). In general therefore, once k_z has been fixed, not any arbitrary a will be permissible but, if the problem has a solution, there will be particular values of a such that both (8) and (11) can be solved. Before studying the anisotropic cases it is interesting to see how equations (8) and (11) become for the well known case of the guide filled with lossless isotropic homogeneous medium.

3. SOLUTION FOR THE ISOTROPIC CASE

In this case the modes of the parallel plate guide possess reflection symmetry along the x-direction and k_x^2 is real. By virtue of these properties the orthogonality conditions (4) become

$$\int_0^b \mathbf{H}_s \times \mathbf{x}_0 \,.\, \mathbf{E}_r^* \mathrm{d}y = 0 \qquad \text{if } k_{xs}^2 \neq k_{xr}^2 \tag{14}$$

where b is the height of the guide.

The conditions (14) will make zero all the coefficients ϕ_{rs} when $k_{xs} \neq \pm k_{xr}$, and therefore both systems (8) and (11) will break into an infinite number of systems each one constituted by two equations in two unknowns.

In view of the reciprocity of the structure it is now convenient to use equal and opposite subscripts to refer to modes having equal and opposite propagation constants k_x. Such modes will be put consecutively in the set. The sums of equations (8) and (11) shall be therefore extended from ± 1 up to $\pm\infty$.

Operating in this way, the system (8) becomes

$$\begin{aligned}
&\phi_{11}P_1 + \phi_{1-1}P_{-1} &&= 0\\
&\phi_{-11}P_1 + \phi_{-1-1}P_{-1} &&= 0\\
&\qquad\qquad \phi_{22}P_2 + \phi_{2-2}P_{-2} &&= 0\\
&\qquad\qquad \phi_{-22}P_2 + \phi_{-2-2}P_{-2} &&= 0\\
&\qquad\qquad\qquad\qquad \phi_{33}P_3 + \phi_{3-3}P_{-3} &&= 0\\
&\qquad\qquad\qquad\qquad \phi_{-33}P_3 + \phi_{-3-3}P_{-3} &&= 0
\end{aligned}$$

and similarly for the system (11).

We have now, therefore, pairs of independent modes. In order to study singularly each pair, we have to consider two cases: 1st, $k_x^2 > 0$; 2nd, $k_x^2 < 0$ (the case $k_x^2 = 0$, when corresponding to a solution for a parallel plate guide, it is a TE_{on}-solution; the boundary conditions (5)–(6) are automatically satisfied and all the relative coefficients in the systems (8) and (11) vanish).

1*st case*: $k_x^2 > 0$.

In this case $r^* = r$ and from (8) and (11) we obtain

$$\left.\begin{aligned} \phi P_r - \phi^* P_{-r} = 0 \\ \phi^* P_r - \phi P_{-r} = 0 \end{aligned}\right\} \tag{15}$$

$$\left.\begin{aligned} \phi P_r - \phi^* \chi^{-1} P_{-r} = 0 \\ \phi^* \chi P_r - \phi P_{-r} = 0 \end{aligned}\right\} \tag{16}$$

where

$$\phi = \int_0^b \mathbf{E}_r \times \mathbf{x}_0 \,.\, \mathbf{H}_r^* \mathrm{d}y \tag{17}$$

$$\chi = \exp\,[2jk_x a]\,, \tag{18}$$

and a is the width of the guide along x; k_x is the positive value of k_x^2. Since in this case ϕ is real, from (15) we obtain

$$P_{-r} = P_r,$$

while (16) is simultaneously satisfied only if

$$\chi = 1\,,$$

i.e.

$$k_x = \frac{m\pi}{a} \qquad m = 1, 2, \ldots$$

2nd case: $k_x^2 < 0$.

In this case $r^* = -r$ and from (8) and (11) we obtain

$$\left.\begin{aligned} \phi^* P_r + \phi P_{-r} = 0 \\ \phi P_r + \phi^* P_{-r} = 0 \end{aligned}\right\} \tag{19}$$

$$\left.\begin{aligned} \phi^* \chi P_r + \phi P_{-r} = 0 \\ \phi P_r + \phi^* \chi^{-1} P_{-r} = 0 \end{aligned}\right\} \tag{20}$$

where ϕ is given by (17) and

$$\chi = \exp\left[-2\,|k_x|\,a\right]. \tag{21}$$

Since in this case ϕ is purely imaginary, from (19) we obtain again

$$P_{-r} = P_r,$$

but it is now impossible to simultaneously satisfy (20), since χ cannot be equal to 1.

It is therefore impossible, as it is well known, to obtain a mode for the isotropic rectangular waveguide with imaginary transverse wavenumber. It is, however, possible to obtain a solution for a semi-infinite rectangular waveguide, i.e. without the metallic wall for $x = a$, since in this case only (19) must be satisfied.

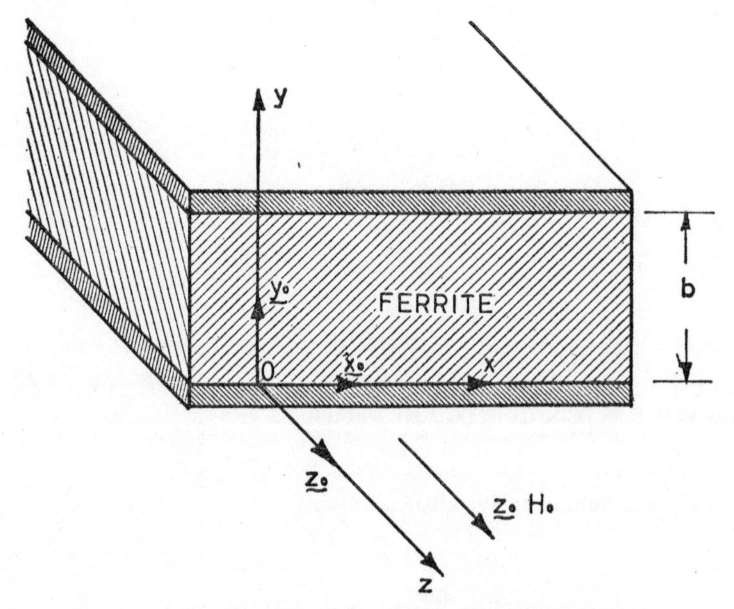

FIG. 3. The parallel plate guide completely filled with ferrite.

4. CASE OF THE RECTANGULAR GUIDE COMPLETELY FILLED WITH LONGITUDINALLY MAGNETIZED FERRITE

We shall consider now in details the case of the rectangular guide completely filled with longitudinally magnetized lossless ferrite. To this case will refer the numerical example carried out in the next section.

We must consider the spectrum of modes for the structure represented in Fig. 3. Such a spectrum has been discussed in a previous work[1].

We shall report here the analytical expression for the modes and the relative characteristics equation.

Let us assume for the ferrite a scalar dielectric constant $\epsilon_0\epsilon$ and a tensor magnetic permeability $\boldsymbol{\mu}$ given by

$$\boldsymbol{\mu} = \mu_0 \begin{vmatrix} \mu_1 & j\mu_2 & 0 \\ -j\mu_2 & \mu_1 & 0 \\ 0 & 0 & 1 \end{vmatrix}$$

where

$$\mu_1 = 1 + \frac{\rho}{1-\tau^2}, \qquad \mu_2 = \frac{\tau\rho}{1-\tau^2},$$

$$\rho = \frac{M_0}{\mu_0 H_0}, \qquad \tau = \frac{\omega}{\omega_0};$$

and M_0 is the intensity of the saturation magnetization, H_0 is the internal d.c. magnetic field of sufficient intensity to saturate the ferrite, ω and $\omega_0 = -\gamma H_0$ are the applied and the resonant circular frequencies, γ is the gyromagnetic ratio for the electron, μ_0 and ϵ_0 are the permeability and the dielectric constant of the vacuum.

The vector functions $\mathbf{E}(y)$ and $\mathbf{H}(y)$ are given by

$$\left.\begin{aligned} \begin{Bmatrix} \mathbf{E}(y) \\ \mathbf{H}(y) \end{Bmatrix} = \begin{Bmatrix} \sqrt{\left(\dfrac{\mu_0}{\epsilon_0\epsilon}\right)} \\ 1 \end{Bmatrix} \Bigg[A_1^+ \begin{Bmatrix} \mathbf{e}_1^+ \\ \mathbf{h}_1^+ \end{Bmatrix} \exp[jk_{y1}y] + \\ + A_1^- \begin{Bmatrix} \mathbf{e}_1^- \\ \mathbf{h}_1^- \end{Bmatrix} \exp[-jk_{y1}y] + \\ + A_2^+ \begin{Bmatrix} \mathbf{e}_2^+ \\ \mathbf{h}_2^+ \end{Bmatrix} \exp[jk_{y2}y] + \\ + A_2^- \begin{Bmatrix} \mathbf{e}_2^- \\ \mathbf{h}_2^- \end{Bmatrix} \exp[-jk_{y2}y] \Bigg]. \end{aligned}\right\} \quad (22)$$

The propagation constants k_x, $k_{y1,2}$, k_z, normalized with respect to

$\omega\sqrt{(\mu_0\epsilon_0\epsilon)}$ must satisfy the following characteristic equation,

$$2\mu_2^2 k_{y1}k_{y2}k_z^2\,(1-\cos k_{y1}b\cos k_{y2}b)+ \\ +(N_1k_x^2+N_2)\sin k_{y1}b\sin k_{y2}b=0, \qquad (23)$$

where

$$N_1=-(\mu_1-1)^2\frac{\mu_1^2-\mu_2^2}{\mu_1}k_z^4+ \\ +\left[2\mu_1^2(\mu_1-1)^2-2\mu_1\mu_2^2(2\mu_1-1)+2\frac{\mu_2^4}{\mu_1}(\mu_1+1)\right]k_z^2+ \\ -\mu_1^3(\mu_1-1)^2+\mu_1\mu_2^2(\mu_1-1)(3\mu_1-1)-\mu_2^4(3\mu_1-2)+\frac{\mu_2^6}{\mu_1},$$

$$N_2=-(\mu_1-1)^2k_z^6+[3\mu_1(\mu_1-1)^2-\mu_2^2(3\mu_1-1)]k_z^4+ \\ +\left[-3\mu_1^2(\mu_1-1)^2+\mu_2^2(3\mu_1-1)(2\mu_1-1)-\frac{\mu_2^4}{\mu_1}(3\mu_1+1)\right]k_z^2+ \\ +\mu_1^3(\mu_1-1)^2-\mu_1\mu_2^2(\mu_1-1)(3\mu_1-1)+\mu_2^4(3\mu_1-2)-\frac{\mu_2^6}{\mu_1}.$$

k_{y1} and k_{y2} are given by

$$k_{y1,\,2}^2=t_{1,\,2}^2-k_x^2 \qquad (24)$$

where $t_{1,\,2}^2$ are the solutions of the following equation,

$$\mu_1t^4+[(\mu_1+1)\,k_z^2-(\mu_1^2-\mu_2^2+\mu_1)]\,t^2+ \\ +k_z^4-2\mu_1k_z^2+\mu_1^2-\mu_2^2=0. \qquad (25)$$

The **e** and **h** are constant adimensional vectors whose components are given by the following expressions:

$$\left.\begin{aligned}
e_{x1,\,2}^{\pm}&=-k_z[\pm(\mu_1-1)\,k_xk_{y1,\,2}+j\mu_2(1-k_x^2)]\\
e_{y1,\,2}^{\pm}&=-k_z[k_z^2-\mu_1(1-t_{1,\,2}^2)-(\mu_1-1)\,k_x^2\mp j\mu_2k_xk_{y1,\,2}]\\
e_{z1,\,2}^{\pm}&=j\mu_2k_x(1-t_{1,\,2}^2)\pm k_{y1,\,2}[k_z^2-\mu_1(1-t_{1,\,2}^2)]\\
h_{x1,\,2}&=k_x^2[k_z^2-(1-t_{1\,2}^2)]-[k_z^2-\mu_1(1-t_{1,\,2}^2)]\\
h_{y1,\,2}^{\pm}&=\pm k_xk_{y1,\,2}[k_z^2-(1-t_{1,\,2}^2)]+j\mu_2(1-t_{1,\,2}^2)\\
h_{z1,\,2}^{\pm}&=-k_z[k_x(\mu_1-k_z^2-t_{1,\,2}^2)\pm j\mu_2k_{y1,\,2}].
\end{aligned}\right\} \qquad (26)$$

The coefficients A are given by the following expressions:

$$\left.\begin{aligned}
A_1^+&=N[e_{x1}^-e_{x2}^-e_{z2}^+(a_1^--a_2^-)+e_{x1}^-e_{x2}^+e_{z2}^-(a_2^+-a_1^-)+\\
&\qquad+e_{x2}^+e_{x2}^-e_{z1}^-(a_2^--a_2^+)]\\
A_1^-&=N[e_{x1}^+e_{x2}^-e_{z2}^+(a_2^--a_1^+)+e_{x1}^+e_{x2}^+e_{z2}^-(a_1^+-a_2^+)+\\
&\qquad+e_{x2}^+e_{x2}^-e_{z1}^+(a_2^+-a_2^-)]\\
A_2^+&=N[e_{x1}^+e_{x2}^-e_{z1}^-(a_1^+-a_2^-)+e_{x1}^+e_{x1}^-e_{z2}^-(a_1^--a_1^+)+\\
&\qquad+e_{x1}^-e_{x2}^-e_{z1}^+(a_2^--a_1^-)]\\
A_2^-&=N[e_{x1}^+e_{x2}^+e_{z1}^-(a_2^+-a_1^+)+e_{x1}^+e_{x1}^-e_{z2}^+(a_1^+-a_1^-)+\\
&\qquad+e_{x1}^-e_{x2}^+e_{z1}^+(a_1^--a_2^+)]
\end{aligned}\right\} \qquad (27)$$

where

$$\alpha^{\pm}_{1,\,2} = \exp\,[\pm\, jk_{y1,\,2}b] \tag{28}$$

and N is a proportionality factor discussed later.

In what follows we shall consider the set of modes having the same real value of k_z.

Since equation (23), taking into account (24), is a relation between k_z^2 and k_x^2 the set of modes is reciprocal, i.e. to a solution $+k_x$ always corresponds a solution $-k_x$. However there is not reflection symmetry along the x-direction, but the solution fields corresponding to $+k_x$ and $-k_x$ present specular symmetry with respect to the plane $y = b/2$; in this specular symmetry the x and y-field components change their sign, while the z-component does not change its sign, or vice versa.

For the particular case under consideration the coefficients (9) become

$$\phi_{rs} = \sum_{[\,]=1}^{4}\ \sum_{(\,)=1}^{4} \eta_{[\overset{\pm}{1,2}]\,(\overset{\pm}{1,2})rs} \tag{29}$$

where

$$\eta_{[\overset{\pm}{1,2}]\,(\overset{\pm}{1,2})rs} = A_{(\overset{\pm}{1,2})s}A^{*}_{[\overset{\pm}{1,2}]r}\Big[h_{z(\overset{\pm}{1,2})s}e^{*}_{y[\overset{\pm}{1,2}]r} + h_{y(\overset{\pm}{1,2})s}e^{*}_{z[\overset{\pm}{1,2}]r}\Big]\,\psi_{[\overset{\pm}{1,2}]\,(\overset{\pm}{1,2})rs} \quad \text{if } s \neq r^{*},$$

$$\eta_{[\overset{\pm}{1,2}]\,(\overset{\pm}{1,2})rr^{*}} = A_{(\overset{\pm}{1,2})r^{*}}A^{*}_{[\overset{\pm}{1,2}]r}\Big[e_{z(\overset{\pm}{1,2})r^{*}}h^{*}_{y[\overset{\pm}{1,2}]r} + e_{y(\overset{\pm}{1,2})r^{*}}h^{*}_{z[\overset{\pm}{1,2}]r}\Big]\,\psi_{[\overset{\pm}{1,2}]\,(\overset{\pm}{1,2})rr^{*}}$$

and

$$\psi_{[\overset{\pm}{1,2}](\overset{\pm}{1,2})rs} = \begin{cases} \dfrac{\exp\,[j(\pm\, k_{y(1,2)s} \mp k^{*}_{y[1,2]r})\,b] - 1}{j(\pm\, k_{y(1,2)s} \mp k^{*}_{y[1,2]r})} & \text{if } (\pm\, k_{y(1,2)s} \mp k_{y[1,2]r}) \neq 0 \\ b & \text{if } (\pm\, k_{y(1,2)s} \mp k^{*}_{y[1,2]r}) = 0\,. \end{cases}$$

It should be noted that the indexes in brackets ( ) and [ ] can assume four values as follows $\overset{+}{1}$; $\overset{-}{1}$; $\overset{+}{2}$; $\overset{-}{2}$. The relative sums are therefore extended from 1 up to 4.

To judge the importance of the various component modes in the expansion (2)–(3), it is convenient to choose the proportionality coefficient N introduced in (27) in such a way that for similar amplitude P there corresponds for the various modes, maximum e.m. fields of the same order of magnitude for $0 \leq x \leq a$.

This can be approximately accomplished by choosing for each mode the proportionality coefficient N in such a way that the maximum between the following 96 terms

$$\left| A_{\substack{\pm \\ 1,2}} \begin{Bmatrix} e \\ \\ h \end{Bmatrix}_{\begin{Bmatrix} x \\ y \\ z \end{Bmatrix} \substack{\pm \\ 1,2}} \exp \left[j \left(\pm k_{y1,\,2} \begin{Bmatrix} y=0 \\ \\ y=b \end{Bmatrix} + k_x \begin{Bmatrix} x=0 \\ \\ x=a \end{Bmatrix} \right) \right] \right| \tag{30}$$

is of the same order of magnitude for all the modes. In (30) the quantities into { } brackets refer to different cases to be considered.

Operating in this way it can be seen from (29) that, if $|k_{xr}^2| \to \infty$ and k_{xs}^2 remains finite, the coefficients ϕ_{rs}, $\phi_{rs}\chi_{rs}$, ϕ_{sr} and $\phi_{sr}\chi_{sr}$ go to zero at least as $1/|k_{xr}|$.

In view of the reciprocity of the structure of Fig. 3, it is convenient to use equal and opposite subscripts to refer to modes having equal and opposite propagation constants k_x. Such modes will be put consecutively in the set. The systems (8), (11) become therefore

$$\sum_{s=\pm 1}^{\pm \infty} \phi_{rs} P_s = 0 \qquad r = \pm 1, \pm 2, \ldots \tag{31}$$

$$\sum_{s=\pm 1}^{\pm \infty} \phi_{rs} \chi_{rs} P_s = 0 \qquad r = \pm 1, \pm 2, \ldots \tag{32}$$

5. NUMERICAL EXAMPLE

Let us now apply the preceding considerations to a numerical case.

The parameters characterizing the modal spectrum for the completely filled parallel plate guide are ρ, τ and b. We have chosen the following values, $\rho = 3$ $\tau, = 3{\cdot}21482$, $b = 0{\cdot}596$, which may be taken to correspond to the following numerical values for the actual physical parameters:

$$\epsilon = 10; \; H_0 = \frac{10^6}{4\pi} \; A/m; \; M_0 = 0{\cdot}3 \; Wb/m^2; \; f = \frac{\omega}{2\pi} = 9000 \; Mc/s;$$

$$\frac{b}{\omega\sqrt{(\mu_0 \epsilon_0 \epsilon)}} = 10^{-3} \text{ m.}$$

The small-value of b has been chosen in order to consider a rectangular guide which will be well under cut-off without the d.c. magnetization.

For the parallel plate guide considered we have solved the characteristic equation for real values of k_x^2. The solutions are represented in Fig. 4, which is an extension of Fig. 4 of the quoted work[1]. The numerical values of k_x^2 have been computed with five significant figures for $k_z^2 = 98{\cdot}7286$, which corresponds to a positive value of k_x^2 equal to $4\pi^2/b^2$. The modal curves for the attenuated modes, not represented in the figure, are approximately parallel to the curves $k_x^2 = t_{1,\,2}^2$ which belong to an hyperbola.

For the numerical case considered we have, with reference to the x-direc-

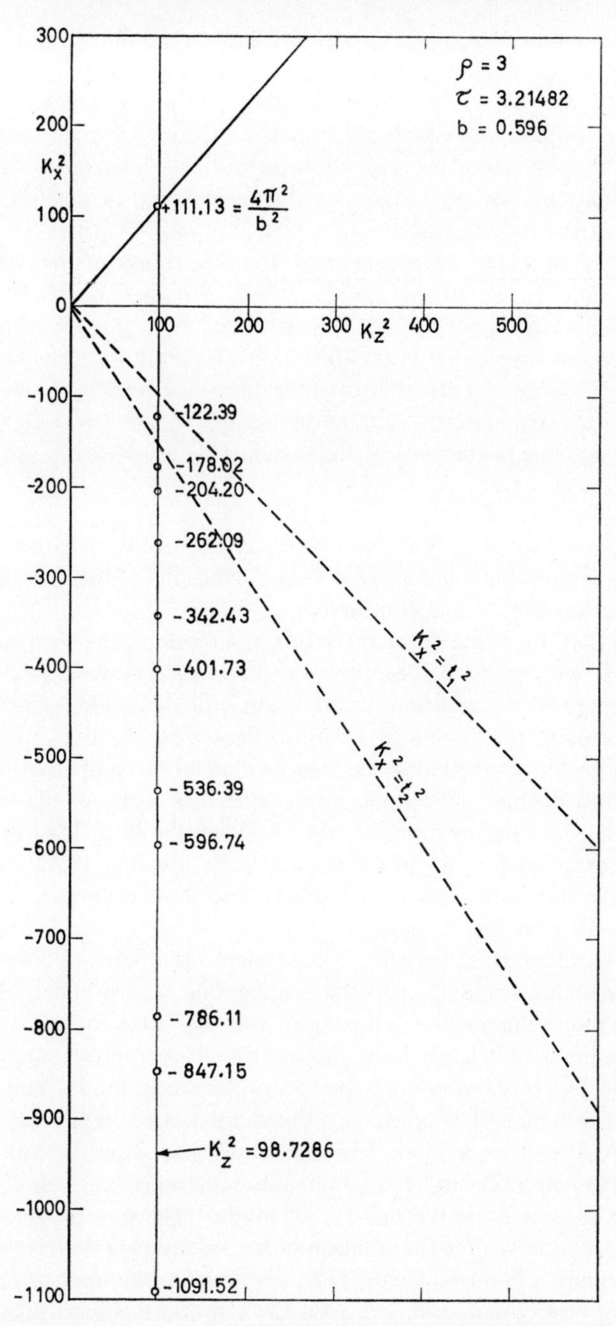

FIG. 4. Modal spectrum for the parallel plate guide completely filled with ferrite magnetized parallel to the walls and for the indicated numerical values of the parameters.

tion, only one pair of propagation modes and an infinite number of pairs of attenuated modes. Only twelve roots have been computed.

We shall order the set of modes according to decreasing values of k_x^2. The modes corresponding to equal and opposite values of k_x will be put consecutively in the set, and a positive subscript will be associated with the mode with propagation constant k_x lying on the positive real or positive imaginary half-axis of the complex plane.

The modes associated with equal and opposite subscripts possess specular symmetry with respect to the plane $y = b/2$. We shall choose the proportionality coefficient N for the modes associated with positive subscripts by referring to the value $x = 0$ in the form. (30); for each mode associated with a negative subscript $-s$ we shall assume the same amplitude as the mode associated with the subscript $+s$, multiplied by $\left| \exp(-jk_{x\,-s}a) \right|$. Operating in this way the coefficients of the system (31) must satisfy the following property

$$\phi_{-r,\,-s} = -\left| \chi_{rs} \right| \phi_{rs}. \tag{33}$$

In (33) the subscripts r and s may be either positive or negative.

Let us now examine the matrices of the coefficients of the systems (31) and (32) for the numerical case considered.

We note that, by virtuc of the preceding assumptions, the coefficients of the systems (31) with positive subscripts approach zero moving away from the principal diagonal. In addition, if the width a of the guide is not too small, the coefficients of the system (31) with at least one negative subscript, corresponding to an attenuated mode, are very small in comparison to the coefficients with positive subscripts. The coefficients $\chi_{rs}\phi_{rs}$ of the system (32) behave in an opposite manner, i.e. the coefficients with at least one positive subscript corresponding to an attenuated mode are very small with respect to the coefficients with negative subscripts, and these approach zero moving away from the principal diagonal.

We can conclude therefore that in the system (31) there are only significant the coefficients corresponding to the propagating modes and to the modes attenuated along the positive x-direction, and vice versa for the system (32).

By assuming a sufficiently large, we can therefore approximately solve the system (31) by considering only the two propagating modes and a certain number of attenuated modes, i.e. those associated with the subscripts $+1, -1, +2, +3, +4, \ldots$. We shall show later how we can choose a so that the system (32) can be automatically satisfied.

In order to determine the number of modes necessary to obtain a satisfactory approximation for the solution of the system (31) we have considered a set of systems obtained from (31) by limiting the number of modes considered respectively to 4, 5, 6, ..., 13. Each system has been solved and the values of the unknowns recorded in Fig. 5 against the number n of modes; we have assumed $P_1 = 1$.

From the figure it is apparent that by increasing the number n of equations the various amplitudes P approach a definite limit. For instance the real and imaginary parts P_2^R and P_2^J of P_2 become practically constant for $n \geq 8$. From

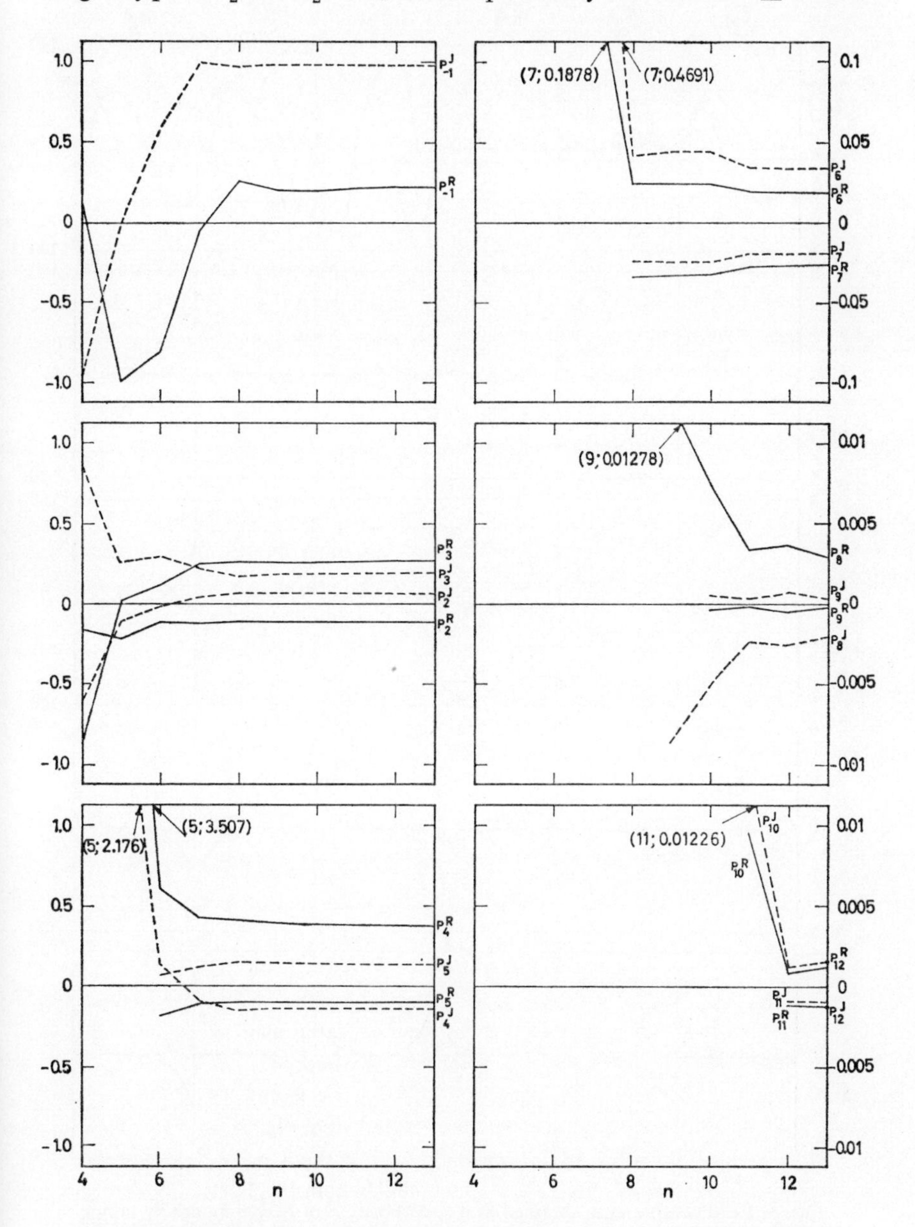

FIG. 5. Amplitudes of the various modes obtained from the system (31) by taking an increasing number n of equations. R and J indicate real and imaginary parts.

the diagrams it appears that for $n = 13$ all the mode amplitudes up to P_{12} are obtained with sufficient accuracy. It also appears that for any number of

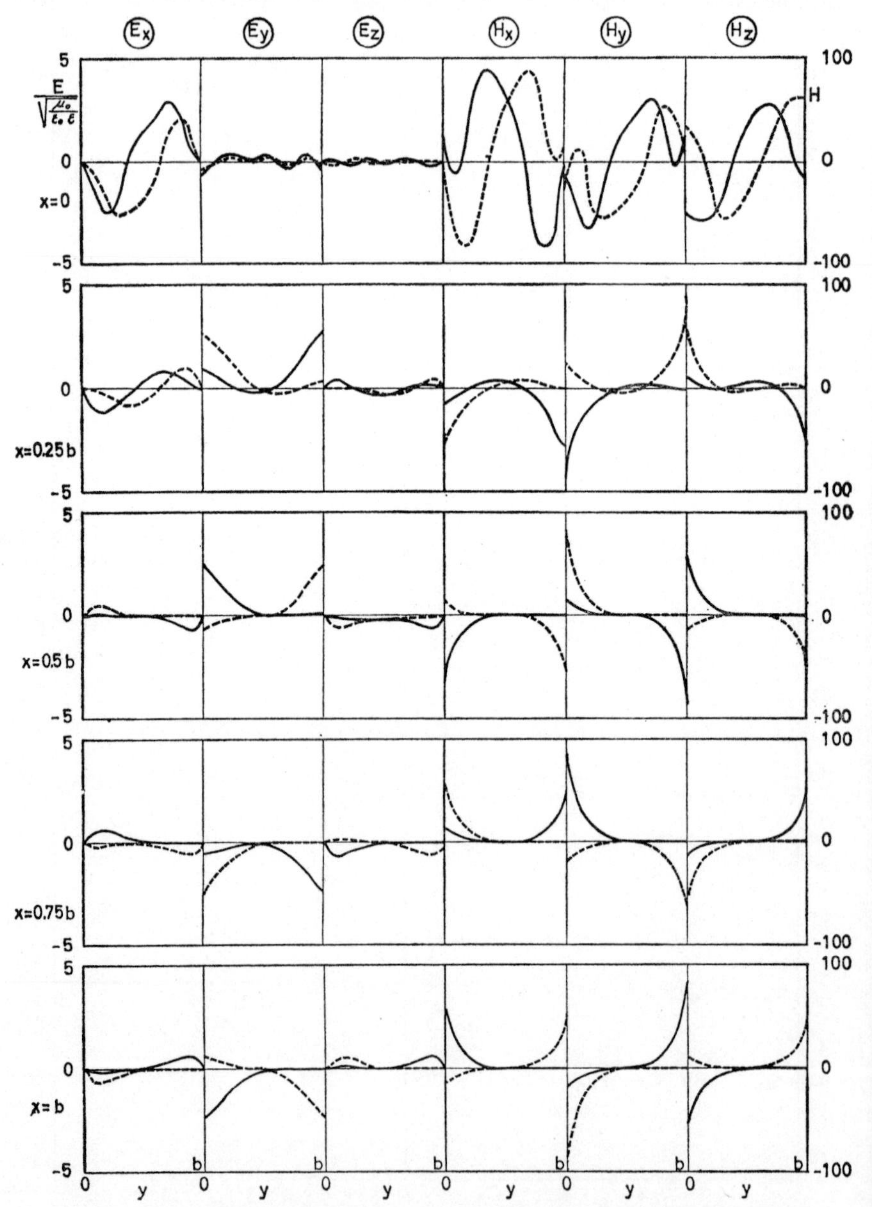

FIG. 6. Field components obtained as a superposition of twelve modes by taking into account only the discontinuity at $x = 0$ (continuous and dotted lines refer respectively to the real and imaginary parts).

equations n considered $(P^R_{-1})^2 + (P^J_{-1})^2 = 1$, as it should be, since the only two modes able to carry power are those associated with the subscripts $+1$ and -1; we have in fact neglected in our expansion the attenuated modes with

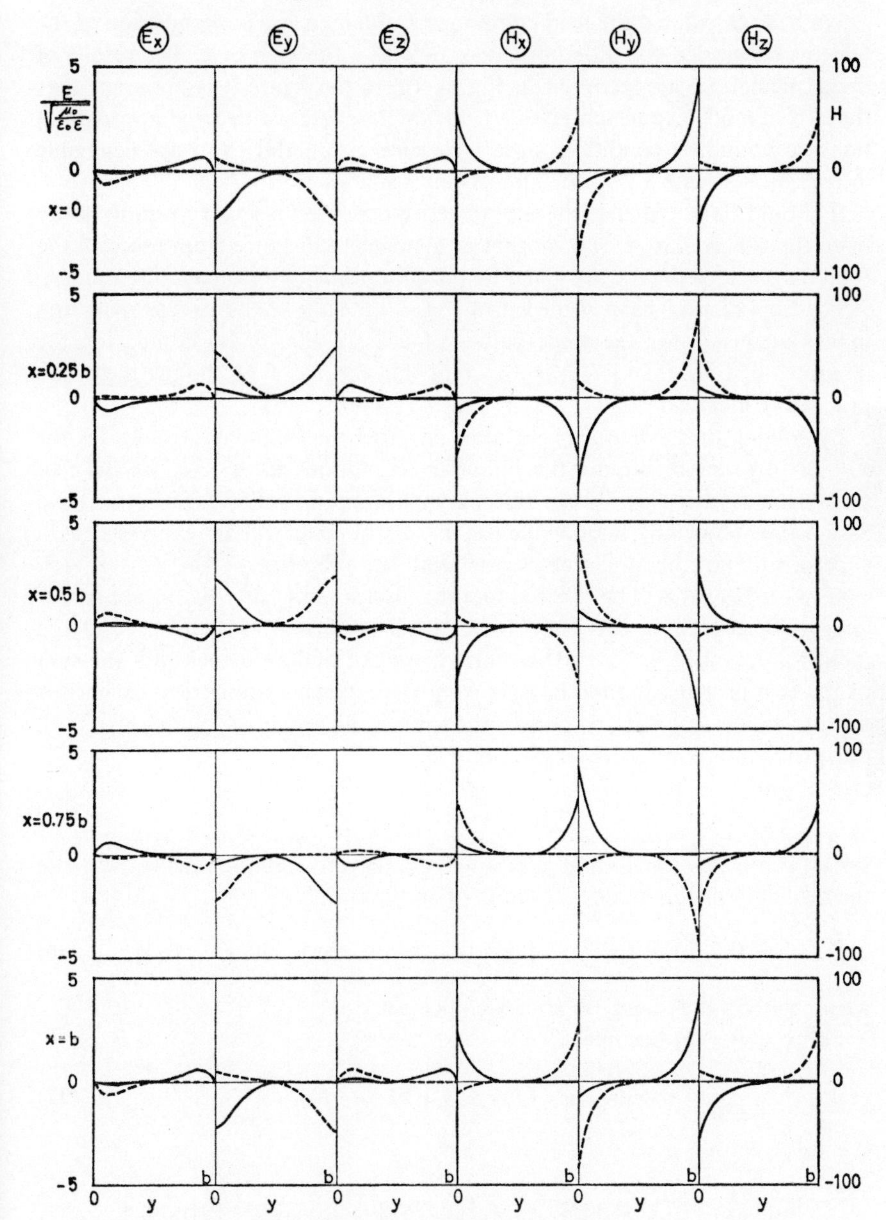

FIG. 7. Field components obtained as a superposition of only the two propagating modes for the same case as in Fig. 6.

negative subscripts. It is therefore

$$P_{-1} = \exp [j\phi]. \tag{34}$$

We have calculated the field components obtained as a superposition of the first twelve modes, for different values of x as a function of y. The results of these calculations are recorded in Fig. 6. From the figure it is apparent that the twelve mode expansion gives a field which satisfies with good approximation the boundary conditions at $x = 0$, since $|E_y|$ and $|E_z|$ are negligible in comparison with $|E_x|$ for almost any value of y.

It should be noted that the attenuated modes decrease very rapidly away from the side wall at $x = 0$, so that at a sufficient distance from the wall the total e.m. field is practically given by the sum of the two propagating modes.

To show this we have recorded in Fig. 7 the sum of the two propagating modes against y for the same values of x as in Fig. 6. By comparing the diagrams of Figs. 6 and 7, it is apparent that for $x \geq 0{\cdot}5b$ the two fields are practically identical.

By solving the system (31) we have imposed the boundary conditions for $x = 0$. In order to impose the boundary conditions for $x = a$, we have to satisfy the system (32). To do this we shall argue as follows. Since we have assumed a sufficiently large value for $x = a$ the field will be expressed as a superposition of the modes associated with the subscripts $+1, -1, -2, -3, -4, \ldots$. Now it will be recalled that the modes $-2, -3, -4, \ldots$ are specularly symmetrical (in the sense indicated in section 4) with respect to the modes $+2, +3, +4, \ldots$. Therefore, if we can find an a such that the sum of the two propagating modes at $x = a$ is specularly symmetrical except for a complex factor k with respect to the sum of the same two modes at $x = 0$, our problem will be solved by assuming

$$P_{-2} = kP_2, \qquad P_{-3} = kP_3. \tag{35}$$

To see if it is possible to find an a satisfying the above requirements, we write the field sum of the modes $+1$ and -1 in the form

$$\mathbf{F}(x, y) = \mathbf{f}_1(y) \exp [jk_{x_1}x] + \mathbf{f}_{-1}(y) \exp [-j(k_{x_1}x - \phi)] \tag{36}$$

where the symbols have obvious significance.

For $x = 0$, (36) becomes

$$\mathbf{F}(0, y) = \mathbf{f}_1(y) + \mathbf{f}_{-1}(y) \exp [j\phi]. \tag{37}$$

We must choose a in such a way that

$$\left.\begin{aligned} \mathbf{F}(a, y) &= \mathbf{f}_1(y) \exp [jk_{x_1}a] + \mathbf{f}_{-1}(y) \exp [-j(k_{x_1}a - \phi)] \\ &= k\{\mathbf{f}_{-1}(y) + \mathbf{f}_1(y) \exp [j\phi]\}. \end{aligned}\right\} \tag{38}$$

From (38) we obtain

$$a = \frac{m\pi + \phi}{k_{x_1}} \qquad m = 0, 1, 2, \ldots \tag{39}$$

$$k = \exp\,[-\,jm\pi] = \pm\, 1. \tag{40}$$

It can be verified that, if (39)–(40) are satisfied, (35) is a solution of the system (32).

In our numerical case

$$a = m\,0{\cdot}298 + 0{\cdot}128. \tag{41}$$

From (41) and from the diagrams of Figs. 6 and 7 it appears that for $m = 0$, a is not sufficiently large to fulfil the preceding assumptions; however, for $m \geq 1$ the procedure followed is justified.

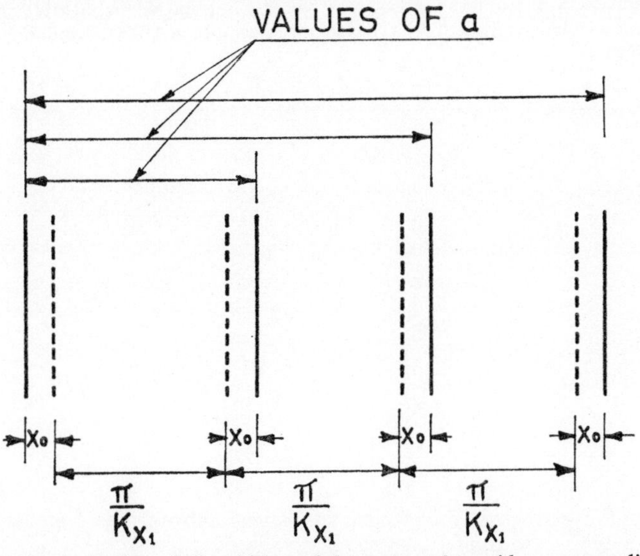

FIG. 8. Determination of the width a of the rectangular guide corresponding to prescribed b and k_z.

From eqn (39) it appears that the lack of reflection symmetry introduces in the *reflected wave* from the wall at $x = 0$ an additional phase shift, in comparison with the reflection symmetrical case having the same k_{x_1}. This is equivalent to say that the *reflection* takes place from a wall at

$$x = x_0 = \frac{\phi}{2k_{x_1}}$$

instead of at $x = 0$.

Therefore, in order to satisfy the boundary conditions we have to choose a width a of the guide larger of an amount $2x_0$ with respect to the reflection symmetrical case (see Fig. 8).

REFERENCES

1. BARZILAI, G. and GEROSA, G. Modes in rectangular guides filled with magnetized ferrite, *L'Onde électrique*, 38e Année, No. 376ter, pp. 612–617, Supplément Spécial—Congrès International Circuits et Antennes Hyperfréquences, Paris 21–26 October 1957.
2. BARZILAI, G. and GEROSA, G. Modes in rectangular guides partially filled with transversely magnetized ferrite, *I.R.E. Trans. on Antennas and Propagation*, **AP–7**, Special Supplement, pp. S471–S474; December 1959—For details: Barzilai G. and Gerosa, G. Modes in rectangular guides partially filled with transversely magnetized ferrite, Istituto Elettrotecnico dell'Università di Roma, Techn. Note No. 1, Contract No. AF61(052)–101; June 3, 1959.
3. BARZILAI, G. and GEROSA, G. Modes in rectangular guides loaded with a transversely magnetized slab of ferrite away from the side walls, *I.R.E. Trans. on Microwave Theory and Techniques*,**MTT–9**, pp.403–408; September 1961. Istituto di Elettronica dell'Universita di Roma—Techn. Note No. 2, Contract No. AF61(052)–101; July 30, 1960—presented orally at the XIII General Assembly of URSI, London 5–15 September 1960.

UNIDIRECTIONAL WAVES IN ANISOTROPIC MEDIA†

AKIRA ISHIMARU

Department of Electrical Engineering, University of Washington,
Seattle 5, Washington, U.S.A.

ABSTRACT

This paper presents the resolution of the so-called thermodynamic paradox, which was first pointed out by Lax and Button. It is shown that Bresler's resolution is insufficient, and that there can exist a single unidirectional mode in a lossless medium. Poynting's theorem is investigated for a discontinuity in a one-way system, and it is shown that the energy equations exhibit marked differences depending on whether the conductivity is zero or approaches zero.

It is shown that the problem of solving Maxwell's equations with a completely lossless medium which leads to the thermodynamic paradox is in fact an "Improperly-posed problem", which does not correspond to physical reality.

1. INTRODUCTION

The first one-way transmission system was proposed by Lord Rayleigh (1901), (Hogan, 1956) using Faraday rotation in an optical system. Since then, through the work of McMillian (1946), Gamo (1959) and others (Carlin, 1954), it is well established from general energy considerations that a one way transmission system must include resistive elements.

In 1955, however, Lax and Button (1955, 1956) pointed out the possibility of the existence of a one–way propagating mode in a *lossless* ferrite loaded waveguide, thus apparently violating the basic laws of thermodynamics. In an attempt to resolve this so-called "thermodynamic paradox", it was argued that the power flow in reverse direction takes place via cut-off modes (Kales, 1956). That this does not provide a satisfactory resolution was shown by Bresler (1960) and Seidel (1959). At present, there are two schools of thought on the resolution of this paradox. One approach (Seidel, 1957) is that a ferrite medium possesses an "intrinsic loss" and this approach is based on a consideration of the atomic model from which the ferrite properties are

† The research reported in this document has been sponsored by the Electronics Research Directorate of The Air Force Cambridge Research Laboratories, Office of Aerospace Research, United States Air Force, Bedford, Massachusetts.

deduced. The other approach was advanced by Bresler (1960), who rejected the intrinsic loss approach. On the basis of the lossless permeability dyadic, Bresler showed that while a single unidirectional mode can be obtained for a ferrite slab placed at a rectangular waveguide wall, if a gap of width d between the ferrite and the wall is considered, a different secular equation results, and that in the limit $d \rightarrow 0$, there are always an even number of propagating modes, half of which are the forward waves and the other half the backward waves. Thus, Bresler stated that this clears the way for the resolution of the thermodynamic paradox.

This paper first shows that Bresler's resolution is not valid for the general case of a single unidirectional propagating mode. There exists a unidirectional mode in a lossless medium.

In order to resolve this thermodynamic paradox, we first note that there are no thermodynamic difficulties involved in an infinite lossless system, and that the difficulty occurs only when there are some discontinuities in the waveguide, which include input and output terminations.

Next, Poynting's theorem is studied in detail, in particular, its integral form for the case of a one-way system terminated with a lossless short. It is shown that the solutions and the power relations show marked discontinuities whether conductivity is zero or approaches zero.

From this discontinuous behavior, it is shown that the solution of Maxwell's equations with the conductivity approaching zero is a "Properly-posed problem" which corresponds to physical reality and which satisfies three basic requirements. On the other hand, the solution of Maxwell's equations with zero conductivity is an "Improperly-posed problem" which does not correspond to physical reality and satisfies only two of the three basic requirements.

Thus, the thermodynamic paradox is resolved by stating that only an "Improperly-posed problem" presents any thermodynamic difficulties, and that if the problem is properly-posed, there is no thermodynamic difficulty.

2. BRESLER'S MODES AND THE EXISTENCE OF A SINGLE UNIDIRECTIONAL MODE

In this section, it is shown that the discontinuous behavior of the solution discussed by Bresler is not valid for the general case of a single unidirectional mode, and a single unidirectional mode can exist in a lossless medium.

Instead of the ferrite loaded waveguide, a simpler model shown in Fig. 1 is chosen to illustrate the point.

Consider a semi-infinite lossless ferrite region $y > d$ which is bounded by an air gap $d > y > 0$ and an electric wall at $y = 0$. The d.c. magnetic field is directed in the z-direction and it is assumed that the geometry and the

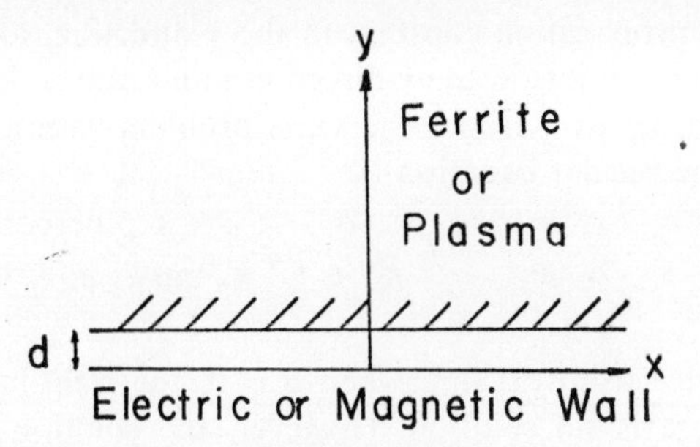

FIG. 1. Unidirectional wave in anisotropic medium bounded by an electric or magnetic wall.

field quantities are independent of z. The permeability tensor is given by

$$\bar{\bar{\mu}} = \begin{bmatrix} \mu & jK & 0 \\ -jK & \mu & 0 \\ 0 & 0 & \mu_z \end{bmatrix} \tag{1}$$

where

$$\mu = \mu_0 \left(1 + \frac{\omega_M \omega_0}{\omega_0^2 - \omega^2}\right) \tag{2}$$

$$K = \mu_0 \frac{\omega \omega_M}{\omega_0^2 - \omega^2}$$

$\omega_M = -\gamma \mu_0 M_0$ saturation magnetization frequency

$\omega_0 = -\gamma \mu_0 H_i$ gyromagnetic resonance frequency.

The differential equation for E_z in the ferrite is

$$\left(\frac{\partial^2}{\partial x^2} + \frac{\partial^2}{\partial y^2} + k_f^2\right) E_z = 0 \tag{3}$$

where

$$k_f^2 = \omega^2 \epsilon_f \frac{\mu^2 - K^2}{\mu}.$$

H_x is given in terms of E_z,

$$H_x = -\frac{1}{j\omega\mu_T}\left(\frac{\partial}{\partial y} E_z + j\frac{K}{\mu}\frac{\partial}{\partial x} E_z\right) \tag{4}$$

where $\mu_T = \dfrac{\mu^2 - K^2}{\mu}$.

Now, matching the tangential electric field E_z and the magnetic field H_x at the boundaries, we obtain a secular equation

$$-jk_{fy} + \frac{K}{\mu} K_x = \frac{\mu_T}{\mu_0} \frac{k_y}{\tan k_y d} \tag{5}$$

where k_x is the propagation constant in the x-direction, and k_y and k_{fy} are the propagation constants in the y-direction in air and in ferrite respectively. For comparison, let us consider the same problem with a magnetic wall at $y = 0$. Then, the secular equation is

$$-jk_{fy} + \frac{K}{\mu} K_x = \frac{\mu_T}{\mu_0} k_y \tan k_y d. \tag{6}$$

First, we notice from (5) that when $d = 0$, the right-hand side becomes infinite and there is no solution. However, the solution of (6) exhibits a continuous behavior irrespective of whether $d = 0$ or $d \to 0$.

The same situation arises when a lossless plasma is bounded by an air gap and an electric wall. It can be shown that in a frequency range

$$\sqrt{(\omega_p^2 + \omega_c^2)} < \omega < \tfrac{1}{2}[\sqrt{(4\omega_p^2 + \omega_c^2)} + \omega_c] \tag{7}$$

where ω_p is the plasma frequency and ω_c is the cyclotron frequency, a unidirectional surface wave can propagate in a lossless opaque plasma (Ishimaru, 1961 and 1962). In this range the dielectric constant is negative, and the plasma is opaque, thus, there is no radiation in the y-direction. In such a case, the only power flow is by means of a single unidirectional surface wave mode and this mode is not accompanied by any backward waves.

It is important to note that this single unidirectional wave mode is a surface wave type mode, which means that the field decays exponentially in transverse direction. As will be shown later this behavior is important in the discussion of the thermodynamic paradox.

3. POYNTING THEOREM AND THERMODYNAMIC PARADOX

In the previous section, it was shown that a unidirectional surface wave can exist in a lossless medium.

We note first of all that no thermodynamic paradox occurs for an infinite uniform (or endless) lossless one-way system. For example an infinite lossless waveguide which carries the one-way mode does not violate any thermodynamic law, because energy is being transferred from minus infinity to plus infinity and there is no source or load which may be cooled or heated. The same can be said about a ring type waveguide which propagates a one-way mode. Also, even though a source or a load (a dipole) is introduced in this infinite waveguide, the amount of energy received by the dipole is equal to the amount of energy transmitted by the dipole, and there is no accumulation of energy on the dipole.

From the above considerations, we note that the so-called thermodynamic paradox occurs only when there are some discontinuities in the waveguide. The energy consideration which lead to McMillian's conclusion of resistive elements in the one-way system was based on the fact that the system is terminated at both input and output ends, where the voltage or current can be specified. This is only possible when there are some discontinuities in the system.

Since there must be discontinuities in the waveguide if the thermodynamic difficulty is to occur, let us consider a typical problem of terminating the one-

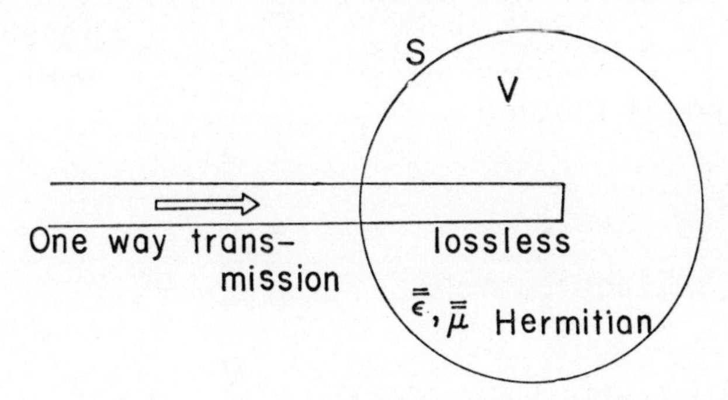

FIG. 2. Poynting theorem applied to a one-way system terminated by a lossless short.

way system by a lossless short circuit. This is shown in Fig. 2. In order to investigate this problem, it is necessary to examine Poynting's theorem.

From Maxwell's equations, we get

$$\nabla \cdot \mathbf{E} \times \mathbf{H}^* = j\omega[\mathbf{E} \cdot \mathbf{D}^* - \mathbf{H}^* \cdot \mathbf{B}].$$

Taking the volume integral of both sides of the equation we get

$$\int_S \mathbf{E} \times \mathbf{H}^* \cdot \mathrm{d}\mathbf{S} = \int_V j\omega[\mathbf{E} \cdot \mathbf{D}^* - \mathbf{H}^* \cdot \mathbf{B}]\, \mathrm{d}V \tag{8}$$

provided that such an integral exists.

It is significant to note that this well known integral form is valid only when the integration can be performed to yield a finite value.

Let us first recognize that for the case of a lossless medium, $\bar{\bar{\mu}}$ and $\bar{\bar{\epsilon}}$ are Hermitian, and thus the integrand in the right side of (8) is pure imaginary as is well known. We note first the apparent difficulty associated with this problem. From the integral form of Poynting's theorem, provided that this form is valid, the left side of (8) has a real part representing the real power carried by the one-way mode. The right-hand side, however, has the pure imaginary integrand, and it appears that the integral may also be pure imaginary. Thus the real power carried by the one-way mode must be equal to the imaginary power, which is contradictory.

To investigate this difficulty more closely let us consider a simple example.

Let us consider an opaque plasma bounded by an electric wall as shown in Fig. 3. The one-way surface wave mode propagates towards the right. Suppose that the magnetic wall is placed as shown and does not support the propagating mode. Thus, the real power carried by the one-way mode is being stopped by a lossless wall.

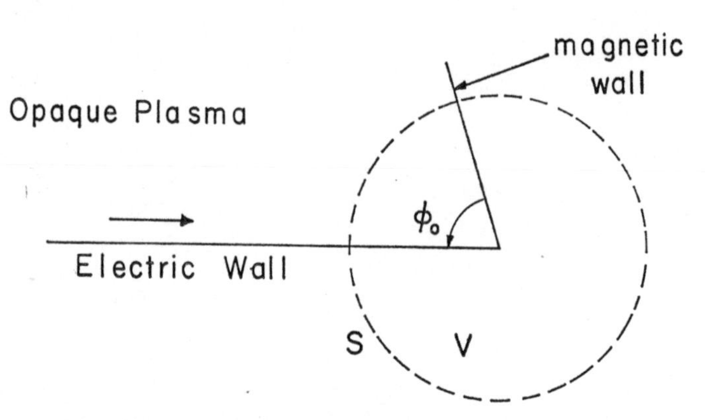

FIG. 3. Unidirectional wave along an electric wall terminated by a magnetic wall.

First, consider the behavior of the field components near this corner. H_z satisfies the wave equation

$$\left(\frac{\partial^2}{\partial x^2}+\frac{\partial^2}{\partial y^2}+k_p^2\right) H_z=0, \quad k_p^2=\omega^2\mu\epsilon_T. \tag{9}$$

The boundary condition on the electric wall is

$$\left(\frac{\partial}{\partial y}+j\lambda\frac{\partial}{\partial x}\right) H_z=0, \quad \lambda=\frac{q}{\epsilon}. \tag{10}$$

The boundary condition on the magnetic wall is

$$H_z=0. \tag{11}$$

The dielectric tensor of the plasma is given by

$$\bar{\bar{\epsilon}}=\begin{bmatrix}\epsilon & jq & 0\\ -jq & \epsilon & 0\\ 0 & 0 & \epsilon_z\end{bmatrix}, \tag{12}$$

where

$$\epsilon=1+\frac{\omega_p^2}{\omega^2-\omega^2} \tag{13}$$

$$q = \pm \frac{\omega_c \omega_p^2}{\omega(\omega_c^2 - \omega^2)} \tag{14}$$

$$\epsilon_z = 1 - \frac{\omega_p^2}{\omega^2} \tag{15}$$

$$\epsilon_T = \frac{\epsilon^2 - q^2}{\epsilon}.$$

The plus sign in (14) applies when the d.c. magnetic field is directed in the $+z$–direction, and the minus sign is for the $-z$–direction.

We note that in a frequency range given by (7),

$$\epsilon_T < 0 \quad \text{and} \quad \left| \frac{q}{\epsilon} \right| > 1. \tag{16}$$

H_z can be represented by

$$H_z = \sum_i a_{\nu_i} Z_{\nu_i}(k_p \rho) \sin \nu_i \phi \tag{17}$$

where Z_{v_i} is a Bessel function of order ν_i and a_{v_i} is a constant.

Consider a small distance from the corner $| k_p \rho | \ll 1$.

Then the first dominant term of H_z is

$$H_z = \rho^\nu \sin \nu\phi \qquad \nu \neq 0. \tag{18}$$

From the boundary condition,

$$\tan \nu\phi_0 = j \frac{1}{\lambda}. \tag{19}$$

Since $| \lambda | > 1$ there is at least one root of (19) which is pure imaginary. This imaginary root represents the exponential decay in transverse direction and this corresponds to the one-way mode. The other field components are

$$\begin{aligned} E_\rho &= \frac{1}{j\omega\epsilon_T} \rho^{\nu-1} \nu(\cos \nu\phi + j\lambda \sin \nu\phi) \\ E_\phi &= \frac{1}{j\omega\epsilon_T} \rho^{\nu-1} \nu(j\lambda \cos \nu\phi - \sin \nu\phi) \\ D_\rho &= \frac{1}{j\omega} \rho^{\nu-1} \nu \cos \nu\phi \\ D_\phi &= \frac{1}{j\omega} \rho^{\nu-1} \nu \sin \nu\phi. \end{aligned} \tag{20}$$

Let us calculate the right-hand integral of (8),

$$\int_v j\omega \mathbf{E} \,.\, \mathbf{D}^* \,\mathrm{d}V = \frac{1}{j\omega\epsilon_T} \int_0^\rho \nu(\nu + \nu^*) \rho^{\nu+\nu^*-1} \,\mathrm{d}\rho \int_0^{\varphi_0} (j\lambda \cos \nu\phi - \sin \nu\phi) \sin \nu^*\phi \,\mathrm{d}\phi \,. \tag{21}$$

First, we note that if the medium and the boundaries are completely lossless, ν is pure imaginary and thus $\nu + \nu^* = 0$. However, the integrand contains $\rho^{\nu+\nu^*-1}$ which becomes ρ^{-1}. Thus, this integrand is indeterminate, and moreover this integral is not defined because the behavior of the integrand at $\rho = 0$ cannot be definitely known.

However, if the integration is performed before the medium is allowed to become lossless, then

$$\int_v j\omega \mathbf{E} \,.\, \mathbf{D}^* \,\mathrm{d}V = \frac{1}{j\omega\epsilon_T} \rho^{v+v^*} \frac{\nu}{2}$$

$$\left[\frac{\sin \nu_r\phi_0}{\nu_r}(j\lambda \sin \nu_r\phi_0 + \cos \nu_r\phi_0) + \frac{\sinh \nu_i\phi_0}{\nu_i}(\lambda \sinh \nu_i\phi_0 - \cosh \nu_i\phi_0)\right] \quad (22)$$

provided that $\nu + \nu^* > 0$, and $\nu = \nu_r + j\nu_i$. As the medium becomes lossless $\nu_r \to 0$.

Thus

$$\int_v j\omega \mathbf{E} \,.\, \mathbf{D}^* \,\mathrm{d}V = \frac{\nu_i}{2\omega\epsilon_T} \phi_0. \quad (23)$$

The left side of the integral of (8) is

$$\int_S \mathbf{E} \times \mathbf{H}^* \,.\, \mathrm{d}\mathbf{S} = \frac{1}{j\omega\epsilon_r} \rho^{v+v^*} \frac{\nu}{2}$$

$$\left[\frac{\sin \nu_r\phi_0}{\nu_r}(j\lambda \sin \nu_r\phi_0 + \cos \nu_r\phi_0) + \frac{\sinh \nu_i\phi_0}{\nu_i}(\lambda \sinh \nu_i\phi_0 - \cosh \nu_i\phi_0)\right] \quad (24)$$

irrespective of whether ν is real, imaginary or complex.

The above results can be summarized as follows:

The power relation represented by the integral form of Poynting's theorem shows a marked discontinuous behavior depending on whether the conductivity of the dielectric medium and the resistivity of the conductor is zero or approaches zero, i.e. $\sigma_d = 0$, $1/\sigma_c = 0$ or $\sigma_d \to 0$, $1/\sigma_c \to 0$.

In terms of the integrals in (8), these two cases correspond to the integration being performed *after* or *before* σ_d (and $1/\sigma_c$) is allowed to become zero.

The left side of (8)

$$\int_S \mathbf{E} \times \mathbf{H}^* \,.\, \mathrm{d}\mathbf{S}$$

yields the real power, irrespective of whether the integration is performed before or after the limit $\sigma \to 0$.

$$\lim_{\sigma \to 0} \int_S \mathbf{E} \times \mathbf{H}^* \,.\, \mathrm{d}\mathbf{S} = \int_S \lim_{\sigma \to 0} [\mathbf{E} \times \mathbf{H}^*] \,.\, \mathrm{d}\mathbf{S} = \frac{\nu_i\phi_0}{2\omega\epsilon_T} \text{ real.} \quad (25)$$

However, the right side of (8) is not defined if the limit is taken before the integration, and this is the case which yields the thermodynamic difficulty. If the limit is taken after the integration this yields the same result as (25) and there is no difficulty associated with the power. It is interesting to note that even though the integrand of the right side of (8) is pure imaginary for a lossless medium, when the integration is performed before the limit $\sigma \to 0$ is taken, then the limit is purely real representing the real power dissipation,

$$\lim_{\sigma \to 0} \int_v j\omega(\mathbf{E} \,.\, \mathbf{D}^* - \mathbf{H}^* \,.\, \mathbf{B})\, \mathrm{dv} = \frac{\nu_i \phi_0}{2\omega\epsilon_T} \text{ real,} \tag{26}$$

$$\int_v \lim_{\sigma \to 0} j\omega(\mathbf{E} \,.\, \mathbf{D}^* - \mathbf{H}^* \,.\, \mathbf{B})\, \mathrm{dv} = \text{Not defined.} \tag{27}$$

4. RESOLUTION OF THERMODYNAMIC PARADOX

In the preceding section, it was shown that the solution of Maxwell's equations shows a remarkable discontinuity at $\sigma_d = 0$ for the dielectric and $1/\sigma_c = 0$ for the conductor. It was shown that if $\sigma_d = 0$ and $1/\sigma_c = 0$, then the integral form of Poynting theorem is not valid, and therefore the power relations cannot be meaningfully discussed. Also this case leads to the so-called thermodynamic paradox. On the other hand if σ_d and $1/\sigma_c$ are allowed to approach zero, then Poynting's theorem is valid and there is no thermodynamic difficulty.

Our problem is to investigate why one solution of Maxwell's equations, namely the case $\sigma_d = 0$, $1/\sigma_c = 0$ yields such difficulties, while the other solution, $\sigma_d \to 0$ and $1/\sigma_c \to 0$ offers no difficulty at all.

We may be tempted to say that the latter solution is satisfactory because of the intrinsic loss mechanism as proposed by Seidel. However, we do not wish to rely on the intrinsic loss idea. We wish to resolve this difficulty within the framework of Maxwell's equations itself, as was advocated by Bresler. In other words, we wish to choose one of the two solutions, purely on the basis of mathematical arguments, without employing the atomic model or the idea of intrinsic loss.

In order to do this, it is first necessary to investigate what requirements must be satisfied by mathematical problems if they are to correspond to physical reality. These requirements are explored by Courant (1962). In general there are three requirements which a mathematical problem should satisfy.

They are:

(1) The solution must exist.
(2) The solution must be uniquely determined.
(3) The solution should depend continuously on the data.

The last requirement is most important for our discussion. Data such as time, space coordinates, angle, dielectric constant, permeability, conductivity, etc., can only be given within a certain margin, and these data cannot be measured without a certain amount of error. Therefore, for a small variation of the data, the variation of the solution must be also small if this problem is to describe physical phenomena.

The mathematical problem which satisfies these three requirements is called a "Properly-posed problem", while the problem whose solution exists and is uniquely determined, but does not continuously depend on the data, is called an "Improperly-posed problem".

From this consideration it is now possible to clearly resolve the thermodynamic paradox.

We note that the solution shows a sharp discontinuity at $\sigma_d = 0$ and $1/\sigma_c = 0$. Thus the problem of solving Maxwell's equations for a purely lossless medium constitutes an "Improperly-posed problem", which simply does not correspond to physical reality. This is why this lossless case leads to the so-called "thermodynamic paradox", which however does not exist in reality.

On the other hand, the problem of solving Maxwell's equations, with a medium whose conductivity approaches zero, constitutes a "Properly-posed problem", and indeed there are no thermodynamic difficulties involved.

5. SOME RELATED PROBLEMS

It may be interesting to note that the idea of the "Improperly-posed problem" may be applicable to other physical problems.

For example the edge condition in the usual sense (Heins, 1955), which states that the energy in any small volume containing the edge should be finite, is meaningless in the problem discussed in the previous section. Instead the edge condition should be stated as

$$\lim_{\sigma \to 0} \int_{v} (\mathbf{E} \,.\, \mathbf{D}^* + \mathbf{H}^* \,.\, \mathbf{B})\, \mathrm{d}V = \text{finite}. \tag{28}$$

Felsen (1959) observed that, in the case of a wedge with linearly varying impedance, the edge condition is apparently violated unless a small loss is introduced. This is in fact the same situation, and the above edge condition should be employed.

REFERENCES

1. BRESLER, A. D. (1960) On the TE_{n0}-modes of a ferrite slab loaded rectangular waveguide and the associated thermodynamic paradox, *I.R.E. Trans.* **MTT–8**, 81 (January).
2. BUTTON, K. J. and LAX, B. (1956) Theory of ferrites in rectangular waveguides, *I.R.E. Trans.* **PGAP 4**, 531–537 (July).

3. Carlin, H. J. (1954) Principles of gyrator networks, Symp. on Modern Advances in Microwave Techniques, P.I.B.
4. Courant, R. and Hilbert, D. (1962) Methods of mathematical physics, *Partial Differential Equations*, Vol. II, 227–229. Interscience.
5. Felsen, L. B. (1959) Electromagnetic properties of wedge and cone surfaces with a linearly varying surface impedance, *I.R.E. Trans.* **PGAP 7**, 231–243, Special Supplement (December).
6. Gamo, Hideya (1959) On passive one-way systems, *I.R.E. Trans. on Circuit Theory* **Ct–6**, 283–298, Special Supplement (May).
7. Heins, A. and Silver, S. (1955) The edge conditions and field representation theorems in the theory of electromagnetic diffraction, *Proc. Camb. Phil. Soc.* **51**, 149–161.
8. Hogan, C. L. (1956) The elements of nonreciprocal microwave devices, *Proc. I.R.E.* 1345–1368 (October).
9. Ishimaru, A. (1961) Unidirectional surface wave along a perfectly conducting plane in a plasma in the presence of a d.c. magnetic field, presented at the URSI–IRE meeting, Austin, Texas (October).
10. Ishimaru, A. (1962) The effect of a unidirectional surface wave along a perfectly conducting plane on the radiation from a plasma sheath, presented at the Second Symposium on the Plasma Sheath, Boston, Mass (April 10–12).
11. Kales, M. L. (1956) Topics in guided wave propagation in magnetized ferrites, *Proc. I.R.E.* **44**, 1403–1409 (October).
12. Lax, B. and Button, K. J. (1955) New ferrite mode configurations and their applications, *J.A.P.* 1186 (September).
13. McMillian, E. M. (1946) Violation of the reciprocity theorem in linear passive electromechanical systems, *J. Acoust. Soc. Amer.* **18**, 344.
14. Rayleigh, Lord (1901) On the magnetic rotation of light and the second law of thermodynamics, *Nature*, **64**, 577.
15. Seidel, H. (1957) Ferrite slabs in transverse electric modes waveguide, *J.A.P.* **28**, 218–226 (February).
16. Seidel, H. and Fletcher, R. C. (1959) Gyromagnetic modes in waveguide partially loaded with ferrite, *BSTJ*, 1427–1456 (November).

BACKWARD WAVES IN LONGITUDINALLY MAGNETIZED FERRITE FILLED GUIDES

G. H. B. THOMPSON†

ABSTRACT

Computed curves are presented to show the propagation characteristics of longitudinally magnetized ferrite filled circular guide of various diameters as a function of frequency. The frequency regions in which backward waves of the volume and surface type occur are related to the characteristic frequencies appearing in the resonance behaviour of plane waves in an infinite ferrite. The backward wave character of both types is shown to be associated with the turning away of a resonance direction of propagation from the magnetic field direction as the frequency increases. Some experimental results are presented.

INTRODUCTION

Exact characteristic equations can in general be formulated for the propagation coefficient in ferrite loaded waveguide, but they are usually of a complicated transcendental form and require numerical methods for their solution.[1–4] They therefore can seldom yield useful results by inspection only and must be evaluated for a particular set of conditions. The propagation coefficient-frequency characteristic obtained in this way for longitudinally magnetized ferrite filled waveguide (see Fig. 1) shows the presence of backward waves over a range of frequencies close to the ferrite resonance frequency. It is the purpose of this paper to show the qualitative relationship between these backward waves and the plane waves which propagate in an infinite ferrite or over an infinite ferrite conductor interface.

WAVES IN FERRITE FILLED CYLINDRICAL GUIDE

Figure 1 shows the relation, for circular guide, between frequency ω' normalized to the gyroresonance frequency of the ferrite, and propagation coefficient, normalized to the plane wave propagation coefficient β_0 in a dielectric of the same permittivity as the ferrite at the frequency of gyroresonance, for various guide radii normalized to $1/\beta_0$. Modes of a single circumferential variation only are shown and the two directions of circular polarization are distinguished. The curves apply to a value of $B/\mu_0 H (=q)$ of 2, but the main characteristics are qualitatively unaltered by changing q. Figure 1

† Standard Telecommunication Laboratories Limited, Harlow, Essex.

shows two frequency regions of backward waves (curves with negative gradient) for small guide diameters, one lying between $\omega' = 1$ and $\omega' = \sqrt{q}$, and the other between $\omega' = \sqrt{q}$ and $\omega' = q$. The infinite manifold of modes in the first region are volume modes, and the single mode in the second is a

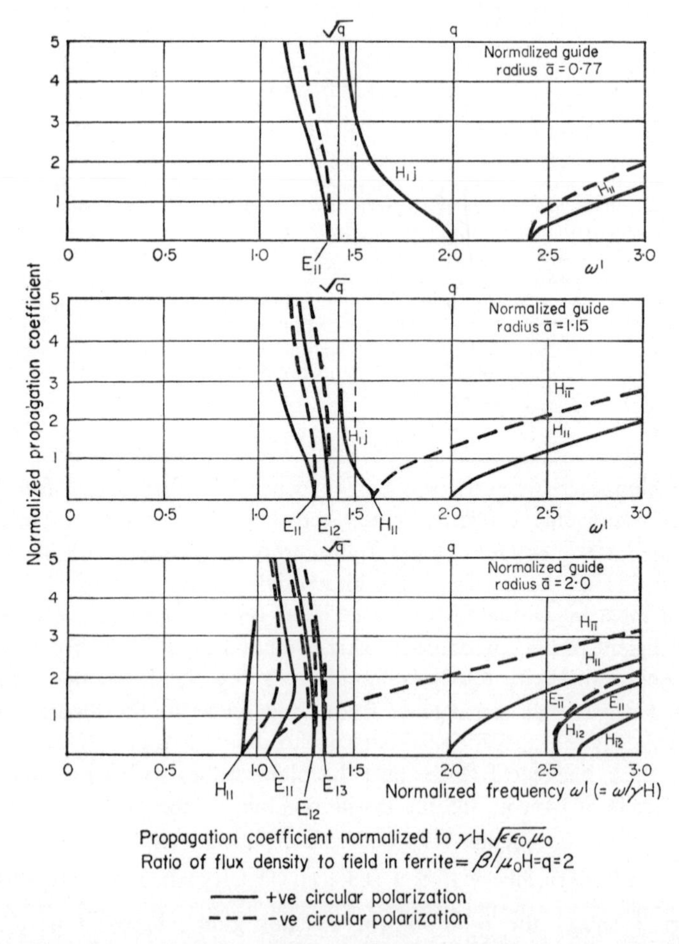

FIG. 1. Relation between propagation coefficient and frequency in ferrite filled guide of various radii.

surface mode and only occurs for positive circular polarization (rotation of the r.f. field vector clockwise about the d.c. magnetic field). Increasing the guide radius reduces the frequency region in which these modes propagate, eventually converts the lower order volume modes into forward waves and suppresses the surface mode entirely.

PLANE WAVES IN FERRITE

The frequency values which divide the guided modes also represent transition points for plane waves. A plane wave is split by a ferrite into two components of in general complementary elliptical polarization which propagate with different velocities and of which only one is resonant as a function of frequency.[5] The shape of the polarization ellipse, the value of the propagation coefficient and, for the extraordinary wave, the resonant frequency depend on the angle θ between the propagation direction and the d.c. magnetic field. Figure 2 shows a plot of the effective permeability of the extraordinary wave

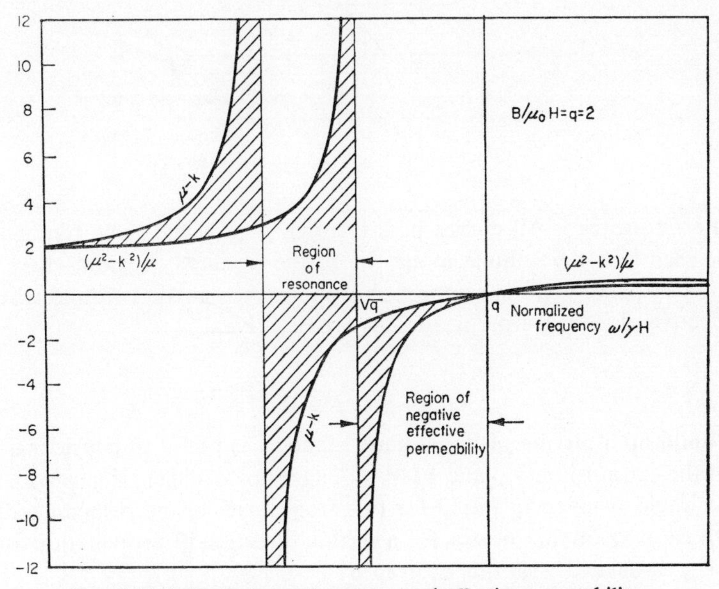

FIG. 2. Relation between frequency and effective permeability.

as a function of frequency for propagation both along the field direction (the $(\mu - k)$ curve) and at right angles (the $(\mu^2 - k^2)/\mu$ curve). (μ and k are components of the permeability tensor.) The effective permeability for intermediate angles is given by the relation[5]

$$\mu_{\text{eff}} = \frac{2\mu - (\mu + k^2 - \mu^2)\sin^2\theta \mp \{(\mu + k^2 - \mu^2)^2 \sin^4\theta + 4k^2 \cos^2\theta\}^{1/2}}{2(\cos^2\theta + \mu \sin^2\theta)} \tag{1}$$

or in terms of normalized frequency ω'

$$\mu_{\text{eff}} = \frac{2(\omega'^2 - q) + (q - 1)\{\mp (q^2 \sin^4\theta + 4\omega'^2 \cos^2\theta)^{1/2} - q \sin^2\theta\}}{2[\cos^2\theta\,(\omega'^2 - 1) + \sin^2\theta\,(\omega'^2 - q)]} \tag{2}$$

where in both cases the negative sign of the pair refers to the extraordinary wave. The corresponding curves of permeability against frequency lie between those given for the extreme cases in Fig. 3 and show resonance at inter-

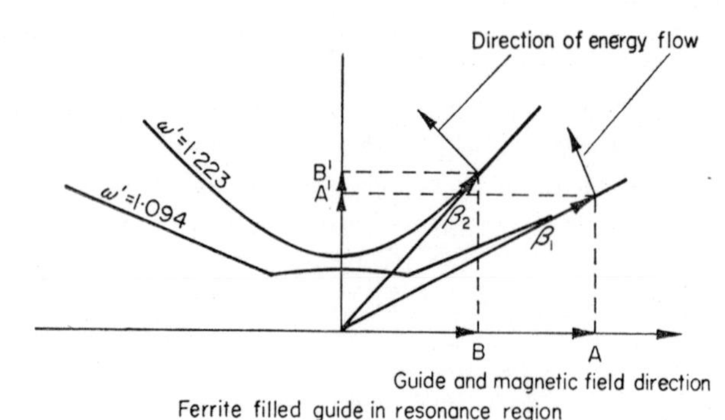

FIG. 3. Effect of frequency on guide wavelength.

mediate frequencies. All curves pass through zero at the same frequency. It can be seen that the volume modes lie in the frequency region where some direction of plane wave is resonant, but the surface mode is found where all the effective plane wave permeabilities are negative.

VOLUME MODES

To build up a picture of the volume modes it is useful to consider a polar plot of the extraordinary plane wave propagation coefficient, i.e. wave vector, against angle as given in Fig. 3 for two frequencies in the resonance region. In both cases resonance occurs for a certain direction of propagation and the angle θ_{res} associated with this is given by

$$\cot^2\theta_{res} = -\mu = \frac{q^2 - \omega'^2}{\omega'^2 - 1} \quad (3)$$

and increases with frequency. The propagation coefficient is infinite for this value of θ, and it decreases for increasing θ to a minimum value when θ is 90°. For θ less than the resonant value the propagation coefficient is imaginary and the wave is evanescent. The direction of energy flow is normal to the wave vector characteristic, and is therefore in general not parallel to the wave vector itself, as indicated in Fig. 3 for two particular wave vectors β_1 and β_2.

Guided waves can be synthesized from plane waves propagating at an angle to the guide axis, the angle being determined by the boundary conditions. Since plane waves in a ferrite contain both transverse components of E it is normally necessary in a ferrite loaded guide for both the ordinary and extraordinary waves to be present to satisfy the condition of zero tangential electric field at the conducting boundary. However, an exception occurs when the waveguide diameter is small. To produce the necessary large transverse variation short wavelength component plane waves are required, which can be obtained from the extraordinary wave if the direction of propagation is close to the resonance direction. The magnetic field of the extraordinary wave under these conditions (see Appendix) is almost entirely longitudinal and very large compared with the electric field. The magnetic field of the ordinary wave which is comparable with the electric field may be ignored. Hence the boundary condition at the conductor requires a zero of the normal component of flux density due to the extraordinary wave alone; giving

$$B_r = \mu H_r - jkH_\theta = 0. \tag{4}$$

When the plane wave vector is close to the resonance direction it may increase without limit (in the lossless case) with negligible change in direction, giving it in effect an arbitrary value but a fixed angle. In terms of the components β_z along the field direction and β_t transversely one must simply satisfy the relation

$$\beta_t/\beta_z = \tan\theta = \left(\frac{-1}{\mu}\right)^{1/2}. \tag{5}$$

The transverse variation of r.f. field in the guide along the radius r, with only the extraordinary wave contributing, is the same as in the isotropic case, or as calculated using Walker's[6] magnetostatic approximation, being of the form

$$\begin{aligned} H_r &\simeq AJ_n'(x), \\ H_\theta &\simeq njAJ_n(x)/x \end{aligned} \tag{6}$$

where $x = \beta_t r$ and the sign of n determines the direction of circular polarization. Substituting these values into eqn (4) one obtains an equation for x_0 at the boundary of the guide:

$$x_0 \frac{J_1'(x_0)}{J_1(x_0)} = \frac{nk}{\mu} \tag{7}$$

and β_z is given in terms of x_0 and r_0, the guide radius, by

$$\beta_z \simeq x_0 \cot\theta/r_0. \tag{8}$$

cot θ dominates the behaviour of this equation, and since θ increases with ω, β_z decreases, and the propagation is of the backward wave type. It may be explained physically by saying that increasing frequency causes the propagation direction of the short wavelength resonant plane waves to turn away from the guide axis and hence reduce the component of the propagation coefficient along the axis. This effect is illustrated in Fig. 3. The pair of plane wave vectors β_1 and β_2 at two different frequencies must give transverse components A' and B' to satisfy the boundary conditions of a particular circular guide. The small change of boundary condition with frequency is due to the frequency dependence of the quantity x_0 in eqn (7) but this is a minor effect. The major effect is imposed by the shape of the wave vector characteristic which requires that the wave vector rotate away from the magnetic field direction with increasing frequency. This reduces the longitudinal component of propagation from vector A to vector B and illustrates the negative gradient in the propagation coefficient-frequency curve.

Figure 3 can also be used to demonstrate directly that the phase velocity and the energy flow in the guide are opposed. It is immediately obvious from the diagram that the longitudinal components of the phase vector, e.g. β_1, and of the corresponding energy flow vector are of opposite sign, due only to the concave shape of the characteristic. The conclusion that the guided wave is backward only applies when the extraordinary wave plays the dominant part, i.e. for small size guides.

Curves of the type shown in Fig. 3 can normally only be used when the wave vectors are large compared with those of the ordinary wave if boundary eqn (7) is to be accurate. In the example given the values are too small for good accuracy. However, an exceptional case occurs at cut-off. The component plane waves then propagate transversely to the Z axis of the guide and the proper polarizations are linear (see Appendix). The ordinary wave has only a Z component of magnetic field whilst the extraordinary has only radial and circumferential components. The boundary provides no coupling between them. Hence only the ordinary wave is present at an H-mode cut off and the extraordinary at an E-mode cut off. Figure 3 can therefore be used exactly at an E-mode cut–off and demonstrates how the backward volume modes stop propagating at a frequency slightly less than $\omega' = \sqrt{q}$ which would make θ a right angle and cot θ zero in eqn (8). The transverse wave vector required at cut-off is the same as for an E_{1n} mode in isotropic guide, i.e. u_{1n}/r_0 where u_{1n} is the nth solution of $J_1(u) = 0$. This value should be plotted on the vertical axis of Fig. 3. Cut-off will then occur at the frequency where the minimum point on the curve has risen to intersect this point. The exact behaviour can be seen by referring to Fig. 1 which shows how the cut-off frequencies decrease as the guide size is made larger and the transverse wave vector is correspondingly reduced.

SURFACE MODES

The occurrence of propagating surface waves in a region of frequency where all the plane waves are evanescent may be loosely ascribed to the fact that the boundary condition at a ferrite-conductor or ferrite–air interface allows a strongly evanescent decay in the normal direction to be associated with real propagation along the surface. It will be shown below that, with the d.c. magnetic field direction in the plane of the surface, a resonant direction of propagation arises along the surface, the angle θ between the two directions depending on frequency. An approximate solution may be obtained using the Walker magnetostatic approach[6] where curl **H** is assumed zero and **H** is taken as grad A. Applying this to the parallel plate configuration of Fig. 4, taking A of the form $A_0\, e^{j\omega t-\beta_x x-j(\beta_y y+\beta_z z)}$ to give an exponential variation normal to the surface, using the boundary condition

$$B_x = \mu H_x - jkH_y = 0 \tag{9}$$

to produce zero normal flux density at both conductor surfaces and substituting

$$H_x = \frac{\delta A}{\delta x}, \quad H_y = \frac{\delta A}{\delta y}$$

in (9) one obtains

$$\beta_x/\beta_y = -k/\mu \tag{10}$$

It should be noted that this relation between β_y and β_x, the real and imaginary components of the transverse wave vector respectively, implies that as indicated in Fig. 4, if the direction of β_y is reversed then the direction of the

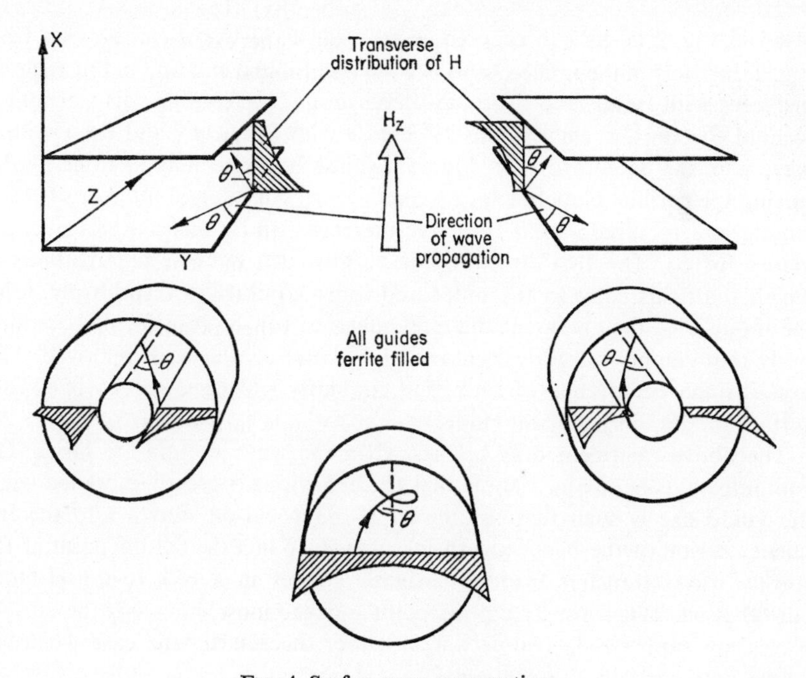

FIG. 4. Surface wave propagation.

normal evanescent decay must also be reversed. This effect precludes the simple reflection of β_y at an XZ plane.

Using div $\mathbf{B} = 0$ one may introduce β_z with the further relation

$$\text{div } \mathbf{B} = \mu\left(\frac{\delta^2 A}{\delta x^2} + \frac{\delta^2 A}{\delta y^2}\right) + \frac{\delta^2 A}{\delta z^2} = 0,$$
$$\mu(-\beta_x^2 + \beta_y^2) + \beta_z^2 = 0, \tag{11}$$

which imposes no stipulation on the sign of β_z.

Eliminating β_x between eqns (10) and (11) gives

$$\frac{\beta_z}{\beta_y} = \left(\frac{k^2 - \mu^2}{\mu}\right)^{1/2} = \cot\theta \tag{12}$$

and since the ratio β_z/β_y depends only on frequency there arises a resonant direction of propagation at an angle θ to the field direction. This effect is illustrated in Fig. 4 and the various orientations of θ with respect to the direction of evanescent decay are shown.

The effective permeability for the normal plane wave,

$$\left(\frac{\mu^2 - k^2}{\mu}\right),$$

given in Fig. 2 is, by eqn (12), equal to $-\cot^2\theta$. Hence it can be seen from Fig. 2 that only in the range $\sqrt{q} < \omega' < q$ is $\cot\theta$ real and that in this region θ increases with frequency. If an additional boundary condition is introduced to hold β_y constant then β_z must decrease with frequency and the resulting wave is of the backward type. Figure 4 shows how this may be done by distorting the parallel plate line into a coaxial line. An integral number of wavelengths are required round the circumference, which sets a series of eigen values for β_y. The two directions of β_y give two circular polarizations in which the fields cling to the outer and inner conductors respectively. Only the mode with the energy at the circumference can be transformed continuously to an empty cylindrical guide as the centre conductor is removed. This explains the occurrence of only one circularly polarized set of backward surface modes in cylindrical guide.

The above treatment only applies when the wave vectors are large. The computed curves of Fig. 1 show that the behaviour is greatly modified when the guide size is such that the lowest H-mode cut-off moves into the frequency region of the backward surface waves. In fact the cut-off point of the surface waves transfers from the original cut-off at $\omega' = q$ to the H-mode cut-off itself. When the H cut-off point moves below $\omega' = \sqrt{q}$ the surface waves are suppressed. This is not, however, necessarily the case when the ferrite only partially fills the guide.

EXPERIMENTAL RESULTS

Figure 5 shows the relationship between measured and calculated results for two diameters of ferrite filled guide, the smaller of which supports backward surface modes. The calculated results are based on the d.c. magnetic

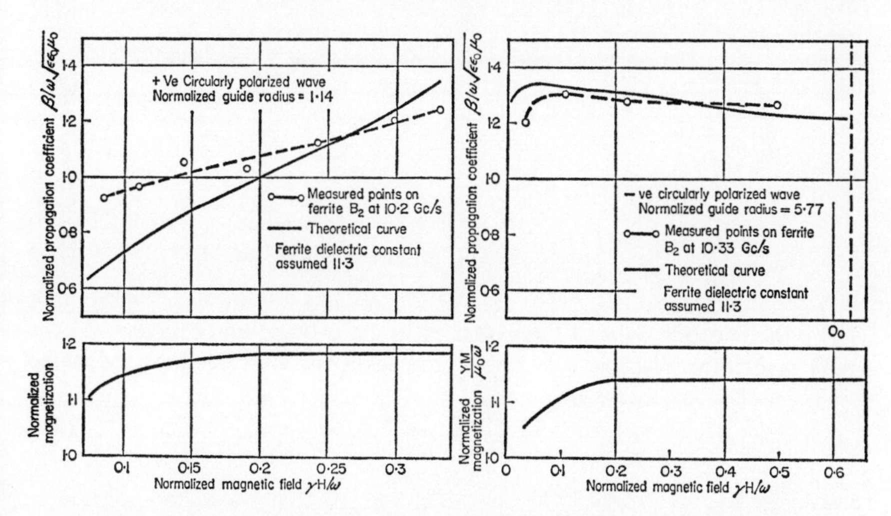

FIG. 5.—Comparison of measured and calculated propagation coefficient (a) in small diameter guide H_{1j} mode (b) in large diameter guide, H_{11} limit mode.

properties of the ferrite and on an adjusted figure for the dielectric constant arranged to give equally good agreement for the two guide sizes. The agreement is seen to be reasonable, particularly in the region where the ferrite is saturated. The negative phase velocity in the smaller guide was confirmed separately.

ACKNOWLEDGEMENT

The author would like to thank Standard Telecommunication Research Laboratories Ltd. for permission to publish the paper.

APPENDIX

Ratio of Field Components for Plane Waves in Ferrites

The ordinary and extraordinary plane waves in a ferrite have complementary elliptical polarization of transverse E and H fields which may be expressed in terms of the axial ratio. They also have a longitudinal component of magnetic field. The ratio of the various field quantities is given in two forms below as a function of either the components of the permeability tensor or the normalized frequency, and of the angle of propagation relative to the magnetic field. The field components are distinguished by subscripts l for the longitudinal components, t for the transverse components in the plane of the magnetic field and t^* for transverse components at right angles to the magnetic field. (t, t^*, l form a right-handed set.)

The axial ratio of the polarization ellipse is given by

$$\begin{aligned}\frac{H_t}{H_{t*}} &= \frac{-j}{2k\cos\theta}[(\mu+k^2-\mu^2)\sin^2\theta \mp \{(\mu+k^2-\mu^2)^2\sin^4\theta+4k^2\cos^2\theta\}^{1/2}] \\ &= \frac{-E_t^*}{E_t}\end{aligned} \tag{A1}$$

or in terms of frequency ω' by

$$\frac{H_t}{H_{t*}} = \frac{-j}{2\omega'\cos\theta}\{q\sin^2\theta \mp (q^2\sin^4\theta + 4\omega'^2\cos^2\theta)^{1/2}\} \tag{A2}$$

where the negative signs of the pair refer to the extraordinary wave.

At resonance where $\omega'^2 = q\sin^2\theta + 2\cos^2\theta$ these expressions reduce to

$$\frac{H_t}{H_{t*}} = \frac{-j\omega'}{\cos\theta} \tag{A3}$$

and

$$\frac{H_t}{H_{t*}} = \frac{j\cos\theta}{\omega'} \tag{A4}$$

for the ordinary and extraordinary waves respectively.

The ratio of the longitudinal to the transverse magnetic field is given by

$$\frac{H_l}{H_{t*}} = \frac{j\sin\theta}{k}\{(\mu-1)\,\mu_{\text{eff}} + \mu + k^2 - \mu^2\} \tag{A5}$$

or in terms of frequency by

$$\frac{H_l}{H_{t*}} = \frac{-j\sin\theta}{\omega'}\{\mu_{\text{eff}} - q\} \tag{A6}$$

The characteristic impedance, relating transverse components of E and H, is of the same form for both waves, namely

$$\begin{aligned}E_t/H_{t*} &= \sqrt{(\mu_0\mu_{\text{eff}}/\epsilon_0\epsilon)} \\ &= \frac{E_{t*}}{H_t}\end{aligned} \tag{A7}$$

These relationships show that, for comparable electric field of the ordinary and extraordinary waves in the resonance region of the latter, the transverse magnetic field in the extraordinary wave is reduced relative to the ordinary wave in inverse ratio to the square root of the corresponding effective permeabilities, whereas the longitudinal component of the extraordinary wave is increased relative to the transverse field of the ordinary directly as the square root of the effective permeabilities, and therefore constitutes the dominant effect.

REFERENCES

1. KALES, M. L. (1953) Modes in waveguides containing ferrites, *J. Appl. Phys.* **24**, 604.
2. VAN TRIER, A. A. Th. M. (1953) Guided electromagnetic waves in anisotropic media, *Appl. Sci. Res.* **B3**, 305.
3. SUHL, H. and WALKER, L. R. (1954) Topics in guided wave propagation through gyromagnetic media, Part I: Completely filled cylindrical guide, *Bell. Syst. Tech. J.* **34**, 5.
4. SEIDEL, H. (1957) The character of waveguide modes in gyromagnetic media, *Bell Syst. Tech. J.* **36**, 409.
5. POLDER, D. (1949) On the theory of ferromagnetic resonance, *Phil. Mag.* **40**, 99.
6. WALKER, L. R. (1957) Magnetostatic modes in ferromagnetic resonance, *Phys. Rev.* **105**, 390.

ON THE PENETRATION OF A STATIC HOMOGENEOUS FIELD IN AN ANISOTROPIC MEDIUM INTO AN ELLIPSOIDAL INCLUSION CONSISTING OF ANOTHER ANISOTROPIC MEDIUM

V. FRANK†

IT IS shown that a homogeneous field in an anisotropic medium will penetrate into an ellipsoidal inclusion of another anisotropic medium in such a manner that the field in the inclusion is also homogeneous. The relation between the outer and the inner field is established, and a method for constructing the tensor of depolarizing coefficients for the system is given. This tensor depends both on the shape of the ellipsoid and on the properties of the symmetrical part of the tensor of permittivity of the outer medium. The case of an elliptic cylinder, which presents some special features, is also considered.

1. INTRODUCTION

The "problem of the ellipsoid" is an old one in field theory, and as far as isotropic media are concerned its solution is found in most standard text-books. In Maxwell's *Treatise on Electricity and Magnetism* (1891) the case of an anisotropic (magnetic) medium inside the ellipsoid is considered; this case presents no special difficulty in comparison with the isotropic case. Recently Eshelby (1957) has given the solution of the problem in connection with *elastic* fields in isotropic solids (the complexity of elastic fields in an isotropic solid and of electrical fields in an anisotropic solid seems to be comparable). The solution of the problem in the case of electric fields in anisotropic media may, beside its academic interest, also have certain applications, e.g. in the theory of the properties of random mixtures of crystallites.

2. THEORY

2.1. *Ellipsoidal Case*

The field in question is described by two time-independent vectors, denoted by $\mathbf{E}$ and $\mathbf{D}$, which are supposed to satisfy everywhere:

$$\text{curl}\,\mathbf{E} = 0 \tag{1}$$

$$\text{div}\,\mathbf{D} = 0 \tag{2}$$

† Physics Department, Technical University of Denmark, Copenhagen.

Further we assume that there exists a linear connection between **D** and **E**:

$$D_i = (\epsilon_{ij} + a_{ij})\, E_j \tag{3}$$

where

$$\epsilon_{ij} = \epsilon_{ji} = \text{real numbers} \tag{4a}$$

and

$$a_{ij} = -\, a_{ji} \tag{4b}$$

In these equations the indices refer to the components of the quantities involved with respect to an arbitrarily chosen system of rectangular axes of reference x_1, x_2, x_3. Summation over repeated indices is implied.

We shall also assume

$$\epsilon_i \;(= \text{principal values of } \tilde{\epsilon}) > 0 \tag{4c}$$

This condition is, for physical reasons, always fulfilled in the static case; but we may note, in passing, that for the problem under consideration it need not be fulfilled as regards the permittivity of the medium in the inclusion. The splitting of the tensor connecting **D** and **E** in a symmetric and an antisymmetric part will prove to be essential.

The field thus specified may represent:

(a) An electrostatic field:

$$\mathbf{E} \to \mathbf{E}, \qquad \mathbf{D} \to \mathbf{D}, \qquad \tilde{\epsilon} \to \tilde{\epsilon}, \qquad \tilde{a} \equiv 0.$$

(b) A magnetostatic field:

$$\mathbf{E} \to \mathbf{H}, \qquad \mathbf{D} \to \mathbf{B}, \qquad \tilde{\epsilon} \to \tilde{\mu}, \qquad \tilde{a} \equiv 0.$$

(c) A stationary distribution of current:

$$\mathbf{E} \to \mathbf{E}, \qquad \mathbf{D} \to \mathbf{i}, \qquad \tilde{\epsilon} \to \tilde{\sigma} \qquad \text{and}$$

$\tilde{a} \neq 0$ (connected with the Hall-effect).

The ellipsoidal surface S dividing the whole space in an *outer region I* and an *inner region II* is specified by

$$b_{ij}x_i x_j = 1 \tag{5a}$$

where

$$b_{ij} = b_{ji} = \text{real numbers} \tag{5b}$$

and

$$b_i \;(= \text{principal values of } \tilde{b}) > 0 \tag{5c}$$

The sources of the field (placed at infinity) are such as to create, in the absence of the ellipsoidal inclusion, a constant value of the field $\mathbf{E}^I$ throughout all space; and at present we *assume* that when the inclusion is introduced a homogeneous field $\mathbf{E}^{II}$ results inside the inclusion, while a dipole-like field is superimposed on $\mathbf{E}^{I}$ in the region I.

Formulated in terms of potentials we thus get:

$$U^I = -E_p^I x_p + \Omega, \qquad x \in I \tag{6a}$$

$$U^{II} = -E_p^{II} x_p, \qquad x \in II \tag{6b}$$

where the potential Ω must satisfy:

$$\epsilon_{ij}^I \frac{\partial^2 \Omega}{\partial x_i\, \partial x_j} = 0, \qquad x \in I \tag{7}$$

(terms in $\tilde{a}_{ij}^I$ are absent because of (4b))
together with the following boundary conditions:

$$\Omega \text{ dipole-like at infinity} \tag{8}$$

$$\Omega = (E_p^I - E_p^{II})\, x_p, \qquad x \text{ on } S \tag{9}$$

$$x_i b_{ip}(\epsilon_{ps}^I + a_{ps}^I)\left(E_s^I - \frac{\partial \Omega}{\partial x_s}\right) = x_i b_{ip}(\epsilon_{ps}^{II} + a_{ps}^{II})\, E_s^{II}, \qquad x \text{ on } S \tag{10}$$

Equation (9) expresses the continuity of U across S; (10) the continuity of the normal component of $\mathbf{D}$ across S; note that a vector normal to S may be chosen as $x_i b_{ip}$.

A solution of (7) with the boundary conditions (8) to (10) does not, of course, exist for arbitrary $\mathbf{E}^{II}$ ($\mathbf{E}^{I}$ regarded as fixed). It will, however, turn out, that a unique $\mathbf{E}^{II}$ exists, which permits *one* solution for Ω, thus demonstrating the correctness of our initial assumption about the qualitative features of the solution.

Next we make the following *real, linear* and *non-singular* coordinate substitution:

$$x_i = \gamma_{is} y_s \triangleq [(\epsilon^I)^{1/2}]_{ij} O_{js} y_s \tag{11a}$$

Here $\tilde{O}$ is an *orthogonal* matrix (to be determined); the "square-root" matrix $(\tilde{\epsilon}^I)^{1/2}$ is in view of (4a) and (4c) uniquely defined as the *real symmetric* matrix, which has as principal values the real positive quantities $\sqrt{\epsilon_i^I}$, the two matrices $\tilde{\epsilon}^I$ and $(\tilde{\epsilon}^I)^{1/2}$ being diagonalized simultaneously by an orthogonal transformation.

The inverse of the substitution (11a) reads:

$$\left.\begin{aligned} y_q &= (\gamma^{-1})_{qs} x_s = (O^{-1})_{qp}[(\epsilon^I)^{-1/2}]_{ps} x_s \\ \text{or} \quad y_q &= O_{pq}[(\epsilon^I)^{-1/2}]_{qs} x_s \end{aligned}\right\} \tag{11b}$$

where the matrix $(\tilde{\epsilon}^I)^{-1/2}$ is defined in analogy with the definition of $(\tilde{\epsilon}^I)^{1/2}$.†

The equations (7) to (10) read as follows in the y-system

$$\frac{\partial^2 \Omega}{\partial y_i \, \partial y_i} = 0, \qquad y \text{ outside } S' \tag{7A}$$

(note: independent of the choice of $\tilde{O}$)

$$\Omega \text{ dipole-like at infinity} \tag{8A}$$

$$\Omega = (E_p^I - E_p^{II}) \, \gamma_{ps} y_s, \qquad y \text{ on } S' \tag{9A}$$

$$\gamma_{ij} y_j b_{ip} (\epsilon_{ps}^I + a_{ps}^I)(\gamma^{-1})_{qs} \frac{\partial \Omega}{\partial y_q}$$

$$= \gamma_{ij} y_j b_{ip} [(\epsilon_{ps}^I + a_{ps}^I) \, E^I - (\epsilon_{ps}^{II} + a_{ps}^{II}) \, E_s^{II}], \qquad y \text{ on } S' \tag{10A}$$

where the surface S' is given by

$$y_s (O^{-1})_{sr} \, [(\epsilon^I)^{1/2}]_{ri} \, b_{ij} [(\epsilon^I)^{1/2}]_{jq} O_{qp} y_p = 1 \tag{5A}$$

The matrix

$$\tilde{A} = (\tilde{\epsilon}^I)^{1/2} \, \tilde{b} (\tilde{\epsilon}^I)^{1/2}$$

is according to (4a) and (5b) a real symmetric matrix. $\tilde{A}$ therefore has three real eigenvalues, which are all positive because a real linear coordinate substitution can only change a closed surface of the second degree into another closed surface (of the same degree).

We are thus able to choose the orthogonal matrix $\tilde{O}$ so that (5A) is on diagonal form:

$$S' : y_i^2 L_i^{-2} = 1 \tag{5B}$$

where we have denoted the (positive) eigenvalues of $\tilde{A}$ by L_i^{-2}. L_i are thus the "lengths" of the semiprincipal axes of the ellipsoid S'.

The problem stated in (7A), (8A), (9A) and (10A) (where $\tilde{\gamma}$ is now a definite matrix and where S' is given by (5B)) is, however, sufficiently close to the known isotropic case to suggest the following representation for Ω (compare with the exposition of the isotropic case in Becker–Sauter (1957)):

$$\Omega = (E_r^I - E_r^{II}) \, \gamma_{rp} N_p^{-1} y_p \, \frac{D(0)}{2} \int_{\mu}^{\infty} \frac{d\lambda}{(L_p^2 + \lambda) \, D(\lambda)} \tag{13}$$

where

$$D(\lambda) = \Big[\prod_i (L_i^2 + \lambda) \Big]^{1/2} \tag{13a}$$

† If the system of coordinate axes is chosen along the principal axes of $\tilde{\varepsilon}^I$, (11a) would read:

$$x_i = \sum_s \sqrt{(\varepsilon_i^I)} \, O_{is} y_s \; (\textit{no sum over } i)$$

Thus in numerical calculations this special coordinate system may be preferred, but in the formalism it seems easier to stick to an arbitrary coordinate system in order to take advantage of the usual rules of matrix algebra.

and u is the *positive* root of the equation:

$$\sum_i \frac{y_i^2}{L_i^2 + u} = 1 \tag{13b}$$

($u = 0$ thus gives the surface S').

The quantities N_p are the *depolarizing coefficients for the ellipsoid* S', defined in the usual way:

$$N_p = \frac{D(0)}{2} \int_0^\infty \frac{d\lambda}{(L_p^2 + \lambda)\, D(\lambda)} \tag{14}$$

Ω, given by (13), is (see R. Becker, *loc. cit.*) formed as the scalar product of a constant vector with the gradient of a Coulomb-like potential; it follows that (7A) and (8A) are satisfied. By inspection it is seen that (9A) is satisfied also. Further, *as shown in the appendix*, (13) also satisfies the only remaining boundary condition (10A), provided the following relation between $\mathbf{E}^I$ and $\mathbf{E}^{II}$ is fulfilled:

$$E_s^I = E_s^{II} + [(\epsilon^I)^{-1/2}]_{sj} N_{jq} [(\epsilon^I)^{-1/2}]_{qr} \; [(\epsilon_{rp}^{II} + a_{rp}^{II}) - (\epsilon_{rp}^I + a_{rp}^I)]\, E_p^{II} \tag{15}$$

where $\tilde{N}$, the tensor of depolarizing coefficients, is given by:

$$N_{jq} = O_{ji}\, N_i (O^{-1})_{iq} \tag{16}$$

We note:

$$\mathrm{Tr}\{\tilde{N}\} = N_{ss} = \sum_i N_i = 1 \tag{17}$$

The elements of $\tilde{N}$ are real numbers, depending on the shape of the ellipsoidal surface S and on the properties of the tensor $\tilde{\epsilon}^I$, but *not* depending on $\tilde{\epsilon}^{II}$, $\tilde{a}^{II}$ or $\tilde{a}^I$. Thus these last mentioned properties of the media I and II only enter the relation between $\mathbf{E}^I$ and $\mathbf{E}^{II}$ through the terms explicitly appearing in (15). Therefore, if we write (15) in the form:

$$\mathbf{E}^I = \mathbf{E}^{II} + (\tilde{\epsilon}^I)^{-1/2}\, \tilde{N} (\tilde{\epsilon}^I)^{-1/2} [\mathbf{D}^{II} - (\tilde{\epsilon}^I + \tilde{a}^I)\, \mathbf{E}^{II}] \tag{18}$$

(the meaning of the symbolic form should be self-evident)
it follows that this formula is correct for an arbitrary (*linear or non-linear, single- or multi-valued*) *connection between* $\mathbf{D}^{II}$ *and* $\mathbf{E}^{II}$. In the case of a multivalued (hysteretic) connection between $\mathbf{D}^{II}$ and $\mathbf{E}^{II}$ (18), of course, only applies to states of the system, which have been established in such a manner, that the inner field *is* homogeneous.

As a special case of (18) we may mention the problem of a metallic ellipsoid in a dielectric medium. In this case we put

$$\mathbf{E}^{II} = 0$$

and

$$\mathbf{D}^{II} = \mathbf{P}^{II} = \frac{1}{V} \mathscr{P}^{II};$$

here $\mathscr{P}^{II}$ and V are the *total* dipole moment and the volume of the ellipsoid S respectively. (18) now gives:

$$\mathbf{E}^I = \frac{1}{V} (\widetilde{\epsilon^I})^{-1/2} \tilde{N} (\widetilde{\epsilon^I})^{-1/2} \mathscr{P}^{II} \tag{19}$$

2.2. *The case of an elliptic cylinder*

This case deserves special consideration because one of the eigenvalues of the matrix $\tilde{b}$ is now zero.

Because (see (12))

$$\det (\tilde{A}) = \det (\tilde{\epsilon}^I) \, . \, \det (\tilde{b}) = 0$$

one of the eigenvalues of $\tilde{A}$ is also zero, or one of the L's is infinite. As this is true independently of $\tilde{\epsilon}^I$, there is a fixed characteristic axis in the problem, which is of course the axis of the cylinder. To simplify matters we choose the coordinate system in the following way:

The third axis (coordinate x_3) is along the axis of the cylinder, while the mutually perpendicular x_1- and x_2-axes are arbitrary in a plane normal to the x_3-axes. Further, the convention will be used, that *greek indices* take the values 1 and 2 and that repeated greek indices are to be summed over these values.

We shall only state the problem and give the solution; the proofs are quite parallel to those in 2.1.

The cylindrical surface C and the potential U are given by (region I and II outside and inside C respectively):

$$C\colon b_{\alpha\beta} x_\alpha x_\beta = 1 \tag{20}$$

$$U^I = - E^I_\alpha x_\alpha - E^I_3 x_3 + \Omega(x_1, x_2), \qquad x \in I \tag{21}$$

$$U^{II} = - E^{II}_\alpha x_\alpha - E^I_3 x_3, \qquad x \in II \tag{22}$$

That Ω depends only on x_1 and x_2 follows from the translational invariance along the x_3-axes, and therefore

$$E^{II}_3 = E^I_3 \tag{23}$$

follows from the continuity of U across C.

The potential Ω must satisfy:

$$\epsilon^I_{\alpha\beta} \frac{\partial^2 \Omega}{\partial x_\alpha \, \partial x_\beta} = 0 \tag{24}$$

$$\Omega \text{ two-dimensional dipole-like at infinity} \tag{25}$$

$$\Omega = (E^I_\alpha - E^{II}_\alpha) \, x_\alpha, \qquad x \text{ on } C \tag{26}$$

$$\left.\begin{aligned} & x_\alpha b_{\alpha\beta}(\epsilon^I_{\beta\gamma} + a^I_{\beta\gamma}) \left(E^I_\gamma - \frac{\partial \Omega}{\partial x\gamma}\right) \\ & \quad = x_\alpha b_{\alpha\beta}(\epsilon^{II}_{\beta\gamma} + a^{II}_{\beta\gamma})\, E^{II}_\gamma + \\ & \qquad + x_\alpha b_{\alpha\beta}[\epsilon^{II}_{\beta 3} + a^{II}_{\beta 3} - (\epsilon^I_{\beta 3} + a^I_{\beta 3})]\, E^I_3, \qquad x \text{ on } C \end{aligned}\right\} \tag{27}$$

In (27), (23) has been used.

We now make the substitution

$$x_\alpha = \gamma_{\alpha\beta} y_\beta \triangleq [(\epsilon^{I*})^{1/2}]_{\alpha\mu} O_{\mu\beta} y_\beta \tag{28}$$

Here the matrix $\sqrt{\epsilon^{I*}}$ is formed from the elements of the 2×2 sub-matrix $\widetilde{\epsilon^{I*}}$ of $\tilde{\epsilon}^I$, which have indices (11), (22) and (12).

$\tilde{O}$ is a 2×2 orthogonal matrix, which is chosen such that C is transformed into:

$$C': y_\alpha^2 L_\alpha^{-2} = 1 \tag{29}$$

where L_α^{-2} are the eigenvalues of the matrix $\sqrt{(\widetilde{\epsilon^{I*}})}\ \tilde{b}\ \sqrt{(\widetilde{\epsilon^{I*}})}$. Then Ω is given by:

$$\Omega = (E^I_\alpha - E^{II}_\alpha)\, \gamma_{\alpha\beta} N_\beta^{-1} y_\beta \frac{D(0)}{2} \int_u^\infty \frac{d\lambda}{(L \ + \lambda)\, D(\lambda)} \tag{30}$$

where

$$D(\lambda) = [(L_1^2 + \lambda)(L_2^2 + \lambda)]^{1/2} \tag{31}$$

and u is the *positive root* of

$$\sum_\alpha \frac{y_\alpha^2}{L_\alpha^2 + u} = 1 \tag{32}$$

For N_β we get

$$N_\beta = \frac{D(0)}{2} \int_0^\infty \frac{d\lambda}{(L_\beta^2 + \lambda)\, D(\lambda)} = \begin{cases} \dfrac{L_2}{L_1 + L_2} & \text{for } \beta = 1 \\[2ex] \dfrac{L_1}{L_1 + L_2} & \text{for } \beta = 2 \end{cases} \tag{33}$$

(the integrals in (30) may also be expressed in terms of elementary functions).

The connection between $\mathbf{E}^I$ and $\mathbf{E}^{II}$ is:

$$\left.\begin{aligned} & E^I_\alpha = E^{II}_\alpha + [(\epsilon^{I*})^{-1/2}]_{\alpha\beta} O_{\beta\gamma} N_\gamma (O^{-1})_{\gamma\theta} [(\epsilon^{I*})^{-1/2}]_{\theta\psi} \\ & \qquad \times [(\epsilon^{II}_{\psi j} + a^{II}_{\psi j}) - (\epsilon^I_{\psi j} + a^I_{\psi j})]\, E^{II}_j \\ & \text{where } j \text{ is summed over } j = 1, 2, \textit{and } 3, \\ & \text{and where } E^{II}_3 = E^I_3. \end{aligned}\right\} \tag{34}$$

The discussion given at the end of Section 2.1 also applies to (34); in the case of a metallic cylinder, however, the further condition $E^I_3 = 0$ must be satisfied in order that the problem is well defined.

APPENDIX

Proof of solution

From (13) and (13b) we get:

$$\left.\frac{\partial\Omega}{\partial y_q}\right|_{u=0} = (E_r^I - E_r^{II})\,(\gamma_{rq} - \gamma_{rp}N_p^{-1}L_p^{-2}L_q^{-2}y_p y_q\,M^{-1}), \quad \textit{no}\text{ sum over } q \qquad \text{(A1)}$$

where $M = y_i^2 L_i^{-4}$

The left-hand side of (10A) then becomes:

$$\gamma_{ij}y_j b_{ip}(\epsilon_{ps}^I + a_{ps}^I)(E_s^I - E_s^{II}) - \{(E_r^I - E_r^{II})\,\gamma_{rl}N_l^{-1}L_l^{-2}y_l\}.\{\gamma_{ij}b_{ip}(\epsilon_{ps}^I + a_{ps}^I)(\gamma^{-1})_{qs}L_q^{-2}y_j y_q\}.M^{-1}$$

Now, from (11a), (5A) and (5B) we have:

$$\gamma_{ij}b_{ik}\gamma_{kn} = L_j^{-2}\delta_{jn} \qquad (\textit{no}\text{ sum over } j)$$

or

$$\gamma_{ij}b_{ip} = L_j^{-2}(\gamma^{-1})_{jp} \qquad (\textit{no}\text{ sum over } j) \qquad \text{(A2)}$$

The last curled bracket in the expression above thus becomes, with help of (11a) and (4a):

$$\begin{aligned} &L_j^{-2}(O^{-1})_{jm}[(\epsilon^I)^{-1/2}]_{mp}(\epsilon_{ps}^I + a_{ps}^I)(O^{-1})_{qn}[(\epsilon^I)^{-1/2}]_{ns}L_q^{-2}y_j y_q \\ &\quad = L_j^{-2}(O^{-1})_{jm}\left\{\delta_{mn} + [(\epsilon^I)^{-1/2}]_{mp}\,a_{ps}[(\epsilon^I)^{-1/2}]_{sn}\right\}O_{nq}L_q^{-2}y_j y_q \\ &\quad = L_j^{-2}\delta_{jq}L_q^{-2}y_j y_q = M \text{ (according to (A1))} \end{aligned}$$

[the matrix $(\tilde{\epsilon^I})^{-1/2}\,\tilde{a}^I(\tilde{\epsilon^I})^{-1/2}$ is antisymmetric, so that the corresponding terms in total amount to zero]. Therefore (10A) becomes

$$\begin{aligned} &\gamma_{ij}y_j b_{ip}(\epsilon_{ps}^I + a_{ps}^I)(E_s^I - E_s^{II}) - (E_r^I - E_r^{II})\,\gamma_{rj}N_j^{-1}L_j^{-2}y_j \\ &\qquad = \gamma_{ij}y_j b_{ip}[(\epsilon_{ps}^I + a_{ps}^I)\,E_s^I - (\epsilon_{ps}^{II} + a_{ps}^{II})\,E_s^{II}] \end{aligned}$$

This is satisfied for all y's on S' if and only if:

$$\begin{aligned} (E_r^I - E_r^{II})\,\gamma_{rj}N_j^{-1}L_j^{-2} &= \gamma_{ij}b_{ip}(\Delta\epsilon)_{ps}E_s^{II} \\ &= L_j^{-2}(\gamma^{-1})_{jp}\,(\Delta\epsilon)_{ps}\,E_s^{II} \\ &\qquad \text{(according to (A2))} \end{aligned} \qquad \textit{no}\text{ sum over } j$$

Here

$$\widetilde{\Delta\epsilon} = \tilde{\epsilon}^{II} + \tilde{a}^{II} - (\tilde{\epsilon}^I + \tilde{a}^I)$$

Thus

$$E_i^I - E_i^{II} = (\gamma^{-1})_{ji}\,N_j(\gamma^{-1})_{jp}\,(\Delta\epsilon)_{ps}\,E_s^{II}$$

Use of (11a) and (4a) directly leads to the result (15).

REFERENCES

BECKER, R. and SAUTER, F. (1957) *Theorie der Elektrizität*, vol. I, §30, 16 Aufl.

ESHELBY, J. D. (1957) Determination of the elastic field of an ellipsoidal inclusion and related problems, *Proc. Roy. Soc.* **A241**, 376–396.

MAXWELL, J. C. (1954) *Electricity and Magnetism*, 3rd edition 1891. Dover Reprint, ii. 66 (1954).

A NOTE ON THE OBLIQUE INCIDENCE OF ELECTROMAGNETIC WAVES UPON AN ABSORBING SLAB

KIYOSHI MORITA

Oki Electric Industry Co., Ltd., Tokyo, Japan

ABSTRACT

R. F. Hiatt and others reported that the scattering cross-section for a metallic sphere covered with a thin absorber sheath is larger in some cases than that of a bare metallic sphere, a fact appearing somewhat strange from the ordinary point of view. The author aims to give reasons for this, and proposes the theory that the large back scattering cross-section comes from the reflection of the partial waves impinging obliquely on the target. He describes how obliquely incident waves are absorbed well by a lossy sheath, when the sheath is composed not of a single homogeneous layer but of multi-layer absorbers having stepwisely increasing lossy characteristic. He also calculated the case of a cylindrical target, and suggests that Hiatt's result would be modified if the absorber were composed of suitably designed multi-layers.

1. PLANE ABSORBER

In the succeeding two papers about absorbing coating on a conducting object, written by three authors, R. F. Hiatt, K. M. Siegel, and H. Weil, there is a description on ineffectiveness of coating in relation to forward scattering. It sounds very curious to hear that the forward scattering is augmented by coating when the wavelength is short. Since the conducting object was spherical in shape, and the wave will impinge on the coated surface generally in an oblique angle, and wave energy would not be absorbed effectively in surface film, especially at the location where the angle of incidence is very large, this may result in the ineffectiveness of coating as reported. The author calculated the reflection coefficient of a multi-layer lossy dielectric slab and showed the effectiveness in adsorbing action as compared to the mono-layer film for the case of oblique incidence.

Figure 1 shows the construction of lossy slabs. Slab No. 1 is largest both in propagation constant and in loss factor and slabs numbering greater than one have larger characteristic impedance, so that matching can be attained between the impedances of air and slab.

Calculation for reflection coefficient is given by considering that the tangential propagation constant is continuous for adjacent layer.

For each layer, when θ_r represents the incident angle to rth layer, we have

Propagation constant in Z-direction:

$$\Gamma_{zr} = j\beta_0\sqrt{\dot{\epsilon}_r}\, f(\theta_r)$$

Characteristic impedance:

$$K_r = (120\pi/\sqrt{\dot{\epsilon}_r})\, f(\theta_r)$$

when E is tangential to layer

$$= (120\pi/\sqrt{\dot{\epsilon}_r})\, [(1/f(\theta_r)]$$

when H is tangential
where

$$f(\theta_r) = \sqrt{[1 - (\sin^2 \theta_r/\dot{\epsilon}_r)]}$$

$\dot{\epsilon}_r$ = complex dielectric constant of the rth layer.

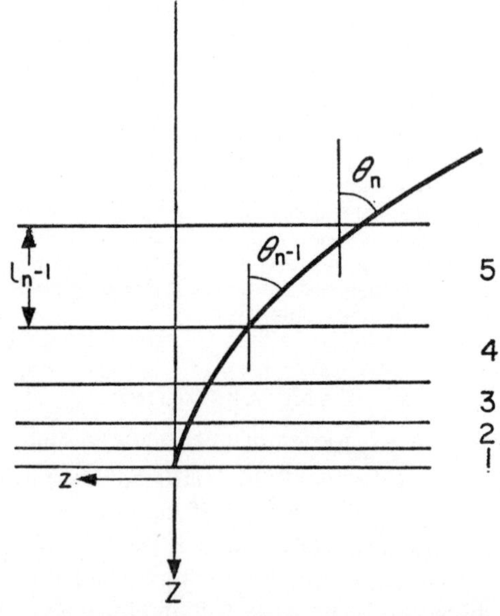

FIG. 1. Oblique incidence of waves upon stratified layers.

The multi-layer absorber is so designed that the phase angle shift in each layer is nearly equal to $\pi/2$ and the logarithm of K_r is a linear function of $(\pi/2)r$.

Moreover, it is required to get matching for the wave vertically incident on the slab, and attenuation in each slab is so designed as to satisfy this requirement. Also the last one of these multi-layer slabs is backed by a metallic plate so as to make an absorbing wall to fit for a shielding room. Voltage reflection coefficients for the waves with the electric field tangential to the surface are calculated, with the results shown in Fig. 2, Table 1 represents material constants for each slab. The dotted and cut lines, curves No. 1′ and 1″ respectively in Fig. 2, represent the case of a single-layer absorber with thickness and characteristic impedance as follows.

No. 1′ $1/\lambda = 0{\cdot}325$ $\dot{K}/K_0 = 0{\cdot}866 < 26° \, 39'$

No. 1″ $1/\lambda = 0{\cdot}058$ $\dot{K}/K_0 = 0{\cdot}225 < 8°$

We see that the thin slab No. 1″ is more effective than the thick one No. 1′, and yet it is inferior to the multi-layer slabs for oblique incidence. After these results, the author wants to suggest using multi-layer absorbing coating to avoid forward scattering.

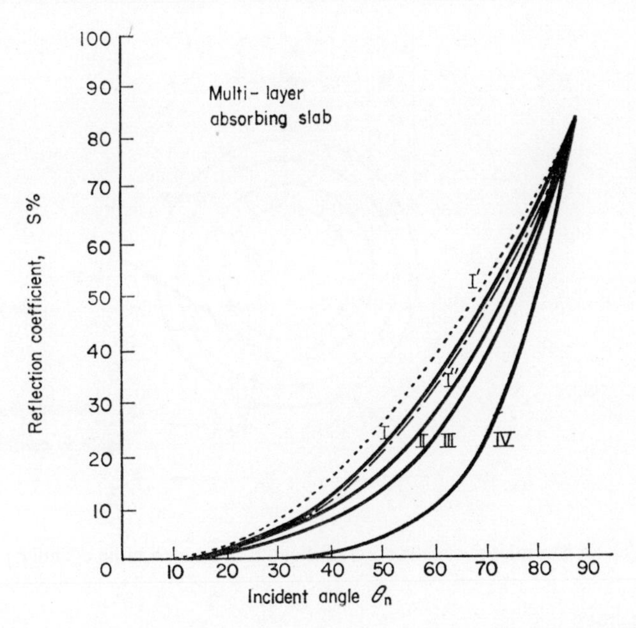

FIG. 2. Incident angle and reflection coefficients numbers of layers as parameter.

2. ABSORBER ON METALLIC CYLINDER

The problem of scattering electromagnetic waves by infinite cylinders coated with thin absorbing layers is solved by the following procedure. Here we assume that the layers are composed of several absorbing sheets having different electrical properties. The accompanying figure represents one typical example of such cylindrical construction (Fig. 3).

In Fig. 3 the innermost circle with radius a_0 represents a metallic cylinder, whereas circles characterized by their radii a_1, a_2, a_3, show that there is a series of absorbing layers, having peculiar material constants $\dot{k}_1, \dot{k}_2, \dot{k}_3$ respectively, where $\dot{k}$ means

$$\dot{K} = \frac{2\pi}{\lambda_0}\sqrt{(\epsilon' + i\epsilon'')}$$

ϵ' = dielectric constant,
ϵ'' = lossy constant.

When plane waves denoted by wave potential ϕ_1 such as

$$\phi_1 = e^{i(kx-\omega t)}$$

impinge on the cylinder, the waves will be scattered.

For analysis, we express ϕ_1 in cylindrical coordinates Z, θ and r, taking the center line of the cylinder as Z-axis.

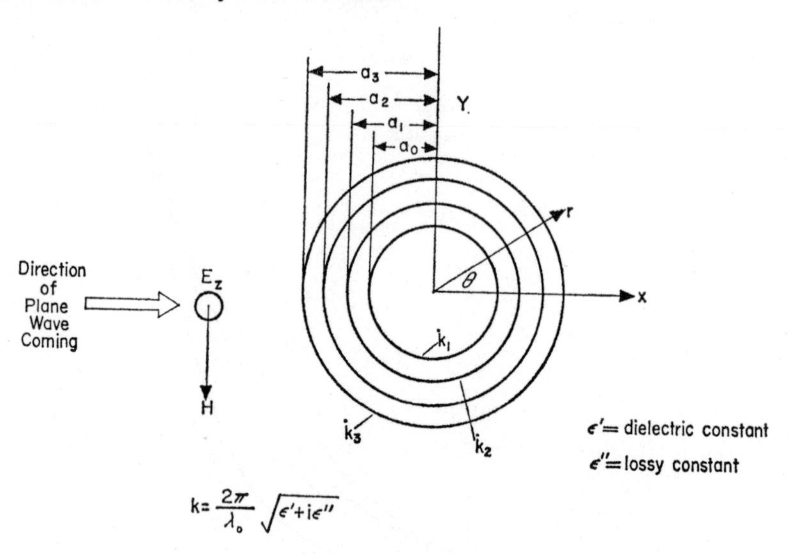

FIG. 3. Multi-layer construction of dielectrics on a metallic cylinder.

Thus we have

$$\phi_1 = \sum_{m=0}^{\infty} {}_1\phi_m$$

$${}_1\phi_m = \epsilon_m i^m \cos m\theta J_m(kr) \tag{1}$$

where $\epsilon_m = 1$ When $m = 0$
$\epsilon_m = 2$ $m > 0$

This is the expression for plane waves. Now we take the Neumann function to represent secondary waves, being excited and emerging from the cylinder, i.e.

$$\phi_2 = \sum_{m=0}^{\infty} {}_2\phi_m$$

$${}_2\phi_m = \epsilon_m i^m \cos m\theta Y_m(kr) \tag{2}$$

Linear combinations of these two functions will give the potential of scattered waves, and the coefficients should be determined by boundary conditions.

In general, the potential function ϕ relates to the electric field E_z and the magnetic field H_θ by the formulas

$$E_z = i\omega\mu\phi; \qquad H_\theta = \frac{\partial\phi}{\partial r} \tag{3}$$

2.1. *Single-Layer Absorber*

Figure 4 shows a metallic cylinder coated with a single-layer absorber.

In the region of open air, we have the following combination of Bessel and Neumann functions,

$$\phi = \sum_{m-0}^{\infty} \epsilon_m i^m \cos m\theta[J_m(kr) \cos \eta_m + Y_m(Kr) \sin \eta_m] \,.\, e^{-i\eta_m} \,.\, e^{-i\omega t} \tag{4}$$

after considering the ultimate expression for the case of r tending to infinity. In this expression η_m is independent of radius r or angle θ.

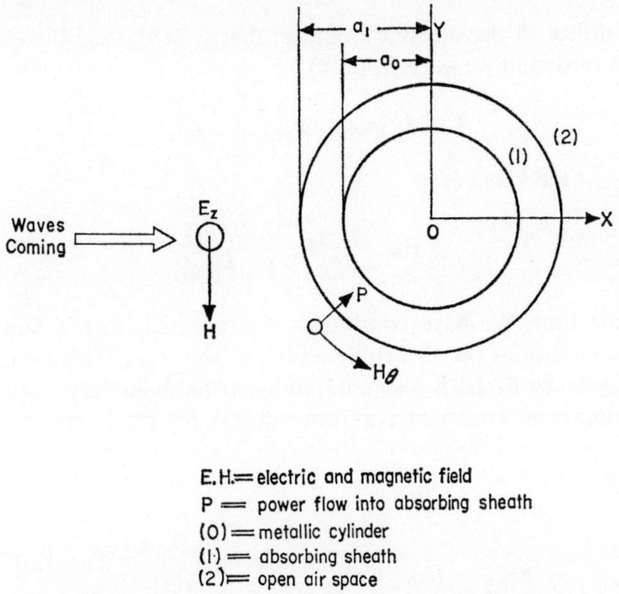

FIG. 4. Single layer dielectric on a metallic cylinder.

For the proof, we have to remember the asymptotic expression of J_m and Y_m.

$$J_m(K_r) \underset{r\to\infty}{=} \sqrt{\left(\frac{2}{\pi kr}\right)} \cos\left[k_r - \frac{\pi}{2}\left(m + \frac{1}{2}\right)\right]$$

$$Y_m(K_r) \underset{r\to\infty}{=} \sqrt{\left(\frac{2}{\pi kr}\right)} \sin\left[k_r - \frac{\pi}{2}\left(m + \frac{1}{2}\right)\right] \tag{5}$$

and inserting these into equation (3) will give

$$\phi = \sqrt{\left(\frac{1}{2\pi kr}\right)} \sum \epsilon_m \cos m\theta(-1)^m e^{-ikr} + e^{ikr} \times e^{-i(2\eta_m + \frac{\pi}{2})} \tag{6}$$

Here we notice, after considering the case of $\theta = 0$ or π, that the term $e^{-i2\eta_m}$ expresses the reflection coefficient of the waves originating from the presence of the cylinder. Yet the value of η_m is not yet determined. It should be evaluated from the boundary condition at the surface of the cylinder. Equation (5) shows that the linear combination of Bessel and Neumann functions in eqn (3) is quite correct.

In the interior part of the absorbing layer, the range denoted as (1) in Fig. 4, we have the expression of the wave potential as follows:

$$\phi = \sum_{m=0}^{\infty} \epsilon_m i^m \cos m\theta [A_m J_m(\dot{k}_1 r) + B_m Y_m(\dot{k}_1 r)] \times e^{-i\omega t} \tag{7}$$

and electric field E_z and magnetic field H_θ can be deduced from the equation already cited.

At the surface of the metallic cylinder, that is at $r = a_0$, the electric field E_z must be zero, and we have from (6),

$$A_m J_m(\dot{k}_1 a_0) + B_m Y_m(\dot{k}_1 a_0) = 0$$

irrespective of the value of m

$$B_m = -A_m \frac{J_m(\dot{k}_1 a_0)}{Y_m(\dot{k}_1 a_0)} \tag{8}$$

At the same time, we have continuity of tangential electric and magnetic fields at the surface of the absorbing layer, i.e. at $r = a_1$. This means that the surface impedance E_z/H_θ is also continuous at the boundary, independently of mode number m. Thus we have from eqns (3), (6), (7) and (8)

In sheath

$$Z_m = \left(\frac{Ez}{H\theta}\right) = \frac{i\omega\mu}{\dot{k}_1} \times \frac{J_m(\dot{k}_1 a_1) - \dfrac{J_m(\dot{k}_1 a_0)}{Y_m(\dot{k}_1 a_0)} Y_m(\dot{k}_1 a_1)}{J'_m(\dot{k}_1 a_1) - \dfrac{J_m(\dot{k}_1 a_0)}{Y_m(\dot{k}_1 a_0)} Y^1_m(\dot{k}_1 a_1)} \tag{9}$$

In air

$$\begin{aligned} Z_m = \left(\frac{Ez}{H\theta}\right) &= \frac{i\omega\mu}{k} \times \frac{J_m(ka_1)\cos\eta_m + Y_m(ka_1)\sin\eta_m}{J'_m(ka_1)\cos\eta_m + Y^1_m(ka_1)\sin\eta_m} \\ &= iZ_{co} \cdot \frac{J_m(ka_1) + \tan\eta_m Y_m(ka_1)}{J^1_m(ka_1) + \tan\eta_m Y^1_m(ka_1)} \end{aligned} \tag{10}$$

$$Z_{co} = 120\pi \text{ ohm.} \tag{11}$$

Equality of the two expressions (9) and (10) leads us to conclude that the value η_m can be calculated, and henceforth the reflection coefficient can also be computed.

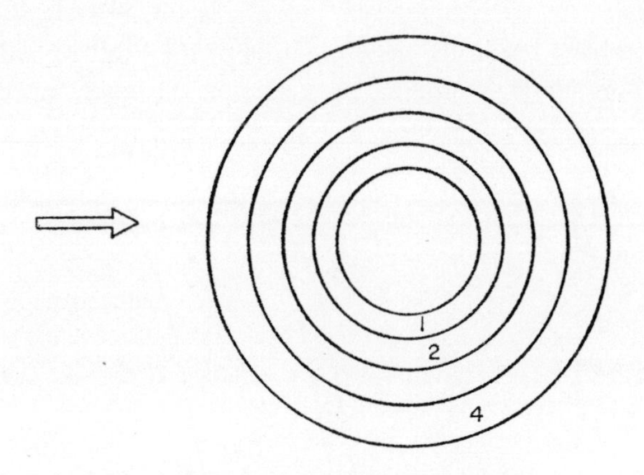

FIG. 5. Multi-layer construction of dielectrics.

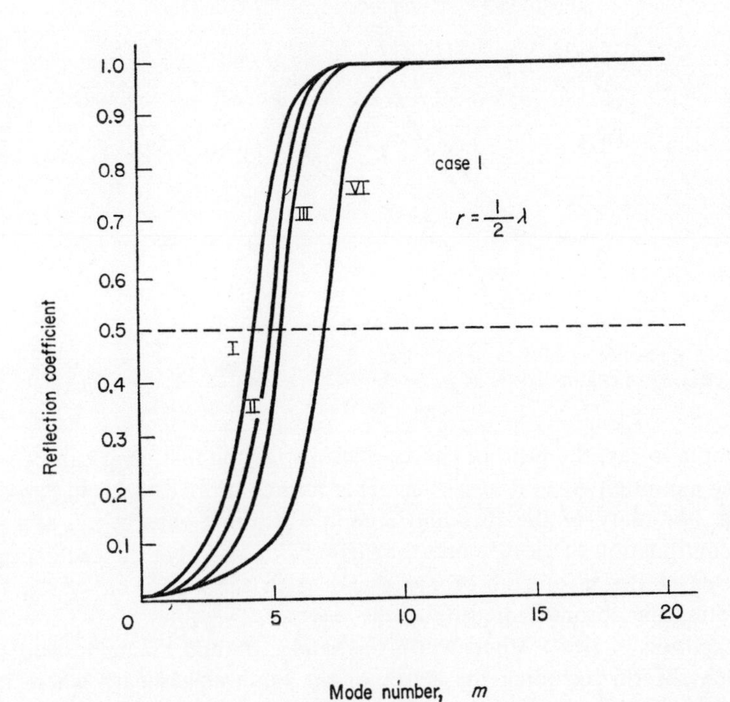

FIG. 6. Reflection coefficient of cylindrical waves as a function of their mode number with number of coating layers as parameter (when radius of metallic cylinder is equal to $\frac{1}{2}\lambda$).

2.2. *Multi-Layer Absorber*

In the case of a multi-layer absorber, as shown in Fig. 5, the solutions for electric and magnetic fields are obtained by nearly the same process as before.

Here the problem lies in the determination of the value of two coefficients in eqn (7). They can be determined in succession starting from the values for the innermost absorbing sheath.

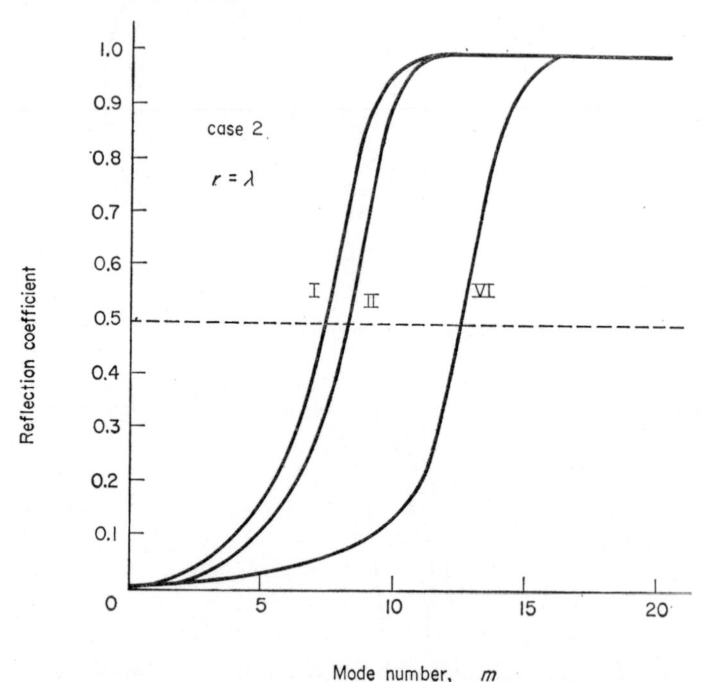

FIG. 7. Reflection coefficient of cylindrical waves as a function of their mode number with numer of coating layers as parameter (when radius of metallic cylinder is equal to λ).

That is to say, the ratio of two coefficients for the first layer can be settled by the condition $E_z = 0$ at $r = a_0$. This result will give the field expression at the boundary of the first and second absorbing sheath, i.e. at $r = a_1$, and continuation of electric and magnetic fields will give the full expression of fields in the second absorbing sheath with the exception of the factor indicating the absolute intensity of the waves.

Repetition of this method will give the electric and magnetic field status in each sheath, including the status in the outermost sheath where it will touch the outside air extending to infinity.

Here again we take into consideration the continuation of both electric and magnetic fields. We can then decide the absolute intensity of the waves in the interior part of the absorber after taking into account the intensity of plane waves impinging on the cylinder.

The reflection coefficients η_m can also be determined by the same procedure as already mentioned in eqn (10).

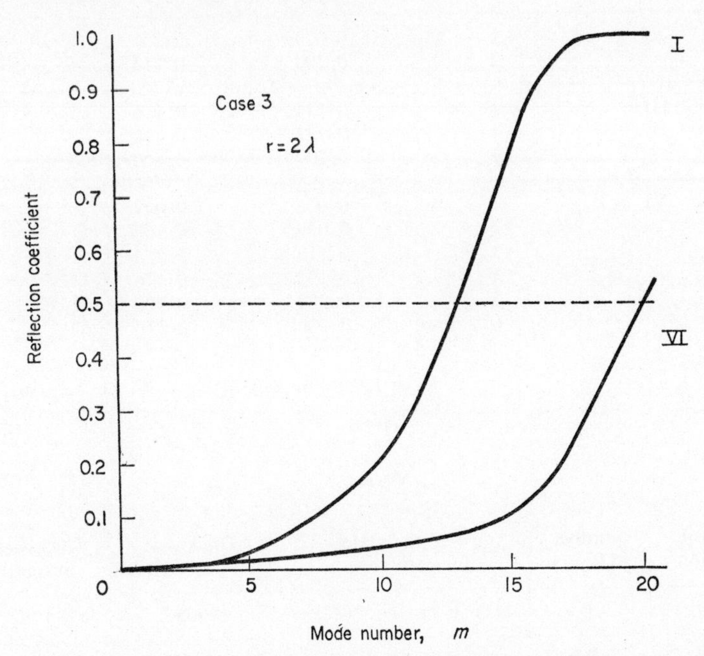

FIG. 8. Reflection coefficient of cylindrical waves as a function of their mode number with number of coating layers as parameter (when radius of metallic cylinder is equal to $1\frac{1}{2}\lambda$).

2.3. *Calculated Result*

We calculated the absolute value of $e^{-i2\eta m}$, that is the reflection coefficient for m mode excitation, for the case of a multi–layer absorbing sheath on conducting cylinders of various sizes. Table 2 gives a list of the various objects under calculation.

The radius of a conducting cylinder is varied for three values, i.e.

Case 1, $r = \frac{1}{2}\lambda$

Case 2, $r = \lambda$

Case 3, $r = 2\lambda$

The following figures give the result of calculation, which shows that the multi-layer absorber consisting of the material with graded lossy characteristics works well for attenuating the secondary radiation from the metallic cylinder.

TABLE 1

Specimen		l/λ	$(K\angle\phi)/K_0$
I	1	0·178	0·584 $\angle 18°$
II	1 2	0·078 0·200	0·256 $\angle 18°$ 0·714 $\angle 5°$
III	1 2 3	0·034 0·122 0·207	0·116 $\angle 15°$ 0·436 $\angle 5°$ 0·800 $\angle 3°$
VI	1 2 3 4 5 6	0·055 0·069 0·116 0·195 0·325 0·300	0·087 $\angle 20° 10'$ 0·168 $\angle 11° 50'$ 0·283 $\angle 11° 50'$ 0·476 $\angle 11° 50'$ 0·800 $\angle 11° 50'$ 0·910 $\angle 3°$

$K_0 = 120\pi$ ohm.

TABLE 2

Classification	Number of layers	$\sqrt{\dot{\varepsilon}}$ of the material for each sheath	Thickness of each sheath	Thickness in total
I	1	(1) 1·72 $\angle -18°$	0·178λ	0·178λ
II	2	(1) 3·91 $\angle -18°$ (2) 1·40 $\angle -5°$	0·078λ 0·200λ	0·278λ
III	3	(1) 8·63 $\angle -15°$ (2) 2·30 $\angle -5°$ (3) 1·24 $\angle -3°$	0·034λ 0·122λ 0·207λ	0·363λ
VI	6	(1) 11·50 $\angle -20° 10'$ (2) 5·95 $\angle -11° 50'$ (3) 3·54 $\angle -11° 50'$ (4) 2·10 $\angle -11° 50'$ (5) 1·25 $\angle -11° 50'$ (6) 1·10 $\angle -3°$	0·033λ 0·069λ 0·116λ 0·195λ 0·325λ 0·300λ	1·038λ

ACKNOWLEDGMENT

The author wishes to thank Mr. Yoshiyuki Naito at the Tokyo Institute of Technology for his valuable contribution to this paper.

PROPAGATION OF ELECTROMAGNETIC WAVES IN AN ANISOTROPIC STRATIFIED MEDIUM

R. W. HOUGARDY† and D. S. SAXON‡

SUMMARY

In the following we discuss approximation methods for describing the propagation of electromagnetic waves in an anisotropic stratified medium. By a stratified medium we mean one with electrical properties which are functions of only a single rectangular coordinate, say the z-coordinate for definiteness. Our methods are most simply applied when the properties of the medium vary "slowly" and continuously with z, and we shall assume this to be the case. Aside from these limitations on their spatial dependence, the constitutive tensors of the medium are taken to be unrestricted.

To furnish a background for our approach, consider first plane wave propagation in an *homogeneous* anisotropic medium. Normally, this problem is treated by introducing plane waves travelling in a specified direction with an unknown wave number and state of polarization, to be determined from Maxwell's equations. Instead, because of the special role of the z-direction in our deliberations, we pose the problem in a different but equivalent way. Let us suppose the x, y components of the wave vector to be given. The z-component of the wave vector, and the state of polarization, are then required in order to specify a solution. Because of the divergence conditions, only four of the six field components are independent. Hence, Maxwell's equations are equivalent to four linear homogeneous equations whose solutions yield four modes, two travelling towards positive z, two towards negative z, each with its characteristic state of polarization.

If the medium is not homogeneous, these four modes become coupled together; transitions between polarization states occur as well as reflections. We shall give a method for treating, in a stratified medium, the situation when the four modes can be separated into two pairs, the members of each pair being more strongly coupled to each other than to either member of the other pair. This situation is common and of great interest; it occurs, for example, in the neighbourhood of turning points (real or complex) or in

† Hughes Aircraft Co., Los Angeles, California, U.S.A.

‡ Faculty of Sciences, University of Paris, Orsay, France. Guggenheim Fellow and Fulbright Lecturer, 1961–1962, on leave from University of California, Los Angeles 24, California, U.S.A.

transition regions. Our method consists simply of the treatment of each pair of more strongly coupled modes together to obtain two second order scalar wave equations, weakly cross-coupled. If this cross-coupling is neglected in first approximation, the two independent solutions of each scalar wave equation generate a corresponding pair of "modes" in which full account is taken of the internal coupling. The weak cross-coupling can then be handled by iteration methods. For the practicability of our approach, it must be supplemented by a method for dealing effectively with the one–dimensional scalar wave equation. Such a method has recently been developed.[1, 2]

Propagation in a stratified medium can be analysed by considering solutions of Maxwell's equations analogous to the plane wave solutions for a homogeneous medium discussed above. Thus we write the fields in the form

$$\left.\begin{aligned} E(x, y, z, t) &= E(z)\, e^{ipx+iqy-iwt} \\ H(x, y, z, t) &= H(z)\, e^{ipx+iqy-iwt} \end{aligned}\right\} \tag{1}$$

where $E(z)$ and $H(z)$ are vector functions to be determined. The transverse components (p, q) of the wave vector characterize the obliquity of the propagation; for $p = q = 0$, propagation proceeds along the z-axis alone.

With the assumed form for the solution, the z-components of E and H are readily eliminated from Maxwell's equations and the resulting four coupled first order differential equations in the transverse components of E and H completely determine the propagation properties of the medium. These equations can be compactly written as follows[3]:

$$\frac{dW}{dz} = i\, M(z)\, W, \tag{2}$$

where $W(z)$ is a four component column vector whose elements are (proportional to) the four transverse components of E and H taken in some order. The quantity $M(z)$ is a four by four matrix with elements which are a complicated combination of the elements of the constitutive tensors and which depend on the obliquity of propagation through p and q. We shall call M the *propagation matrix* of the medium; we shall assume that it can always be diagonalized by a similarity transformation.

For a homogeneous medium, the four eigenvalues of M are the propagation constants of the medium, two of which describe propagation toward positive z, two toward negative z. The polarization modes are the corresponding eigenvectors of M. A familiar example occurs in homogeneous ionospheric models where, under appropriate conditions, there exist the well-known rising and downcoming ordinary and rising and downcoming extraordinary waves.

For an inhomogeneous medium, we transform to local principal axes by introducing the matrix $B(z)$ which diagonalizes $M(z)$. That is, B is defined by

$$B^{-1}(z)\, M(z)\, B(z) = K(z), \tag{3}$$

where $K(z)$ is diagonal. With $W = BX$, eqn (2) becomes

$$\frac{dX}{dz} = i\, K(z)\, X - B^{-1} \frac{dB}{dz} X. \tag{4}$$

The second term on the right determines the coupling between the locally defined propagation modes. In general, it includes both the coupling between waves of similar polarization type travelling in opposite directions (reflection) and between waves of different polarization type (transitions).[3] Also, in general, if the coupling term is large, there is no particular advantage to transforming to local principal axes, one must in any case treat the fourth order equation, equivalent to eqns (2) or (4), in its full complexity. No suitable analytic methods for doing so are known to us.† However, as we have stated, our interest is mainly in the case when the properties of the medium vary only slowly with z. Specifically B^{-1} (dB/dz) is to be regarded as small compared to K, except in the neighbourhood of turning points or transition points, if any. If the off–diagonal terms in eqn (4) are neglected, the resulting uncoupled equations yield the standard *WKB* solution for each mode. When the off-diagonal terms are small, reflection and transition coefficients can be obtained iteratively by the method of variation of constants. Such methods have been well studied and we shall not discuss them further, except to emphasize that they break down whenever the coupling between a pair of modes becomes large, as near a turning point or in a transition region.‡ As we have stated, it is this interesting case we want to treat.

Suppose that, in the notation of eqn (4), X_1 and X_2 are strongly coupled to each other and also, for generality, that X_3 and X_4 are strongly coupled. To handle the strong coupling, one might proceed optimally as follows: First neglect the cross-coupling. Then construct Y_1 and Y_3 as appropriate independent linear combinations of X_1 and X_2. Elimination of Y_3, say, then

† See, however, H. B. Keller, N.Y.U. Inst. Math. Sci. Res. Rep. EM–56 (1953) where exact formal solutions are obtained. The solutions involve an infinite series of integrals and hence appear to be practically useful only for weak coupling.

‡ In general, turning and transition points actually occur for complex values of z. Their importance increases with their nearness to the real axis. We remark that, so far as we know, no *general* study of the conditions for strong coupling between a pair of modes has been given. In reference 3 it is stated merely that the coupling between two modes is strong when their propagation constants become equal. This condition is satisfied at a turning or transition point. It is clearly not a necessary condition. That it is also not a sufficient condition follows from consideration of propagation in an isotropic medium. See reference 1 for a consideration of strong coupling in some special cases.

yields a second order equation in Y_1, the "scalar wave equation", the two independent solutions of which are new "modes" with the internal coupling taken fully into account. Similarly for Y_2 and Y_4 in terms of X_3 and X_4. The weak cross-coupling can then be handled by iteration methods.

The particular linear combinations which are most convenient depend on the detailed nature of the propagation and coupling matrices. We now give an explicit recipe[1] which ignores such fine points. Write

$$Y_1 = X_1 + X_2, \qquad Y_2 = X_3 + X_4, \qquad Y_3 = X_2, \qquad Y_4 = X_4 \,,$$

or in matrix notation

$$Y = \gamma^{-1} X \,.$$

Substitution into eqn (4), followed by elimination of Y_3 and Y_4, yields the *two* component second order equation, in matrix form,

$$\frac{\mathrm{d}^2 Y^0}{\mathrm{d}z^2} + Q\frac{\mathrm{d}Y^0}{\mathrm{d}z} + SY^0 = 0; \qquad Y^0 = \begin{pmatrix} Y_1 \\ Y_2 \end{pmatrix}, \tag{5}$$

where

$$Q = P_{11} + P_{12}P_{22}P_{12}^{-1} - P_{12}\frac{\mathrm{d}P_{12}}{\mathrm{d}z},$$

$$S = P_{12}P_{22}P_{12}^{-1}P_{11} - P_{12}P_{21} - \frac{\mathrm{d}P_{11}}{\mathrm{d}z} - P_{12}\frac{\mathrm{d}P_{12}}{\mathrm{d}z}P_{11},$$

and where the P_{ij} are two by two matrices defined by

$$P \equiv \begin{Bmatrix} P_{11} & P_{12} \\ P_{21} & P_{22} \end{Bmatrix} \equiv \gamma^{-1}\left(iK - B^{-1}\frac{\mathrm{d}B}{\mathrm{d}z}\right)\gamma \,.$$

To reduce (5) to canonical form, write,

$$Y_1 = Z_1 \mathrm{e}^{-1/2\int Q_{11}\mathrm{d}z}, \qquad Y_2 = Z_2 \mathrm{e}^{-1/2\int Q_{22}\mathrm{d}z}$$

whence finally

$$\frac{\mathrm{d}^2 Z_1}{\mathrm{d}z^2} + \left(S_{11} - \tfrac{1}{4}Q_{11}^2 - \tfrac{1}{2}\frac{\mathrm{d}Q_{11}}{\mathrm{d}z}\right)Z_1$$

$$= \left[(S_{12} - \tfrac{1}{2}Q_{12}Q_{22})\,Z_2 + Q_{12}\frac{\mathrm{d}Z_2}{\mathrm{d}z}\right]\mathrm{e}^{-1/2\int(Q_{22}-Q_{11})\mathrm{d}z} \tag{6}$$

and similarly for Z_2. To understand the structure of this equation, we remark that if the medium were homogeneous, the right side would vanish and Z_1 would satisfy

$$\frac{\mathrm{d}^2 Z_1}{\mathrm{d}z^2} + \left(\frac{K_1 - K_2}{2}\right)^2 Z_1 = 0 \,,$$

with $Y_1 = Z_1 e^{i(K_1+K_2)z/2}$. The more complicated coefficient of Z_1 in equation (6) is similar but is supplemented by terms involving the derivative of the electrical parameters of the medium, while the right side involves only such derivatives and is, of course, taken to be small. Note that, regardless of the algebraic signs of the K's, the solutions of eqn (6) always involve rising and descending waves; the strong coupling effects are expressed in the familiar language of reflection and transmission coefficients. Thus, for example, a transition between two rising polarization modes finds expression in the existence of a turning point in eqn (6).

If solutions of the homogeneous equations (that is, eqn (6) with the right side set equal to zero) can be found, the cross–coupling can then be taken into account by iteration methods. Some specific examples for ionospheric propagation have been worked out in reference (1) and a detailed account will be published.

REFERENCES

1. Hougardy, R. W., Ph.D. Thesis, Dept. of Physics, UCLA, Los Angeles 24, California.
2. Hougardy, R. W. and Saxon, D. S., to be published.
3. Clemmow, P. C. and Heading, J. (1954) *Proc. Camb. Phil. Soc.* **50**, Pt. 2, 319.

DIFFRACTION AT HIGH FREQUENCIES IN A STRATIFIED MEDIUM

D. S. JONES

Department of Mathematics, University of Keele, Keele, Staffordshire

ABSTRACT

The field produced at high frequencies by a line source in a stratified medium above a plane is found to a first approximation when there is an impedance boundary condition on the plane and the refractive index increases steadily with distance from the plane. The formula is cast into a form which is invariant under a conformal mapping and is then applicable to all media and boundaries which can be obtained by conformal mapping. In particular it applies to the circular cylinder in a medium which is homogeneous or radially stratified.

NONUNIFORM TRANSMISSION LINES OR STRATIFIED LAYERS

G. LATMIRAL, G. FRANCESCHETTI and R. VINCIGUERRA†

ABSTRACT

An investigation has been carried out on the behaviour of a nonuniform transmission line, or stratified layer, used as broad band matching junction or termination, from both analysis and synthesis points of view.
For the inhomogeneous or stratified matching junctions the scattering matrix and the longest transmitted wavelength are deduced. Many numerical calculations (performed by means of both analog and digital computers) are reported.

1. The design of a nonuniform line or stratified layer, to be used as a broadband matching junction or termination, may be done by following either analysis or synthesis techniques. Let n be the relative refraction index, Z_0 the intrinsic impedance of the line, or layer,

$$\rho(x) = [Z(x) - Z_0(x)]/[Z(x) + Z_0(x)]$$

the voltage reflection coefficient, defined in the "Schelkunoff form", and $\rho_0(\eta)$ the reflection coefficient at the beginning of the line ($\eta = 4\pi/\lambda$).

In the first case, i.e. analysis, for given functions $n(x)$ and $Z_0(x)$ (defined in the interval 0, L in which the inhomogeneity is confined), the $\rho_0(\eta)$ function should be evaluated. The behaviour of $\rho_0(\eta)$ and, particularly, the value of η, say $\bar{\eta}$, such that, for $\eta \geq \bar{\eta}$, $\rho_0(\eta) \leqslant 10$ per cent shows, *a posteriori*, whether or not the chosen $n(x)$ and $Z_0(x)$ have favourable characteristics. Conversely, in the case of synthesis, for a given $\rho_0(\eta)$ spectrum, two functions $n(x)$ and $Z_0(x)$ have to be found, to give the $\rho_0(\eta)$ reflection when examined by the analysis procedure.

2. In the case of matching, the analysis may be performed by solving (mostly numerically) the following Riccati differential equation

$$\rho' - j\eta n\rho + \frac{Z_0'}{2Z_0}(1 - \rho^2) = 0 \tag{1}$$

($\rho' = \mathrm{d}\rho/\mathrm{d}x$; $Z_0' = \mathrm{d}Z_0/\mathrm{d}x$), where Z_0 and n are real quantities, under the boundary condition $\rho(L) = 0$.

† Istituto Universitario Navale, Naples, Italy. This research has been sponsored by the European Office, Office of Aerospace Research, USAF, under the Contract AF61(052)–589.

Obviously, when $\mu_r = 1$ (non-magnetic junctions), (1) can be written

$$\rho' - j\eta n\rho - \frac{n'}{2n}(1 - \rho^2) = 0 \tag{2}$$

Ten different matching $n(x)$ functions were tested, solving (2) by means of a PACE-TR-10 Analog Computer for "transformation rates" $n(L)/n(0)=2$; 5; 10.

The results are summarized in Table 1, where the ratio $\xi = \lambda/L$ is such that, for $\lambda \leqslant L\xi$, $|\rho_0(\eta)| \leqslant 10$ per cent.

TABLE 1

VALUES OF $\xi = \lambda/L$, SUCH THAT $|\rho_0(L/\lambda)| = 10\%$ (MATCHING CASE)

Functions $n(x)$	$\exp[k\,x/L]$	$\exp[k(x/L)^2]$	$\exp[k(x/L)^3]$	$1 + k\,x/L$	$1 + k\,(x/L)^2$	$1 + k\,(x/L)^3$	$\frac{1}{1 - k\,x/L}$	$\exp[e^{kx/L} - 1]$	$\cos h\,[k\,x/L]$	$1 + k\,(1 - \cos \pi x/L)$
$n(L) = 2$	3·85	2·60	2·04	3·78	3·35	2·62	3·57	3·64	3·08	2·94
$n(L) = 5$	1·36	2·0	2·82	<1·0	2·50	1·99	1·58	2·41	2·41	1·33
$n(L) = 10$	1·44	1·89	2·18	<1·0	1·61	1·46	1·45	1·80	1·77	<1·0

To carry out the synthesis, we should remark that (1) can be linearized by assuming $|\rho(x)|^2 \ll 1$ (as it is for a matching junction and for wavelengths not too long in comparison with the optical length). Solving for $\rho_0(\eta)$, we get:

$$\rho_0(\eta) = \int_0^L \frac{Z_0'}{2Z_0} \exp[-j\eta \int_0^x n(x)\,\mathrm{d}x].\mathrm{d}x \tag{3}$$

When $n = \text{const.} = 1$, eq. (3) shows that $\rho_0(\eta)$ is the Fourier transform of the function (of x), $Z_0/2Z_0$, which is zero outside the interval 0, L within which the inhomogeneity is confined (Bolinder, 1950, 1956). We can deduce, using Fourier transform techniques, that it is possible for $n = \text{const.}$ (TEM guided waves) to synthetize $Z_0(x)$, starting from a specified $\rho_0(\eta)$.

When $\mu = \text{const.}$, we can get a transform pair, even when n depends on x. Introducing in (3) the new variable of integration

$$y = \int_0^x n(x)\,\mathrm{d}x$$

(which is steadily increasing and admits, consequently, the inverse function

$x = x(y)$), it is possible to show that $\rho_0(\eta)$ and $\phi(y) = -n'[x(y)]/2n^2[x(y)]$ are a Fourier transform pair (Latmiral *et al.*, 1962).

If the $\rho_0(\eta)$ function is specified, the $\phi(y)$ function results determined by means of the inverse Fourier transform; and $n(x)$, which is the datum of practical interest, can be obtained by solving the 2nd order D.E.

$$-\frac{y''}{2y'^2} = \phi(y) \tag{4}$$

where the derivatives are with respect to x.

In certain special cases (4) can be solved analytically; but in general, only numerical solutions are possible.

For five different $\phi(y)$ functions (to which favourable spectra of reflection are related) the corresponding $n(x)$ functions have been synthesized by solving (4). This calculation has been done by means of the IBM 1620 Digital Computer of the Faculty of Science of the University of Naples (for the first four cases) and by means of a PACE-TR-10 Analog Computer (for the last one). The results are summarized in Table 2.

TABLE 2

VALUES OF $\xi = \lambda/L$, SUCH THAT $|\rho_0 (L/\lambda)| = 10\%$ (MATCHING CASE)

Functions $\phi(y)$	$A\,[yy_L - y^2]$	$A\,[yy_L - y^2]^2$	$A\,(1 - \cos 2\pi\, y/y_L)$	$A \sin \pi\, y/y_L$	Triangular pulse between $0, y_L$
$n(L) = 2$	2·90	2·08	2·02	2·90	2·27
$n(L) = 5$	3·20	2·50	2·50	3·20	2·63
$n(L) = 10$	3·70	2·80	2·63	3·80	3·03

3. For a lossless junction, and provided that the conditions under which (3) is fulfilled are satisfied, it is easy to write the corresponding scattering matrix (Franceschetti, 1962). Really, writing (3) not only for side 1 of the junction, but also for side 2, and using of the matricial equation $(S)(\tilde{S})^* = (I)$, by rather tedious, but simple calculations, we get

$$(S) = \left\| \begin{matrix} \rho_0 & \sqrt{(1-|\rho_0|^2)} \exp\left[-j\frac{\eta}{2}\int_0^L n(x)\,dx\right] \\ \sqrt{(1-|\rho_0|^2)} \exp\left[-j\frac{\eta}{2}\int_0^L n(x)\,dx\right] & -\rho_0^* \exp[-j\eta \int_0^L n(x)\,dx] \end{matrix} \right\| \tag{5}$$

where $\rho_0^*(\eta)$ is the complex conjugate of $\rho_0(\eta)$.

Furthermore, let us define an interval $\Delta\eta$ such that

$$\Delta\eta^2 = \frac{1}{I}\int_{-\infty}^{+\infty} \eta^2 \mid \rho_0(\eta) \mid^2 \mathrm{d}\eta \tag{6}$$

where

$$I = \int_{-\infty}^{+\infty} \mid \rho_0(\eta) \mid^2 \mathrm{d}\eta \tag{7}$$

and in a similar way an interval Δy. Obviously the function $\rho_0(\eta)$ will differ considerably from zero only in the interval $\pm\Delta\eta$.

But, since $\phi(y)$ and $\rho_0(\eta)$ are a Fourier transform pair, the following relation must hold

$$\Delta\eta \;.\; \Delta y \geqslant 2\pi \tag{8}$$

If we put $\Delta y \simeq y(L)$, we obtain from (8) the value of the longest transmitted wavelength $\lambda_{\max}$

$$\lambda_{\max} \leqslant 4\bar{n}L \tag{9}$$

where $\bar{n}$ is the mean value of $n(x)$ in the interval $(0, L)$.

4. In the case of absorption, the analysis can be performed solving the same Riccati D.E. (1) or (2), $Z_0(x)$ and $n(x)$ being now complex quantities. Ten different functions $n(x)$ were tested (both nonmagnetic and magnetic), and their results summarized in Tables 3 and 4.†

In some cases the conditions for physical realizability (i.e. the Kramers–Krönig relations) were taken into account.

Perhaps the most useful results are those listed under Table 3.

In order to apply the synthesizing procedure developed under Sec. 2, we must remark that, in this case, it is not possible to linearize (2), since we have, at least at the far end, $\rho(L) = -1$.

However, a Riccati D.E. can be transformed in a Bernouilli D.E., once a particular integral is known. We may consider as such the solution of the Riccati D.E. for the condition which corresponds to perfect matching at the far end, i.e. $\rho(L) = 0$.

Calling $\sigma_0(\eta)$ such a solution, this may be computed by means of the linearized equation (2).

Solving the Bernouilli equation, after long and tedious computations, we

† When there is an initial discontinuity in $n(x)$ or $Z_0(x)$, the Riccati D.E.

$$Z' = j\frac{1}{2}\eta\frac{n}{Z_0}(Z^2 - Z_0^2)$$

must be solved instead of (1) and ρ_0 calculated once $Z(0)$ is known.

TABLE 3

POWER REFLECTION COEFFICIENT $|\rho_0(\eta)|^2\%$, FOR SEVERAL VALUES OF λ/L (ABSORBING CASE)

$$\varepsilon = A\ [\cos 2\phi \frac{x}{L} - j \frac{\lambda}{\bar{\lambda}} \sin 2\phi\, x/L]\ \exp\ [2m\, x/L] \quad \mu = B \exp\ [2(n - j\psi)\, x/L]$$

A, m, ϕ B, n, ψ $\bar{\lambda}$	1·5 1·0 0·7 1·0 0·0 0·0 5	1·5 1·5 0·3 1·0 0·4 0·4 3	1·5 1·5 0·3 1·0 0·4 0·4 5	1·5 1·5 0·3 1·0 0·6 0·4 3	1·5 1·5 0·3 1·0 0·6 0·4 5
$\lambda/L = 1$	1·70	1·83	1·56	1·75	1·58
2	0·46	2·13	2·24	1·91	2·01
3	5·90	2·99	3·26	2·92	2·12
4	2·90	3·87	3·76	3·12	3·34
5	0·50	5·46	5·51	3·94	3·76
6	4·50	7·47	7·52	5·36	5·20
7	12·4	9·35	8·21	7·03	6·45
8	22·3	11·0	8·26	8·74	7·03

TABLE 4

POWER REFLECTION COEFFICIENT $|\rho_0(\eta)|^2\%$, FOR SEVERAL VALUES OF λ/L (ABSORBING CASE)

$$\varepsilon = A \exp\ [2(m - j\phi)\, x/L] \qquad \mu = B \exp\ [2(n - j\psi)\, x/L]$$

A, m, ϕ B, n, ψ	1·0 1·0 0·7 1·0 0·0 0·0	1·5 1·0 0·7 1·0 0·0 0·0	1·5 1·5 0·6 1·0 0·6 0·6	1·5 1·0 0·7 1·0 1·0 0·6	1·0 1·0 0·7 1·0 0·5 0·5
$\lambda/L = 1·0$	0·30	0·75			0·054
2·0	1·20	1·0	2·31	0·61	0·29
3·0	2·30	2·70	2·92	0·60	0·45
4·0	0·80	2·90	3·07	0·82	0·59
5·0	8·2	0·50	4·12	0·80	0·66
6·0	23·8	3·75	4·95	0·80	1·06
7·0		14·8	5·60	0·72	1·99
8·0		30·16	6·04	0·59	3·45

can write approximatively

$$\rho_0(\eta) = \sigma_0(\eta) - \frac{\exp[-j\eta \int_0^L n(x)\, dx]}{1 + \bar{\sigma}(L)} \tag{10}$$

where

$$\sigma_0(\eta) = \int_0^L \frac{Z_0'}{2Z_0} \exp[-j\eta \int_0^L n(x)\, dx]\ .\ dx \tag{11}$$

and $\bar{\sigma}(L)$ is the reflection coefficient in $x = L$ for a wave which proceeds from right to left, in the layer matched at $x = L$.

When $|\bar{\sigma}(L)| \ll 1$, (10) can be further simplified as follows

$$\rho_0(\eta) \simeq \sigma_0(\eta) - \exp[-j\eta \int_0^L n(x)\,dx] \tag{12}$$

Equation (12) shows, as a matter of fact, that minimizing $|\rho_0(\eta)|$ is equivalent to minimizing $|\sigma_0(\eta)|$, provided that the following relation is fulfilled

$$\text{Re}\,[\eta \int_0^L n(x)\,dx] > K \tag{13}$$

where K is a positive suitable constant.

The synthesizing procedure developed under Sec. 2, for the case $\mu = $ const., can then be applied to $\sigma_0(\eta)$, by making recourse to the Fourier transform $\Sigma_0(\eta)$ of the function:

$$\phi(y) = \frac{1}{2}\frac{d}{dy}\{\ln\,[n_1(y) - jn_2(y)]\} \tag{14}$$

where

$$y = \int_0^x n_1(x)\,dx$$

It can be shown (Latmiral *et al.*, 1962) that Σ_0 is a function larger, on the mean, than σ_0.

Once $\Sigma_0(\eta)$ has been chosen, the $\phi(y)$ function results determined by means of the inverse transform, and from (14) the two functions $n_1(x)$ and $n_2(x)$ can be deduced.

A numerical checking is obviously always necessary.

For two different $\phi(y)$ functions (to which favourable spectra of reflection coefficient are related), the corresponding $n(x) = n_1(x) - jn_2(x)$ functions were synthesized, solving (14) numerically.

TABLE 5

POWER REFLECTION COEFFICIENT $|\rho_0(\eta)|^2\%$ FOR SEVERAL VALUES OF L/λ (ABSORBING CASE)

Functions	$\ln\sqrt{n_1^2 + n_2^2} = A \sin \frac{2\pi}{T_1 y_L} y$ $\arctan \frac{n_2}{n_1} = B \sin \frac{2\pi}{T_2 y_L} y$	(Triangular + j rectangular) pulse between 0, y_L
$L/\lambda = 0{\cdot}1$	59·9	76·9
0·2	31·2	12·5
0·3	0·55	3·3
0·4	3·82	0·25
0·5	1·14	0·02
0·6	0·65	
0·7	0·68	

The results of the numerical checking carried out by solving numerically (2), are summarized in Table 5.

CONCLUSION

In the case of matching, the linearization of the basic Riccati D.E. allows an almost rigorous synthesis of a matching device, starting from a given reflection spectrum.

When dealing with absorbers, on the contrary, the synthesis procedure can be used only for orientative purposes and under a number of rather severe restrictions. Only the analysis gives exact results.

REFERENCES

BOLINDER, E. F., Correspondence on PIRE, November 1950, April 1956.

FRANCESCHETTI, G., "The inhomogeneous line, or layer, as a matching device ", *Alta Frequenza*, **31**, N. 11, 1962.

LATMIRAL, G., FRANCESCHETTI G. and VINCIGUERRA, R. "Synthesization of matching or absorbing layers or lines, starting from given reflection spectra ", Technical Report, Contract AF61(052)–36, 31st January 1962.

EFFECTS OF MUTUAL COUPLING IN A SYSTEM OF MAGNETIZED FERRITE SAMPLES

A. L. MICAELJAN and V. I. ANTONJANZ

State's Committee of Radioelectronics, U.S.S.R.

ABSTRACT

If two or more ferrite samples are placed in the immediate vicinity of each other some new interesting effects will occur at ferromagnetic resonance. The reason is that closely located ferrite samples behave as a coupled system and not as a system of independent resonance circuits. This is caused by the fact that, besides the external alternating field, each sample is affected by secondary (reradiated) fields created by adjacent samples.

In order to make clear these effects of coupling the most elementary case of two ferrite samples located in a free space is considered.

1. NATURAL FREQUENCIES OF A SYSTEM OF FERRITE SAMPLES

In each of the two samples, to be denoted by indexes 1 and 2, the relations between the magnetic moment and the effective field are described by the equation of motion:

$$\frac{\partial \overline{M}_{1,2}}{\partial t} = -\mu_0 \gamma [\overline{M}^i_{1,2} \bar{H}_{1,2}] \tag{1}$$

The field applied to the first sample consists of the external field $\bar{H}^e_1$ and of the field $\bar{h}_2$ radiated by the second second sample

$$\bar{H}_1 = \bar{H}^e_2 + \bar{h}_2 \tag{2}$$

Similarly, the field applied to the second sample is given by

$$\bar{H}_2 = \bar{H}^e_2 + \bar{h}_1 \tag{3}$$

where $\bar{h}_1$ is the field radiated by the first sample.

Let us consider the case of free oscillations in the absence of the external alternating field.

To simplify the computation procedure we assume dimensions of the samples to be small in comparison with the distance between them. Besides, field $\bar{h}_1$ around the second sample and field $\bar{h}_2$ around the first one are assumed to be uniform.

Under these conditions the magnetization of the samples will be uniform in volume while the samples can be presented as magnetic dipoles with moments $\bar{m}_1$ and $\bar{m}_2$ if v_1 and v_2 are the sample volumes. Radiation fields can be found with the use of the Hertz vector.

For a magnetic dipole

$$\bar{\Pi} = \frac{1}{4\pi} \sqrt{\left(\frac{\mu}{\epsilon}\right)} [\bar{m}_m \bar{\tau}_0] \left(\frac{1}{z} - \frac{j}{k_0 \tau^2}\right) e^{j(\omega t - k_0 \tau)} \tag{4}$$

where $\bar{\tau}_0$—unit-vector;

k_0—propagation wave number;

$\bar{m}_m$—amplitude of the sample's magnetic moment.

The origin of the coordinate system coincides with the center of the first sample (see Fig. 1).

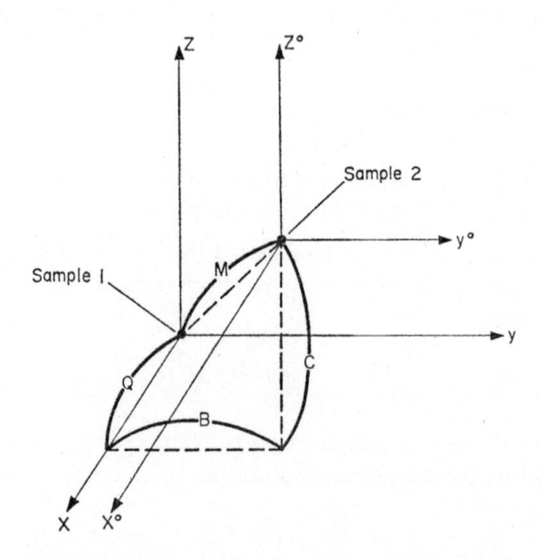

FIG. 1. The disposition of samples in coupling system.

Let us assume that the magnetization field H^e_{01} is directed along the z-axis and that the magnetic moment $\bar{m}_1$ of the sample is located in the plane XOY. In this case

$$h_{x_1} = \left[m_{xm_1}\left(R_1 \frac{x^2}{\tau^2} - R \frac{yz + z^2}{\tau^2}\right) + m_{ym_1}(R_1 + R_2) \frac{xy}{\tau^2}\right] e^{j\omega t} \tag{5}$$

$$h_{y_1} = \left[m_{xm_1}(R_1 + R_2) \frac{xy}{\tau^2} + m_{ym_1}\left(R_1 \frac{y^2}{\tau^2} - R_2 \frac{x^2 + z^2}{\tau^2}\right)\right] e^{j\omega t} \tag{6}$$

$$h_{z_1} = \left[m_{xm_1}(R_1 + R_2) \frac{xz}{\tau^2} + m_{ym_1}(R_1 + R_2) \frac{yz}{\tau^2} \right] e^{j\omega t} \tag{7}$$

$$R_1 = \frac{2}{4\pi\tau^3}(1 + jk_0\tau)\, e^{-jk_0\tau}, \qquad R_2 = \frac{1}{4\pi\tau^3}(1 - k_0^2\tau^2 + jk_0\tau)\, e^{-jk_0\tau} \tag{8}$$

Now, if we place the second sample at the origin of another coordinate system x^0, y^0 and z^0 the sample's radiation field in this coordinate system can be written just in the same manner as the field of the first sample in the x, y, z coordinate system. Besides, if the coordinate axes of the x^0, y^0, z^0-system are parallel to the axes of the x, y, z-system we will have

$$x^0 = x - a; \qquad y^0 = y - b; \qquad z^0 = z - c$$

where a, b, c, are the coordinates of the second sample in the x, y, z-system.

Let us consider the case when the samples are located in the plane $z = 0$ which is normal to the direction of the constant magnetic field. In this plane all the positions of the second sample are equivalent, therefore it can be positioned at a point with coordinates $x = a$ and $y = 0$ (the first sample is always at the origin of the coordinate system).

Then the field of the first sample at the point $(a, 0, 0)$ will be given by

$$h_{x_1} = m_{xm_1} R_1\, e^{j\omega t} \tag{9}$$

$$h_{y_1} = -\, m_{ym_1} R_2\, e^{j\omega t} \tag{10}$$

The field of the second sample at the location of the first one (at the origin of the coordinate system) will be given by

$$h_{x_2} = m_{xm_2} R_1\, e^{j\omega t} \tag{11}$$

$$h_{y_2} = -\, m_{ym_2} R_2\, e^{j\omega t} \tag{12}$$

Substitution of these field expressions into the equations of motion gives

$$\frac{1}{v_1}\dot{m}_{x_1} = -\,\omega_M m_{y_2} R_2 - \omega f_1 \frac{1}{v_1} m_{y_1} \tag{13}$$

$$\frac{1}{v_1}\dot{m}_{y_1} = -\,\omega_M m_{x_2} R_1 + \omega f_1 \frac{1}{v_1} m_{x_1} \tag{14}$$

$$\frac{1}{v_2}\dot{m}_{x_2} = -\,\omega_M m_{y_1} R_2 - \omega f_2 \frac{1}{v_2} m_{y_2} \tag{15}$$

$$\frac{1}{v_2}\dot{m}_{y_2} = -\,\omega_M m_{x_1} R_1 + \omega f_2 \frac{1}{v_2} m_{x_2} \tag{16}$$

Here $\omega_{f1} = \mu_0\gamma H_{01} - (N_z - N_T)\,\omega_M$ and $\omega_{f2} = \mu_0\gamma H_{02} - (N_t - N_T)\,\omega_M$ are the resonance frequencies of the first and the second samples

$N_x = N_y = N_T =$ demagnetization factors

H_{01} and $H_{02} =$ constant fields in the region of the first and the second samples directed along the Z-axis.

Solving the system (13)–(16) we obtain the characteristic equation

$$(\omega^2 - \omega_{f1}^2)(\omega^2 - \omega_{f2}^2) + \omega^2 2\omega_M{}^2 v_1 v_2 R_1 R_2 - \\ - \omega_{f1}\omega_{f2}\omega_M^2 v_1 v_2 (R_1^2 + R_2^2) + \omega_M^4 v_1^2 v_2^2 R_1^2 R_2^2 = 0 \tag{17}$$

If R_1 and $R_2 \to 0$ (with $z \to \infty$) or if v_1 and $v_2 \to 0$ the equation (17) is divided into two independent parts $\omega^2 - \omega_{f1}^2 = 0$ and $\omega^2 - \omega_{f_2}^2 = 0$, which determine the resonance frequencies of each of the samples.

In the simple case when the natural frequencies of the samples are equal we have $\omega_{f_1} = \omega_{f_2} = \omega_f$

$$\omega_{I,\,II} \approx \omega_f \pm \tfrac{1}{2}\,\omega_M \sqrt{(v_1 v_2)}(R_1 - R_2). \tag{18}$$

Assuming the natural frequencies to be complex in nature it is possible to determine the damping of free oscillations in a coupled system $\omega = \omega' + j\omega''$ $\omega_f = \omega_f' + j\omega_f''$, where ω_f'' and ω'' characterize the losses of each of the samples and the losses of the coupled system respectively.

From (18) we obtain

$$\omega_{I,\,II}' \cong \omega_f' \pm \tfrac{1}{2}\,\omega_M \sqrt{(v_1 v_2)}(R_1' - R_2') \tag{19}$$

$$\omega_{I,\,II}'' \cong \omega_f'' \pm \tfrac{1}{2}\,\omega_M \sqrt{(v_1 v_2)}(R_1'' - R_2'') \tag{20}$$

where $R_1 = R_1' + jR_1''$ and $R_2 = R_2' + jR_2''$

$$R_1' = \frac{1}{2\pi\tau^3}(\cos k_0\tau + k_0\tau \sin k_0\tau)$$

$$R_1'' = \frac{1}{2\pi\tau^3}(k_0\tau \cos k_0\tau - \sin k_0\tau)$$

while

$$\left.\begin{aligned} R_2' &= \frac{1}{4\pi\tau^3}[k_0\tau \sin k_0\tau + (1 - k_0^2\tau^2)\cos k_0\tau] \\ R_2'' &= \frac{1}{4\pi\tau^3}[k_0\tau \cos k_0\tau - (1 - k_0^2\tau^2)\sin k_0\tau] \end{aligned}\right\} \tag{21}$$

Since R_1 and R_2 are complex quantities the value of frequency ω will also be complex even in the absence of heat losses in the ferrite samples. The imaginary part of the value of ω is due, in this case, to the radiation losses. The problem of losses will be discussed in detail somewhat later.

The results obtained so far indicate that two ferrite samples located closely enough to each other manifest mutual coupling the amount of which can be given by a coupling ratio k just as it is done in a system of two coupled resonance circuits.

The dependance of the coupling ratio and of the separation of the natural frequencies of the samples upon the distance between them is shown in Fig. 2 where the values τ of the distance between the samples are fixed along the ordinate.

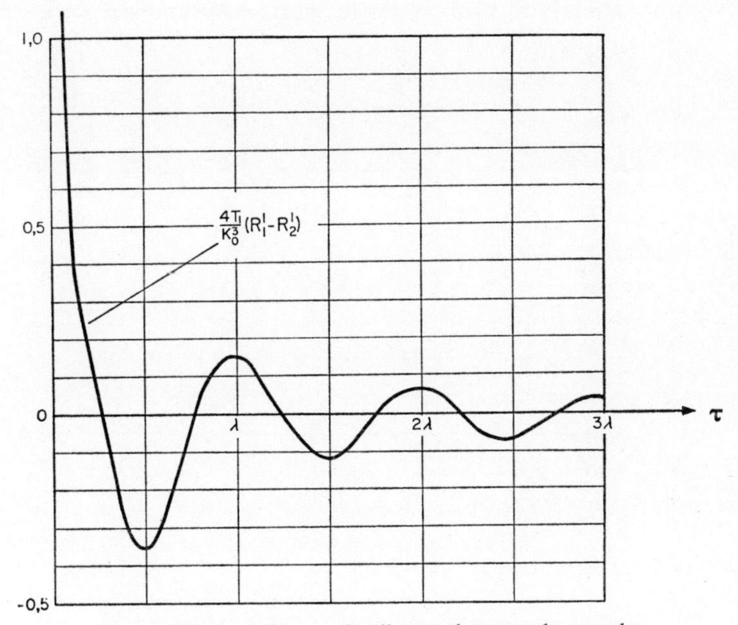

FIG. 2. The coupling vs. the distance between the samples.

It is interesting to note that these values at certain distances become zero and in the oscillating section they decrease very slowly with the increase of the distance.

With the decrease of the distance it is necessary to take into account that the value of $\sqrt{(v_1 v_2)}/\tau^3$ which goes into the expressions for $\omega_{I,\,II}$ remains finite since z can in no case be less than the sum of the samples' radii.

2. LOSSES IN A COUPLED SYSTEM AND RADIATION LOSSES OF AN INDIVIDUAL SAMPLE

Now let us consider in detail the problem of losses, the problem of the imaginary part of the natural frequency. From eqn. (20) it is evident that the imaginary part of the natural frequency depends upon the losses of an individual sample ω_f'' and upon "the mutual" losses which are summed at one natural frequency and are subtracted at the other.

The mutual losses are characteristic only of a coupled system; with the increase of the distance between the samples the value of these losses approaches zero. Their dependance upon the distance is shown in Fig. 3 for two cases of different positions of the samples with respect to each other. The value of losses in an individual sample can be obtained experimentally through the half-width of the absorption curve $\omega_f'' = \mu_0 \gamma \Delta H$. The value of ω_f'' takes into account all kinds of losses: heat, radiation, etc.

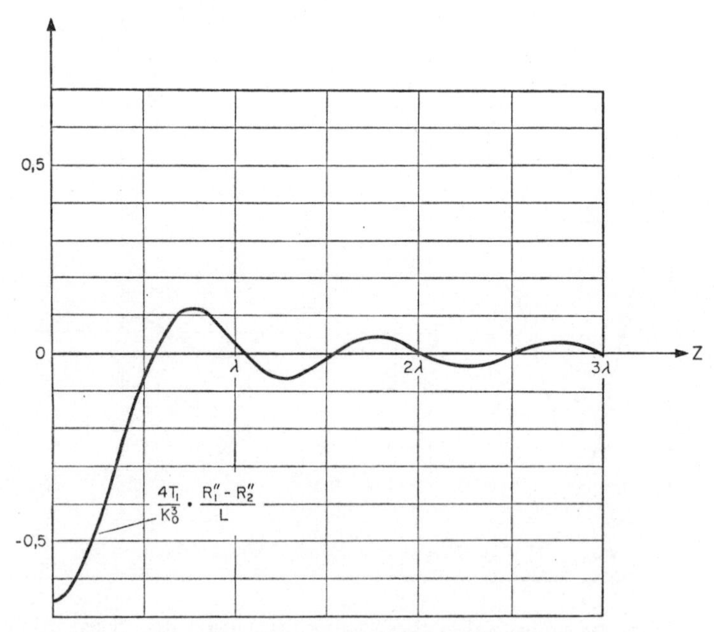

FIG. 3. The mutual losses vs. the distance between samples.

Usually it is assumed, however, that all the energy absorbed by a ferrite is converted into heat. But in the present case we can discriminate between the losses, and estimate the value of radiation losses P_r separately.

Assume that the whole of the energy absorbed by ferrite P is radiated and that there are no other types of losses. Then

$$P = P_r \tag{22}$$

As is known, the energy when a ferrite is excited by a circularly polarized field with an amplitude of H is given by

$$P = \frac{\mu_0 \omega v \omega_M \omega''_{fr} H^2}{(\omega_f - \omega)^2 + \omega''^2_{fr}} \tag{23}$$

where ω''_{fr} characterizes the radiation losses.

The amount of radiation losses can be obtained with the use of the previously derived expressions for the fields.

$$P_r = \tfrac{1}{2} \int_S \mathrm{Re}\,[\bar{E}\bar{h}^*]\,\mathrm{d}\bar{S} \tag{24}$$

Taking into account that when a sample is magnetized by a circularly polarized field the magnetization of the former is polarized in a circle and that $m_y = -jm_x$ we obtain

$$P_r = \frac{k_0^4}{6\pi}\sqrt{\left(\frac{\mu_0}{\epsilon_0}\right)}\,|\,\bar{m}\,|^2 = \frac{k_0^4 v^2}{6\pi}\sqrt{\left(\frac{\mu_0}{\epsilon_0}\right)}\,|\,\bar{M}\,|^2 \tag{25}$$

or

$$P_r = \frac{v_2 k_0^4}{6\pi}\sqrt{\left(\frac{\mu_0}{\epsilon_0}\right)}\,\frac{\omega_M^2 H^2}{(\omega_0 - \omega)^2 + \omega''_{fr}} \tag{26}$$

Assuming $P = P\tau$ we have

$$\omega''_{fr} = \frac{\omega_M v k_0^3}{6\pi} \tag{27}$$

A similar expression can be deduced in a more rigorous manner.

This value is always higher than the value of "mutual" losses (they are equal only in the extreme case when $r = 0$).

If there are also heat losses ω''_{ft}

$$\omega''_f = \omega''_{fr} + \omega''_{ft} \tag{28}$$

Thus, the obvious conclusion is that in a coupled system there are constant heat and radiation losses of each sample ω''_f, as well as "mutual" losses peculiar only for coupled system.

This "mutual" component of the losses can cause either a considerable decrease of losses at one of the natural frequencies when it is subtracted from ω''_f, or it can cause an increase of losses at the other frequency when "0" it is summed up with ω''_f.

By way of illustration let us consider the composition of losses in a coupled system of two samples which have the form of spheres 2 mm in diameter and which are located in a plane $Z = 0$ at a distance $\tau = \frac{3}{4}\lambda$ from each other.

The losses will be determined through the half-width of the resonance curve ΔH as was shown above:

$$\Delta H_{I,\, II} = \Delta H_{ft} + \Delta H_{fr} \pm \tfrac{1}{2} M_0 v(R_1'' - R_2'') \tag{29}$$

where the values of ΔH_{ft} and ΔH_{fr} depend upon the heat and radiation losses respectively.

$$\Delta H_{fr} = \frac{\mu_0 v k_0^3}{6\pi} = 2{\cdot}96 \text{ oersteds}$$

The amount of the "mutual" losses is found with the use of the curve given in Fig. 3.

Consequently

$$\Delta H_{I,\, II} = (\Delta H_{ft} + 2{\cdot}96 \pm 0{\cdot}5) \text{ oersted} \tag{30}$$

It is evident that the radiation losses cause considerable expansion of the resonance curve and can greatly exceed the amount of heat losses. This effect makes it possible to use ferrite samples as effective radiators whose parameters could be controlled by means of regulating the strength of the magnetic field.

But this problem must be treated separately.